Third Edition

A Simplified Text in
Electrical
Machine
Design

for BE/BTech EEE Course

Other CBS Titles by *A Nagoor Kani*

- Advanced Control Theory, 3/e
- Circuit Analysis, 2/e (*Anna University/ECE*)
- Circuit Theory, 2/e (*Anna University/EEE*)
- Control Systems, 5/e
- Control Systems (*Anna University/EEE*)
- Control Systems Engineering (*Anna University/ECE*)
- Control Systems Engineering, 2/e
- Design of Electrical Apparatus (*Anna University/EEE*)
- Microprocessors and Microcontrollers, 2/e (*Anna University/ECE*)
- Microprocessors and Microcontrollers (*Anna University/EEE*)
- Power System Analysis
- Signals and Systems—Simplified 2/e (*Anna University/ECE*)
- 8085 Microprocessor and its Applications, 4/e
- 8086 Microprocessor and its Applications, 3/e
- Digital Electronics (*in Press*)
- Digital Logic Circuits (*in Press*)
- Digital Electronics (*Anna University/ECE*) (*in Press*)
- Digital Logic Circuits (*Anna University/EEE*) (*in Press*)
- Digital Principles and System Design (*in Press*)
- Digital Principles and System Design (*Anna University/CSE/IT*) (*in Press*)
- Digital Signal Processing, 3/e (*in Press*)
- Digital Signal Processing (*Anna University/EEE*) (*in Press*)
- Discrete Time Signal Processing (*Anna University/ECE*) (*in Press*)
- Electric Circuits, 2/e (*in Press*)
- Microprocessors and Microcontrollers, 3/e (*in Press*)
- Signals and Systems, 2/e (*in Press*)

Third Edition

A Simplified Text in
Electrical
Machine
Design

for BE/BTech EEE Course

A Nagoor Kani

Founder, RBA Educational Group
Chennai

CBS

CBS Publishers & Distributors Pvt Ltd

New Delhi • Bengaluru • Chennai • Kochi • Kolkata • Mumbai
Hyderabad • Jharkhand • Nagpur • Patna • Pune • Uttarakhand

Third Edition

A Simplified Text in
**Electrical
Machine
Design**

ISBN: 978–93-90709-92-2

Third Edition: 2022
First Edition: 1998
Second Edition: 2000

Published by Satish Kumar Jain and produced by Varun Jain for
CBS Publishers & Distributors Pvt Ltd
4819/XI Prahlad Street, 24 Ansari Road, Daryaganj, New Delhi 110 002, India
Ph: 011–23289259, 23266861, 23266867   Fax: 011–23243014
Website: www.cbspd.com   e-mail: delhi@cbspd.com; cbspubs@airtelmail.in

Corporate Office: 204 FIE, Industrial Area, Patparganj, Delhi 110 092, India
Ph: 011–49344934   Fax: 011–49344935   e-mail: publishing@cbspd.com; publicity@cbspd.com

Branches

• **Bengaluru:** Seema House 2975, 17th Cross, K.R. Road, Banasankari 2nd Stage, Bengaluru 560 070, Karnataka, India
  Ph: +91-80-26771678/79   Fax: +91-80-26771680   e-mail: bangalore@cbspd.com

• **Chennai:** 7, Subbaraya Street, Shenoy Nagar, Chennai 600 030, Tamil Nadu, India
  Ph: +91-44-26680620, 26681266   Fax: +91-44-42032115   e-mail: chennai@cbspd.com

• **Kochi:** 42/1325, 1326, Power House Road, Opposite KSEB, Power House, Ernakulum-682018, Kochi, Kerala, India
  Ph: +91-484-4059061–67   Fax: +91-484-4059065   e-mail: kochi@cbspd.com

• **Kolkata:** 147, Hind Ceramics Compound, 1st Floor, Nilgunj Road, Belghoria, Kolkata 700056, West Bengal, India
  Ph: +91-9096713055/7798394118, 9836841399   e-mail: kolkata@cbspd.com

• **Mumbai:** PWD Shed, Gala No. 25/26, Ramchandra Bhatt Marg, Next JJ Hospital Gate No. 2
  Opp. Union Bank of India, Noorbaug, Mumbai-400009, Maharashtra, India
  Ph: +91-22-66661880/89   e-mail: mumbai@cbspd.com

Representatives

• **Hyderabad**	0-9885175004	• **Jharkhand**	0-9811541605	• **Nagpur**	0-9421945513
• **Patna**	0-9334159340	• **Pune**	0-9623451994	• **Uttarakhand**	0-9716462459

Printed at: Mudrak, Noida, UP, India.

to

---

my mother
Late Rohia Beevi Allaudeen

PREFACE

Electrical machine design deals with various concepts involved in machines and equipment operated by electrical energy and it is one of the core subjects for BE/BTech EEE students and for those who study postgraduate in electrical machines. The reader is expected to have prior knowledge in theory of electrical machines. This text is designed for EEE branch students studying in various universities in India.

This book is organized with 9 chapters. The various concepts of design of electrical machines are presented in a very easiest and elaborative manner. Throughout the book, carefully chosen examples are presented so that the reader will have a clear understanding of the concepts discussed. The analysis-based computer aided design of electrical machines has been presented in this book which includes simple C-programs for design of various components of electrical machines.

Chapter 1 presents the basic concepts and general problems of electrical machine design. Specific electrical and magnetic loadings, output equation and separation of D and L for all rotating machines are discussed in this chapter. An introduction to modern trend and computer aided design of electrical machine is also included in this chapter.

Chapter 2 is devoted to discussion on various materials used in construction of electrical machines. Brief discussion on properties and types of various conductive, resistive, magnetic and insulating materials are presented. Chapter 3 focuses on field and armature electro-magnetic systems. The various concepts of magnetic circuits, leakage flux and reactance are explained with examples.

Chapter 4 explains the concept of heating and cooling in electrical machines. The various types of ventilation and cooling method are discussed. Types of enclosures and machine rating are also presented. Estimation of short time rating of machine is explained with better clarity. Carefully chosen examples will help to understand the heating and cooling concepts clearly.

Chapter 5 deals with the design of armature winding for DC and AC machine armature with detailed examples and solved problems. Single and double layer and integral slot and fractional slot AC machine windings are also explained with examples. Every winding example is supported by various pitch calculations for better understanding.

Chapter 6 is concerned with design of DC machines. The design of main dimensions of armature and field system are presented with clear explanation and number of solved problems. The design of armature core, winding and commutator are also explained with appropriate examples.

Chapter 7 focuses on design of core and shell type transformers. This chapter deals with various types of cores for constructing transformers and design of windings. The concepts of heat generation and temperature rise in transformers and cooling of transformer to reduce temperature rise are also presented with clear examples.

Chapter 8 introduces the concepts of design of 3-phase squirrel cage and wound rotor induction motor. The design of various dimensions of rotor and stator are discussed with clear examples. The choice of stator and rotor slots to minimize harmonics are also discussed. The prediction of performance characteristics of motor from the design data using circle diagram is discussed with clear example.

Chapter 9 deals with design of synchronous machines. The design of main dimensions of salient pole and non-salient pole synchronous machines are explained with examples. The design of field system and damper winding are also presented with examples.

Several concepts and procedures are illustrated through simple examples rather than generalized formulation. Each chapter provides the foundations and practical implications of their own topic with number of solved problems for better understanding.

Special care has been taken in presenting the concepts in simple manner and choosing examples and solved problems. I hope that the teaching and student community will welcome the book. The readers can feel free to convey their suggestions or criticism for further improvement of the book.

A Nagoor Kani

ACKNOWLEDGEMENTS

I express my heartfelt thanks to my wife Ms C Gnanaparanjothi Nagoor Kani and my sons N Bharath Raj alias Chandrakani Allaudeen and N Vikram Raj, for the support, encouragement and cooperation they have extended to me throughout my career. I thank Ms T A Benazir, Manager, RBA Group, and all my office-staff for their cooperation in carrying out my day-to-day activities.

It's my pleasure to acknowledge the contributions of our technical editors, Ms S Saranya and Ms P Kanimozhi and Ms K G Sathyapriya, for editing, proof-reading and typesetting of the manuscript and preparing the layout of the book.

My sincere thanks to all reviewers for their valuable suggestions and comments which helps me to explore the subject to greater depth.

I am also grateful to Mr SK Jain, CMD, CBS Publishers & Distributors, for his keen interest in publishing this work in CBS banner. My sincere thanks to all team members of CBS Publishers & Distributors, for their concern and care in publishing this work.

Finally, a special note of appreciation is due to my sisters, brothers, relatives, friends, students and the entire teaching community for their overwhelming support and encouragement to my writing.

A Nagoor Kani

CONTENTS

CHAPTER 2: ELECTRICAL ENGINEERING MATERIALS   2.1–2.14

CHAPTER 5: ARMATURE WINDING   5.1–5.116

CHAPTER 7: DESIGN OF TRANSFORMERS   7.1–7.90

CHAPTER 8: DESIGN OF THREE-PHASE INDUCTION MOTOR   8.1–8.90

CHAPTER 9: DESIGN OF SYNCHRONOUS MACHINE   9.1–9.66

> Chapter 1

BASIC PRINCIPLES OF ELECTRICAL MACHINE DESIGN

List of symbols

Symbol	Meaning	Unit
a	Number of parallel paths in armature winding	-
a_z	Area of cross section of conductor	mm^2
ac	Specific electric loading	$amp.cond./m$
B_{av}	Specific magnetic loading	Wb/m^2 or $Tesla$
B_{gm}	Maximum air-gap flux density under load conditions	Wb/m^2 or $Tesla$
b	Pole arc	m
b_p	Width of the pole body	m
C_o	Output coefficient	$kVA/m^3 - rps$
c	Cooling coefficient	$°C\ W\text{-}m^2$
D	Armature diameter or stator bore	m
d_s	Depth of slot	mm
E	Generated emf or back emf	V
E_{cm}	Maximum voltage between adjacent segments	V
E_{ph}	Induced emf per phase	V
f	Frequency	Hz
I	Rated current	A
I_a	Armature current	A
I_z	Current in each conductor	A
I_{ph}	Current per phase	A
L	Armature length or stator core length	m
N	Speed	rpm
n	Speed	rps
n_s	Synchronous speed	rps
P	Rating of machine (Rated output power)	kW
P_a	Power developed by armature	kW

Symbol	Meaning	Unit
Q	kVA rating of machine	kVA
Q_l	Loss dissipated	kW
q	Loss dissipated per unit area	kW/m^2
R	Resistance	Ω
S	Dissipating surface	m^2
T_c	Turns per coil	-
T_{ph}	Turns per phase	-
V_a	Peripheral speed	m/sec
w_s	Width of the slot	mm
y_s	Slot pitch	mm
Z	Total number of armature or stator conductors	-
ϕ	Magnetic flux	Wb
Ψ	Ratio of pole arc to pole pitch	-
τ	Pole pitch	m
η	Efficiency	-
θ	Temperature rise	$°C$
δ	Current density	A/mm^2
ρ	Resistivity	$\Omega\text{-}m$

1.1 INTRODUCTION

Electrical machine design involves application of science and technology to produce cost-effective, durable, quality and efficient machines. Also the machines should be designed as per standard specifications. The requirements like low cost and high quality will be conflicting in nature and so a compromise should be made between them.

The electrical machines can be classified into static and dynamic machines. The transformer is a static (stationary) machine. The motors and generators are dynamic (rotating) machines. The transformer converts electrical energy from one voltage level to another voltage level. The rotating machines converts electrical energy to mechanical energy or vice-versa.

The conversion in any electrical machine takes place through magnetic field. The required magnetic field is produced by an electromagnet which requires a core and winding. The basic principle of operation of all electrical machine is governed by Faraday's law of electromagnetic induction.

1.1.1 Constructional Elements of Transformer

The transformer is a static electromagnetic device used to transfer electrical energy from a high potential (voltage) circuit to low potential (voltage) circuit or vice-versa. It consists of two or more windings which link with a common magnetic field. An iron core serves as a path for magnetic flux.

The constructional elements of the transformer are windings, core, tank and cooling tubes or radiators. A simple transformer has two windings and they are called high voltage winding and low voltage winding. One of the winding is connected to supply and it is called *primary*. The other winding is connected to load and it is called *secondary*.

The two different types of transformer constructions are core type and shell type. In core type transformer the windings surround the core and in shell type transformer the core surround the windings. The core and winding assembly is housed in the tank. Cooling tubes or radiators are provided around the tank surface in order to increase the effective cooling surface.

1.1.2 Constructional Elements of Rotating Machines

The rotating electrical machines converts electrical energy to mechanical energy or vice-versa. The energy conversion takes place through magnetic field. Every rotating machine have the following three quantities. The presence of any two quantities, will produce the third quantity.

1.   Magnetic field - I (Field)

2.   Magnetic field - II (Armature)

3.   Mechanical force

In generator, the armature is rotated by a mechanical force inside a magnetic field or the magnetic field is rotated by keeping armature stationary. By Faraday's law of induction, an emf is induced in the armature. When the generator is loaded, the armature current flows, which produce another magnetic field (armature magnetic field). Hence in a generator, by the presence of a magnetic field and mechanical force, an another magnetic field is produced.

The mechanical force developed by the motor is due to the reaction of two magnetic fields. A current carrying conductor has a magnetic field around it. When it is placed in another magnetic field it experiences a mechanical force due to the reaction of two magnetic field. Hence in a motor by the presence of two magnetic fields a mechanical force is developed.

From the above discussion it is clear that any rotating machine requires two magnetic field and one of the field is rotating. Hence a rotating machine will have a stationary and rotating electromagnet, each consisting of a core and winding. The stationary electromagnet is called *stator* and the rotating electromagnet is called *rotor*.

The basic constructional elements of a rotating electrical machine are stator and rotor. In DC machines the stator consists of field core and winding. The rotor comprises of armature core and winding. In AC machines the stator has armature core and winding. The rotor consists of field core and winding. The constructional elements of various electrical machines are listed here.

Constructional elements of DC machine

Stator	-	Yoke or Frame	Rotor	-	Armature core
	-	Field pole		-	Armature winding
	-	Pole shoe		-	Commutator
	-	Field winding	Others	-	Brush
	-	Interpole		-	Brush holder

Constructional elements of salient pole synchronous machine

Stator	-	Frame	Rotor	-	Field pole
	-	Armature core		-	Pole shoe
	-	Armature winding		-	Field winding
				-	Damper winding

Constructional elements of cylindrical rotor synchronous machine

Stator	-	Frame	Rotor	-	Solid rotor
	-	Armature core		-	Field conductors or bars
	-	Armature winding			

Constructional elements of squirrel cage induction motor

Stator	-	Frame	Rotor	-	Rotor core
	-	Stator core		-	Rotor bars
	-	Stator winding		-	End ring

Constructional elements of slip ring induction motor

Stator	-	Frame	Rotor	-	Rotor core
	-	Stator core		-	Rotor winding
	-	Stator winding		-	Slip rings

1.1.3 Classification of Design Problems

The design of an electrical machine involves solution of many complex and diverse engineering problems. The design problems may be classified under the following four major headings.

1. Electromagnetic design
2. Mechanical design
3. Thermal design
4. Dielectric design

Each major problem may be solved separately and the results are combined to give overall solution. Each major problem may be further divided into simple problems and solutions of individual problem are combined to give the solution of a major design problem.

The electromagnetic design problem in rotating machines involves the design of stator and rotor core dimensions, stator and rotor teeth dimensions, air-gap length, stator and rotor windings. In transformer it is the problem of designing the core and the windings.

The mechanical design in rotating machine involves the design of frame (enclosure), shaft and bearings. In transformer it is the design of tank (i.e., housing for core and winding assembly).

The thermal design in rotating machine involves the design of cooling ducts in core and cooling fans. In case of large machines coolants like air or hydrogen may be forced to circulate in the ducts and air-gap. In transformer it involves the design of cooling tubes or radiators.

Another important design problem, that may require great attention in the design of insulations (dielectric design). Dielectric materials are used to insulate one conductor from other and also the windings from the core. The dielectric materials are designed to withstand high voltage stresses. The breakdown of dielectric materials may lead to failure of machine.

1.1.4 Standard Specifications   *(AU, Apr'17, 8 M)*

Every country has a standards organisation to fix standard specifications for the manufacturers. The specifications are guidelines for the manufacturers to produce economic products without compromising quality. The manufacturers who are compiling with the standards will be issued a certification for their products. The quality of the certified products will be periodically monitored by the standards organisation.

The standard specifications issued for electrical machines includes the following,

1. Standard ratings of machines
2. Types of enclosure
3. Standard dimensions of conductors to be used
4. Method of marking ratings and name plate details
5. Performance specifications to be met
6. Types of insulation and permissible temperature rise
7. Permissible loss and range of efficiency
8. Procedure for testing of machine parts and machines
9. Auxiliary equipments to be provided
10. Cooling methods to be adopted.

In India, Indian Standards Institution (ISI) was started in the year 1947 to lay down specification for various products. ISI was renamed as Bureau of Indian Standards (BIS) in the year 1986.

The standard specifications of a product (or part of a product) will be framed and released with a prefix IS (Indian standard) followed by number and year of publication.

The standards will be amended time to time, in order to include the latest developments in technology. Recently they have released revised standards ISO 9002, to comply with international standards.

The name plate of a rotating machine has to bear the following details as per ISI specifications.

1. kW or kVA rating of machine
2. Rated working voltage
3. Operating speed
4. Full load current
5. Class of insulation
6. Frame size
7. Manufacturers name
8. Serial number of the product

Some of the Indian standard specifications numbers along with year of issue for electrical machines are listed here.

IS 325-1966:   Specifications of three-phase induction motor.

IS 1231-1974:   Specifications of foot mounted induction motor.

IS 4029-1967:   Guide for testing three-phase induction motor.

IS 996-1979:   Specifications of single-phase AC and universal motor.

IS 1885-1993:   Specifications of electric and magnetic circuits.

IS 9499-1980:   Conventions concerning electric and magnetic circuits.

IS 7538-1996:   Specifications of three-phase induction motor for centrifugal pumps and agricultural applications.

IS 12615-1986: Specifications of energy efficient induction motor.

IS 9320-1979:   Guide for testing DC machines.

IS 4722-1992:   Specifications of rotating electrical machines.

IS 12802-1989:	Temperature rise measurement of rotating electrical machines.
IS 4889-1968:	Method of determination of efficiency of rotating electrical machines.
IS 13555-1993:	Guide for selection and application of three-phase induction motor for different types of driven equipment.
IS 7132-1973:	Guide for testing synchronous machines.
IS 5422-1996:	Specifications of turbine type generators.
IS 7572-1974:	Guide for testing single-phase AC and universal motors.
IS 8789-1996:	Values of performance characteristics for three-phase induction motors.
IS 12066-1986:	Specifications of three-phase induction motors for machine tools.
IS 1180-1989:	Specifications of outdoor 3-phase distribution transformer upto $100\,kVA$. (Sealed and Non-sealed type)
IS 2026-1994:	Specifications of power transformers.
IS 11171-1985:	Specifications of dry type power transformers.
IS 5142-1969:	Continuously variable voltage auto transformers.
IS 10028-1985:	Code of practice for selection, installation and maintenance of transformers.
IS 10561-1983:	Application guide for power transformers.
IS 13956-1994:	Testing transformers.
IS 9678-1980:	Methods of measuring temperature rise of electrical equipment.
IS 12063-1987:	Classification of degree of protections provided by enclosures of electrical equipment.
IS 3855-1966:	Standard dimensions of rectangular enamelled copper conductor.
IS 449-1962:	Standard dimensions of enamelled round copper conductor (oleo resinous enamel).
IS 1595-1960:	Standard dimensions of enamelled round copper conductor (synthetic enamel).
IS 1897-1962:	Standard dimensions of bare copper strip.
IS 1666-1961:	Standard dimensions of paper covered rectangular copper conductor for transformer windings.
IS 2068-1962:	Standard dimensions of cotton covered rectangular copper conductor for transformer windings.
IS 3454-1966:	Standard dimensions of paper covered round conductors used for transformer windings.
IS 450-1964:	Standard dimensions of cotton covered round conductors used for transformer windings.

1.1.5 Major Considerations in Electrical Machine Design   *(AU, Apr' 17, 8 M)*

The major considerations in electrical machine design are the following,

1. Cost
2. Durability
3. Performance as per specifications.

Most of the design aspects will be oriented towards reducing the cost of a machine but sometimes the low cost and durability cannot be achieved easily. Generally, a compromise is made between low cost and durability. The cost of a machine can be greatly reduced by choosing low cost materials without compromising performance specifications.

In rotating electrical machines, the volume of active part (and hence size of machine) is inversely proportional to speed for a specified output. Therefore, the machine can be designed to operate at the highest permissible speed so that the volume of active part will be less which results in low cost. Similarly, in transformers the volume of core is inversely proportional to flux density. Therefore, transformers can be designed to operate at higher flux densities so that the volume of core will be less which results in low cost.

The durability of a machine can be enhanced by choosing good quality materials and adapting latest design techniques and this naturally leads to higher cost. Mostly electrical machines are designed for longer life (typically more than 20 years) and continuous running.

The components of electrical machines are designed to have satisfactory performance as per specifications. The major components of rotating electrical machines are stator, rotor and windings. The major components of transformers are core and windings. The major components are designed to meet specifications like voltage, current and power rating, power factor, losses and efficiency and temperature rise.

1.2 GENERAL DESIGN PROCEDURE

In general any electrical machine has two windings. The transformer has primary and secondary winding. The DC machine and synchronous machine has armature and field winding. The induction machine has stator and rotor winding. The basic principle of operation of all electrical machine is governed by Faraday's law of induction. Also in every electrical machine the energy is transferred through the magnetic field. Hence a general design procedure can be developed for the design of electrical machines.

The general design procedure is to relate the main dimensions of the machine to its rated power output. An electrical machine is designed to deliver a certain amount of power called rated power. The rated power output of a machine is defined as the maximum power that can be delivered by the machine safely. In DC machine the power rating is expressed in kW and in AC machine in kVA. In case of motor the output power is expressed in HP.

In electrical machines the core and winding of the machine are together called **active part** (because the energy conversion takes place only in the active part of the machine). The active part of rotating machine is cylindrical in shape. If L is length and D is diameter of active part then volume of active part is $D^2 L$. A general output equation can be developed for DC machine which relates the power output to volume of active part ($D^2 L$), speed, magnetic and electric loading. Similarly a general output equation can be developed for AC machine which relates kVA rating to volume of active part ($D^2 L$), speed, magnetic and electric loading.

1.2.1 Main Dimensions of Rotating Machines   *(PTU, May'19, 2 M)*

In rotating machines the active part is cylindrical in shape. The volume of the cylinder is given by the product of area of cross section and length. If D is the diameter and L is the length of cylinder, then the volume is given by $\pi D^2 L/4$. Therefore D and L are specified as main dimensions.

In case of DC machine, D represent the diameter of armature and L represent the length of armature. The Fig. 1.1 shows the main dimensions of DC machine.

Here,   D = Diameter of armature

l_g = Length of air-gap

L = Length of armature

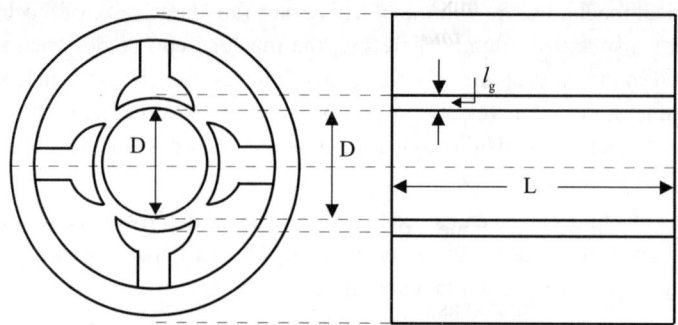

Fig. a: Main dimensions of DC machines.

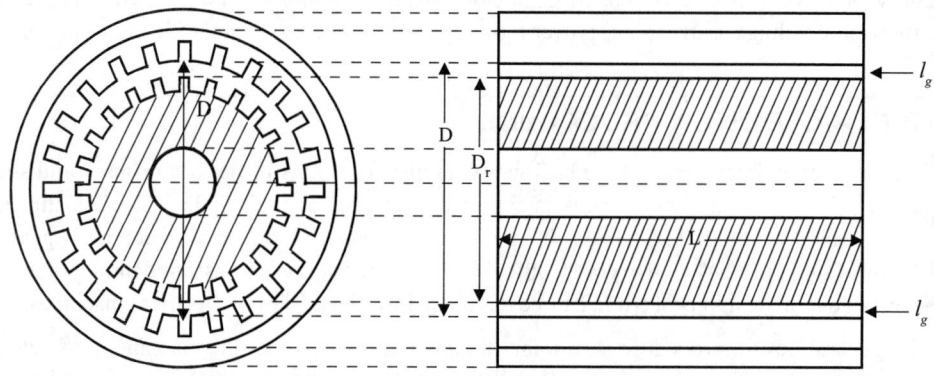

Fig. b: Main dimensions of AC machines.

Fig. 1.1: Main dimensions of rotating machines.

1.3 MAGNETIC AND ELECTRIC LOADINGS

Consider a conductor of length L, carrying a current of I_z amperes. If the conductor is moved in a uniform magnetic field of flux density, B_{av} *Wb/m²* then the work done in moving the conductor through a distance X is given by,

$$\text{Work} = X\, B_{av}\, L\, I_z \qquad\qquad\qquad .....(1.1)$$

When the conductor is moved through a distance X, the conductor cuts through a magnetic flux of $\phi = X\, B_{av} L$ webers.

$$\therefore \text{Work} = \phi\, I_z$$

In rotating machines the conductors are placed in armature. In one revolution of the armature, each conductor moves through a total flux of $p\phi$ webers, where ϕ is flux per pole and p is number of poles. If Z is the total number of armature conductors then the work done in one revolution is given by,

$$\text{Work} = p\, \phi \times I_z\, Z \qquad\qquad\qquad .....(1.2)$$

In equation (1.2) the term $p\phi$ represent the total flux entering and leaving the armature and so it is called **total magnetic loading** (or total flux). The term $I_z Z$ represents the sum of currents in all the conductors on the armature and so it is called **total electric loading** (or total current volume or total ampere conductors on the armature).

$$\therefore \text{ Total magnetic loading } = p\phi \qquad \qquad \qquad \qquad \qquad .....(1.3)$$

$$\text{Total electric loading } = I_z Z \qquad \qquad \qquad \qquad \qquad .....(1.4)$$

Therefore we can say that the work done in one complete revolution is given by the product of total magnetic loading and total electric loading.

The **total magnetic loading** is defined as the total flux around the armature periphery at the air-gap. The **total electric loading** is defined as the total number of ampere conductors around the armature periphery.

1.3.1 Specific Magnetic Loading   *(PU, Nov'19, 8 M)*

Each unit area of armature surface is capable of receiving a certain magnetic flux. Hence the flux per unit area is an important parameter to estimate the intensity of magnetic loading and it is also a criterion to decide the volume of active material. This flux per unit area is expressed as the average value of the flux density at the armature surface or specific magnetic loading (by assuming that the armature is smooth). It is denoted by B_{av}.

The **specific magnetic loading** or average flux density, B_{av} is given by the ratio of flux per pole and area under a pole.

$$\therefore \text{ Specific magnetic loading, } B_{av} = \frac{\text{Flux per pole}}{\text{Area under a pole}} = \frac{\text{Flux per pole}}{\text{Pole pitch} \times \text{Length of armature}}$$

$$= \frac{\phi}{\frac{\pi D}{p} \times L} = \frac{p\phi}{\pi DL} \qquad \qquad \qquad .....(1.5)$$

From equation (1.5) we can say that the **specific magnetic loading** is also given by the ratio of total flux around the air-gap and the area of flux path at the air-gap.

$$\therefore \text{ Specific magnetic loading, } B_{av} = \frac{\text{Total flux around the air-gap}}{\text{Area of flux path at the air-gap}} = \frac{p\phi}{\pi DL} \qquad .....(1.6)$$

The typical values of specific magnetic loading for various types of rotating machines are listed in Table 1.1.

1.3.2 Choice of Specific Magnetic Loading   *(HTU, Dec'18, 10 M)*

The specific magnetic loading is determined by,

1. Maximum flux density in iron parts of machine

2. Magnetizing current

3. Core losses.

Maximum flux density in iron

The maximum flux density in any iron part of machine must be below a certain limiting value. The maximum flux density occurs in the teeth of the armature (or stator core). [Teeth are the portion of the core in between slots].

The flux density in the teeth is directly proportional to specific magnetic loading. Hence the choice of specific magnetic loading should be such that the maximum value of flux density in the teeth is not exceeded. The maximum value of flux density in the teeth is between 1.7 to 2.2 Wb/m^2.

Magnetizing Current

The magnetizing current of a machine is directly proportional to mmf. The mmf is directly proportional to specific magnetic loading. Hence a large value of specific magnetic loading results in increased values of magnetizing mmf and magnetizing current.

The value of magnetizing current is not usually a serious design consideration in dc machines. But in induction motors an increased value of magnetizing current results in low power factor. Hence specific magnetic loading in induction motors is lower than in DC machines. For synchronous machines the magnetizing current is not so critical and the value of specific magnetic loading is intermediate between that of DC and induction machines.

Core loss

The core loss in any part of the magnetic circuit is directly proportional to flux density for which it is going to be designed. The flux density is directly proportional to the specific magnetic loading. Hence the core loss in a machine varies directly as the specific magnetic loading. Thus a large value of specific magnetic loading results in increased core loss and consequently a decreased efficiency and an increased temperature rise.

With a given specific magnetic loading, the core loss increases as the frequency of flux reversals is increased. This is because the hysteresis loss is directly proportional to the frequency and eddy current loss is proportional to the square of the frequency. It follows that for high speed dc machines, or high frequency AC machines, specific magnetic loadings must be reduced in order to achieve lower iron loss.

1.3.3 Specific Electric Loading

Every section of armature is capable of carrying certain amount of current. Hence ampere-turn per unit section of armature periphery (circumference) is an important parameter to estimate the intensity of electric loading and it is also a criterion to decide the volume of active material. This ampere-turn per unit section of armature periphery is expressed as the specific electric loading. It is denoted by **ac**.

The *specific electric loading* is given by the ratio of total armature ampere conductors and armature periphery (circumference) at air-gap.

$$\therefore \text{ Specific electric loading, } \mathbf{ac} = \frac{\text{Total armature conductors}}{\text{Armature periphery at air-gap}} = \frac{I_z Z}{\pi D} \qquad .....(1.7)$$

The value of specific electric loading for various types of rotating machines are listed in Table 1.1.

Table 1.1: Specific Magnetic and Electric Loadings

Machine	Specific magnetic loading B_{av} in Wb/m²	Specific electric loading ac in amp.cond./m
DC machine	0.4 to 0.8	15000 to 50000
Induction motor	0.3 to 0.6	5000 to 45000
Synchronous machine	0.52 to 0.65	20000 to 40000
Turbo alternator	0.52 to 0.65	50000 to 75000

1.3.4 Choice of Specific Electric Loading

The choice of specific electric loading depends on,

1. Permissible temperature rise
2. Voltage rating of machine
3. Size of machine
4. Current density

Permissible temperature rise

Let, θ = Temperature rise c = Cooling Coefficient

S = Dissipating surface δ = Current density

Q_l = Loss dissipated ρ = Resistivity

Loss dissipated per unit area of armature surface $\left.\right\}$ $q = \dfrac{(\text{Current})^2 \times \text{No.of conductors} \times \text{Resistance}}{\text{Surface area of armature}}$

$$= \frac{I_z^2 \times Z \times \rho\, L/a_z}{\pi D L} = \frac{I_z Z}{\pi D} \times \frac{I_z}{a_z} \times \rho = \mathbf{ac}\,\delta\,\rho \qquad .....(1.8)$$

where, $\delta = I_z / a_z$

Also, $q = Q_l / S$ $\qquad\qquad\qquad .....(1.9)$

The temperature rise, $\theta = \dfrac{Q/c}{S} = qc$ $\qquad .....(1.10)$

From equation (1.10), $q = \theta/c$ $\qquad\qquad .....(1.11)$

On equating the equations (1.8) and (1.11),

$$\mathbf{ac}\,\delta\,\rho = \frac{\theta}{c}$$

$\therefore$ Maximum allowable specific electric loading, $\mathbf{ac} = \dfrac{\theta}{\rho\,\delta\,c}$ $\qquad .....(1.12)$

From equation (1.8) it can be inferred that the heat dissipated per unit area of armature is proportional to specific electric loading.

From equation (1.12) it is clear that allowable specific electric loading is fixed by allowable temperature rise and the cooling coefficient. A high value of **ac** can be used in a machine when a high temperature rise is allowed. The maximum allowable temperature rise of a machine is determined by the type of insulating materials used in it. When better quality insulating materials which can withstand high temperature rises are used in the machines, increased values of specific electric loading can be used. This results in reduction in the size of the machine.

A high value of electric loading may be used if the cooling coefficient of the machine is small. The value of cooling coefficient depends upon the ventilation conditions in the machine. High speed machines will have better ventilation and so higher value of **ac** can be used.

Voltage

Let, w_s = Width of the slot y_s = Slot pitch
      d_s = Depth of the slot δ = Current density
      S_f = Slot space factor

The specific electric loading can be related to the above terms by the equation,

$$ac = d_s \, (w_s/y_s) \, \delta \, S_f \qquad\qquad .....(1.13)$$

From equation (1.13) it is clear that the specific electric loading is directly proportional to slot space factor S_f. In high voltage machines, greater insulation thickness is required and therefore the space factor for these machines is lower. Hence an increase in voltage will, in general, necessitate a reduction in specific electric loading **ac**.

Size of machine

From the equation (1.13) it is clear that **ac** depends on the dimension of the slot. For large machines the depth of the slot will be greater and so higher values of **ac** can be used. Actually if the current density and the slot space factor are assumed constant, then specific electric loading is proportional to the diameter as slot depth usually depends upon the diameter.

Current density

From the equation, $q = ac \, \delta \, \rho$ it is clear that a higher value of specific electric loading can be used in a machine which employs lower current density in its conductors. (because $ac = q/\delta \, \rho$).

Typical values of current density are in the range of 2 to 5 A/mm^2. The temperature rise is usually 40°C (above ambient) for normal applications and cooling coefficient is between 0.02 and 0.035°C $W\text{-}m^2$.

1.4 OUTPUT EQUATION

The output of a machine can be expressed in terms of its main dimensions, specific magnetic and electric loadings and speed. The equation which relates the power output to D, L, B_{av}, **ac** and n of the machine is known as output equation.

Output equation and Output coefficient of DC machine

The following equations are used to derive the output equation.

$$\text{Induced emf in armature, } E = \frac{\phi ZN}{60} \, \frac{p}{a} \quad \Rightarrow \quad E = \frac{\phi Znp}{a} \qquad \boxed{n = \frac{N}{60}} \; .....(1.14)$$

$$\text{Current through each conductor, } I_z = \frac{I_a}{a} \quad \Rightarrow \quad I_a = a \, I_z \qquad\qquad\qquad .....(1.15)$$

Specific magnetic loading, $B_{av} = \dfrac{p\phi}{\pi DL}$ $\Rightarrow$ $p\phi = \pi DL\, B_{av}$ .....(1.16)

Specific electric loading, $ac = \dfrac{I_z Z}{\pi D}$ $\Rightarrow$ $I_z Z = \pi D\, ac$ .....(1.17)

where, n = Speed in rps   Z = Number of armature conductors

I_a = Armature current   D = Diameter of rotor

p = Number of poles   L = Length of rotor

a = Number of parallel paths ϕ = Flux per pole

In DC generator the electrical power generated in the armature is given by the product of induced emf and armature current. In case of DC motor the mechanical equivalent of electrical power in armature is given by the product of induced emf (back emf) and armature current.

$\therefore$ Power developed in armature, $P_a = E\,I_a \times 10^{-3}$ in kW .....(1.18)

$$= \frac{\phi Znp}{a} \times aI_z \times 10^{-3}$$

> Using equations (1.14) and (1.15)

$$= p\phi \times I_z Z \times n \times 10^{-3}$$

$$= \pi DL\, B_{av} \times \pi D\, ac \times n \times 10^{-3}$$

> Using equations (1.16) and (1.17)

$$= \pi^2 B_{av}\, ac \times 10^{-3} \times D^2\, L\, n$$

$$= C_0\, D^2\, L\, n$$ .....(1.19)

where, $C_0 = \pi^2 B_{av}\, ac \times 10^{-3}$ .....(1.20)

The equation, $P_a = C_0\, D^2\, Ln$ is called output equation of DC machine and the term C_0 is called *output coefficient* of DC machine.

The output coefficient, C_0 in terms of maximum gap density, B_g is given by,

$C_0 = \pi^2\, \psi B_g\, ac \times 10^{-3}$ .....(1.21)

where, $B_g = \dfrac{B_{av}}{\psi}$ .....(1.22)

$\psi = \dfrac{b}{\tau}$ = Ratio of pole arc to pole pitch .....(1.23)

The term $D^2 L$ in the output equation is proportional to volume of active part. Therefore if C_0 is constant then we can say the power output is directly proportional to the product of volume of active part and speed.

i.e., P_a α Volume of active part × Speed

If C_0 is varied then power output is directly proportional to the four quantities and they are B_{av}, ac, volume of active part and speed.

i.e., P_a α $B_{av} \times ac \times$ Volume of active part × Speed.

Power developed by the armature, P_a is different from the rated power output P, of the machine. The relationship between the two are,

$$P_a = \frac{P}{\eta} \text{ - For DC generator} \quad\quad .....(1.24)$$

$$P_a = P \text{ - For DC motors} \quad\quad .....(1.25)$$

where, η = Efficiency of DC generator

Alternative expression for power developed in armature

Consider the equation (1.20), $P_a = p\phi \times I_z Z \times n \times 10^{-3}$.

Here $p\phi$ = Total magnetic loading

$I_z Z$ = Total electric loading

n = Speed in rps

Hence power developed in the armature can be expressed as shown in equation (1.26)

$$\left.\begin{array}{c}\text{Power developed in}\\ \text{armature in KW}\end{array}\right\} = \begin{pmatrix}\text{Total}\\ \text{Magnetic}\\ \text{loading}\end{pmatrix} \times \begin{pmatrix}\text{Total}\\ \text{electric}\\ \text{loading}\end{pmatrix} \times \begin{pmatrix}\text{Speed}\\ \text{in}\\ \text{rps}\end{pmatrix} \times 10^{-3} \quad .....(1.26)$$

In DC generator the power developed in armature has to supply for copper losses and load. Therefore the power developed in the armature of DC generator can also be expresseds shown in equation (1.27).

$$\left.\begin{array}{c}\text{Power developed by}\\ \text{armature of a DC}\\ \text{generator}\end{array}\right\} = \begin{pmatrix}\text{Output}\\ \text{power}\end{pmatrix} + \begin{pmatrix}\text{Armature}\\ \text{copper loss}\end{pmatrix} + \begin{pmatrix}\text{Field}\\ \text{copper loss}\end{pmatrix} \quad .....(1.27)$$

If we consider the input mechanical power from the prime mover then the electrical power developed in the armature of DC generator can be expressed as shown in equation (1.28).

$$\left.\begin{array}{c}\text{Power developed by}\\ \text{armature of a DC}\\ \text{generator}\end{array}\right\} = \begin{pmatrix}\text{Input}\\ \text{power}\end{pmatrix} - \begin{pmatrix}\text{Friction, Windage and}\\ \text{Iron losses}\end{pmatrix} \quad .....(1.28)$$

If P = output, η = efficiency then, input power = P/η

In DC motor the power developed in armature has to supply for constant losses and the remaining power is delivered to load.

$$\left.\begin{array}{c}\text{Power developed by}\\ \text{armature of a DC motor}\end{array}\right\} = \begin{pmatrix}\text{Output}\\ \text{power}\end{pmatrix} + \begin{pmatrix}\text{Friction, Windage and}\\ \text{Iron losses}\end{pmatrix} \quad .....(1.29)$$

In case of large machines the friction, windage and iron losses can be neglected.

$\therefore P_a = P/\eta$ for generators and $P_a = P$ for motors

In case of small machines the friction, windage and iron losses can be taken as one-third of total losses.

$$\text{Total losses} = \begin{pmatrix}\text{Input}\\ \text{power}\end{pmatrix} - \begin{pmatrix}\text{Output}\\ \text{power}\end{pmatrix} = \frac{P}{\eta} - P = P\left(\frac{1-\eta}{\eta}\right) \quad .....(1.30)$$

Power developed by armature of a DC motor (small motor) $= \left(\dfrac{\text{Output}}{\text{power}}\right) + \left(\dfrac{\text{Friction, Windage and}}{\text{Iron losses}}\right)$

$$= P + \frac{1}{3} P\left(\frac{1-\eta}{\eta}\right) = P\left[\frac{3\eta + 1 - \eta}{3\eta}\right] = P\left(\frac{1+2\eta}{3\eta}\right) \qquad .....(1.31)$$

Power developed by armature of a DC generator (small generator) $= \left(\dfrac{\text{Input}}{\text{power}}\right) - \left(\dfrac{\text{Friction, Windage and}}{\text{Iron losses}}\right)$

$$= \frac{P}{\eta} - \frac{1}{3} P\left(\frac{1-\eta}{\eta}\right) = P\left[\frac{3 - 1 + \eta}{3\eta}\right] = P\left(\frac{2+\eta}{3\eta}\right) \qquad .....(1.32)$$

Output equation and output coefficient of AC machine

The equations of induced emf, frequency, current through each conductor and total number of armature conductors of an AC machine are given below. These equations are obtained from the knowledge of machine theory.

Induced emf per phase, $E_{ph} = 4.44\, f\, \phi\, T_{ph}\, K_{ws}$ .....(1.33)

The frequency of induced emf, $f = \dfrac{p n_s}{2}$ .....(1.34)

Current through each conductor, $I_z = \dfrac{I_{ph}}{a}$ .....(1.35)

where, I_{ph} = Current per phase

and a = Number of parallel circuits or paths per phase.

Total number of armature conductors, Z = Number of phases $\times 2\, T_{ph}$

$$= 3 \times 2\, T_{ph} = 6\, T_{ph} \qquad .....(1.36)$$

Consider a 3-phase machine having one circuit (one parallel path) per phase. The volt-ampere rating of one phase is given by the product of voltage per phase and current per phase. Hence the kVA rating of 3-phase AC machine can be written as shown in equation (1.37).

kVA rating of 3-phase machine, $Q = 3\, E_{ph}\, I_{ph} \times 10^{-3}$ .....(1.37)

On substituting for E_{ph}, I_{ph} and f from equations (1.33), (1.35) and (1.34) respectively in equation (1.37) we get,

$$Q = 3 \times 4.44\, f\, \phi\, T_{ph}\, K_{ws} \times I_z \times 10^{-3} = 3 \times 4.44 \times \frac{p n_s}{2} \times \phi\, T_{ph}\, K_{ws}\, I_z \times 10^{-3}$$

$$= 6.66\, p\, n_s\, \phi\, T_{ph}\, K_{ws}\, I_z \times 10^{-3} \qquad .....(1.38)$$

$$= 1.11 \times p\phi \times I_z\, 6\, T_{ph} \times n_s \times K_{ws} \times 10^{-3}$$

On substituting for $6T_{ph}$ from equation (1.36) in equation (1.38) we get,

$$Q = 1.11 \times p\phi \times I_z Z \times n_s \times K_{ws} \times 10^{-3} \qquad .....(1.39)$$

On substituting for $p\phi$ and $I_z Z$ from equations (1.16) and (1.17) in equation (1.40) we get,

$$Q = 1.11 \times \pi\, D\, \textbf{ac} \times n_s \times K_{ws} \times 10^{-3}$$

$$= 1.11\, \pi^2\, B_{av}\, \textbf{ac}\, K_{ws} \times 10^{-3} \times L\, n_s$$

$$= 11\, B_{av}\, \textbf{ac}\, K_{ws} \times 10^{-3} \times D^2\, L\, n_s$$

$$= C_o\, D^2\, L\, n_s \qquad\qquad\qquad\qquad\qquad .....(1.40)$$

where, $C_o = 11\, B_{av}\, \textbf{ac}\, K_{ws} \times 10^{-3}$ $\qquad\qquad\qquad\qquad\qquad$ .....(1.41)

The equation, $Q = C_o\, D^2\, L\, n_s$ is called **output equation** and C_o is called **output coefficient**. The term $D^2\, L$ in the output equation is proportional to volume of active part. Therefore, if C_o is constant then we can say that the kVA rating is directly proportional to the product of volume of active part and speed.

i.e. $Q\ \alpha$ Volume of active part $\times$ Speed

If C_o is varied then power output is directly proportional to the four quantities : B_{av}, **ac**, volume of active part and speed.

i.e. $Q\ \alpha\ B_{av} \times \textbf{ac} \times$ Volume of active part $\times$ Speed

1.5 SEPARATION OF D AND L

From equations (1.19) and (1.40) we can say that the output of a rotating electrical machine is directly proportional to the term $D^2\, L$. The active part of the rotating electrical machine is cylindrical in shape and the volume of cylinder is $\pi D^2\, L$. Hence the term $D^2\, L$ is related to volume of active part. The seperation of D and L refers to the selection of an appropriate values for D and L for a given volume of active part.

For a given volume of active part there are various choice of D and L. The choice of D and L and the ratio between them depends on a number of factors in various types of rotating machines. In the following sections a brief discussion about various factors that decides the choice of D and L in various types of rotating machines are presented. In general a ratio of L/τ or L/D is assumed where τ = pole pitch = $\pi D/p$. Using the output equation and from the knowledge of kW or kVA rating, specific loadings and speed, the value of $D^2\, L$ is estimated. Then by solving the two equations (i.e., the equation of L/τ or L/D and $D^2\, L$) the values of D and L are estimated.

1.5.1 Separation of D and L for DC Machines

In DC machines the separation of D and L depends on,

1. Pole proportions
2. Peripheral speed
3. Moment of inertia
4. Voltage between adjacent commutator segments

Pole proportions

The dimensions of the machine are decided by the square pole criterion. This states that for a given flux and cross-section area of pole, the length of mean turn of field winding is minimum, when the periphery forms a square. This means that the length, L must be approximately equal to pole arc or $L = b = \psi\, \tau$. The value of Ψ is usually between 0.64 to 0.72. ($\therefore$ The ratio $L/\tau = 0.64$ to 0.72). However in practice L is slightly greater than pole arc, b and L/τ is usually between 0.7 to 0.9. For square pole criterion choose L/τ as 0.7.

b_p = Width of pole body

b = Pole arc

L = Core length

τ = Pole pitch = $\pi D/p$

Ψ = b/τ = Ratio of pole arc to Pole pitch

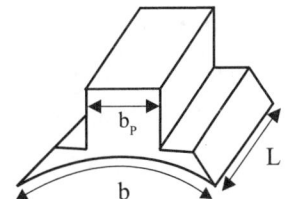

Peripheral speed

Fig. 1.2: Pole Dimensions.

The peripheral speed of armature (V_a) is sometimes a limiting factor to the value of diameter. The peripheral speed should not exceed about 30 *m/s*.

Maximum peripheral of armature, V_{am} = 30 *m/s*

Moment of inertia

For machines used in control systems, a small moment of inertia is desirable. For low moment of inertia the diameter should be made as small as possible. Conversely a high inertia machine may be required for impact load applications and such machines are designed with larger diameter.

Voltage between adjacent commutator segments

The maximum core length is fixed by the maximum voltage that can be allowed between adjacent segments. It can be shown that the maximum voltage between adjacent commutator segments in DC machines is given by equation (1.42) (using equations (3.47) and (3.50) of section 3.6.5).

Maximum voltage between adjacent segments, $E_{cm} = 2\,B_{gm}\,L\,V_a\,T_c$ .....(1.42)

where, B_{gm} = Maximum air-gap flux density under load conditions.

T_c = Turns per coil

The limiting values of B_{gm}, V_a, E_{cm} are B_{gm} = 1.2 *Wb/m²*, V_{am} = 30 *m/s*, E_{cm} = 30 *V*. With these limiting values, for T_c = 1, $L \approx 0.4$ *m*. This is only an indication of limiting value of core length, but it must be clear that large DC machines should have large diameters rather than large core lengths.

1.5.2 Separation of D and L for Induction Motors

The operating characteristics of an induction motor are mainly influenced by the ratio L/τ. The ratio of L/τ for various design features are listed below.

For minimum cost, L/τ = 1.5 to 2

For good power factor, L/τ = 1.0 to 1.25

For good efficiency, L/τ = 1.5

For good overall design, L/τ = 1.0

Generally L/τ lies between 0.6 to 2. It can be shown that for best power factor the pole τ pitch is given by the equation, $\tau = \sqrt{0.18\,L}$.

1.5.3 Separation of D and L for Synchronous Machines

In synchronous machines the separation of D and L depends on,

1. Pole proportions
2. Peripheral speed
3. Number of poles
4. Short circuit ratio.

Pole proportions

In salient pole synchronous machines the choice of diameter (D) depends on the type of pole and the permissible peripheral speed. The two-types of poles used in salient pole machines are round poles and rectangular poles.

When round poles are used the ratio of L/τ is between 0.6 to 0.7, where τ = pole pitch = $\pi D/p$. When rectangular poles are employed the ratio of L/τ is between 1 to 5.

Peripheral speed

For large high speed machines D is fixed by the limiting peripheral speed. The output equation for synchronous machine can be expressed in terms of peripheral speed (V_a) as shown below.

The output equation is, $Q = C_o \, D^2 \, L \, n_s$

$$\text{where, } C_o = 11 \, B_{av} \, ac \, K_{ws} \times 10^{-3}$$

$$\text{But } V_a = \pi \, D \, n_s ; \qquad \therefore \ D = \frac{V_a}{\pi \, n_s} \left[\begin{array}{c} \text{Here length of air - gap } (l_g) \\ \text{is neglected.} \end{array} \right]$$

On substituting the expression for C_o and D in the output equation,

$$Q = 11 \, B_{av} \, ac \, K_{ws} \times 10^{-3} \left(\frac{V_a}{\pi \, n_s} \right)^2 L n_s$$

$$Q = 1.11 \, B_{av} \, ac \, K_{ws} \times 10^{-3} \, \frac{V_a^2 \, L}{n_s} \qquad\qquad\qquad .....(1.43)$$

From equation (1.43) we can say that an increase in machine rating will necessitate an increase in V_a which in turn is achieved by increasing D, until the maximum permissible peripheral speed is reached. Once this happens, the value of D cannot be increased further and the only way to get increased output is to increase the length L. In general the value of D is calculated using the limiting value of peripheral speed V_a for cylindrical rotor synchronous machines. Then using the values of $D^2 L$ and D, the value of L is estimated.

Number of poles

The small diameter and large number of poles results in the small pole pitch and so less space for field coils. Hence a large diameter is advisable for machines having large number of poles. The empirical relationship used to decide the number of poles is

$$\tau/L = 0.5 + (6/p) \qquad\qquad\qquad\qquad .....(1.44)$$

Short circuit ratio

A major factor influencing the design of synchronous machines is their short circuit ratio (SCR).

$$SCR = \frac{\text{Field current required to produce rated voltage on open circuit}}{\text{Field current required to produce rated current at short circuit}}$$

Higher values of SCR results in higher stability limit and a low value of inherent regulation. For high values of SCR, the length of core (L) should be large.

1.6 FACTORS AFFECTING THE SIZE OF ROTATING MACHINES

The factors affecting the size of rotating machines are speed and output coefficient. The output coefficient in turn depends on specific electric and magnetic loadings.

Speed

The power developed in armature of a rotating machine is directly proportional to speed. Also the speed is inversely proportional to volume of active parts.

Power developed in armature, $P_a = C_o D^2 L n$

∴ Power, $P_a \quad \alpha \quad n$ (Provided C_o and D^2L is constant)

Speed, $n \quad \alpha \quad \dfrac{P_a}{D^2 L}$ (Provided C_o is constant)

Since D^2L is propotional to volume of active parts, for same volume with increase in speed the output will increase. For a given output, a high speed machine will have less volume and costs less. Therefore for reducing the cost highest possible speed may be selected. The maximum speed is limited by mechanical stresses of the rotating parts.

Output coefficient

The power developed in armature of rotating machine is directly proportional to output coefficient. Also the volume of active parts is inversely proportional to output coefficient.

Power developed in armature, $P_a = C_o D^2 L n_s$

∴ $P_a \quad \alpha \quad C_o$ (Provided D^2L and n are constant)

$C_o \quad \alpha \quad \dfrac{1}{D^2 L}$ (Provided P_a and n are constant)

Hence higher value of C_o results in higher output. With high C_o, the volume of active parts decreases and the machine costs less.

The output coefficient, C_o depends on specific loadings, B_{av} and **ac**. Hence for higher C_o, higher specific loadings are chosen.

$C_o = \pi^2 B_{av} \, \textbf{ac} \times 10^{-3}$

∴ $C_o \propto B_{av} \, \textbf{ac}$

The high values of specific loadings may affect some of the performance characteristics of the machines like temperature rise, efficiency, commutation conditions, etc. So, specific loadings are chosen such that they give best performance and minimum cost.

1.7 VARIATION OF OUTPUT AND LOSSES WITH LINEAR DIMENSIONS

Consider two machines of the same type with all their linear dimensions in the ratio x :1 and having the same speed, flux density and current density. Let the machine with linear dimensions x times be called A and the other machine B.

Output

$$\text{Output} = C_0 \, D^2 \, Ln$$

$$= \pi^2 \, B_{av} \, ac \, D^2 \, Ln \times 10^{-3}$$

Using equations (1.19) and (1.20)

Let, B_{av} and n are constants,

$\therefore$ Output $\propto$ **ac** D^2 L

$$ac = \frac{I_z Z}{\pi D} = \frac{\delta a_z Z}{\pi D}$$

$$\boxed{\delta = \frac{I_z}{a_z}}$$

Here, $D^2 \propto x^2$ and $L \propto x$

Since δ, Z and π are constants,

Since, Output $\propto$ **ac** D^2 L,

$$ac \propto \frac{a_z}{D}$$

Output $\propto x \times x^2 \times x$

Here, $a_z \propto x^2$ and $D \propto x$

$\therefore$ **Output** $\propto x^4$

$$\therefore \textbf{ac} \propto \frac{x^2}{x}$$

$$\therefore \textbf{ac} \propto x$$

Therefore the output of machine A is x^4 times the output of machine B.

Losses

I^2 R loss = Number of conductors × Copper loss in each conductor

$$= Z \, I_z^2 \left(\rho \, \frac{L}{a_z} \right)$$

$$\frac{\rho l}{a} = \text{Resistance}$$

$$\therefore I^2 R \text{ loss} = Z \, (\delta a_z)^2 \, \frac{\rho L}{a_z} = \delta^2 \, \rho \times (Z \, a_z \, L)$$

$$\boxed{\delta = \frac{I_z}{a_z}}$$

$$= \delta^2 \, \rho \times \text{Volume of active portion of conductors}$$

$\therefore I^2 R$ loss α Volume of active portion of conductors

$\therefore I^2 R$ loss α x^3

Thus the copper loss of machine A is x^3 times those of B.

Total iron loss = Loss per unit volume × Volume of iron

$\therefore$ Total iron loss $\propto x^3$

Both I^2 R loss and iron losses vary as the third power of linear dimensions. **Hence total losses of machine A is x^3 times those of B.**

Efficiency

The efficiency increases with increase in x.

$$\text{Efficiency, } \eta = \frac{\text{Output}}{\text{Output} + \text{Losses}}$$

$$\therefore \eta \propto \frac{x^4}{x^4 + k\,x^3} = \frac{1}{1 + k/x}$$

With increase in x (i.e., as the size of machine increases) the term k/x becomes smaller and smaller. Hence the efficiency increases, with increase in the linear dimensions of the machine.

1.8 LIMITATIONS IN DESIGN
(VTU, Dec' 19, 6 M)

The specifications imposed by consumers and standards organisation are major limitations in design. Some of the specifications to be met are allowable temperature rise, losses, efficiency, power factor, voltage rating, torque requirement, etc.

The following factors impose limitations on design of electrical machines.

1. Saturation
2. Temperature rise
3. Stress on insulation
4. Efficiency
5. Stress and strain on rotating parts and bearings
6. Mechanical precision of air-gap
7. Commutation
8. Power factor
9. Specifications.

For maximum utilization of active material the specific magnetic and electric loading can be kept as high as possible. The value of specific magnetic loading is limited by the saturation of magnetic materials used in the machine.

The value of specific electric loadings is limited by the allowable temperature rise in the machine, which in turn depends on the insulating material. If an insulating material is subjected to temperatures above its limit, then its life is drastically reduced.

The heat developed in the electrical machines (due to losses) impose thermal stress on the insulating materials. The operating voltage impose electrical stress on the insulating materials. The short circuit currents that may flow in the windings create mechanical stresses on the insulating materials. The type of insulating material is decided by the operating temperature and the size (dimension) of insulation is decided by the electrical and mechanical stresses.

The capital and running cost of machine depends on efficiency. If the losses in the machine are low then running cost will be less. In order to keep the electric and magnetic losses to a low value, the specific electric and magnetic loadings should be as low as possible. This requires larger volume of active material (iron and copper/aluminium) which results in higher capital cost. Hence, for a machine with high value of efficiency the capital cost will be higher and running cost will be lesser.

The dimension of rotor and the central shaft are limited by the mechanical stress and strain on them. In general the rotor should be stiff and there should not be any significant deflection due to strain. In high speed machines the rotor slot dimensions are selected such that the mechanical stresses at the bottom of rotor teeth do not exceed the limit.

The type of bearings to be used in rotating machines are decided by the rotor weight, external loads, forces due to unbalanced rotors and unbalanced magnetic pull.

In induction motors, the length of air-gap is kept as small possible for high power factor. The length of air-gap is mainly limited by the precision (accuracy) of mechanical fabrication technology. By employing modern CNC machines and dynamic balancing of rotors smaller air-gaps can be achieved.

The commutation conditions in DC machines limit the maximum power output that can be delivered by the machine. The maximum power output that can be obtained from a single DC machine is 10 MW.

In general the power factor should be high in order to reduce the current level for the same power. For high value of power factor, the specific magnetic loading should be less and the length of air-gap should be as small as possible. These requirements will increase the cost of the machine. (Because when magnetic loading is reduced the volume of active material has to be increased and for small values of air-gap the fabrication cost will be high).

1.9 MODERN MACHINES MANUFACTURING TECHNIQUES

The modern electrical machines are characterized by a very wide range of power outputs. The power range varies from a fraction of a watt to several hundreds of megawatts. The range of rotational speeds of electrical machines is also very wide. The range of speed may vary from few revolutions per second to several thousand revolutions per second. The large varied fields of applications and wide range of both power output and speed of operation of electrical machines has led to a variety of types of construction. The type of construction to be adopted in mostly influenced by the operating speed of the machine.

Some of the modern trend in electrical machine manufacturing techniques are given below:

1. Low-speed machines (below 250 rpm) are built with large diameter and small axial length and high-speed machines (3000 rpm and above) with small diameter and a long core length.

2. The size machine is designed as small in size as possible. This leads to use of lesser material with same efficiency and overload capacity. The increase in power ratings using smaller size coupled with good overall performance has been possible due to technology advancement in loss reduction techniques and design of better cooling systems.

3. Use of magnetic materials having permeability, a low iron loss and a high mechanical strength. These characteristics permit use of the high values of flux density and therefore result in reduction in the size of the machine and help to achieve higher power output.

4. The modern trend is to use improved insulating materials and high quality newer insulating materials that can withstand much higher temperatures. This leads to design small size machines with higher power output.

5. Modern machine manufacturing techniques are equipped with use of higher electro-magnetic loadings for active parts and increased mechanical loadings for construction materials.

6. The modern trend is to analyze and improve the design of individual parts for better performance and cost reduction.

7. Also, modern design trend takes care of satisfactory operation under the desired environmental conditions.

1.10 MODERN TRENDS IN DESIGN OF ELECTRIC MACHINES   *(JNTKU, Apr'19, 6 M)*

The modern trend in electrical machine design is considering all types of electrical machine as electromechanical energy network and use optimization technique to search best design for a specified objective like low cost, small size, highly efficient, etc.

Another, modern trend is to design a series of machines having different ratings to fit into a single frame size. In this case, the finished designs of machines are in groups, in which all designs within a group are interdependent. This again is an optimization problem because frame sizes have to be optimum giving due weightage to designs of all the machines in a series or a group.

The design of electrical machines by optimization techniques is an iterative process in which the assumed data may have to be varied a number of times to arrive at the desired design. Therefore the final design data to meet the specified optimum criteria is a matter of long and tedious iterations, This fact has led to the application of fast digital computers to the design of electrical machines.

The digital computer has completely revolutionized the field of electrical machine design. The computer aided design has the advantages of eliminating tedious and time-consuming hand calculations thereby allowing the designer to concentrate on physical and logical ideas.

Also, the use of computer makes possible more trial designs and enables sophisticated calculations to be made with in tolerable limits in lesser time. When computers are used it is easy to check data at every design stage, easy to handles non-linearities and incorporates the designer's ideas. Computer aided design permits implementation of more detailed and precise functional relationships which give rise to possibility of new and comprehensive design procedures.

1.11 COMPUTER AIDED DESIGN OF ELECTRICAL MACHINES   *(KTU, Feb'18, 4 M)*

(AKTU, Dec'19, 5 M)

The various computer aided design techniques available for electrical machine design can be broadly classified into Analysis method and Synthesis method. In the analysis method the design will be finalized by human and computer interactions, whereas in synthesis method the design is finalized by computer itself by employing optimization techniques to fit the design for a specified performance.

1.11.1 Analysis Method of Design of Electrical Machines   *(UTU, Dec'13, 5 M)*

The various steps in analysis method of electrical machine design are shown as a flowchart in Fig. 1.3. In analysis method the human or a designer select the choice of dimensions, materials and types of construction to be employed and input these details as input data to a computer program. The computer does all the design calculations and then using the design data the computer estimate the performance of the machine. The design data and estimated performance of the machine are given as output to the designer. The designer examines the performance and if the performance is not satisfactory then the designer modifies the input design parameters and make another run of the computer program with modified data and this process is repeated until a satisfactory design is achieved. The design is finalized by the designer.

The analysis method of design is good for a beginner using digital computers for the design of electrical machines and in fact most of the designer engineers initially start machine design by analysis method.

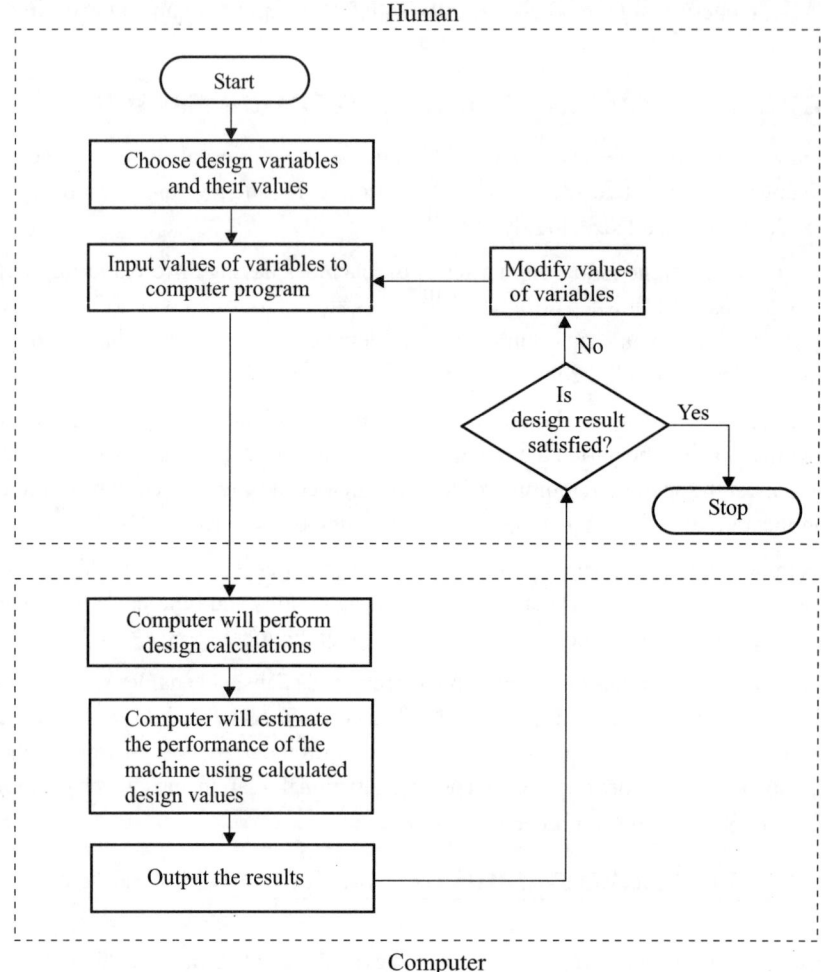

Fig. 1.3: *Analysis method of computer aided design of electrical machines.*

The advantages of analysis method are,

1. It is easy and straight forward to program the design equations.

2. It results in considerable time saving in performing calculations.

3. Analysis method programs are simple and they become the foundations for later larger and sophisticated programs.

4. The interaction of designer and computer yields highly acceptable results.

1.11.2 Synthesis Method of Design of Electrical Machines

(AKTU, Dec'19, 7 M)

(UTU, Dec'13, 5 M)

The various steps in synthesis method of electrical machine design are shown as a flowchart in Fig. 1.4. The major difference between analysis and synthesis method is that the desired performance is also given as input along with design variables and data to the computer. The logical decisions required to modify the variables to achieve the desired performance are implemented in the program as a set of instructions or an algorithm. Therefore, human or designer interaction is not necessary in synthesis method and the computer itself will decide the final design. Most of the synthesis method employ an optimization algorithm to achieve a best design. Sometimes, the design is carried by different optimization algorithms and the results of various optimization techniques are compared to choose better design.

Synthesis method of design is designing a machine that satisfies a set of specifications or performance indices. A large number of solutions are possible for a given set of specifications and it will be difficult to choose a particular design. Therefore, the synthesis design includes optimum design techniques with an objective for good design. It will be very difficult to arrive at final design in synthesis method without optimization technique.

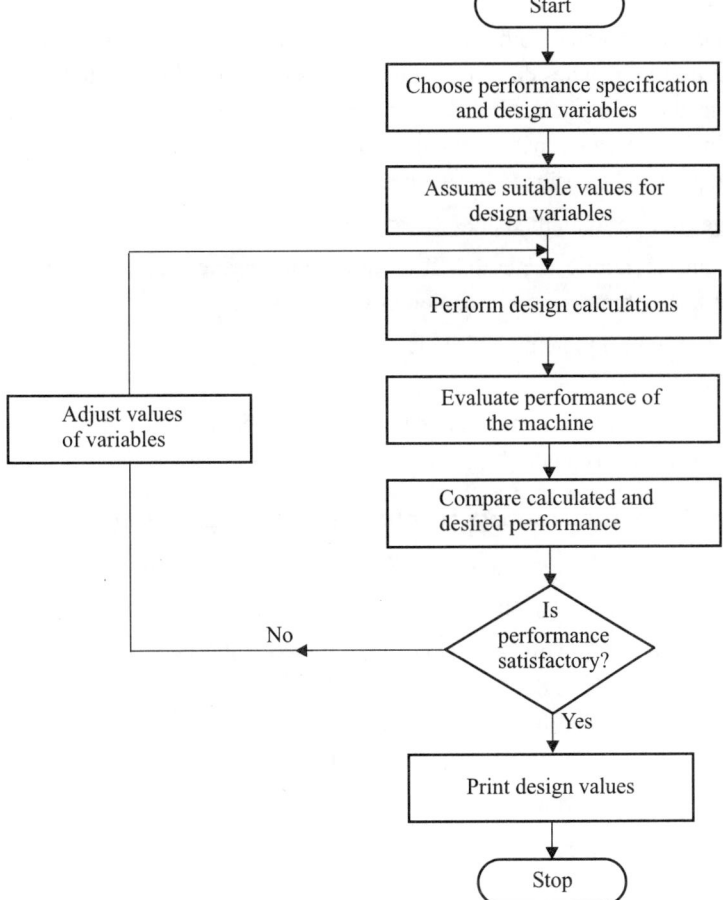

Fig. 1.4: *Synthesis method of computer aided design of electrical machine.*

The advantages of synthesis method of design is the savings in time and in engineering man hours due to the decision making left to the computer.

The disadvantages in synthesis method of design are,

1.   The synthesis method involves too much of logic and before incorporating these logical decisions in the program they have to be accepted by team of design engineers. In a team, different members may give different suggestions to produce an optimum design and it becomes really hard to formulate an unique logical decisions.

2.   The synthesis method programs are too complex which lead to high software development cost and require computers with high configurations involving huge expenditure.

3.   The synthesis program developed at high cost needs periodic updating to incorporate latest development in materials, manufacturing techniques, performance standards, costs of materials and market conditions. Also, require changes in the design logic itself. Therefore, a synthesis program requires both additional effort and cost.

1.11.3 Hybrid Method of Design of Electrical Machines

Hybrid method incorporates both the analysis as well as the synthesis method in the program. Normally a machine design includes design of number parts of the machine. In hybrid method of design some parts are designed by analysis method and other parts are designed by synthesis method. Since the synthesis methods involve greater cost, a classification is needed to decide the parts to be designed by analysis method and parts to be designed by synthesis method.

The advantages of hybrid method are,

1.   The results of hybrid method will be more appropriate, due to combination of designer involvement in finalizing design and optimization of vital parts.

2.   The cost of hybrid systems will be lesser than synthesis systems, due to design of non vital parts by analysis method.

3.   Since the design of some parts are carried by analysis method, the time required for design will also be lesser than synthesis method.

1.11.4 Optimization Techniques in Electrical Machine Design   (UTU, Dec'13, 10 M)

The aim of optimization in the design of electrical machine is to achieve a best design solution for a given set of specifications by satisfying the performance requirements and constraints.

The optimization process, involves possible range of values of varies variables and a logic to change the values of variables in order to satisfy all limitations or restraints imposed on the performance of the machine. Therefore, optimization is a process of finding a best solution for a given set of conditions.

In order to achieve the best possible solution, it is necessary to define the objective of optimization process. The objective may vary from one problem to another and in general the objective may be either economical or technical. In electrical machine design, some of the objectives are low cost, small size, high efficiency, etc.

The design problem of electrical machines has conflicting conditions or constraints. For example, the cost of active materials in induction motors can be reduced by using high values of specific electric and magnetic loadings but these high values of specific loadings will result in unsatisfactory performance like high temperature rise and poor power factor. The choice of low values of specific loading has the undesirable effect of high cost and better performance. Therefore, the cost and performance are conflicting in nature. The best design can be obtained by making a compromise between cost and performance.

1.11.5 General Procedure for Optimization in Design of Electrical Machines

1.   Choose a design problem and define an objective for the design problem under consideration. For example, design of small DC motor with high efficiency. In this example design of small DC motor is design problem and high efficiency is objective.

2.   Determine the restrictions or constraints in design the problem. For example, limitation in speed, maximum values of main dimensions, etc.

3.   Identify the physical system of the design problem and examine the structure of the system and the inter-relationship of the system elements. In design of small DC motor, physical system is DC machine and components of DC machine are system elements.

5.   Construct a mathematical model of the system using design equations and variables.

6.   Determine the range of system variables. Also, define the various restrictions imposed on the system variables.

7.   Choose an optimization algorithm and develop software codes in a programming language to perform design calculations and estimate the performance.

8.   The optimization involves a number of iterations or repetitions of design calculations and performance estimation to achieve the objective.

9.   In each iteration the estimated performance is compared with specified objective and difference is estimated. In next iteration the variables are modified and algorithm is applied to get next best solution. This process is repeated until we get a solution very close to specified and estimated performance.

1.12 SUMMARY OF DESIGN EQUATIONS

1.   Total magnetic loading $= p\phi$

2.   Total electric loading $= I_z Z$

3.   Specific magnetic loading, $B_{av} = \dfrac{p\phi}{\pi DL}$

4.   Specific electric loading, $\mathbf{ac} = \dfrac{I_z Z}{\pi D}$

5.   Power developed in the armature of DC machine, $P_a = C_o D^2 L\, n$

6.   Output coefficient of DC machine, $C_o = \pi^2 B_{av}\, \mathbf{ac} \times 10^{-3}$

7.   kVA rating of 3-phase AC machine, $Q = C_o D^2 L\, n_s$

8.   Output coefficient of AC machine, $C_o = 11 B_{av}\, \mathbf{ac}\, K_{ws} \times 10^{-3}$

9.   Pole pitch, $\tau = \pi D/p$

10.   Choice of L/τ for DC machines
      i)   In general, $L/\tau = 0.7$ to 0.9
      ii)   For square pole criterion, $L/\tau = 0.7$

11.   Choice of L/τ for induction motors

   i)   For minimum cost, L/τ = 1.5 to 2

   ii)   For good power factor, L/τ = 1.0 to 1.25

   iii)   For good efficiency, L/τ = 1.5

   iv)   For good overall design, L/τ = 1.0

   v)   For best power factor, $\tau = \sqrt{0.18\,L}$

12.   Choice of L/τ for salient pole synchronous machines

   i)   When round poles are employed, L/τ = 0.6 to 0.7

   ii)   When rectangular poles are employed, L/τ = 1 to 5

13.   In cylindrical rotor synchronous machines,

   i)   Diameter of rotor, $D_r = \dfrac{V_a}{\pi n_s}$

   ii)   Stator inner diameter, $D = D_r + 2\,l_g$

   > **Note :** If l_g (length of air-gap) is neglected then $D = D_r$

14.   Maximum allowable specific electric loading, $\mathbf{ac} = \dfrac{\theta}{\rho\,\delta\,c}$

1.13 SOLVED PROBLEMS   *(KTU, Feb' 18, 10 M) (AU, Apr' 18, 8 M)*

EXAMPLE 1.1

   A 350 *kW*, 500 *V*, 450 *rpm*, 6 pole DC generator is built with an armature diameter of 0.87 *m* and core length of 0.32 *m*. The lap wound armature has 660 conductors. Calculate the specific electric and magnetic loadings.

GIVEN DATA

P = 350 *kW*	D = 0.87 *m*	V = 500 *V*	L = 0.32 *m*
n = 450/60 *rps*	Z = 660	p = 6	Lap wound

SOLUTION

Specific electric loading, $\mathbf{ac} = \dfrac{I_z Z}{\pi D}$

Specific magnetic loading, $B_{av} = \dfrac{p\phi}{\pi DL}$

Power output of the generator, P = VI × 10⁻³ in *kW*

Full load current, $I = \dfrac{P}{V \times 10^{-3}} = \dfrac{350}{500 \times 10^{-3}} = 700\,A$

$\left.\begin{array}{c}\text{Current through each}\\ \text{armature conductor}\end{array}\right\}\,I_z = \dfrac{\text{Armature current}}{\text{Number of parallel paths}} = \dfrac{I_a}{a}$

$= \dfrac{700}{6} = 116.67\,A$

Specific electric loading, $\mathbf{ac} = \dfrac{I_z Z}{\pi D} = \dfrac{116.67 \times 660}{\pi \times 0.87} = 28173\,amp.cond./m$

Induced emf in DC generator, $E = \phi Z n \dfrac{P}{a} = \phi Z n$

Flux per pole, $\phi = \dfrac{E}{Zn} \approx \dfrac{V}{Zn} = \dfrac{500}{660 \times 450/60} = 0.101 \, Wb$

Specific magnetic loading, $B_{av} = \dfrac{p\phi}{\pi DL} = \dfrac{6 \times 0.101}{\pi \times 0.87 \times 0.32} = 0.6929 \, Wb/m^2$

RESULT

Specific electric loading,   ac = 28173 $amp.cond./m$

Specific magnetic loading, B_{av} = 0.6929 Wb/m^2.

EXAMPLE 1.2

The output coefficient of 1250 kVA, 300 rpm, synchronous generator is 200 kVA/m^3-rps.

Determine the main dimensions D and L for following three cases.

Case (i):   The ratio of length to diameter is 0.2.

Case (ii):   Specific loadings are decreased by 10% each with speed remaining the same as in case (a).

Case (iii): Speed is decreased to 150 rpm with specific loadings remaining the same as in case (a).

Assume the same ratio of length to diameter in all the three cases. Comment upon the results.

GIVEN DATA

C_o = 200 kVA/m^3-rps Q = 1250 kVA

$\dfrac{L}{D} = 0.2$ N_s = 300 rpm

SOLUTION

Case (i)

Given that, $\dfrac{L}{D} = 0.2$

     $\therefore$ Length of stator core, L = 0.2 D   .....(1)

Synchronous speed, $n_s = \dfrac{300}{60} = 5 \, rps$

kVA rating, $Q = C_o \, D^2 \, L \, n_s$

     $\therefore D^2 L = \dfrac{Q}{C_o \, n_s} = \dfrac{1250}{200 \times 5} = 1.25 \, m^3$

     $\therefore$ D²L = 1.25   &boxed; Using equation (1)

     D² (0.2D) = 1.25

     $\therefore$ Diameter of stator bore, $D = \left(\dfrac{1.25}{0.2}\right)^{1/3} = 1.842 \, m$

                                                         Using equation (1)

     $\therefore$ Length of stator core, L = 0.2 D = 0.2 × 1.842 = 0.368 m

Case (ii)

Let, Output coefficient of machine-II, $C_{o2} = \pi^2 B_{av2} \, ac_2 \times 10^{-3}$

In case (ii) the specific loadings are reduced (decreased) by 10%.

$$\therefore B_{av2} = 0.9 \, B_{av} \text{ and } ac_2 = 0.9 \, ac$$

$$\left. \begin{array}{l} \therefore \text{Output coefficient} \\ \text{of machine} - \text{II} \end{array} \right\} C_{o2} = 11 \, B_{av2} \, ac_2 \times 10^{-3}$$

$$= 11 \times 0.9 \, B_{av} \times 0.9 \, ac \times 10^{-3}$$

$$= 0.9 \times 0.9 \times (11 \, B_{av} \, ac \times 10^{-3}) = 0.9 \times 0.9 \times C_o$$

$$= 0.9 \times 0.9 \times 200 = 162 \, kVA/m^3\text{-}rps$$

kVA rating of machine-II, $Q_2 = C_{o2} \, D_2^2 \, L_2 \, n_s$

$$\therefore D_2^2 \, L_2 = \frac{Q_2}{C_{o2} \, n_s}$$

$$= \frac{1250}{162 \times 5} = 1.543 \, m^3$$

$\boxed{Q_2 = 1250 \ kVA \text{ (same as case (a))}}$

$$\therefore D_2^2 L_2 = 1.543$$

$$D_2^2 \, (0.2 \, D_2) = 1.543$$

$$\therefore \text{Diameter of stator bore, } D_2 = \left(\frac{1.543}{0.2} \right)^{1/3} = 1.976 \, m$$

$\boxed{\begin{array}{c} \text{Using equation (1)} \\ L_2 = 0.2 \, D_2 \end{array}}$

$$\therefore \text{Lenght of stator core, } L_2 = 0.2 \, D_2 = 0.2 \times 1.976 = 0.395 \, m$$

Comment : Here $D_2^2 \, L_2 = 1.543 \, m^3$ but $D^2 L = 1.25 \, m^3$, therefore volume of machine-II is more than machine-I and so the size of the machine increases with decrease in specific loadings.

Case (iii)

In case (iii) the speed is reduced to 150 *rpm*.

$$\therefore \text{Synchronous speed in rps, } n_{s3} = \frac{150}{60} = 2.5 \, rps$$

kVA rating of machine-III, $Q_3 = C_o \, D_3^2 \, L_3 \, n_{s3}$

$$\therefore D_3^2 \, L_3 = \frac{Q_3}{C_o \, n_{s3}} = \frac{1250}{200 \times 2.5} = 2.5 \, m^3$$

$\boxed{\begin{array}{l} Q \text{ and } C_o \text{ same as case (a)} \\ \therefore Q_3 = 1250 \ kVA \text{ and} \\ \therefore C_o = 200 \ kVA/m^3\text{-}rps. \end{array}}$

$$\therefore D_3^2 \, L_3 = 2.5$$

$$D_3^2 (0.2 \, L_3) = 2.5$$

$\boxed{\begin{array}{c} \text{Using equation (1)} \\ L_3 = 0.2 \, D_3 \end{array}}$

$$\therefore \text{Diameter of stator bore, } D_3 = \left(\frac{2.5}{0.2} \right)^{1/3} = 2.32 \, m$$

$$\therefore \text{Length of stator core, } L_3 = 0.2 \, D_3 = 0.2 \times 2.32 = 0.464 \, m$$

Comment : Here $D_3^2 L_3 = 2.5 \, m^3$ but $D^2 L = 1.25 \, m^3$, therefore volume of machine-III is more than machine-I and so the size of machine increases with decrease in operating speed.

RESULT

Case (i)	Case (ii)	Case (iii)
$D = 1.842\ m$	$D_2 = 1.976\ m$	$D_3 = 2.32\ m$
$L = 0.368\ m$	$L_2 = 0.395\ m$	$L_3 = 0.464\ m.$

EXAMPLE 1.3

A 20 HP, 440 V, 4 pole, 50 Hz, 3-phase induction motor is built with a stator bore of 0.25 m and core length of 0.16 m. The specific electric loading is 23000 $amp.\ cond./m$. Find the specific magnetic loading of the machine. Assume full load efficiency of 84% and a power factor of 0.82. Using the data of the above machine determine the main dimensions, number of stator slots and stator conductors for a 15 HP, 460 V, 6 pole, 50 Hz motor. Take $K_{ws} = 0.955$.

GIVEN DATA

Machine-I		Machine-II
20 HP	4 pole	15 HP
440 V	50 Hz	460 V
$D = 0.25\ m$	3-phase	50 Hz
$L = 0.16\ m$		6 pole
$ac = 23000\ amp.cond./m$		$K_{ws} = 0.955$

SOLUTION

Machine-I

$$\text{kVA input, } Q_1 = \frac{HP_1 \times 0.746}{\eta \times pf} = \frac{20 \times 0.746}{0.84 \times 0.82} = 21.66\ kVA$$

$$\text{Synchronous speed, } n_{s1} = \frac{2f}{p} = \frac{2 \times 50}{4} = 25\ rps$$

Also, kVA input, $Q_1 = C_o\ D^2\ L\ n_s,$

where, $C_o = 11\ B_{av}\ ac\ K_{ws} \times 10^{-3}$

$$\therefore\ Q_1 = 11\ B_{av}\ ac\ K_{ws} \times 10^{-3} D^2\ L\ n_s$$

$$\therefore\ B_{av} = \frac{Q_1}{11\ ac\ K_{ws} \times 10^{-3} D^2\ L\ n_s}$$

$\boxed{K_{ws} = 0.955}$

$$= \frac{21.66}{11 \times 23000 \times 0.955 \times 10^{-3} \times 0.25^2 \times 0.16 \times 25} = 0.3586\ Wb/m^2$$

Machine-II

$$\text{kVA input, } Q_2 = \frac{HP_2 \times 0.746}{\eta \times pf} = \frac{15 \times 0.746}{0.84 \times 0.82} = 16.246\ kVA$$

$$\text{Synchronous speed, } n_{s2} = \frac{2f}{p} = \frac{2 \times 50}{6} = 16.667\ rps$$

The value of ac (specific electric loading) decreases when voltage rating is increased. Hence the ratio of specific electric loading can be expressed as shown below.

$$\frac{ac_2}{ac_1} = \frac{V_{L2}}{V_{L1}}$$

The ratio of voltage rating $= \frac{V_{L2}}{V_{L1}} = \frac{460}{440} = 1.0455$

$$\therefore ac_2 = ac_1 \times \frac{V_1}{V_2} = \frac{ac_1}{V_2/V_1}$$

$$= \frac{ac_1}{1.0455} = \frac{23000}{1.0455} = 21999.044 \approx 22000 \; amp.cond./m$$

Let us assume same L/τ ratio for both the machines.

For machine-I, $\quad \dfrac{L_1}{\tau_1} = \dfrac{L_1}{\dfrac{\pi D_1}{p_1}} = \dfrac{L_1 \, p_1}{\pi \, D_1} = \dfrac{0.16 \times 4}{\pi \times 0.25} = 0.8149$

For machine-II, $\quad \dfrac{L_2}{\tau_2} = 0.8149$

$$\therefore L_2 = 0.8142 \; \tau_2 = 0.8149 \times \frac{\pi D_2}{p_2} = \frac{0.8149 \times \pi}{6} D_2 = 0.4267 \, D_2$$

> $\tau = \dfrac{\pi D}{p}$

$\therefore$ Length of stator core, $L_2 = 0.4267 \, D_2$

.....(1)

kVA input, $Q_2 = C_{o2} \, D_2^2 \, L_2 \, n_{s2}$

where, $C_{o2} = 11 \, B_{av} \, ac_2 \, K_{ws} \times 10^{-3}$

> Taking same B_{av} for both machines.

$\therefore Q_2 = 11 \, B_{av} \, ac_2 \, K_{ws} \times 10^{-3} \times D_2^2 \, L_2 \, n_{s2}$

$$\therefore D_2^2 \, L_2 = \frac{Q_2}{11 \, B_{av} \, ac_2 \, K_{ws} \times 10^{-3} \, n_{s2}}$$

> $K_{ws} = 0.955$

$$= \frac{16.246}{11 \times 0.3586 \times 22000 \times 0.955 \times 10^{-3} \times 16.667} = 0.0118 \; m^3$$

$\therefore D_2^2 \, L_2 = 0.0118 \; m^3$

> Using equation (1)

$$D_2^2 \, (0.4267 \, D_2) = 0.0118$$

$\therefore$ Diameter of stator bore, $D_2 = \left(\dfrac{0.0118}{0.4267}\right)^{1/3} = 0.3024 \; m = 0.3 \; m$

$\therefore$ Length of stator core, $L_2 = 0.4267 \times 0.3 = 0.128 \; m$

Maximum flux per pole, $\phi_m = \dfrac{B_{av} \, \pi \, D_2 \, L_2}{p}$

$$= \frac{0.3586 \times \pi \times 0.3 \times 0.128}{4} = 0.011 \; Wb$$

Stator turns per phase, $T_s = \dfrac{E_s}{4.44 \, f \, \phi_m \, K_{ws}}$

> Delta connected motor.
> $\therefore E_s = V_L$

$$= \frac{440}{4.44 \times 50 \times 0.011 \times 0.955} = 188$$

The stator slot pitch should be lie between 15 mm to 25 mm.

Stator slot, $S_s = \dfrac{\pi D}{y_{ss}}$

When, $y_{ss} = 15\ mm$, $S_s = \dfrac{\pi \times 0.3}{15 \times 10^{-3}} = 62$

When, $y_{ss} = 25\ mm$, $S_s = \dfrac{\pi \times 0.3}{25 \times 10^{-3}} = 38$

The stator slot should be lie between 38 to 62.

The stator slot be multiple of q, where q is slots per pole per phase.

Stator slot, S_s = Number of phase × Poles × q = 3 p q

When, q = 2 $S_s = 3 \times 4 \times 2 = 24$

When, q = 3 $S_s = 3 \times 4 \times 3 = 36$

When, q = 4 $S_s = 3 \times 4 \times 4 = 48$

The S_s value of 48 lie in the range of 38 to 64.

Conductors per slot, $Z_{ss} = \dfrac{6\ T_s}{S_s}$

$$= \dfrac{6 \times 188}{48} = 23.5 \simeq 24$$

∴ Total stator conductors, $Z = S_s \times Z_{ss} = 48 \times 24 = 1152$

∴ New value of turns per phase, $T_s = \dfrac{Z_{ss} \times S_s}{6}$

$$= \dfrac{48 \times 24}{6} = 192$$

RESULT

Machine-I	Specific magnetic loading,	B_{av}	=	0.3586 Wb/m^2
Machine-II	Diameter of stator bore,	D_2	=	0.2972 m
	Length of stator core,	L_2	=	0.127 m
	Conductors per slot,	Z_{ss}	=	24
	Stator slot,	S_s	=	48
	Total stator conductors,	Z	=	1152
	Stator turns per phase,	T_s	=	192

EXAMPLE 1.4

Calculate the main dimensions of a 20 HP, 1000 rpm, 400 V, DC motor. Given that, B_{av} = 0.37 Wb/m^2 and ac = 16000 $amp.cond./m$. Assume an efficiency of 90%.

GIVEN DATA

P = 20 HP B_{av} = 0.37 Wb/m^2 N = 1000 rpm

V = 400 V ac = 16000 $amp.cond./m$ η = 90%

SOLUTION

Power input, $P_i = \dfrac{P}{\eta} = \dfrac{20 \times 0.746}{0.9} = 16.58 \, kW$

Also power input, $P_i = VI \times 10^{-3}$

$\therefore$ Load current, $I = \dfrac{P_i}{V \times 10^{-3}} = \dfrac{16.58}{400 \times 10^{-3}} = 41.45 \, A$

Armature current, $I_a \approx I = 41.45 \, A$

Since the armature current is less than 200 A, the current per parallel path will not exceed the upper limit of 200 A.

Let, $p = 2$, $f = \dfrac{pN}{120} = \dfrac{2 \times 1000}{120} = 16.67 \, Hz$

$p = 4$, $f = \dfrac{pN}{120} = \dfrac{4 \times 1000}{120} = 33.33 \, Hz$

$p = 6$, $f = \dfrac{pN}{120} = \dfrac{6 \times 1000}{120} = 50 \, Hz$

The frequency of flux reversals should lie in the range of 25 to 50 Hz. For minimum cost the highest possible choice of poles should be chosen.

Hence, the number of poles, $p = 6$

Let us assume a square pole face.

For square pole face, $\dfrac{L}{\tau} = 0.7$

$\therefore L = 0.7 \, \tau = 0.7 \times \dfrac{\pi D}{p} = \dfrac{0.7 \times \pi}{6} \times D = 0.3665 \, D$

$\boxed{\tau = \dfrac{\pi D}{p} \\ \quad\quad .....(1)}$

$\therefore$ Length of armature, $L = 0.3665 \, D$

Output coefficient, $C_o = \pi^2 \, B_{av} \, ac \times 10^{-3}$

$= \pi^2 \times 0.37 \times 16000 \times 10^{-3}$

$= 58.428 \, kW/m^3\text{-}rps$

For DC motor, power developed in armature $\Big\}$ $P_a \approx P = 20 \, HP = 20 \times 0.746 = 14.92 \, kW$

Also, power developed in armature, $P_a = C_o \, D^2 \, L \, n$

$\therefore D^2 L = \dfrac{P_a}{C_o \, n}$

$= \dfrac{14.92}{58.428 \times (1000 / 60)} = 0.0153 \, m^3$

$\therefore D^2 L = 0.0153$

$D^2 (0.3665D) = 0.0153$

$\therefore$ Diameter of armature, $D = \left(\dfrac{0.0153}{0.3665}\right)^{1/3} = 0.3469 \, m \approx 0.35 \, m$

$\therefore$ Length of armature, $L = 0.3665 \, D = 0.3665 \times 0.35 = 0.128 \, m$

$\boxed{\text{Using equation (1)}}$

RESULT

Diameter of armature, D = 0.35 m

Length of armature,   L = 0.128 m.

EXAMPLE 1.5

A 600 *rpm*, 50 *Hz*, 10000 *V*, 3-phase, synchronous generator has the following design data. B_{av} = 0.48 *Wb/m²*, δ = 2.7 *A/mm²*, slot space factor = 0.35, number of slots = 144, slot size = 120 × 20 *mm*, D = 1.92 *m* and L = 0.4 *m*. Determine the kVA rating of the machine.

GIVEN DATA

3-phase	B_{av} = 0.48 *Wb/m²*	D = 1.92 *m*
600 *rpm*	δ = 2.7 *A/mm²*	L = 0.4 *m*
10000 *V*	Slots = 144	Slot size = 120 × 20 *mm*
		Slot space factor = 0.35

SOLUTION

We know that,

Current density, $\delta = \dfrac{I_z}{a_z}$

∴ Current in each armature conductor, $I_z = \delta\, a_z$ .....(1)

Conductors area in a slot = Slot area × Slot space factor = 120 × 20 × 0.35 = 840 *mm²*

Total number of armature conductors $\left. \right\}$ $Z = \dfrac{\text{Conductors area in a slot} \times \text{Number of slots}}{\text{Area of cross-section of each conductor}}$

$$= \frac{840 \times 144}{a_z} \qquad .....(2)$$

Specific electric loading, $\mathbf{ac} = \dfrac{I_z Z}{\pi D}$

> Using equations (1) and (2).

$$= \frac{\delta\, a_z \dfrac{840 \times 144}{a_z}}{\pi D} = \frac{\delta \times 840 \times 144}{\pi D}$$

$$= \frac{2.7 \times 840 \times 144}{\pi \times 1.92} = 54144\ amp.cond./m$$

kVA rating, $Q = C_o\, D^2\, L n_s$

> $C_o = 11\, B_{av}\, K_{ws} \times 10^{-3}$

$$= 11\, B_{av}\, \mathbf{ac}\, K_{ws} \times 10^{-3}\, D^2\, L\, n_s$$

$$= 11 \times 0.48 \times 54144 \times 0.955 \times 10^{-3} \times 1.92^2 \times 0.4 \times \frac{600}{60}$$

$$= 4025\ kVA$$

RESULT

kVA rating, Q = 4025 *kVA*.

1.14 SHORT-ANSWER QUESTIONS

Q1.1 What are the constructional elements of a transformer?

The constructional elements of a transformer are core, high and low voltage windings, cooling tubes or radiators and tank.

Q1.2 List the constructional elements of a DC machine.

The major constructional elements of a DC machine are stator, rotor, brushes and brush holders. The various parts of stator and rotor are listed below,

Stator	-	Yoke (or) Frame	**Rotor**	-	Armature core
	-	Field pole		-	Armature winding
	-	Pole shoe		-	Commutator
	-	Field winding	**Others**	-	Brush
	-	Interpole		-	Brush holder

Q1.3 List the constructional elements of salient pole synchronous machine.

The various constructional elements of salient pole synchronous machine are,

Stator	-	Frame	**Rotor**	-	Field pole
	-	Armature core		-	Pole shoe
	-	Armature winding		-	Field winding
				-	Damper winding

Q1.4 What are the constructional elements of cylindrical rotor synchronous machine?

The constructional elements of cylindrical rotor synchronous machine are,

Stator	-	Frame	**Rotor**	-	Solid rotor
	-	Armature core		-	Field conductors or bars
	-	Armature winding			

Q1.5 List the constructional elements of squirrel cage induction motor.

The constructional elements of squirrel cage induction motor are,

Stator	-	Frame	**Rotor**	-	Rotor core
	-	Stator core		-	Rotor bars
	-	Stator winding		-	End ring

Q1.6 List the constructional elements of slip ring induction motor.

The constructional elements of slip ring induction motor are,

Stator	-	Frame	**Rotor**	-	Rotor core
	-	Stator core		-	Rotor winding
	-	Stator winding		-	Slip rings

Q1.7 How the design problems of an electrical machine can be classified?

The design problems of electrical machine can be classified as,

1. Electromagnetic design
2. Mechanical design
3. Thermal design
4. Dielectric design.

Q1.8 *What are the major considerations to evolve a good design of electrical machines?*   *(AU, Apr' 18, 2 M)*

The major consideration to achieve a good electrical machine are,

1. Cost
2. Durability
3. Performance as per specifications.

Q1.9 *Write a short note on standard specifications.*

The standard specifications are the specifications issued by the standards organisation of a country. The standard specifications serve as guideline for the manufacturers to produce quality products at economical prices. The standard specifications for electrical machines include ratings, types of enclosure, dimensions of conductors, name plate details, performance indices, permissible temperature rise, permissible loss, efficiency, etc.

Q1.10 *What are the items to be mentioned in the rating plates of rotating machinery?*

The name plate of a rotating machine has to bear the following details as per ISI specifications.

1. kW or kVA rating of machine
2. Rated working voltage
3. Operating speed
4. Full load current
5. Class of insulation.
6. Frame size
7. ISI specification number
8. Manufacturers name
9. Serial number of product

Q1.11 *List the Indian Standard specifications for transformer.*   *(AU, Nov' 17, 2 M)*

i)   IS 1180 - 1989:   Specifications for out door 3-phase distribution transformer upto 100 *kVA*.

ii) IS 2026 - 1972:   Specifications of power transformers.

Q1.12 *List the Indian Standard Specifications for induction motor.*

i)   IS 325 - 1966:   Specifications of three-phase induction motor.

ii)   IS 1231 - 1974:   Specifications of foot mounted induction motor.

iii) IS 4029 - 1967:   Guide for testing three-phase induction motor.

iv)   IS 996 - 1979:   Specifications of single-phase AC and universal motor.

Q1.13 *What is meant by general design procedure?*

The general design procedure is to relate the main dimensions of the machine to its rated power output. A general output equation can be developed for electrical machines which relates the power output to volume of active part (D^2L), speed, magnetic and electric loadings.

Q1.14 *What is active part?*

In electrical machines the core and winding of the machine are together called active part. Because, the energy conversion takes place only in the active part of the machine.

Q1.15 *What are the main dimensions of a rotating machine?*

The main dimensions of a rotating machine are the armature diameter or stator bore, D and armature or stator core length, L.

Q1.16 Define total magnetic loading.

The total magnetic loading is defined as the total flux around the armature (or stator inner) periphery at the air-gap.

$$\text{Total magnetic loading} = p\phi$$

where p = Number of poles ; ϕ = Flux per pole

Q1.17 Define total electric loading.

The total electric loading is defined as the total number of ampere conductors around the armature (or stator) periphery.

$$\text{Total electric loading} = I_z Z$$

where I_z = Current through one armature conductor ; Z = Total number of armature conductors.

Q1.18 Define specific magnetic loading.   *(VTU, Dec'19, 8 M)*

The specific magnetic loading is defined as the average flux density over the air-gap of a machine.

$$\text{Specific magnetic loading, } B_{av} = \frac{\text{Total flux around the air-gap}}{\text{Area of flux path at the air-gap}} = \frac{p\phi}{\pi DL}$$

Q1.19 Define specific electric loading.

The specific electric loading is defined as the number of armature (or stator) ampere conductors per metre of armature (or stator) periphery at the air-gap.

$$\text{Specific electric loading, } \mathbf{ac} = \frac{\text{Total armature ampere conductors}}{\text{Armature periphery at air-gap}} = \frac{I_z Z}{\pi D}$$

Q1.20 Why total loadings are not used to determine the output of a rotating machine?

Each unit area of armature surface is capable of receiving a certain magnetic flux. Similarly every section of armature is capable of carrying certain amount of current. Hence specific loadings indicates the intensity of loading and the utility of active materials. Therefore specific loadings are used to determine the output rather than total loadings.

Q1.21 Give typical values of specific electric and magnetic loading.

Machine	Specific magnetic loading, B_{av} in Wb/m²	Specific electric loading, ac in amp.cond./m.
DC machine	0.4 to 0.8	15000 to 50000
Induction motor	0.3 to 0.6	5000 to 45000
Synchronous machine	0.52 to 0.65	20000 to 40000
Turbo-alternator	0.52 to 0.65	50000 to 75000

Q1.22 What is output equation.

The equation which relates the kVA input to the main dimensions (D and L), Specific loadings (B_{av} and **ac**) and speed (n) of a machine is known as output equation.

The output equation of DC machine is, $P_a = C_o D^2 L n$, in kW

The output equation of AC machine is, $Q = C_o D^2 L n_s$, in kVA

where, P_a = Power developed in armature of DC machine.

$\quad\quad Q$ = kVA rating of AC machine.

$\quad\quad C_o$ = Output coefficient.

Q1.23 Write the expression for output coefficient.

For DC machine, Output coefficient, $C_o = \pi^2 \, B_{av} \, \textbf{ac} \times 10^{-3}$, in kW/m^3- rps

For AC machine, Output coefficient, $C_o = 11 \; B_{av} \, \textbf{ac} \, K_{ws} \times 10^{-3}$, in kVA/m^3- rps.

Q1.24 How the power developed by armature is taken for a generator and for a motor?

Let   P = Rated power output of a DC machine

η = Efficiency.

P_a = Power developed by the armature of DC machine.

Now, $P_a = P/\eta$ for DC generators

$P_a = P$ for DC motors

> **Note :** The above concepts are applicable for large DC machines (above 1 kW), because in these machines the friction, windage and iron losses can be neglected.

Q1.25 How fixed losses are accounted in small DC machines?

The fixed losses in dc machines includes the friction, windage and iron losses. For design purpose, the fixed losses can be taken as one-third of total losses.

Q1.26 What are the factors that can be varied to vary the power output of a rotating electrical machine?

The power output of a rotating electrical machine depends of specific electric loading, specific magnetic loading, volume of active part and speed. Hence by varying these four quantities the power output of a machine can be varied.

Q1.27   Give the expression for the torque developed by a DC motor in terms of main dimensions of armature.

Let, T_a = Torque developed in the armature

We know that,

Power = Torque × Angular velocity $\Rightarrow$ $P_a = T_a \times 2\pi n$

$$\boxed{\begin{aligned} P_a &= C_o \, D^2 \, L \, n \\ &= \pi^2 \, B_{av} \, \textbf{ac} \times 10^{-3} \, D^2 \times L \, n \end{aligned}}$$

$\therefore$ Torque, $T_a = \dfrac{1}{2\pi n} P_a$

$\qquad\qquad = \dfrac{1}{2\pi n} \pi^2 \, B_{av} \, \textbf{ac} \times 10^{-3} \, D^2 L n$

$\qquad\qquad = \dfrac{\pi}{2} \, B_{av} \, \textbf{ac} \, D^2 \, L \times 10^{-3}$

Q1.28   Calculate the main dimensions for a 500 kW, 1kV, 600 rpm, 6 pole DC machine. Take $L/\tau = 1$ and C_o = 220 kW/m³-rps.

Solution

Given that, $L / \tau = 1$,

$\therefore L = \tau = \dfrac{\pi D}{p} = \dfrac{\pi}{6} D = 0.5236 \, D$ .....(1)

We know that, $P_a = C_o \, D^2 \, L \, n$

$\therefore D^2 L = \dfrac{P_a}{C_o n} = \dfrac{500}{220 \times 600 / 60} = 0.2273 \, m^3$

$\therefore D^2 L = 0.2273$

$\therefore D^2 (0.5236 \, D) = 0.2273$ Using equation (1)

$\therefore D = \left(\dfrac{0.2273}{0.5236}\right)^{1/3} = 0.76 \, m$

$\therefore L = 0.5236 \, D = 0.5236 \times 0.76 = 0.4 \, m$

Q1.29 How to separate D and L for rotating machines?

The seperation of D and L refers to the selection of an appropriate values for D and L for a given volume of active part. There are various choice of D and L for a given volume of active part.

In general a ratio of L/τ or L/D is assumed to form an equation relating D and L, where τ = pole pitch = $\pi D/p$.

Using the output equation and from the knowledge of kW/kVA rating, specific loadings and speed, the value of D^2L is estimated which gives another equation relating D and L. Then by solving the two equations the values of D and L are estimated.

Q1.30 What is the significance of the ratio of core length and pole pitch in induction motor?

In induction motors the operating characteristics are mainly influenced by the ratio of core length and pole pitch, L/τ. The factors influencing this choice are,

For minimum cost,	L/τ = 1.5 to 2	For good efficiency, L/τ = 1.5
For good power factor,	L/τ = 1.0 to 1.25	For good overall design, L/τ = 1.0

Q1.31 List the factors that influences the separation of D and L of a DC machine.

In DC machines the separation of D and L depends on,

1. Pole proportions
2. Peripheral speed
3. Moment of inertia
4. Voltage between adjacent commutator segments

Q1.32 What is square pole criterion?

The square pole criterion states that for a given flux and cross section area of pole, the length of mean turn of field winding is minimum, when the periphery forms a square. This implies that the length L must be approximately equal to pole arc b, i.e. **L = b = $\psi\,\tau$.**

Q1.33 In a DC machine, What are the limiting values of armature peripheral speed and voltage between adjacent commutator segments?

Maximum armature peripheral speed, $V_{a\,max}$ = 30 *m/sec.*
Maximum voltage between commutator segments, E_{cm} = 30 *V.*

Q1.34 List the various values of L/τ used for separation of D and L in induction motor.

The operating characteristics of an induction motor are mainly influenced by the ratio L/τ. The various values of L/τ used are listed here.

For minimum cost,   L/τ = 1.5 to 2
For good power factor,   L/τ = 1.0 to 1.25
For good efficiency,   L/τ = 1.5
For overall design,   L/τ = 1.0

Q1.35 What is the factor that is used to design an induction motor for best power factor?

For best power factor the separation of D and L is performed using the equation,

Pole pitch, $\tau = \sqrt{0.18\,L}$

where $\tau = \pi D/p$.

Q1.36 What are the factors to be considered for the separation of D and L of synchronous machine?

In synchronous machine the separation of D and L depends on pole proportions peripheral speed, number of poles and short circuit ratio.

Q1.37 List the various values of L/τ used for separation of D and L in synchronous machine.

In salient pole synchronous machines, when round poles are used the ratio of L/τ is between 0.6 to 0.7, and when rectangular poles are employed the ratio of L/τ is between 1 to 5.

Note : *In cylindrical rotor synchronous machines, the ratio L/τ is not used for separation of D and L. In this type of machine, the permissible peripheral speed is used to calculate D. Then L is estimated using output equation and this value of D.*

Q1.38 What is peripheral speed? Write the expression for peripheral speed of a rotating machine.

The peripheral speed is a translational speed that may exist at the surface of the rotor, while it is rotating. (It is a translational speed equivalent to the angular speed at the surface of the rotor).

$$\text{Peripheral speed, } V_a = \pi D_r n \text{ in } m/sec$$

where, D_r = Diameter of rotor   ;   n = Speed of the rotor.

Q1.39 High speed alternators have very long armature. Why?

In high speed alternators, the peripheral speed will be high and so the diameter has to be kept low to limit the peripheral speed. For a given volume of active part, if the diameter is kept low then the length has to be increased. Therefore the high speed alternators have very long armature.

Q1.40 What are the factors that affect the size of rotating machine?

The factors affecting the size of rotating machines are speed and output coefficient. The output coefficient in turn depends on specific electric and magnetic loadings.

Q1.41 What is the effect of speed on the size of the machine?

For a given output power the speed is inversely proportional to volume of active parts. Hence for a given output a high speed machine will have less volume and costs less. Therefore for reducing the cost highest possible speed may be selected. The maximum speed is limited by mechanical stresses of the rotating parts.

Q1.42 What is the importance of output coefficient?

The power output of a machine is directly proportional to output coefficient. Hence higher value of C_o results in higher power output.

For a given power output and speed, the volume of active part is inversely proportional to output coefficient. Hence with high values of C_o, the volume of active parts decreases and the machine costs less.

Q1.43 Why the output coefficient changes with size and type of machines?

For a given power output and speed, the volume of active part is inversely proportional to output coefficient. Hence if the output coefficient is higher then size of the machine will be small and vice versa.

The output coefficient is directly proportional to specific electric and magnetic loadings. Therefore for high values of C_o, high values of specific loadings are chosen, which affects the performance characteristics like temperature rise, efficiency, commutation conditions, etc. The type of machine (for e.g., forced air cooled) is decided by the performance requirements and so by output coefficient.

Q1.44 What are the factors that decide the choice of specific magnetic loading?

The value of magnetic loading is determined by

1. Maximum flux density in iron parts of machine
2. Magnetizing current and
3. Core losses.

Q1.45 The effect of magnetizing current is considered as important in case of induction motor. Why?

In induction motor the magnetizing current decides the power factor of the motor. When the magnetizing current is high, the power factor is low and vice versa. If power factor is low, then to deliver the same power output the current rating will be higher, which increases the cost of winding and motor.

Q1.46 Give the relation between core losses and frequency.

The hysteresis and eddy current losses are called core losses. The hysteresis loss is directly proportional to the frequency and eddy current loss is proportional to the square of the frequency.

Q1.47 What are the factors that decide the choice of specific electric loading?

The choice of specific electric loading depends on the following factors

　　1.　Permissible temperature rise　　　　　3.　Voltage rating of machine

　　2.　Size of machine　　　　　　　　　　　4.　Current density

Q1.48 Calculate the maximum allowable ac (specific electric loading), given that maximum allowable temperature rise = 55°C, resistivity = 2.7 ×10⁻⁸ Ω-m, current density = 2.75 A/mm² and cooling coefficient = 0.025.

Solution

$$\left.\begin{array}{l}\text{Maximum allowable specific}\\ \text{electric loading}\end{array}\right\} \mathbf{ac} = \frac{\theta}{\rho\,\delta\,c} = \frac{55}{2.7\times10^{-8}\times2.75\times10^{6}\times0.025}$$

$$= 29629.6 \approx 29630\ amp.cond./m$$

Q1.49 Give the typical values of current density, temperature rise and cooling coefficient for the best choice of specific electric loading.

For the best choice of specific electric loading the current density can be in the range of 2 to 5 *A/mm²*, the temperature rise can be in the range of 40° to 60°C above ambient and cooling coefficient can be in the range of 0.02 to 0.035 °C *W-m²*.

Q1.50 Give typical values for specific electric and magnetic loading for a 3.7 kW, 1440 rpm squirrel cage induction motor.

One of the typical choice of specific loadings for a 3.7 *kW*, 1440 *rpm* squirrel cage induction motor are,

Specific electric loading = 28,000 *amp.cond./m*　；　Specific magnetic loading = 0.56 *Wb/m²*.

Q1.51 Smaller machines have low specific magnetic loadings. Why?

A higher value of specific magnetic loading results in increased core loss and higher temperature rise. Consequently the efficiency of the machine will be low. In small machines, the losses has to be kept low in order to get higher power output and so a low value of specific magnetic loading is preferred in small machines.

Q1.52 In 2 DC machines running at same speed and having same number of poles, the physical dimensions are in the ratio 105 : 1. Compare their output.

Let linear dimensions are in the ratio 105 : 1 = x : 1

The output of dc machine, $P_a = C_o\, D^2\, L\, n$

The output coefficient, $C_o = \pi^2\, B_{av}\, \mathbf{ac} \times 10^{-3}$ | Also $D^2 \propto x$ and $L \propto x$

∴ Output, $P_a = \pi^2\, B_{av}\, \mathbf{ac} \times 10^{-3}\, D^2\, L\, n$ | ∴ Output $\propto x \times x^2\, x$

Here B_{av} and n are constants. | Output $\propto x^4$

$$\mathbf{ac} = \frac{I_z\, Z}{\pi D} = \frac{\delta\, a_z\, Z}{\pi D}$$ | Since x = 105, Output $\propto 105^4$

$a_z \propto x^2$ and $D \propto x$ | ∴ $\mathbf{ac} \propto \dfrac{x^2}{x} \propto x$

Q1.53 To what value the output of rotating machine will be reduced if the dimensions are scaled down to 75% of their original values?

If x is the linear dimension of the rotating machine, then the output of a rotating machine is proportional to x^4. The linear dimensions are reduced to 0.75 (i.e., 75%). Let the linear dimensions of original machine be 1(100%). The ratio of linear dimensions is 1 : 0.75. Hence the ratio of output is $1^4 : 0.75^4$.

$$0.75^4 = 0.3164 \Rightarrow 31.64\,\%$$

The output of the machine reduces by 31.64% if the linear dimensions are reduced by 75%.

Q1.54 Prove that large machines are more efficient than small machines.   *(PTU, May'19, 2 M)*

If x is the linear dimension of a machine then we can show that the output is proportional to x^4 and total loss is proportional to x^3.

$$\text{Efficiency, } \eta = \frac{\text{Output}}{\text{Input}} = \frac{\text{Output}}{\text{Output} + \text{Total losses}}$$

$$\therefore \eta \, \alpha \, \frac{x^4}{x^4 + k\,x^3} \quad \Rightarrow \quad \eta \, \alpha \, \frac{1}{1 + k/x}$$

where, k is a constant.

With increase in x (i.e., as the size of machine increases) the term k/x becomes smaller and smaller. Hence the efficiency increases, with increase in the linear dimensions of the machine. Therefore, large machines are more efficient than small machines.

Q1.55 What are the factors that impose technical limitations on the design?

The factors that impose technical limitations on design of electrical machines are

1. Saturation
2. Temperature rise
3. Stress on insulation
4. Efficiency
5. Stress and strain on rotating parts and bearings
6. Mechanical precision of air-gap
7. Commutation
8. Power factor
9. Specifications

Q1.56 Write approximate efficiency of static and dynamic devices.

The efficiency of static devices will be in the range of 90 to 98%. The efficiency of dynamic devices will be in the range of 85 to 95%. The static electrical devices will have copper and iron losses, and there is no friction losses. Hence the efficiency of static devices will be more than that of dynamic devices.

Q1.57 How does power factor influence the design aspect?

For same power output when power factor is less, the current level will be high. The higher current rating is achieved by increasing the specific electric loading which increases the size of conductor and so the cost of winding and machine.

On the other hand, if power factor is high then the current level will be less. The higher power factor is achieved by decreasing the specific magnetic loading. For low value of specific magnetic loading the volume of active material should be large or the air-gap should be as small as possible. These requirements will increase the cost of the machine.

Q1.58 What are the various methods of computer aided design of electrical machines?

The methods of computer aided design for electrical machines are,

      1. Analysis method

      2. Synthesis method

      3. Hybrid method.

Q1.59 What are the advantages of analysis method of computer aided design?

The advantages of analysis method are,

1. It is easy and straight forward to program the design equations.

2. It results in considerable time saving in performing calculations.

3. Analysis method programs are simple and they become the foundations for later larger and sophisticated programs.

4. The interaction of designer and computer yields highly acceptable results.

Q1.60 Give any two disadvantages of synthesis method of computer aided design.

1. The synthesis method involves too much of logic and before incorporating these logical decisions in the program they have to be accepted by team of design engineers. In a team, different members may give different suggestions to produce an optimum design and it becomes really hard to formulate an unique logical decisions.

2. The synthesis method programs are too complex which lead to high software development cost and require computers with high configurations involving huge expenditure.

Q1.61 Mention the advantages of hybrid method of computer aided design.   *(AKTU, Dec' 19, 3 M)*

The advantages of hybrid method are,

1. The results of hybrid method will be more appropriate, due to combination of designer involvement in finalizing design and optimization of vital parts.

2. The cost of hybrid systems will be lesser than synthesis systems, due to design of non vital parts by analysis method.

3. Since the design of some parts are carried by analysis method, the time required for design will also be lesser than synthesis method.

1.15 EXERCISES

I.   Fill in the blanks

1.   The electrical machines can be classified into .... and .... machines.

2.   The basic constructional elements of a rotating electrical machine are .... and ....

3.   The general design procedure is to relate the .... of the machine to its rated ....

4.   In electrical machines the core and the winding of the machine are together called ....

5.   The .... is defined as the total flux around the armature periphery.

6.   The .... is defined as the total number of ampere conductors around the armature periphery.

7.   The .... and .... of armature are called main dimensions.

8.   The unit of output coefficient is ....

9.   In small DC machines, the friction, windage and iron losses are approximately equal to .... of the ....

10.   In large DC machines, if P is the kW rating of the machine and η is efficiency then the power developed in the armature of generator is .... and that of motor is ....

11.   The main dimensions of a machine .... with .... in speed and output coefficient.

12.   In induction motors, low power factor is due to .... in the value of magnetizing current.

13.   The value of specific magnetic loading in synchronous machines is .... than DC machines and .... than induction machines.

14.   If specific magnetic loading is increased then the core loss .... and efficiency ....

15.   Space factor is the ratio of .... to total slot area.

16.   In high .... machines, the space factor is less due to .... thickness of insulation.

17.   The organic and inorganic insulating materials can withstand the temperatures of .... and .... respectively.

18.   In rotating machines if .... is the linear dimension then the output is proportional to ....

19.   In rotating machines, if .... is the linear dimension then the losses are proportional to ....

20.   The .... loss is proportional to frequency and .... loss is proportional to the square of the frequency.

Answers

1.	static, dynamic	11.	decreases, increase
2.	stator, rotor	12.	increase
3.	main dimensions, power output	13.	lower, higher
4.	active part	14.	increases, decreases
5.	total magnetic loading	15.	bare conductor area
6.	total electric loading	16.	voltage, higher
7.	diameter, length	17.	105°C, 180°C
8.	kW/m^3-rps or kVA/m^3-rps	18.	x, x^4
9.	one-third, total losses	19.	x, x^3
10.	P/η, P	20.	hysteresis, eddy current

II.   State whether the following statements are True/False

1.   The basic principle of operation of all electrical machine is Faraday's law of electromagnetic induction.

2.   In DC machines the power rating is expressed in kVA and in AC machines in kW.

3.   The work done per revolution is given by the product of total electric and magnetic loading.

4.   Total loadings are used to determine the output of a machine.

5.   The winding factor is accounted only for AC machines.

6.   Specific electric and magnetic loadings are directly proportional to each other.

7.   In small DC machines the friction, windage and iron losses are very small, hence they can be neglected.

8.   In a rotating machine, the volume of active part is inversely proportional to speed.

9.   The size and cost of the machine increases with increase in output coefficient.

10.   Two machines having different speed, can have same output power.

11.   The maximum flux density occurs in the teeth.

12.   In general a ratio of L/τ is assumed for separation of D and L.

13. In induction machines, higher value of magnetizing current is preferred in order to achieve good power factor.

14. Core loss is observed only when the iron parts are subjected to alternating magnetization.

15. Air-gap flux density is directly proportional to frequency of flux reversals.

16. Higher values of ac are used for machines having round conductors, because space factor is less for them.

Answers

1.	True	5.	True	9.	False	13.	False
2.	False	6.	False	10.	True	14.	True
3.	True	7.	False	11.	True	15.	False
4.	False	8.	True	12.	True	16.	False

III. Unsolved problems

E1.1 Estimate the main dimensions of a 4-pole, 100 kW, 1500 rpm DC generator assuming specific electric and magnetic loadings as 19000 $amp.conductors\ per\ metre$ and 0.4 T respectively. Assume that the length of armature is equal to the pole pitch.

$$(D = 0.41\ m\ ;\ L = 0.32\ m)$$

E1.2 Find the suitable values for number of poles, diameter and length of armature core of a 400 kW, 500 V, 180 rpm DC generator. Assume B_{av} = 0.6 Wb/m^2 and ac = 35000. Choose L/τ = 1.2.

$$(p = 20\ ;\ D = 1.5\ m\ ;\ L = 0.28\ m)$$

E1.3 Determine the main dimensions of a DC shunt generator with the following specifications: 5 kW, 220 V, 1500 rpm, 4-pole, ac = 200 $amp.cond./cm$, B_{av} = 0.6 T, η = 90% and pole arc = 0.7τ. Choose $L=\tau$.

$$(D = 0.15\ m\ ;\ L = 0.08\ m)$$

E1. 4 A 400 kW, 440 V, 600 rpm, 6 pole DC generator is built with an armature diameter of 0.9 m and core length of 0.45 m. The lap wound armature has 660 conductors. Calculate the specific electric and magnetic loadings.

$$(ac\ =\ 35364\ amp.cond./m,\ B_{av}\ =\ 0.3159\ Wb/m^2)$$

E1. 5 Calculate the main dimensions of a 40 HP, 1400 rpm, 500 V, DC motor. Given that, B_{av} = 0.5 Wb/m^2 and ac = 18000 $amp.cond./m$. Assume an efficiency of 90%.

$$(D\ =\ 0.34\ m,\ L\ =\ 0.125\ m)$$

E1. 6 A 400 rpm, 50 Hz, 8000 V, 3-phase, synchronous generator has the following design data. B_{av} = 0.32 Wb/m^2, δ = 1.9 A/mm^2, slot space factor = 0.18, number of slots = 36, slot size = 120 × 20 mm, D = 1.12 m and L = 0.2 m. Determine the kVA rating of the machine.

$$(Q = 47\ kVA)$$

> Chapter 2

ELECTRICAL ENGINEERING MATERIALS

2.1 MATERIALS FOR ELECTRICAL APPARATUS

(JNTKU, Apr' 19, 9 M)

Material for electrical apparatus can be broadly classified as follows.

1. Electrical conducting materials
2. Electrical carbon materials
3. Insulating materials
4. Magnetic materials.

Electrical conducting materials can be classified into two main groups.

1. High conductivity materials: The materials having low resistivity are highly conductive and these materials are used as conductors for all types of windings required in electrical machines, apparatus and devices. The main purpose of choosing highly conductive materials is to minimize the power loss due to resistance of the conductors.

2. High resistivity material (Alloy): Highly resistive materials will have large power loss and if the loss is released in the form of heat then these materials are used for constructing heating devices. These materials are also used for making wide range of resistances.

2.2 HIGH CONDUCTIVITY MATERIALS

The fundamental requirements of high conductivity materials are,

1. Highest possible conductivity (or least resistivity)
2. Least possible temperature coefficient of resistance
3. High tensile strength and absence of brittleness
4. Rollability and drawability
5. Good weldability and solderability
6. Low immunity to oxidation and corrosion.

The materials that are used as high conductivity materials are copper, aluminium, iron and steel and also alloys of copper such as bronze, beryllium copper, cadmium copper, brass and copper mixed with silver.

2.2.1 Copper as Conductive Materials

Copper is the best conductor and widely used electrical conductor due to high electrical conductivity with excellent mechanical properties and very low immunity to oxidation and corrosion. It is highly malleable and ductile metal and so it can be easily cast, forged, rolled, drawn and machined. Mechanical working hardens it but annealing restores it to soft state.

2.2.2 Aluminium as Conductive Materials

Aluminium is another best choice for conducting wires whose properties are very close to copper. Aluminium is available in abundant in nature when compared to copper. Therefore, cost of aluminium is lower when compared to copper but aluminium has higher resistivity than copper.

The comparison of copper and aluminium conductors are listed in Table 2.1 and 2.2 for an equal resistance per unit length.

Table 2.1: Properties of Copper and Aluminium   (VTU, Dec' 19, 4 M)

Characteristic	Copper	Aluminium
(i) Density, kg/m^3	8900	2700
(ii) Melting point, ^oC	1083	660
(iii) Thermal conductivity, W/m-^oC	350	200
(iv) Resistivity, Ω-m	0.01724×10^{-6}	0.0287×10^{-6}
(v) Resistivity temperature coefficient at 20^oC, $\Omega/^oC$	0.00393	0.00429
(vi) Coefficient of thermal expansion at 20^oC, $(^oC)^{-1}$	16.7×10^{-6}	25.5×10^{-6}
(vii) Specific heat, J/kg-^oC	390	890
(viii) Tensile strength, MN/m^2	220-250	90

Table 2.2: Comparison of Aluminium and Copper wires

Characteristic	Copper	Aluminium
Cost	1	0.49 K
Cross-section	1	1.62
Diameter	1	1.27
Volume	1	2.04
Weight	1	0.49
Breaking strength	1	0.64
K = Ratio of cost of copper and aluminium per kg.		

2.2.3 Iron and Steel as Conductive Materials

Pure iron is highly immune to oxidation. But the two forms of iron: cast iron and steel are very low immune to oxidation. The cast iron is mixture of iron, carbon and silicon. In cast iron, carbon is added upto 4% and silicon is added upto 3%. The amount of carbon and silicon in iron will give rise to wide variety of cast iron. Cast iron is used in the manufacture of resistance grids used in the starters of large motors.

Steel is an alloy of iron, carbon, nickel or chromium. Steel is used for making starter rheostats where lightness combined with robustness and good heat dissipation are important considerations.

2.2.4 Alloys of Copper as Conductive Materials

1. Bronze: It is a group of copper alloy containing tin, cadmium, berryllium and certain other metals with high conductivity. All bronzes posses high mechanical strength as compared with copper, but have higher resistivities. Bronze is less immune to corrosion than pure copper.

2. Beryllium copper: It has been found that the addition of 1 to 2.5% beryllium to copper makes a hard alloy which is capable of being rolled and formed into springs and contact strips. Therefore, it is used for current carrying springs, brush holders, sliding contacts and knife switch blades. Its resistivity is 3 to 6 times that of copper.

3. Cadmium copper: Alloys containing 1.1% cadmium gives wires which are stiffer, harder and of higher tensile strength than hard-drawn copper. It is used for making contact wires and commutator segments. Cadmium copper is also used for cage windings because it can be flame brazed without deterioration.

4. Brass: It is an alloy made with 66% copper and 34% zinc. It has greater mechanical strength and lesser wear and tear than copper, but with lower conductivity. Brass is easily shaped by press forming methods, lends itself to deep drawing, has good weldability and solderability and is fairly resistant to corrosion. Therefore, it has gained wide use in the manufacture of electrical apparatus.

5. Copper silver alloy: This alloy contains 99.10% copper and 0.06 to 0.1% silver. It has a resistivity 0.01814×10^{-6} Ω-m. Silver bearing copper is used in turbo alternators due to its resistance to thermal shortening and creep.

Table 2.3 shows peroperties of copper alloys.

Table 2.3: Properties of Copper Alloys

Material	State	Resistivity (Multiple of resistivity of copper)	Maximum tensile strength MN/m²
Cadmium copper	Annealed	1.05	300
(0.9% Cd)	Hard	1.1 to 1.2	715
Bronze	Annealed	1.65 to 1.8	280
(Cd 0.8%, Sn 0.6%)	Hard	1.8 to 2	715
Beryllium copper	Annealed	6.0	590
(Be 2.25%)	Ageing at 350 °C	3.3	1080
Brass	Annealed	4.0	510
(Cu 70%, Zn 30%)	Hard	4.0	860

2.3 HIGH RESISTIVITY MATERIALS

Materials of high resistivity are alloys of different metals. They can be classified into three catagories based on applications.

1. Materials used in precision measuring instruments and in making standard resistances and resistance boxes.

2. Materials from which resistance elements are made for all kinds of rheostats and similar control devices.

3. Materials suitable for making high temperature elements for electric furnaces, heating devices and loading rheostats.

2.3.1 Resistivity Materials used for Precision Work

The requirement of high resistivity materials used in precision electrical instruments and for making standard resistances is stability of resistance. Hence, the material should have a low resistance temperature coefficient. The thermo-electro-motive force resulting from contact of material with copper should be minimum to minimize errors in measurements.

The popular material used in instruments is manganin. Its an alloy of 86% copper, 12% manganese and 2% nickel. This alloy has a resistivity of 0.43×10^{-6} Ω-m and a resistance temperature coefficient of the order of 1×10^{-5} per °C.

2.3.2 Resistivity Materials used for Rheostats

The resistance materials used in making rheostats can have large thermo-emf and a large resistance temperature coefficient. But this material should have a high permissible working temperature and low cost.

The popular material used for rheostats is constantan, consisting of 60 to 65% copper and 40 to 35 % nickel. Sometimes small amounts of manganese and iron are also included. Constantan wire has resistivity of 0.46 to 0.53×10^{-6} Ω-m. Constantan can be safely used upto a temperature of 500 °C.

2.3.3 Resistivity Materials used for Heating Devices

The primary requirement for high temperature resistance alloys for use in electric furnaces and heating devices is a high working temperature. The most extensively used high working temperature resistance materials are alloys of nickel, chromium and iron called nichrome and alloy of aluminium, iron and chromium. The working temperature of these alloys depends upon the chromium content. The resistivity of nichrome varies from 1.1 to 1.27×10^{-6} Ω-m.

The optimum working temperature for nichrome wire is 900° to 1000 °C. Platinum is an incorrodable material with a high melting point (1710 °C) with a resistivity 0.117×10^{-6} Ω-m. Platinum is used in laboratory electric furnaces with a working temperature of 1300 °C.

2.4 ELECTRICAL CARBON MATERIALS

Electrical carbon materials are manufactured from various types of carbon and graphite and combination of carbon or graphite with other materials. Graphite is one of the crystalline forms of carbon.

Various applications of carbon in electrical apparatus are,

1. Carbon filaments of incandescent lamp
2. Carbon electrical contacts
3. Carbon resistors
4. Carbon brushes for DC machines and alternators
5. Carbon electrodes for battery
6. Carbon electrodes for electric furnaces
7. Carbon electrodes for arc lighting and welding.

In electrical machines carbon brushes are used as electrical contacts for current transfer between moving and stationary parts. Based on the composition and manufacturing process, carbon brushes can be classified into following five groups.

1. Hard carbon: It is made from amorphous carbons like retort coke or petroleum coke bonded with pitch or resin. Then the mixture of materials is pressed and kilned at temperatures up to 1200 °C.

The kilning process carbonises the binder and a hard carbon is produced with much better polishing qualities than electrographite. Hard carbon brushes are mostly used on small machines which require strongly polishing brushes. The permissible electrical and thermal load for hard carbon brushes is low and their use is limited to machines with peripheral speeds of up to 20 *m/s*.

2. Carbon graphite: They are manufactured similar to that of hard carbon but graphite is used instead of amorphous carbon. Its property is intermediate between hard carbon and electrographite. Carbon graphite brushes are mainly used in FHP (Fractional Horse Power) motors and thyristor fed DC machines.

3. Electrographite: It is made from amorphous carbons like retort coke or petroleum coke bonded with pitch or resin. Then mixture of materials is pressed and baked at temperatures upto 2400 °C. This converts the carbon material into a micro-crystalline form of graphite, called electrographite, a material having good electrical and thermal conductivity. High temperature treatment improves the elasticity and refractoriness but polishing properties are lower than hard carbon or graphite brushes. Electrographite brushes are used on large commutator machines with high electrical, thermal and mechanical requirements.

4. Graphite: It is made from graphite bonded with resin, which is pressed and heat treated in a special process. The advantage of graphite brushes is their high contact drop and high internal resistance. The disadvantage is that the high contact resistance causes high electrical loss due to circulating current. The main field of application for graphite brushes are in machines with high commutating requirements, but with relatively low brush current like three-phase commutator motors and small machines with mica undercut and their use is limited to machines with peripheral speeds of up to 35 *m/s*.

5. Metal graphite: Metal graphite brushes are made from natural graphite and fine metal powders. Copper is the most common metallic constituent, but silver, tin, lead and other metals are sometimes used. Metal graphite are ideal for a variety of applications due to their low resistivity. Since the contact and internal resistance of metal graphite brushes are low, they are used in machines with high brush loads. Due to their low elasticity, metal graphite brushes can be used with peripheral speeds up to 30 *m/s*.

2.5 MAGNETIC MATERIALS

The magnetic properties of materials depends on their value of relative permeability. The magnetic materials can be broadly classified as ferromagnetic, paramagnetic and diamagnetic materials based on the permeability.

1. Ferromagnetic materials: The relative permeabilities of these materials are much greater than unity and permeability of the material depends upon the magnetizing force.

2. Paramagnetic materials: These materials have relative permeabilities only slightly greater than unity and permeability is independent of magnetizing force. The value of susceptibility is thus positive for these materials.

3. Diamagnetic materials: These materials have their relative permeabilities slightly less than unity and permeability is independent of the mgnetizing force.

The ferromagnetic materials can produce large magnetic forces and so they are widely used in electrical apparatus. Iron, nickel and cobalt are natural ferromagnetic materials. The magnetic properties of these metals can be modified by making alloys of these materials with aluminium, chromium, manganese, copper and silver.

2.6 TYPES OF FERROMAGNETIC MATERIALS   *(VTU, Dec' 19, 5 M) (MU, May' 19, 5 M)*

Ferromagnetic materials can be broadly classified into soft and hard magnetic materials based on the hysteresis loop of the B-H characteristics of the material.

1. Soft magnetic materials: The ferromagnetic materials with small or narrow hysteresis loop are called soft magnetic materials. The hysteresis loss depends upon the area of hysteresis loop. Magnetic cores used in alternating magnetic fields are made from materials whose hysteresis loops are narrow to reduce hysteresis loss.

Soft magnetic materials are mainly used in the manufacture of magnetic core of electrical rotating machines and transformers.

2. Hard magnetic materials: Materials with broad hysteresis loops are called hard magnetic materials. Hard magnetic materials are used in certain types of electrical machines of low power rating, and in all kinds of instruments and devices requiring permanent magnets.

2.7 SOFT MAGNETIC MATERIALS

Soft magnetic materials that are employed in magnetic cores of electrical apparatus can be classified into following three groups.

1. Solid core magnetic materials.
2. Electrical steel sheets and strips used as laminations of magnetic core.
3. Special purpose alloys.

2.7.1 Solid Core Magnetic Materials   *(VTU, Dec' 19, 5 M)*

Solid core materials are normally used for parts of magnetic circuits carrying steady flux such as cores of DC electromagnets, relays and field frames of DC machines. The basic requirements of solid core material are high permeability, at high values of flux densities and small hysteresis loop. The materials used for solid magnetic cores are iron, low carbon iron, silicon steel, soft steel, cast steel, cast iron and ferro-cobalt.

1. Iron, low carbon iron and silicon steel: Iron is widely used in many kinds of electrical apparatus and instruments as cores and pole shoe for electromagnets. Iron in pure form is difficult to make and so iron will always have some amount of carbon and silicon. Carbon and silicon content in iron will improve mechanical strength at the expense of resistivity. Therefore, carbon and silicon content is minimized as far as possible to get qualities of pure iron.

2. Cast iron: It is magnetically inferior to iron or steel but casting complex shapes is much easier with cast iron. It usually contains 2.7 to 3.6% carbon, 2.0 to 2.7% silicon, 0.15 to 0.5% phosphorous and 0.15% sulphur. It has a low relative permeability and it is used in field frames where low cost is of primary importance and extra weight is not objectionable.

3. Cast steel: Cast steel is used to construct magnetic circuit which carry steady flux and need superior mechanical qualities. The composition of cast steel are carbon 5.5%, silicon 0.2%, manganese 2.5%, phosphorous 0.08% and sulphur 0.05%.

4. Soft steel: The soft steel has lower carbon content than cast steel. The soft steel has superior mechanical properties than cast steel. Rolled and welded frames of soft steel plates are now widely used in place of cast steel.

5. Ferro-cobalt: It is an alloy of iron with 35% cobalt. It is characterized by very high permeability. It has its saturation flux density 10% higher than that of pure iron. Its cost is relatively high and its use is limited to pole pieces where a high value of magnetic induction is desired.

2.7.2 Sheet Steels as Magnetic Materials   *(VTU, Dec' 19, 4 M)*

Sheet steels are used to make laminated iron cores to reduce eddy current and hysteresis losses. Sheet steels are available in various thickness from 0.23 *mm* to 0.35 *mm*. Sheet steels are cut into required laminations of required shape and then stacked together to form iron cores. Sheet steel are basically classified into two major types: Hot rolled silicon steel and Cold rolled silicon steel.

1. Hot rolled silicon steel: Hot rolled sheet steel are formed by heating large rectangular iron billet at 1700 °F and then flattened into a large roll. Then at a high temperature of 800 to 1200 °F they are run through a series of compression rollers to achieve its finished dimensions and spin the rolled steel into coils and leave it to cool.

Since hot rolled steel is cooled after processing, there is less control over its final shape, making it less suitable for precision applications. Hot rolled steel will have scaling on its surface. Scaling can be removed by grinding, sand blasting or acid-bath pickling. Once scaling is removed, various brush or mirror finishes can be applied. In hot rolled steel the iron crystals are randomly oriented and as a result the permeability will be low and so cannot be used for cores with high flux densities. Hot rolled sheet steel is cheaper when compared to cold rolled sheet steel and so hot rolled sheet steel is used for low cost machines.

The permeability and resistivity of hot rolled steel can be varied by altering the silicon content. If silicon content is less than 0.5% then the sheet steel will have high saturation flux densities and low resistivity and they are used in cores of rotating machines and called dynamo grade steels.

If silicon content is high (up to 4%) then the steel is called transformer grade steel having high resistivity and so loss is minimal and efficiency will be high.

2. Cold rolled steel: Cold rolled steel is essentially hot rolled steel that has gone through additional cold rolling processes. To get cold rolled steel, manufacturers generally take cooled-down hot rolled steel and perform one or more cold roll process to get more exact dimensions and better surface qualities. The cold rolling is performed to improve the grain orientation and hence to achieve higher permeability. Therefore, cold rolled grain-oriented sheet steel can be used to construct cores working at high flux densities especially in large power transformers and alternators.

2.7.3 Special Purpose Alloys for Magnetic Materials

In order to obtain high permeability in week magnetic fields special magnetic alloys have been developed. A group of iron alloys containing nickel between 30 to 90% with addition of molybdenum and chromium have very high permeabilities at low flux densities and much lower losses than iron. The important alloys are permalloy and mumetal.

1. Mumetal: It is an alloy of nickel, iron, copper and chromium or molybdenum. A popular composition of metals in this alloy is 77% nickel, 16% iron, 5% copper and 2% chromium or molybdnum. It has a lower reluctance but has high electrical resistance so that eddy current losses are lower.

2. Permalloys: It is an alloy of nickel and iron according to the nickle content, permalloys can be divided into two groups: high nickel and low nickel.

High Nickel Permalloy: It is an alloy of 80% nickel and 20% iron. Small amounts of molybdenum or molybdenum with copper or molybdenum with chromium are also added. It has high initial and maximum permeability and high resistivity. This makes it suitable for magnetic amplifier, current transformer cores, induction coils for communication and control equipments.

Low Nickel Permalloy : It contains nickel from 38 to 50% and iron with additions of manganese, silicon and chromium. It has lower permeability than high nickel variety but has higher resistivity. It is used for making cores of instrument transformers, induction coils, chokes and communication equipment.

3. Super-permalloy: It consist of 20% iron, 75% nickel and 5% molybdenum. This alloy is distinguished by its high purity. It has a very high initial relative permeability of upto 1,00,000.

4. Perminvar: Perminvar is an alloy of iron, nickel and cobalt. In perminvar the permeabilty is independemt of the field strength. Such material find applications in certain kinds of chokes and current transformers. Its use is limited by high cost and difficulties in manufacture.

5. Permendur: It contains 49% cobalt, 2% vanadium, and 49% cast iron. It is used in cores and poles of magnetic circuits to provide a strong magnetic field in the air-gap of apparatus such as electromagnets, oscillographs, microphones, etc.

2.8 INSULATING MATERIALS   *(HTU, Dec' 18, 10 M)   (VTU, Dec' 19, 8 M)   (AU, Nov' 18, 13 M)*

Insulating materials are used to electrically isolate conducting parts in electrical machines and hence they are high resistive non-conducting materials. There are large varities of insulating materials. Some of the natural insulating materials are paper, cloth, paraffin wax and natural resins. Some of the inorganic insulating materials are glass, ceramics and mica. Many of the insulating materials are man-made products manufactured in the form of resins, insulating films, etc.

Electrical properties of insulating materials are,

1. High dielectric strength.

2. High resistivity or specific resistance.

3. Low dielectric hysteresis.

4. Good thermal conductivity.

5. High degree of thermal stability.

2.8.1 Classification of Insulating Materials   (AU, Apr' 17, 8 M) (AU, Apr' 18, 5 M)

Basically insulation materials are classified into seven classes of insulating materials in relation to their thermal stability as shown in Table 2.4.

Table 2.4: Classification of Insulation Materials

Class	Temperature
Y	90 °C
A	105 °C
E	120 °C
B	130 °C
F	155 °C
H	180 °C
C	above 180 °C

Class Y: This class of insulation are made by materials or combinations of materials, such as cotton, silk and paper without impregnation. Example: Cotton, silk, paper, cellulose, wood, etc., neither impregnated nor immersed in oil.

Class A: This class of insulation are made by materials of class Y impregnated with natural resins cellulose esters, insulating oil, etc. Also included in this class are laminated wood, varnished paper.

Class E: This class of insulation are made by materials or combinations of materials such as synthetic resin enamles, cotton and paper laminates with formaldehyde bonding, etc., which are capable of operating at class E temperature.

Class B: This class of insulation are made by materials or combinations of materials such as mica, glass fibre, asbestos, etc., with suitable bonding substances.

Class F: This class of insulation are made by materials of class B with bonding materials of higher thermal stability.

Class H: This class of insulation are made by materials, such as silicon elastomer and combinations of materials such as mica, glass fibre, asbestos, etc., with suitable bonding substances like silicon resins.

Class C: This class of insulation are made by materials or combinations of materials such as mica, porcelain, glass and quartz with or without an inorganic binder of higher thermal stability.

2.8.2 Insulating Electrical Materials used in Modern Electrical Machines

Mica: In original form mica is brittle and so it is used in the form of sheets of splittings with shellac, bitumin or synthetic or polyester binding.

Micafolium: It is a wrapping consisting of mica splittings wrapped in paper and air dried.

Fibrous glass: It is glass in the form of fibre and made from material which is free from alkali metal oxides. Tapes and clothes woven from continuous filament yarns of glass have a high resistivity, thermal conductivity and tensile strength and form a good class B insulation.

Asbestos: Laminates of asbestos with synthetic resins have good mechanical strength and thermal resistivity. Asbestos in the form of wire and strip coverings have resilience and abrasion resistance.

Cotton fibre: Fibre cotton woven from acetylated cotton have remarkable resistance to heat. They are much less hygroscopic than ordinary cotton materials.

Polyamides: Polyamides in the form of nylon tapes have high mechanical strength and have a good space factor due to thinness. Nylon film is one of plastic films having adequate resistance to temperature and can withstand tearing.

Synthetic-resin materials: These are enamels of the vinyl-acetate or nylon types have an excellent smooth finish and have been used for coating windings. They also give good binding to windings.

Slot-lining materials: The materials used for slot lining are mica composites or two-ply or three-ply varnished cotton cloth.

Wood: Synthetic-resin-impregnated and compressed laminations of wood are robust and accurate materials for packing blocks, coil supports and spacers.

Silicones: Silicones are semi-inorganic materials with a basic structures of alternate silicon and oxygen atoms. They are extremely resistant to heats. They act as binders in Class H insulation and permit their continuous operations at 180 °C. Silicones are water-repellent and anti-corrosive. They have a high thermal conductivity and improved heat transfer coefficient.

Epoxide thermosetting resins: These materials have assumed importance in casting potting, laminating-adhesive and varnishing applications and in the encapsulation of small transformers.

Synthetic resin: It is used as bonding materials to make laminates of bonded paper, cotton and glass fibre.

Petroleum-based mineral oils: They are extensively used in the cooling and insulation of transformers. Their dielectric strength is good when they are clean and moisture-proof.

Askarels: They are synthetic non-flammable insulating liquids. When they are decomposed by an electric arc, they produce non-explosive gases. The commonest askarel is a 60/40 mixture of hexachlorodiphenyl and trichlorobenzene.

2.8.3 Temperature Rise and Insulating Materials   *(KTU, Feb' 18, 4 M) (AU, Apr' 19, 13 M)*

The various losses produced in the electrical machine are converted into heat energy, which leads to heating of various parts of the electrical machine. Accumulated heat energy will increase the temperature of various parts of the machine. The losses are mainly produced in the active parts of the electrical machine which constitutes the core made of iron and winding made of copper. Therefore, the heat energy appears mainly in the active parts resulting in increase in temperature of iron and copper.

For safe operation of electrical machines, the heating of every part must be maintained within certain limits. This rise in operating temperature mainly affects the insulating materials used to isolate the windings from the iron parts. There is a safe maximum permissible temperature for every insulating material up to which it will not lose its insulating property. Since the losses and hence temperature rise depends on load, the maximum allowable load also depends on insulating materials.

The life of class A insulating materials can be expressed by an empirical relation,

$$\text{INSL} = 72 \times 10^3 \times e^{-0.09\,\theta}$$

where, INSL = Life of insulating material, years,

θ = Maximum temperature to which the material can be continuously subjected, °C.

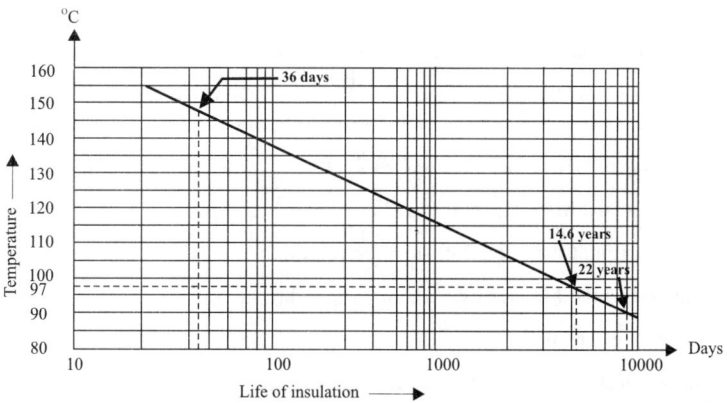

Fig. 2.1: *Temperature rise life curve for class A insulation.*

Some estimation of insulation life using the above equation is given below:

Let, $\theta = 91$ °C, $\text{INSL} = 72 \times 10^3 \times e^{-0.09\theta} = 72 \times 10^3 \times e^{-0.09 \times 91} = 19.97 = 20$ years

Let, $\theta = 98.6$ °C, $\text{INSL} = 72 \times 10^3 \times e^{-0.09\theta} = 72 \times 10^3 \times e^{-0.09 \times 98.6} = 10.0786 = 10$ years

Let, $\theta = 150$ °C, $\text{INSL} = 72 \times 10^3 \times e^{-0.09\theta} = 72 \times 10^3 \times e^{-0.09 \times 150} = 0.0987$ years

$$= 0.0987 \times 365 = 36 \text{ days}$$

Let, $\theta = 200$ °C, $\text{INSL} = 72 \times 10^3 \times e^{-0.09\theta} = 72 \times 10^3 \times e^{-0.09 \times 200} = 1.0966 \times 10^{-3}$ years

$$= 1.0966 \times 10^{-3} \times 365 = 0.4 \text{ days}$$

$$= 0.4 \times 24 = 9.6 \text{ hours}$$

From the above examples we can infer that the life of insulation reduces drastically with increase in temperature. Therefore, even if we operate the machine a little above the maximum allowable temperature, it seriously affects the life of insulating materials. In electrical machines the condition of insulation is of prime importance and so the insulating materials must be worked within safe temperature limits. Fig. 2.1 shows the relationship between temperature and life of class A insulating materials.

The heat produced in a machine depends upon the losses and this heat is removed by the presence of ambient air, a cooling or ventilating system. The operating temperature of the machine is reached when the rates of heat production and dissipation become equal.

The temperature rise in a machine can be maintained within safe limits by properly designing a cooling and ventilation system. A higher output can be obtained from a given machine frame by a proper choice of ventilation system and insulating materials.

2.9 SHORT-ANSWER QUESTIONS

Q2.1 *What are the desirable properties of high conductivity materials?* *(VTU, Dec' 19, 5 M)*

The properties of high conductivity materials are,

1. Highest possible conductivity (or least resistivity)
2. Least possible temperature coefficient of resistance
3. High tensile strength and absence of brittleness
4. Rollability and drawability
5. Good weldability and solderability
6. Low immunity to oxidation and corrosion.

Q2.2 *What are the different conducting materials used in rotating machines?* *(AU, Nov' 18, 2 M)*

The materials that are used as high conductivity materials are copper, aluminium, iron and steel and also alloys of copper such as bronze, beryllium copper, cadmium copper, brass and copper mixed with silver.

Q2.3 *Name the types of magnetic materials based on hysterisis loops? Give examples.*

1. Soft magnetic materials: Materials with small or narrow hysteresis loop.

Examples: Cast iron, Cast steel, Soft steel, Ferro-cobalt.

2. Hard magnetic materials: Materials with large or broad hysteresis loop.

Examples: Nickle, Cobalt.

Q2.4 *What are the electrical properties of insulating materials?* *(AU, Apr' 17, 2 M) (AU, Apr' 19, 2 M)*
(MU, May' 19, 5 M)

Electrical properties of insulating materials are,

1. High dielectric strength
2. High resistivity or specific resistance
3. Low dielectric hysteresis
4. Good thermal conductivity
5. High degree of thermal stability.

Q2.5 *What are the categories of high resistivity materials based on applications?*

1. Materials used in precision measuring instruments and in making standard resistances and resistance boxes.
2. Materials from which resistance elements are made for all kinds of rheostats and similar control devices.
3. Materials suitable for making high temperature elements for electric furnaces, heating devices and loading rheostats.

Q2.6 *How magnetic materials are classified based on permeability?*

1. Ferromagnetic materials
2. Paramagnetic materials
3. Diamagnetic materials.

Q2.7 *How the soft magnetic materials employed in magnetic cores of electrical apparatus are classified?*

1. Solid core magnetic materials
2. Electrical steel sheets and strips used as laminations of magnetic core
3. Special purpose alloys.

Q2.8 *What are the materials used for solid magnetic materials?*

   1. Iron, low carbon iron and silicon steel

   2. Cast iron

   3. Cast steel

   4. Soft steel

   5. Ferro-cobalt.

Q2.9 *Name the classification of sheet steels as magnetic materials?*

   1. Hot rolled silicon steel

   2. Cold rolled steel.

Q2.10 *What are the important special purpose alloys for magnetic materials?*

   1. Mumetal

   2. Permalloys

      i) High Nickel Permalloy

      ii) Low Nickel Permalloy

   3. Super-Permalloy

   4. Perminvar

   5. Permendur.

Q2.11 *How insulating materials are classified?*

The insulation materials are classified based on thermal stability to withstand high temperatures. Conventionally the maximum withstanding temperature is referred to as class of insulation and listed in the following table.

Class	Temperature
Y	90 °C
A	105 °C
E	120 °C
B	130 °C
F	155 °C
H	180 °C
C	above 180 °C

2.10 EXERCISES

I. Fill in the blanks

1. An insulating material should have .... dielectric loss.

2. By adding sillicon to iron .... properties are improved.

3. An ideal insulating material should have .... thermal conductivity.

4. .... material has the highest permeability.

5. The percentage of silicon in transformer grade steel is usually limited to ....

6. .... steel can be worked with higher flux densities.

7.   Fibre glass insulation can be used upto temperature of ....

8.   Bronze is less immune to corrosion than ....

9.   Silver bearing copper is used in ....

10.   Materials with broad hysteresis loops are called ....

11.   Cotton insulation can be used upto temperature of ....

12.   .... materials are used for making wide range of resistances.

13.   Steel is used in ....

14.   The resistivity of nichrome varies from ....

15.   Carbon graphite are mainly used in ....

Answers					
1.	high	6.	cold rolled silicon	11.	90°C
2.	magnetic	7.	130 °C	12.	high resistivity
3.	good	8.	pure copper	13.	starter rheostats
4.	permalloy	9.	turbo alternator	14.	1.1 to 1.27×10^{-6} Ωm
5.	4%	10.	hard magnetic materials	15.	FHP motors

II.   State whether the following statements are True/False

1.   The high resistive materials have small power loss.

2.   The high resistive materials are used for wide range of resistance.

3.   In high conductive materials the cost of aluminium is lower when compared to copper.

4.   In electrical machines carbon brushes are used as electrical contacts for current transfer between moving and stationary parts.

5.   The resistivity of nichrome varies from 1.8 to 2.27×10^{-6} Ω-m.

6.   For graphite brushes the limiting valus of peripheral speed is 40 *m/sec*.

7.   The ferromagnetic materials can produce large magnetic force.

8.   Hot rolled sheet steel is used for high cost machines.

9.   The commonest askarel is a 60/30 mixture of hexachlorodiphenyl and trichlorobenzene.

10.   The resistivity of copper silver alloy is 0.01814×10^{-6} Ω-m.

Answers			
1.	False	6.	False
2.	True	7.	True
3.	True	8.	False
4.	True	9.	False
5.	False	10.	True

> # Chapter 3

MAGNETIC CIRCUITS IN MACHINES

List of symbols

Symbol	Meaning	Unit
A	Area of cross-section of the flux path	m^2
A_a	Area of cross-section of air path	m^2
A_g	Actual area of air-gap per pole	m^2
A_g'	Area of contracted air-gap	m^2
A_i	Area of cross-section of iron path	m^2
A_p	Area per pole	m^2
A_t	Area of cross-section of tooth at the desired section	m^2
AT	Total mmf	AT
AT_g	mmf for air-gap	AT
AT_i	mmf for iron parts	AT
AT_t	Total mmf for teeth	AT
AT_m	mmf for the peak flux density B_m	AT
at	mmf per metre of magnetic path	AT/m
at$_g$	mmf per metre for air-gap	AT/m
at$_{mean}$	Mean value of **at**	AT/m
at$_{real}$	mmf per metre across the tooth for tooth density B_{real}	AT/m
B	Flux density	Wb/m^2
B_a	Flux density in air-gap	Wb/m^2
B_g	Flux density at the centre of the pole	Wb/m^2
B_m	Maximum flux density	Wb/m^2
B_t	Flux density of tooth corresponding to A_t	Wb/m^2
B_{av}	Average gap density or specific magnetic loading	Wb/m^2
B_{app}	Apparent flux density in tooth	Wb/m^2

Symbol	Meaning	Unit
B_{real}	Real flux density in tooth	Wb/m^2
b	Pole arc	m
C_l	Leakage coefficient	-
D	Stator inner diameter	mm
d_s	Depth of slot	mm
H	Magnetizing force	AT/m
h	Height of the field	mm
I_m	Magnetizing current	A
I_{m_nsd}	Magnetizing current (Non-sinusoidal flux distribution)	A
I_{m_sd}	Magnetizing current (Sinusoidal flux distribution)	A
I_{m_rms}	Rms value of magnetizing current	A
I_{m_max}	Maximum value of magnetizing current	A
I_z	Current in each conductor	A
K_f	Field form factor	-
K_g	Total gap contraction factor	-
K_{cd}	Carter's coefficient for ducts	-
K_{cs}	Carter's coefficient for slots	-
K_{css}	Carter's coefficient for stator slots	-
K_{csr}	Carter's coefficient for rotor slots	-
K_{gd}	Gap contraction factor for ducts	-
K_{gs}	Gap contraction factor for slots	-
K_{gsal}	Gap contraction factor for saliency of pole	-
K_{gsr}	Gap contraction factor for rotor slots	-
K_{gss}	Gap contraction factor for stator slots	-
K_p	Peak factor	-
K_{w1}	Winding factor for fundamental	-
L	Length of core	m
L'	Effective axial length	m
L_i	Net iron length	m
l	Length of magnetic path	m
l_g	Air-gap length	mm
l_t	Length of tooth	mm
l_{ge}	Effective air-gap length	mm
N	Number of turn in winding	-
n_d	Number of ducts	-
n_t	Number of teeth under a pole	-
p	Number of poles	-
S	Reluctance	AT/Wb

Symbol	Meaning	Unit
S_g	Reluctance of air-gap	AT/Wb
S_{g_cas}	Reluctance of air-gap in machines with smooth armature	AT/Wb
S_{g_oas}	Reluctance of air-gap in machines with open armature slots	AT/Wb
S_{g_oasf}	Reluctance of air-gap in machines with open armature slots (including the effect of fringing)	AT/Wb
S_{g_d}	Reluctance of air-gap in machine with radial ventilating ducts	AT/Wb
S_r	Number of rotor slots	-
S_s	Number of stator slots (neglecting the effect of fringing)	-
T	Total number of turns	-
T_s	Number of stator turns per phase	-
w_d	Width of duct	mm
w_o	Slot opening	mm
w_{os}	Stator slot opening	mm
w_{or}	Rotor slot opening	mm
w_s	Width of slot	mm
w_t	Width of tooth	mm
X_s	Slot leakage reactance	Ω
y_s	Slot pitch	mm
y_s'	Effective slot pitch	mm
y_{ss}	Stator slot pitch	mm
y_{sr}	Rotor slot pitch	mm
Z_s	Number of conductors per slot	-
τ	Pole pitch	m
λ	Specific permeance	$Wb/AT\text{-}m$
λ_c	Specific permeance for conductor portion of slot	$Wb/AT\text{-}m$
λ_{sa}	Specific permeance for non-conductor portion of slot	$Wb/AT\text{-}m$
λ_{sc}	Specific permeance for conductor portion of slot	$Wb/AT\text{-}m$
λ_w	Specific permeance for slot opening	$Wb/AT\text{-}m$
Λ	Permeance	Wb/AT
Λ_a	Permeability of air-path	Wb/AT
Λ_s	Slot permeability	Wb/AT
Λ_{sa}	Permeance of non-conductor portion of slot	Wb/AT
Λ_{sc}	Permeance of conductor portion of slot	Wb/AT
ϕ	Magnetic flux	Wb
ϕ_a	Flux through air	Wb
ϕ_i	Flux through iron	Wb
ϕ_s	Flux over one slot pitch	W

Symbol	Meaning	Unit
ϕ_{sl}	Slot leakage flux	Wb
Ψ	Ratio of pole arc to pole pitch	-
μ	Absolute permeability of the magnetic material	H/m
μ_o	Permeability of free-space or non-magnetic material ($4\pi \times 10^{-7}$ H/m)	H/m
μ_r	Relative permeability	-

3.1 DESIGN OF MAGNETIC CIRCUITS

The magnetic circuit is the path of magnetic flux. The mmf of the circuit creates flux in the path by overcoming the reluctance of the path. The various elements in the flux path of salient pole machines are poles, pole shoes, air-gap, armature teeth, armature core and yoke. The various elements in the flux path of non-salient pole machines are stator core, stator teeth, air-gap, rotor teeth and rotor core. In transformers the flux is set-up through the core. The magnetic circuit of various electrical machines are shown in Fig. 3.1.

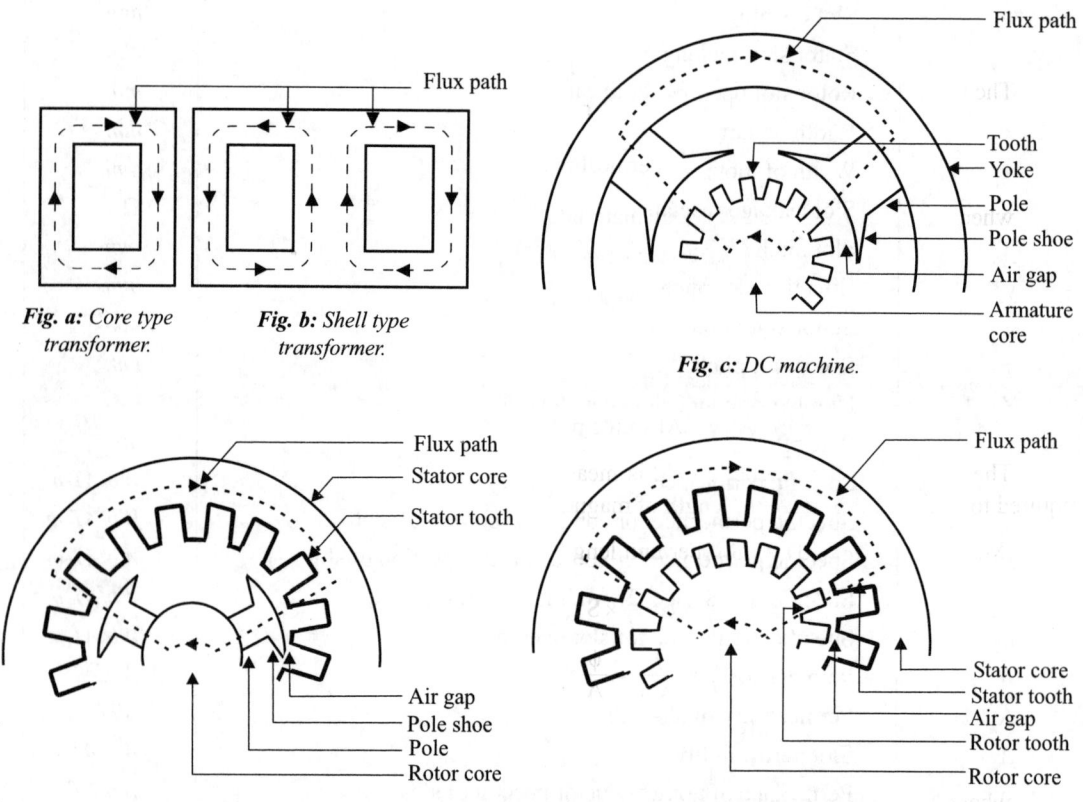

Fig. a: Core type transformer.

Fig. b: Shell type transformer.

Fig. c: DC machine.

Fig. d: Salient pole electrical machine.

Fig. e: Induction motor or Non-salient pole synchronous machine.

Fig. 3.1: Magnetic circuits in electrical machines.

A magnetic circuit is analogous to an electric circuit. In electric circuit the emf circulates current against resistance when a closed path is provided. Similarly, in a magnetic circuit the mmf creates flux in a closed path against reluctance of the path. A coil wound on an iron core with N turns and carrying a current, I will produce an mmf of NI ampere-turns.

The equation which relates flux, mmf and reluctance is given by,

$$\text{Flux} = \frac{\text{mmf}}{\text{Reluctance}} \quad \text{(Similar to Ohm's of electric circuit)}$$

$$\therefore \text{Flux, } \phi = \frac{\text{AT}}{\text{S}} \qquad \qquad .....(3.1)$$

Since the inverse of reluctance is permeance, the flux is also given by.

$$\text{Flux, } \phi = \text{AT} \times \Lambda \qquad \qquad .....(3.2)$$

where, AT = mmf

S = Reluctance

Λ = Permeance

The reluctance of the magnetic material can be estimated using the following equation.

$$\text{Reluctance, } S = \frac{\text{Length}}{\text{Area}} \times \frac{1}{\text{Permeability}} = \frac{l}{A\,\mu} = \frac{l}{A\,\mu_r\,\mu_o} \qquad .....(3.3)$$

where, l = Length of magnetic material

A = Area of cross-section of magnetic material

μ = Permeability of the magnetic material

μ = $\mu_r\,\mu_o$

μ_r = Relative permeability

μ_o = $4\pi \times 10^{-7}$ *H/m* = Absolute permeability of free space

The strength of the magnetic field is measured by the term magnetizing force, H. It is the mmf required to establish flux in a unit length of magnetic path.

Magnetizing force, H = mmf per unit length = Flux × Reluctance per unit length

$$= \phi \times \frac{S}{l} = \phi \times \frac{1}{l} \times S$$

$$= \phi \times \frac{1}{l} \times \frac{l}{A\,\mu} = \frac{\phi}{A} \times \frac{1}{\mu} \qquad \boxed{S = \frac{l}{A\,\mu}} \quad \boxed{B = \frac{\phi}{A}}$$

$$= \frac{B}{\mu} \qquad \qquad .....(3.4)$$

where, B = Flux density

In the case of a material of length, l and carrying a uniform flux, the total mmf AT is,

$$\text{AT} = \text{H} \times l = \textbf{at} \times l \qquad \boxed{H = \textbf{at}} \quad .....(3.5)$$

In a series magnetic circuit, the total reluctance is the sum of reluctances of individual parts. [Fig. 3.1, a, c, d and e are examples of series magnetic circuit].

Therefore, in series magnetic circuits,

$$\text{Total Reluctance, } S = S_1 + S_2 + S_3 + \dots\dots\dots \qquad \dots\dots(3.6)$$

where, $S_1, S_2, S_3, \dots\dots$ are reluctances of individual parts connected in series.

The total mmf acting around a complete magnetic circuit is given by,

$$\text{Total mmf, } AT = \phi \, S \qquad \dots\dots(3.7)$$

$$= \phi \, (S_1 + S_2 + S_3 + \dots\dots\dots) \qquad \boxed{\text{Using equation (3.6)}}$$

$$= AT_1 + AT_2 + AT_3 + \dots\dots\dots$$

$$= \textbf{at}_1 l_1 + \textbf{at}_2 l_2 + \textbf{at}_3 l_3 + \dots\dots\dots$$

$$= \int \textbf{at} \, l \qquad \dots\dots(3.8)$$

The equation (3.8) represent the circuital law for magnetic circuits where $\textbf{at}_1$, $\textbf{at}_2$, $\textbf{at}_3$, etc., are the mmf's per metre for individual part and l_1, l_2, l_3, etc., are lengths of parts connected in series.

In parallel circuits, the same mmf is applied to each of the parallel paths and the total flux divides between the paths in inverse proportion to their reluctances. [Fig. 3.1, b is an example of parallel magnetic circuit.]

Therefore, in parallel magnetic circuits,

$$\text{Total flux, } \quad \phi = \phi_1 + \phi_2 + \phi_3 + \dots\dots\dots \qquad \dots\dots(3.9)$$

where, ϕ_1, ϕ_2, ϕ_3, $\dots\dots$ are flux in paths connected in parallel.

On dividing the equation (3.9) by AT,

$$\frac{\phi}{AT} = \frac{\phi_1}{AT} + \frac{\phi_2}{AT} + \frac{\phi_3}{AT} + \dots\dots\dots$$

$$\frac{1}{S} = \frac{1}{S_1} + \frac{1}{S_2} + \frac{1}{S_3} + \dots\dots\dots \qquad \boxed{S = \dfrac{AT}{\phi}}$$

$$\Lambda = \Lambda_1 + \Lambda_2 + \Lambda_3 \qquad \dots\dots(3.10)$$

where, S   = Total reluctance of magnetic circuit

$S_1, S_2, S_3, \dots\dots$ = Reluctance of individual parts.

Λ = Total permeance of magnetic circuit.

$\Lambda_1, \Lambda_2, \Lambda_3, \dots\dots$ = Permeance of individual parts.

The similarities and the differences between electric and magnetic circuits are presented in Table 3.1 and 3.2 respectively.

Table 3.1: Similarities in Electric and Magnetic Circuits

Electric Circuit	Magnetic Circuit
1. The emf circulates current in a closed path.	1. The mmf creates or establish flux in a closed path.
2. Flow of current is opposed by resistance of the circuit.	2. The creation of flux is opposed by reluctance of the circuit.
3. The path of current is called electric circuit.	3. The path of flux is called magnetic circuit.
4. Resistance, $R = \dfrac{\rho l}{A} = \dfrac{l}{\sigma A}$ $\left(\sigma = \dfrac{1}{\rho} = \text{conductivity}\right)$	4. Reluctance, $S = \dfrac{l}{\mu A}$
5. $\text{Current} = \dfrac{\text{emf}}{\text{Resistance}}$	5. $\text{Flux} = \dfrac{\text{mmf}}{\text{Reluctance}}$
6. Current density, $\delta = \dfrac{\text{Current}}{\text{Area of cross-section}}$	6. Flux density, $B = \dfrac{\text{Flux}}{\text{Area of cross-section}}$

Table 3.2: Differences between Electric and Magnetic Circuits

Electric Circuit	Magnetic Circuit
1. Current actually flow in the electric circuit.	1. Flux does not flow, but it is only assumed to flow.
2. When current flows, the energy is spent continuously.	2. Energy is needed only to create the flux but not to maintain it.
3. Resistance of the electric circuit in independent of current strength.	3. Reluctance of the magnetic circuit depends on total flux or flux density in the material.
4. Resistance changes with temperature.	4. Reluctance changes with saturation of flux or flux density.

3.2 MAGNETIC CURVE

In magnetic materials the magnetizing force required to establish a given flux density depends on the saturation of the material. If the material is not saturated then a small increase in magnetizing force will result in a proportional increase in flux density. But when the material is saturated a large increase in magnetizing force will result in a small increase in flux density (or there will not be any increase in flux density). Therefore the permeability of the magnetic material is not constant.

In non-magnetic materials like air, copper, etc., there is no such phenomena of saturation. Hence the permeability of non-magnetic material is constant and the relation between B and H is linear. Therefore the B-H curve will be a straight line passing through origin.

In magnetic materials the relation between the flux density, B and the magnetizing force, H is non-linear. Hence it is difficult to express the relation in terms of mathematical equation. Therefore to calculate the mmf per metre of flux path for a given flux density the B-**at** curve is employed.

The manufacturers of stampings or laminations for transformer, induction motor, DC machines, etc., will supply B-H curve (or B-**at** curve). Also, the manufacturers will supply a loss curve, since the core loss is also non-linear and depends on magnetizing force. A typical B-H and loss curve for the laminations used for induction motor are shown in Fig. 3.2.

These curves are used to estimate magnetizing force, H and core loss for a given flux density (or for a required flux density) in any part of a machine.

3.3 MAGNETIC CIRCUIT CALCULATIONS

The magnetic circuit calculations involves estimation of reluctance, flux density and mmf for various sections of magnetic circuit. The ultimate aim of magnetic circuit calculation is to estimate the total mmf required to establish the desired flux in a magnetic circuit.

The magnetic circuit is split into convenient parts (sections) which may be connected in series or parallel. Then the reluctance, flux density and mmf for every section of the magnetic circuit is estimated. The summation of mmf of all sections in series gives the total mmf for the magnetic circuit.

The following procedure can be used to estimate the mmf of a section of magnetic circuit.

1. Determine the flux in the concerned section from the knowledge of flux per pole.

2. Calculate the area of cross-section of the section from the specified dimensions.
    The ratio of flux and area of cross-section will give the flux density, B in this section.

3. For the calculated flux density, B, determine the mmf per unit length from the B-H (or B-**at**) curve.

4. The mmf for the concerned section is given by the product of length of the section and mmf per unit length.

The method looks quite simple but there are some parts in the magnetic circuit like air-gap and tapered teeth which present complex magnetic problems. The reluctance of the air-gap is modified or affected due to slots, radial ventilating ducts, and non-uniform air-gaps. Hence the calculation of mmf for air-gap cannot be generalized and so the calculation of reluctance should be attempted for each type of machine.

The dimensions of a tooth depends on the type of slot. Also, the dimensions of a tooth is not uniform, hence the reluctance of a tooth is non-uniform. Hence special methods are needed for estimating the mmf for teeth.

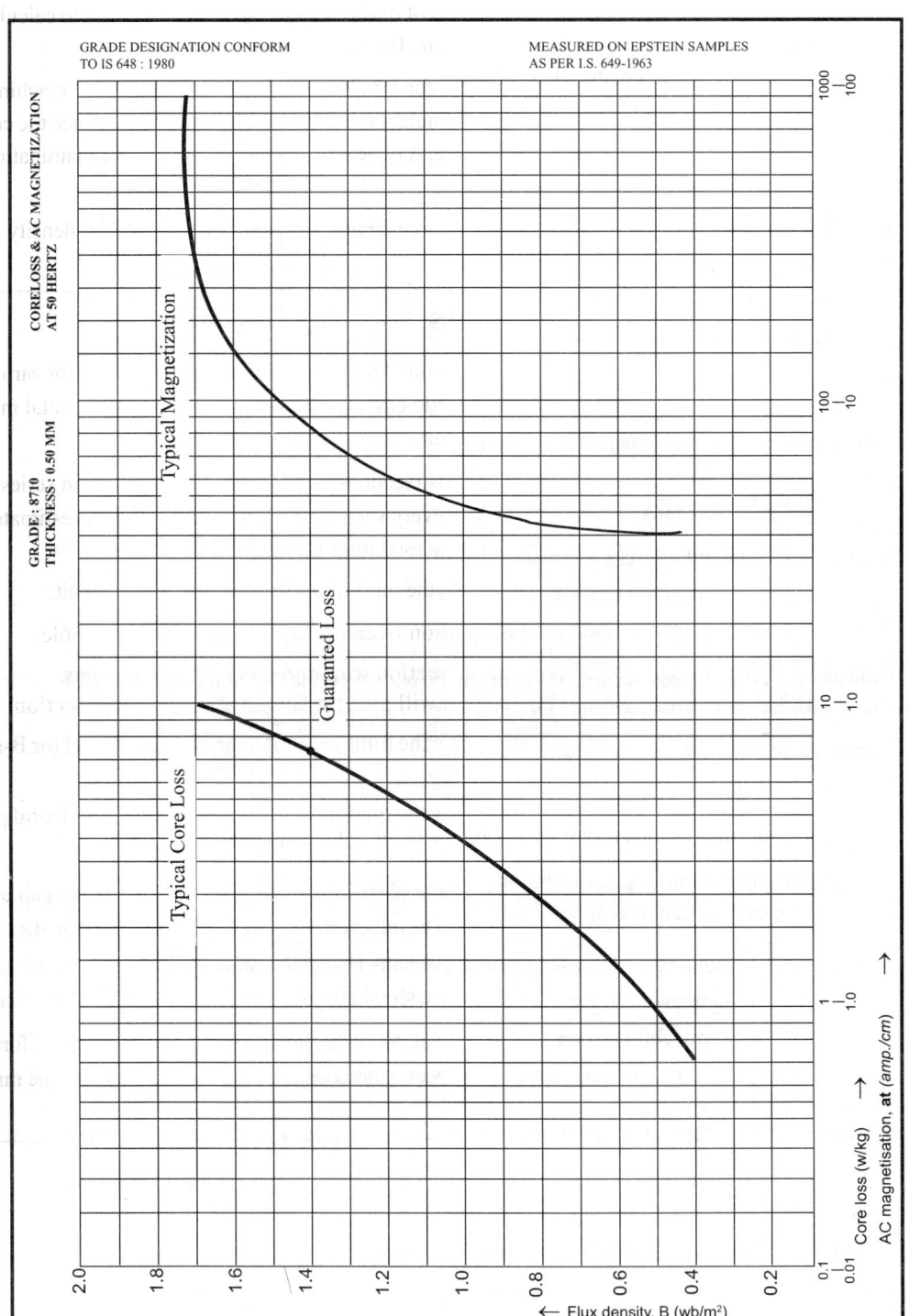

Fig. 3.2: B-H (B-at) curve and loss curve.

3.4 RELUCTANCE OF AIR-GAP IN MACHINES WITH SMOOTH ARMATURE

(AU, Apr' 17, 6 M)

The rotating machines will have a small air-gap between armature and pole surface. Smooth armature surfaces are possible only if the armature has closed slots. Consider the iron surfaces on the two sides of the air-gap to be smooth as shown in Fig. 3.3. The flux is uniformly spread over the entire slot pitch and goes straight across the air-gap.

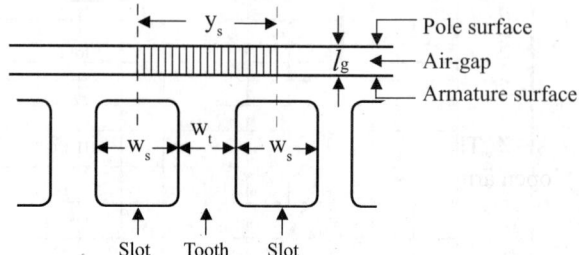

L = Length of armature

l_g = Air-gap length

y_s = Slot pitch

w_s = Width of slot

w_t = Width of tooth

Fig. 3.3: Smooth armature surface.

The reluctance of a magnetic path is given by,

$$\text{Reluctance, } S = \frac{l}{\mu A}$$

where, l = Length of magnetic path

μ = Permeability of the medium

A = Area of cross-section of the magnetic path

Consider the area of cross-section of the magnetic path over one slot of the armature. It is given by the product of the length of armature and slot pitch.

Therefore, in this case,

$$l = l_g, \quad \mu = \mu_o, \quad A = L\,y_s$$

Hence the reluctance of the air-gap in machine with closed armature slots can be written as,

$$\left.\begin{array}{l}\text{Reluctance of air-gap in machine} \\ \text{with closed armature slots}\end{array}\right\} S_{g\_cas} = \frac{l_g}{\mu_o\,L\,y_s} \qquad\qquad .....(3.11)$$

where, S_{g_cas} = Reluctance of air-gap in machine with closed armature slots

l_g = Length of air-gap

μ_o = Permeability of air

$L\,y_s$ = Area of cross-section of air-gap over one slot.

3.5 RELUCTANCE OF AIR-GAP IN MACHINES WITH OPEN ARMATURE SLOTS

In armature with open and semienclosed slots, the flux will flow through the teeth of the armature. Hence the effective area of flux path is decreased, which results in increased reluctance of air-gap.

Case (i) : Reluctance of air-gap neglecting fringing effect

Consider the armature with open type of slots as shown in Fig. 3.4. Here the flux is only confined to the tooth width. Hence the area of cross-section of the air-gap through which the flux passes is $L\,(y_s - w_s)$.

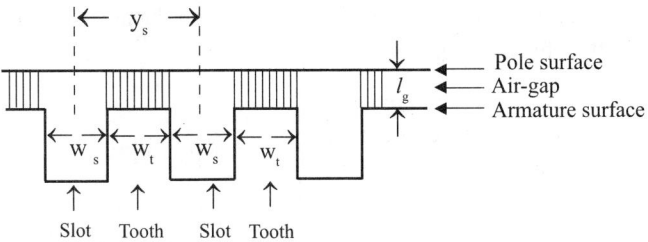

Fig. 3.4: *Slotted armature surface.*

Therefore, y_s is replaced by, $y_s - w_s$ in equation (3.11) for reluctance of air-gap in machine with open armature slots.

$$\left.\begin{array}{l}\text{Reluctance of the air-gap in}\\\text{machines with open armature slots}\end{array}\right\} S_{g\_oas} = \frac{l_g}{\mu_o\, L\,(y_s - w_s)} \qquad .....(3.12)$$

Case (ii) : Reluctance of air-gap including the effect of fringing

The equation (3.12) is applicable only if the fringing of the flux is neglected. In armatures with open slots the flux would fringe around the tooth and this fringing would increase the area of cross-section of flux path.

Consider the armature with open type slot shown in Fig. 3.5. Here the fringing of flux can be accounted by increasing the area of cross-section of flux path by δw_s as shown in Fig. 3.6.

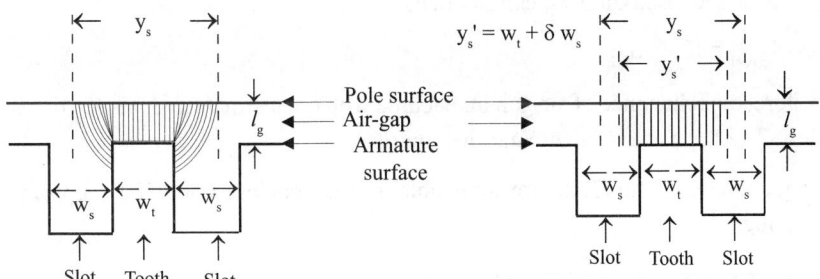

Fig. 3.5: *Slot with magnetic fringing.*

Fig. 3.6: *Contracted slot pitch to account for magnetic fringing.*

The reluctance in this case is more than that of a air-gap in smooth armature but lesser than that of the case where the whole flux is assumed to be confined over the tooth width.

A simple method to calculate reluctance in this case is to assume that the air-gap flux is uniformly distributed over the whole of slot pitch except for a fraction of slot width, as shown in Fig. 3.6. This fraction depends on the ratio of slot width to air-gap length. Thus the flux of one slot is distributed over $w_t + \delta\, w_s$.

Effective or contracted slot pitch, $y_s' = w_t + \delta\, w_s$ .....(3.13)

$$= w_t + w_s + \delta\, w_s - w_s$$

$$= y_s + \delta\, w_s - w_s$$

$$= y_s - (1 - \delta)\, w_s$$

$$= y_s - K_{cs}\, w_s \qquad .....(3.14)$$

where, $K_{cs} = (1 - \delta) =$ Carter's gap coefficient for slots.

Therefore, y_s is replaced by y_s' in equation (3.11) for reluctance of air-gap in armature with open slots including the effect of fringing.

$$\left.\begin{array}{l}\text{Reluctance of air-gap in machine}\\\text{with open armature slot including}\\\text{fringing effect}\end{array}\right\} S_{g\_oasf} = \frac{l_g}{\mu_o\, L\, y_s'}$$

$$= \frac{l_g}{\mu_o L(y_s - K_{cs} w_s)} \qquad\qquad .....(3.15)$$

The **gap contraction factor for slots,** K_{gs} is defined as the ratio of reluctance of air-gap in machine with open armature slot to reluctance of air-gap in machine with smooth armature.

$$\therefore K_{gs} = \frac{\text{Reluctance of air-gap in machine with open armature slot including fringing effect}}{\text{Reluctance of air-gap in machine with closed armature slot}}$$

$$= \frac{S_{g\_oasf}}{S_{g\_cas}}$$

$$= \frac{l_g/\mu_o L(y_s - K_{cs} w_s)}{l_g/\mu_o L(y_s)} = \frac{y_s}{y_s - K_{cs} w_s} = \frac{y_s}{y_s'} \qquad\qquad .....(3.16)$$

In equation (3.16) if fringing effect is neglected then $K_{cs} = 1$.

From the above discussion we can say that,

$$S_{g\_oasf} = K_{gs} \times S_{g\_cas}$$

Therefore, the reluctance of air-gap in machine with open armature slot is K_{gs} times that with closed armature slots. The K_{gs} has a value greater than unity.

The equations (3.12) to (3.16) are applicable for semi enclosed slots if we replace, w_s by w_o, where w_o is slot opening.

Estimation of Carter's gap coefficient for slots

The carter's gap coefficient, K_{cs} depends on the ratio of slot opening to gap length. The value of K_{cs} can be determined from the standard curve shown in Fig. 3.7. The curve shows relation between the w_o/l_g and K_{cs} where w_o = slot opening. ($w_o = w_s$ in open type slots).

The K_{cs} can also be calculated from the empirical formula,

$$\text{Carter's gap coefficient for slots, } K_{cs} = \frac{1}{1 + 5\, l_g/w_o} \qquad\qquad .....(3.17)$$

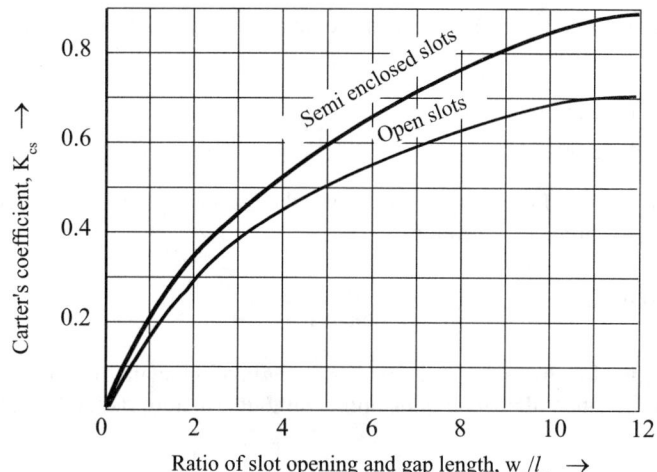

Fig. 3.7: Carter's air-gap coefficient.

Another useful relationship which can be used for calculation of Carter's coefficient for parallel sided open slots is,

$$K_{cs} = \frac{2}{\pi} \left[\tan^{-1} y - \frac{1}{\pi} \log \sqrt{(1+y^2)} \right] \qquad \text{.....(3.18)}$$

where, $y = \dfrac{w_o}{2l_g}$

3.6 EFFECT OF VENTILATING DUCTS ON RELUCTANCE OF AIR-GAP

(PU, Nov' 19, 6 M) (AU, Nov' 17, 8 M)

When the length of the armature is higher than the diameter or when the length is greater than 0.1 *m*, radial ventilating ducts are provided for better cooling of the core. The radial ventilating ducts are small gaps of width, w_d in between the stacks of armature core as shown in Fig. 3.8. The core is normally divided into stacks of 40 to 80 *mm* thick, with ventilating ducts of width 10 *mm* in between two stacks.

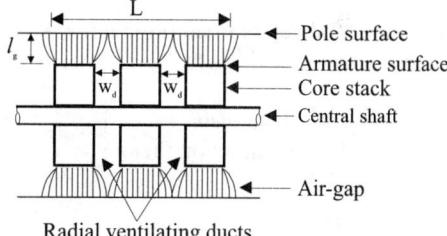

Fig. 3.8: Cross-section of armature with radial ventilating ducts.

The provision of radial ventilating ducts results in contraction of flux in the axial direction. Due to this the effective axial length of the machine is reduced and this results in an increase in the reluctance of air-gap.

We can derive an expression for gap contraction factor for ventilating ducts by treating stacks of laminations as teeth and the ducts as slots.

Contracted or effective axial length, $L' = L - K_{cd} \, n_d \, w_d$ .....(3.19)

where, K_{cd} = Carter's coefficient for ducts

       w_d = Width of the duct

       n_d = Number of the ducts

On replacing L by L' in equation (3.11) we get the reluctance of air-gap in machine with radial ventilating ducts.

$$\left.\begin{array}{l}\therefore \text{ Reluctance of air-gap in machine} \\ \text{with radial ventilating ducts}\end{array}\right\} S_{gd} = \frac{l_g}{\mu_o \, L' \, y_s} = \frac{l_g}{\mu_o \, (L - K_{cd} \, n_d \, w_d) \, y_s}$$

The values of K_{cd} can be obtained from Fig. 3.7 by using the ratio w_d/l_g in place of ratio w_o/l_g. The K_{cd} can also be calculated from the empirical formula,

$$\text{Carter's gap coefficient for ducts, } K_{cd} = \frac{1}{1 + 5l_g/w_d}$$

The **gap contraction factor for ducts,** K_{gd} is defined as the ratio of reluctance of air-gap in machine with armature radial ducts to reluctance of air-gap in machine without armature radial ducts.

$$\text{Gap contraction factor for ducts, } K_{gd} = \frac{S_{gd}}{S_{g\_oas}} = \frac{l_g/\mu_o L' y_s}{l_g/\mu_o L y_s} = \frac{L}{L'}$$

$$= \frac{L}{L - K_{cd} n_d w_d} \qquad\qquad .....(3.20)$$

3.7   TOTAL GAP CONTRACTION FACTOR

The effect of both open slots and ventilating ducts can be allowed for in a single expression.

$$\left.\begin{array}{l}\text{Reluctance of air-gap in machines} \\ \text{with open armature slots and without ducts}\end{array}\right\} S_{g\_cas} = \frac{l_g}{\mu_o L y_s}$$

$$\left.\begin{array}{l}\text{Reluctance of air-gap in machines} \\ \text{with open armature slots and ducts}\end{array}\right\} S_{g\_oasd} = \frac{l_g}{\mu_o L' y_s'}$$

The **total gap contraction factor,** K_g is defined as the ratio of reluctance of air-gap in machines with slotted armature and ducts to the reluctance of air-gap in machines with smooth armature and without ducts.

$$\left.\begin{array}{l}\text{Total gap} \\ \text{contraction factor}\end{array}\right\} K_g = \frac{S_{g\_oasd}}{S_{g\_cas}} = \frac{l_g/\mu_o L' y_s'}{l_g/\mu_o L y_s} = \frac{y_s}{y_s'} \times \frac{L}{L'}$$

$$= K_{gs} \times K_{gd} \qquad\qquad .....(3.21)$$

The total gap contraction factor is equal to the product of gap contraction factor for slots and ducts.

Gap contraction factor for Induction motor

In induction motor both the rotor and the stator has slots. Hence the gap contraction factor should be computed for both the stator and the rotor. The total gap contraction factor is given by the product of rotor and stator gap contraction factor.

$$K_{gs} = K_{gss} \, K_{gsr} \qquad\qquad .....(3.22)$$

K_{gs} = Total gap contraction factor for slots

K_{gss} = Gap contraction factor for stator slots

K_{gsr} = Gap contraction factor for rotor slots

The formulae used to calculate the Carter's coefficient for stator and rotor slots and the gap contraction factor for stator and rotor slots are given below.

$$\text{Carter's coefficient for stator slots, } K_{css} = \frac{1}{1 + 5l_g/w_{os}}$$

$$\text{Carter's coefficient for rotor slots, } K_{csr} = \frac{1}{1 + 5l_g/w_{or}}$$

$$\text{Gap contraction factor for stator slots, } K_{gss} = \frac{y_{ss}}{y_{ss} - K_{css} w_{os}}$$

$$\text{Gap contraction factor for rotor slots, } K_{gsr} = \frac{y_{sr}}{y_{sr} - K_{csr} w_{or}}$$

where, y_{ss} = Stator slot pitch

w_{os} = Stator slot opening

y_{sr} = Rotor slot pitch

w_{or} = Rotor slot opening

$$\text{Stator slot pitch, } y_{ss} = \frac{\text{Stator inner diameter}}{\text{Number of stator slots}} = \frac{D}{S_s}$$

$$\text{Rotor slot pitch, } y_{sr} = \frac{\text{Rotor outer diameter}}{\text{Number of rotor slots}} = \frac{D - 2l_g}{S_r}$$

where, D = Stator inner diameter

S_r = Number of rotor slots

S_s = Number of stator slots

3.8 MMF FOR AIR-GAP
(AU, Nov' 18, 6 M)

Non-magnetic materials (like air, copper, etc.,) have a constant value of permeability and so the B-H curve for them is a straight line passing through the origin.

$$\left.\begin{array}{l}\text{mmf per meter of the path} \\ \text{in non-magnetic material}\end{array}\right\} at_{nm} = \frac{B}{\mu_o} = \frac{B}{4\pi \times 10^{-7}} \approx 800,000 \, B \text{ in } AT/m \qquad .....(3.23)$$

where, B = Flux density in the non-magnetic material

μ_o = Permeability of non-magnetic material

mmf for air-gap in rotating machines

The iron surfaces around the air-gap are not smooth and so the calculation of mmf for the air-gap requires special attention. The non-uniform air-gaps are due to the following,

1. In machines with open or semi-enclosed slots, the flux concentrates on teeth, i.e., the flux is not uniformly distributed in the air-gap.

2. There may be radial ventilating ducts in the machine for cooling purposes. This results in contraction of flux in the axial direction.

3. In salient pole machines, the air-gap dimensions are not constant over whole of the pole pitch.

mmf of air-gap in machines with smooth armature (or closed armature slots)

mmf per metre for air-gap, $\quad at_g = \dfrac{B_{av}}{\mu_o} = \dfrac{B_{av}}{4\pi \times 10^{-7}} \approx 800,000\, B_{av}$ .....(3.24)

where, B_{av} = Average flux density in the air-gap

         μ_o = $4\pi \times 10^{-7}\ H/m$ = Permeability of air-gap

If l_g is the length of air-gap, then

$\left.\begin{array}{l}\text{mmf required for air-gap of length } l_g \\ \text{in machines with closed armature slots}\end{array}\right\}$ $AT_g = at_g \times l_g = 800,000\, B_{av} l_g$ .....(3.25)

mmf of air-gap in machines with open armature slot and radial ventilating ducts

The reluctance of air-gap in machines with open armature slots is higher than with closed armature slots. The ratio of the two reluctances is equal to K_g, the gap contraction factor. Hence the mmf required for air-gap in machines with open armature slot is K_g times the mmf required for air-gap with closed armature slots.

$\left.\begin{array}{l}\text{mmf required for air-gap in machines} \\ \text{with open armature slots and ducts}\end{array}\right\}$ = $K_g \times AT_g$ for air-gap in machines with smooth armature surface

$$= K_g \times 800,000\, B_{av}\, l_g$$

$$= 800,000\, B_{av}\, K_g\, l_g \qquad\qquad .....(3.26)$$

Note :   1. K_g is also called gap expansion factor

          2.   Effective air-gap, $l_{ge} = K_g\, l_g$

          3.   Area of contracted air-gap, $A_g' = A_g / K_g$

               where, A_g = Actual area of air-gap per pole, m^2

          4.   If the armature with open slot does not have ducts, then $K_g = K_{gs}$

          5.   If the armature has smooth surface and radial ventilating ducts then $K_g = K_{gd}$.

Effect of Saliency on the mmf for air-gap

In case of salient pole machines, the length of air-gap is not constant over the whole pole pitch. To find the mmf in this case, we can consider the length of air-gap as an effective gap given by $K_{gsal}\, l_g$, where K_{gsal} is the gap contraction factor for salient poles.

For calculating AT_g (mmf for air-gap) in salient pole machines, the maximum gap density, B_g at the centre of the pole is considered instead of average gap density.

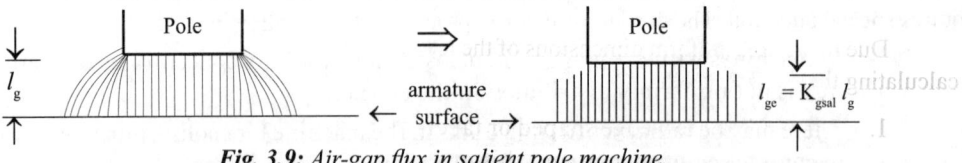

Fig. 3.9: Air-gap flux in salient pole machine.

$\left.\begin{array}{l}\therefore \text{ mmf for air-gap in salient}\\ \text{pole machines (DC/AC machines)}\end{array}\right\} = 800,000 \, B_g \, K_g \, l_g$ ....(3.27)

where, $K_g = K_{gs} \, K_{gd} \, K_{gsal}$

Here K_g is the total gap contraction factor including the effect of saliency.

The maximum flux density can be estimated from the knowledge of field form factor, K_f or using the ratio of pole arc to pole pitch, Ψ as shown below.

$$\text{Field form factor, } K_f = \frac{\text{Average gap density over the pole pitch}}{\text{Maximum flux density in the air-gap}} = \frac{B_{av}}{B_g}$$

$$\text{Also, } K_f \approx \Psi = \frac{\text{Pole arc}}{\text{Pole pitch}} = \frac{b}{\tau}$$

On equating the above two equations of K_f we get,

$$\frac{b}{\tau} = \frac{B_{av}}{B_g}$$

$\therefore$ Maximum flux density in air-gap, $B_g = \dfrac{B_{av}}{b/\tau}$

$$= B_{av} \times \frac{\tau}{b} \qquad\qquad \boxed{B_{av} = \frac{p\phi}{\pi DL}}$$

$$= \frac{p\phi}{\pi DL} \times \frac{\pi D}{p} \times \frac{1}{b} \qquad\qquad \boxed{\tau = \frac{\pi D}{p}}$$

$$= \frac{\phi}{Lb} \qquad\qquad\qquad .....(3.28)$$

3.9 MMF FOR TEETH

(AKTU, Dec' 19, 10 M)

The mmf required for teeth depends on area of cross-section of a tooth and flux passing through it. The area of cross-section depends on the dimensions of a tooth which in turn depends on the type of slot. The Fig. 3.10 shows some typical cross-section of teeth and it can be observed that the width of the teeth is not constant.

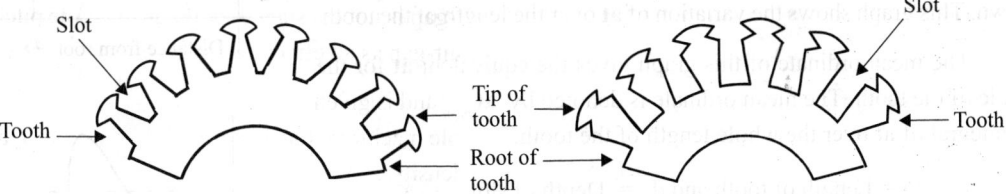

Fig. 3.10: Section of armature lamination showing the non-uniform dimension of teeth.

Due to the non-uniform dimensions of the teeth, the following problems may be encountered while calculating the mmf for teeth.

1.   The teeth are wedge shaped or tapered. This means that the area presented to the path of flux is not constant and this gives different values of flux density over the length a tooth.

2.   The slot provides another parallel path for the flux, shunting the tooth. The teeth are normally worked in the saturation region and therefore their permeability is low, and as a result an appreciable portion of the flux goes down the depth of the slots. The presence of two parallel paths and the reluctance of one part depending upon the degree of saturation in the other, makes the problem intricate.

The mmf required for teeth can be easily calculated whatever may be their shape, if the flux going down the slot is neglected. The correction, to take slot flux into account, can be incorporated later on.

There are three methods employed for the calculation of mmf required for teeth. They are :

     1.   Graphical Method
     2.   Three Ordinate Method (Simpson's rule)
     3. $B_{t1/3}$ Method.

Graphical Method

In this method, first the flux density at various sections of a tooth are determined.

The flux density at any section of a tooth can be estimated using the following equation.

$$\left.\begin{array}{l}\text{Flux density at a section of tooth}\\ \text{with area of cross-section } A_t\end{array}\right\} B_t = \frac{\phi}{n_t A_t} \qquad .....(3.29)$$

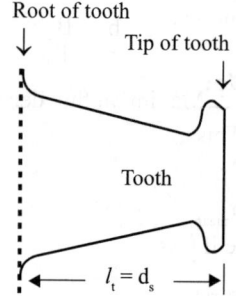

where, A_t =   Area of cross-section of tooth at the desired section

       B_t =   Flux density of tooth corresponding to A_t

       n_t =   Number of teeth under a pole

       ϕ =   Flux per pole

A graph between flux density and distance from the root of a tooth is drawn. This graph shows the variation of flux density over the tooth length, l_t. Then for each point of tooth the mmf per metre, **at** is found from the knowledge of **B-at** curve. A graph between **at** and distance is drawn. This graph shows the variation of **at** over the length of the tooth.

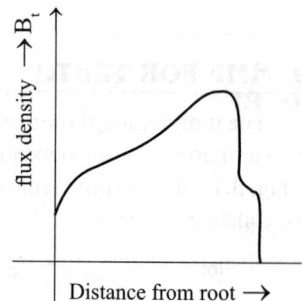

The mean ordinate of this graph gives the equivalent **at** for the whole of the tooth. The mean ordinate is denoted by **at**$_{mean}$ and is given by integral of **at** over the whole length of the tooth.

Let, l_t = Length of tooth and d_s = Depth of slot.

Here, $l_t = d_s$

$$\therefore \text{ at}_{mean} = \int \text{ at } dl_t$$

Total mmf for tooth, $AT_t = \text{at}_{mean} \times l_t$

$$= \text{at}_{mean} \times d_s \qquad .....(3.30)$$

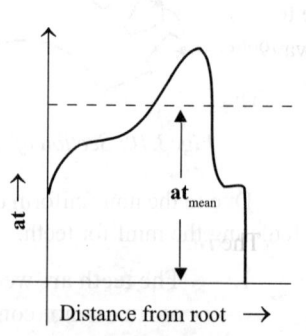

Fig. 3.11

Three Ordinate Method (Simpson's rule)

This method can be applied to teeth of very simple form and having small taper. This method is based on the assumption that the curve relating mmf per metre, **at** with flux density is a parabola.

In this method the flux density and the corresponding values of mmf per metre, **at** are obtained at three equidistant points. The three points chosen are root, centre and tip of a tooth. The flux density at these three points are estimated using the equation (3.29) and the corresponding **at** are determined from **B-at** curve.

Let at_1 = at for the root of tooth.

        at_2 = at for the centre of tooth.

        at_3 = at for the tip of tooth.

Mean value of **at**, $at_{mean} = \dfrac{at_1 + 4at_2 + at_3}{6}$

mmf required for the tooth, $AT_t = at_{mean} \times l_t = at_{mean} \times d_s$

$B_{t1/3}$ Method

This method is applied to teeth of small taper. This method is based upon the assumption that value of mmf per metre, **at** obtained for flux density at a section one-third of tooth height from the narrow end is the at_{mean}. This method is the most simple of all the methods. The results in this method are sufficiently accurate if the teeth are worked at low saturation.

First calculate the flux density at one-third height from the narrow end using equation (3.29). Then from **B-at** curve, find the value of **at** for this flux density. Let this **at** be denoted as $at_{1/3}$.

Total mmf for tooth, $AT_t = at_{1/3} \times l_t = at_{1/3} \times d_s$ .....(3.31)

3.10 REAL AND APPARENT FLUX DENSITIES   (KTU, Feb' 18, 8 M)

The flux entering an armature from the air-gap flows in teeth. If the flux density in the teeth is very high then the mmf acting on the teeth is high. Since the slots are in parallel with teeth, this mmf will act on the slots also. Thus some of the fluxes pass through slots. At higher flux densities the flux passing through the slots becomes large and cannot be neglected.

Hence the real flux passing through the teeth is always less than the total or apparent flux. As a result, the real flux density in the teeth is always less than the apparent flux density.

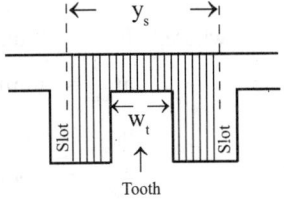

The **apparent flux density** is defined as,

$B_{app} = \dfrac{\text{Total flux in a slot pitch}}{\text{Tooth area}}$

Fig. 3.12: Slot flux due to saturation in teeth.

The **real flux density** is defined as,

$B_{real} = \dfrac{\text{Actual flux in a tooth}}{\text{Tooth area}}$

In an actual machine, there are two parallel paths for the flux over one slot pitch. They are iron path of tooth and air and conductor path of slot.

Let, ϕ_i = Flux passing through iron (tooth) over a slot pitch.

        ϕ_a = Flux passing through slot (air and conductor) over a slot pitch.

y_s = Slot pitch		w_t = Width of tooth
L = Length of slot		μ_o = Permeability of air = $4\pi \times 10^{-7}$ H/m
L_i = Net iron length		A_i = Area of iron path of tooth
A_a = Area of air path of slot		ϕ_s = Flux over one slot pitch

Apparent flux density, $B_{app} = \dfrac{\text{Total flux over a slot pitch}}{\text{Iron area over a slot pitch}}$

$$= \frac{\phi_s}{A_i} = \frac{\phi_i + \phi_a}{A_i} \qquad \boxed{\phi_s = \phi_i + \phi_a}$$

$$= \frac{\phi_i}{A_i} + \frac{\phi_a}{A_i} = B_{real} + \frac{\phi_a}{A_a} \times \frac{A_a}{A_i} \qquad \boxed{\begin{array}{l}\text{Multiply and} \\ \text{divide by } A_a.\end{array}}$$

$$= B_{real} + B_a K \qquad \boxed{\dfrac{\phi_i}{A_i} = B_{real}} \quad \boxed{\dfrac{\phi_a}{A_a} = B_a} \quad .....(3.32)$$

where, $K = \dfrac{A_a}{A_i} = \dfrac{\text{Air area (over slot)}}{\text{Iron area (over tooth)}}$

Let, $K_s = \dfrac{\text{Total area}}{\text{Iron area}} = \dfrac{A_i + A_a}{A_i}$

$$= 1 + \frac{A_a}{A_i} = 1 + K \qquad .....(3.33)$$

From equation (3.32) we can write,

Real flux density, $B_{real} = B_{app} - B_a K$ $\boxed{B_a = \mu_o \, at_{real}}$

$$= B_{app} - \mu_o \, at_{real} K \qquad \boxed{\text{Using equation (3.33)}}$$

$$= B_{app} - \mu_o \, at_{real} (K_s - 1) \qquad .....(3.34)$$

where, B_a =   Flux density in air = $\mu_o H = \mu_o \, at_{real}$

      at_{real} =   mmf per metre across the tooth for tooth density B_{real}

      μ_o = $4\pi \times 10^{-7}$ H/m

      A_i =   Tooth width × Net iron length = $w_t L_i$

      A_a =   Total area – Net iron area = $L y_s - w_t L_i$

      K_s =   Total area / Iron area = $L y_s / L_i w_t$

      L_i = $S_f L$

      S_f =   0.9

3.11 MAGNETIZING CURRENT

The magnetizing current is decided by the total mmf required to establish the required flux in the magnetic circuit of a machine. The magnetizing current also depends on the number of turns in the exciting winding and the way in which the winding is distributed.

3.11.1 Magnetizing Current for Concentrated Windings

In a concentrated winding, it is assumed that whole of the flux links with all the turns.

$$\therefore \text{ Magnetizing current, } I_m = \frac{AT}{T} \qquad \qquad .....(3.35)$$

where, AT = Total mmf

N = Number of turn in winding

In transformers, AT is always calculated on the basis of maximum flux density.

$$\left. \begin{array}{l} \therefore \text{ Maximum value} \\ \text{of magnetizing current} \end{array} \right\} I_{m\_max} = \frac{AT_m}{T} \qquad \qquad .....(3.36)$$

$$\left. \begin{array}{l} \therefore \text{ Rms value of} \\ \text{magnetizing current} \end{array} \right\} I_{m\_rms} = \frac{I_{m\_max}}{K_p} \qquad \qquad .....(3.37)$$

where, K_p = Peak factor

3.11.2 Magnetizing Current for Distributed Windings

In a distributed winding, the flux may be sinusoidal or non-sinusoidal. When the flux distribution is sinusoidal the magnetizing current can be directly calculated from total mmf. But when the flux distribution is non-sinusoidal then the harmonics in flux wave has to be analysed and the magnetizing current is calculated using harmonics having sufficient strength.

(i) Sinusoidal flux distribution

In a winding, if the flux is steady and is sinusoidally distributed in space then,

$$\text{Magnetizing current, } I_m = \frac{AT_m}{T} \qquad \qquad .....(3.38)$$

$$\left. \begin{array}{l} \text{Rms value of} \\ \text{magnetizing current} \end{array} \right\} I_{m\_rms} = \frac{AT_m}{\sqrt{2}\,T} \qquad \qquad .....(3.39)$$

where, AT_m = mmf for the peak flux density B_m.

In a three-phase winding uniformly distributed in space with $120°$ phase spread and excited with balanced sine wave currents the magnetizing mmf and current are given by following equations.

$$\left. \begin{array}{l} \text{Amplitude of fundamental} \\ \text{magnetizing mmf} \end{array} \right\} AT_{m1} = \frac{2.7\, I_m\, T_s\, K_{w1}}{p} \qquad \qquad .....(3.40)$$

$$\therefore \text{ Magnetizing current per phase, } I_{m\_nsd} = \frac{p}{2.7 \, T_s \, K_{w1}} \, AT_{m1}$$

(Non-sinusoidal flux distribution)

$$= \frac{0.37 \, p \, AT_{m1}}{T_s \, K_{w1}} \qquad \qquad .....(3.41)$$

where, T_s = Number of stator turns per phase

K_{w1} = Winding factor for fundamental

p = Number of poles

(ii) Non-sinusoidal flux distribution

The flux distribution becomes non-sinusoidal due to saturation in iron parts. Under such situation, the mmf produces a flat topped flux wave, which may be represented by sum of fundamental and odd harmonics flux wave.

Let, B_o = Flux density at an angle, θ from neutral axis

Now, $B_o = B_{m1} \sin \theta + B_{m3} \sin 3\theta + B_{m5} \sin 5\theta + ...$

where, B_{m1}, B_{m3} and B_{m5} are the maximum values of fundamental, 3rd harmonic and 5th harmonic flux densities respectively.

Generally, the harmonics above the third are small in magnitude and so the flux density wave shape can be considered to consist of a fundamental sine wave with a super-imposed third harmonic as shown in Fig. 3.13.

In order to calculate the required mmf any closed path can be chosen. The knowledge of flux density along that path is required to calculate mmf. If mmf is calculated along pole centre then knowledge of fundamental and third harmonic are required. On the otherhand if the mmf is calculated along 60° from interpolar axis then knowledge of fundamental mmf wave is sufficient and this is sinusoidal too.

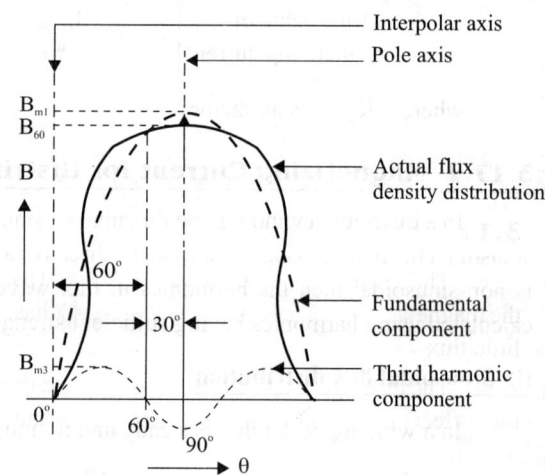

Fig. 3.13: *Flux density distribution in space with its fundamental and third-harmonic components.*

Let, B_{60} = Flux density at 60° from interpolar axis.

$$\therefore B_{60} = B_{m1} \sin 60° + B_{m3} \sin (3 \times 60°)$$

$$= B_{m1} \sin 60° \qquad \qquad \boxed{\sin 180° = 0}$$

$$= \frac{\sqrt{3}}{2} \, B_{m1} \qquad \qquad .....(3.42)$$

It is clear from the above expression that B_{60} is the same whether third harmonic component is present or not.

Here, B_{m1} is sinusoidally distributed.

$$\therefore \ B_{m_1} = \frac{\pi}{2} \ B_{av_1} = \frac{\pi}{2} \ \frac{\phi_1}{A_1} \qquad\qquad \text{.....(3.43)}$$

where, ϕ_1 = Fundamental flux per pole

A_1 = Effective area of each fundamental pole

B_{av1} = Average flux density in air-gap = $\frac{\phi_1}{A_1}$ (fundamental component)

From equations (3.42) and (3.43) we can write,

$$B_{60} = \frac{\sqrt{3}}{2} \ B_{m1} = \frac{\sqrt{3}}{2} \ \frac{\pi}{2} \ B_{av_1}$$

$$= 1.36 \ B_{av1}$$

Let, AT_{60} = mmf acting at 60° from interpolar axis

$$\therefore \ AT_{60} = AT_{m_1} \sin 60° = \frac{\sqrt{3}}{2} \ AT_{m_1}$$

$$= \frac{\sqrt{3}}{2} \times 2.7 \ \frac{I_{m\_sd} \ T_s \ K_{w_1}}{p} \qquad \boxed{\text{Using equation (3.40)}} \ \text{.....(3.44)}$$

Magnetizing current per phase, $I_{m\_sd} = \dfrac{2}{\sqrt{3} \times 2.7} \ \dfrac{p \ AT_{60}}{K_{w1} \ T_s} = \dfrac{0.427 \ p \ AT_{60}}{K_{w_1} \ T_s} \qquad \text{.....(3.45)}$
(Sinusoidal flux distribution)

3.12 FLUX LEAKAGE   *(PU, Nov' 19, 6 M)*

The flux which passes through unwanted path is called the leakage flux. It is impossible to confine all the magnetic flux to a given path. The designer has to provide a path of low reluctance so that comparatively little flux leaks away from the desired path.

The leakage flux does not contribute to either transfer or conversion of energy. However, the leakage flux affects the performance of transformers and rotating machines. The leakage flux affects the following performance indices of various machines.

1. Excitation demand of salient pole machines
2. Performance of AC machines depends on the leakage reactance
3. Forces between the windings under short circuit conditions
4. Voltage regulation of generators and transformers
5. Commutation conditions in DC machines
6. Stray load losses
7. Circulating currents in transformer tank walls.

For magnetic circuit calculations, a term leakage coefficient is introduced in order to take into account the leakage flux. The *leakage coefficient* is defined as the ratio of total flux to useful flux.

$$\text{Leakage coefficient, } C_l = \frac{\text{Total flux}}{\text{Useful flux}} = \frac{\text{Useful flux + Leakage flux}}{\text{Useful flux}}$$

Types of leakage flux

The ***armature leakage fluxes*** affects most of the performances of rotating machines. Hence the different types of armature leakage fluxes are discussed in this section. The different types of armature leakage fluxes are.

1. Slot leakage flux
2. Tooth top leakage flux
3. Zigzag leakage flux
4. Overhang leakage flux
5. Harmonic or differential leakage flux
6. Skew leakage flux
7. Peripheral leakage flux.

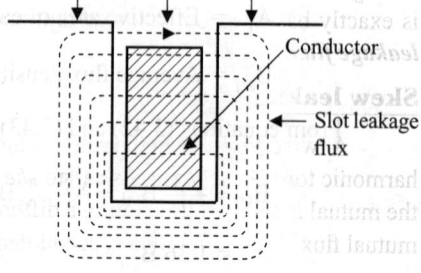

Fig. 3.14: Slot leakage flux.

Slot leakage flux

The fluxes that crosses the slot from one tooth to the next and returning through iron are called ***slot leakage flux***. They link the conductors below them, as shown in Fig. 3.14.

Tooth top leakage flux

The flux flowing from top of one tooth to the top of another tooth as shown in Fig. 3.15 is called ***tooth top leakage flux***. This leakage flux is considered only in machines having large air-gap length like DC machines and synchronous machines. Since in induction machines the air-gap length is very small the tooth top leakage flux is negligible.

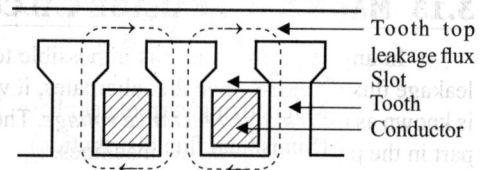

Fig. 3.15: Tooth top leakage flux.

Zigzag leakage flux

The flux passing from one tooth to another in a zigzag fashion across the air-gap as shown in Fig. 3.16 is called ***zigzag leakage flux***. The magnitude of this flux depends on the length of air-gap and the relative positions of the tips of rotor and stator teeth.

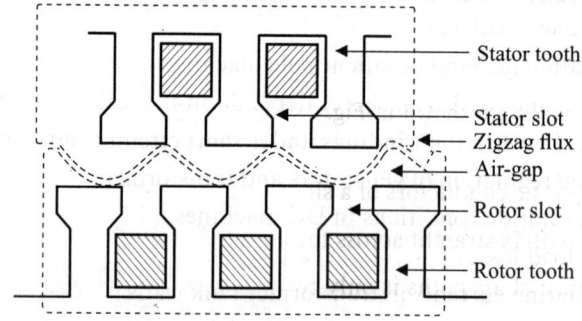

Fig. 3.16: Zigzag leakage flux.

Overhang leakage flux

The end connections (The conductor which connects the two sides of a coil) are called ***overhang***. The fluxes produced by the overhang portion of the armature winding are called ***overhang leakage flux***. It depends on the arrangement of overhang and the nearby metal parts (for example core stiffners and end covers).

Harmonic (or Differential or Belt) leakage flux

The harmonic leakage flux are due to dissimilar mmf distribution in the stator and rotor. Actually the difference in the harmonic contents of stator and rotor mmfs produces harmonic leakage fluxes. In squirrel cage induction motor the rotor current is exactly balanced by stator current and so there is no *harmonic leakage flux.*

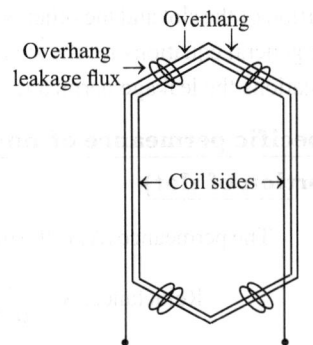

Skew leakage flux

A twist provided in the rotor of induction motors to eliminate harmonic torques and noise is called *skewing*. The skewing reduces the mutual flux and thus creating a difference between total flux and mutual flux. This difference is accounted as *skew leakage flux*.

Fig. 3.17: A three turn armature coil.

Peripheral leakage flux

The fluxes flowing circumferentially round the air-gap without linking with any of the windings are called *peripheral leakage flux*. Usually this leakage flux is negligible in most of the machines.

3.13 MAGNETIC LEAKAGE CALCULATIONS

In any magnetic path it is impossible to confine all the flux in the useful paths. There is always a leakage flux. If the leakage flux alternates, it will induce voltage in any winding with which it links. This is known as the *leakage reactance voltage*. The reactance corresponding to this voltage plays an important part in the performance of AC machines.

In DC machines, the leakage flux passing in non-useful paths affects the field excitation of the machines. The excitation has to be increased to compensate for loss of flux. This leakage flux also induces reactance voltage in the commutator, which opposes the reversal of current and makes the commutation difficult.

The geometry of leakage path is highly complex and so exact mathematical estimation of leakage flux is difficult. Usually the theoritical values are compared with experimental data to verify the accuracy of theoritical estimation. The estimation of slot leakage permeance for various types of slots are attempted in this book.

Slot Leakage

The slot leakage fluxes are shown in Fig. 3.14. The slot leakage can be calculated by making the following assumptions.

1. The current in the conductors of a slot is uniformly distributed over their cross-section.

2. The leakage path is straight across the slot and around the iron at the bottom.

3. The permeance of air paths is only considered. The reluctance of iron paths is assumed as zero.

The slot leakage permeance depends upon the shape of the slot and the arrangement of the conductors in the slot.

In general the total height (or depth) of the slot is divided into smaller sections and the specific permeance of each section is estimated. Then the specific permeance of all the sections are added to get the total specific permeance of the slot.

While dividing the slot into smaller sections we may come across two general cases, one is conductor portion of the slot and the other is non-conductor portion (or air-path or insulation portion) of the slot. First the general equations for the specific permeance of these two cases has been derived and then using these equations the leakage permeance of different types of slots are estimated.

Specific permeance of non-conductor portion of slot (or air-path or insulation portion of slot)

The permeance (Λ) is the inverse of reluctance (S). The reluctance of a magnetic material is given by,

$$\text{Reluctance, } S = \frac{l}{\mu A} \qquad \qquad .....(3.46)$$

where, l = Length of flux path.

            A = Area of cross-section of flux path.

            μ = Permeability of the medium.

$$\therefore \text{Permeability of air-path, } \Lambda_a = \frac{1}{S} = \frac{\mu_o A}{l} = \mu_o \times \frac{\text{Area of flux path}}{\text{Length of flux path}} \qquad .....(3.47)$$

where, μ_o = Permeability of free space

Let the area of cross-section of the non-conductor portion of the slot be rectangular in shape as shown in Fig 3.18.

Let,   y = Length of field

        L = Depth of field

        h = Height of field

Also, L = Length of slot or length of armature

$\therefore$ Area of cross-section of flux path, $A = L\,h$

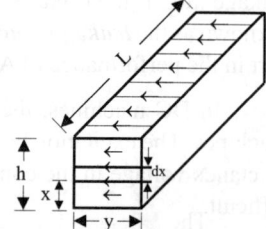

Fig. 3.18: Leakage flux in non-conductor portion of a slot.

Consider a small height, dx in the total height, h of the field.

Now, $h = dx$, $\therefore A = L\,dx$

Also, in this case $L = y$.

$\therefore$ Permeance of the section with height dx, $\delta\Lambda_{sa} = \mu_o \dfrac{L\,dx}{y}$ $\boxed{\text{Using equation (3.47)}}$ .....(3.48)

The permeance of the non-conductor portion of slot can be obtained by integrating the equation (3.48) between limits 0 to h as shown below.

$$\left.\begin{array}{l}\text{Permeance of non-conductor}\\ \text{portion of slot}\end{array}\right\} \Lambda_{sa} = \int_o^h \delta\Lambda_{sa} = \int_o^h \frac{L}{y}\,dx = L\int_o^h \frac{1}{y}\,dx \qquad .....(3.49)$$

The specific permeance is defined as the permeance per unit length of slot or depth of field.

$$\left.\begin{array}{l}\therefore \text{Specific permeance for}\\ \text{non-conductor portion of slot}\end{array}\right\} \lambda_{sa} = \frac{\text{Permeance}}{\text{Depth of field}} = \frac{\Lambda_{sa}}{L} = \mu_o \int_o^h \frac{dx}{y} \qquad .....(3.50)$$

Specific permeance for conductor portion of slot

In the conductor portion of the slot, only a part of the leakage flux will link with all the conductors. (But in non-conductor portion of the slot, the leakage flux will link the entire non-conductor portion of a slot.)

Hence an effective flux is assumed for conductor portion of slot. The effective flux is a hypothetical flux that is assumed to link with all the conductors of a slot producing same amount of flux linkages as are produced by actual arrangement of winding and field.

$$\text{Effective flux} = \frac{\text{Total flux linkages}}{\text{Total number of conductors}} \quad .....(3.51)$$

$$\text{Permeance} = \frac{\text{Effective flux}}{\text{Total mmf}}$$

Using the above concepts, a separate equation for specific permeance of conductor portion of slot is derived as shown below.

Let the area of cross-section of the conductor portion of the slot be rectangular in shape as shown in Fig. 3.18.

Consider a small section of the conductor portion of slot of height dx at a distance x from the bottom of slot. Let Z_x be the number of conductors upto a height of x from the bottom of slot. The flux linkages associated with this small number of conductors can be estimated as shown below.

Let,   dx   = Height of small section in the total height h of the field

        Z_x = Number of conductors below the section of height dx

        $d\phi_x$ = Leakage flux linking with Z_x conductors

        I_z = Current in each conductor

        mmf_x = mmf required to produce the flux $d\phi_x$

Now, $mmf_x = I_z Z_x$ .....(3.52)

The flux $d\phi_x$ is given by the product of the mmf required to produce this flux and the permeance of the section with height dx.

$$\therefore d\phi_x = mmf_x \times \text{Permeance}$$

Using equations (3.52) and (3.48).

$$= I_z Z_x \mu_o L \frac{dx}{y} \quad ....(3.53)$$

$$\text{Flux linkages associated with } Z_x \text{ conductors} = Z_x d\phi_x \quad .....(3.54)$$

On substituting for $d\phi_x$ from equation (3.53) in equation (3.54) we get,

$$\left.\begin{array}{l}\text{Flux linkages associated} \\ \text{with } Z_x \text{ conductors}\end{array}\right\} = Z_x \times I_z Z_x \mu_o L \frac{dx}{y} = \mu_o L I_z Z_x^2 \frac{dx}{y} \quad .....(3.55)$$

The total flux linkages in the conductor portion with height **h** is obtained by integrating the equation (3.55) between limits 0 to h as shown below.

$$\therefore \text{Total flux linkages} = \int_0^h \mu_o L I_z Z_x^2 \frac{dx}{y} = \mu_o L I_z \int_0^h Z_x^2 \frac{dx}{y} \quad .....(3.56)$$

$$\text{Effective flux} = \frac{\text{Total flux linkages}}{\text{Total conductors per slot}} = \frac{\mu_o L I_z}{Z_s} \int_0^h Z_x^2 \frac{dx}{y} \quad .....(3.57)$$

where, Z_s = Total number of conductors in a slot

       Total mmf $= I_z Z_s$ .....(3.58)

$$\left.\begin{array}{l}\text{Permeance of conductor}\\\text{portion of slot}\end{array}\right\} \Lambda_{sc} = \frac{\text{Effective flux}}{\text{Total mmf}} = \frac{\dfrac{\mu_0 L I_z}{Z_s} \displaystyle\int_0^h Z_x^2 \, \dfrac{dx}{y}}{I_z Z_s}$$

$$= \frac{\mu_0 L}{Z_s^2} \int_0^h Z_x^2 \, \frac{dx}{y} = \mu_0 L \int_0^h \left(\frac{Z_x}{Z_s}\right)^2 \frac{dx}{y} \qquad\qquad .....(3.59)$$

Specific permeance is defined as the permeance per unit length of slot or depth of field.

$$\left.\begin{array}{l}\therefore \text{Specific permeance for}\\\text{conductor portion of slot}\end{array}\right\} \lambda_{sc} = \frac{\text{Permeance}}{\text{Depth of field}} = \frac{\Lambda_{sc}}{L} = \mu_0 \int_0^h \left(\frac{Z_x}{Z_s}\right)^2 \frac{dx}{y} \qquad .....(3.60)$$

3.14 LEAKAGE IN ARMATURE

3.14.1 Leakage Permeance of Parallel Sided Slot   *(HTU, Dec' 18, 10 M)*

The slot with parallel sides is shown in Fig. 3.19.

Let, Z_s = Total number of conductors per slot

      w_0 = Slot opening

      w_s = Slot width

      h = Depth of slot

Also,   h = Height of field

The total depth of the slot is divided into different sections of height h_1, h_2, h_3 and h_4 as shown in Fig. 3.19 and the specific permeance of each section be λ_1, λ_2, λ_3 and λ_4.

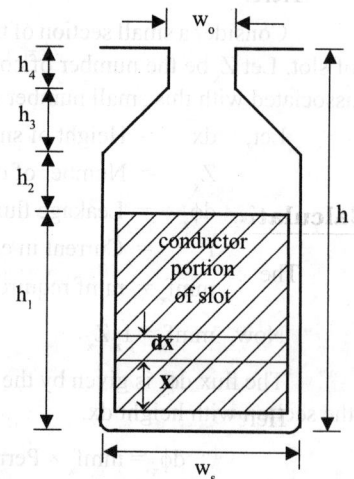

Fig. 3.19: Parallel sided slot.

Total specific slot permeance, $\lambda_s = \lambda_1 + \lambda_2 + \lambda_3 + \lambda_4$

Calculation of λ_1

Here h_1 is the depth over which conductors occupy the slot. The permeance of this section can be obtained by considering a strip of height **dx** at a distance x from the bottom.

The equation (3.60) can be used to calculate the permeance of the conductor portion of the slot.

$$\lambda_{sc} = \mu_0 \int_0^h \left(\frac{Z_x}{Z_s}\right)^2 \frac{dx}{y} \qquad\qquad\qquad \boxed{\text{Equation (3.60)}}$$

Here, $\lambda_{sc} = \lambda_1$; $y = w_s$; $h = h_1$

Z_x = Number of conductors below the section of height dx.

$$= \frac{x}{h_1} Z_s$$

> Flux at a distance x from bottom of conductors links with x/h$_1$ of the conductors.

$$\therefore \text{ Specific permeance of section-1, } \lambda_1 = \mu_o \int_0^{h_1} \left(\frac{\frac{X}{h_1} Z_s}{Z_s}\right)^2 \frac{dx}{w_s}$$

$$= \mu_o \frac{1}{h_1^2 w_s} \int_0^{h_1} x^2 \, dx = \mu_o \frac{1}{h_1^2 w_s} \left[\frac{x^3}{3}\right]_0^{h_1}$$

$$= \mu_o \frac{1}{h_1^2 w_s} \left[\frac{h_1^3}{3} - 0\right] = \mu_o \frac{h_1}{3w_s} \qquad .....(3.61)$$

Calculation of λ_2

The equation (3.50) can be used to calculate the specific permeance of section with height h_2.

$$\lambda_{sa} = \mu_o \int_0^h \frac{dx}{y} \qquad \boxed{\text{Equation (3.50)}}$$

Here, $\lambda_{sa} = \lambda_2$; $\quad h = h_2$; $\quad y = w_s$

$$\therefore \text{ Specific permeance of section-2, } \lambda_2 = \mu_o \int_0^{h_2} \frac{dx}{w_s} = \frac{\mu_o}{w_s} \int_0^{h_2} dx$$

$$= \frac{\mu_o}{w_s} [h_2 - 0] = \frac{\mu_o h_2}{w_s} \qquad .....(3.62)$$

Calculation of λ_3

The equation (3.50) can be used to calculate the specific permeance of the section with height h_3.

$$\lambda_{sa} = \mu_o \int_0^h \frac{dx}{y} \qquad \boxed{\text{Equation (3.50)}}$$

Here, $\lambda_{sa} = \lambda_3$; $h = h_3$; $\quad y = $ Mean of the widths w_s and w_o.

$$\therefore y = \frac{w_s' + w_o}{2}$$

$$\left.\begin{array}{l}\therefore \text{ Specific permeance} \\ \text{of section-3}\end{array}\right\} \lambda_3 = \mu_o \int_0^{h_3} \frac{dx}{y} = \mu_o \int_0^{h_3} \frac{dx}{(w_s + w_o)/2}$$

$$= \mu_o \frac{2}{w_s + w_o} \int_0^{h_3} dx = \mu_o \frac{2}{w_s + w_o} [x]_0^{h_3}$$

$$= \mu_o \frac{2}{w_s + w_o} [h_3 - 0] = \mu_o \frac{2 h_3}{w_s + w_o} \qquad ....(3.63)$$

Calculation of λ_4

The equation (3.50) can be used to calculate the specific permeance of the section with height h_4.

$$\lambda_{sa} = \mu_o \int_0^h \frac{dx}{y} \qquad \boxed{\text{Equation (3.50)}}$$

Here, $\lambda_{sa} = \lambda_4$; $h = h_4$; $y = w_o$.

$$\left.\begin{array}{c} \therefore \text{ Specific permeance} \\ \text{of section - 4} \end{array}\right\} \lambda_4 = \mu_o \int_o^{h_4} \frac{dx}{w_o} = \frac{\mu_o}{w_o} \int_o^{h_4} dx$$

$$= \frac{\mu_o}{w_o} [x]_0^{h_4} = \frac{\mu_o}{w_o} [h_4 - 0]$$

$$= \mu_o \frac{h_4}{w_o} \qquad\qquad .....(3.64)$$

Total specific permeance

> Using equations (3.61) to (3.64).

$$\left.\begin{array}{c} \text{Total specific slot permeance} \\ \text{of parallel sided slots} \end{array}\right\} \lambda_s = \lambda_1 + \lambda_2 + \lambda_3 + \lambda_4$$

$$= \mu_o \left[\frac{h_1}{3w_s} + \frac{h_2}{w_s} + \frac{2h_3}{w_s + w_o} + \frac{h_4}{w_o} \right] \qquad .....(3.65)$$

3.14.2 Specific Permeance of Tapered Slot

The slot with tapered sides is shown in Fig. 3.20.

Let Z_s = Total number of conductors per slot

    w_o = Slot opening

    w_s = Slot width at the bottom

    h = Depth of slot

Also, h = Height of field

The total slot depth is divided into various heights h_1, h_2, h_3 and h_4 as shown in Fig. 3.20, and the specific permeance of each section be λ_1, λ_2, λ_3 and λ_4 respectively.

Total specific slot permeance, $\lambda_s = \lambda_1 + \lambda_2 + \lambda_3 + \lambda_4$

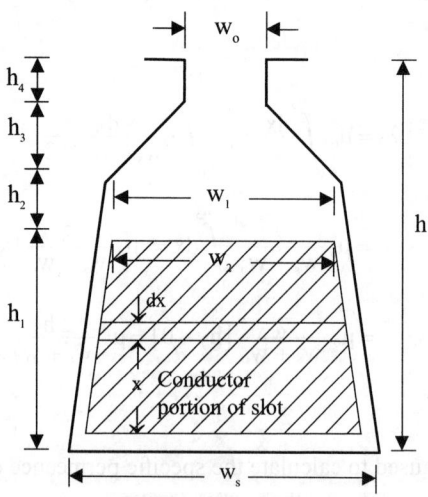

Fig. 3.20: Tapered slot.

Calculation of λ_1

Here h_1 is the depth over which conductors occupy the slot. The permeance of this section can be obtained by considering a strip of height **dx** at a distance x from the bottom.

The equation (3.60) can be used to calculate the permeance of the conductor section of the slot.

$$\lambda_{sc} = \mu_o \int_0^h \left(\frac{Z_x}{Z_s}\right)^2 \frac{dx}{y} \qquad \boxed{\text{Equation (3.60)}}$$

Here, $\lambda_{sc} = \lambda_1$; $\qquad h = h_1$; $\qquad y = $ Mean of the widths w_s and w_2

$$\therefore y = \frac{w_s + w_2}{2}$$

Z_x = Number of conductors below the section of height dx.

$$= \frac{x}{h_1} Z_s \qquad \boxed{\begin{array}{l}\text{Flux at a distance x from bottom of conductors} \\ \text{links with x/h}_1 \text{ of the total conductors.}\end{array}}$$

$\therefore$ Specific permeance
of section-1 $\left.\begin{array}{l}\\\\\end{array}\right\} \lambda_1 = \mu_o \int_0^{h_1} \left(\frac{\frac{x}{h_1} Z_s}{Z_s}\right)^2 \frac{dx}{\frac{w_s + w_2}{2}}$

$$= \mu_o \frac{2}{h_1^2 (w_s + w_2)} \int_0^{h_1} x^2 \, dx = \mu_o \frac{2}{h_1^2 (w_s + w_2)} \left[\frac{x^3}{3}\right]_0^{h_1}$$

$$= \mu_o \frac{2}{h_1^2 (w_s + w_2)} \left[\frac{h_1^3}{3} - 0\right]$$

$$= \mu_o \frac{2h_1}{3 (w_s + w_2)} \qquad\qquad .....(3.66)$$

Calculation of λ_2

The equation (3.50) can be used to calculate λ_2.

$$\lambda_{sa} = \mu_o \int_0^h \frac{dx}{y} \qquad \boxed{\text{Equation (3.50)}}$$

Here, $\lambda_{sa} = \lambda_2$; $\qquad h = h_2$; $\qquad y = $ Mean of the widths w_2 and w_1.

$$\therefore y = \frac{w_2 + w_1}{2}$$

$\therefore$ Specific permeance
of section-2 $\left.\begin{array}{l}\\\\\end{array}\right\} \lambda_2 = \mu_o \int_0^{h_2} \frac{dx}{\frac{w_2 + w_1}{2}} = \mu_o \frac{2}{w_2 + w_1} \int_0^{h_2} dx = \mu_o \frac{2}{w_2 + w_1}[x]_0^{h_2}$

$$= \mu_o \frac{2}{w_2 + w_1}[h_2 - 0]$$

$$= \mu_o \frac{2h_2}{w_2 + w_1} \qquad\qquad .....(3.67)$$

Calculation of λ_3

The equation (3.50) can be used to calculate λ_3.

$$\lambda_{sa} = \mu_o \int_0^h \frac{dx}{y}$$

Equation (3.50)

Here, $\lambda_{sa} = \lambda_3$; $h = h_3$; $y = $ Mean of the widths w_1 and w_0.

$$\left.\begin{array}{l}\therefore \text{Specific permeance} \\ \text{of section} - 3\end{array}\right\} \lambda_3 = \mu_o \int_0^{h_3} \frac{dx}{\dfrac{w_1 + w_o}{2}} = \mu_o \frac{2}{w_1 + w_o} \int_0^{h_3} dx$$

$$= \mu_o \frac{2}{w_1 + w_o} [x]_0^{h_3} = \mu_o \frac{2}{w_1 + w_o} [h_3 - 0]$$

$$= \mu_o \frac{2h_3}{w_1 + w_o} \qquad\qquad .....(3.68)$$

Calculation of λ_4

The equation (3.50) can be used to calculate λ_4.

$$\lambda_{sa} = \mu_o \int_0^h \frac{dx}{y}$$

Equation (3.50)

Here, $\lambda_{sa} = \lambda_4$; $h = h_4$; $y = w_o$.

$$\left.\begin{array}{l}\therefore \text{Specific permeance} \\ \text{of section} - 4\end{array}\right\} \lambda_4 = \mu_o \int_0^{h_4} \frac{dx}{w_o} = \mu_o \frac{1}{w_o} \int_0^{h_4} dx$$

$$= \mu_o \frac{1}{w_o} [x]_0^{h_4} = \mu_o \frac{1}{w_o} [h_4 - 0]$$

$$= \mu_o \frac{h_4}{w_o} \qquad\qquad .....(3.69)$$

Total specific permeance

$$\left.\begin{array}{l}\text{Total specific slot permeance} \\ \text{of tapered slot}\end{array}\right\} \lambda_s = \lambda_1 + \lambda_2 + \lambda_3 + \lambda_4$$

Using equations (3.66) to (3.69).

$$= \mu_o \left[\frac{2h_1}{3(w_s + w_2)} + \frac{2h_2}{w_2 + w_1} + \frac{2h_3}{w_1 + w_o} + \frac{h_4}{w_o} \right]$$

$$= \mu_o \left[\frac{2h_1}{3(w_s + w_2)} + \frac{2h_2}{w_2 + w_1} + \frac{2h_3}{w_1 + w_o} + \frac{h_4}{w_o} \right] \qquad .....(3.70)$$

3.14.3 Specific Permeance of Circular slot

The circular slot is shown in Fig. 3.21.

Total specific slot permeance, $\lambda_s = \lambda_c + \lambda_w$.

where, λ_c = Specific permeance for conductor portion of slot

λ_w = Specific permeance for slot opening

The specific permeance for conductor portion of slot can be estimated as 66% of absolute permeability.

$$\therefore \lambda_c = 0.66 \, \mu_o$$

The specific permeance for slot opening can be estimated from following equation.

$$\lambda_w = \mu_o \, \frac{h}{w_o}$$

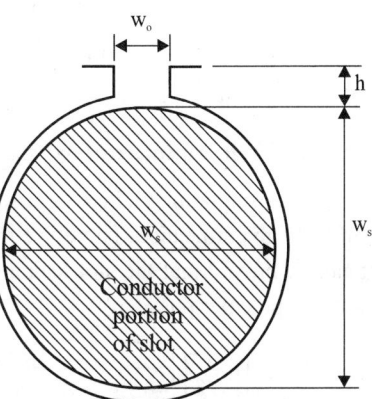

$$\therefore \lambda_s = 0.66 \, \mu_o + \mu_o \, \frac{h}{w_o}$$

Fig. 3.21: *Circular slot.*

$$= \mu_o \left[0.66 + \frac{h}{w_o} \right] \qquad \qquad .....(3.71)$$

where, μ_o = Absolute permeability = $4\pi \times 10^{-7}$ *H/m*

w_o = Slot opening

h = Height of slot above circular portion.

3.15 SLOT LEAKAGE REACTANCE

A reactive voltage is produced, when the leakage flux is associated with a winding carrying alternating current. This reactive voltage may be considered as a voltage drop due to a leakage reactance, which in turn due to the property of inductance.

The slot leakage flux, ϕ_{sl} is given by the product of mmf and slot permeance.

$$\therefore \text{ Slot leakage flux, } \phi_{sl} = \text{mmf} \times \text{Permeance} = Z_s \, I_z \, \Lambda_s \qquad .....(3.72)$$

where, Λ_s = Slot permeance

The *inductance* is defined as flux linkages per unit current. Let us consider a slot with Z_s conductors and each conductor carrying a current of I_z.

$$\text{Slot inductance, } L_s = \frac{\text{Flux linkages}}{\text{Current}} = \frac{Z_s \, \phi_{sl}}{I_z}$$

$$= \frac{Z_s}{I_z} \, Z_s \, I_z \, \Lambda_s = Z_s^2 \, \Lambda_s \qquad \boxed{\text{Using equation (3.72)}} \; .....(3.73)$$

$\therefore$ Slot leakage reactance, $X_s = 2\pi f \times$ Slot inductace

$$= 2\pi f\, Z_s^2\, \Lambda_s = 2\pi f\, Z_s^2 L\, \lambda_s \qquad \boxed{\Lambda_s = L\lambda_s} \quad .....(3.74)$$

where,   f = Frequency of current

        L = Length of slot

The equation (3.74) can be used to estimate the slot leakage reactance of one slot.

Slot leakage reactance of poly phase machine (AC machine)

From equation (3.74) it is observed that the slot leakage reactance is directly proportional to specific slot leakage permeance. Therefore the slot leakage reactance is modified by the slots per pole per phase. It can be proved that the slot leakage reactance is inversely proportional to the square of the number of slots per pole per phase, q.

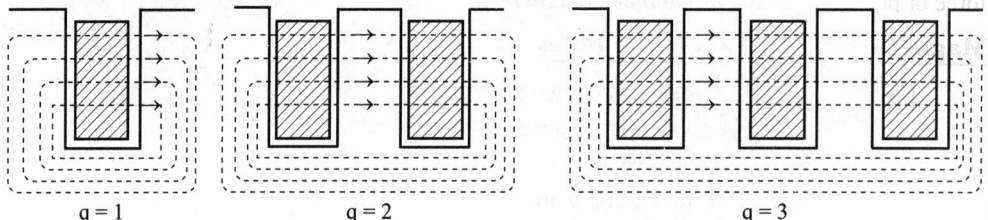

Fig. 3.22: *Slot leakage fluxes in slots of a phase under one pole.*

The slot leakage fluxes in the slots of a phase under one pole are shown in Fig. 3.22. Let q be the number of slots per pole per phase. Here all the conductors in the q number of slots carry the same current. Hence the leakage fluxes in these slots links the conductors of all the q slots. Therefore if λ_s is the specific slot permeance, then the effective specific slot permeance for slots of a phase under one pole is λ_s/q, because the permeance of individual slots are in series.

The slot leakage reactance due to the conductors of a phase under one pole can be estimated using the equation (3.74) if we make the following replacements.

1.   Replace Z_s by $Z_s\, q$

2.   Replace λ_s by λ_s / q

$$\left.\begin{array}{l}\therefore \text{Slot leakage flux due to conductors} \\ \text{of a phase under a pole}\end{array}\right\} = 2\pi f\,(Z_s q)^2\, L \times \frac{\lambda_s}{q} = 2\pi f\, Z_s^2\, q\, L\, \lambda_s \qquad .....(3.75)$$

In polyphase machines the conductors of a phase under all the poles are connected in series. Hence the slot leakage reactance per phase is given by the product of number of poles and the slot leakage flux due to conductors of a phase under a pole.

$$\therefore \text{Slot leakage reactance per phase, } X_s = 2\pi f\, Z_s^2\, q\, L\, \lambda_s \times p = 2\pi f\, p\, q\, Z_s^2 L\, \lambda_s \qquad .....(3.76)$$

where,   p = Number of poles.

If T_s is stator turns per phase then the conductors per slot is given by,

$$\text{Conductors per slot, } Z_s = \frac{\text{Conductors per phase}}{\text{Slots per phase}} = \frac{2\,T_s}{q\,p} \quad\quad .....(3.77)$$

On substituting for Z_s from equation (3.77) in equation (3.76) we get,

$$\left.\begin{array}{l}\text{Slot leakage reactance}\\ \text{per phase}\end{array}\right\} X_s = 2\pi fpq \left(\frac{2\,T_s}{q\,p}\right)^2 L\,\lambda_s = \frac{8\pi f\,T_s^2\,L\,\lambda_s}{q\,p} \quad\quad .....(3.78)$$

3.16 UNBALANCED MAGNETIC PULL

In rotating machines if the air-gap around the armature periphery is non-uniform then radial forces are developed in the rotor. The radial forces will act perpendicular to rotor axis. This force or pull is called the *unbalanced magnetic pull*.

Magnetic force between two poles

Let	F	=	Force between two poles
	B	=	Flux density in air-gap
	A_P	=	Area of each pole
	μ_o	=	Permeability of air
	l_g	=	Length of air-gap
	AT	=	Total mmf
	AT_g	=	mmf for air-gap
	AT_i	=	mmf for iron parts

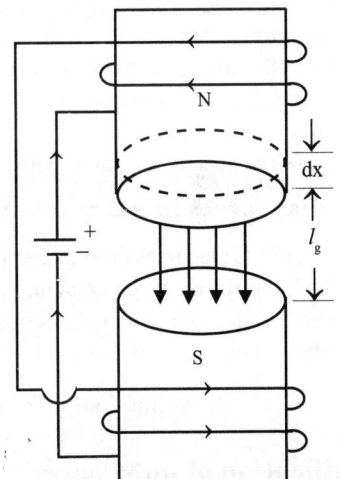

Fig. 3.23 : Electromagnet

Let an electromagnet be arranged as shown in Fig 3.23. If one of the poles is moved by a distance dx, then the work done is equal to the change of energy stored in magnetic field.

Now, Work done = Force × Distance = Fdx $\quad\quad .....(3.79)$

Change in energy = Energy density × Change in volume

$$= \frac{1}{2}\frac{B^2}{\mu_o} \times A_p\,dx = \frac{1}{2}\frac{B^2}{\mu_o}\,A_p\,dx \quad\quad .....(3.80)$$

On equating the expressions for work done and change in energy we get,

$$F\,dx = \frac{1}{2}\frac{B^2}{\mu_o}\,A_p\,dx$$

$$\therefore \text{Force, } F = \frac{1}{2}\frac{B^2}{\mu_o}\,A_p \quad\quad .....(3.81)$$

The flux density in the air-gap depends upon the mmf of the exciting winding. The total mmf is equal to the sum of mmf for air-gap and iron parts. If the iron parts are not saturated, then mmf for iron parts is negligible and total mmf is equal to mmf for air-gap. The force between the two magnet can be estimated from the knowledge of mmf.

If the mmf for iron parts are neglected then the flux density is given by,

$$\text{Flux density, } B = \frac{\mu_o\,AT_g}{l_g} \quad\quad .....(3.82)$$

On substituting for B from equation (3.82) in equation (3.81) we get,

$$\text{Force, } F = \frac{1}{2} \mu_o \left(\frac{AT_g}{l_g}\right)^2 A_p \qquad\qquad .....(3.83)$$

If the saturation of iron parts are not neglected, then

$$\text{Force, } F = \frac{1}{2} \mu_o \left(\frac{AT}{l_g}\right)^2 A_p \qquad\qquad .....(3.84)$$

Radial forces in DC machine

Radial forces are developed in DC machines due to attraction or repulsion of field poles and armature poles. If the rotor is perfectly concentric with stator and the air-gap is uniform then the radial forces developed under north pole will be exactly opposite to that under south pole. Hence the resultant radial force will be zero. In DC machines the flux distribution is uniform and so the equations (3.81), (3.83) and (3.84) can be used to calculate the radial force of a pole. Here area per pole is given by armature periphery per pole. If the rotor is slightly eccentric the radial forces under opposite poles will not be exactly balanced and a net radial force acts on the rotor, which is called unbalanced magnetic pull.

Radial forces in AC machine

In AC machines the flux distribution is sinusoidal. Hence the radial forces are estimated by considering the radial force on a elemental angular surface and integrating over the entire surface. It can be shown that the net radial force in a perfect concentric rotor is zero. If there is a small eccentricity then there will be a resultant radial force which is called unbalanced magnetic pull.

$$\text{In AC machines, Force, } F = \frac{B_m^2}{3\mu_o} DL \qquad\qquad .....(3.85)$$

Estimation of unbalanced magnetic pull

The unbalanced magnetic pull in DC and AC machines can be estimated from the following formulae. In ac machines, B represents the rms value of flux density. If the flux distribution is sinusoidal and B_m is the maximum value of flux density then rms value of flux density is $B_m / \sqrt{2}$.

Let, e = Displacement of rotor (or eccentricity of rotor)

$$\text{Magnetic pull per unit area, } P_m = \frac{1}{2} \frac{B^2}{\mu_o} \qquad\qquad .....(3.86)$$

$$\text{Area per pole, } A_p = \frac{\pi D L}{p} \qquad\qquad .....(3.87)$$

$$\left.\begin{array}{l}\text{Unbalanced magnetic pull}\\\text{due to pair of poles}\end{array}\right\} P_p = 2 A_p P_m \frac{e}{l_g} \qquad\qquad .....(3.88)$$

$$\left.\begin{array}{l}\text{Unbalanced magnetic pull}\\\text{due to p number of poles}\end{array}\right\} = \text{pole pairs} \times P_p.$$

$$= \frac{p}{2} \times 2 A_p P_m \frac{e}{l_g} = p A_p P_m \frac{e}{l_g} \qquad\qquad ....(3.89)$$

When the poles are located at an angle θ with horizontal axis, the unbalanced magnetic pull acting downwards due to pair of poles, can be estimated using the following equation.

$$\left.\begin{array}{l}\text{Unbalanced magnetic pull acting}\\ \text{downwards due to pair of poles}\end{array}\right\} = 4 \times A_p\, P_m\, \frac{e}{l_g}\, \sin^2\theta \qquad\qquad .....(3.90)$$

Undesirable effects of unbalanced magnetic pull

The unbalanced magnetic pull produces the following undesirable effects on the rotating machines.

1. Saturation of magnetic materials due to reduction in air-gap.
2. Excessive vibration and noise due to unbalanced radial forces.

Design guidelines to reduce unbalanced magnetic pull

The following techniques are employed to overcome unbalanced magnetic pull in rotating machines.

1. The length of the rotor can be kept small and diameter can be made higher.
2. Ball bearings are employed and rotor is dynamically balanced.
3. The combination of rotor and stator slots which produce vibrations are avoided.

3.17 SUMMARY OF DESIGN EQUATIONS

1. Flux, $\phi = AT \times \Lambda = \dfrac{AT}{S}$

2. Reluctance, $S = \dfrac{l}{A\,\mu}$

3. Magnetizing force, $H = \dfrac{B}{\mu}$ $\boxed{H = at}$

4. Total mmf, $AT = \phi S = $ at l

5. Reluctance of air-gap in machines with smooth armature, $S_{g\_cas} = \dfrac{l_g}{\mu_o\, L\, y_s}$

6. $\left.\begin{array}{l}\text{Reluctance of air-gap in machines with open}\\ \text{armature slots (neglecting the effect of fringing)}\end{array}\right\} S_{g\_oas} = \dfrac{l_g}{\mu_o\, L\, (y_s - w_s)}$

7. $\left.\begin{array}{l}\text{Reluctance of air-gap in machines with open}\\ \text{armature slots (including the effect of fringing)}\end{array}\right\} S_{g\_oasf} = \dfrac{l_g}{\mu_o\, L\, (y_s - K_{cs} w_s)}$

8. Gap contraction factor for slots, $K_{gs} = \dfrac{y_s}{y_s - K_{cs} w_s}$

> **Note :** In equations 6, 7 and 8 for semienclosed slots replace w_s by w_o.

9. Carter's coefficient for slots, $K_{cs} = \dfrac{1}{1 + 5\, l_g/w_o}$

(or) $K_{cs} = \dfrac{2}{\pi}\left[\tan^{-1} y - \dfrac{1}{\pi}\log\sqrt{1+y^2}\right]$; where, $y = \dfrac{w_o}{2l_g}$

> **Note :** For open type slots replace w_o by w_s.

10. Reluctance of air-gap in machine with radial ventilating ducts $\left.\right\}$ $S_{g\_d} = \dfrac{l_g}{\mu_o \left(L - K_{cd}\, n_d\, w_d\right) y_s}$

11. Gap contraction factor for ducts, $K_{gd} = \dfrac{L}{L - K_{cd}\, n_d\, w_d}$

12. Carter's coefficient for ducts, $K_{cd} = \dfrac{1}{1 + 5\, l_g/w_d}$

13. Total gap contraction factor, $K_g = K_{gs} \times K_{gd}$ or $K_g = K_{gs} \times K_{gd} \times K_{gsal}$

14. Total gap contraction factor for slots in induction motor, $K_{gs} = K_{gss}\, K_{gsr}$

15. Gap contraction factor for stator slots, $K_{gss} = \dfrac{y_{ss}}{y_{ss} - K_{css}\, w_{os}}$

16. Gap contraction factor for rotor slots, $K_{gsr} = \dfrac{y_{sr}}{y_{sr} - K_{csr}\, w_{or}}$

17. Carter's coefficient for stator slots, $K_{css} = \dfrac{1}{1 + 5\, l_g/w_{os}}$

18. Carter's coefficient for rotor slots, $K_{csr} = \dfrac{1}{1 + 5\, l_g/w_{or}}$

19. Stator slot pitch, $y_{ss} = \dfrac{D}{S_s}$

20. Rotor slot pitch, $y_{sr} = \dfrac{D - 2\, l_g}{S_r}$

21. mmf per meter of the path in non-magnetic material $\left.\right\}$ $at_{nm} = \dfrac{\beta}{\mu_o} \approx 800,000\, B$

22. mmf required for air-gap of length l_g in non-salient pole machine with closed armature slots $\left.\right\}$ $AT_g = 800,000\, B_{av}\, l_g$

23. mmf required for air-gap in non-salient pole machines with open armature slots and radial ventilation ducts $\left.\right\}$ $K_g\, AT_g = 800,000\, B_{av}\, K_g\, l_g$

> **Note :** *For salient pole machines (or DC machines) replace B_{av} by B_g.*

24. Field form factor, $K_f = \dfrac{B_{av}}{B_g}$

> Also, $K_f \approx \Psi = \dfrac{\text{Pole arc}}{\text{Pole pitch}}$

25. Maximum flux density in air-gap, $B_g = B_{av} \times \dfrac{\tau}{b}$

26. Flux density at a section of a tooth with area of cross-section A_t $\left.\right\}$ $B_t = \dfrac{\phi}{n_t\, A_t}$

27. Total mmf for tooth, $AT_t = at_{mean} \times l_t = at_{mean} \times d_s$

28. Mean value of **at**, $\text{at}_{\text{mean}} = \dfrac{\text{at}_1 + 4\text{at}_2 + \text{at}_3}{6}$ $\boxed{\text{Simpson's rule/ Three ordinate method}}$

29. Total mmf for teeth, $\text{AT}_t = \text{at}_{1/3} \times l_t = \text{at}_{1/3} \times d_s$ $\boxed{B_{t1/3} \text{ method}}$

30. $B_{\text{app}} = B_{\text{real}} + B_a K$, where, $K = \dfrac{A_a}{A_i}$ or $K = K_s - 1$ and $K_s = \dfrac{L\, y_s}{L_i\, w_t}$

31. Magnetizing current per phase, $I_{m\_nsd} = \dfrac{0.37\, p\, \text{AT}_{m1}}{K_{w1}\, T_s}$
    (Non-sinusoidal flux distribution)

32. Magnetising current per phase, $I_{m\_sd} = \dfrac{0.427\, p\, \text{AT}_{60}}{K_{w1}\, T_s}$
    (Sinusoidal flux distribution)

33. Leakage coefficient, $C_l = \dfrac{\text{Total flux}}{\text{Useful flux}} = \dfrac{\text{Useful flux} + \text{Leakage flux}}{\text{Useful flux}}$

34. Specific permeance for non-conductor portion of slot $\left.\right\}$ $\lambda_{sa} = \mu_o \displaystyle\int_0^h \dfrac{dx}{y}$

35. Specific permeance for conductor portion of slot $\left.\right\}$ $\lambda_{sc} = \mu_o \displaystyle\int_0^h \left(\dfrac{Z_x}{Z_s}\right)^2 \dfrac{dx}{y}$

36. Total specific permeance of parallel sided slot $\left.\right\}$ $\lambda_s = \mu_o\left[\dfrac{h_1}{3w_s} + \dfrac{h_2}{w_s} + \dfrac{2h_3}{w_s + w_o} + \dfrac{h_4}{w_o}\right]$

37. Total specific permeance of tapered slot $\left.\right\}$ $\lambda_s = \mu_o\left[\dfrac{2h_1}{3(w_s + w_2)} + \dfrac{2h_2}{(w_2 + w_1)} + \dfrac{2h_3}{(w_1 + w_o)} + \dfrac{h_4}{w_o}\right]$

38. Total specific permeance of circular slot $\left.\right\}$ $\lambda_s = \mu_o\left[0.66 + \dfrac{h}{w_o}\right]$

39. Slot leakage reactance of one slot, $X_s = 2\pi f\, Z_s^2\, L\, \lambda_s$

40. Slot leakage reactance per phase in AC machines, $X_s = 2\pi f\, pq\, Z_s^2\, L\, \lambda_s$

    (or) $X_s = \dfrac{8\,\pi\, f\, p\, q\, T_s^2\, L\, \lambda_s}{q\, p}$

41. Force due to a pole, in DC machine, $F = \dfrac{1}{2}\dfrac{B_2}{\mu_o}A_p$

42. Force due to a pole, in AC machine, $F = \dfrac{B_m^2}{3\mu_o}DL$

43. Area per pole, $A_p = \dfrac{\pi\, DL}{p}$

44. Magnetic pull per unit area, $P_m = \dfrac{1}{2}\dfrac{B^2}{\mu_o}$

45. Unbalanced magnetic pull due to pair of poles, $P_p = 2A_p\,P_m\,\dfrac{e}{l_g}$

46. Unbalanced magnetic pull due to p number of poles, $\text{UMP} = p\,A_p\,P_m\,\dfrac{e}{l_g}$

47. When poles are located at an angle θ with horizontal axis $\Big\}$ $= 4\,A_p\,P_m\,\dfrac{e}{l_g}\,\sin^2\theta$
     the UMP acting downwards due to a pair of poles

48. When poles are located at an angle θ with horizontal $\Big\}$ $= 2\,p\,A_p\,P_m\,\dfrac{e}{l_g}\,\sin^2\theta$
     axis the UMP acting downwards due to p poles

3.18 SOLVED PROBLEMS

EXAMPLE 3.1

Calculate the mmf required for air-gap of a DC machine with an axial length of 20 *cm* (no ducts) and a pole arc of 18 *cm*. The slot pitch = 27 *mm*, slot opening = 12 *mm*, air-gap = 6 *mm* and the useful flux per pole = 25 *mWb*. Take carter's coefficient for slot as 0.3.

GIVEN DATA

$L = 20\ cm$	$b_s = 18\ cm$	$l_g = 6\ mm$
$y_s = 27\ mm$	$w_o = 12\ mm$	$K_{cs} = 0.3$
$\phi = 25\ mWb$		

SOLUTION

Gap contraction factor for slots, $K_{gs} = \dfrac{y_s}{y_s - K_{cs}\,w_o}$

$$= \frac{27}{27 - 0.3 \times 12} = 1.1538$$

Flux density at the $\Big\}$ $B_g = \dfrac{B_{av}}{\Psi} = \dfrac{p\phi\,/\,\pi DL}{b\,/\,\tau} = \dfrac{p\phi\tau}{b \times \pi DL}$
centre of the pole

$\qquad\qquad\qquad\qquad\boxed{B_{av} = \dfrac{p\phi}{\pi DL}}\ \ \boxed{\Psi = \dfrac{b}{\tau}}$

$$= \frac{p\phi \times \pi D\,/\,p}{b \times \pi DL}$$

$\qquad\qquad\qquad\qquad\boxed{\tau = \dfrac{\pi D}{p}}$

$$= \frac{\phi}{b \times L}$$

$$= \frac{25 \times 10^{-3}}{18 \times 10^{-2} \times 20 \times 10^{-2}} = 0.6944\ Wb/m^2$$

Since there are no ducts, $K_g = K_{gs}$

$\therefore$ mmf required for air-gap, $AT_g = 800{,}000\,K_g\,B_g\,l_g$

$$= 800{,}000 \times 1.1538 \times 0.6944 \times 6 \times 10^{-3}$$

$$= 3846\ AT$$

RESULT

mmf required for air-gap, $AT_g = 3846\ AT$

EXAMPLE 3.2

A 15 kW, 230 V, 4 pole DC machine has the following data : armature diameter = 0.25 m, armature core length = 0.125 m, length of air-gap at pole centre = 2.5 mm, flux per pole = 11.7 × 10⁻³ Wb, pole arc/pole pitch = 0.66

Calculate the mmf required for air-gap (i) if the armature surface is treated as smooth (ii) if the armature is slotted and the gap contraction factor is 1.18.

GIVEN DATA

15 kW	D = 0.25 m	ϕ = 11.7 × 10⁻³ Wb	l_g = 2.5 mm	K_g = 1.18
230 V	L = 0.125 m	Ψ = b/τ = 0.66	4 pole	

SOLUTION

$$\text{Specific magnetic loading, } B_{av} = \frac{p\phi}{\pi DL} = \frac{4 \times 11.7 \times 10^{-3}}{\pi \times 0.25 \times 0.125} = 0.4767 \ Wb/m^2$$

$$\text{Flux density at the centre of the pole, } B_g = \frac{B_{av}}{\Psi} = \frac{0.4767}{0.66} = 0.7223 \ Wb/m^2$$

mmf required for air-gap with smooth armature $\left.\right\}$ $AT_g = 800,000 \ B_g \ l_g = 800,000 \times 0.7223 \times 2.5 \times 10^{-3}$

$$= 1445 \ AT$$

mmf required for air-gap with slotted armature $\left.\right\}$ $AT_g = 800,000 \ K_g B_g \ l_g$

$$= 800,000 \times 1.18 \times 0.7223 \times 2.5 \times 10^{-3}$$

$$= 1705 \ AT$$

RESULT

mmf for air-gap with smooth armature = 1445 AT

mmf for air-gap with slotted armature = 1705 AT

EXAMPLE 3.3

Determine the air-gap length of a DC machine from the following particulars: gross-length of core = 0.12 m, number of ducts = 1, width of duct = 10 mm, slot pitch = 25 mm, slot width = 10 mm, carter's coefficient for slots and ducts = 0.32, gap density at pole centre = 0.7 Wb/m^2, field mmf/pole = 3900 AT, mmf required for iron parts of magnetic circuit = 800 AT.

GIVEN DATA

L = 0.12 m	y_s = 25 mm	B_g = 0.7 Wb/m^2
n_d = 1	w_t = 10 mm	mmf per pole = 3900 AT
w_d = 10 mm	K_{cs} = K_{cd} = 0.32	mmf for iron = 800 AT

SOLUTION

mmf for air-gap, AT_g = mmf per pole − mmf for iron parts

$$= 3900 - 800 = 3100 \ AT$$

$$\text{Gap contraction factor for slots, } K_{gs} = \frac{y_s}{y_s - K_{cs} \ w_t} = \frac{25}{25 - 0.32 \times 10} = 1.1468$$

Gap contraction factor for ducts, $K_{gd} = \dfrac{L}{L - K_{cd}\, n_d\, w_d}$

$$= \dfrac{0.12}{0.12 - 0.32 \times 1 \times 10 \times 10^{-3}} = 1.0274$$

Gap contraction factor, $K_g = K_{gs} \times K_{gd} = 1.1468 \times 1.0274 = 1.1782$

mmf for air-gap, $AT_g = 800{,}000\, K_g\, B_g\, l_g$

$\therefore$ Length of air-gap, $l_g = \dfrac{AT_g}{800{,}000\, B_g\, K_g} = \dfrac{3100}{800{,}000 \times 0.7 \times 1.1782}$

$$= 4.7 \times 10^{-3} = 4.7 \ mm$$

RESULT

Length of air-gap, $l_g = 4.7 \ mm$

EXAMPLE 3.4

Find the permeability at the root of a tooth in a DC machine armature from the following data : slot pitch = 2.1 cm, tooth width at the root = 1.07 cm, gross length = 32 cm, stacking factor = 0.9, real flux density at the root of the tooth = 2.25 Wb/m², apparent flux density at the root = 2.36 Wb/m².

GIVEN DATA

$y_s = 2.1 \ cm$ $B_{real} = 2.25 \ Wb/m^2$ $S_f = 0.9$

$w_t = 1.07 \ cm$ $B_{app} = 2.36 \ Wb/m^2$ $L = 32 \ cm = 0.32 \ m$

SOLUTION

$K_s = \dfrac{L y_s}{L_i\, w_t} = \dfrac{L y_s}{S_f\, L\, w_t} = \dfrac{0.32 \times 0.021}{0.9 \times 0.32 \times 0.0107} = 2.181$ $\boxed{L_i = S_f\, L}$

$B_{app} = B_{real} + 4\pi \times 10^{-7} \ at\ (K_s - 1)$

$\therefore at = \dfrac{B_{app} - B_{real}}{4\pi \times 10^{-7}\, (K_s - 1)} = \dfrac{2.36 - 2.25}{4\pi \times 10^{-7}\, (2.181 - 1)} = 74119.6 = 74120 \ AT/m$

The permeability at the root of the tooth at real flux density $\left.\right\} \ \mu = \dfrac{B_{real}}{at} = \dfrac{2.25}{74120} = 30.356 \times 10^{-6} \ H/m$

RESULT

Permeability at the root of the tooth at real flux density, $\mu = 30.356 \times 10^{-6} \ H/m$

EXAMPLE 3.5   (UTV, Dec' 13, 5 M)

Calculate the apparent flux density at a section of a tooth in an armature of a DC machine from the following data at that section: slot pitch = 24 mm, slot width = tooth width = 12 mm, length of armature core including 5 ducts of 10 mm width = 0.38 m, iron stacking factor = 0.92. True flux density in tooth at that section is 2.2 Wb/m² for which the mmf is 70,000 AT/m.

GIVEN DATA

$y_s = 24\ mm$ $L = 0.38\ m$ $B_{real} = 2.2\ Wb/m^2$

$w_s = 12\ mm$ $n_d = 5$ $at = 70,000\ AT/m$

$w_t = 12\ mm$ $w_d = 10\ mm$ $S_f = 0.92$

SOLUTION

L_i = Stacking factor × (Core length − Total width of ducts)

$$= S_f\,(\,L - 5w_d\,) = 0.92 \times (\,0.38 - 5 \times 10 \times 10^{-3}\,) = 0.3036$$

$$K_s = \frac{L\,y_s}{L_i\,w_t} = \frac{0.38 \times 0.024}{0.3036 \times 0.012} = 2.5$$

$$B_{app} = B_{real} + [\mu_o\,at\,(K_s - 1)]$$

$$= 2.2 + [4\pi \times 10^{-7} \times 70,000 \times (2.5 - 1)] = 2.332\ Wb/m^2$$

RESULT

Apparent flux density, B_{app} = 2.332 Wb/m^2

EXAMPLE 3.6

Calculate the mmf for air-gap in a three-phase induction motor from the following data. Stator bore = 500 mm, core length = 220 mm, stator slots = 76, rotor slots = 94, slot opening = 2 mm, air-gap length = 0.9 mm. Take K_{gd} = 1.15 and air-gap flux density = 0.54 Wb/m^2.

GIVEN DATA

$D = 500\ mm$ $w_{os} = w_{or} = 2\ mm$

$L = 220\ mm$ $l_g = 0.9\ mm$

$S_s = 76$ $K_{gd} = 1.15$

$S_r = 94$ $B_{av} = 0.54\ Wb/m^2$

SOLUTION

Carter's coefficient for stator slots, $K_{css} = \dfrac{1}{1 + 5\,l_g/w_{os}} = \dfrac{1}{1 + 5 \times 0.9/2} = 0.3077$

Since, $w_{os} = w_{or}$, $K_{css} = K_{csr} = 0.3077$

Stator slot pitch, $y_{ss} = \dfrac{\pi D}{S_s} = \dfrac{\pi \times 500}{76} = 20.67\ mm$

Rotor slot pitch, $y_{sr} = \dfrac{\pi\,(D - 2l_g)}{S_r} = \dfrac{\pi \times (500 - 2 \times 0.9)}{94} = 16.65\ mm$

Gap contraction factor for stator slots, $K_{gss} = \dfrac{y_{ss}}{y_{ss} - K_{css}\,w_{os}}$

$$= \frac{20.67}{20.67 - (0.3077 \times 2)} = 1.0307$$

Gap contraction factor for rotor slots, $K_{gsr} = \dfrac{y_{sr}}{y_{sr} - K_{csr}\, w_{or}}$

$$= \dfrac{16.65}{16.65 - (0.3077 \times 2)} = 1.0384$$

Gap contraction factor for slots, $K_{gs} = K_{gss} \times K_{gsr} = 1.0307 \times 1.0384 = 1.0703$

Total gap contraction factor, $K_g = K_{gs} \times K_{gd} = 1.0703 \times 1.15 = 1.2308$

mmf for air-gap, $AT_g = 800{,}000\, B_{av}\, K_g\, l_g$

$$= 800{,}000 \times 0.54 \times 1.2308 \times 0.9 \times 10^{-3}$$

$$= 478.5 \; AT$$

RESULT

mmf for air-gap, AT_g = 478.5 AT

EXAMPLE 3.7   *(UTV, Dec' 13, 10 M)*

A DC machine has the following data: pole arc = 32 *cm*, length of armature = 40 *cm*, length of air-gap = 0.8 *cm*, slot pitch = 2.6 *cm*, width of slot = 1.2 *cm*, number of ventilating ducts in armature = 5, width of each ventilating duct = 1 *cm*, flux per pole = 0.75 × 10⁻³ *Wb*. Find the mmf required for the air-gap. Given,

$$\dfrac{\text{Slot width}}{\text{Gap length}} = 5, \qquad \text{Carter's coefficient for slots} = 0.21$$

$$\dfrac{\text{Duct width}}{\text{Gap length}} = 1.25, \qquad \text{Carter's coefficient for ducts} = 0.18$$

GIVEN DATA

b = 32 *cm* = 0.32 *m*	y_s = 2.6 *cm* = 0.026 *m*	n_d = 5
L = 40 *cm* = 0.4 *m*	w_s = 1.25 *cm* = 0.0125 *m*	K_{cs} = 0.21
l_g = 0.8 *cm* = 0.08 *m*	w_d = 1 *cm* = 0.01 *m*	K_{cd} = 0.18

SOLUTION

Gap density at the centre of the pole, $B_g = \dfrac{\text{Flux per pole}}{\text{Pole arc} \times \text{Length of core}}$

$$= \dfrac{0.75 \times 10^{-3}}{0.32 \times 0.4} = 5.8594 \times 10^{-3} \; Wb/m^2$$

Gap contraction factor for slots, $K_{gs} = \dfrac{y_s}{y_s - K_{cs}\, w_s}$

$$= \dfrac{2.6}{2.6 - 0.21 \times 1} = 1.0879$$

Gap contraction factor for ducts, $K_{gd} = \dfrac{L}{L - K_{cd}\, n_d\, w_d}$

$$= \dfrac{40}{40 - 0.18 \times 5 \times 1} = 1.023$$

Total gap contraction factor, $K_g = K_{gs} K_{gd}$

$$= 1.0879 \times 1.023 = 1.1129$$

mmf for air-gap, $AT_g = 800{,}000 \ K_g \ B_g \ l_g$

$$= 800{,}000 \times 1.1129 \times 5.8594 \times 10^{-3} \times 0.08$$

$$= 417.3393 \ AT \approx 418 \ AT$$

RESULT

mmf for air-gap, $AT_g = 418 \ AT$

EXAMPLE 3.8

Calculate the unbalanced magnetic pull of 4 pole DC machine having air-gap flux density of 0.85 Wb/m^2 and a gap length of 2 mm. The area of each pole is $24 \times 10^{-3} \ m^2$ and the poles are symmetrically mounted. The eccentricity is 15% in the air-gap.

GIVEN DATA

$p = 4$	$e = 15\%$	$A_p = 24 \times 10^{-3} \ m^2$
$l_g = 2 \ mm$	$B = 0.85 \ Wb/m^2$	

SOLUTION

Magnetic pull per unit area, $P_m = \dfrac{1}{2} \dfrac{B^2}{\mu_0}$

$$= \frac{1}{2} \times \frac{0.85^2}{4\pi \times 10^{-7}} = 287.4736 \times 10^3 \ N/m^2$$

Eccentricity, $e = 15 \ \%$ of $l_g = \dfrac{15}{100} \times 2 = 0.3 \ mm$

Unbalanced magnetic pull due to four poles $= p \ A_p \ P_m \ \dfrac{e}{l_g}$

$$= 4 \times 24 \times 10^{-3} \times 287.4736 \times 10^3 \times \frac{0.3}{2}$$

$$= 4139.6 = 4140 \ N$$

RESULT

Unbalanced magnetic pull due to four poles $= 4140 \ N$

EXAMPLE 3.9

The following data refers to a 20 kW, 2 pole DC motor. Air-gap $= 2.5 \ mm$, Area under each pole $= 20 \times 10^{-3} \ m$, air-gap flux density $= 0.8 \ Wb/m^2$, vertical displacement of rotor $= 0.4 \ mm$. Calculate the unbalanced magnetic pull acting downwards if the poles are located 45° with horizontal axis.

GIVEN DATA

$p = 2$	$B = 0.8 \ Wb/m^2$	$A_p = 20 \times 10^{-3} \ m^2$	A_3'
$l_g = 2.5 \ mm$	$\theta = 45°$		
$e = 0.4 \ mm$	$P = 20 \ kW$		

SOLUTION

Magnetic pull per unit area $P_m = \dfrac{1}{2}\dfrac{B^2}{\mu_0}$

$$= \frac{1}{2} \times \frac{0.8^2}{4\pi \times 10^{-7}} = 254.648 \times 10^3 \; N/m^2$$

Unbalanced magnetic pull $\Big\}= 2\,p\,A_p\,P_m\,\dfrac{e}{l_g}\sin^2\theta$
due to 2 poles

$$= 2 \times 2 \times 20 \times 10^{-3} \times 254.648 \times 10^3 \times \frac{0.4}{2.5} \times (\sin 45)^2$$

$$= 1629.7 \approx 1630 \; N$$

RESULT

Unbalanced magnetic pull due to 2 poles = 1630 N

3.19 SHORT-ANSWER QUESTIONS

Q3.1 What is magnetic circuit?

The magnetic circuit is the path of magnetic flux. The mmf of the circuit creates flux in the path against the reluctance of the path. The equation which relates flux, mmf and reluctance is given by,

$$\text{Flux} = \frac{\text{mmf}}{\text{Reluctance}}$$

Q3.2 What are the constituents of magnetic circuit in rotating machines?

The various elements in the flux path of salient pole machines are poles, pole shoes, air-gap, armature teeth, armature core and yoke. The various elements in the flux path of non-salient pole machines are stator core, stator teeth, air-gap, rotor teeth and rotor core.

Q3.3 Draw the magnetic circuit of transformer.

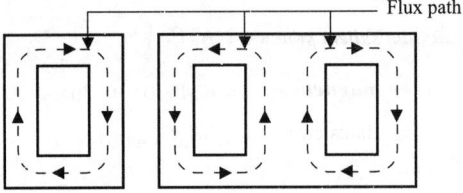

Fig. a: Core   *Fig. b: Shell type transformer.*
type transformer.

Fig. Q3.3: Magnetic circuit of trasformer.

Q3.4 Draw the magnetic circuit of DC machine.

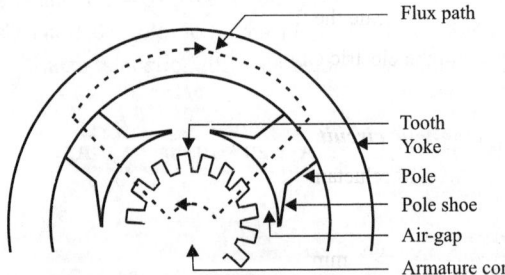

Fig. Q3.4: Magnetic circuit of DC machine.

Q3.5   What are the components of magnetic circuit of an induction motor?

The components of magnetic circuit of an induction motor are stator core, stator teeth, air-gap, rotor core and rotor teeth. A typical magnetic circuit of an induction motor is shown below.

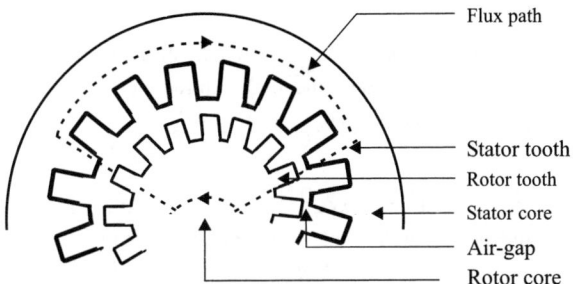

Fig. Q3.5: Magnetic circuit of induction motor.

Q3.6   Draw the magnetic circuit of salient pole synchronous machine.

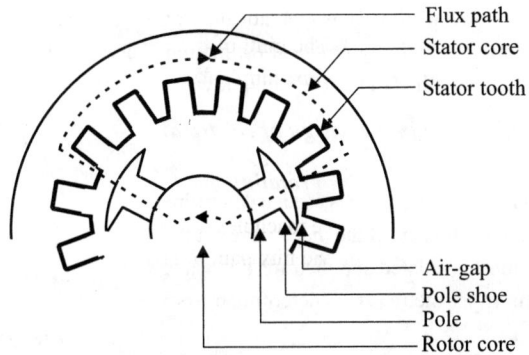

Fig. Q3.6: Magnetic circuit of salient pole synchronous machine.

Q3.7   Write any two similarities between magnetic and electric circuits.

i)   In electric circuit the emf circulates current in a closed path. Similarly in a magnetic circuit the mmf creates flux in a closed path.

ii)   In electric circuit the flow of current is opposed by resistance of the circuit. Similarly in a magnetic circuit the creation of flux is opposed by reluctance of the circuit.

Q3.8   Write any two essential differences between magnetic and electric circuits.

i)   When the current flows in electric circuit the energy is spent continuously, whereas in magnetic circuit the energy is needed only to create the flux but not to maintain it.

ii)   Current actually flows in the electric circuit; whereas the flux does not flow in a magnetic circuit but it is only assumed to flow.

Q3.9   Write the Ohm's law of a magnetic circuit.

The equation relating mmf, flux and reluctance of a magnetic circuit can be called Ohm's law of magnetic circuit. It is given by,

$$\text{Flux} = \frac{\text{mmf}}{\text{Reluctance}} \quad \Rightarrow \quad \text{mmf} = \text{Flux} \times \text{Reluctance}$$

The Ohm's law of magnetic circuit can be stated as follows.

The mmf of a magnetic circuit is directly proportional to flux established in it provided no part of the magnetic circuit is saturated. The constant of proportionality is the reluctance of the magnetic circuit.

i.e., mmf α flux   or   mmf = Flux × Reluctance.

Q3.10 What is magnetization curve?

The magnetization curve is a graph showing the relation between the magnetic field intensity, H and the flux density, B of a magnetic material. It is used to estimate the mmf required for flux path in the magnetic material and it is supplied by the manufacturers of stampings or laminations.

Q3.11 What is loss curve?

The loss curve is a graph showing the relation between iron loss and magnetic field intensity, H. It is used to estimate the iron loss of the magnetic materials and it is supplied by the manufacturers of stampings or laminations.

Q3.12 For magnetic circuit calculations, B-at curves are generally used. Why?

The relation between the flux density (B) and the magnetizing force (H) or mmf per metre (**at**) is non-linear. Hence it is difficult to express the relation in terms of mathematical equation. The manufacturers of laminations will supply the B-**at** curves which they obtain by practical estimation. Therefore to calculate the mmf per metre of flux path for a given flux density the B-**at** curve is employed.

Q3.13 How magnetisation curve is made use of in the design of electrical machines?

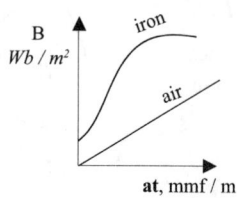

Fig. Q3.13: B-at curve.

In electrical machine design the magnetisation curve is used to determine the mmf per metre of magnetic path in any part of the machine. First the flux density in any part of machine is calculated and then for this value of flux density (B) the value of mmf per metre (**at**) is determined from magnetisation curve (B-**at** curve).

Q3.14 Draw the B-at curve for air. Give its H value in terms of B.

The air is a non-magnetic material. Hence it will have a constant value of permeability and so the B-**at** curve for air will be a straight line passing through the origin. The B and H of air is related by μ_o, where μ_o is the permeability of free space.

Here, $H = B/\mu_o$;   where, $\mu_o = 4\pi \times 10^{-7}$ H/m.

Q3.15 What is the difference in permeability of magnetic and non-magnetic materials?

In magnetic materials the permeability is not constant and it depends on the saturation of the magnetic material. But in non-magnetic materials the permeability is constant.

Q3.16 What is meant by magnetic circuit calculations?

The calculations of reluctance, flux density and mmf for various sections of magnetic circuit are commonly referred to as magnetic circuit calculations.

Q3.17 How the mmf of a section of magnetic circuit is determined? or How to find total mmf in series circuit?

The various steps in estimation of mmf of a section of magnetic circuit are
1.   Determine the flux in the concerned section.
2.   Calculate the area of cross-section of the section.
3.   Calculate the flux density in the section.
4.   From B-**at** curve of the magnetic material, determine the mmf per meter (**at**) for the calculated flux density.
5.   The mmf of the section is given by the product of length of the section and mmf per metre.

Q3.18 How the mmf of a magnetic circuit is determined?   *(PTU, May' 19, 2 M)*

The magnetic circuit is split into convenient parts (sections) which may be connected in series or parallel. Then the reluctance, flux density and mmf for every section of the magnetic circuit is estimated. The summation of mmf of all sections in series gives the total mmf for the magnetic circuit.

Q3.19 Why the calculation of mmf for air-gap cannot be generalized?

The reluctance of the air-gap is modified or affected by the slot opening, radial ventilating ducts and non uniform air-gaps. The machines may have different type of slots. Some machines are provided with ventilating ducts and some may not. Hence the calculations of reluctance of air-gap should be attempted individually for each type of machine. Therefore the calculation of mmf for air-gap cannot to generalized.

Q3.20 What are the factors which modify the reluctance of air-gap?

The reluctance of air-gap is modified by slots, radial ventilating ducts and non-uniform air-gaps.

Q3.21 Define gap contraction factor for slots.   *(HTU, Dec' 18, 2 M)*

The gap contraction factor for slots, K_{gs} is defined as the ratio of reluctance of air-gap in machines with slotted armature to reluctance of air-gap in machines with smooth armature.

$$K_{gs} = \frac{\text{Reluctance of air-gap in machines with slotted armature}}{\text{Reluctance of air-gap in machines with smooth armature}}$$

Q3.22 Define gap contraction factor for ducts.

The gap contraction factor for ducts, K_{gd} is defined as the ratio of reluctance of air-gap in machines with ducts to reluctance of air-gap in machines without ducts.

$$K_{gd} = \frac{\text{Reluctance of air-gap in machines ducts}}{\text{Reluctance of air-gap in machines without ducts}}$$

Q3.23 Define total gap contraction factor, K_g.   *(UTV, Dec' 13, 2 M)*

The total gap contraction factor, K_g is defined as the ratio of reluctance of air-gap in machines with slotted armature and ducts to the reluctance of air-gap in machines with smooth armature and without ducts. The total gap contraction factor is equal to the product of gap contraction factor for slots and ducts.

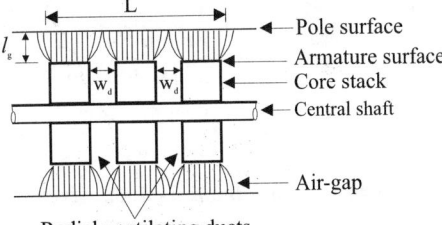

Q3.24 What are ventilating ducts?

The radial ventilating ducts are small gaps of width, w_d in between the stacks of armature core as shown in figure below. They are provided for better cooling of the core when the length of the core is greater than $0.1m$.

Fig. Q3.23: Cross-section of armature with radial ventilating ducts.

(UTV, Dec' 13, 2 M)

Q3.25 What is Carter's coefficient? What is its usefulness in the design of DC machine?   *(AU, Apr' 19, 2 M)*

The Carter's coefficient is a parameter that can be used to estimate the contracted or effective slot pitch in case of armatures with open or semienclosed slots. It is a function of the ratio of w_o/l_g, where w_o is slot opening and l_g is air-gap length.

Carter's coefficient is also used to estimate the effective length of armature when ducts are employed. In this case it will be a function of w_d/l_g, where w_d is the width of duct.

In electrical machine design the Carter's coefficient is used to estimate the air-gap expansion (or contraction) factor for slots and ducts. The increase in reluctance of air-gap due to slotting and ducts is accounted as additional air-gap, which in turn is given by air-gap expansion factor.

Q3.26 Write down the importance of gap contraction factor for slots and ducts in the design of magnetic circuits.

The slots and ducts in the armature (or stator and rotor) of electrical machines increases the reluctance of air-gap, which in turn increases the mmf required for air-gap. The gap contraction factor represents the increase in reluctance as an increase in air-gap length. Hence, with the knowledge of gap contraction factor the mmf required for air-gap can be estimated without calculating the increase in reluctance due to slots and ducts.

Q3.27 Write the expressions for gap contraction factor for slots and ducts.

$$\text{Gap contraction factor for slots, } K_{gs} = \frac{y_s}{y_s - K_{cs}w_s}$$

(HTU, Dec, 18, 2 M)

$$\text{Gap contraction factor for ducts, } K_{gd} = \frac{L}{L - K_{cd}\, n_d\, w_d}$$

where, y_s = Slot pitch w_d = Width of duct

          K_{cs} = Carter's coefficient for slots K_{cd} = Carter's coefficient for ducts

          w_s = Slot opening in case of semienclosed slots n_d = Number of ducts

                (or slot width in case of open type slots)

          L = Length of armature

Q3.28 Write the expressions for reluctance of air-gap in machines with smooth armature and slotted armature.

Reluctance of air-gap in machines with closed armature slots $\left\}\; S_{g\_cas} = \dfrac{l_g}{\mu_0\, L\, y_s}\right.$

Reluctance of air-gap in machines with open armature slots including fringing effect $\left\}\; S_{g\_oasf} = \dfrac{l_g}{\mu_0\, L\, y'_s}\right.$

(HTU, Dec, 13, 2 M)

where, l_g = Length of air-gap

          μ_0 = Permeability of free space $y'_s = y_s - K_{cs}w_s$ = Effective slot pitch

          L = Length of armature w_s = Width of slot

          y_s = Slot pitch K_{cs} = Carter's gap coefficient for slots.

Q3.29 Find the width of the duct. Given that L_i = 0.25 m, L = 0.33 m, K_i = 0.9 and number of ducts = 5.

Solution

$$\text{Length of armature excluding ducts} = \frac{\text{Net iron length}}{\text{Stacking factor}} = \frac{L_i}{K_i} = \frac{0.25}{0.9} = 0.2778\ m$$

Length of armature including ducts = L = 0.33 m

$$\text{Total width of all the ducts} = L - \frac{L_i}{K_i} = 0.33 - 0.2778 = 0.0522\ m$$

$$\text{Width of one duct} = \frac{\text{Total width of all the ducts}}{\text{Number of ducts}} = \frac{0.0522}{2} = 0.0104\ m = 10.4\ mm$$

Q3.30 Define air-gap expansion factor K_g.

The air-gap expansion factor is defined as the ratio of reluctance of air-gap in machine with slotted armature and duct to the reluctance of air-gap in machine with smooth armature and without ducts. It is same as gap contraction factor.

Q3.31 How ventilating ducts are provided in the armature of DC machine ?

The armature core is divided into stacks of 40 to 80 mm thick, with ventilating ducts of width 10 mm in between two stacks. Normally ventilating radial ducts are provided if the length of the core is more than 0.1 m.

Q3.32 Write an expression to find the total mmf required to be produced by each pole.

Total mmf to be produced by a pole (in DC machines) $\left.\right\}$ = mmf for air gap + mmf for teeth + mmf for pole and pole shoe + mmf for yoke + mmf for armature core

Q3.33 Write down the formula for computing the mmf for air-gap.

mmf for air-gap = 800,000 B K_g l_g

where, B = Flux density in air-gap

$\qquad$ = B_{av} - For non-salient pole machines

$\qquad$ = B_g - For salient pole machines

K_g = Gap contraction factor

$\qquad$ = $K_{gs} K_{gd}$ - For non-salient pole machines

$\qquad$ = $K_{gs} K_{gd} K_{gsal}$ - For salient pole machines

l_g = Length of air-gap

Q3.34 What is the effect of salient poles on the air-gap mmf?

In salient pole machines the length of air-gap is not constant over the whole pole pitch. Hence the effective air-gap length is given by $K_g l_g$ where K_g is the gap contraction factor. Also for calculating mmf, the maximum gap density, B_g at the centre of the pole is considered instead of average gap density.

mmf for air-gap in salient pole machine = 80,000 $B_g K_g l_g$

Q3.35 Define field form factor. *(UTV, Dec' 13, 2 M)*

The field form factor, K_f is defined as the ratio of average gap density over the pole pitch to maximum flux density in the air-gap.

$$\text{Field from factor, } K_f = \frac{B_{av}}{B_g} \quad ; \quad \text{Also, } K_f \approx \psi = \frac{\text{Pole arc}}{\text{Pole pitch}}$$

Q3.36 What are the problems encountered in estimating the mmf for teeth?

The problems encountered in estimating the mmf for teeth are,

1. The flux density in different section of a tooth is not uniform.
2. The slot provides another parallel path for the flux, shunting the tooth.

Q3.37 List the methods used for estimating the mmf for teeth (tapered teeth) *(HTU, Dec' 18, 2 M)*

The following are the three methods used for estimating the mmf for teeth.

1. Graphical method
2. Three ordinate method (simpson's rule)
3. $B_{t1/3}$ method

Q3.38 What is real and apparent flux density? or Distinguish between real and apparent flux densities in the tooth section of slot. *(AU, Apr' 17, 2 M)*

The real flux density is due to the actual flux through a tooth. The apparent flux density is due to total flux that has to be pass through the tooth. Since some of the flux passes through slot, the real flux density is always less than the apparent or total flux density.

$$B_{app} = \frac{\text{Total flux in a slot pitch}}{\text{Tooth area}} \quad ; \quad B_{real} = \frac{\text{Actual flux in a tooth}}{\text{Tooth area}}$$

Q3.39 Define the terms tapered slot and tapered teeth.

If the width of the slot gradually increases (or decreases) from top of slot to bottom of the slot, then the slot is called tapered slot.

If the width of the tooth gradually increases (or decreases) from tip of tooth to root of tooth, then the tooth is called tapered tooth.

Q3.40 In which way the air-gap length influence the design of machines.

The total mmf to be produced by a pole is approximately equal to mmf for air-gap. Hence the design of field system primarily depends on length of air-gap.

Q3.41 Calculate the mmf per metre for a flux density of 1.7 Wb/m² and a permeability of 23.5 × 10⁻⁶ H/m.

Solution

$$\text{mmf per metre, } \mathbf{at} = \frac{B}{\mu} = \frac{1.7}{23.5 \times 10^{-6}} = 72340 \ AT$$

Q3.42 Give the simpon's rule for calculation of mmf for tooth.

$$\text{mmf for tooth} = \mathbf{at}_{mean} \times l_t = \mathbf{at}_{mean} \times d_s$$

$$\mathbf{at}_{mean} = \frac{\mathbf{at}_1 + 4\mathbf{at}_2 + \mathbf{at}_3}{6}$$

where, $l_t = d_s =$ Length of tooth $\mathbf{at}_2 = $ at for the centre of the tooth
          $\mathbf{at}_1 = $ at for the root of tooth $\mathbf{at}_3 = $ at for the tip of tooth

Q3.43 Write the rule for calculation of mmf for tooth by $B_{t\,1/3}$ method.

$$\text{mmf for tooth} = \mathbf{at}_{1/3} \times l_t = \mathbf{at}_{1/3} \times d_s$$

where, $\mathbf{at}_{1/3} = $ at for flux density at one-third height from the narrow end
          $l_t = d_s =$ Length of tooth

Q3.44 How the variable cross-section of the tooth is accounted for in determining ampere turns for tooth?

When the tooth has variable cross-section the mmf can be estimated either by simpon's rule or by $B_{t\,1/3}$ method. In simpon's method the "**at**" at the root, centre and tip are estimated.

$$\text{Now, } \mathbf{at}_{mean} = \frac{\mathbf{at}_1 + 4\mathbf{at}_2 + \mathbf{at}_3}{6}$$

$$\text{mmf for tooth} = \mathbf{at}_{mean} \times l_t$$

where, $\mathbf{at}_1 = $ at for the root of tooth $\mathbf{at}_3 = $ at for the tip of the tooth
          $\mathbf{at}_2 = $ at for the centre of tooth $l_t = $ Length of tooth

In $B_{t\,1/3}$ method the "**at**" for the flux density at one-third height from the narrow end of tooth is determined.

$$\text{mmf for tooth} = \mathbf{at}_{1/3} \times l_t$$

where, $\mathbf{at}_{1/3} = $ at for flux density at one-third height from narrow end of tooth
          $l_t = $ length of tooth

Q3.45 What is fringing flux?

The bulging of magnetic path at the air-gap is called fringing. The fluxes in the bulged portion are called fringing flux.

Q3.46 State the relation between apparent flux density and real flux density.

The real and apparent flux density are related by the equation,

Real flux density, $B_{real} = B_{app} - \mu_o \, at_{real} \, (K_s - 1)$ and $K_s = \dfrac{L \, y_s}{L_i \, w_t}$

where, B_{app} =   Apparent flux density L_i =   Net iron length of armature

         μ_o = $4\pi \times 10^{-7}$ = Permeability of free space y_s =   Slot pitch

         at_{real} =   at for B_{real} of tooth w_s =   Width of tooth

         L =   Length of armature

Q3.47 What is magnetic leakage (or leakage flux) and leakage coefficient ?

The leakage flux is the flux passing through unwanted path. The leakage flux will not help either for transfer or conversion of energy.

The leakage coefficient is defined as the ratio of total flux to useful flux.

$$\text{Leakage coefficient, } C_l = \frac{\text{Total flux}}{\text{Useful flux}}$$

Q3.48 Write down the importance of leakage coefficient in the design of magnetic circuits.

The leakage coefficient is used to estimate the leakage flux. Since the leakage flux affects the performance of transformers and rotating machines, the knowledge of leakage flux is essential.

The leakage flux affects the excitation demand, regulation, forces on the winding under short circuit conditions, commutation, stray load losses and circulating currents in transformer tank walls.

Q3.49 What are the differences between leakage flux and fringing flux?

The leakage flux is not useful for energy transfer or conversion. But the fringing flux is useful flux. The leakage flux flows in the unwanted path. But the fringing flux flows in the magnetic path. The effect of leakage flux on machine performance is accounted by leakage reactance. The fringing flux increases the slot reactance.

Q3.50 List the different types of armature or stator leakage fluxes.

The different types of armature or stator leakage fluxes are,

1.   Slot leakage flux   5.   Tooth top leakage flux
2.   Zig-zag leakage flux   6.   Overhang leakage flux
3.   Harmonic or differential leakage flux   7.   Skew leakage flux
4.   Peripheral leakage flux.

Q3.51 What is slot leakage flux ?

The fluxes that crosses the slot from one tooth to the next and returning through iron are called slot leakage flux.

Q3.52 What is tooth top leakage flux ?

The flux flowing from top of one tooth to the top of another tooth is called tooth top leakage flux.

Q3.53 What is Zig-Zag leakage flux ?

The flux passing from one tooth to another in a zig-zag fashion across the air-gap is called zig-zag leakage flux.

Q3.54 What is overhang leakage flux ?

The fluxes produced by overhang portion of the armature winding are called overhang leakage flux.

Q3.55 What is harmonic leakage flux ?

The fluxes produced due to the difference in stator and rotor harmonic contents are called harmonic leakage flux.

Q3.56 What is skew leakage flux ?

The reduction in the mutual flux due to skewing of rotor in induction motor is accounted as skew leakage flux.

Q3.57 What is pheripheral leakage flux ?

The fluxes flowing circumferentially around the air-gap without linking with any of the windings are called peripheral leakage flux.

Q3.58 How will you minimize the magnetic leakage ?

Most of the leakage fluxes are flowing through air-gap of the machine. If the air-gap of the machine is kept as low as possible, these leakage fluxes can be minimized. The harmonic leakage can be minimized by balancing the stator and rotor currents. The slot leakage can be minimized if the width of the slot is more than the width of tooth.

Q3.59 List the factor that are affected by the leakage flux.

The leakage flux affects the following performance indices of electrical machines.

1. Excitation demand of salient pole machines.
2. Performance of AC machines depends on the leakage reactance.
3. Forces between the windings under short circuit condition.
4. Voltage regulation of generators and transformers.
5. Commutation condition in DC machines.
6. Stray load losses.
7. Circulating currents in transformer tank walls.

Q3.60 Define slot leakage reactance.

The slot leakage reactance is the reactance offered by slot leakage flux. The slot leakage flux will offer a reactive voltage drop which is equivalent to an inductive effect and so the effect can be measured in terms of reactance.

Q3.61 What is reactance voltage? Is it beneficial ?

The reactance voltage is the induced voltage due to leakage flux. It is not beneficial. The reactance voltage will affect commutation and produce sparking in DC machines. The reactance voltage will increase the regulation in case of transformers and rotating machines.

Q3.62 Write an expression for slot reactance.

$$\text{Slot leakage reactance, } X_s = 2\pi\, f\, Z_s^2\, L\, \lambda_s$$

where,   f   = Frequency   L   = Length of armature

Z_s = Conductors per slot λ_s = Specific slot permeance

Q3.63 Write the equation for slot leakage reactance per phase in poly phase machine.

$$\text{Slot leakage reactance per phase, } X_s = 2\pi\, f\, pq\, Z_s^2\, L\, \lambda_s$$

$$\text{or } X_s = 8\pi\, f\, T_{ph}^2\, L\, \lambda_s / pq$$

Q3.64 What is permeance ?

The permeance is inverse of reluctance. The reluctance of a magnetic path is given by,

$$\text{Reluctance, } S = \frac{l}{A\mu} \quad ; \quad \therefore \text{Permeance, } \Lambda = \frac{A\mu}{l}$$

where,   l   = Length of flux path

A = Area of cross-section of flux path

μ = Permeability of the medium

Q3.65 Define specific permeance of a slot.

Specific permeance of a slot is defined as the permeance per unit length of slot or depth of field.

Q3.66 What is unbalanced magnetic pull?   *(HTU, Dec' 18, 2 M) (AU, Nov' 17, 2 M)*

The unbalanced magnetic pull is the radial force acting on the rotor due to non uniform air-gap around armature periphery.

Q3.67 List the various techniques employed to overcome unbalanced magnetic pull (or to reduce unbalanced magnetic pull).

The techniques employed to overcome the unbalanced magnetic pull are the following.

1. The length of the rotor is kept small and diameter is made higher.

2. Ball bearings are employed and rotor is dynamically balanced.

3. The combinations of rotor and stator slots which produce vibrations are avoided.

Q3.68 Write an expression to determine unbalanced magnetic pull.

$$\left.\begin{array}{l}\text{Unbalanced magnetic pull}\\ \text{due to p number of poles}\end{array}\right\} = p\, A_p\, P_m\, \frac{e}{l_g}$$

$$\text{Magnetic pull per unit area, } P_m = \frac{1}{2}\frac{B^2}{\mu_o}$$

where, B = Flux density in air-gap   p = Number of poles

μ_o = $4\pi \times 10^{-7}$ H/m = Permeability of air   e = Eccentricity

A_p = Area per pole l_g = Length of air-gap

Q3.69 Mention the importance of conductor dimensions.

The dimensions of the conductors directly affects the following factors in rotating machines.

1. Allowable temperature rise   3. Current density

2. Resistivity   4. Specific electric loading.

Q3.70 List the different types of slots that are used in rotating machines.

The different types of slots are,

1. Parallel sided slots with flat bottom   4. Tapered slots with circular bottom

2. Tapered slots with flat bottom   5. Circular slots.

3. Parallel sided slots with circular bottom

Q3.71 Mention the undesirable effects of unbalanced magnetic pull.

The undesirable effects of unbalanced magnetic pull are

1. Saturation of magnetic materials due to reduction in air-gap

2. Excessive vibration and noise due to unbalanced radial forces.

Q3.72 Write an expression to determine the unbalanced magnetic pull when the poles are located at an angle θ with horizontal axis.

$$\left.\begin{array}{l}\text{Unbalanced magnetic pull acting}\\ \text{downwards due to a pair of poles}\end{array}\right\} = 4\, A_p\, P_m\, \frac{e}{l_g}\sin^2\theta$$

3.20 EXERCISES

I.   Fill in the blanks

1.   The magnetic circuit is the path of ....

2.   The mmf creates flux in a closed path against the .... of the path.

3.   The value of permeability of non-magnetic material is ....

4.   In a series magnetic circuit .... is same but .... is different in every section.

5.   In a parallel magnetic circuit .... is same but........is different in every section.

6.   The mmf per metre of any non-magnetic material is .... times the flux density B.

7.   In magnetic materials the B-H relationship is ....

8.   The .... curve is used to estimate .... for a given flux density.

9.   The .... curve is used to estimate .... for a given flux density.

10.   The reluctance of the air-gap in machines with smooth armature is given by the equation......

11.   The carter's coefficient for slots depends on the ratio of ....

12.   The carter's coefficient for ducts depends on the ratio of ....

13.   The bulging of magnetic path at the air-gap is called ....

14.   The mmf required for air-gap with slotted armature is .... than that of smooth armature.

15.   The equation for at_{mean} in three ordinate method is ....

16.   The two components of iron losses are .... and .... losses.

17.   In .... method the at_{mean} is calculated using the value of "at" at three equidistant points.

18.   The induced voltage due to the alternating leakage flux is called ....

19.   The permeance per unit length of slot or depth of field is called ....

20.   The harmonic or differential leakage flux is also called ....

21.   The two parallel flux paths over one .... are iron path of tooth and conductor path of slot.

22.   In .... the "at" obtained for flux density at a section one-third of tooth height from the narrow end is the ....

23.   The flux which passes through unwanted path is called the ....

24.   The .... is defined as the ratio of total flux to useful flux.

Answers

1. magnetic flux	9. loss, core loss/kg	17. simpson's
2. reluctance	10. $S_g = l/\mu_o \, L \, y_s$	18. leakage reactance voltage
3. $4\pi \times 10^{-7}$ *H/m*	11. slot opening to gap length	19. specific permeance
4. flux, mmf	12. duct width to gap length	20. belt leakage flux
5. mmf, flux	13. fringing	21. slot pitch
6. 800,000	14. K_g times	22. $B_{t1/3}$ method, at_{mean}
7. non-linear	15. $at_{mean} = (at_1 + 4at_2 + at_3)/6$	23. leakage flux
8. magnetization, mmf/m	16. hysteresis, eddy current	24. leakage coefficient

II.   State whether the following statements are True/False

1.   Energy is needed only to create the flux but not to maintain it.
2.   The reluctance of a magnetic material is constant.
3.   The permeability of non-magnetic material is constant.
4.   The B-**at** curve for air will be a straight line passing through origin.
5.   The permeability of the air-gap is not modified by slots and ducts.
6.   The reluctance of air-gap in machines with smooth armature is higher than with slotted armature .
7.   The value of carter's gap coefficient for semienclosed slots is higher than that for open slots.
8.   The gap contraction factor represents the increase in reluctance as an increase in air-gap length.
9.   The fringing flux is not an useful flux.
10.   The mmf for air-gap in machines with slotted armature is lesser than that for smooth armature.
11.   The armature slots and ducts increases the mmf required for air-gap.
12.   In salient pole machines the air-gap dimensions are uniform over whole of the pole pitch.
13.   The permeability of teeth is low because they normally work in the saturated region.
14.   The reluctance of slot portion is dependent on the degree of saturation of teeth.
15.   The mmf across the slot and tooth are same, because they are parallel in the magnetic circuit.
16.   Tooth top leakage flux is important in dc and synchronous machines, because they are having large air-gaps.
17   Skew leakage flux is present only when the slots are skewed.
18.   The real flux passing through the tooth is always higher than the apparent flux.
19.   The slot leakage reactance is inversely proportional to specific slot leakage permeance.
20.   The slot leakage reactance is inversely proportional to the square of the number of slots per pole per phase.
21.   The unbalanced magnetic pull produces excessive vibrations.
22.   The saturation of iron parts decreases due to unbalanced magnetic pull.
23.   The slot space factor depends on thickness of the insulation of the conductors.
24.   The dimensions of the conductor directly affects the specific electric loading.
25.   To reduce unbalanced magnetic pull the diameter of rotor is kept small and length is made higher.

Answers									
1.	True	6.	False	11.	True	16.	True	21.	True
2.	False	7.	True	12.	False	17.	True	22.	False
3.	True	8.	True	13.	True	18.	False	23.	True
4.	True	9.	False	14.	True	19.	False	24.	True
5.	False	10.	False	15.	True	20.	True	25.	False

III.   Unsolved problems

E3.1   Calculate the mmf required for the air-gap of a machine having core length = 0.25 m, including 4 ducts of 10 mm each, pole arc = 0.19 m, slot pitch = 64 mm, slot opening = 3 mm, air-gap length = 0.95 mm, flux per pole = 52 mWb, Carter's co-efficient for slot = 0.46, Carter's coefficient for duct = 0.65.

(**Note :** For semi enclosed slots, replace w_s by w_o in the expression for k_{gs} where w_o is slot opening).

$$(AT_g = 949\ AT)$$

E3.2 Estimate the mmf for air-gap of a three phase slip ring induction motor from the following data, stator bore = 637 *mm*, core length = 250 *mm*, number of stator slots = 90, slot opening = 2 *mm*, rotor slots = 120, rotor slot opening = 2 *mm*, air-gap length = 1 *mm*, K_{gd} = 1.07, air-gap density = 0.62 Wb/m^2.

(**Note** : $K_{css} = \dfrac{1}{1 + 5\,l_g / w_{os}}$ and $K_{csr} = \dfrac{1}{1 + 5\,l_g / w_{or}}$).

$$(AT_g = 564.1\ AT)$$

E3.3 Determine the air-gap length of a DC machine from the following particulars: gross-length of core = 0.16 *m*, number of ducts = 3, width of duct = 12 *mm*, slot pitch = 28 *mm*, slot width = 12 *mm*, carter's coefficient for slots and ducts = 0.36, gap density at pole centre = 0.9 Wb/m^2, field mmf/pole = 3900 *AT*, mmf required for iron parts of magnetic circuit = 900 *AT*.

$$(l_g = 3.5\ mm)$$

E3.4 Calculate the mmf required for air-gap of a DC machine with an axial length of 25 *cm* (no ducts) and a pole arc of 20 *cm*. The slot pitch = 29 *mm*, slot opening = 15 *mm*, air-gap = 7 *mm* and the useful flux per pole = 28 *mWb*. Take carter's coefficient for slot as 0.5.

$$(AT_g = 4229\ AT)$$

E3.5 A 20 *kW*, 330 *V*, 4 pole DC machine has the following data : armature diameter = 0.29 *m*, armature core length = 0.25 *m*, length of air-gap at pole centre = 3.1 *mm*, flux per pole = 12.5 × 10⁻³ *Wb*, pole arc/pole pitch = 0.72

Calculate the mmf required for air-gap (i) if the armature surface is treated as smooth (ii) if the armature is slotted and the gap contraction factor is 1.29.

(mmf for air-gap with smooth armature = 756 *AT*
mmf for air-gap with slotted armature = 975 *AT*)

E3.6 Find the permeability at the root of the teeth of a DC machine armature from the following data, slot pitch = 28 *mm*, tooth width =18 *mm*, gross core length = 0.35 *m*, stacking factor = 0.9, real flux density = 2.15 Wb/m^2, app. flux density = 2.2156 Wb/m^2.

$$(\mu = 3 \times 10^{-5}\ H/m)$$

E3.7 Calculate the apparent flux density at a particular section of a tooth from the following data. Tooth width = 4 *mm*, slot width = 19 *mm*, gross core length = 75 *mm*, number of ventillating ducts = 4, width of duct = 10 *mm*, real flux density = 1.1 Wb/m^2, permeability of teeth = 31.4 x 10⁻⁶ *H/m*, stacking factor = 0.95.

$$(B_{app} \approx 1.6269\ Wb/m^2)$$

E3.8 Calculate the unbalanced magnetic pull of 2 pole DC machine having an air-gap flux density of 0.9 Wb/m^2 and a gap length of 4 *mm*. The area of each pole is 32 ×10⁻³ m^2 and the poles are vertically mounted. The eccentricity is 10% in the air-gap.

$$(2063\ N)$$

E3.9 The following data refers to a 50 *kW*, 4 pole DC generator, air-gap = 3 *mm*, area under each pole face = 28 × 10⁻³ m^2, air-gap flux density = 0.7 Wb/m^2, vertical displacement of rotor = 0.8 *mm*. Calculate the unbalanced magnetic pull acting downwards if the poles are centred 45° with horizontal axis.

$$(5823\ N)$$

> # Chapter 4

HEATING AND COOLING

List of symbols

Symbol	Meaning	Unit
a	Length of copper per metre of heat path in winding	m
c_p	Specific heat capacity	J/kg-°C
c_{pa}	Specific heat capacity of air	J/kg-°C
c_{ph}	Specific heat capacity of hydrogen	J/kg-°C
c_{pw}	Specific heat capacity of water	J/kg-°C
c_{po}	Specific heat capacity of oil	J/kg-°C
e	Coefficient of emissivity	-
G	Weight of active parts of machine	kg
H	Barometric pressure	mm Hg
H_a	Barometric pressure of air at inlet of cooling system	mm Hg
I_{eq}	Equivalent current	A
I_{nom}	Nominal load current	A
P_a	Pressure of air	N/m^2
P_c	Variable loss at nominal or full load	W
P_{csh}	Variable loss on short time load	W
P_{eq}	Equivalent power	W
P_i	Constant loss	W
P_{sh}	Load power of short time rating	W
P_{nom}	Load power of continuous rating	W
P_{ho}	Heating overload ratio	-
P_{mo}	Mechanical overload ratio	-
Q_{con}	Heat dissipated by conduction	J/s or W
Q_{rad}	Heat dissipated by radiation	J/s or W
Q_L	Power loss or heat to be dissipated	J/s or W
Q_{av}	Average losses	J/s or W

Symbol	Meaning	Unit
Q_{sh}	Permissible losses on short time duty	J/s or W
Q_{nom}	Nominal power loss	J/s or W
q	Heat produced per unit volume	W/m^3
q_{rad}	Heat transfer by radiation per unit surface	W/m^2
q_{conv}	Heat dissipated by convection per unit surface	W/m^2
R_c	Thermal resistance of heat conduction	$°C/W$ or Ω
S	Surface area of hot body	m^2
S_f	Space factor	-
T_h	Heating time constant	s or min
T_c	Cooling time constant	s or min
T_{eq}	Equivalent torque	N-m
T	Absolute temperature	$°K$
t	Length of heat path	m
t_h	Heating period	min or h
t_{sh}	Time duration of short time load	min or h
t_{shm}	Maximum permissible time on short time rating	min or h
V_a	Volume of coolant air per unit time	m^3/s
V_{a_NTP}	Volume of air at NTP per unit time	m^3/s
V_c	Volume of coolant per unit time	m^3/s
V_h	Volume of coolant hydrogen gas per unit time	m^3/s
V_w	Volume of coolant water per unit time	l/s
V_o	Volume of coolant oil per unit time	l/s
W_c	Weight of coolant per unit time	kg/s
θ_a	Temperature of ambient medium	$°C$
θ_e	Temperature of emitting surface	$°C$
θ_i	Initial temperature over ambient medium	$°C$
θ_r	Temperature rise while heating	$°C$
θ_{ra}	Temperature rise of coolant air	$°C$
θ_{rh}	Temperature rise of coolant hydrogen	$°C$
θ_{rsh}	Temperature rise with short time rated load	$°C$
θ_{ro}	Temperature rise of coolant oil	$°C$
θ_{rw}	Temperature rise of coolant water	$°C$
θ_f	Temperature fall while cooling	$°C$
θ_d	Temperature difference	$°C$
θ_m	Final steady temperature rise while heating / Maximum temperature rise limit / Hot spot temperature	$°C$
θ_{msh}	Maximum temperature with short time rated load	$°C$
θ_n	Final steady temperature rise while cooling	$°C$

Symbol	Meaning	Unit
θ_{sh}	Power loss with short time rating	$^{\circ}C$
θ_s	Surface temperature	$^{\circ}C$
λ	Specific heat dissipation	$W/m^2\text{-}^{\circ}C$
λ_{conv}	Specific heat dissipation by convection or emissivity	$W/m^2\text{-}^{\circ}C$
ρ	Thermal resisitivity of material	$^{\circ}C\text{-}m/W \text{ or } \Omega\text{-}m$
ρ_x	Thermal resistivity across the laminations	$\Omega\text{-}m$
ρ_y	Thermal resistivity along the laminations	$\Omega\text{-}m$
ρ_c	Density of coolant	kg/m^3
ρ_{ca}	Density of coolant air	kg/m^3
ρ_{ch}	Density of coolant hydrogen	kg/m^3
ρ_{cw}	Density of coolant water	kg/m^3
ρ_{co}	Density of coolant oil	kg/m^3
ρ_i	Thermal resistivity of insulation	$\Omega\text{-}m$
ρ_e	Effective thermal resistivity of winding	$\Omega\text{-}m$
σ	Thermal conductivity	$W/^{\circ}C\text{-}m$
η	Efficiency	-
η_f	Fan efficiency	-

4.1 INTRODUCTION

In electrical machines power or energy loss occur in winding, core and in rotating parts and the various losses appear in the form of heat. The continuous operation of the machines results in continuous heat generation and the generated heat dissipates in the surrounding medium. The various modes of heat dissipation are conduction, convection and radiation. If the heat generation is in excess of heat dissipation then heat accumulates in the machine and lead to increase in temperature above the ambient temperature. Every machine will have a maximum temperature rating and if the temperature increases above this maximum limit then machine will be damaged. Therefore, machines are designed with proper cooling system to prevent temperature rise.

In small capacity machines the power loss per unit volume will be less and so no special arrangements are needed for cooling. Mostly small capacity machines are provided with cooling fins on the outer surface of the machine which will increase the surface area and cooled by natural convection and radiation. But in large capacity machines the power loss per unit volume will be high and so special arrangements are needed for cooling. The cooling is provided in large capacity machines by providing ventilators through which air or water or hydrogen gas is circulated to remove excess heat.

4.1.1 Heat Conduction   *(UTU, Dec' 13, 4 M) (PU, Nov' 19, 6 M)*

Heat conduction is one of the modes of heat transfer (or transfer of internal thermal energy) in a body by the collisions of microscopic particles and movement of electrons within a body. In conduction mode of heat transfer, the heat flow is within and through the body itself. The microscopic particles in the heat conduction can be molecules, atoms and electrons.

Let, Q_{con} = Heat dissipated in conduction

θ_1, θ_2 = Temperature of two boundary surfaces

R_c = Thermal resistance for heat conduction

ρ = Thermal resistivity of material

$\sigma = \dfrac{1}{\rho}$ = Thermal conductivity of material

t = Length of heat path

S = Surface area of the body

The thermal resistance for heat conduction in a hot body is defined as,

Thermal resistance for heat conduction, $R_c = \dfrac{\rho\, t}{S}$ .....(4.1)

Since, $\rho = \dfrac{1}{\sigma}$

Thermal resistance for heat conduction, $R_c = \dfrac{t}{\sigma\, S}$ .....(4.2)

Now, the heat dissipated in conduction is given by,

Heat dissipated in conduction, $Q_{con} = \dfrac{\theta_1 - \theta_2}{R_c}$ .....(4.3)

Using Equation (4.1) in Equation (4.3) we get,

Heat dissipated in conduction, $Q_{con} = \dfrac{\theta_1 - \theta_2}{\rho\, t / S} = \dfrac{S(\theta_1 - \theta_2)}{\rho\, t}$

$$= \dfrac{S\, \theta_d}{\rho\, t} \quad .....(4.4)$$

where, $\theta_d = \theta_1 - \theta_2$ = Temperature difference across medium

From Equation (4.4) we get,

Temperature difference across medium, $\theta_d = Q_{con} \dfrac{\rho\, t}{S}$ .....(4.5)

Table 4.1: Thermal Resistivities of Materials used in Electrical Machines

Material	Thermal Resistivity (Ohm metre)	Material	Thermal Resistivity (Ohm metre)
Air	20	Asbestos	4
Cotton cloth	14	Empire cloth	4
Micanite	8	Mica	3
Compressed paper	8	Laminated core	
Paper	7.5	– along laminations	0.02
Transformer coil	6.25	– across laminations	0.05 to 0.1
Pressboard	6	Brass	0.01
Varnished cloth	5	Aluminium	0.005
Mica tape	2.6 to 6.6	Copper	0.026

4.1.2 Heat Radiation

Heat transfer through radiation (or thermal radiation) takes place in form of electromagnetic waves mainly in the infrared region. Radiation emitted by a body is due to thermal agitation of the constituent of the body like molecules, atoms and electrons.

Radiation heat transfer can be described by reference to the black body. The black body is defined as a body that absorbs all radiation that falls on its surface. Actual black bodies don't exist in nature-though its characteristics are approximated by a hole in a box filled with highly absorptive material.

A black body is a hypothetical body that completely absorbs all wavelengths of thermal radiation incident on it and so the black body do not reflect light. When heated, all black bodies emit thermal radiation.

Let, q_{rad} = Heat transfer by radiation

The radiation energy from a black body is proportional to the fourth power of the absolute temperature and can be expressed with Stefan-Boltzmann Law,

Heat transfer by radiation, $q_{rad} = \sigma\, T^4\, S$ .....(4.6)

where, σ = Stefan-Boltzmann constant = 5.6703×10^{-8}

T = Absolute temperature in Kelvin

S = Surface area of the emitting body

If a hot object is radiating energy to its cooler surroundings the net radiation heat loss rate can be expressed as,

Heat transferred by radiation per unit surface, $q_{rad} = \sigma\, e\, (T_e^4 - T_a^4)$ .....(4.7)

where, θ_e = Temperature of emitting medium in centigrade

θ_a = Temperature of ambient medium in centigrade

$T_e = 273 + \theta_e$ = Absolute temperature of emitting medium in Kelvin

$T_a = 273 + \theta_a$ = Absolute temperature of ambient medium in Kelvin

e = Coefficient of emissivity

For electrical machines the equation (4.7) can be simplified as shown below.

$$q_{rad} = \sigma\, e\, (T_e^4 - T_a^4) = \sigma\, e\, (T_e^2 + T_a^2)(T_e^2 - T_a^2) \qquad \boxed{a^2 - b^2 = (a+b)(a-b)}$$

$$= \sigma\, e\, (T_e^2 + T_a^2)(T_e + T_a)(T_e - T_a)$$

$$= \sigma\, e\, (T_e - T_a)(T_e^3 + T_e T_a^2 + T_a T_e^2 + T_a^3)$$

The term $T_e^3 + T_e T_a^2 + T_a T_e^2 + T_a^3$ will have small changes for the temperature range in electrical machines and so can be considered as constant.

Let, $k_{rad} = T_e^3 + T_e T_a^2 + T_a T_e^2 + T_a^3$

$\therefore q_{rad} = \sigma\, e\, k_{rad}\, (T_e - T_a)$

The temperature difference in °K will be same as temperature difference in °C, and so $T_e - T_a = \theta_e - \theta_a$.

$\therefore q_{rad} = \sigma\, e\, k_{rad}\, (\theta_e - \theta_a)$

$$= \lambda_{rad}\, \theta_d \qquad .....(4.8)$$

Here, $\lambda_{rad} = \sigma \, e \, k_{rad} = $ Specific heat dissipation due to radiation

$\theta_d = \theta_e - \theta_a = $ Temperature difference between emitting medium and ambient

$\therefore$ Total heat dissipation by radiation, $Q_{rad} = q_{rad} \, S = \lambda_{rad} \, \theta_d \, S$ .....(4.9)

Here, the specific heat dissipation by radiation is directly propotional to emissivity. From Table 4.2 we can observe that emissivity of dull metallic paints like white and grey are higher than metals. Therefore, the electrical machines are painted with dull metallic paints to keep better heat dissipation by radiation.

Table 4.2: Emissivity of Materials

Surface	Emissivity e
Aluminium	0.10
Copper	0.15
Steel	
- Rough	0.24
- Sheet	0.55
Metal Paint Aluminium	0.55
Lead Paints	
- White	0.90
- Grey	0.95

4.1.3 Heat Convection

Heat convection is a mode of heat transfer occurs on the surface of a hot body when the surrounding fluid move heat energy away from the hot body. Convection heat transfer occurs when the surface temperature of a hot body differs from that of surrounding fluid.

Let, $q_{conv} = $ Heat dissipated per unit surface by convection.

For electrical machines,

Heat dissipated per unit surface by convection, $q_{conv} = K_c \, (\theta_e - \theta_a)^n$ .....(4.10)

where, K_c = Constant (Depend on dimensions of hot body)

        n = Constant (Depend on dimensions of hot body)

        θ_e = Temperature of emitting surface

        θ_a = Temperature of ambient

Let, $n = 1$

    $\theta_d = (\theta_e - \theta_a) = $ Temperature difference between emitting surface and ambient

    $K_c = \lambda_{conv} = $ Specific heat dissipation due to convection.

$\therefore$ Heat dissipated per unit surface by convection, $q_{conv} = \lambda_{conv} \, \theta_d$ .....(4.11)

Total heat dissipation due to convection, $Q_{conv} = q_{conv} \, S = \lambda_{conv} \, \theta_d \, S$ .....(4.12)

4.2 NEWTON'S LAW OF COOLING

Newton's law of cooling states that the rate of heat loss of a body is directly proportional to the difference in the temperature between the body and its surroundings.

$$\text{Heat dissipation, } Q = (\lambda_{rad} + \lambda_{conv})\,\theta_d\,S = \lambda\,\theta_d\,S \qquad \qquad .....(4.13)$$

where, $\lambda = \lambda_{rad} + \lambda_{conv}$ = Sum of specific heat dissipation due to radiation and convection

θ_d = Temperature difference between body and ambient

S = Cooling surface

λ = Specific heat dissipation

The Newton's law of cooling, is applicable when the temperature difference is small and the nature of heat transfer is constant. This condition is generally met in heat conduction. In convective heat transfer, Newton's Law is followed for forced air or pumped fluid cooling, where the properties of the fluid do not vary strongly with temperature. In the case of heat transfer by thermal radiation, Newton's law of cooling holds only for very small temperature differences.

When stated in terms of temperature differences, Newton's law with several further simplifying assumptions, results in a simple differential equation expressing temperature-difference as a function of time. The solution to that equation will be an exponential decrease of temperature-difference over time. This characteristic decay of the temperature-difference is also associated with Newton's law of cooling.

Table 4.4: Specific Heat Dissipation of Various Surfaces at 40°C

Surface	λ
	$W/m^2\text{-}°C$
Polished metal	8.2
Tarnished metal	9.1
Aluminium paint	10.8
Oil paint	13.0
Shellac varnish	13.5
Cotton tape	12.4
Varnished tape	15.0

4.3 INTERNAL TEMPERATURES IN CORE AND WINDING

In electrical machines power loss occur in core and winding. The power loss in core is called iron loss and the power loss in winding is called copper loss. The power loss is released in the form of heat and accumulated heat will increase the temperature of the core and winding. In order to prevent high temperature rises the heat is removed by cooling system. The power loss is distributed throughout the length and breadth of the core and winding and so the heat generation and dissipation are also distributed in nature. Therefore, estimation of temperature rise is a complex problem. But for analysis purpose the heat distribution is assumed in a particular direction.

Since the external surface of core and winding are in contact with surrounding environment, they will have better cooling than internal surface. Due to this, the internal temperature of the core and winding will be higher than the surface temperature of the core and winding. Sometimes hot spots are developed inside the core and winding that may damage or deteriorate the insulation. Therefore, estimation of internal and hotspot temperature of the core and winding is essential. The cooling system should be designed to keep the temperature of hotspots within safe limits.

4.3.1 Internal Temperature of a Hot Body

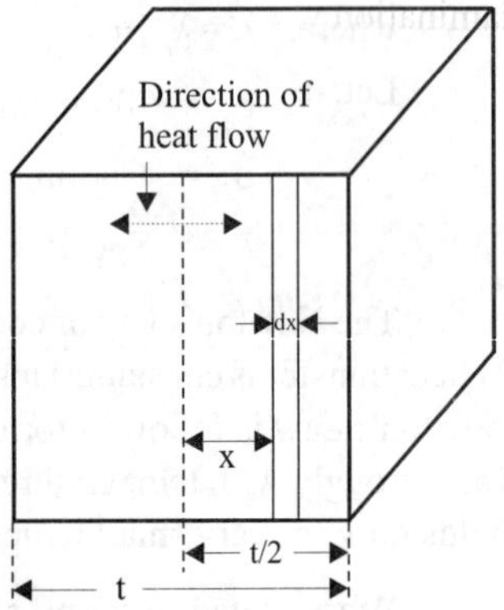

Consider the hot body shown in Fig. 4.1 with thickness t.

Now, the distance from center of the body to outer surface is t/2.

Let, x = Distance of a surface from center of the body

Q = Total heat

q = Q/St = Heat produced per unit volume

S = Surface area of hot body

t = Length of heat path

ρ = Thermal resistivity of hot body

θ_{dx} = Temperature difference between center and surface at a distance x

θ_{do} = Temperature difference between centre and outer surface

Fig. 4.1: Hot body.

Let us assume that the heat flows from centre of body to outer surface as shown in Fig. 4.1. Now the temperature difference between centre of body and the surface at a distance x can be estimated using following equation.

Temperature between center and x, $\theta_{dx} = \dfrac{q \rho x^2}{2}$ .....(4.14)

The temperature difference between centre and outer surface can be estimated from equation (4.14) by substituting, x = t/2.

∴ Temperature between center and outer surface, $\theta_{do} = \dfrac{q \rho (t/2)^2}{2}$

$$= \dfrac{q \rho t^2}{8}$$ .....(4.15)

Let, θ_s = Temperature at surface of the body.

The maximum temperature occur at the centre of the body. Therefore, the sum of θ_{do} and θ_s will give the temperature of the hottest spot or hot spot temperature.

Hot spot temperature, $\theta_m = \theta_{do} + \theta_s$

$$= \dfrac{q \rho t^2}{8} + \theta_s$$ .....(4.16)

4.3.2 Internal Temperature in Cores

The cores are constructed using laminated sheet steel. For analysis of heat conduction, two directions are chosen as shown in Fig. 4.2. One direction is across the lamination and another direction is along the lamination. The value of thermal resistivity across lamination will be higher than along lamination.

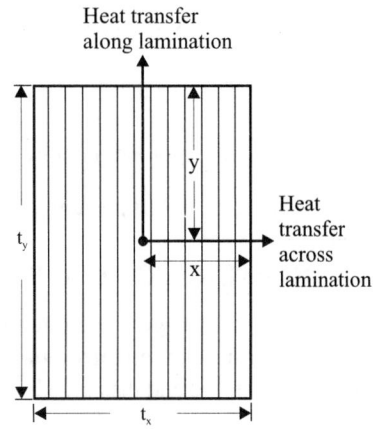

Fig. 4.2: Laminated core.

Let, t_x = Length of heat transfer path across lamination

t_y = Length of heat transfer path along lamination

$x = t_x/2$ = Distance from core centre to outer surface across lamination

$y = t_y/2$ = Distance from core centre to outer surface along lamination

ρ_x = Thermal resistivity across lamination

ρ_y = Thermal resistivity along lamination

Across Lamination,

$$\left.\begin{array}{l}\text{Temperature between center}\\ \text{and outer surface}\end{array}\right\} \theta_{dx} = \frac{q\,\rho_x\,x^2}{2}$$

$$= \frac{q\,\rho_x\,(t_x/2)^2}{2} = \frac{q\,\rho_x\,t_x^2}{8} \qquad .....(4.17)$$

Along Lamination,

$$\left.\begin{array}{l}\text{Temperature between centre}\\ \text{and outer surface}\end{array}\right\} \theta_{dy} = \frac{q\,\rho_y\,y^2}{2}$$

$$= \frac{q\,\rho_y\,(t_y/2)^2}{2} = \frac{q\,\rho_y\,t_y^2}{8} \qquad .....(4.18)$$

4.3.3 Internal Temperature in Winding

The winding copper is coated with insulation varnish and so the thermal resistivity of winding is sum of resistivity of copper and insulation.

Let, ρ_e = Effective resistivity of copper and insulation

Effective resistivity of winding, $\rho_e = \rho_i\,(1 - a)$ 　　　　.....(4.19)

Where, ρ_i = Thermal resistivity of insulation

a = Length of copper per meter of heat path in winding

Note: Length of heat path in winding will have copper and insulation.

Also, $a = \sqrt{S_f}$

Where, S_f = Space factor

Let, x = Distance of a surface from centre of winding

t = Length of heat path in winding

Temperature between center and x, $\theta_{dx} = \dfrac{q \, \rho_e \, x^2}{2}$

Temperature between center and outer surface, $\theta_{do} = \dfrac{q \, \rho_e \, (t/2)^2}{2} = \dfrac{q \, \rho_e \, t^2}{8}$

Hot spot temperature in winding, $\theta_m = \theta_{do} + \theta_s$

Where, θ_s = Surface temperature of winding.

4.4 HEATING EQUATION   *(HTU, Dec' 18, 10 M) (AU, Nov' 17, 16 M) (AU, Apr' 18, 13 M)*

When heat is generated in a body, a part of heat is stored in the body and another part of heat is dissipated to surrounding medium. The heat equation can be developed by equating the heat generated to sum of heat stored and dissipated.

Let,   G   =   Weight of body

      c_p =   Specific heat capacity of the body

      λ =   Specific heat dissipation of the body

      S   =   Surface area of the body

      Q   =   Heat energy generated per second

      θ_r =   Temperature rise while heating

Consider a small-time duration dt. Let, $d\theta_r$ be the raise in temperature of the body in the time dt.

Now, Heat generated in time dt = Heat energy generated per second × dt

$$= Q \, dt \qquad\qquad .....(4.20)$$

Heat stored in time dt

$$= \text{Weight} \times \text{Specific heat capacity} \times \text{Temperature rise}$$

$$= G \times c_p \times d\theta_r$$

$$= G \, c_p \, d\theta_r \qquad\qquad .....(4.21)$$

Heat dissipated in time dt

$$= \text{Specific heat dissipation} \times \text{Surface area} \times \text{Temperature rise} \times \text{Time}$$

$$= \lambda \times S \times \theta_r \times dt$$

$$= \lambda \, S \, \theta_r \, dt \qquad\qquad .....(4.22)$$

Let, Heat generated = Heat stored + Heat dissipated

$$\therefore \; Q \, dt = G \, c_p \, d\theta_r + \lambda \, S \, \theta_r \, dt$$

$$Q \, dt - \lambda \, S \, \theta_r \, dt = G \, c_p \, d\theta_r$$

$$\left(\dfrac{Q}{G \, c_p} - \dfrac{\lambda \, S}{G \, c_p} \, \theta_r \right) dt = d\theta_r \qquad\qquad \boxed{\text{Using equations (4.20) to (4.22).}}$$

$$dt = \dfrac{d\theta_r}{\left(\dfrac{Q}{G \, c_p} - \dfrac{\lambda \, S}{G \, c_p} \, \theta_r \right)} \qquad\qquad .....(4.23)$$

The solution of above differential equation is,

$$t = -\frac{G\,c_p}{\lambda\,S}\,\log_e\left(\frac{Q}{G\,c_p} - \frac{\lambda\,S}{G\,c_p}\,\theta_r\right) + K \qquad\qquad .....(4.24)$$

where, K is integral constant.

The constant K can be solved by using initial conditions.

The initial conditions are,

At time, $t = 0$, Temperature, $\theta_r = \theta_i$

On substituting the initial conditions in the equation (4.24), we get,

$$0 = -\frac{G\,c_p}{\lambda\,S}\,\log_e\left(\frac{Q}{G\,c_p} - \frac{\lambda\,S}{G\,c_p}\,\theta_i\right) + K$$

$$K = \frac{G\,c_p}{\lambda\,S}\,\log_e\left(\frac{Q}{G\,c_p} - \frac{\lambda\,S}{G\,c_p}\,\theta_i\right)$$

On substituting the equation of K in equation (4.24) we get,

$$t = -\frac{G\,c_p}{\lambda\,S}\,\log_e\left(\frac{Q}{G\,c_p} - \frac{\lambda\,S}{G\,c_p}\,\theta_r\right) + \frac{G\,c_p}{\lambda\,S}\,\log_e\left(\frac{Q}{G\,c_p} - \frac{\lambda\,S}{G\,c_p}\,\theta_i\right)$$

$$= -\frac{G\,c_p}{\lambda\,S}\left(\log_e\left(\frac{Q}{G\,c_p} - \frac{\lambda\,S}{G\,c_p}\,\theta_r\right) + \log_e\left(\frac{Q}{G\,c_p} - \frac{\lambda\,S}{G\,c_p}\,\theta_i\right)\right)$$

$$= -\frac{G\,c_p}{\lambda\,S}\,\log_e\frac{\dfrac{Q}{G\,c_p} - \dfrac{\lambda\,S}{G\,c_p}\,\theta_r}{\dfrac{Q}{G\,c_p} - \dfrac{\lambda\,S}{G\,c_p}\,\theta_i}$$

$$= -\frac{G\,c_p}{\lambda\,S}\,\log_e\frac{\dfrac{\lambda\,S}{G\,c_p}\left(\dfrac{Q}{\lambda\,S} - \theta_r\right)}{\dfrac{\lambda\,S}{G\,c_p}\left(\dfrac{Q}{\lambda\,S} - \theta_i\right)}$$

$$= -\frac{G\,c_p}{\lambda\,S}\,\log_e\frac{\left(\dfrac{Q}{\lambda\,S} - \theta_r\right)}{\left(\dfrac{Q}{\lambda\,S} - \theta_i\right)} \qquad\qquad .....(4.25)$$

The equation (4.25) can be modified using final conditions.

The final conditions are,

At time, $t = \infty$, Temperature, $\theta_r = \theta_m$ and $d\theta_r = 0$

On substituting, $\theta_r = \theta_m$ and $d\theta_r = 0$, in equation (4.22) we get,

$$Q\,dt = G\,c_p\,d\theta_r + \lambda\,S\,\theta_r\,dt$$

$$Q\,dt = \lambda\,S\,\theta_m\,dt$$

$$\theta_m = \frac{Q}{\lambda\,S}$$

On substituting, $\theta_m = \dfrac{Q}{\lambda S}$ in the equation (4.25) we get,

$$t = -\frac{G\,c_p}{\lambda\,S}\,\log_e\frac{(\theta_m - \theta_r)}{(\theta_m - \theta_i)} \qquad\qquad .....(4.26)$$

Let us examine the units of the term, $\dfrac{G\,c_p}{\lambda\,S}$

$$\frac{kg\,\dfrac{J}{kg\,{}^\circ C}}{\dfrac{W}{m^2\,{}^\circ C}\,m^2} = \frac{J}{W} = \frac{J}{J/s} = s$$

From the above we can infer that the unit of $\dfrac{G\,c_p}{\lambda\,S}$ is same as unit of time seconds.

Therefore, the term $\dfrac{G\,c_p}{\lambda\,S}$ can be considered as heating time constant, T_h.

On substituting, $\dfrac{G\,c_p}{\lambda\,S} = T_h$ in the equation (4.26) we get,

$$t = -\,T_h\,\log_e\frac{(\theta_m - \theta_r)}{(\theta_m - \theta_i)}$$

$$-\frac{t}{T_h} = \log_e\frac{(\theta_m - \theta_r)}{(\theta_m - \theta_i)}$$

On taking natural logarithm of above equation we get,

$$e^{-t/T_h} = \frac{(\theta_m - \theta_r)}{(\theta_m - \theta_i)}$$

$$(\theta_m - \theta_i)\,e^{-t/T_h} = (\theta_m - \theta_r)$$

$$\theta_m\,e^{-t/T_h} - \theta_i\,e^{-t/T_h} = \theta_m - \theta_r$$

$$\therefore\ \theta_r = \theta_m - \theta_m\,e^{-t/T_h} + \theta_i\,e^{-t/T_h}$$

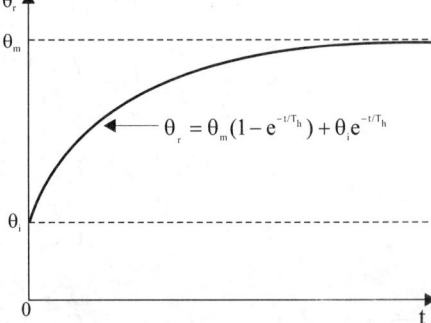

$$\theta_r = \theta_m(1 - e^{-t/T_h}) + \theta_i\,e^{-t/T_h}$$

Fig. 4.3: Heating curve when $\theta_i \neq 0$.

$$\therefore\ \text{Temperature rise, } \theta_r = \theta_m(1 - e^{-t/T_h}) + \theta_i\,e^{-t/T_h} \qquad .....(4.27)$$

The equation (4.27) is the general equation to estimate the rise in temperature while heating at any time, t. The initial and final conditions of temperature can be calculated as shown below. The variation of temperature rise, θ_r with respect to time, t is exponential as shown in Fig. 4.3.

In equation (4.27), when $t = 0$, $\theta_r = \theta_m(1 - e^0) + \theta_i\,e^0 = \theta_m(1 - 1) + \theta_i \times 1 = \theta_i$

In equation (4.27), when $t = \infty$, $\theta_r = \theta_m(1 - e^{-\infty}) + \theta_i\,e^{-\infty} = \theta_m(1 - 0) + \theta_i \times 0 = \theta_m$

When electrical machine started from cold or initial temperature is ambient temperature then, θ_i is condidered as zero.

$$\therefore\ \text{When } \theta_i = 0,$$

$$\text{Temperature rise, } \theta_r = \theta_m(1 - e^{-t/T_h}) \qquad\qquad .....(4.28)$$

The equation (4.28) is the equation to estimate the rise in temperature when machine started from cold condition. The initial and final values of temperature rise when machine started from cold are estimated as shown below. The variation of temperature rise, θ_r when started from cold with respect to time, t is shown in Fig. 4.4. Also, the value of temperature rise at $t = T_h$ is calculated and found to be 0.632 times θ_m. Therefore, the heating time constant can also be estimated using equation (4.28) by taking θ_r equal to 0.0632 θ_m. Also from Fig. 4.4 we can observe that the heating time constant is given by the slope of the curve at $t = 0$.

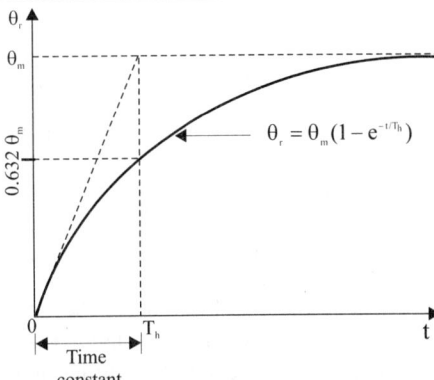

$$\theta_r = \theta_m(1 - e^{-t/T_h})$$

Fig. 4.4: Heating curve when $\theta_i = 0$.

In equation (4.28), when $t = 0$, $\theta_r = \theta_m (1 - e^0)$

$$= \theta_m (1 - 1) = 0$$

In equation (4.28), when $t = \infty$, $\theta_r = \theta_m (1 - e^{-\infty}) = \theta_m (1 - 0) = \theta_m$

In equation (4.28), when $t = T_h$, $\theta_r = \theta_m (1 - e^{-T_h/T_h}) = \theta_m (1 - e^{-1}) = 0.632\, \theta_m$

Where, t = Time

T_h = Heating time constant

θ_m = Final steady temperature while heating

θ_i = Initial temperature rise over ambient

Also, Maximum temperature rise, $\theta_m = \dfrac{Q}{S\,\lambda}$ .....(4.29)

Also, Heating time constant, $T_h = \dfrac{G\,c_p}{\lambda\,S}$ .....(4.30)

4.5 COOLING EQUATION

The equation to estimate fall in temperature when the cooling process starts in an electrical machine can be directly obtained from heating equation by replacing θ_m by θ_n and T_h by T_c in equation (4.27).

Here, θ_n = Final steady temperature while cooling

T_c = Cooling time constant

Let, θ_f = Fall in temperature while cooling

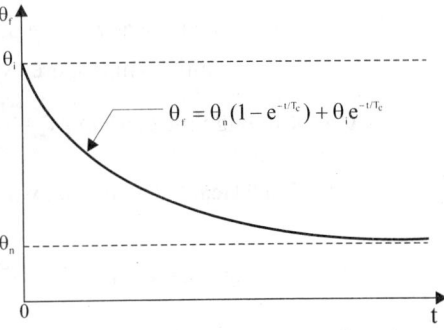

$$\theta_r = \theta_n(1 - e^{-t/T_c}) + \theta_i\, e^{-t/T_c}$$

Fig. 4.5: Cooling curve when $\theta_n \neq 0$.

Temperature fall, $\theta_f = \theta_n (1 - e^{-t/T_c}) + \theta_i\, e^{-t/T_c}$ .....(4.31)

The equation (4.31) is the general equation to estimate the fall in temperature while cooling at any time, t. The initial and final conditions of temperature can be calculated as shown below. The variation of temperature while cooling, θ_f with respect to time, t is exponential as shown in Fig. 4.5.

In equation (4.31), when $t = 0$, $\theta_f = \theta_n (1 - e^0) + \theta_i e^0 = \theta_n (1 - 1) + \theta_i \times 1 = \theta_i$

In equation (4.31), when $t = \infty$, $\theta_f = \theta_n (1 - e^{-\infty}) + \theta_i e^{-\infty} = \theta_n (1 - 0) + \theta_i \times 0 = \theta_n$

During cooling process if the final steady temperature is ambient temperature then, θ_n is condidered as zero.

∴ When $\theta_n = 0$,

Temperature fall, $\theta_f = \theta_i\, e^{-t/T_c}$ .....(4.32)

The equation (4.32) is the equation to estimate the fall in temperature when machine is cooled till ambient temperature is reached. The initial and final values of temperature fall when machine the machine is cooled to ambient temperature are estimated as shown below. The variation of temperature fall, θ_f when the machine is cooled to ambient temperature with respect to time, t is shown in Fig. 4.6. Also, the value of temperature fall at $t = T_c$ is calculated and found to be 0.368 times θ_i. Therefore, the cooling time constant can also be estimated using equation (4.32) by taking θ_f equal to 0.0368 θ_i. Also from Fig. 4.6 we can observe that the cooling time constant is given by the slope of the curve at $t = 0$.

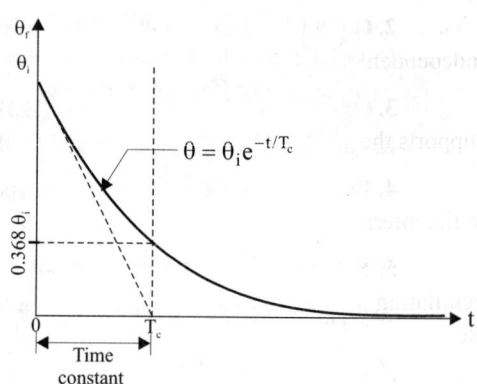

Fig. 4.6: *Cooling curve when* $\theta_n = 0$.

In equation (4.32), when $t = 0$, $\theta_f = \theta_i e^0 = \theta_i \times 1 = \theta_i$

In equation (4.32), when $t = \infty$, $\theta_f = \theta_i e^{-\infty} = \theta_i \times 0 = 0$

In equation (4.32), when $t = T_c$, $\theta_f = \theta_i\, e^{-T_c/T_c} = \theta_i\, e^{-1} = 0.368\, \theta_i$

where,   t   = Time

       T_c = Cooling time constant

       θ_n = Final steady temperature while cooling

       θ_i = Initial temperature over ambient

Also, Cooling time constant, $T_c = \dfrac{G\, c_p}{\lambda\, S}$

Also, Final steady temperature while cooling, $\theta_n = \dfrac{Q}{S\,\lambda}$ .....(4.33)

Also, Cooling time constant, $T_c = \dfrac{G\, c_p}{\lambda\, S}$ ......(4.34)

where,   G   =   Weight of active part of machine

       c_p =   Specific heat capacity

       S   =   Cooling surface

       λ   =   Specific heat dissipation

4.6 TYPES OF ENCLOSURES FOR ROTATING ELECTRICAL MACHINES

(AKTU, Dec' 19, 5 M)

Electrical machine parts have to be protected from water, dust, etc, with a proper enclosure. The enclosure should also provide cooling or ventilation for machine parts. The various types of enclosures of electrical machines and a brief description about the enclosure are listed here.

1. Open Machine: In these machines most of the machine parts are open and so will have good ventilation. The enclosure mainly depends on good mechanical strength.

2. Open Pedestal Machine (OP): It is an open machine provided with pedestal bearing supported independently on the machine frame.

3. Open End-Bracket Machine (OED): It is an open machine provided with end-brackets which supports the end bearings.

4. Protected Machine (P): In this type of machines, the enclosure is designed to prevent contact to the internal parts without obstructing ventilation.

5. Screen Protected Machine (SP): In this type of machines, the enclosure is designed with ventilating openings which are covered by screens made of wire mesh, expanded metal, perforated metal, etc.

6. Drip-Proof Machine (DP): In this type of machines, the ventilators and its openings are designed to prevent the entry of water or dirt through the ventilation openings when water falls on the top of the machine.

7. Splash-Proof Machine (SPLP): In this type of machines, the ventilators and its openings are designed such water splash at any angle not greater than 100° from the vertical cannot enter the machine.

8. Hose-Proof Machine (HSP): In this type of machine, the enclosure is designed to prevent entry of water when the machine is cleaned using water from a hose.

9. Pipe-Ventilated or Duct-Ventilated Machine: In this type of machines, the enclosure is designed to support ventilating pipes or ducts through which ventilation is provided by circulating air or gas or liquid.

10. Totally Enclosed Machine (TE): In this type of machines, the enclosure is designed to completely cover the machine so that external or ambient air has no contact with internal parts of the machine.

11. Totally Enclosed Fan-Cooled Machine (TEFC): A totally enclosed machine provided with ventilating ducts and cooling fan driven by the motor itself in order to provide forced circulation of air to remove the internal heat generated in the machine.

12. Totally Enclosed Separately Air-Cooled Machine (TESAC): A totally enclosed machine provided with ventilating ducts and separately driven cooling fan in order to provide forced circulation of air to remove the internal heat generated in the machine.

13. Totally Enclosed Water or Liquid-Cooled Machine (TEWC): A totally enclosed machine provided with external surface cooling arrangements using water or liquid.

14. Totally Enclosed Closed Air Circuit Machine: A totally enclosed machine provided with external surface cooling arrangements using air and an external cooling circuit to cool the hot air using external cooler.

15. Totally Enclosed Closed Gas Circuit Machine (CGGW): A totally enclosed machine provided with external surface cooling arrangements using gas and an external cooling circuit to cool the hot gas using external cooler.

16. Weather Proof Machine (WP): A totally enclosed machine constructed to prevent damage from adverse weather conditions specified by the consumer.

17. Watertight Machine (WT): A totally enclosed machine constructed to prevent entry of water in to the machine up to a specified water pressure. Normally tested in water at a depth of not less than 1 m, or at an external water pressure of 0.1 kg/cm^2 for a period of one hour.

18. Submersible Machine: A totally enclosed machine designed to work permanently in water wells when submerged under a specified head of water.

19. Flame Proof Machine (FLP): The enclosures are designed such that any flame generated inside the machine (sparking in commutators and slip rings) is contained in the machine itself. The ventilating openings in the machine are made long enough for the flame to be contained within the frame. These complies with the requirements of IS: 2148 - 1962 "Specification for flame-proof enclosure of electrical apparatus".

4.7 VENTILATION IN ELECTRICAL MACHINES   *(PTU, May' 19, 10 M)*

Ventilation is the process of circulation of air inside the machine so that the external air will pass through the heated parts of the machine and cool the heated parts in order to maintain the temperature with in limits.

Electrical machines require proper ventilation arrangements for better cooling. In one way of classification the ventilation can be classified into,

1. Induced ventilation

2. Forced ventilation.

In another way of classification, the ventilation in electrical machines can be classified into

1. Radial ventilation system

2. Axial ventilation

3. Combined radial and axial ventilation system.

4.7.1 Induced Ventilation

The induced ventilation is provided either by an internal fan mounted at one end of the shaft within the end shield or by an external fan. The ventilation of the machine is induced when the fan produces a decreased pressure inside the machines causing the air to be sucked inside the machine as air flows from high pressure to low pressure. Then the air is pushed out by the fan into atmosphere. Figures 4.7 and 4.8 shows a simple sketch of the induced ventilation scheme respectively by internal and externals fan. Induced ventilation is commonly provided in machines of small and medium power outputs.

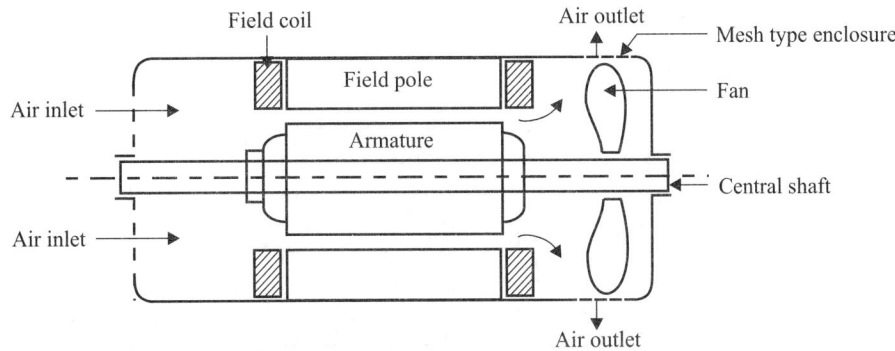

Fig. 4.7: Induced ventilation with internal fan.

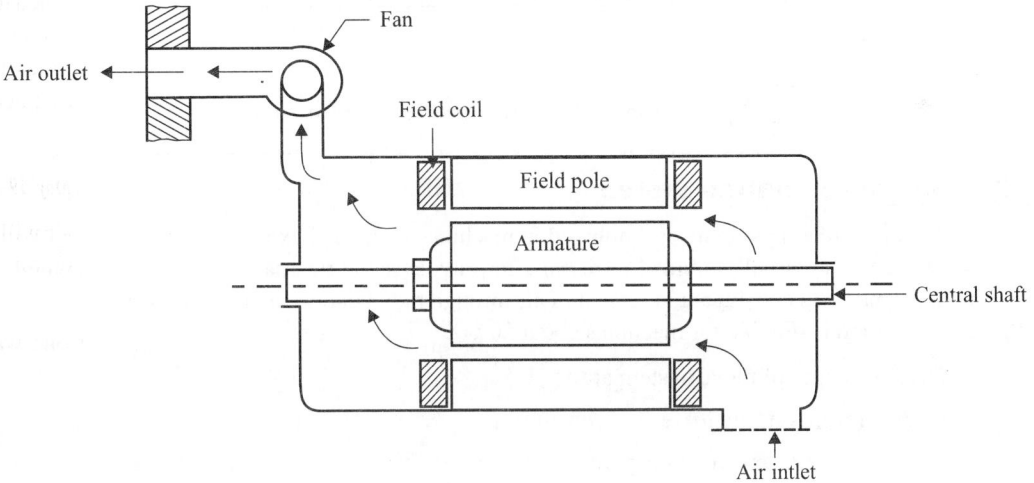

Fig. 4.8: Induced ventilation with external fan.

4.7.2 Forced Ventilation

The forced ventilation is also provided by an internal or external fan but in this arrangement the fan sucks the air from the atmosphere and forces it into the hot parts of machine, and then heated air pushed out to the atmosphere. Figures 4.9 and 4.10 shows a simple sketch of the forced ventilation scheme respectively by internal and externals fans. Forced ventilation is provided in large capacity machines.

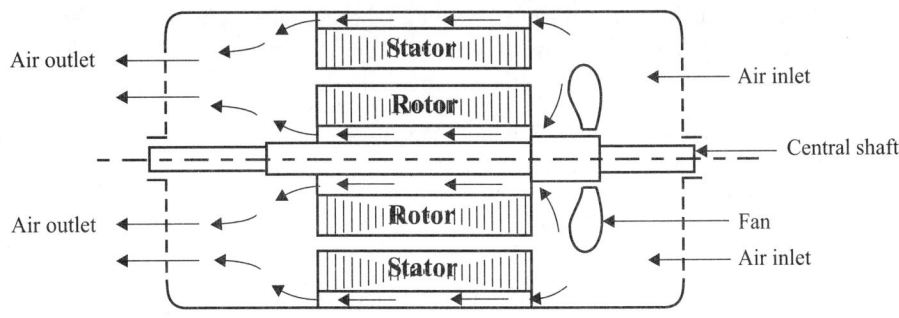

Fig. 4.9: Forced ventilation with internal fan.

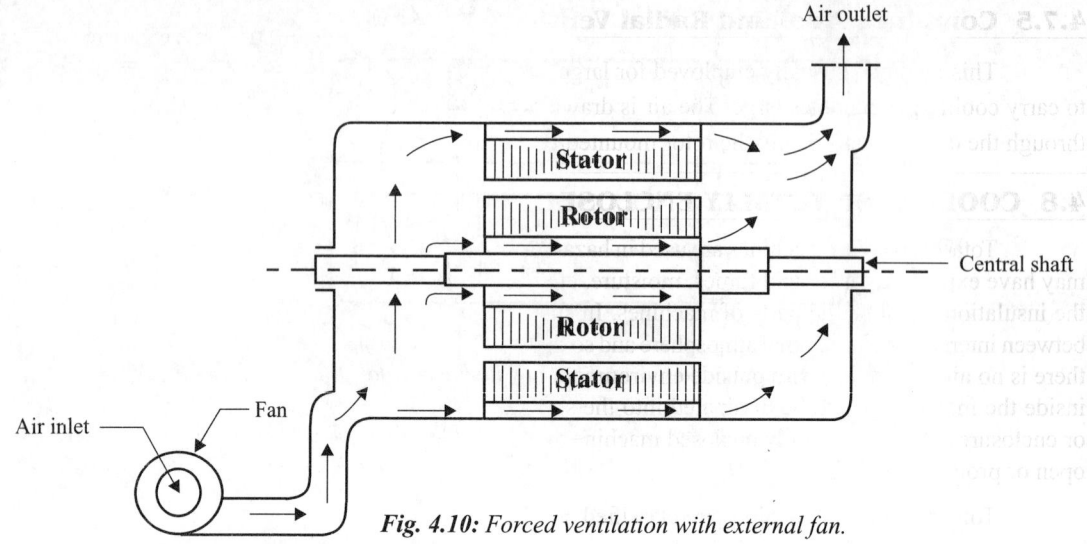

Fig. 4.10: *Forced ventilation with external fan.*

4.7.3 Radial Ventilation System

Radial ventilating systems is employed in machines with large axial length and capacity. In such machines the core is normally divided in to parts to provide radial ventilating ducts. The movement of rotor induces a natural centrifugal movement of air through radial ducts which is augmented by provision of fan. This method is suitable for machines about 20 *kW*.

The advantages of radial system are,

1. Minimum energy losses for ventilation.
2. Sufficiently uniform temperature rise of machines in the axial direction.

The disadvantages of radial system are,

1. It makes the machines lengthy as space for ventilating ducts has to be provided along the core length.
2. It may become unstable in respect to quantity of cooling air flowing.

4.7.4 Axial Ventilation System

The axial ventilating system is employed in small and medium capacity induction machines. This system of ventilation is suitable for high-speed machines, because the high-speed machines have a solid rotor and reduced core length in order to avoid centrifugal stresses. In order to increase the air flow, holes are punched on the laminations to provide axial air circulation path.

The advantages and disadvantages of axial system are,

1. It improves cooling but requires a large core diameter for the increased core depth.
2. Non uniform heat transfer.
3. Increased iron loss.
4. In large number of cases this loss is more than compensated by improved cooling.

4.7.5 Combined Axial and Radial Ventilating System

This method is usually employed for large motors and small turbo alternators. The area of ducts to carry cooling air becomes large. The air is drawn to the machines from one end is encouraged to pass through the ducts by and finally the rotor mounted fan forces out the air.

4.8 COOLING OF TOTALLY ENCLOSED MACHINES

Totally enclosed machines are used in hazardous environments where the surrounding atmosphere may have explosive gases, acid fumes, moisture, etc., that may create fire accidents as well as deteriorate the insulation and metallic parts of machines. In such environments there should not be any connection between internal air and external atmosphere and so the use of open and protected machines is avoided. Since there is no air connection with outside environment in a totally enclosed machine, all the heat developed inside the machine should be dissipated into the surroundings through the external surface of the frame or enclosure. Therefore, totally enclosed machines will have heavier and more expensive enclosure than open or protected machines.

Totally enclosed machines are classified into two types.

1. Ventilated frame machines

2. Ventilated radiator machines

Ventilated frame machines: In this type of machines, the machines are protected by two frames: inner and outer frame. A fan mounted on the shaft and protected by a separate enclosure, will force air in the gap between two frames. The inner frame is also provided with cooling fins to enhance the cooling. This ventilation arrangement is effective for machines with power rating up to 25 kW.

In machines with power rating above 25 kW, internal circulation of air is necessary to prevent development of hot spots and so an additional fan is mounted inside the inner frame to force circulation of air in air-gap and ducts. In this cooling arrangement, the air inside the machine acts as primary coolant and the air outside the machine acts as secondary coolant.

Ventilated radiator machines: For machines with power rating above 200 kW, the circulation of air by fans alone will not be sufficient to remove the heat developed. In large capacity machines, the enclosures will be same as ventilated frame machines but in addition, an external heat exchanger or radiator is provided to remove the heat from circulating air. The radiator will have an external fan to suck the hot air from machine and force through heat exchanger to dissipate heat and then push the cooled air to machine. This type of cooling can be provided for machines up to 5 MW rating.

4.9 COOLING OF TURBO ALTERNATORS   *(AKTU, Dec' 19, 5 M)*

Turbo alternators are used for electrical power generation in thermal and gas turbine power plants and constructed for MW rating and operated at high speeds. The size of Turbo alternators will be very large with longer core length when compared to diameter of the core. Due to larger axial length and high-speed operation, the problem of cooling of turbo alternator is a complex problem.

Based on the cooling system, the turbo alternators can be classified in to following three types.

1. Air cooled turbo alternators

2. Hydrogen cooled turbo alternators

3. Direct cooled turbo alternators.

4.9.1 Air Cooled Turbo Alternators

Turbo alternators with rating less than 60 *MW* are air-cooled turbo alternators and they are used as emergency generator in large power plants or peak load power plants. The following three methods of ventilation are employed in air cooled Turbo alternators to maintain air circulation.

1. One-sided axial ventilation: In this method of cooling, the air is blown in axial direction from one end of the machine through the air-gap using propeller fan and so the cool air enters the machine from one side and the hot air leave from the other side. The disadvantage in this method is that it cannot provide sufficient cooling in machines with large axial length and so this method of ventilation is limited to machines up to 3 *MW* rating.

2. Two-sided axial ventilation: In this method of cooling, the air is blown in axial direction from both end of the machine using propeller fan. The advantage in two-sided ventilation is that the temperature rise at both end of the machine will be same. This method of ventilation is limited to machines up to 12 *MW* rating.

3. Multiple inlet system: In axial ventilation system, the central portion of the core will not have sufficient cooling and so the temperature rise at the center of the core will more than that at ends. Therefore, radial ventilation with multiple inlet and outlet air champers. The cooling air is forced radially in machines with rating above 12 *MW*.

In multiple inlet ventilation system, the stator outer casing is divided into compartments to provide alternate inlet and outlet air chambers. Cooling air is circulated under pressure radially through inlet chamber and the hot air, leave from outlet chamber. A part of cooling air will also flow in the axial direction through airgap. The hot air from outlet chambers is cooled using heat exchangers and recirculated through inlet chamber. This method of ventilation is limited to machines up to 60 *MW* rating.

4.9.2 Hydrogen Cooled Turbo Alternators   *(HTU, Dec' 18, 10 M)*

In Turbo alternators with rating above 60 *MW* it will be difficult to cool inner parts of the machine with air cooling system. Also, the volume of air and the fan power required to circulate air will be high in alternators with large power rating. Therefore, hydrogen gas is used as coolant in Turbo alternators with rating above 60 *MW*.

In hydrogen cooling system, the cool hydrogen gas at a pressure higher than that of atmosphere is circulated through ventilating ducts and airgap with the help of fans mounted on the rotor. The hot hydrogen gas is collected from outlet ventilating ducts and cooled in external heat exchanger for recirculation. The higher pressure is maintained to prevent external air entry inside the machine which may create explosion. When hydrogen gas is used at high pressure any leakage will be from machine to external atmosphere and so internal hydrogen-air mixture that may create explosion is avoided.

The following are the advantage of hydrogen cooling.

Increased efficiency: Turbo alternators cooled by hydrogen gas will have very high efficiency up to 99.2% and this efficiency cannot be achieved in any other cooling system. The higher efficiency is due to low density and higher specific heat capacity of hydrogen gas. The density of hydrogen is 1/7 times density of air and specific heat capacity of hydrogen gas is 12 times that of air. Therefore lesser volume of hydrogen gas will remove more heat than air which results in lesser power for circulation of hydrogen gas.

Increase in rating: Heat conducted by hydrogen gas will be 7 times higher than that of air and so for a given size of machine the power output can be higher and hence machine rating will be higher than that of air-cooled machine.

Increase in life: The life of machine mainly depends on life of insulation which deteriorates at high temperatures. Since hydrogen gas more effectively remove heat than air, the temperature rise in insulation can be limited to a lesser value which enhance the life of insulation and hence life of machine.

Elimination of fire hazard: Hydrogen gas will not support fire and so machine filled with hydrogen gas will not have fire hazard inside the machine. (Air contains oxygen which supports fire).

Smaller size of coolers: For a given power rating the volume of hydrogen gas coolant required will be lesser when compared to air and so the size of coolers will also be smaller.

Less noise: Since the density of hydrogen gas is lesser than air the rotor will generate lesser noise when hydrogen gas is used as coolant.

4.9.3 Direct Cooled Turbo Alternators

The process of circulating the coolant inside the hollow conductors to directly remove the heat from conductors is called direct cooling. The other names for direct cooling are supercharged cooling, inner cooling and conductor cooling. The purpose of direct cooling is to remove more heat and to enhance the machine rating for a given size of machine.

In direct cooled alternators, the coolant is circulated inside the stator and rotor core and conductors directly to remove the heat. The stator and rotor conductors will be made using tubular conductors so that coolant can be circulated inside the tubes. The rotor and stator cores will have radial and axial ventilating ducts for circulation of coolant. In conventional cooling the coolant will not be in contact with conductor whereas in direct cooling, the coolant is passed inside the slots or ducts in the stator and rotor and through central hole in the conductors.

The coolants used for direct cooling are hydrogen gas, water and oil.

Hydrogen gas: The rotor and stator conductors are made from tubular conductors and hydrogen gas at a higher pressure than atmosphere is circulated inside the central hole of tubular conductors. The conductors are made using hard drawn silver bearing copper. Direct hydrogen cooling is employed in machines up to 300 *MW* rating.

Water: In very high-power rating turbo alternators, water has replaced hydrogen gas for direct cooling. In direct water-cooled alternators, the conductors are constructed similar to that of hydrogen cooled machine, but water is used as coolant instead of hydrogen gas. Water has superior heat transfer capacity and so lesser amount of coolant is required to remove head. Also, water has low viscosity and so can be circulated with lesser pressure than hydrogen.

The temperature of water is maintained at 60 to 70 °C. The water is circulated at pressure of 2200 *mm Hg* and velocity of water is between 1,5 *m/s* and 2.5 *m/s*. Pure water is used as coolant to avoid corrosion and cavitation.

Oil: Some of the manufacturers employ high grade transformer oil instead of water due to lesser viscosity of oil than water. But oil has a flash point which is development of spark at high temperature rise. But in normal working condition the flash point temperature will not be reached. There is a possibility of flash point temperature rise during fault conditions and it can be avoided by limiting fault current to safe value.

4.9.4 Estimation of Coolant

The various coolants used in cooling system of electrical machines are air, hydrogen gas, water and transformer oil. The estimation of coolant required to maintain a specific temperature in an electric machine are presented in this section.

Lower loss or heat to be dissipated

θ_r = Temperature rises of cooling medium

c_p = Specific heat capacity

Now, the weight of coolant required to maintain a temperature rise, θ_r is given by,

$$\text{Weight of coolant, } W_c = \frac{Q_L}{c_p \, \theta_r} \qquad\qquad .....(4.35)$$

The volume of coolant can be obtained from weight and density.

Let, ρ_c = Density of coolant

Now, volume of coolant is given by,

Using equation (4.35)

$$\text{Volume of coolant, } V_c = \frac{W_c}{\rho_c} = \frac{Q_L}{c_p \, \theta_r} \frac{1}{\rho_c} \qquad\qquad .....(4.36)$$

For air or hydrogen gas,

$$\text{Volume of coolant, } V_c = \frac{W_c}{\rho_c} = \frac{Q_L}{c_p \, \theta_r} \frac{1}{\rho_c} \qquad\qquad .....(4.37)$$

In case liquid coolants, the volume should be expressed in litres per second.

For water or oil,

$1\ m^3 = 1000\ l$

$$\text{Volume of coolant, } V_c = \frac{W_c}{\rho_c} = \frac{Q_L}{c_p \, \theta_r} \frac{1}{\rho_c} \times 1000 \qquad\qquad .....(4.38)$$

Estimation of volume of coolant air

The volume of air or gas depends on temperature and pressure. Normally, the circulating coolant in the machine will be at a higher temperature and pressure than NTP. The volume of air or gas at two different temperature and pressure are related by the following relation (also known as Kelvin's law or Boyle's law).

$$\frac{H_1 \, V_1}{T_1} = \frac{H_2 \, V_2}{T_2} \qquad\qquad .....(4.39)$$

where, H_1, H_2 = Barometric pressure (Atmospheric pressure)

V_1, V_2 = Volume of gas

T_1, T_2 = Temperature of gas

The equation (4.39) can be used to estimate the volume of coolant at working temperature from the knowledge of volume at NTP.

> **Note:** *NTP refer to Normal Temperature and Pressure, which is 0 °C (or 273 °K) at 760 mm Hg.*

Let, $V_1 = V_{a\_NTP}$ = Volume of air at NTP

T_1 = Temperature at NTP = 273 °K

H_1 = Barometric pressure at NTP = 760 *mm Hg*

θ_i = Inlet temperature of coolant air

T_2 = Inlet temperature of coolant air in Kelvin = $\theta_i + 273$ °K

$H_2 = H_a$ = Barometric pressure of air at inlet of cooling system

Let, V_a = Volume of coolant air at inlet temperature of cooling system

Now, using equation (4.39) we can write,

$$\text{Volume of coolant air, } V_a = V_2 = V_1 \frac{T_2}{T_1} \frac{H_1}{H_2}$$

$$= V_{a\_NTP} \frac{273 + \theta_i}{273} \frac{760}{H_a}$$

On substituting for V_{a_NTP} from equation (4.37) we get,

$$\text{Volume of coolant air, } V_a = \frac{Q_L}{c_{pa} \, \theta_{ra}} \frac{1}{\rho_{ca}} \frac{273 + \theta_i}{273} \frac{760}{H_a} \qquad .....(4.40)$$

where, Q_L = Power loss or heat to be dissipated

θ_{ra} = Temperature rises of coolant air

c_{pa} = Specific heat capacity of air = 995 J/kg-°C

ρ_{ca} = Density of coolant air at NTP = 1.2903 kg/m³

The coolant air is circulated using fan and the power required to operate the cooling fan is given by the product of volume and pressure of air.

Let, P_a = Pressure of air

η_f = Efficiency of fan

Now, Power required by fan, $P_f = P_a \dfrac{V_a}{\eta_f}$ $\qquad .....(4.41)$

Estimation of volume of coolant hydrogen gas

The equation (4.40) is also applicable for hydrogen gas if specific heat capacity and density of hydrogen gas are considered instead of air.

Let, V_h = Volume of coolant hydrogen gas at inlet temperature of cooling system

H_h = Barometric pressure of hydrogen gas at inlet of cooling system

Now, using equation (4.40) we can write,

$$\text{Volume of coolant hydrogen gas, } V_h = \frac{Q_L}{c_{ph} \, \theta_{rh}} \frac{1}{\rho_{ch}} \frac{273 + \theta_i}{273} \frac{760}{H_h} \qquad .....(4.42)$$

where, Q_L = Power loss or heat to be dissipated

θ_{rh} = Temperature rises of coolant hydrogen gas

c_{ph} = Specific heat capacity of hydrogen gas = 12540 J/kg-°C

ρ_{ch} = Density of coolant hydrogen gas at NTP = 0.0893 kg/m³

Estimation of volume of coolant water

The equation (4.38) can be used to estimate the volume of coolant water by using specific heat capacity and density of water.

Let, V_w = Volume of coolant water at inlet temperature of cooling system

Now, using equation (4.38) we can write,

$$\text{Volume of coolant water, } V_w = \frac{Q_L}{c_{pw} \, \theta_{rw}} \frac{1}{\rho_{cw}} \times 1000 \qquad \qquad .....(4.43)$$

where, Q_L = Power loss or heat to be dissipated

θ_{rw} = Temperature rises of coolant water

c_{pw} = Specific heat capacity of water = 4180 $J/kg\text{-}°C$

ρ_{cw} = Density of water = 1000 kg/m^3

Estimation of volume of coolant oil

High grade transformer oil refined from crude petroleum is used as coolant oil. The equation (4.38) can be used to estimate the volume of coolant oil by using specific heat capacity and density of oil.

Let, V_o = Volume of coolant oil at inlet temperature of cooling system

Now, using equation (4.38) we can write,

$$\text{Volume of coolant oil, } V_o = \frac{Q_L}{c_{po} \, \theta_{ro}} \frac{1}{\rho_{co}} \times 1000 \qquad \qquad .....(4.44)$$

where, Q_L = Power loss or heat to be dissipated

θ_{ro} = Temperature rises of coolant oil

c_{po} = Specific heat capacity of oil = 2060 $J/kg\text{-}°C$

ρ_{co} = Density of oil = 890 kg/m^3

4.10 RATING OF ELECTRICAL MACHINES

The values of electrical quantities like power, voltage, current, power factor, etc., and type of insulation and enclosure, etc., are called rating of electrical machines. The ratings may also include temperature rise and load duty cycle.

The ratings of the electrical machines are provided by the manufacturers of the electrical machines in order to provide information to users regarding the machine, so that the user can select the machines that suit for their specific applications. Sometimes, the manufactures may also construct electrical machines for a specified rating by the user for a specific application.

The ratings of electrical machines will help the user to select proper choice of machine for a particular application. The overloading capacity, power factor and temperature rise are of prime importance while selecting a machine. The overloading capability can be used to select 10 to 20 % lesser capacity machine provided the temperature rise is within limits. Also, overloading capability can be used to select lesser capacity machine for intermittent loads. Most of the electrical machines are capable of taking overload but with increase in temperature rise. The temperature rise will deteriorate the quality of insulation and hence will decide the life of a machine. Therefore, the machines have to be operated with in safe temperature limits.

4.10.1 Types of Duties and Rating

Electrical motors can be operated continuously or intermittently depending on the type of load and as well as consumer requirements. The time duration of operation of motors is referred to as duty. Also, most of the motors are capable of operating on overload for short time. Therefore, when motors are operated intermittently then there is a possibility of operating the motor at rating above the designed full load rating.

Electrical motors can be classified into following eight types depending on operating time or duty.

1. Duty type S1: Continuous duty

Electrical motors operated continuously for a long-time duration at a constant temperature are referred to as continuous duty motors. These motors are operated at a load lesser than or equal to full load with temperature rise not exceeding maximum permissible value. The continuous duty motors are used in pumps, paper mill drives, compressors, conveyors, etc.

2. Duty type S2: Short time duty

Electrical motors operated intermittently for a short-time duration with permissible temperature rise at loads higher than full-load are referred to as short time duty motors. These motors are operated at a load higher than full load with temperature rise not exceeding maximum permissible value.

The period of operation is so short that the temperature rise of the motor does not exceed the maximum permissible value and the period of rest is long enough for machine temperature to cool down to near ambient condition. Standard short time ratings are: 10, 30, 60 and 90 minutes. The short time duty motors are used in cranes, toll-gates, home appliances, etc.

3. Duty type S3: Intermittent periodic duty

Electrical motors that run alternatively on load for a short time duration and no-load for another short time duration are referred to as intermittent periodic duty motors. These motors will run continuously but load alone will be disconnected periodically to keep temperature rise within maximum permissible value. The intermittent periodic duty motors are encountered in continuously operated cranes, lifts, drilling machines, etc.

4. Duty type S4: Intermittent periodic duty with starting

Electrical motors that alternatively ON and OFF for a short time duration are referred to as intermittent periodic duty motors with starting. The motor ON duration will include a starting time and run time with load. The OFF duration is called rest time. These motors will not run continuously and so will not attain thermal equilibrium in one duty cycle but will attain a steady thermal state after several duty cycles. The motors used in metal cutting and drilling tool drives encounter intermittent periodic duty with starting.

5. Duty type S5: Intermittent periodic duty with starting and braking

Electrical motors that alternatively ON and OFF for a short time duration with ON duration having a starting time and braking time are referred to as intermittent periodic duty motors with starting and braking. The motor ON duration will include a starting time, run time with load and a braking time. The OFF duration is called rest time. These motors will not run continuously and so will not attain thermal equilibrium in one duty cycle but will attain a steady thermal state after several duty cycles.

6. Duty type S6: Continuous duty with intermittent periodic duty

Electrical motors that continuously running with a sequence of identical duty cycles with each duty cycle consisting of a period of operation at constant load and a period of operation at no load are referred to as continuous duty with intermittent periodic duty.

7. Duty type S7: Continuous duty with starting and braking

Electrical motors that continuously run with periodic starting and braking are referred to as continuous duty with starting and braking. Motors with this type of duty cycle will not have a rest time and so motor will be started immediately after breaking.

8. Duty type S8: Continuous duty with periodic speed changes

Electrical motors that run with periodic speed and load changes are referred to as continuous duty with periodic speed changes.

4.10.2 Determination of Motor Rating for Variable Load Drives   *(AU, Apr' 17, 8 M)*

Motors are manufactured at standard ratings for continuous duty and the specifications of motors will be available from the catalogue of the manufacturers of motors. The selection of right capacity of motor for continuous duty and constant load is straight forward and the motors can be directly selected from the manufacturers catalogue.

In case of variable load drives the following methods can be used for selection of motor rating.

1. Average loss method
2. Equivalent current method
3. Equivalent torque method
4. Equivalent power method

Average loss method

In this method the motor rating is selected such that, the nominal power loss of motor at rated load is equal to average loss for a specified load pattern or a load cycle.

Let, $Q_1, Q_2, Q_3, \ldots, Q_n$, be the power loss at various loads and $t_1, t_2, t_3, \ldots$, tn be the corresponding time duration of the load in a load cycle with n different value of loads.

Let, Q_{av} = Average power loss of the motor for the load pattern

$$\text{Average loss, } Q_{av} = \frac{Q_1 t_1 + Q_2 t_2 + Q_3 t_3 + \ldots + Q_n t_n}{t_1 + t_2 + t_3 + \ldots + t_n} \quad\quad \ldots(4.45)$$

Let, Q_{nom} = Nominal power loss of motor at full load

The motor rating is selected such that Q_{nom} of the motor is equal to or slightly higher lesser than Q_{av}.

The disadvantage in this method is that the peak temperature rise at variable load conditions are not taken into account. Therefore, for safety factor 10 to 30 % higher capacity motor is selected.

Equivalent current method

In this method the motor rating is selected such that, the nominal load current of motor at rated load is equal to equivalent load current for a specified load pattern or a load cycle.

Let, $I_1, I_2, I_3, \ldots\ldots, I_n$ be the load current at various loads and $t_1, t_2, t_3, \ldots\ldots, t_n$ be the corresponding time duration of the load in a load cycle with n different value of loads.

Let, I_{eq} = Equivalent load current of the motor for the load pattern

$$\text{Equivalent current, } I_{eq} = \sqrt{\frac{I_1^2 t_1 + I_2^2 t_2 + I_3^2 t_3 + \ldots\ldots + I_n^2 t_n}{t_1 + t_2 + t_3 + \ldots\ldots + t_n}} \qquad \ldots\ldots(4.46)$$

Let, I_{nom} = Nominal load current of motor at full load

The motor rating is selected such that I_{nom} of the motor is equal to or slightly higher lesser than I_{eq}.

Since iron loss of motor is constant and copper loss is directly proportional to current, the equivalent current method will give best choice of motor rating among all methods. The disadvantage in this method is that the current may not be constant in a duration during starting, breaking and changes in load, but this can be overcome by integrating the current in a every load duration to find average currents.

Equivalent torque method

When flux and power factor of motor is constant the torque is directly proportional to current. Hence, equivalent torque can be estimated similar to equivalent current. In equivalent torque method the motor rating is selected such that, the nominal load torque of motor at rated load is equal to equivalent torque for a specified load pattern or a load cycle.

Let, $T_1, T_2, T_3, \ldots\ldots, T_n$ be the torque at various loads and $t_1, t_2, t_3, \ldots\ldots, t_n$ be the corresponding time duration of the load in a load cycle with n different value of loads.

Let, T_{eq} = Equivalent load torque of the motor for the load pattern

$$\text{Equivalent torque, } T_{eq} = \sqrt{\frac{T_1^2 t_1 + T_2^2 t_2 + T_3^2 t_3 + \ldots\ldots + T_n^2 t_n}{t_1 + t_2 + t_3 + \ldots\ldots + t_n}} \qquad \ldots\ldots(4.47)$$

The equivalent torque method should not be applied to motors with variable flux like DC series motor.

Equivalent power method

When speed of the motor is constant the power is directly proportional to torque. Hence, equivalent power can be estimated similar to equivalent torque. In equivalent power method the motor rating is selected such that, the nominal load power of motor at rated load is equal to equivalent power for a specified load pattern or a load cycle.

Let, $P_1, P_2, P_3, \ldots\ldots, P_n$ be the load power at various loads and $t_1, t_2, t_3, \ldots\ldots, t_n$ be the corresponding time duration of the load in a load cycle with n different value of loads.

Let, P_{eq} = Equivalent power of the motor for the load pattern

$$\text{Equivalent power, } P_{eq} = \sqrt{\frac{P_1^2 t_1 + P_2^2 t_2 + P_3^2 t_3 + \ldots\ldots + P_n^2 t_n}{t_1 + t_2 + t_3 + \ldots\ldots + t_n}} \qquad \ldots\ldots(4.48)$$

The equivalent power method should not be applied to motors operated at variable speeds.

4.10.3 Temperature Rise in Short Time Rating

Let, θ_m = Maximum temperature rise with continuous rated load

θ_{msh} = Maximum temperature rise with short time rated load

θ_{rsh} = Temperature rise with short time rated load

t_{sh} = Time duration of short time load

Now the temperature rise with short time rating can be estimated using the following equation.

$$\theta_{rsh} = \theta_{msh} \left(1 - e^{-t_{sh}/T_h}\right) \qquad \boxed{\text{Using equation (4.28)}} \quad .....(4.49)$$

For safe operation of machine at short time rating the temperature rise on short time rating should not exceed the maximum temperature rise with continuous rated load. Therefore, when $\theta_{rsh} = \theta_m$, the equation (4.49) can be written as shown below.

$$\theta_m = \theta_{msh} \left(1 - e^{-t_{sh}/T_h}\right) \qquad\qquad .....(4.50)$$

$$\frac{\theta_{msh}}{\theta_m} = \frac{1}{\left(1 - e^{-t_{sh}/T_h}\right)} \qquad\qquad .....(4.51)$$

$$\therefore \theta_{msh} = \theta_m \frac{1}{\left(1 - e^{-t_{sh}/T_h}\right)} \qquad\qquad .....(4.52)$$

In equation (4.52), the term $\dfrac{1}{\left(1 - e^{-t_{sh}/T_h}\right)}$ will have a value greater than unity and so the maximum temperature rise and hence permissible losses and load on short time rating will be higher than maximum temperature rise on continuous load.

The heating overload ratio is defined as ratio of power loss with short time load and power loss with nominal load.

Let, Q_{sh} = Power loss with short time rating

Q_{nom} = Power loss with nominal load

Now, Heating overload ratio, $p_{ho} = \dfrac{Q_{sh}}{Q_{nom}}$

Since, the power loss is proportional to temperature rise the heating overload ratio can be written as shown below.

$$\text{Heating overload ratio, } p_{ho} = \frac{Q_{sh}}{Q_{nom}} = \frac{\theta_{msh}}{\theta_m} \qquad \boxed{\text{Using equation (4.51)}}$$

$$= \frac{1}{\left(1 - e^{-t_{sh}/T_h}\right)} \qquad\qquad ......(4.53)$$

Let, t_{shm} = Maximum permissible time on short time rating

Now, from equation (4.53) we can write,

$$p_{ho} = \frac{1}{\left(1 - e^{-t_{shm}/T_h}\right)}$$

$$p_{ho} = \left(1 - e^{-t_{shm}/T_h}\right) = 1$$

$$p_{ho} - p_{ho} \, e^{-t_{shm}/T_h} = 1$$

$$e^{-t_{shm}/T_h} = \frac{p_{ho} - 1}{p_{ho}}$$

$$t_{shm} = -T_h \, \ln\left(\frac{p_{ho} - 1}{p_{ho}}\right)$$

Maximum permissible time on short time rating, $t_{shm} = -T_h \, \ln\left(\frac{p_{ho} - 1}{p_{ho}}\right)$ .....(4.54)

Like heating overload ratio, an overload ratio can be defined for mechanical load.

The mechanical overload ratio is defined as ratio of short time power rating and nominal power rating of the motor.

Let, P_{sh} = Load power of short time rating

P_{nom} = Load Power of continuous rating

Now, Mechanical overload ratio, $p_{mo} = \dfrac{P_{sh}}{P_{nom}}$ .....(4.55)

The heating and mechanical overload ratios can be expressed in terms of ratio of constant and variable losses of motor as shown below.

Let, P_i = Constant loss

P_c = Variable loss at nominal or full load

The constant loss is same for any value of loss but the variable loss (copper loss) will depend on load. Therefore, the variable loss on short time load can be expressed as shown below.

Variable loss on short time load, $P_{csh} = \left(\dfrac{P_{sh}}{P_{nom}}\right)^2 P_c = p_m^2 \, P_c$ $\boxed{\text{Using equation (4.55)}}$

Now, Power loss with short time rating, $Q_{sh} = P_i + \left(\dfrac{P_{sh}}{P_{nom}}\right)^2 P_c = P_i + p_m^2 \, P_c$

Power loss with nominal load, $Q_{nom} = P_i + P_c$

Now, Heating overload ratio, $p_{ho} = \dfrac{Q_{sh}}{Q_{nom}} = \dfrac{P_i + p_{mo}^2 \, P_c}{P_i + P_c}$

Let us divide the numerator and denominator of above equation by P_c.

Heating overload ratio, $p_{ho} = \dfrac{\dfrac{P_i}{P_c} + p_{mo}^2}{\dfrac{P_i}{P_c} + 1}$

$$= \frac{K + p_{mo}^2}{K + 1}$$ .....(4.56)

where, K = Ratio of constant and variable loss at full load = $\dfrac{P_i}{P_c}$

On rearranging equation (4.56) we can write, $p_{ho}(K+1) = K + p_{mo}^2$

$\therefore$ Mechanical overload ratio, $p_{mo} = \sqrt{(K+1)p_{ho} - K}$ ......(4.57)

4.11 SUMMARY OF DESIGN EQUATIONS

1. Thermal resistance for heat conduction, $R_c = \dfrac{t}{\sigma S}$

2. Heat dissipated in conduction, $Q_{con} = \dfrac{S \theta_d}{\rho t}$

3. Temperature difference across medium, $\theta_d = Q_{con} \dfrac{\rho t}{S}$

4. Heat transferred per unit surface by radiation, $q_{rad} = \sigma e (T_e^4 - T_a^4)$

5. Heat dissipated per unit surface by convection, $q_{conv} = \lambda_{conv} \theta_d$

6. Total heat dissipation due to convection, $Q_{conv} = q_{conv} S = \lambda_{conv} \theta_d S$

7. Total heat dissipation by radiation and convection, $Q = (\lambda_{rad} + \lambda_{conv}) \theta_d S = \lambda \theta_d S$

8. Temperature between centre and outer surface, $\theta_{do} = \dfrac{q \rho_e t^2}{8}$

9. Hot spot temperature, $\theta_m = \dfrac{q \rho t^2}{8} + \theta_s$

10. Temperature difference across lamination, $\theta_{dx} = \dfrac{q \rho_x x^2}{2}$

11. Temperature difference along lamination, $\theta_{dy} = \dfrac{q \rho_y y^2}{2}$

12. Effective resistivity of winding, $\rho_e = \rho_i (1 - a)$

13. Temperature rise while heating, $\theta_r = \theta_m (1 - e^{-t/T_h}) + \theta_i e^{-t/T_h}$

14. Maximum temperature rise, $\theta_m = \dfrac{Q}{S \lambda}$

15. Heating time constant, $T_h = \dfrac{G c_p}{\lambda S}$

16. Temperature fall while cooling, $\theta_f = \theta_n (1 - e^{-t/T_c}) + \theta_i e^{-t/T_c}$

17. Cooling time constant, $T_c = \dfrac{G c_p}{\lambda S}$

18. Final steady temperature while cooling, $\theta_n = \dfrac{Q}{S \lambda}$

19. Weight of coolant per unit time, $W_c = \dfrac{Q_L}{c_p \theta_r}$

20. Volume of coolant air per unit time, $V_a = \dfrac{Q_L}{c_{pa} \theta_{ra}} \dfrac{1}{\rho_{ca}} \dfrac{273 + \theta_i}{273} \dfrac{760}{H_a}$

21. Power required by fan, $P_f = P_a \dfrac{V_a}{\eta_f}$

22. Volume of coolant hydrogen gas per unit time, $V_h = \dfrac{Q_L}{c_{ph}\,\theta_{rh}}\,\dfrac{1}{\rho_{ch}}\,\dfrac{273+\theta_i}{273}\,\dfrac{760}{H_h}$

23. Volume of coolant water per unit time, $V_w = \dfrac{Q_L}{c_{pw}\,\theta_{rw}}\,\dfrac{1}{\rho_{cw}}\times 1000$

24. Volume of coolant oil per unit time, $V_o = \dfrac{Q_L}{c_{po}\,\theta_{ro}}\,\dfrac{1}{\rho_{co}}\times 1000$

25. Average loss, $Q_{av} = \dfrac{Q_1\,t_1 + Q_2\,t_2 + Q_3\,t_3 + \ldots + Q_n\,t_n}{t_1 + t_2 + t_3 + \ldots + t_n}$

26. Equivalent current, $I_{eq} = \sqrt{\dfrac{I_1^2\,t_1 + I_2^2\,t_2 + I_3^2\,t_3 + \ldots + I_n^2\,t_n}{t_1 + t_2 + t_3 + \ldots + t_n}}$

27. Equivalent torque, $T_{eq} = \sqrt{\dfrac{T_1^2\,t_1 + T_2^2\,t_2 + T_3^2\,t_3 + \ldots + T_n^2\,t_n}{t_1 + t_2 + t_3 + \ldots + t_n}}$

28. Equivalent power, $P_{eq} = \sqrt{\dfrac{P_1^2\,t_1 + P_2^2\,t_2 + P_3^2\,t_3 + \ldots + P_n^2\,t_n}{t_1 + t_2 + t_3 + \ldots + t_n}}$

29. (i) Heating overload ratio, $p_{ho} = \dfrac{Q_{sh}}{Q_{nom}}$

    (ii) Heating overload ratio, $p_{ho} = \dfrac{1}{\left(1 - e^{-t_{sh}/T_h}\right)}$

    (iii) Heating overload ratio, $p_{ho} = \dfrac{K + p_{mo}^2}{K + 1}$; $K = \dfrac{P_i}{P_c}$

30. Maximum permissible time on short time rating, $t_{shm} = -\,T_h\,ln\left(\dfrac{p_{ho}-1}{p_{ho}}\right)$

31. Variable loss on short time load, $P_{csh} = p_{mo}^2\,P_c$

32. Power loss with short time rating, $Q_{sh} = P_i + p_{mo}^2\,P_c$

33. Power loss with nominal load, $Q_{nom} = P_i + P_c$

34. (i) Mechanical overload ratio, $p_{mo} = \dfrac{P_{sh}}{P_{nom}}$

    (ii) Mechanical overload ratio, $p_{mo} = \sqrt{(K+1)p_{ho} - K}$; $K = \dfrac{P_i}{P_c}$

4.12 SOLVED PROBLEMS

EXAMPLE 4.1

In a winding, a copper loss of is 0.5 W is dissipated through a layer of paper having a thermal resistivity of 6 Ω-m. Find the temperature difference between the two sides of the paper. The thickness of paper is 0.5 mm and its surface area is 250 mm^2.

GIVEN DATA

ρ = 6 Ω-m t = 0.5 mm

Q_{con} = 0.5 W S = 250 mm^2

SOLUTION

Let, θ_d = Temperature difference between two sides of the paper

$\therefore$ Temperature difference, $\theta_d = Q_{con} \dfrac{\rho\, t}{S}$

$$= 0.5 \times \frac{6 \times 0.5}{250} = 6\,°C$$

RESULT

Temperature difference, θ_d = 6 °C

EXAMPLE 4.2

The thermal resistivity of laminations in the direction of lamination is 40 times lesser than the resistivity in the direction perpendicular to laminations. Calculate the loss that will be conducted across the laminations in a stack 0.1 m thick and 20 × 10^{-3} m^2 in cross-section with a temperture difference 15 °C. A temperature difference of 5 °C will cause 64 W to be conducted through a stack with surface area 6 × 10^{-3} m^2 and 25 mm thick measured along the laminations.

GIVEN DATA

$\rho_y = \rho_x / 40$ $S_y = 6 \times 10^{-3}\ m^2$

$t_x = 0.1\ m$ $Q_{con\_y} = 64\ W$

$t_y = 25\ mm$ $\theta_{dx} = 15\ °C$

$S_x = 20 \times 10^{-3}\ m^2$ $\theta_{dy} = 5\ °C$

SOLUTION

Temperature rise along lamination, $\theta_{dy} = \dfrac{q_y\, \rho_y\, y^2}{2}$ $\boxed{Q_y = q_y\, S_y\, t_y}$

$$= \frac{(Q_y / S_y\, t_y)\, \rho_y\, (t_y / 2)^2}{2}$$ $\boxed{y = t_y / 2}$

$$= \frac{Q_y\, \rho_y\, t_y^2}{8\, S_y\, t_y}$$

$\therefore$ Resistivity along lamination, $\rho_y = \dfrac{8\, S_y\, \theta_{dy}}{Q_y\, t_y}$

$$= \frac{8 \times 6 \times 10^{-3} \times 5}{64 \times 25 \times 10^{-3}} = 0.15\ \Omega\text{-}m$$

Given that, $\rho_y = \rho_x / 40 \implies \rho_x = 40\,\rho_y$

$\therefore$ Resistivity across lamination, $\rho_x = 40\,\rho_y = 40 \times 0.15 = 6\ \Omega\text{-}m$

Temperature rise across lamination, $\theta_{dx} = \dfrac{q_x\,\rho_x\,x^2}{2}$

$$= \dfrac{(Q_x / S_x\,t_x)\,\rho_x\,(t_x/2)^2}{2}$$

$$= \dfrac{Q_x\,\rho_x\,t_x^2}{8\,S_x\,t_x}$$

$\boxed{Q_x = q_x\,S_x\,t_x}$

$\boxed{x = t_x / 2}$

$\therefore$ Heat dissipated across lamination, $Q_x = \dfrac{8\,S_x\,\theta_{dx}}{\rho_x\,t_x}$

$$= \dfrac{8 \times 20 \times 10^{-3} \times 15}{6 \times 0.1} = 4\ W$$

RESULT

Heat dissipated across lamination, $Q_x = 4\ W$

EXAMPLE 4.3

A field coil has a length of mean turn of 0.8 m, a space factor of 0.545, a current of 2.55 A at a terminal voltage of 55 V, and an overall cross-section of 105 × 60 mm^2. Estimate the hot spot temperature if the surface of the coil is 35 °C. Thermal resisitivity of insulating materials = 8 $\Omega\text{-}m$.

GIVEN DATA

$\rho_i = 8\ \Omega\text{-}m$ $S_f = 0.545$

$I_f = 2.55\ A$ Overall cross-section = 105 × 60 mm^2

$V_f = 55\ V$ Length of mean turn = 0.8 m

$\theta_s = 35\ °C$

SOLUTION

Effective thermal resistivity of winding and insulation $\Big\} \; \rho_e = \rho_i\,(1 - \sqrt{S_f})$

$$= 8\,(1 - \sqrt{0.545})$$

$$= 2.0941\ \Omega\text{-}m$$

Volume of coil = Overall cross-section × Length of mean turn

$$= 105 \times 60 \times 10^{-6} \times 0.8$$

$$= 5.04 \times 10^{-3}\ m^3$$

Heat per unit volume, $q = \dfrac{\text{Total heat}}{\text{Volume of coil}} = \dfrac{V_f \times I_f}{5.04 \times 10^{-3}}$

$\boxed{\text{Power = Total heat}}$

$$= \dfrac{55 \times 2.55}{5.04 \times 10^{-3}}$$

$$= 27827.38\ W/m^3$$

Temperature difference between coil centre and outer surface $\Big\} \; \theta_{do} = \dfrac{q\,\rho_e\,t^2}{8}$

The hot spot temperature is given by sum of θ_{do} and θ_s.

$$\text{Hot spot temperature, } \theta_m = \theta_{do} + \theta_s = \frac{q\, \rho_e\, t^2}{8} + \theta_s$$

Take, lenght of heat path as 60 mm.

$$= \frac{27827.38 \times 2.0941 \times (60 \times 10^{-3})^2}{8} + 35° = 61.2 °C$$

RESULT

Hot spot temperature, $\theta_m = 61.2 °C$

EXAMPLE 4.4

The full load efficiency of a 100 MW hydrogen cooled turbo-alternator is 99 %. The hydrogen enters the machine with a temperature of 25 °C and leaves the machine with a temperature of 55 °C. Determine the volume of hydrogen required if the hydrogen pressure is 1500 mm above a gauge pressure of 760 mm of mercury. Specific heat of hydrogen is 12600 J/kg-°C. Volume of 1 kg of hydrogen at NTP is 11.2 m^3.

GIVEN DATA

P = 100 MW

H_h = 1500 mm + 760 = 2260 mm

V = 1 kg of NTP = 11.2 m^2

c_{ph} = 12600 J/kg-°C

$\eta = 99 \% = \dfrac{99}{100}$

$\theta_i = 25° C$

$\theta_{rh} = 55 - 25 = 30 °C$

$\rho_{ch} = \dfrac{1}{11.2} = 0.0893 \ kg/m^3$

SOLUTION

$$\text{Power loss, } Q_L = P\,(1 - \eta) = P \times \left(1 - \frac{99}{100}\right)$$

$$= 100 \times (1 - 0.99) = 1 \ MW = 10^6 \ W$$

$$\text{Barometric pressure of hydrogen, } H_h = 1500 + 760 = 2260 \ mm \ Hg$$

$$\therefore \text{ Volume of hydrogen, } V_h = \frac{Q_L}{c_{ph}\, \theta_{rh}}\, \frac{1}{\rho_{ch}}\, \frac{273 + \theta_i}{273}\, \frac{760}{H_h}$$

$$= \frac{10^6}{12600 \times 30} \times 11.2 \times \frac{273 + 25}{273} \times \frac{760}{2260}$$

$$= 10.8764 \ m^3/s$$

RESULT

Volume of hydrogen, V_h = 10.8764 m^3/s.

EXAMPLE 4.5

The full load losses in a 16 MVA transformer are : iron loss, 82 kW; copper loss, 134 kW. The tank is 3.6 m high × 2.9 m long × 1.35 m wide and contains a cooling coil through which 2 litres of water per second are passed. Assuming that the tank sides dissipate 7.5 W/m^2-°C and the average temperature rise of cooling water is 25 °C, estimate the temperature rise of the tank.

GIVEN DATA

16 MVA

V_w = 2 l/s

λ = 7.5 W/m^2 -°C

Iron loss = 82 kW

θ_{rw} = 25 °C

Copper loss = 134 kW

SOLUTION

Volume of water, $V_w = \dfrac{Q_L}{c_{pw}\,\theta_{rw}}\dfrac{1}{\rho_{cw}} \times 1000$

∴ Power loss dissipated by water, $Q_L = \dfrac{V_w\,c_{pw}\,\theta_{rw}\,\rho_{cw}}{1000}$

$$= \dfrac{2 \times 4180 \times 25 \times 1000}{0.24}$$

$$= 209000 \ W = 209 \ kW$$

Take, $c_{pw} = 4180 \ J/kg\text{-}°C$

$\rho_{cw} = 1000 \ kg/m^3$

Total loss = 82 + 134 = 216 kW

Loss dissipated by walls by convection, Q_w = Total loss – Loss dissipated by water

$$= 216 - 209 = 7 \ kW$$

Area of tank walls, $S = 2 \times h \times l + 2 \times h \times w$ (neglecting top and bottom surface)

$$= 2 \times h \times (l + w)$$

$$= 2 \times 3.6 \times (2.9 + 1.35) = 30.6 \ m^2$$

Temperature rise of tank, $\theta_t = \dfrac{Q_w}{S\,\lambda} = \dfrac{7 \times 1000}{30.6 \times 7.5} = 30.5\,°C$

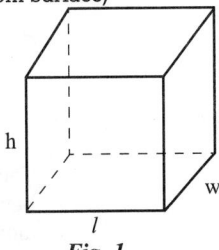

Fig. 1

RESULT

Temperature rise of tank, θ_t = 30.5 °C

EXAMPLE 4.6

An induction motor is running on load and it is shut down when it is heated to a temperature of 62 °C. Calculate temperature of motor at a time of 20 minutes after the shutdown if the cooling time constant is 60 minutes. Take ambient temperature as 30 °C.

GIVEN DATA

θ_m = 60 °C $\qquad\qquad$ θ_a = 30 °C

T_c = 60 min $\qquad\qquad$ t_c = 20 min

SOLUTION

Given that, maximum temperature rise, θ_m = 62 °C

Temperature rise over ambient, $\theta_r = \theta_m - \theta_a = 62 - 30 = 32$ °C

The temperature, θ_r is the initial temperature when the cooling process is started after shutdown of motor.

Let, t_c = Cooling time = 20 min

$\left.\begin{array}{l}\text{Fall in temperature at 20} \\ \text{minutes after shutdown}\end{array}\right\}\theta_{20} = \theta_i \times e^{-t_c/T_c}$

$$= 32 \times e^{-20/60}$$

$$= 22.9\ °C$$

$\boxed{\theta_i = \theta_r}$

$\left.\begin{array}{l}\text{Temperature of motor at 20} \\ \text{minutes after shutdown}\end{array}\right\}\theta_{m\_20} = \theta_a + \theta_{20}$

$$= 30 + 22.9$$

$$= 52.9\ °C$$

RESULT

$\left.\begin{array}{l}\text{Temperature of motor at 20} \\ \text{minutes after shutdown}\end{array}\right\}\theta_{m\_20} = 52.9\,°C$

EXAMPLE 4.7

The width of a transformer core-plate is 500 *mm*. The plate has a core loss of 4 *W/kg*, density of 7500 *kg/m³*, and a thermal resistivity of 0.01 *Ω-m*. Assume that the core loss is uniformly distributed and space factor is 0.9 and the heat flow is along the lamination. Compute the maximum core temperature when the two surfaces are maintained at a temperature of 40 °C.

GIVEN DATA

Core loss = 4 *W/kg*   t = 500 *mm* = 0.5 *m*

$S_f = 0.91$ $\theta_s = 40$ °C

Density = 7500 *kg m⁻³*   ρ = 0.01 *Ω-m*

SOLUTION

Heat per unit volume, q = Core less × Density × Space factor

$$= 4 \times 7500 \times 0.91$$

$$= 27300 \ W/m^3$$

Temperature difference between centre and surface along lamination $\Bigg\} \theta_{dy} = \dfrac{q \rho y^2}{2}$

$$= \dfrac{q \rho (t/2)^2}{2} \qquad \boxed{y = t/2}$$

$$= \dfrac{q \rho t^2}{8}$$

$$= \dfrac{27300 \times 0.01 \times 0.5}{8} = 17\,°C$$

Maximum core temperature, $\theta_m = \theta_{dy} + \theta_s$

$$= 17 + 40 = 57 \ °C$$

RESULT

Maximum core temperature, $\theta_m = 57$ °C

EXAMPLE 4.8

A 15 *kW* squirrel-cage induction motor having maximum efficiency of 90% when continuously operated at full-load. The temperature-rise of motor is 41.8 °C after 30 minutes and 50 °C after one hour under the above operating conditions.

a) Compute its final steady-state temperature rise on continuous load, and the heating time constant.

b) If the motor is operated on a short-time load for 6 minutes, estimate the temperature rise.

c) Compute the maximum overload that can be appiled on a short-time 6 minutes rating so that the temperature rise is the same as the final steady-state temperature rise on continuous load.

GIVEN DATA

When $t = t_1 = 30$ *min* or 1/2 h, $\theta_r = \theta_{r1} = 41.8$ °C

When $t = t_2 = 1$ h, $\theta_r = \theta_{r2} = 50$ °C

SOLUTION

a)   Temperature rise at anytime t, $\theta_r = \theta_m(1 - e^{-t/T_h})$

When $t = t_1$ and $\theta_r = \theta_{r1}$, $\theta_{r1} = \theta_m(1 - e^{-t_1/T_h})$

$\therefore 41.8 = \theta_m(1 - e^{-1/2T_h})$ .....(1)

When $t = t_2$ and $\theta_r = \theta_{r2}$, $\theta_{r2} = \theta_m(1 - e^{-t_2/T_h})$

$\therefore 50 = \theta_m(1 - e^{-1/T_h})$ .....(2)

On dividing equation (2) by (1) we get,

$$\frac{50}{41.8} = \frac{1 - e^{-1/T_h}}{1 - e^{-1/2T_h}}$$

$$\frac{50}{41.8} = \frac{(1 - e^{-1/2T_h})(1 + e^{-1/2T_h})}{(1 - e^{-1/2T_h})}$$

$\boxed{a^2 - b^2 = (a + b)(a - b)}$

$1.1962 = 1 + e^{-1/2T_h}$ $\Rightarrow$ $1.1962 - 1 = e^{-1/2T_h}$

$0.1962 = e^{-1/2T_h}$

$\ln(0.1962) = -\dfrac{1}{2T_h}$

$\therefore$ Heating time constant, $T_h = -\dfrac{1}{2 \ln 0.1962} = 0.307\ h$

From equation (2) we get,

         Final steady temperature, $\theta_m = \dfrac{41.8}{1 - e^{-1/2T_h}}$

$$= \frac{41.8}{1 - e^{-1/(2 \times 0.307)}} = 52\,°C$$

b)   Temperature rise on short time load for 6 *min.*

Here, $t = 6\ min = \dfrac{6}{60} = 0.1\ h$

$\therefore$ Temperature rise, $\theta_r = \theta_m(1 - e^{-0.1/T_h})$

$$= 52(1 - e^{-0.1/0.307}) = 14.46\ °C$$

c)   Maximum steady state temperature rise on short time rating so as to obtain 52 °C after 6 *min.*

Here, $\theta_s = \theta_{msh}(1 - e^{-t/T_h})$

$\therefore \theta_{msh} = \dfrac{\theta_s}{(1 - e^{-0.1/T_h})}$

$$= \frac{52}{1 - e^{-0.1/0.307}} = 187\,°C$$

Total loss $= P \times (1 - \eta)$ $\boxed{\text{Given that, } \eta = 90\%}$

$$= 15 \times 10^3 \times \left(1 - \frac{90}{100}\right) = 1500\ W$$

$$\therefore P_i + P_c = 1500 \ W$$

Here, $P_i = P_c = \dfrac{1500}{2} = 750 \ W$ | At maximum efficiency, $P_i = P_c$

Total loss at full load $= P_i + P_c$

Total loss at short time rating $= x^2 \ P_i + P_c$

where, $x = $ Overload at short time rating.

Now, $\dfrac{\theta_{msh}}{\theta_s} = \dfrac{x^2 \ P_c + P_i}{P_c + P_i}$

$\dfrac{187}{52} = \dfrac{x^2 \ 750 + 750}{1500} \quad \Rightarrow \quad \dfrac{187}{52} \times 1500 - 750 = x^2 \ 750 \quad \Rightarrow \quad 4644 = 750 \ x^2$

$\therefore$ Overload at short time rating, $x = \sqrt{\dfrac{4644}{750}} = 2.48$ times full load

RESULT

Final steady state temperature rise, θ_s = 52 °C

Temperature rise on short time load, θ = 14.46 °C

Maximum steady state temperature rise on short-time rating, θ_{msh} = 187 °C

Overload at short time rating,   x = 2.48 times full load.

EXAMPLE 4.9

When started from cold condition, the temperature rise of a loaded transformer is 26 °C after one hour and 37.7 °C after 2 hours. When the transformer is switched off the temperature falls from the final steady temperature to 35 °C in 2.8 hours. Taking the ambient temperature as 25 °C, calculate the following for the transformer (i) heating time constant (ii) final steady full load temperature rise and (iii) cooling time constant.

GIVEN DATA

When, $t = t_1 = 1 \ h,$ $\theta_r = \theta_{r1} = 26$ °C $\theta_a = 25$ °C

When, $t = t_2 = 2 \ h,$ $\theta_r = \theta_{r2} = 37.7$ °C

When, $t = t_3 = 2.8 \ h,$ $\theta_c = 35$ °C

SOLUTION

(i)   Heating time constant,

Temperature rise at any time, t during heating process is given by the equation,

Temperature rise, $\theta_r = \theta_m \left(1 - e^{-t/T_h}\right)$

Where, θ_m = Maximum temperature rise

T_h = Heating time constant

Temperature rise at time t_1, $\theta_{r1} = \theta_m \left(1 - e^{-t_1/T_h}\right)$

$\therefore 26 = \theta_m \left(1 - e^{-1/T_h}\right)$ .....(1)

Temperature rise at time t_2, $\theta_{r2} = \theta_m \left(1 - e^{-t_2/T_h}\right)$

$\therefore 37.7 = \theta_m \left(1 - e^{-2/T_h}\right)$ .....(2)

On dividing equation (2) by (1) we get,

$$\frac{37.7}{26} = \frac{(1 - e^{-2/T_h})}{(1 - e^{-1/T_h})}$$

$$1.45 = \frac{(1 - e^{-1/T_h})(1 + e^{-1/T_h})}{(1 - e^{-1/T_h})}$$

$\boxed{a^2 - b^2 = (a + b)(a - b)}$

$$\therefore 1.45 = 1 + e^{-1/T_h} \quad \Rightarrow \quad 1.45 - 1 = e^{-1/T_h}$$

$$0.45 = e^{-1/T_h} \quad \Rightarrow \quad \ln 0.45 = -\frac{1}{T_h}$$

$$\therefore T_h = -\frac{1}{\ln 0.45} = 1.25 \ h$$

(ii) Final steady temperature rise,

From equation (1) we get,

$$\text{Final steady temperature, } \theta_m = \frac{26}{1 - e^{-1/T_h}} = \frac{26}{1 - e^{-1/1.25}}$$

$$= 47.2 \ °C$$

(iii) Cooling time constant,

Initial temperature during cooling process, $\theta_i = \theta_m = 47.2 \ °C$

Let, $t = t_c = $ Cooling time $= 2.8 \ h$

$\theta = \theta_c = $ Temperature at time, t_c

When $t = 2.8 \ h$, Fall in temperature, $\theta_f = \theta_c - \theta_a = 35 - 25 = 10 \ °C$

Temperature fall at any time, t during the cooling process is given by,

Temperature fall, $\theta_f = \theta_i \ e^{-t/T_c}$

$$10 = 47.2 \ e^{-2.8/T_c} \quad \Rightarrow \quad \frac{10}{47.2} = e^{-2.8/T_c} \quad \Rightarrow \quad 0.2119 = e^{-2.8/T_c}$$

$$\ln 0.2119 = -\frac{2.8}{T_c}$$

$$\therefore \text{Cooling time constant, } T_c = -\frac{2.8}{\ln 0.2119} = 1.8 \ h$$

RESULT

Heating time constant,	T_h	$= 1.25 \ h$
Final steady temperature,	θ_m	$= 47.2 \ °C$
Cooling time constant,	T_c	$= 1.8 \ h$

EXAMPLE 4.10

The rate of temperature rise as measured from a temperature time curve of a DC motor is 0.0803 °C per minute, and 0.0605 °C per minute when the temperaturer rise is 20.5 °C and 28.5 °C respectively. Estimate (i) Final steady temperature rise and (ii) heating time constant of DC motor.

GIVEN DATA

When, $\theta_r = \theta_{r1} = 20.5 \ °C$, $d\theta_{r1}/dt = 0.0803 \ °C/min$

When, $\theta_r = \theta_{r2} = 28.5 \ °C$, $d\theta_{r2}/dt = 0.0605 \ °C/min$

SOLUTION

Temperature rise at any time t, $\theta_r = \theta_m(1 - e^{-t/T_h})$ ......(1)

On differentiating the above equation we get,

$$\frac{d\theta_r}{dt} = \theta_m \frac{e^{-t/T_h}}{T_h}$$ .....(2)

From equation (1) we can write,

$$e^{-t/T_h} = \frac{\theta_m - \theta_r}{\theta_m}$$ .....(3)

From equations (2) and (3) we can write,

$$\frac{d\theta_r}{dt} = \theta_m \frac{\theta_m - \theta_r}{\theta} \times \frac{1}{T_h}$$

$$\therefore \frac{d\theta_r}{dt} = \frac{\theta_m - \theta_r}{T_h}$$ .....(4)

When, $\theta_r = \theta_{r1} = 20.5\ °C$, $\quad \dfrac{d\theta_{r1}}{dt} = \dfrac{\theta_m - \theta_{r1}}{T_h} \Rightarrow 0.0803 = \dfrac{\theta_m - 20.5}{T_h}$ .....(5)

When, $\theta_r = \theta_{r2} = 28.5\ °C$, $\quad \dfrac{d\theta_{r2}}{dt} = \dfrac{\theta_m - \theta_{r2}}{T_h} \Rightarrow 0.0605 = \dfrac{\theta_m - 28.5}{T_h}$ .....(6)

On dividing equation (5) by (6) we get,

$$\frac{0.0803}{0.0605} = \frac{\theta_m - 20.5}{\theta_m - 28.5} \quad \Rightarrow \quad 1.3273 = \frac{\theta_m - 20.5}{\theta_m - 28.5}$$

$$1.3273\ \theta_m - 28.5 \times 1.3273 = \theta_m - 20.5$$

$$1.3273\ \theta_m - \theta_m = 28.5 \times 1.3273 - 20.5$$

$$0.3273\ \theta_m = 17.3281$$

$\therefore$ Final steady temperature, $\theta_m = \dfrac{17.3281}{0.3273} = 52.9\ °C$

From equation (5) we get,

Heating time constant, $T_h = \dfrac{\theta_m - 20.5}{0.0803}$

$$= \frac{52.9 - 20.5}{0.0803} = 403\ min$$

RESULT

Final steady temperature rise, θ_m = 52.9 °C

Heating time constant, $\quad T_h$ = 403 min

EXAMPLE 4.11

Estimate the volume of cooling air required for an alternator with a minimum temperature rise of 40 °C from the following data :

Total weight of active iron = 1500 kg

Losses per kg in iron = 8 W

Total weight of active copper = 700 kg

Barometric pressure of air = 780 $mm\ Hg$

Losses per kg in copper = 20 W

Stray losses = 20% of the total losses

Inlet temperature of air = 25 °C

GIVEN DATA

Total weight of active iron = 1500 kg

Total weight of active copper = 700 kg

Stray losses = 20% of the total losses

Barometric pressure of air, H_a = 780 $mm\ Hg$

Losses per kg in iron = 8 W

Losses per kg in copper = 20 W

Inlet temperature of air, θ_i = 25 °C

Temperature rise of air, θ_{ra} = 40 °C

SOLUTION

Total iron losses = Total weight of iron × Losses per kg in iron

$$= 1500 \times 8 = 12000\ W$$

Total copper losses = Total weight of copper × Losses per kg in copper

$$= 100 \times 20 = 14000\ W$$

Sum of iron and copper losses = 12000 + 14000 = 26000 W

$$\text{Stray losses} = \frac{20}{100} \times \text{Sum of iron and copper losses}$$

$$= \frac{20}{100} \times 26000 = 5200\ W$$

Total losses, Q_L = Iron loss + Copper loss + Stray losses

$$= 26000 + 5200 = 31200\ W$$

Volume of coolant air, $V_a = \dfrac{Q_L}{c_{pa}\ \theta_{ra}}\ \dfrac{1}{\rho_{ca}}\ \dfrac{273+\theta_i}{273}\ \dfrac{760}{H_a}$

Take, ρ_{ca} = 1.2903 kg/m^3
c_{pa} = 995 $J/kg\text{-}°C$

$$= \frac{312000}{995 \times 40} \times \frac{1}{1.2903} \times \frac{273+\theta_i}{273} \times \frac{760}{280}$$

$$= 0.6462\ m^3/s$$

RESULT

Volume of coolant air, V_a = 0.6462 m^3/s

EXAMPLE 4.12

A turbo alternator runs on test at a continuous rated load of 25 MVA with a power factor of 0.88. The following cooling air measurements are taken: volume of cooling air measured at intake = 25 m^3/s, intake air temperature = 10 °C, outlet air temperature = 40 °C, barometric pressure = 700 $mm\ Hg$. (i) Find the efficiency of the machine, taking the specific heat of air at constant pressure as 900 $J/kg\text{-}°C$ and the volume of 1 kg of air at 0 °C and a pressure of 760 $mm\ Hg$ as 0.72 m^3. (ii) Calculate the amount of cooling water in litre per second to cool the air, assuming the temperature rise of water to be 7 °C.

GIVEN DATA

25 MVA

pf = 0.88

c_{pa} = 900 $J/kg\text{-}°C$

$\rho_{ca} = \dfrac{1}{0.72}\ kg/m^3$

V_a = 25 m^3/s

θ_i = 10 °C

θ_o = 40 °C

H_a = 700 $mm\ Hg$

θ_{rw} = 7 °C

SOLUTION

(i)   Let, Temperature rise of air, $\theta_{ra} = \theta_o - \theta_i = 40 - 10 = 30\ °C$

$$\text{Volume of air, } V_a = \frac{Q_L}{c_{pa}\,\theta_{ra}}\ \frac{1}{\rho_{ca}}\ \frac{273+\theta_i}{273}\ \frac{760}{H_a}$$

$$\therefore \text{ Power loss, } Q_L = V_a \times c_{pa}\,\theta_{ra} \times \rho_{ca} \times \frac{273}{273+\theta_i} \times \frac{H_a}{760}$$

$$= 25 \times 900 \times 30 \times \frac{1}{0.72} \times \frac{273}{273+10} \times \frac{700}{760}$$

$$= 832975\ W$$

$$= 832.975\ kW = 833\ kW$$

Given that, Output MVA, $Q_o = 25\ MVA = 25 \times 10^3\ kVA$

$\therefore$ Output kilowatt $= Q_o \times pf = 25 \times 10^3 \times 0.88 = 22 \times 10^3\ kW$

$$\text{Efficiency, } \eta = \frac{Q_o}{Q_o + Q_L}$$

$$= \frac{22000}{22000 + 833} = 96.35\ \%$$

(ii)   Given that, temperature rise of water, $\theta_{rw} = 7\ °C$

$$\text{Volume of water, } V_w = \frac{Q_L}{c_{pw}\,\theta_{rw}}\ \frac{1}{\rho_{cw}} \times 1000$$

$$= \frac{833 \times 10^3}{4180 \times 7} \times \frac{1}{1000} \times 1000 = 28.5\ l/s$$

$\rho_{cw} = 1000\ kg/m^3$

$c_{pw} = 4180\ J/kg\text{-}°C$

RESULT

Efficiency, $\eta = 96.35\ \%$

Volume of water, $V_w = 28.5\ l/s$

EXAMPLE 4.13

Half hour rating of a motor is 31.5 kW. The heating time constant is 85 minutes. Find the heating and mechanical overload ratios and hence find the continuous rating of the motor. The maximum efficiency of the motor occurs at 80% of full load.

GIVEN DATA

$P_h = 31.5\ kW$ $t_h = 30\ min$

$T_h = 85\ min$ $\eta_m = 80\% = \dfrac{80}{100}$

SOLUTION

$$\text{Heating overload ratio, } p_{ho} = \frac{1}{1 - e^{-t_h/T_h}}$$

$$= \frac{1}{1 - e^{-30/85}} = 3.3627$$

$t_h = 30\ min$
(half-an-hour)

Maximum efficiency occurs at 80% full load.

$$\therefore \text{ Constant losses, } P_i = \eta_m^2\,P_c = \left(\frac{80}{100}\right)^2 P_c = 0.64\ P_c$$

Where, P_c = Copper loss or variable loss

Now, $K = \dfrac{P_i}{P_c} = 0.64$

Mechanical overload ratio, $p_{mo} = \sqrt{(K+1)\,p_{ho} - K}$

$$= \sqrt{(0.64+1) \times 3.3627 - 0.64} = 2.1945$$

$\therefore$ Continuous rating of motor, $P_{nom} = \dfrac{P_h}{p_{mo}}$

$$= \dfrac{31.5}{2.1945} = 14.354 \; kW$$

RESULT

Heating overload ratio,	p_{ho}	= 3.3627
Mechanical overload ratio,	p_{mo}	= 2.1945
Continuous rating of motor,	P_{nom}	= 14.354 kW

EXAMPLE 4.14

The total losses in a transformer is 250 kW. The transformer is oil cooled and the heat from oil is removed using water cooled radiator. The oil remove 75% of heat generated and tank surface dissipates the remaining heat. During cooling process the temperature of oil is raised by 25 °C and that of water by 12 °C. Estimate the volume of oil and water required for this cooling system.

GIVEN DATA

Transformer loss = 250 kW

Loss dissipated by cooling system = 75%

Temperature rise of oil, $\theta_{ro} = 25$ °C

Temperature rise of water, $\theta_{rw} = 12$ °C

SOLUTION

Loss dissipated by coolant, $Q_L = 250 \times \dfrac{75}{100} = 187.5 \; kW$

Volume of oil, $V_o = \dfrac{Q_L}{c_p\,\theta_{ro}}\,\dfrac{1}{\rho_{co}} \times 1000$

$\boxed{\begin{aligned} c_{po} &= 2060 \; J/kg\text{-}°C \\ \rho_{co} &= 890 \; kg/m^3 \end{aligned}}$

$$= \dfrac{187.5 \times 1000}{2060 \times 25} \times \dfrac{1}{890} \times 1000$$

$$= 4.09 \; l/s$$

Volume of water, $V_w = \dfrac{Q_L}{c_p\,\theta_{rw}}\,\dfrac{1}{\rho_{cw}} \times 1000$

$\boxed{\begin{aligned} c_{pw} &= 4180 \; J/kg\text{-}°C \\ \rho_{cw} &= 1000 \; kg/m^3 \end{aligned}}$

$$= \dfrac{187.5 \times 1000}{4180 \times 12} \times \dfrac{1}{1000} \times 1000$$

$$= 3.74 \; l/s$$

RESULT

Volume of oil,	V_o	= 4.09 l/s
Volume of water,	V_w	= 3.74 l/s

EXAMPLE 4.15

An induction motor has to perform the following duty cycle indefinitely.

85 kW for 15 minutes,

30 kW for 10 minutes,

55 kW for 14 minutes,

No load for 6 minutes,

Determine a suitable capacity of a continuously rated motor to perform the load duty cycle. Motor of standard (continuous) ratings 50, 65, 90 kW are available. The ratio of maximum torque to nominal torque should be less than 1.7.

GIVEN DATA

P_1 = 85 kW t_1 = 15 min

P_2 = 30 kW t_2 = 10 min

P_3 = 55 kW t_3 = 14 min

P_4 = 0 kW t_4 = 8 min

SOLUTION

$$\text{Equivalent power, } P_{eq} = \sqrt{\frac{P_1^2\, t_1 + P_2^2\, t_2 + P_3^2\, t_3 + P_4^2\, t_4}{t_1 + t_2 + t_3 + t_4}}$$

$$= \sqrt{\frac{(85^2 \times 15) + (30^2 \times 10) + (55^2 \times 14) + (0 \times 6)}{15 + 10 + 14 + 6}} = 59.57\ kW$$

Given that, available motor ratings are 50, 65, 90 kW.

Let, P_{nom} = 65 kW

Now, P_{nom} is slightly higher than P_{eq} and so the motor with 65 kW rating will be capable of performing given load duty cycle.

Therefore, the motor with a standard rating of 65 kW is selected.

Since the induction motor is practically a constant speed motor, the ratio of maximum torque to nominal torque is equal to the ratio maximum power to nominal power.

From load patteren, we get,

P_{max} = 85 kW, $\therefore$ T_{max} α 85

Here,

P_{nom} = 65 kW $\therefore$ T_{nom} α 65

$\therefore$ $\dfrac{T_{max}}{T_{nom}} = \dfrac{85}{65} = 1.31$

For the selected motor rating the ratio T_{max} / T_{nom} is less than the maximum allowable value of 1.7. Therefore, the selected motor satisfy the requirement.

RESULT

Capacity of selected motor, P = 65 kW

4.13 SHORT-ANSWER QUESTIONS

Q4.1 *What are the different modes of heat dissipation in electric machines?*

The different modes of heat dissipation in electric machines are conduction, radiation and convection.

Q4.2 *What is heat conduction?*

Heat conduction is one of the modes of heat transfer (or transfer of internal thermal energy) in a body by the collisions of microscopic particles and movement of electrons within a body. In conduction mode of heat transfer, the heat flow is within and through the body itself. The microscopic particles in the heat conduction can be molecules, atoms and electrons.

Q4.3 *What is heat radiation?*

Heat transfer through radiation (or thermal radiation) takes place in form of electromagnetic waves mainly in the infrared region. Radiation emitted by a body is due to thermal agitation of the constituent of the body like molecules, atoms and electrons.

Q4.4 *What is heat convection?*

Heat convection is a mode of heat transfer occurs on the surface of a hot body when the surrounding fluid move heat energy away from the hot body. Convection heat transfer occurs when the surface temperature of a hot body differs from that of surrounding fluid.

Q4.5 *What is Newton's law of cooling?*

Newton's law of cooling states that the rate of heat loss of a body is directly proportional to the difference in the temperature between the body and its surroundings.

$$\text{Heat dissipation, } Q = (\lambda_{rad} + \lambda_{conv}) \, \theta_d \, S = \lambda \, \theta_d \, S$$

where, $\lambda = \lambda_{rad} + \lambda_{conv}$ = Sum of specific heat dissipation due to radiation and convection

θ_d = Temperature difference between body and ambient

S = Cooling surface

λ = Specific heat dissipation

Q4.6 *What is thermal resistance and its unit?*

In heat conduction the thermal resistance is defined as the thermal resistance which causes a temperature drop of 1°C per watt of heat flow. The thermal resistance is defind as,

$$\text{Thermal resistance, } R_c = \frac{\rho \, t}{S} = \frac{t}{\sigma S}$$

where, ρ = Thermal resistivity of material

$\sigma = 1/\rho$ = Thermal conductivity

t = Length of heat path

S = Surface area

The unit of thermal resistance is *Ω-m* or *°C-m/W*.

Q4.7 *Why the electric machines are usually painted with dull metallic paints?*

The specific heat dissipation due to radiation for surfaces painted with dull metallic paints is large. Hence all electrical machines are painted with dull metallic paints in order to increase the heat dissipation due to radiation.

Q4.8 *Define short time rating.*   *(PTU, May' 19, 2 M) (AU, Nov' 17, 2 M)*

Electrical motors operated intermittently for a short-time duration with permissible temperature rise at loads higher than full-load are referred to as short time duty motors. These motors are operated at a load higher than full load with temperature rise not exceeding maximum permissible value.

Q4.9 Define intermittent rating.

(PTU, May' 19, 2 M)

Electrical motors that run alternatively on load for a short time duration and no-load for another short time duration are referred to as intermittent periodic duty motors. These motors will run continuously but load alone will be disconnected periodically to keep temperature rise within maximum permissible value.

(HTU, Dec' 18, 2 M)

Q4.10 Why short time rating of an electrical machine is much higher than continuous rating?

In short time rating, the period or time duration of operation of motor is very less for machine to attain steady temperature rise. Also, the cooling time is long enough for the motor to cool to ambient temperature. Therefore, short time rating is higher than continuous rating.

Q4.11 Write an expression to find the hotspot temperature ?

The temperature of hottest spot is given by,

$$\text{Hotspot temperature, } \theta_m = \theta_{do} + \theta_s \quad \Rightarrow \quad \theta_m = \frac{q\,\rho\,t^2}{8} + \theta_s$$

where, $\theta_{do} = \dfrac{q\,\rho\,t^2}{8}$ = Temperature rise between centre and outer surface

θ_s = Surface temperature

Q4.12 Write the relation between the effective thermal resistivity of winding, thermal resistivity of insulation and space factor in electrical machines?

A relation between the effective thermal resistivity of winding, thermal resistivity of insulation and space factor in electrical machines is given by,

$$\text{Effective resistivity of winding, } \rho_e = \rho_i(1 - \sqrt{S_f})$$

where, S_f = Space factor

ρ_i = Thermal resistivity of insulation.

Q4.13 Write the equation of temperature rise with time while heating in electric machine. What is heating time constant?

$$\text{Temperature rise while heating, } \theta_r = \theta_m(1 - e^{-t/T_h})$$

where, θ_m = Maximum steady state temperature

T_h = Heating time constant

If the initial rate of change of temperature is maintained, the machine would reach its final steady temperature rise in a time equal to its heating time constant.

Heating time constant is also given by the time for temperature rise to reach $0.632\ \theta_m$.

Q4.14 Write the equation of temperature fall with time while cooling in electric machine. What is cooling time constant?

$$\text{Temperature fall while cooling, } \theta_f = \theta_i\, e^{-t/T_c}$$

where, θ_i = Initial temperature

T_c = Cooling time constant

If the initial rate of change of temperature is maintained, the machine would reach its final steady temperature fall in a time equal to its cooling time constant.

Cooling time constant is also given by the time for temperature fall to reach $0.368\ \theta_i$.

Q4.15 Write a short note on "enclosures of electric machines"?

(PTU, May' 19, 2 M)

The enclosure of electrical machine is a cabinet or box that protects electrical equipment and prevents electrical shock. Enclosures are usually made from rigid plastics or metals such as steel, stainless steel or cast iron. Enclosuers are designed to provide protection against hazardous, non-hazardous and other specific environmental conditions. The enclosure should also provide cooling or ventilation for machine parts.

Q4.16 What are the advantages of hydrogen cooling of turbo-alternator?

1. Increased efficiency

2. Increase in rating

3. Increase in life

4. Elimination of fire hazard

5. Smaller size of coolers

6. Less noise.

Q4.17 What is direct cooling of turbo-alternator?

Direct cooling is the process of dissipating the armature and field winding losses to a cooling medium circulating within the winding insulation wall. Machines cooled in this manner are also called "super-charged", "inner cooled" or "conductor cooled" by various manufacturers.

Q4.18 List the different duties of electrical machines.

(AU, Apr' 17, 2 M) (AU, Nov' 18, 2 M)

1. Continuous duty

2. Short time duty

3. Intermittent periodic duty

4. Intermittent periodic duty with starting

5. Intermittent periodic duty with starting and braking

6. Continuous duty with intermittent periodic loading

7. Continuous duty with starting and braking

8. Continuous duty with periodic speed changes

Q4.19 Define continuous duty.

The continuous rating of a motor may be defined as the load that may be carried by the machine for an indefinite time without the temperature rise of any part exceeding the maximum permissible value.

Q4.20 What are the methods for determination of motor rating for variable load drives?

1. Method of average losses   3. Equivalent torque method

2. Equivalent current method   4. Equivalent power method.

Q4.21 Describe equivalent current method?

The equivalent current method is based upon the assumption that the actual variable current may be replaced by an equivalent current, I_{eq} which produces the same losses in the motor as the actual current. This method also assumes that the constant losses are independent of the load.

$$\text{Equivalent current, } I_{eq} = \sqrt{\frac{I_1^2\, t_1 + I_2^2\, t_2 + I_3^2\, t_3 + ... + I_n^2\, t_n}{t_1 + t_2 + t_3 + ... + t_n}}$$

Q4.22 What are the categories of ventialing systems?

      1.   Radial ventilating system

      2.   Axial ventilating system

      3.   Combined radial and axial ventilating system.

Q4.23 Name the categories of self and seperate ventilation?

      1.   Induced ventilation

      2.   Forced ventilation.

4.14 EXERCISES

I. Fill in the blanks

1. The temperature rise at any time t is given by ....

2. The temperature fall at any time t is given by ....

3. The time taken by the machine to attain 0.632 of its final steady temperature rise is called ....

4. The time taken by the machine to attain 0.368 of its final steady temperature fall is called ....

5. Every machine will have a .... temperature rating.

6. In small capacity machines the power loss per unit volume will be ....

7. In large capacity machines the power loss per unit volume will be ....

8. Heat conduction is one of the modes of ....

9. The range of thermal resistivity of transformer coil across laminations is ....

10. In electrical machines power loss occur in .... and ....

11. The power loss in core is called ....

12. The power loss in winding is called ....

13. In ventilation system the forced ventilation is provided in .... machine.

14. In ventilation system the axial ventilation is suitable for .... machine.

15. The volume of air or gas depends on .... and ....

16. The rating of electrical machines may also include .... and ....

Answers

1.	$\theta_r = \theta_m \left(1 - e^{-t/T_h}\right)$	7.	high	13.	large capacity
2.	$\theta_f = \theta_i\, e^{-t/T_c}$	8.	heat transfer	14.	high speed
3.	heating time constant	9.	0.05 to 0.1 Ω-m	15.	temperature, pressure
4.	cooling time constant	10.	core, winding	16.	temperature rise, load duty cycle
5.	maximum	11.	iron loss		
6.	less	12.	copper loss		

II. State whether the following statements are True/False

1. Every machine will have a minimum temperature rating.

2. The internal temperature of the core and winding will be higher than the surface temperature of the core and winding.

3. Induced ventilation is commonly provided in machines of small and medium power outputs.

4. In ventilation system the forced ventilation is provided in small capacity machines.

5. Axial ventilation system is suitable for high speed machines.

6. In open machine, the enclosure mainly depends on good mechanical strength.

7. In electrical machines, the iron loss is constant and copper loss is directly propotional to power.

8. In equivalent power method the speed of the motor is considered constant and so the power is directly propotional to torque.

Answers			
1.	False	5.	True
2.	True	6.	True
3.	True	7.	False
4.	False	8.	True

III. Unsolved problems

E4.1 In a laminated core, the thermal resistivity in the direction of lamination is 20 times lesser than the resistivity in the direction perpendicular to laminations. A temperature difference of 5 °C will cause 60 W to be conducted through a stack with surface area 4×10^{-3} m^2 and 22 mm thick measured along the laminations. Calculate the loss that will be conducted across the laminations in a stack 0.1 m thick and 10×10^{-3} m^2 in cross-section with a temperture difference 10 °C.

$$(Q_x = 3.3 \ W)$$

E4.2 In a winding, a copper loss of is 0.6 W is dissipated through a layer of insulation having a thermal resistivity of 7 Ω-m. Find the temperature difference between the two sides of the insulation. The thickness of insulation is 0.6 mm and its surface area is 260 mm^2.

$$(\theta_d = 10°C)$$

E4.3 A 80 MW hydrogen cooled turbo-alternator has a full load efficiency of 98.8 %. The hydrogen enters the machine with a temperature of 22 °C and leaves the machine with a temperature of 51 °C. Determine the volume of hydrogen required if the hydrogen pressure is 1800 mm above a gauge pressure of 760 mm of mercury. Specific heat of hydrogen is 12600 J/kg-°C. Volume of 1 kg of hydrogen at NTP is 11.2 m^2.

$$(V_h = 50 \ m^3/s)$$

E4.4 Consider a transformer core of width 300 mm, core loss of 2 W/kg, density of 7600 kg/m^3, and a thermal resistivity of 0.04 Ω-m. The core loss is distributed and space factor is 0.89 and the heat flow is along the lamination. Compute the maximum core temperature when the surface temperature is 30 °C.

$$(\theta_m = 50.3 \ °C)$$

E4.5 In a DC motor, the rate of temperature rise is 0.0908 °C per minute and 0.0703 °C per minute when the temperaturer rise is 21.5 °C and 30.5 °C respectively. Estimate the final steady temperature rise and the heating time constant of DC motor.

$$(\theta_f = 61.4 \text{ °C, } T_h = 442 \text{ minutes})$$

E4.6 In an induction motor on load, the motor is shut down when it is heated to a temperature of 65 °C. Calculate temperature of motor 10 minutes after the shutdown if the cooling time constant is 40 minutes. Take ambient temperature as 28 °C.

$$(\theta_{m\_10} = 56.8 \text{ °C})$$

E4.7 In a load test on a turbo alternator at a rated load of 40 MVA with a power factor of 0.8, the following cooling air measurements are taken: volume of cooling air measured at intake = 28 m^3/s, intake air temperature = 15 °C, outlet air temperature = 40 °C, barometric pressure = 720 mm Hg. Determine the power loss and the efficiency of the machine. Take the specific heat of air at constant pressure as 990 J/kg-°C and the volume of 1 kg of air at 0 °C and a pressure of 760 mm Hg as 0.73 m³. Also, calculate the amount of cooling water in litre per second to cool the air, if the temperature rise of water is 9 °C.

$$(\eta = 97.4 \text{ \%, } V_w = 22.66 \text{ l/s})$$

E4.8 For a motor with half hour rating 39.5 kW, the heating time constant is 60 minutes. Determine the heating and mechanical overload ratios and hence the continuous rating of the motor. The maximum efficiency of the motor occurs at 85% full load.

$$(p_{ho} = 2.5, \ p_{mo} = 1.893, \ P_{nom} = 20.9 \text{ kW})$$

> # Chapter 5

ARMATURE WINDING

List of symbols

Symbol	Meaning	Unit
C	Number of coils	-
k_c	Chording angle	-
k_d	Distribution factor	-
k_w	Winding factor	-
q	Slots pole per phase	-
n	Slots per pole	-
p	Number of poles	-
S	Number of slot	-
β	Slot angle	-
α	Angle of chording	-
Y_b	Back pitch	-
Y_f	Front pitch	-
Y	Winding pitch	-
Y_c	Commutator pitch	-

5.1 BASIC CONCEPTS OF ARMATURE WINDING

The two electric circuits in rotating machines are armature winding and field winding. The armature winding consists of a number of series connected coils accommodated in armature slots. The coils are either multiturn coils or single turn coils. Each turn consists of two conductors. The coils are made of copper. In low cost machines the coils are made of aluminium. In AC machines the armature winding is also called as *stator winding*.

The field winding consists of a number of series connected coils accommodated in field poles (or main poles). Each pole has one coil and the coils are excited by dc supply to produce the required amount of mmf. In case of induction motors the field winding is absent and it is replaced by short circuited rotor bars or winding.

Coils, Turns and Conductors

The armature windings of rotating machines consists of a number of coils and each coil has a number of turns. One turn is made of two conductors. The number of coils and turns are decided based on the following factors,

1. Emf per turn
2. Current density
3. Flux density
4. Type of insulation
5. Number of slots
6. Type of winding
7. Capacity of the machine

The area of cross-section of the armature conductor is decided based on the allowable temperature rise in the machine. The resistivity of the material used for conductor, current density and cooling methods adopted are the factors which determines the temperature rise in machines. Generally copper is used for all types of windings. The resistivity of copper at 20°C is 1.724×10^{-8} ohm-m. The current density in the copper conductor varies from 2.5 to 5 A/mm^2. In bar type conductors a higher current density upto 7 A/mm^2 can be allowed.

In this chapter armature winding design is discussed with inputs like number of slots, conductors per slot and number of poles.

Definition of various terms used in armature winding

Conductor: The active length of copper or aluminium wire in the slot is called **conductor**.

Turn: Two conductors connected for additive emf is called a **turn**. The two conductors of a turn are placed approximately a pole pitch apart.

Coil: A coil consists of a number of turns and it is the principal element of armature winding. The coil with single turn is called **single turn coil** and the coil with several turns is called **multi turn coil**.

Coil side: The active portions of the conductors in a coil are called coil sides. A coil will have two sides as shown in Fig. 5.1 and they are upper (top) coil side and lower (bottom) coil side. Usually the top coil side is represented by solid line and bottom coil side by dotted (or broken) line. The top coil side is placed in the upper portion of a slot and the bottom coil side is placed at the lower portion of another slot. The distance between the two coil sides is kept approximately as one pole pitch.

Overhang: The end portion of the coil connecting the two coil sides is called **overhang**.

Coil span: The distance between the two coil sides of a coil is called coil span. It is expressed in terms of number of slots or in electrical degrees.

Full pitch coil: When the coil span is equal to pole pitch, the coils are called **full pitched coils**.

Short pitched or chorded coil: When the coil span is less than the pole pitch, the coils are called **short pitched or short chorded coils**.

Single layer winding: When the coil sides are arranged in a single layer in a slot, the winding is called **single layer winding**.

Double layer winding: When the coil sides are arranged in two layers in a slot, the winding is called **double layer winding**.

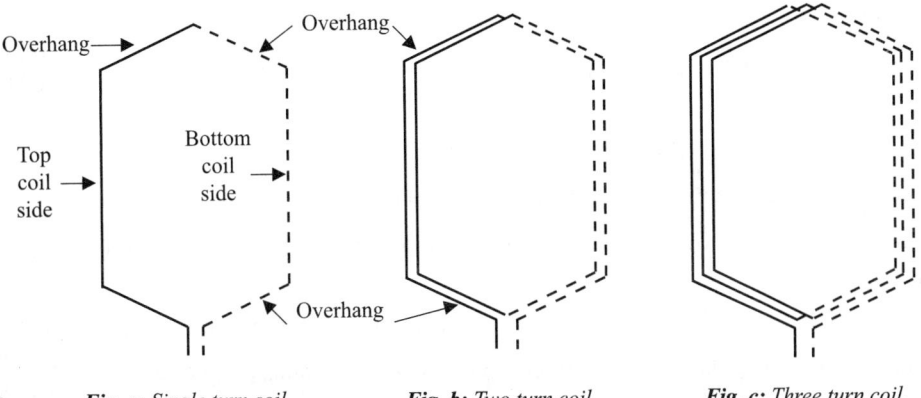

Fig. a: *Single turn coil.*   **Fig. b:** *Two turn coil.*   **Fig. c:** *Three turn coil.*

Fig. 5.1: *Armature winding coil.*

Back Pitch (Y_b)

The distance between top and bottom coil sides of a coil measured around the back of the armature (away from the commutator) is called the ***back pitch***, Y_b. The back pitch is measured in terms of coil sides. Since Y_b is difference between odd and even number, it is always an odd number.

Front Pitch (Y_f)

The distance between two coil sides connected to the same commutator segment is called the ***front pitch***, Y_f.

Winding Pitch (Y)

The distance between the starts of two consecutive coils measured in terms of coil sides is called ***winding pitch***, Y. The winding pitch is always an even integer.

Commutator Pitch (Y_c)

The distance between the two commutator segments to which the two ends (start and finish) of a coil are connected is called the ***commutator pitch***, Y_c and it is measured in terms of commutator segment.

5.2 DC MACHINE ARMATURE WINDING

In general the armature winding consists of a number of coils connected in series and number of such series circuits are connected in parallel. The coils are diamond shaped and are made in special forming machines. The coils may be single turn or multi turn coil, and a turn consists of two conductors.

The coils are placed in the slots on the armature periphery. In full pitched winding the two coil sides of a coil are placed one pole pitch apart. The DC machine armature windings are double layer windings, which means that each slot has two coil sides.

The design of armature winding is discussed in Chapter-3, Section 3.6.7 and it involves the selection of type of winding, estimation of number of armature coils, turns per coil, conductors per slot, total number of armature conductors and dimensions of the conductor.

5.2.1 Types of Armature Winding for DC Machine

DC machines employ two general types of double layer windings. They are

1. Simplex lap winding
2. Simplex wave winding

These two types of windings primarily differ from each other in the following two factors.

1. The number of circuits between the positive and negative brushes, i.e., number of parallel paths.

2. The manner in which the coil ends are connected to the commutator segments.

In simplex lap winding the number of parallel paths is equal to number of poles, whereas in simplex wave winding the number of parallel paths is two.

In simplex lap winding the finish of a coil is connected to start of next coil. In simplex wave winding the finish of a coil is connected to start of a coil which is lying one pitch away from the finish.

The simplex lap or wave windings are suitable for most of the DC machines used for various applications. But occasionally the number of parallel paths have to be increased to a value more than that provided by simplex windings. In such cases the multiplex windings are employed. When the number of parallel paths in a multiplex winding is twice that of simplex winding it is called *duplex winding*. When the number of parallel paths in a multiplex winding is thrice that of simplex winding it is called *triplex winding* and so on.

In general the lap winding and wave winding refers to simplex windings. The discussions and design aspects presented in this book refers to simplex windings. A comparative study of lap and wave winding is presented in Table 5.1.

Table 5.1: Comparison of Lap and Wave Winding

Lap winding	Wave winding
1. The number of parallel paths is equal to number of poles.	1. The number of parallel paths is two.
2. Current through a conductor is I_a/p, where I_a is armature current and p is number of poles.	2. Current through a conductor is $I_a/2$ where I_a is armature current.
3. For a specified voltage rating, the number of armature conductors required is p/2 times that of wave winding.	3. For a specified voltage rating, the number of armature conductors required is 2/p times that of lap winding.
4. For a specified current rating the area of cross-section of conductor is 2/p times that of wave winding.	4. For a specified current rating the area of cross-section of conductor is p/2 times that of lap winding.

Table 5.1: Continued...

Lap winding	Wave winding
5. For a specified power rating large number of conductors with smaller area of cross-section is required. But the volume of copper is same as that of wave winding.	5. For a specified power rating less number of conductors with larger area of cross-section is required. But the volume of copper is same as that of lap winding.
6. Since the lap winding has large number of conductors, the area required for insulation is more and so slot area will be large. Also the number of coils will be large. Therefore the cost will be high.	6. Since the wave winding has less number of conductors, the area required for insulation is less and so slot area required will be lesser. Also the number of coils will be less. Therefore the cost will be less.
7. Equalizer connections have to be employed.	7. Equalizer connections are not needed.
8. The winding is easier and short pitched coils can be made, which results in reduction in overhang length.	8. It is difficult to wind wave winding and short pitched coils can't be used.
9. The lap winding is used for large capacity machines and when the current rating is more than 400 A.	9. The wave winding is used for small and medium capacity machines. Also used for high voltage and slow speed machines.

5.2.2   Winding Pitches for DC Machine Lap and Wave Winding

Back Pitch (Y_b)

The distance between top and bottom coil sides of a coil measured around the back of the armature (away from the commutator) is called the ***back pitch***, Y_b. The back pitch is measured in terms of coil sides. Since Y_b is difference between odd and even number, it is always an odd number.

In the Example shown in Figs. 5.2 and 5.3,

Back pitch, $Y_b = 8 - 1 = 7$ - For lap winding

Back pitch, $Y_b = 8 - 1 = 7$ - For wave winding

The back pitch of a coil determines the size of the coil (coil span) and is nearly equal to coil sides per pole or pole pitch.

Front Pitch (Y_f)

The distance between two coil sides connected to the same commutator segment is called the *front pitch*, Y_f.

In the Example shown in Figs. 5.2 and 5.3,

$\quad\quad\quad$ Front pitch, $Y_f = 8 - 3 = 5$ $\quad$ - $\quad$ For lap winding

$\quad\quad\quad$ Front pitch, $Y_f = 15 - 8 = 7$ $\quad$ - $\quad$ For wave winding

The front pitch determines the type of the winding only and it does not affect the size of the coils.

Winding Pitch (Y)

The distance between the starts of two consecutive coils measured in terms of coil sides is called *winding pitch*, Y. The winding pitch is always an even integer.

$\quad\quad\quad$ Winding pitch, $Y = Y_b - Y_f$ $\quad$ - $\quad$ For lap winding

$\quad\quad\quad$ Winding pitch, $Y = Y_b + Y_f$ $\quad$ - $\quad$ For wave winding

In the Example shown in Figs. 5.2 and 5.3,

$\quad\quad\quad$ Winding pitch, $Y = 7 - 5 = 2$ (always 2) $\quad$ - $\quad$ For lap winding

$\quad\quad\quad$ Winding pitch, $Y = 7 + 7 = 14$ $\quad\quad\quad\quad$ - $\quad$ For wave winding

Commutator Pitch (Y_c)

The distance between the two commutator segments to which the two ends (start and finish) of a coil are connected is called the *commutator pitch*, Y_c and it is measured in terms of commutator segment.

In the Example shown in Figs. 5.2 and 5.3,

$\quad\quad\quad$ Commutator pitch, $Y_c = 2 - 1 = 1$ (always 1) $\quad$ - $\quad$ For lap winding

$\quad\quad\quad$ Commutator pitch, $Y_c = 8 - 1 = 7$ $\quad\quad\quad\quad$ - $\quad$ For wave winding

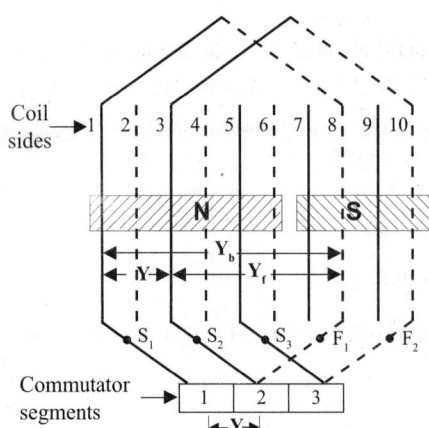

S_1, S_2, S_3:   Start of coils 1, 2, 3 respectively

F_1, F_2:   Finish of coils 1, 2 respectively

(Note: Refer Example 5.3)

Fig. 5.2: Lap winding.

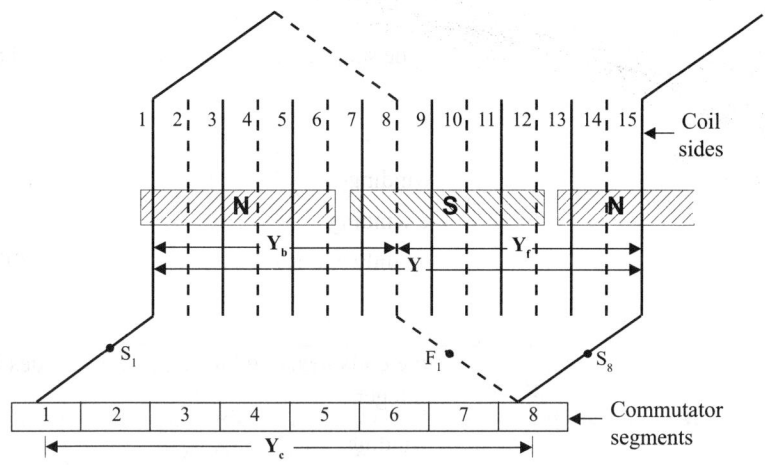

S$_1$, S$_8$: Start of coils 1 and 8 respectively
F$_1$:   Finish of coil 1

*(**Note:** Refer Example 5.7)*

***Fig. 5.3:** Wave winding.*

5.2.3 Simplex Lap Winding for DC Machine

In simplex lap winding the finish of a coil is connected to start of next coil. This winding scheme results in a number of parallel paths which is equal to number of poles. The simplex lap winding is a closed winding. In a closed winding if we trace the winding starting from one point, we will reach the same point after travelling through all the turns. But the electrical circuit closes through external load in case of generator and through external supply in case of motor. The simplex winding has one closed electrical circuit. (i.e., all the parallel paths electrically closes through external load or supply).

The two types of simplex lap winding used are progressive lap winding and retrogressive lap winding.

In the progressive lap winding the joining to the commutator progress around the commutator in the same direction as the coils progress around the armature, as shown in Fig. 5.4.

In the retrogressive lap winding the joining to the commutator segment progresses around the commutator in the opposite direction to the progress of coils around the armature, as shown in Fig. 5.5.

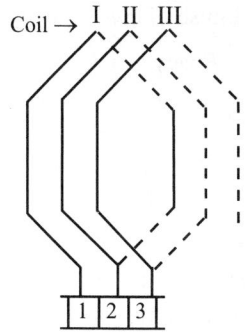

***Fig. 5.4:** Progressive lap.*

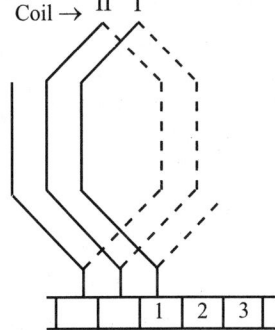

***Fig. 5.5:** Retrogressive lap.*

Table 5.2: Winding Pitches for Lap Winding

Progressive lap	Retrogressive lap
$Y_b = \dfrac{2C}{p} + K$	$Y_b = \dfrac{2C}{p} - K$
$Y = Y_b - Y_f$ and $Y = 2$	$Y = Y_b - Y_f$ and $Y = -2$
$Y_f = Y_b - Y = Y_b - 2$	$Y_f = Y_b - Y = Y_b + 2$
$Y_c = Y/2 = 1$	$Y_c = Y/2 = -1$

Note: K is a number to make Y_b an odd integer nearly equal to 2C/p, i.e., coil sides per pole.

The various winding pitches for simplex lap winding are listed in Table 5.2. In lap winding the back and front pitch are always odd integers. The winding pitch is always two and commutator pitch is always one. Usually the simplex lap winding is wound with two coil sides per slot. But simplex lap winding is possible with 4, 6, etc, (i.e., even number) coil sides per slot.

When the coil sides per slot is more than two, the back pitch, Y_b should be chosen such that all the coils having their top coil sides in one slot should have all their corresponding bottom coil sides together in another slot, which is one pole pitch away. If the back pitch is not properly chosen then the coil sides in the upper layer of one slot will be connected to bottom coil sides of two different slots. For this arrangement split coils have to be used, which is not desirable from practical point of view.

The split coils will have more than two coil sides. When all the top coil sides of a coil are lying in one slot and their corresponding bottom coil sides are accommodated in two different slots then the coil is called **split coil**.

Guide lines for drawing simplex lap winding diagram

The following guidelines will be useful for drawing simplex lap winding with two coil sides per slot. The winding diagram of Example 5.1 have been used to explain the various steps in the guide lines.

1. Determine the number of coil sides and represent the coil sides by parallel straight lines as shown in Fig. 5.6. The top coil sides are shown by solid (or continuous) lines and bottom coil sides are shown by broken (or discontinuous) lines. In the winding diagram, the top and bottom coil sides are shown alternatively because each slot has one top coil side and one bottom coil side.

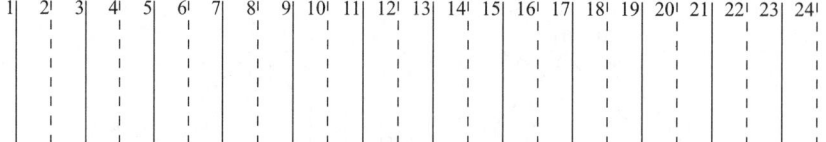

Fig. 5.6: Representation of coil sides in winding diagram.

2. Number the coil sides such that the top coil sides are represented by odd numbers and bottom coil sides by even numbers as shown in Fig. 5.6.

3. Determine the coil sides per pole, which gives the number of coil sides lying under a pole at any instant of time. The enclosure of coil sides by a pole is represented by a shaded rectangle as shown in Fig. 5.7. Different type of shadings are provided for north and south poles. The north and south poles are placed alternatively.

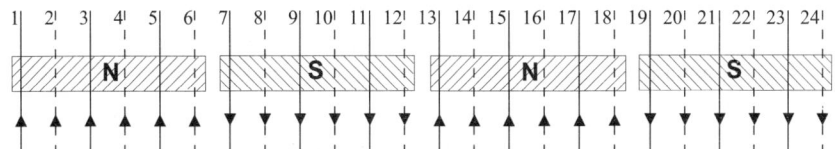

Fig. 5.7: Representation of poles and direction of current in the winding diagram.

4.   The direction of current through conductors under the north pole will be opposite to the direction of current through conductors under the south pole. Mark the direction of current through conductors under north pole as upwards and that of south pole as downwards as shown in Fig. 5.7.

5.   Calculate the back pitch and front pitch using the equations in Table 5.2. The back connection is decided by back pitch and the front connection is decided by front pitch. Connecting the top coil side to bottom coil side of the same coil is called **back connection**. Connecting the bottom coil side of a coil to the start of next coil is called **front connection**.

Bottom coil side of a coil = Top coil side of same coil + Back pitch

Top coil side of next coil = Bottom coil side of previous coil − Front pitch

6.   Determine all the back and front connections. Represent the connections by a winding table. For progressive lap winding the winding table can be prepared as shown below.

Let,   Back pitch, $Y_b = 7$ | Refer Example 5.1 |
          Front pitch, $Y_f = 5$

Winding calculations	**Back connections**	**Front connections**
$1 + 7 = 8$	$1 \leftarrow 8$	$8 \rightarrow 3$
$8 - 5 = 3$	$3 \leftarrow 10$	$10 \rightarrow 5$
$3 + 7 = 10$	$5 \leftarrow 12$	$12 \rightarrow 7$
$10 - 5 = 5$	$7 \leftarrow 14$	$14 \rightarrow 9$
$5 + 7 = 12$	$9 \leftarrow 16$	$16 \rightarrow 11$
$12 - 5 = 7$	$11 \leftarrow 18$	$18 \rightarrow 13$
$7 + 7 = 14$	$13 \leftarrow 20$	$20 \rightarrow 15$
$14 - 5 = 9$	$15 \leftarrow 22$	$22 \rightarrow 17$
$9 + 7 = 16$	$17 \leftarrow 24$	$24 \rightarrow 19$
$16 - 5 = 11$	$19 \leftarrow 2$	$2 \rightarrow 21$
$11 + 7 = 18$	$21 \leftarrow 4$	$4 \rightarrow 23$
$18 - 5 = 13$	$23 \leftarrow 6$	$6 \rightarrow 1$
$13 + 7 = 20$		
$20 - 5 = 15$		
$15 + 7 = 22$	**Winding table**	
$22 - 5 = 17$		
$17 + 7 = 24$		
$24 - 5 = 19$		
$19 + 7 = 26 - 24 = 2$		
$26 - 5 = 21$		
$21 + 7 = 28 - 24 = 4$		
$28 - 5 = 23$		
$23 + 7 = 30 - 24 = 6$		
$6 - 5 = 1$		

Winding table

$1 \leftarrow 8 \rightarrow 3 \leftarrow 10 \rightarrow 5 \leftarrow 12 \rightarrow 7 \leftarrow 14$
$\rightarrow 9 \leftarrow 16 \rightarrow 11 \leftarrow 18 \rightarrow 13 \leftarrow 20 \rightarrow 15 \leftarrow 22$
$\rightarrow 17 \leftarrow 24 \rightarrow 19 \leftarrow 2 \rightarrow 21 \leftarrow 4 \rightarrow 23 \leftarrow 6 \rightarrow 1$

7. After preparing the winding table, draw all the front connections and back connections. One back connection and one front connection are shown in Fig. 5.8. All the back connections are shown in Fig. 5.9 and all the front connections are shown in Fig. 5.10.

8. The meeting points of coil ends formed by the front connections are terminated on the commutator segments. In simplex lap winding with two coil sides per slot, the number of commutator segments is equal to number of coils. The commutator segment connection is represented by placing one segment at the meeting point of one top coil side and one bottom coil side of each front connection as shown in Fig. 5.10.

9. The number of brushes will be equal to number of poles. Half the number of brushes are positive and the remaining half are negative. Brush locations can be shown in the diagram by observing the currents entering and leaving the commutator segments. The current enters or leaves the commutator segment through the conductors connected to them. Two conductors are connected to each commutator segment.

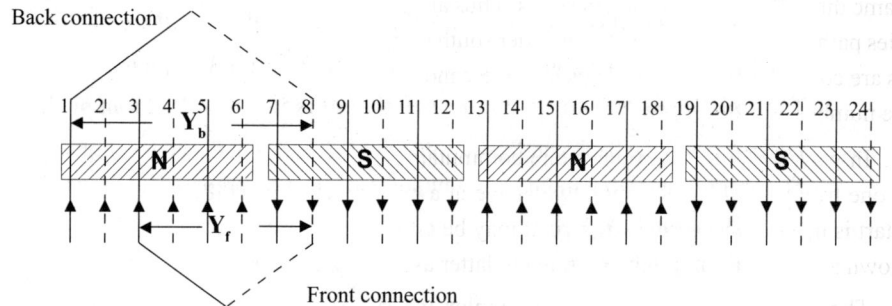

Fig. 5.8: *Back and front connection of simplex lap winding.*

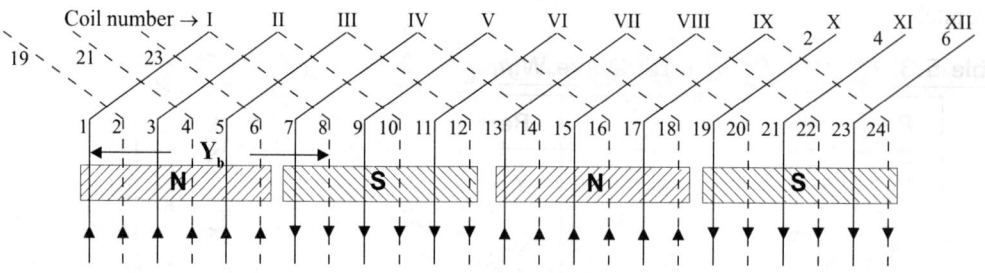

Fig. 5.9: *Back connection of simplex lap winding with two coil sides per slot.*

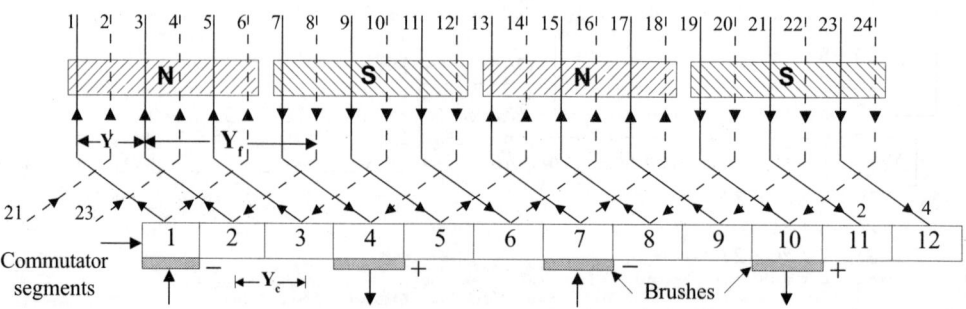

Fig. 5.10: *Front connections, commutator segment connections and location of brushes in simplex lap winding with two coil sides per slot.*

If current enter the segment through one conductor and leaves via another conductor then the brush cannot be located at that segment. If current enters the commutator segment from both the conductors then a positive brush can be placed at that location for a generator. If current leaves the commutator segment through both the conductors then a negative brush can be placed at that location for a generator. The location of brushes are shown in Fig. 5.10.

> *Note : The direction of current through armature conductors for a motoring operation is opposite to that of generator, provided the direction of field (or flux) and speed remaining same as that of generator operation.*

5.2.4 Simplex Wave Winding for DC Machine

In simplex wave winding the finish of a coil is connected to start of a coil which is lying one pole pitch away from it. In the simplex wave winding, the arrangement is such that all coils carrying current in the same direction are connected in series. Thus all the conductors lying under north pole are connected in a series path and the conductors lying under south pole are connected in another series path. The two series paths are connected in parallel. Thus in wave winding there are only two parallel paths, quite independent of the number of poles.

In wave winding, starting at one commutator segment and tracing the winding from coil to coil, after one trip around the armature, we arrive at a commutator segment next to the segment from which the start is made. The segment of arrival may be either ahead or behind the starting segment. The former is known as progressive winding while the latter as retrogressive.

The various winding pitches for simplex wave winding are listed in Table 5.3. In simplex wave winding the back pitch and front pitch are both odd integers. Since the winding pitch is sum of front and back pitch, it is always an even integer.

Table 5.3: Winding Pitches for Wave Winding

Progressive wave	Retrogressive wave
$Y_b = \dfrac{2C}{p} + K$	$Y_b = \dfrac{2C}{p} - K$
$Y = \dfrac{2C + 2}{p/2}$	$Y = \dfrac{2C - 2}{p/2}$
$Y = Y_b + Y_f$ and $Y_f = Y - Y_b$	$Y = Y_b + Y_f$ and $Y_f = Y - Y_b$
$Y_c = \dfrac{C + 1}{p/2} = \dfrac{Y}{2}$	$Y_c = \dfrac{C - 1}{p/2} = \dfrac{Y}{2}$

> *Note : K is a number to make Y_b an odd integer nearly equal to 2C/p, i.e., coil sides per pole.*

Guide lines for drawing simplex wave winding diagram

The following guidelines will be useful for drawing simplex wave winding with two coil sides per slot. The winding diagram of Example 5.4 have been used to explain the various steps in the guide lines.

Steps 1, 2, 3 and 4 are same as the guide lines for drawing simplex lap winding diagram.

5.   Calculate the back pitch and front pitch using the equations in Table 5.4. The back connection is decided by back pitch and the front connection is decided by front pitch. Connecting the top coil side to bottom coil side of the same coil is called back connection. Connecting the bottom coil side of a coil to the start of a coil lying approximately one pole pitch away is called front connection.

Bottom coil side of a coil = Top coil side of same coil + Back pitch

$$\text{Top coil side} = \text{Bottom coil side of a coil lying one pole pitch behind the top coil side} + \text{Front pitch}$$

6.   Determine all the back and front connections. Represent the connections by a winding table. For progressive wave winding the winding table can be prepared as shown below.

Let,   Back pitch, $Y_b = 7$

Front pitch, $Y_f = 7$

> Refer Example 5.4

Winding calculations

$1 + 7 = 8$	$14 + 7 = 21$
$8 + 7 = 15$	$21 + 7 = 28 - 26 = 2$
$15 + 7 = 22$	$2 + 7 = 9$
$22 + 7 = 29 - 26 = 3$	$9 + 7 = 16$
$3 + 7 = 10$	$16 + 7 = 23$
$10 + 7 = 17$	$23 + 7 = 30 - 26 = 4$
$17 + 7 = 24$	$4 + 7 = 11$
$24 + 7 = 31 - 26 = 5$	$11 + 7 = 18$
$5 + 7 = 12$	$18 + 7 = 25$
$12 + 7 = 19$	$25 + 7 = 32 - 26 = 6$
$19 + 7 = 26$	$6 + 7 = 13$
$26 + 7 = 33 - 26 = 7$	$13 + 7 = 20$
$7 + 7 = 14$	$20 + 7 = 27 - 26 = 1$

Back connections	Front connections
1 ← 8	8 → 15
15 ← 22	22 → 3
3 ← 10	10 → 17
17 ← 24	24 → 5
5 ← 12	12 → 19
19 ← 26	26 → 7
7 ← 14	14 → 21
21 ← 2	2 → 9
9 ← 16	16 → 23
23 ← 4	4 → 11
11 ← 18	18 → 25
25 ← 6	6 → 13
13 ← 20	20 → 1

Winding table

```
  1  ←  8  →  15  ←  22  →  3   ←  10  →  17  ←  24
→  5  ←  12 →  19  ←  26  →  7   ←  14  →  21  ←  2
→  9  ←  16 →  23  ←  4   →  11  ←  18  →  25  ←  6
→  13 ←  20 →  1
```

7.   After preparing the winding table, draw all the front connections and back connections. One back connection and one front connection is shown in Fig. 5.11. All the back connections are shown in Fig. 5.12 and all the front connections are shown in Fig. 5.13.

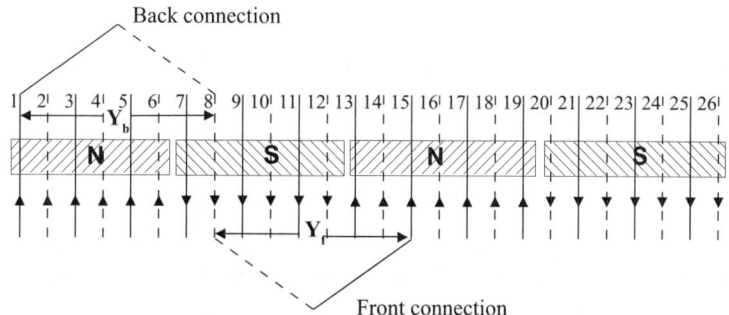

Fig. 5.11: *Back and front connection of simplex wave winding.*

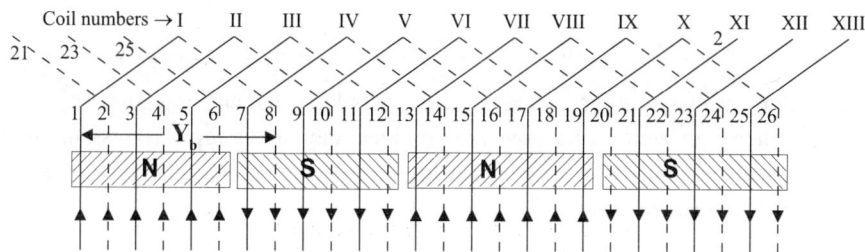

Fig. 5.12: *Back connections of simplex wave winding with two coil sides per slot.*

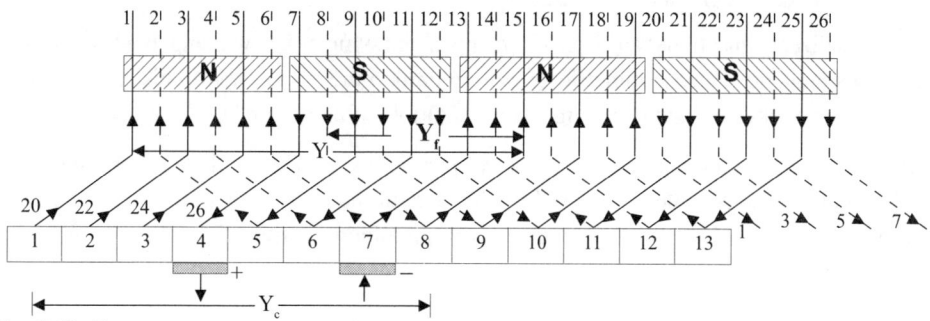

Fig. 5.13: *Front connections, commutator segment connections and location of brushes in simplex wave winding with two coil sides per slot.*

Steps 8 and 9 are same as the guide lines for drawing simplex lap winding diagram. But in wave winding the total number of brushes is always two with one as positive and the other as negative.

Dummy coils

In DC machine armature windings the back pitch and the commutator pitch should be an integer. Sometimes with certain number of coils it may not be possible to get integer value for Y_b or Y_c. In this case, some of the coils have to be left free, i.e., the coils will be placed in the slot for mechanical balance but not connected electrically to the armature winding. Such coils are called *dummy coils*. When dummy coils are provided the winding is also called *forced winding*.

Equalizer connections

The number of parallel paths in simplex lap winding is equal to number of poles. If the induced emf in each parallel path is not equal then there will be circulating currents flowing through brushes in the armature winding. The circulating current act as additional loads on the winding and brushes, thus causing unequal heating in the armature winding and commutation difficulties.

The difference in induced emf may be due to the following reasons.

1. The defects in castings of iron parts may give rise to unequal reluctance in the parallel magnetic circuits.

   | Note : Casting is the process of making metal parts of desired shape. |

2. The asymmetrical arrangement of brushes on the commutator.

3. The defective machining or assembling may give rise to unequal air-gap length under various pole.

4. The error in field winding turns may give rise to difference in ampere-turns of various poles.

The difference in the emf generated in various parallel paths can be equalized by providing equalizer connections. When equalizer, connections are provided, the circulating currents through brushes can be avoided. The equalizer connections are made using copper conductors usually in the form of rings and so they are also called *equalizer rings*.

In armature winding the points at same potential under ideal conditions are connected to one equalizer ring. Hence the conductors which are two pole pitches apart are connected to same equalizer ring. The net emf induced in an equalizer connection is zero.

The equalizer connections can be provided only in the symmetrical winding. For symmetrical winding the conductors must be placed symmetrically with regard to the field system. This is achieved only if the number of slots and the commutator segments are multiple of pair of poles.

In simplex wave winding there are only two parallel paths. The conductors forming a parallel path will be distributed equally under all the poles. Hence both the parallel paths are equally affected by the asymmetry in the magnetic circuit and so there is no circulating current. Therefore there is no necessity for equalizer connections.

5.3 AC MACHINE ARMATURE WINDING

The armature winding of synchronous machine and stator winding of three-phase induction motor are discussed in this section. These windings are wound for three-phase and designed to satisfy the following characteristics of three-phase balanced system.

1. The magnitude of emfs of all the phases should be equal.

2. The emfs of the phases should have indentical waveform and frequency.

3. The phase displacement between any two phases should to 120 electrical degrees.

The above requirements of three-phase winding can be satisfied by forming identical phase groups under each pole pitch (or pole area) and then arranging the phase groups of each phase to have an effective displacement of 120° electrical in space with respect to any other phase group. The phase group is defined as the conductors (or slots) of a phase under a pole pitch (or one pole area).

The pole pitch in terms of mechanical and electrical degrees are,

$$\text{Pole pitch in mechanical degree} = \frac{360°}{p}$$

$$\text{Pole pitch in electrical degree} = \frac{p}{2} \times \frac{360°}{p} = 180°$$

$$\boxed{1°m = \frac{p}{2}\, °e}$$

In terms of slots of armature, the pole pitch is given by slots per pole.

$$\therefore \text{Pole pitch} = \frac{\text{Slots}}{\text{Pole}}$$

Let, S = Number of slots in armature

p = Number of poles

n = Slots per pole (or pole pitch)

q = Slots per pole per phase

Now, for a 3-phase winding,

Number of slots under a pole pitch = 3 q

Total number of slots, S = p n

$$\therefore \text{Slots per pole, } n = \frac{S}{p}$$
(or pole pitch)

Slots per pole per phase, $q = \frac{n}{3} = \frac{S}{3p}$

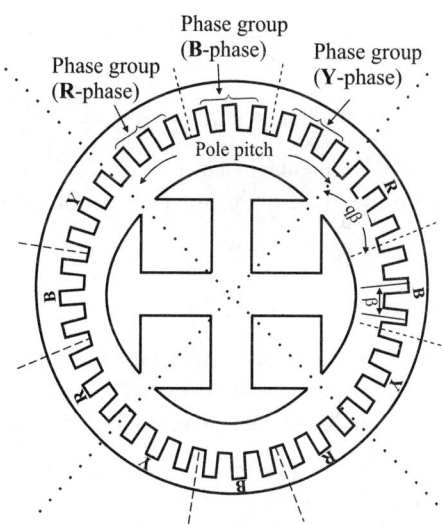

Phase group (B-phase)
Phase group (R-phase)   Phase group (Y-phase)

Pole pitch

Fig. 5.14: Allotment of slots for R, Y, B phases in a 3-phase machine.

When n and q are integers integral slot winding is possible. For integral slot winding coil span is made equal to 180 °e which is same as slots per pole. For fractional slot winding the coil span is made lesser or higher than 180 °e and so coil span is either lesser or higher than slots per pole.

$\therefore$ Coil span, $C_s = 180° = \dfrac{S}{p}$ -   For full pitch winding

$\phantom{\therefore \text{Coil span, } C_s = } < 180° \left(\text{or} < \dfrac{S}{p} \right)$ -   For short pitch winding

$\phantom{\therefore \text{Coil span, } C_s = } > 180° \left(\text{or} > \dfrac{S}{p} \right)$ -   For over pitch winding.

The allotment of slots for a 36 slot, 4 pole, 3-phase machine is shown in Fig. 5.14.

Slots, $S = 36$

Slots per pole, $n = \dfrac{S}{p} = \dfrac{36}{4} = 9$

Slots per pole per phase, $q = \dfrac{n}{3} = \dfrac{9}{3} = 3$

The angular distance covered by one slot is called **slot angle** and it is denoted by β. It is given by the angular distance between the centre of two adjacent slots as shown in Fig. 5.14.

The angular distance of pole pitch is 180° electrical and we have n slots under one pole pitch. Therefore the slot angle can be obtained by dividing 180° by n.

$\therefore$ Slot angle, $\beta = \dfrac{\text{Pole pitch}}{\text{Slots / Pole}} = \dfrac{180°}{n} = \dfrac{180°}{9} = 20°$ electrical $\qquad \boxed{1°\text{m} = \dfrac{p}{2} °\text{e}}$

The slot angle can also be estimated by dividing 360° mechanical by number of slots and then converting it to electrical degrees.

$\therefore$ Slot angle, $\beta = \dfrac{360°}{S} \times \dfrac{p}{2} = \dfrac{360}{36} \times \dfrac{4}{2} = 20°$ electrical

The angular distance of the slots of one phase group is called **phase spread**. It is given by the product of slots per pole per phase and the slot angle.

$\therefore$ Phase spread $= q\beta = 3 \times 20° = 60°$ electrical

5.3.1 Chording, Distribution and Winding Factors

In a three-phase winding, the conductors of a phase are connected in series to form the winding of a phase with two terminals. And so we have three electrically separate windings with six terminals. These six terminals can be connected in star (or delta) to form a star (or delta) connected three-phase winding.

Since the conductors of a phase are connected in series, the emf induced in a phase is given by vector sum of emf induced in all the conductors of a phase. The vector sum of emf induced in two coil sides of a coil will be less than the arithmatic sum if the coil is short pitched or over pitched. Hence to account for this reduction, the arithmatic sum is multiplied by a factor called **chording factor** (or pitch factor), and it is denoted on k_c.

Let, $\propto$ = Angle of chording

Chording factor, $k_c = \cos \dfrac{\propto}{2}$

The chording angle is the difference between pole pitch (180°) and the coil span. For full pitched coils the coil span is equal to pole pitch (180°).

$\therefore \propto = 180° -$ Coil span

Sometimes the chording angle is specified in terms of slot angle. For example, if the coil is short chorded by one slot then chording angle, α is equal to slot angle, β, if the coil is short chorded by two slot then chording angle, α is equal to twice the slot angle (2β) and so on.

In AC machines, the short chording is preferred to eliminate certain harmonics in induced emf. The short chording angle, α needed to eliminate n^{th} harmonic is given by, $\alpha = \dfrac{180°}{n}$

The vector sum of induced emf in the conductors of a phase group will be less than the arithmatic sum due to phase difference between induced emf in the adjacent conductors of a phase group. The phase difference is due to distribution of conductors of a phase group in various slots. To account for the reduction of induced emf due to distribution of conductors, the arithmatic sum is multiplied by a factor called ***distribution factor*** and it is denoted as k_d.

$$\text{Distribution factor, } k_d = \frac{\sin \dfrac{q\beta}{2}}{q \sin \dfrac{\beta}{2}}$$

where, $q\beta$ = Phase spread

　　　　　q = Slots per pole per phase

　　　　　β = Slot angle

The product of chording and distribution factor is called ***winding factor*** and it is denoted by k_w or k_{ws}.

∴ Winding factor, $k_w = k_c \times k_d$

5.3.2 Types of AC Machine Winding

The AC machine windings can be broadly classified into single layer and double layer windings. In single layer winding, one coil side will be placed in each slot, whereas in double layer winding, two coil sides will be placed in each slot. The various types of single and double layer windings are given below.

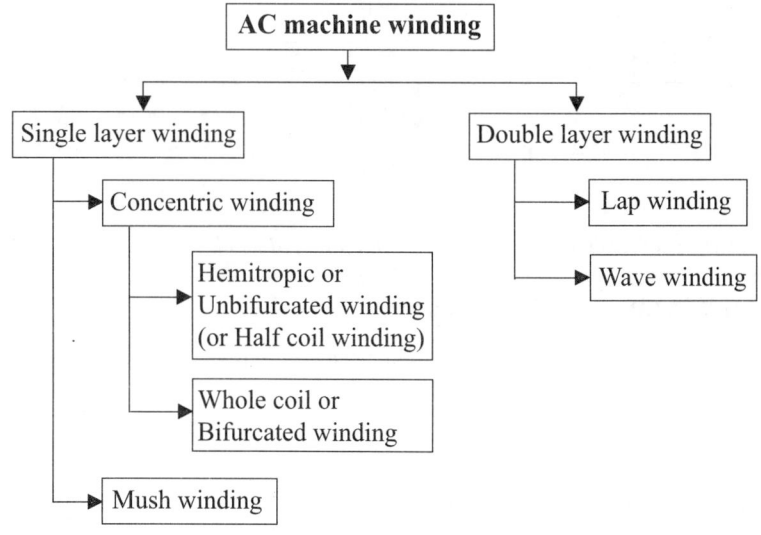

5.3.3 Unbifurcated Winding for AC Machine

In unbifurcated windings, the coil sides of a pair of pole phase groups under adjacent poles are connected to form concentric coils. Here the concentric coils of a coil group will have variable span. The innermost coil will have the smallest span and the outermost coil will have largest span. However the average span of the coils of one coil group will be equal to pole pitch. These coils under two adjacent poles form one coil group and thus there is one coil group per pair of poles.

$$\text{Number of coil groups} = \text{Pole pairs} \times \text{Number of phases} = \frac{3p}{2}.$$

A single layer concentric winding will have to be accomdated in more than one plane. A three-phase unbifurcated winding having 3 plane overhang is shown in Example 5.6.

The salient features of unbifurcated winding are given below:

i)   Coil sides of similar phase under two adjacent poles are connected to form one coil group.

ii) Number of coil groups is given by p/2, where p is number of poles.

iii) The coils will have different span.

iv) The number of coil spans and planes of overhang will be more than that of bifurcated winding.

v)   The number of coil sides will be equal to number of slots.

The connection of conductors of one phase (R-phase) of unbifurcated winding for a 36 slot, 3-phase winding is shown in Fig. 5.15. The detailed winding diagram is presented in Example 5.6.

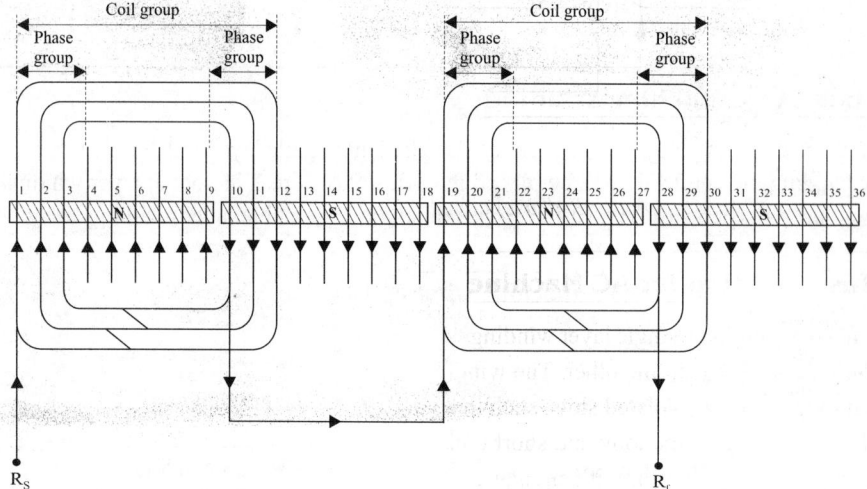

Fig. 5.15: One phase (R-phase) connection of unbifurcated winding.

5.3.4 Bifurcated Winding for AC Machine

In bifurcated windings, the coil sides of each pole phase group is split into two sections. Coil sides of one section is connected to coil sides of previous pole phase group to form concentric coils. Coil sides of another section is connected to coil sides of next pole phase group to form concentric coils.

Therefore, we have one coil group per pole per phase and total number of coil groups is three times the number of poles. The end connections in a 3-phase whole coil winding can be arranged to lie in either 2 or 3 planes. A three-phase bifurcated winding having 3 plane overhang is shown in Example 5.7.

The salient features of bifurcated winding are given below.

i) Section of coil sides of similar phase under two adjacent poles are connected to form coil groups.

ii) Number of coil groups is equal to number of poles.

iii) The coils will have different spans.

iv) The number of coil spans and planes of overhang will be less than that of unbifurcated winding.

v) The number of coil sides will be equal to number of slots.

The connection of conductors of one phase(R-phase) of bifurcated winding for a 48 slot, 4 pole, 3-phase winding is shown in Fig. 5.16. The detailed winding diagram is presented in Example 5.7.

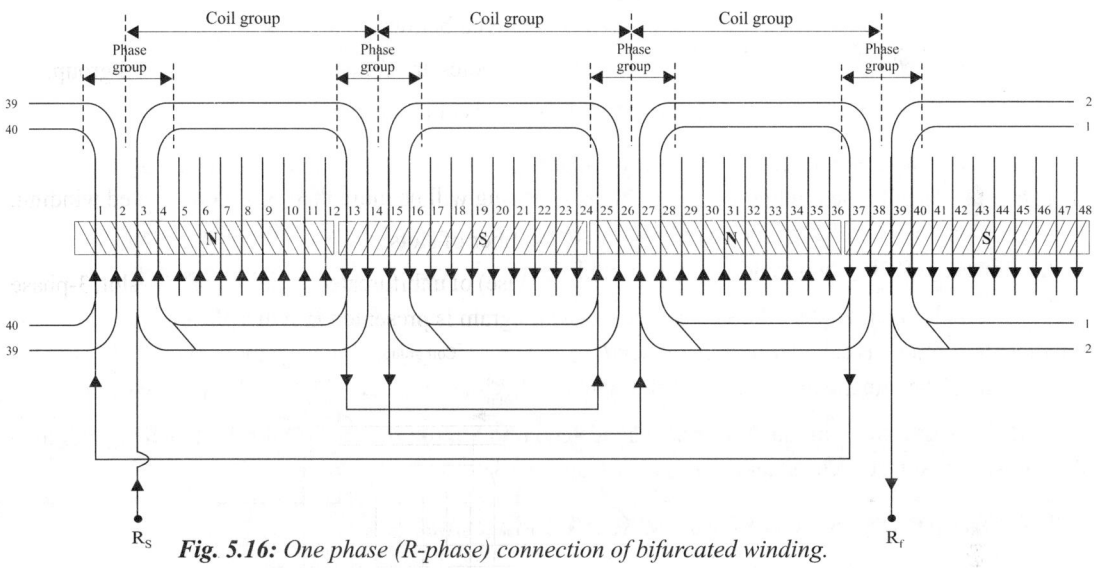

Fig. 5.16: *One phase (R-phase) connection of bifurcated winding.*

5.3.5 Mush Winding for AC Machine

The mush winding is a single layer winding with coils of same span. Each coil is wound on a former, with one coil side shorter than the other. The winding is placed on the core by dropping the conductors, one by one into previously insulated slots, such that the short coil sides are placed first and then the long coil sides. In mush winding the long and short coil sides occupy alternate slots. This type of winding is suitable for small induction motor with circular conductors.

The salient features of mush winding are given below.

i) The coils have a constant span.

ii) There is only one coil side per slot and therefore the number of coil sides is equal to number of slots

iii) Mush winding has one coil group per phase per pole pair and therefore, the maximum number of parallel paths per phase is equal to pole pairs.

The connection of conductors of one phase(R-phase) of mush winding is shown in Fig. 5.17. The detailed winding is presented in Example 5.8.

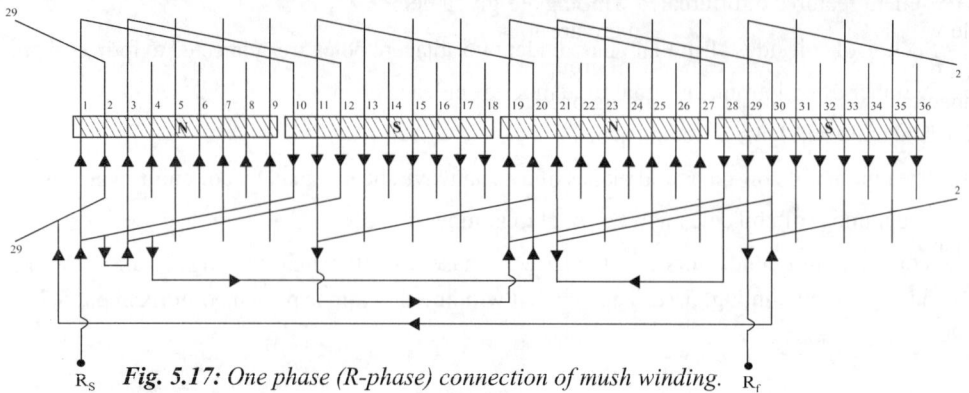

Fig. 5.17: *One phase (R-phase) connection of mush winding.* R_S R_f

5.3.6 Double Layer AC Machine Winding

In double layer winding, two coil sides are placed in each slot. The double layer winding has diamond shaped coils. Usually the double layer winding have short pitched coils. The short chording of coils will reduce the length of overhang (and so reduce the copper requirement) and also eliminate certain harmonics in the induced emf. The two major types of double layer winding are lap and wave winding.

In double layer winding the slots per pole can be an integer or a fraction. when the slots per pole is an interger, the winding is called *integral slot winding* and when it is a fraction the winding is called fractional slot winding. The integral and fractional slot windings are possible in both lap and wave connections.

For integral slot winding coil span is made equal to S/p (or 180 °e). But for fractional slot winding the coil span is selected either lesser or higher than S/p (or 180 °e).

5.3.7 Integral Slot Lap Winding for AC Machine

In lap winding connection the top conductors of a phase group under one pole are connected to bottom conductors of the same phase group under next pole to form a coil group. While connecting the coils of a coil group, the finish of a coil is connected to start of next coil. The coil groups of a phase are connected in series to form the winding of a phase. In lap connection the number of coil groups will be equal to the number of poles.

The concepts discussed for DC machine armature lap winding are applicable for AC machine armature lap winding except the commutator connections.

Therefore, the equations for front pitch, back pitch and winding pitch of AC machine armature lap winding are same as that of DC machine armature lap winding.

Back pitch, $Y_b = \dfrac{2C}{p} + K$ -   For integral slot winding

$\qquad\qquad\qquad = $ Coil span $+ K$ -   For fractional slot winding

Winding pitch, $Y = Y_b - Y_f$ and $Y = 2$

Front pitch, $Y_f = Y_b - Y$

where K is a number to make Y_b an odd integer.

| In integral slot winding, |
| Coil span $= \dfrac{S}{p} = \dfrac{2C}{p}$ and $K = 1$ |

Note : *For fractional slot winding choose coil span as per given specifications.*

The connection of conductors of one phase (R-phase) of lap winding is shown in Fig. 5.18. In the example winding shown in Fig. 5.18, there are four coil groups. The start of coil groups are denoted as R_{s1}, R_{s2}, R_{s3} and R_{s4} and the finish of coil group are denoted as R_{f1}, R_{f2}, R_{f3} and R_{f4}. The four coil groups are connected in series to form the winding of R-phase.

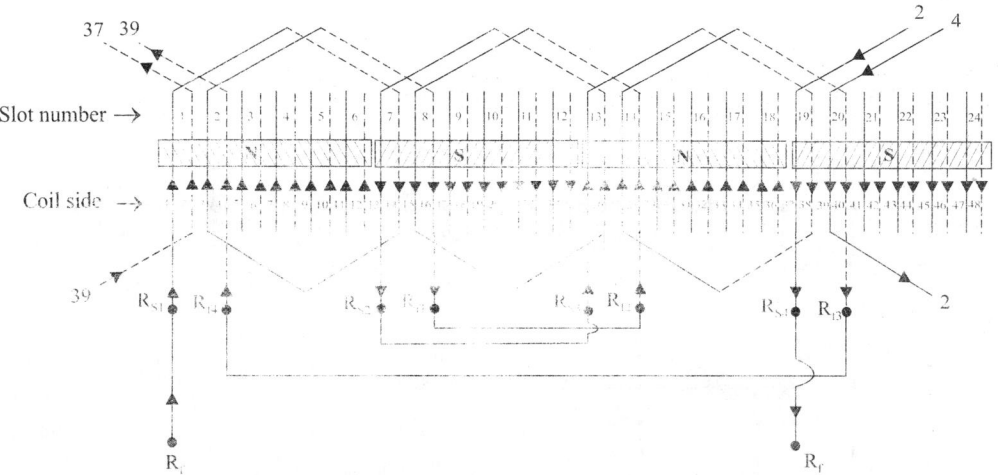

Fig. 5.18: One phase (R-phase) connection of lap winding.

5.3.8 Integral Slot Wave Winding for AC Machine

In wave winding connections the finish of a coil is connected to start of a coil which is lying one pole pitch apart, to form a coil group. In simplex wave connection it is possible to form only 2 coil groups per phase. The two coil groups of a phase are connected in series to form the winding of a phase. In order to connect the two coil groups in series or in the natural direction of current, the direction of winding connection after connecting coils of first coil group should be reversed for next coil group. (Refer winding calculations of Examples 5.8 and 5.9).

The concepts discussed for DC machine armature wave winding are applicable for AC machine armature wave winding except the commutator connections and winding pitch.

Therefore, the equations for front pitch and back pitch of AC machine armature wave winding are same as that of DC machine armature wave winding.

Back pitch, $Y_b = \dfrac{2C}{p} + K$ - For integral slot winding

                $=$ Coil span $+ K$ - For fractional slot winding

Winding pitch, $Y = \dfrac{4C}{p}$

Front pitch, $Y_f = Y - Y_b$

> In integral slot winding,
>
> Coil span $= \dfrac{S}{p} = \dfrac{2C}{p}$ and $K = 1$

where K is a number to make Y_b an odd integer.

> *Note : For fractional slot winding choose coil span as per given specifications.*

The connection of conductors of one phase (R-phase) of wave winding is shown in Fig. 5.19. In the example winding shown in Fig. 5.19, there are two coil groups. The start of coil groups are denoted as R_{s1} and R_{s2} and the finish of coil groups are denoted as R_{f1} and R_{f2}. The natural connection of wave winding provides automatic series connection of coil groups and so extra series connections of coil groups are not required in wave winding.

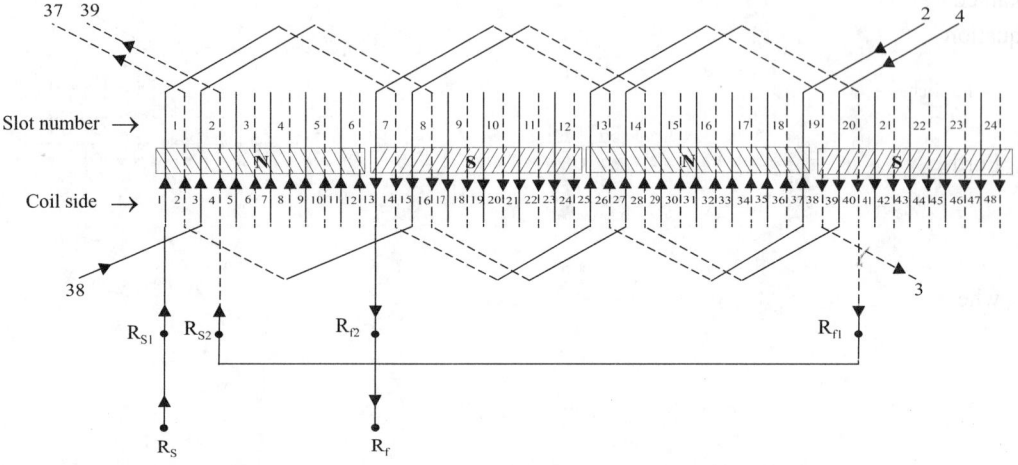

Fig. 5.19: *One phase (R-phase) connection of wave winding*

5.3.9   Fractional Slot Winding

In integral slot winding the slots per pole per phase, q will be an integer. But, in the fractional slot winding the slots per pole per phase, q will not be an integer.

Advantages of fractional slot winding

1.   When the machine has large number of poles the fractional slot winding will give rise to an equivalent infinite distribution winding which results in reducing harmonic mmfs and emfs.

2.   In fractional slot winding only short chorded winding is possible which results in reduced length of overhang and saving in copper and cost.

3.   Due to reduced length of overhang than integral slot winding the leakage reactance in fractional slot winding will be lesser than integral slot winding.

4.   Since the number of slots per pole need not be an integer, the standard armature core laminations with any number of slots can be used to construct the armature of the machine, which save the cost of separate design and manufacturing facility for laminations.

Procedure for fractional slot winding

The poles can be divided into units so that poles per unit is integer.

Let, p = Number of poles

S = Number of slots

m = Number of phases

Now, Slots per pole per phase, $q = \dfrac{S}{m\,p}$

In fractional slot winding the value of q will not be an integer.

Let us express q as ratio of two integers M and d which does not have a common divisor. This can be obtained by dividing the numerator and denominator of above equation of q by a common divisor. Now, the equation of q can be expressed as shown below.

$$\text{Slots per pole per phase, } q = \frac{S}{m\,p} = \frac{S\,/\,Divisor}{m\,p\,/\,Divisor} = \frac{M}{d}$$

Further, M/d can be expressed as a sum of integer and fraction as shown below.

$$\text{Slots per pole per phase, } q = \frac{S}{m\,p} = \frac{M}{d} = I\,\frac{n}{d} = I + \frac{n}{d}$$

where,   d   =   Number of poles in a unit

   M   =   q d   =   Number of slots per phase in each unit

   mM   =   Number of slots in each unit

   p/d   =   Number of pole units

Now, each phase in a pole unit will have d number of coil groups and in this d − n number of coil groups will have I coils each and n number of groups will have I + 1 coils each.

The distribution of slots between various phases in a pole unit can be obtained by calculating the distance between coil sides of a coil using the following equation.

Let, D = Distance between coil sides of a coil in terms of slots.

Now, $D = \dfrac{1 + mMP}{d}$

Where P is a smallest possible value of integer that makes D as an integer.

Now, the slot distribution is given by,

$$1,\ 1 + D,\ 1 + 2D,\ 1 + 3D,\ 1 + 4D,\ \ldots\ldots\ldots$$

In the above calculation, if the sum exceeds the total number of slots in a unit, mM then subtract sum by mM or integral multiple of mM. Therefore, the above equation can be modified as shown below.

$$1,\ 1 + D,\ 1 + 2D,\ 1 + 3D,\ \ldots\ldots ,\ 1 + xD - mM,\ 1 + (x + 1)D - mM\ \ldots\ldots ,\ 1 + yD$$
$$- 2mM,\ 1 + (y + 1)D - 2mM\ \ldots\ldots,$$

$$1 + zD - 3mM,\ 1 + (z + 1)D - 3mM\ \ldots\ldots,\ \text{etc.}$$

The detailed slot allotment are shown in Examples 5.13 to 5.16. The slot distribution for fractional slot lap and wave winding are same and differ only in coil connections.

5.3.10 Fractional Slot Lap Winding for AC Machine

The fractional slot lap winding connections are similar to that integral slot lap winding connections. Therefore, in lap winding connection the top conductors of a phase group under one pole are connected to bottom conductors of the same phase group under next pole to form a coil group. While connecting the coils of a coil group, the finish of a coil is connected to start of next coil. The coil groups of a phase are connected in series to form the winding of a phase. In lap connection the number of coil groups will be equal to the number of poles.

The equations for front pitch, back pitch and winding pitch of fractional slot AC machine armature lap winding are same as that of integral slot AC machine armature lap winding.

$$\text{Back pitch,} \quad Y_b = \frac{2C}{p} + K \qquad - \quad \text{For integral slot winding}$$

$$= \text{Coil span} + K \qquad - \quad \text{For fractional slot winding}$$

$$\text{Winding pitch,} \quad Y = Y_b - Y_f \text{ and } Y = 2$$

$$\text{Front pitch,} \quad Y_f = Y_b - Y$$

In integral slot winding, the coil span is integer and given by $2C/p$ and K is an integer to make Y_b as an odd integer. In fractional slot winding, the coil span may not be an integer and so K is any number to make Y_b as an odd integer.

5.3.11 Fractional Slot Wave Winding for AC Machine

The fractional slot wave winding connections are similar to that of integral slot wave winding connections. Therefore, in wave winding connections the finish of a coil is connected to start of a coil which is lying one pole pitch apart, to form a coil group. In simplex wave connection it is possible to form only 2 coil groups per phase. The two coil groups of a phase are connected in series to form the winding of a phase. In order to connect the two coil groups in series or in the natural direction of current, the direction of winding connection after connecting coils of first coil group should be reversed for next coil group.

The equations for front pitch and back pitch of fractional slot AC machine armature wave winding are same as that of integral slot AC machine armature wave winding.

$$\text{Back pitch,} \quad Y_b = \frac{2C}{p} + K \qquad - \quad \text{For integral slot winding}$$

$$= \text{Coil span} + K \quad - \quad \text{For fractional slot winding}$$

$$\text{Winding pitch,} \quad Y = \frac{4C}{p}$$

$$\text{Front pitch,} \quad Y_f = Y - Y_b$$

In integral slot winding, the coil span is integer and given by $2C/p$ and K is an integer to make Y_b as an odd integer. In fractional slot winding, the coil span may not be an integer and so K is any number to make Y_b as an odd integer.

5.4 SUMMARY OF DESIGN EQUATIONS

1.   Design equations for DC machine armature winding

    i)   Design equations for lap winding (Simplex progressive lap winding)

$$Back\ pitch,\qquad Y_b = \frac{2C}{p} + K$$

$$Winding\ pitch,\qquad Y = 2$$

$$Front\ pitch,\qquad Y_f = Y_b - Y$$

$$Commutator\ pitch,\ Y_c = \frac{Y}{2} = 1\ segment$$

    ii)   Design equations for wave winding (Simplex progressive wave winding)

$$Back\ pitch,\qquad Y_b = \frac{2C}{p} + K$$

$$Winding\ pitch,\qquad Y = \frac{2C + 2}{p/2}$$

$$Front\ pitch,\qquad Y_f = Y - Y_b$$

$$Commutator\ pitch,\ Y_c = \frac{C+1}{p/2} = \frac{Y}{2}$$

2.   Design equations for AC machine armature winding

    i)   Design equations for lap winding (Simplex progressive lap winding)

$$Back\ pitch,\quad Y_b = \frac{2C}{p} + K \qquad -\quad For\ integral\ slot\ winding$$

$$= Coil\ span + K \quad -\quad For\ fractional\ slot\ winding$$

$$Winding\ pitch,\ Y = 2$$

$$Front\ pitch,\quad Y_f = Y_b - Y$$

    ii)   Design equations for wave winding (Simplex progressive wave winding)

$$Back\ pitch,\quad Y_b = \frac{2C}{p} + K \qquad -\quad For\ integral\ slot\ winding$$

$$= Coil\ span + K \quad -\quad For\ fractional\ slot\ winding$$

$$Winding\ pitch,\ Y = \frac{4C}{p}$$

$$Front\ pitch,\quad Y_f = Y - Y_b$$

Note : K *is a number to make* Y_b *an odd integer nearly equal to 2C/p.*

5.5 SOLVED PROBLEMS

EXAMPLE 5.1

Draw the winding diagram in the developed form for a 4 pole, 12 slots, simplex lap connected DC generator with commutator having 12 segments. Indicate the position of brushes.

SOLUTION

In DC machine armature winding the number of coils will be equal to number of commutator segments.

Given that number of commutator segments = 12.

$\therefore$ Number of coils, C = 12

Number of coil sides = 2C = 2 × 12 = 24

$$\text{Coil sides per slot} = \frac{\text{Number of coil sides}}{\text{Number of slots}} = \frac{24}{12} = 2$$

The coil sides are represented by parallel straight lines as shown in Fig. 1. Since two coil sides are accommodated in one slot, each slot will have one top coil side and one bottom coil side. The top coil sides are shown by solid (or continuous) lines and bottom coil sides are shown by broken (or discontinuous) lines. In Fig. 1, the coil sides are numbered such that the top coil sides are represented by odd numbers and bottom coil sides by even numbers.

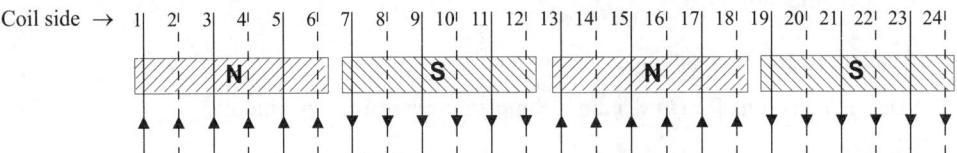

Fig. 1: *Representation of coil sides, poles and direction of current.*

$$\text{Coil sides per pole} = \frac{\text{Number of coil sides}}{\text{Number of poles}} = \frac{24}{4} = 6$$

In this armature winding, six coil sides are enclosed by a pole at any one time instant. The enclosure of coil sides by the poles are represented by shaded rectangles as shown in Fig. 1. The direction of current through conductors lying under north pole are marked upwards and that of south pole are marked downwards.

The winding pitches are calculated as shown below.

Let us choose progressive lap winding.

$$\text{Back pitch, } Y_b = \frac{2C}{p} + K = \frac{2 \times 12}{4} + K = 6 + K$$

Let, K = 1,

$\therefore$ Back pitch, $Y_b = 6 + 1 = 7$ coil sides

Winding pitch, $Y = 2$ (For progressive lap, Y is always +2)

Front pitch, $Y_f = Y_b - Y = 7 - 2 = 5$ coil sides

Commutator pitch, $Y_c = 1$ segment (For progressive lap, Y_c is always +1)

The front and back connections of the coil sides are determined as shown below.

Let us start with top coil side marked 1.

Bottom coil side of the first coil = Top coil side of first coil + Back pitch
(Back connection)

$$= 1 + Y_b = 1 + 7 = 8$$

Top coil side of the second coil = Bottom coil side of first coil − Front pitch
(Front connection)

$$= 8 - Y_f = 8 - 5 = 3$$

Similarly, all the back and front connections are determined. The entire winding calculations, details of back and front connections and the winding table are shown ahead.

Winding calculations	Back connections	Front connections
$1 + 7 = 8$	$1 \leftarrow 8$	$8 \rightarrow 3$
$8 - 5 = 3$	$3 \leftarrow 10$	$10 \rightarrow 5$
$3 + 7 = 10$	$5 \leftarrow 12$	$12 \rightarrow 7$
$10 - 5 = 5$	$7 \leftarrow 14$	$14 \rightarrow 9$
$5 + 7 = 12$	$9 \leftarrow 16$	$16 \rightarrow 11$
$12 - 5 = 7$	$11 \leftarrow 18$	$18 \rightarrow 13$
$7 + 7 = 14$	$13 \leftarrow 20$	$20 \rightarrow 15$
$14 - 5 = 9$	$15 \leftarrow 22$	$22 \rightarrow 17$
$9 + 7 = 16$	$17 \leftarrow 24$	$24 \rightarrow 19$
$16 - 5 = 11$	$19 \leftarrow 2$	$2 \rightarrow 21$
$11 + 7 = 18$	$21 \leftarrow 4$	$4 \rightarrow 23$
$18 - 5 = 13$	$23 \leftarrow 6$	$6 \rightarrow 1$
$13 + 7 = 20$		
$20 - 5 = 15$		
$15 + 7 = 22$		
$22 - 5 = 17$		
$17 + 7 = 24$		
$24 - 5 = 19$		
$19 + 7 = 26 - 24 = 2$		
$26 - 5 = 21$		
$21 + 7 = 28 - 24 = 4$		
$28 - 5 = 23$		
$23 + 7 = 30 - 24 = 6$		
$6 - 5 = 1$		

Winding table

$1 \leftarrow 8 \rightarrow 3 \leftarrow 10 \rightarrow 5 \leftarrow 12 \rightarrow 7 \leftarrow 14$
$\rightarrow 9 \leftarrow 16 \rightarrow 11 \leftarrow 18 \rightarrow 13 \leftarrow 20 \rightarrow 15 \leftarrow 22$
$\rightarrow 17 \leftarrow 24 \rightarrow 19 \leftarrow 2 \rightarrow 21 \leftarrow 4 \rightarrow 23 \leftarrow 6$
$\rightarrow 1$

One back connection and one front connection are shown in Fig. 2. All the back connections are shown in Fig. 3 and all the front connections are shown in Fig. 4. The meeting points of coil ends formed by front connections are terminated on the commutator segments as shown in Fig. 4.

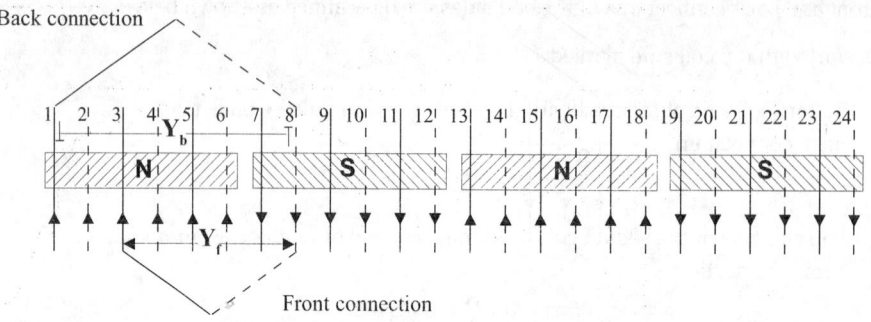

Fig. 2: *The first back and front connection.*

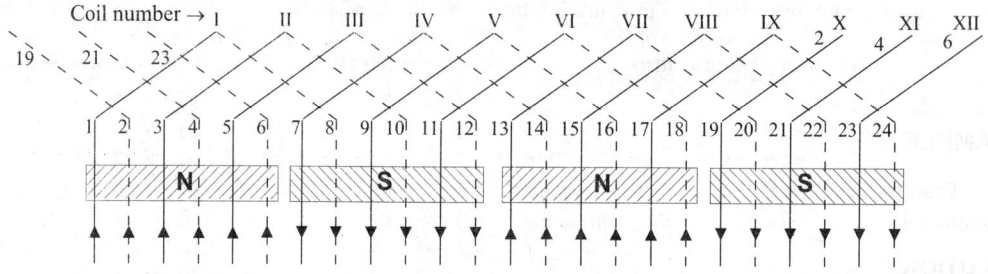

Fig. 3: *Back connections of the winding.*

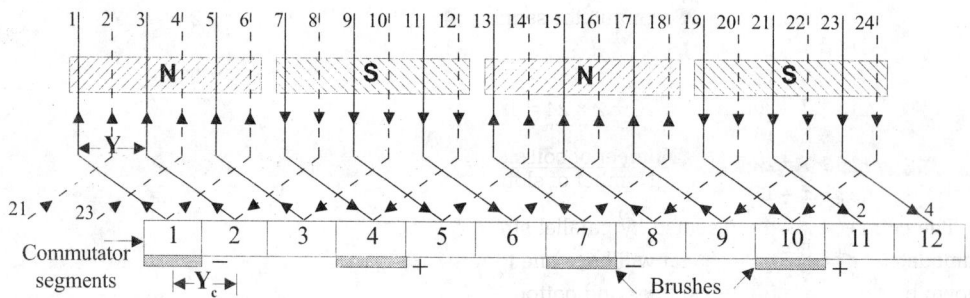

Fig. 4: *Front connections and location of brushes.*

In lap winding the number of brushes will be equal to number of poles. Hence in this winding there will be four brushes with sign of two as positive and two as negative.

On observing the currents entering at the commutator segments we can place positive brushes at segments 4 and 10. Because in segments 4 and 10 the current enter the commutator segment from the two conductors connected to them. In segments 1 and 7 the current leaves the commutator segment through the conductors connected to them. Hence negative brushes can be placed at segments 1 and 7. The complete winding diagram is shown in Fig. 5.

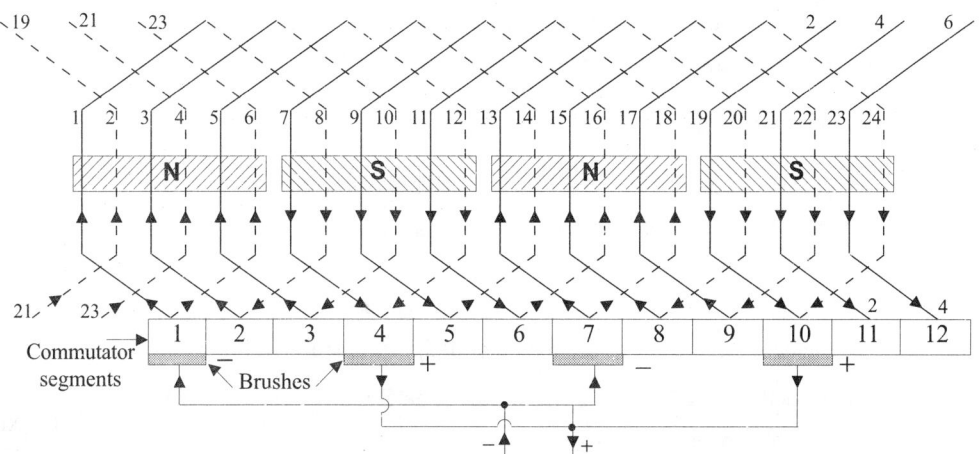

Fig. 5: *Simplex lap winding of a 4 pole DC genertaor armature with 12 slots.*

EXAMPLE 5.2

(JNTKU, Apr' 19, 11 M)

Draw the winding diagram in the developed form for a simplex lap wound 24 slots, 4 pole DC generator armature with 24 commutator segments. Show the position of brushes.

SOLUTION

In DC machine armature winding the number of coils will be equal to number of commutator segments.

Given that number of commutator segments = 24

∴ Number of coils, C = 24

Number of coil sides = 2C = 2 × 24 = 48

Coil sides per slot = $\dfrac{\text{Number of coil sides}}{\text{Number of slots}} = \dfrac{48}{24} = 2$

The coil sides are represented by parallel straight lines as shown in Fig. 1. Since two coil sides are accommodated in one slot, each slot will have one top coil side and one bottom coil side. The top coil sides are shown by solid (or continuous) lines and bottom coil sides are shown by broken (or discontinuous) lines. In Fig. 1, the coil sides are numbered such that the top coil sides are represented by odd numbers and bottom coil sides by even numbers.

Coil sides per pole = $\dfrac{\text{Number of coil sides}}{\text{Number of poles}} = \dfrac{48}{4} = 12$

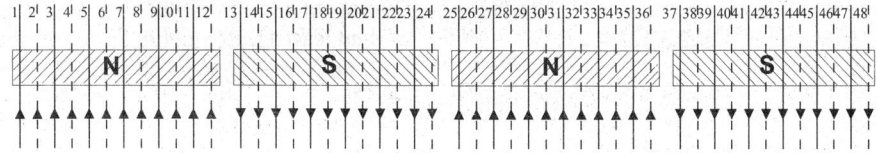

Fig. 1: *Representation of coil sides, poles and direction of current.*

In this armature winding, twelve coil sides are enclosed by a pole at any one time instant. The enclosure of coil sides by the poles are represented by shaded rectangles as shown in Fig. 1. The direction of current through conductors lying under north pole are marked upwards and that of south pole are marked downwards.

The winding pitches are calculated as shown below.

Let us choose progressive lap winding.

$$\text{Back pitch, } Y_b = \frac{2C}{p} + K = \frac{2 \times 24}{4} + K = 12 + K$$

Let, $K = 1$,

    $\therefore$ Back pitch, $Y_b = 12 + 1 = 13$ coil sides

        Winding pitch, $Y = 2$ (For progressive lap, Y is always +2)

        Front pitch, $Y_f = Y_b - Y = 13 - 2 = 11$ coil sides

        Commutator pitch, $Y_c = 1$ segment (For progressive lap, Y_c is always +1)

The front and back connections of the coil sides are determined as shown below.

Let us start with top coil side marked 1.

    Bottom coil side of the first coil = Top coil side of first coil + Back pitch
    (Back connection)

$$= 1 + Y_b = 1 + 13 = 14$$

    Top coil side of the second coil = Bottom coil side of first coil − Front pitch
    (Front connection)

$$= 14 - Y_f = 14 - 11 = 3$$

Similarly, all the back and front connections are determined. The entire winding calculations, details of back and front connections and the winding table are shown ahead.

Winding calculations		Back connections		Front connections	
1 + 13 = 14	25 + 13 = 38	1 ← 14		14 → 3	
14 − 11 = 3	38 − 11 = 27	3 ← 16		16 → 5	
3 + 13 = 16	27 + 13 = 40	5 ← 18		18 → 7	
16 − 11 = 5	40 − 11 = 29	7 ← 20		20 → 9	
5 + 13 = 18	29 + 13 = 42	9 ← 22		22 → 11	
18 − 11 = 7	42 − 11 = 31	11 ← 24		24 → 13	
7 + 13 = 20	31 + 13 = 44	13 ← 26		26 → 15	
20 − 11 = 9	44 − 11 = 33	15 ← 28		28 → 17	
9 + 13 = 22	33 + 13 = 46	17 ← 30		30 → 19	
22 − 11 = 11	46 − 11 = 35	19 ← 32		32 → 21	
11 + 13 = 24	35 + 13 = 48	21 ← 34		34 → 23	
24 − 11 = 13	48 − 11 = 37	23 ← 36		36 → 25	
13 + 13 = 26	37 + 13 = 50 − 48 = 2	25 ← 38		38 → 27	
26 − 11 = 15	50 − 11 = 39	27 ← 40		40 → 29	
15 + 13 = 28	39 + 13 = 52 − 48 = 4	29 ← 42		42 → 31	
28 − 11 = 17	52 − 11 = 41	31 ← 44		44 → 33	
17 + 13 = 30	41 + 13 = 54 − 48 = 6	33 ← 46		46 → 35	
30 − 11 = 19	54 − 11 = 43	35 ← 48		48 → 37	
19 + 13 = 32	43 + 13 = 56 − 48 = 8	37 ← 2		2 → 39	
32 − 11 = 21	56 − 11 = 45	39 ← 4		4 → 41	
21 + 13 = 34	45 + 13 = 58 − 48 = 10	41 ← 6		6 → 43	
34 − 11 = 23	58 − 11 = 47	43 ← 8		8 → 45	
23 + 13 = 36	47 + 13 = 60 − 48 = 12	45 ← 10		10 → 47	
36 − 11 = 25	12 − 11 = 1	47 ← 12		12 → 1	

Winding table

$$1 \leftarrow 14 \rightarrow 3 \leftarrow 16 \rightarrow 5 \leftarrow 18 \rightarrow 7 \leftarrow 20 \rightarrow 9 \leftarrow 22 \rightarrow 11 \leftarrow 24$$
$$\rightarrow 13 \leftarrow 26 \rightarrow 15 \leftarrow 28 \rightarrow 17 \leftarrow 30 \rightarrow 19 \leftarrow 32 \rightarrow 21 \leftarrow 34 \rightarrow 23 \leftarrow 36$$
$$\rightarrow 25 \leftarrow 38 \rightarrow 27 \leftarrow 40 \rightarrow 29 \leftarrow 42 \rightarrow 31 \leftarrow 44 \rightarrow 33 \leftarrow 46 \rightarrow 35 \leftarrow 48$$
$$\rightarrow 37 \leftarrow 2 \rightarrow 39 \leftarrow 4 \rightarrow 41 \leftarrow 6 \rightarrow 43 \leftarrow 8 \rightarrow 45 \leftarrow 10 \rightarrow 47 \leftarrow 12$$
$$\rightarrow 1$$

One back connection and one front connection are shown in Fig. 2. All the back connections are shown in Fig. 3 and all the front connections are shown in Fig. 4. The meeting points of coil ends formed by front connections are terminated on the commutator segments as shown in Fig. 4.

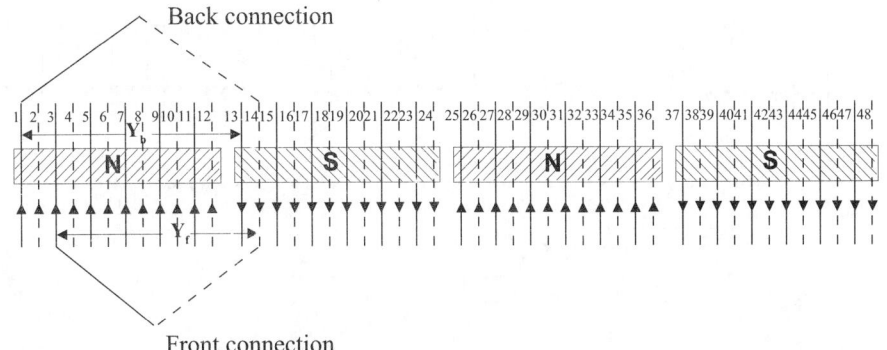

Fig. 2: *The first back and front connection.*

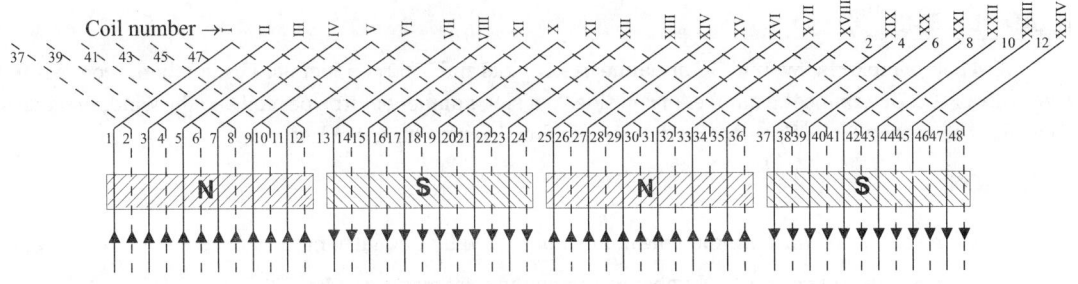

Fig. 3: *Back connections of the winding.*

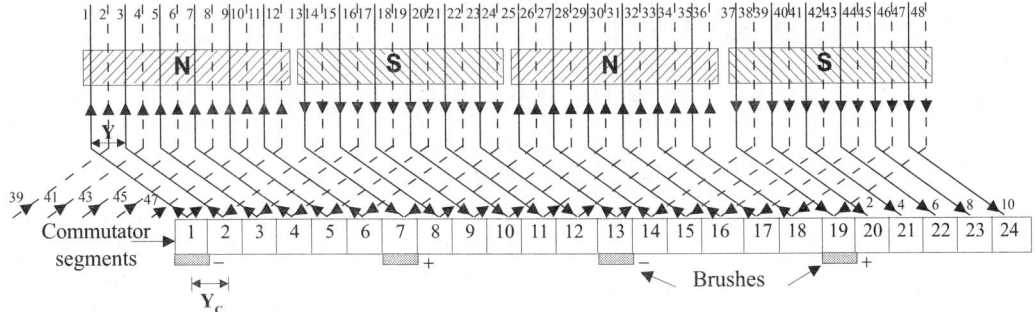

Fig. 4: *Front connections and location of brushes.*

In lap winding the number of brushes will be equal to number of poles. Hence in this winding there will be four brushes and in this four brushes two will be positive and two will be negative. On observing the currents entering at the commutator segments, we can place positive brushes at segments 7 and 19. Because in segments 7 and 19, the current enter the commutator segment from the two conductors connected to them. In segments 1 and 13 the current leaves the commutator segment through the conductors connected to them. Hence negative brushes can be placed at segments 1 and 13. The complete winding diagram is shown in Fig. 5.

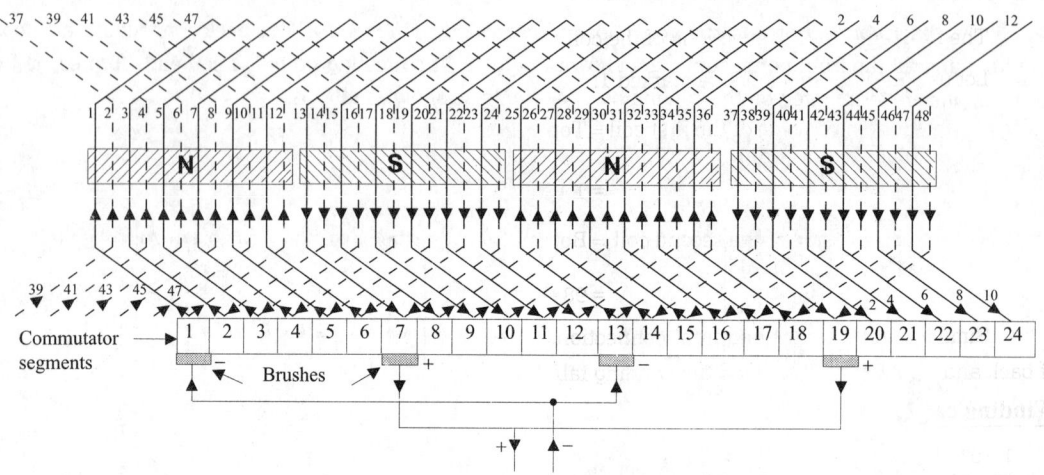

Fig. 5: *Simplex lap winding of a 4 pole DC generator armature with 24 slots.*

EXAMPLE 5.3

A 4 pole, simplex lap wound DC armature has 64 slots and 1152 conductors. The number of commutator segments is 192. Determine the number of coil sides per slot, number of turns per coil and the winding pitches. Draw up the winding table. Specify whether the winding is symmetrical or not.

SOLUTION

In DC machine armature winding the number of coils will be equal to number of commutator segments.

∴ Number of coils, C = Number of commutator segments = 192

Number of coil sides = 2C = 2 × 192 = 384

Conductors per slot = $\dfrac{1152}{64} = 18$

Coil sides per slot = $\dfrac{\text{Number of coil sides}}{\text{Number of slots}} = \dfrac{384}{64} = 6$

Turns per coil = $\dfrac{\text{Conductor per slot}}{\text{Coil sides per slot}} = \dfrac{18}{6} = 3$

The winding pitches are calculated as shown below.

Let us choose progressive lap winding.

Back pitch, $Y_b = \dfrac{2C}{p} + K = \dfrac{2 \times 192}{4} + K = 96 + K$

Let, K = 1,

∴ Back pitch, Y_b = 96 + 1= 97 coil sides

Winding pitch, Y = 2 (For progressive lap, Y is always +2)

Front pitch, Y_f = Y_b − Y = 97 − 2 = 95 coil sides

Commutator pitch, Y_c = 1 segment (For progressive lap, Y_c is always +1)

The front and back connections of the coil sides are determined as shown below.

Let us start with top coil side marked 1.

Bottom coil side of the first coil = Top coil side of first coil + Back pitch
(Back connection)

$$= 1 + Y_b = 1 + 97 = 98$$

Top coil side of the second coil = Bottom coil side of first coil − Front pitch
(Front connection)

$$= 98 − Y_f = 98 − 95 = 3$$

Similarly, all the back and front connections are determined. The entire winding calculations, details of back and front connections and the winding table are shown below.

Winding calculations

1 + 97 = 98	97 + 97 = 194	193 + 97 = 290	289 + 97 = 386 − 384 = 2
98 − 95 = 3	194 − 95 = 99	290 − 95 = 195	386 − 95 = 291
3 + 97 = 100	99 + 97 = 196	195 + 97 = 292	291 + 97 = 388 − 384 = 4
100 − 95 = 5	196 − 95 = 101	292 − 95 = 197	388 − 95 = 293
5 + 97 = 102	101 + 97 = 198	197 + 97 = 294	293 + 97 = 390 − 384 = 6
102 − 95 = 7	198 − 95 = 103	294 − 95 = 199	390 − 95 = 295
7 + 97 = 104	103 + 97 = 200	199 + 97 = 296	295 + 97 = 392 − 384 = 8
104 − 95 = 9	200 − 95 = 105	296 − 95 = 201	392 − 95 = 297
9 + 97 = 106	105 + 97 = 202	201 + 97 = 298	297 + 97 = 394 − 384 = 10
106 − 95 = 11	202 − 95 = 107	298 − 95 = 203	394 − 95 = 299
11 + 97 = 108	107 + 97 = 204	203 + 97 = 300	299 + 97 = 396 − 384 = 12
108 − 95 = 13	204 − 95 = 109	300 − 95 = 205	396 − 95 = 301
13 + 97 = 110	109 + 97 = 206	205 + 97 = 302	301 + 97 = 398 − 384 = 14
110 − 95 = 15	206 − 95 = 111	302 − 95 = 207	398 − 95 = 303
15 + 97 = 112	111 + 97 = 208	207 + 97 = 304	303 + 97 = 400 − 384 = 16
112 − 95 = 17	208 − 95 = 113	304 − 95 = 209	400 − 95 = 305
17 + 97 = 114	113 + 97 = 210	209 + 97 = 306	305 + 97 = 402 − 384 = 18
114 − 95 = 19	210 − 95 = 115	306 − 95 = 211	402 − 95 = 307
19 + 97 = 116	115 + 97 = 212	211 + 97 = 308	307 + 97 = 404 − 384 = 20
116 − 95 = 21	212 − 95 = 117	308 − 95 = 213	404 − 95 = 309
21 + 97 = 118	117 + 97 = 214	213 + 97 = 310	309 + 97 = 406 − 384 = 22
118 − 95 = 23	214 − 95 = 119	310 − 95 = 215	406 − 95 = 311
23 + 97 = 120	119 + 97 = 216	215 + 97 = 312	311 + 97 = 408 − 384 = 24
120 − 95 = 25	216 − 95 = 121	312 − 95 = 217	408 − 95 = 313
25 + 97 = 122	121 + 97 = 218	217 + 97 = 314	313 + 97 = 410 − 384 = 26
122 − 95 = 27	218 − 95 = 123	314 − 95 = 219	410 − 95 = 315
27 + 97 = 124	123 + 97 = 220	219 + 97 = 316	315 + 97 = 412 − 384 = 28
124 − 95 = 29	220 − 95 = 125	316 − 95 = 221	412 − 95 = 317
29 + 97 = 126	125 + 97 = 222	221 + 97 = 318	317 + 97 = 414 − 384 = 30
126 − 95 = 31	222 − 95 = 127	318 − 95 = 223	414 − 95 = 319
31 + 97 = 128	127 + 97 = 224	223 + 97 = 320	319 + 97 = 416 − 384 = 32
128 − 95 = 33	224 − 95 = 129	320 − 95 = 225	416 − 95 = 321
33 + 97 = 130	129 + 97 = 226	225 + 97 = 322	321 + 97 = 418 − 384 = 34
130 − 95 = 35	226 − 95 = 131	322 − 95 = 227	418 − 95 = 323
35 + 97 = 132	131 + 97 = 228	227 + 97 = 324	323 + 97 = 420 − 384 = 36

$132 - 95 = 37$
$37 + 97 = 134$
$134 - 95 = 39$
$39 + 97 = 136$
$136 - 95 = 41$
$41 + 97 = 138$
$138 - 95 = 43$
$43 + 97 = 140$
$140 - 95 = 45$
$45 + 97 = 142$
$142 - 95 = 47$
$47 + 97 = 144$
$144 - 95 = 49$
$49 + 97 = 146$
$146 - 95 = 51$
$51 + 97 = 148$
$148 - 95 = 53$
$53 + 97 = 150$
$150 - 95 = 55$
$55 + 97 = 152$
$152 - 95 = 57$
$57 + 97 = 154$
$154 - 95 = 59$
$59 + 97 = 156$
$156 - 95 = 61$
$61 + 97 = 158$
$158 - 95 = 63$
$63 + 97 = 160$
$160 - 95 = 65$
$65 + 97 = 162$
$162 - 95 = 67$
$67 + 97 = 164$
$164 - 95 = 69$
$69 + 97 = 166$
$166 - 95 = 71$
$71 + 97 = 168$
$168 - 95 = 73$
$73 + 97 = 170$
$170 - 95 = 75$
$75 + 97 = 172$
$172 - 95 = 77$
$77 + 97 = 174$
$174 - 95 = 79$
$79 + 97 = 176$
$176 - 95 = 81$
$81 + 97 = 178$
$178 - 95 = 83$
$83 + 97 = 180$
$180 - 95 = 85$
$85 + 97 = 182$
$182 - 95 = 87$
$87 + 97 = 184$
$184 - 95 = 89$
$89 + 97 = 186$
$186 - 95 = 91$
$91 + 97 = 188$
$188 - 95 = 93$
$93 + 97 = 190$
$190 - 95 = 95$
$95 + 97 = 192$
$192 - 95 = 97$

$228 - 95 = 133$
$133 + 97 = 230$
$230 - 95 = 135$
$135 + 97 = 232$
$232 - 95 = 137$
$137 + 97 = 234$
$234 - 95 = 139$
$139 + 97 = 236$
$236 - 95 = 141$
$141 + 97 = 238$
$238 - 95 = 143$
$143 + 97 = 240$
$240 - 95 = 145$
$145 + 97 = 242$
$242 - 95 = 147$
$147 + 97 = 244$
$244 - 95 = 149$
$149 + 97 = 246$
$246 - 95 = 151$
$151 + 97 = 248$
$248 - 95 = 153$
$153 + 97 = 250$
$250 - 95 = 155$
$155 + 97 = 252$
$252 - 95 = 157$
$157 + 97 = 254$
$254 - 95 = 159$
$159 + 97 = 256$
$256 - 95 = 161$
$161 + 97 = 258$
$258 - 95 = 163$
$163 + 97 = 260$
$260 - 95 = 165$
$165 + 97 = 262$
$262 - 95 = 167$
$167 + 97 = 264$
$264 - 95 = 169$
$169 + 97 = 266$
$266 - 95 = 171$
$171 + 97 = 268$
$268 - 95 = 173$
$173 + 97 = 270$
$270 - 95 = 175$
$175 + 97 = 272$
$272 - 95 = 177$
$177 + 97 = 274$
$274 - 95 = 179$
$179 + 97 = 276$
$276 - 95 = 181$
$181 + 97 = 278$
$278 - 95 = 183$
$183 + 97 = 280$
$280 - 95 = 185$
$185 + 97 = 282$
$282 - 95 = 187$
$187 + 97 = 284$
$284 - 95 = 189$
$189 + 97 = 286$
$286 - 95 = 191$
$191 + 97 = 288$
$288 - 95 = 193$

$324 - 95 = 229$
$229 + 97 = 326$
$326 - 95 = 231$
$231 + 97 = 328$
$328 - 95 = 233$
$233 + 97 = 330$
$330 - 95 = 235$
$235 + 97 = 332$
$332 - 95 = 237$
$237 + 97 = 334$
$334 - 95 = 239$
$239 + 97 = 336$
$336 - 95 = 241$
$241 + 97 = 338$
$338 - 95 = 243$
$243 + 97 = 340$
$340 - 95 = 245$
$245 + 97 = 342$
$342 - 95 = 247$
$247 + 97 = 344$
$344 - 95 = 249$
$249 + 97 = 346$
$346 - 95 = 251$
$251 + 97 = 348$
$348 - 95 = 253$
$253 + 97 = 350$
$350 - 95 = 255$
$255 + 97 = 352$
$352 - 95 = 257$
$257 + 97 = 354$
$354 - 95 = 259$
$259 + 97 = 356$
$356 - 95 = 261$
$261 + 97 = 358$
$358 - 95 = 263$
$263 + 97 = 360$
$360 - 95 = 265$
$265 + 97 = 362$
$362 - 95 = 267$
$267 + 97 = 364$
$364 - 95 = 269$
$269 + 97 = 366$
$366 - 95 = 271$
$271 + 97 = 368$
$368 - 95 = 273$
$273 + 97 = 370$
$370 - 95 = 275$
$275 + 97 = 372$
$372 - 95 = 277$
$277 + 97 = 374$
$374 - 95 = 279$
$279 + 97 = 376$
$376 - 95 = 281$
$281 + 97 = 378$
$378 - 95 = 283$
$283 + 97 = 380$
$380 - 95 = 285$
$285 + 97 = 382$
$382 - 95 = 287$
$287 + 97 = 384$
$384 - 95 = 289$

$420 - 95 = 325$
$352 + 97 = 422 - 384 = 38$
$422 - 95 = 327 \quad 327 + 97 = 424 - 384 = 40$
$424 - 95 = 329$
$329 + 97 = 426 - 384 = 42$
$426 - 95 = 331$
$331 + 97 = 428 - 384 = 44$
$428 - 95 = 333$
$333 + 97 = 430 - 384 = 46$
$430 - 95 = 335$
$335 + 97 = 432 - 384 = 48$
$432 - 95 = 337$
$337 + 97 = 434 - 384 = 50$
$434 - 95 = 339$
$339 + 97 = 436 - 384 = 52$
$436 - 95 = 341$
$341 + 97 = 438 - 384 = 54$
$438 - 95 = 343$
$343 + 97 = 440 - 384 = 56$
$440 - 95 = 345$
$345 + 97 = 442 - 384 = 58$
$442 - 95 = 347$
$347 + 97 = 444 - 384 = 60$
$444 - 95 = 349$
$349 + 97 = 446 - 384 = 62$
$446 - 95 = 351$
$351 + 97 = 448 - 384 = 64$
$448 - 95 = 353$
$353 + 97 = 450 - 384 = 66$
$450 - 95 = 355$
$355 + 97 = 452 - 384 = 68$
$452 - 95 = 357$
$357 + 97 = 454 - 384 = 70$
$454 - 95 = 359$
$359 + 97 = 456 - 384 = 72$
$456 - 95 = 361$
$361 + 97 = 458 - 384 = 74$
$458 - 95 = 363$
$363 + 97 = 460 - 384 = 76$
$460 - 95 = 365$
$365 + 97 = 462 - 384 = 78$
$462 - 95 = 367$
$367 + 97 = 464 - 384 = 80$
$464 - 95 = 369$
$369 + 97 = 466 - 384 = 82$
$466 - 95 = 371$
$371 + 97 = 468 - 384 = 84$
$468 - 95 = 373$
$373 + 97 = 470 - 384 = 86$
$470 - 95 = 375$
$375 + 97 = 472 - 384 = 88$
$472 - 95 = 377$
$377 + 97 = 474 - 384 = 90$
$474 - 95 = 379$
$379 + 97 = 476 - 384 = 92$
$476 - 95 = 381$
$381 + 97 = 478 - 384 = 94$
$478 - 95 = 383$
$383 + 97 = 480 - 384 = 96$
$96 - 95 = 1$

Winding table

```
  1  ←  98  →  3  ← 100 →  5  ← 102 →   7   ← 104 →  9  ← 106 → 11 ← 108
→ 13  ← 110 → 15  ← 112 → 17  ← 114 →  19   ← 116 → 21  ← 118 → 23 ← 120
→ 25  ← 122 → 27  ← 124 → 29  ← 126 →  31   ← 128 → 33  ← 130 → 35 ← 132
→ 37  ← 134 → 39  ← 136 → 41  ← 138 →  43   ← 140 → 45  ← 142 → 47 ← 144
→ 49  ← 146 → 51  ← 148 → 53  ← 150 →  55   ← 152 → 57  ← 154 → 59 ← 156
→ 61  ← 158 → 63  ← 160 → 65  ← 162 →  67   ← 164 → 69  ← 166 → 71 ← 168
→ 73  ← 170 → 75  ← 172 → 77  ← 174 →  79   ← 176 → 81  ← 178 → 83 ← 180
→ 85  ← 182 → 87  ← 184 → 89  ← 186 →  91   ← 188 → 93  ← 190 → 95 ← 192
→ 97  ← 194 → 99  ← 196 → 101 ← 198 → 103   ← 200 → 105 ← 202 → 107 ← 204
→ 109 ← 206 → 111 ← 208 → 113 ← 210 → 115   ← 212 → 117 ← 214 → 119 ← 216
→ 121 ← 218 → 123 ← 220 → 125 ← 222 → 127   ← 224 → 129 ← 226 → 131 ← 228
→ 133 ← 230 → 135 ← 232 → 137 ← 234 → 139   ← 236 → 141 ← 238 → 143 ← 240
→ 145 ← 242 → 147 ← 244 → 149 ← 246 → 151   ← 248 → 153 ← 250 → 155 ← 252
→ 157 ← 254 → 159 ← 256 → 161 ← 258 → 163   ← 260 → 165 ← 262 → 167 ← 264
→ 169 ← 266 → 171 ← 268 → 173 ← 270 → 175   ← 272 → 177 ← 274 → 179 ← 276
→ 181 ← 278 → 183 ← 280 → 185 ← 282 → 187   ← 284 → 189 ← 286 → 191 ← 288
→ 193 ← 290 → 195 ← 292 → 197 ← 294 → 199   ← 296 → 201 ← 298 → 203 ← 300
→ 205 ← 302 → 207 ← 304 → 209 ← 306 → 211   ← 308 → 213 ← 310 → 215 ← 312
→ 217 ← 314 → 219 ← 316 → 221 ← 318 → 223   ← 320 → 225 ← 322 → 227 ← 324
→ 229 ← 326 → 231 ← 328 → 233 ← 330 → 235   ← 332 → 237 ← 334 → 239 ← 336
→ 241 ← 338 → 243 ← 340 → 245 ← 342 → 247   ← 344 → 249 ← 346 → 251 ← 348
→ 253 ← 350 → 255 ← 352 → 257 ← 354 → 259   ← 356 → 261 ← 358 → 263 ← 360
→ 265 ← 362 → 267 ← 364 → 269 ← 366 → 271   ← 368 → 273 ← 370 → 275 ← 372
→ 277 ← 374 → 279 ← 376 → 281 ← 378 → 283   ← 380 → 285 ← 385 → 287 ← 384
→ 289 ←  2  → 291 ←  4  → 293 ←  6  → 295   ←  8  → 297 ← 10 → 299 ← 12
→ 301 ← 14  → 303 ← 16  → 305 ← 18  → 307   ← 20  → 309 ← 22 → 311 ← 24
→ 313 ← 26  → 315 ← 28  → 317 ← 30  → 319   ← 32  → 321 ← 34 → 323 ← 36
→ 325 ← 38  → 327 ← 40  → 329 ← 42  → 331   ← 44  → 333 ← 46 → 335 ← 48
→ 337 ← 50  → 339 ← 52  → 341 ← 54  → 343   ← 56  → 345 ← 58 → 347 ← 60
→ 349 ← 62  → 351 ← 64  → 353 ← 66  → 355   ← 68  → 357 ← 70 → 359 ← 72
→ 361 ← 74  → 363 ← 76  → 365 ← 78  → 367   ← 80  → 369 ← 82 → 371 ← 84
→ 373 ← 86  → 375 ← 88  → 377 ← 90  → 379   ← 92  → 381 ← 94 → 383 ← 96
```

EXAMPLE 5.4

Draw the winding diagram for a 4 pole, 13 slots simplex wave connected DC generator with a commutator having 13 segments. The number of coil sides per slot is 2. Indicate the position of brushes.

SOLUTION

In DC machine armature winding the number of coils will be equal to number of commutator segments.

Given that number of commutator segments = 13

$\therefore$ Number of coils, C = 13

Number of coil sides = 2C = 2 × 13 = 26

Coil sides per slot = $\dfrac{\text{Number of coil sides}}{\text{Number of slots}} = \dfrac{26}{13} = 2$

The coil sides are represented by parallel straight lines as shown in Fig. 1. Since two coil sides are accommodated in one slot, each slot will have one top coil side and one bottom coil side. The top coil sides are shown by solid (or continuous) lines and bottom coil sides are shown by broken (or discontinuous) lines. In Fig. 1 the coil sides are numbered such that the top coil sides are represented by odd numbers and bottom coil sides by even numbers.

$$\text{Coil sides per pole} = \frac{\text{Number of coil sides}}{\text{Number of poles}} = \frac{26}{4} = 6.5$$

Here, the coil sides per pole is not an integer. Hence we can assume that the first two poles enclose six coil sides each and the next two poles enclose seven coil sides each as shown in Fig. 1. The enclosure of coil sides by the poles are represented by shaded rectangles as shown in Fig. 1. The direction of current through conductors lying under north pole are marked upwards and that of south pole are marked downwards.

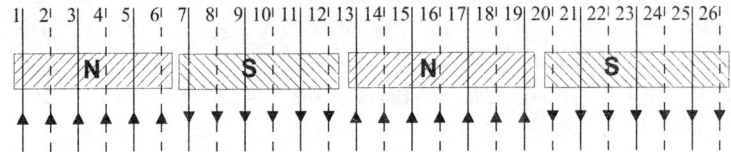

Fig. 1: *Representation of coil sides, poles and direction of current.*

The winding pitches are calculated as shown below.

Let us choose progressive wave winding.

$$\text{Back pitch, } Y_b = \frac{2C}{p} + K = \frac{2 \times 13}{4} + K = 6.5 + K$$

Let, K = 0.5,

Back pitch, $Y_b = 6.5 + 0.5 = 7$ coil sides

Winding pitch, $Y = \frac{2C+2}{p/2} = \frac{(2 \times 13)+2}{4/2} = 14$ coil sides

Front pitch, $Y_f = Y - Y_b = 14 - 7 = 7$ coil sides

Commutator pitch, $Y_c = \frac{Y}{2} = \frac{14}{2} = 7$ coil sides

The front and back connections of the coil sides are determined as shown below.

Let us start with top coil side marked 1.

Bottom coil side of the first coil = Top coil side of first coil + Back pitch
(Back connection)

$$= 1 + Y_b = 1 + 7 = 8$$

Top coil side of the coil lying one pole $\Big\}$ = Bottom coil side of first coil + Front pitch
pitch from first coil (Front connection)$\Big\}$

$$= 8 + Y_f = 8 + 7 = 15$$

Similarly, all the back and front connections are determined. The entire winding calculations, details of back and front connections and the winding table are shown below.

Winding calculations	Back connections	Front connections
$1 + 7 = 8$	$1 \leftarrow 8$	$8 \rightarrow 15$
$8 + 7 = 15$	$15 \leftarrow 22$	$22 \rightarrow 3$
$15 + 7 = 22$	$3 \leftarrow 10$	$10 \rightarrow 17$
$22 + 7 = 29 - 26 = 3$	$17 \leftarrow 24$	$24 \rightarrow 5$
$3 + 7 = 10$	$5 \leftarrow 12$	$12 \rightarrow 19$
$10 + 7 = 17$	$19 \leftarrow 26$	$26 \rightarrow 7$
$17 + 7 = 24$	$7 \leftarrow 14$	$14 \rightarrow 21$
$24 + 7 = 31 - 26 = 5$	$21 \leftarrow 2$	$2 \rightarrow 9$
$5 + 7 = 12$	$9 \leftarrow 16$	$16 \rightarrow 23$
$12 + 7 = 19$	$23 \leftarrow 4$	$4 \rightarrow 11$
$19 + 7 = 26$	$11 \leftarrow 18$	$18 \rightarrow 25$
$26 + 7 = 33 - 26 = 7$	$25 \leftarrow 6$	$6 \rightarrow 13$
$7 + 7 = 14$	$13 \leftarrow 20$	$20 \rightarrow 1$
$14 + 7 = 21$		
$21 + 7 = 28 - 26 = 2$		

Winding table

$$1 \leftarrow 8 \rightarrow 15 \leftarrow 22 \rightarrow 3 \leftarrow 10 \rightarrow 17 \leftarrow 24$$
$$\rightarrow 5 \leftarrow 12 \rightarrow 19 \leftarrow 26 \rightarrow 7 \leftarrow 14 \rightarrow 21 \leftarrow 2$$
$$\rightarrow 9 \leftarrow 16 \rightarrow 23 \leftarrow 4 \rightarrow 11 \leftarrow 18 \rightarrow 25 \leftarrow 6$$
$$\rightarrow 13 \leftarrow 20 \rightarrow 1$$

Winding calculations (cont.)
$2 + 7 = 9$
$9 + 7 = 16$
$16 + 7 = 23$
$23 + 7 = 30 - 26 = 4$
$4 + 7 = 11$
$11 + 7 = 18$
$18 + 7 = 25$
$25 + 7 = 32 - 26 = 6$
$6 + 7 = 13$
$13 + 7 = 20$
$20 + 7 = 27 - 26 = 1$

One back connection and one front connection are shown in Fig. 2. All the back connections are shown in Fig. 3 and all the front connections are shown in Fig. 4. The meeting points of coil ends formed by front connections are terminated on the commutator segments as shown in Fig. 4.

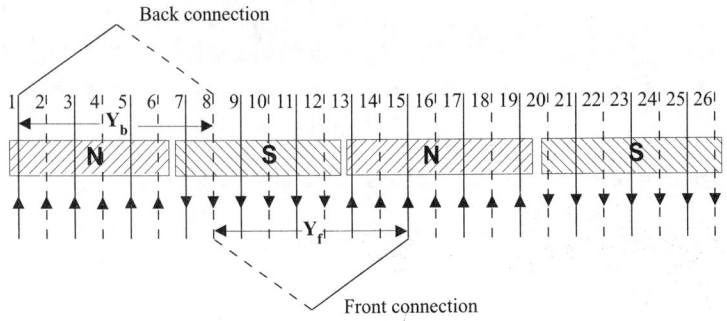

Fig. 2: The first back and front connection.

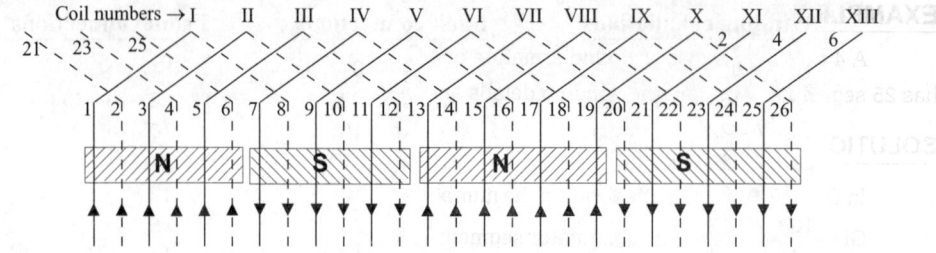

Fig. 3: *Back connections of the winding.*

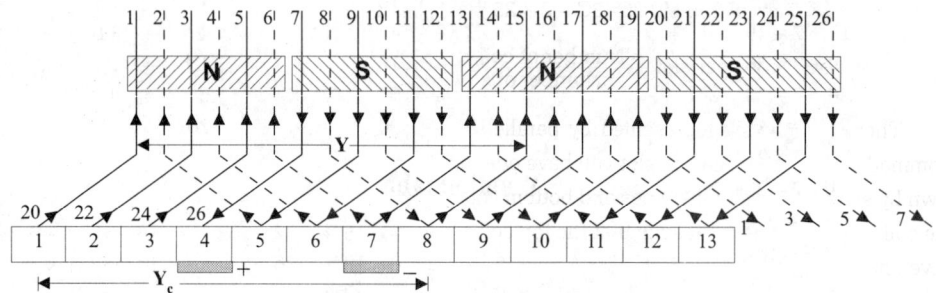

Fig. 4: *Front connections and location of brushes.*

In simplex wave winding the number of brushes are always two. One of the brush will be at positive potential and the other at negative potential. On observing the currents entering at the commutator segments, we can place positive brush at segment-4. Because in segment-4, the current enter the commutator segment from the two conductors connected to them.

In segment-7, the current leaves the commutator segment through the conductors connected to them. Hence negative brush can be placed at segment-7. The complete winding diagram is shown in Fig. 5.

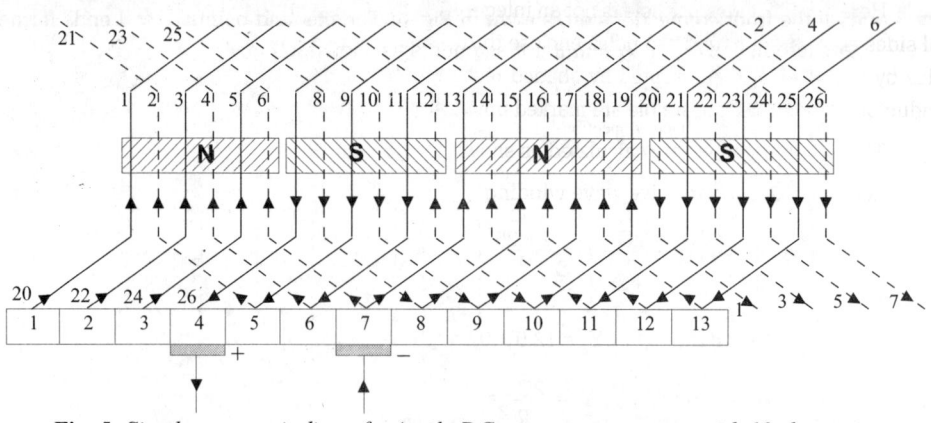

Fig. 5: *Simplex wave winding of a 4 pole DC generator armature with 13 slots.*

EXAMPLE 5.5

A 4 pole simplex wave wound armature of a DC generator has 25 slots and 25 coils. The commutator has 25 segments. Work out the winding details and draw the winding diagram.

SOLUTION

In DC machine armature winding the number of coils will be equal to number of commutator segments.

Given that number of commutator segments = 25

∴ Number of coils, C = 25

Number of coil sides = 2C = 2 × 25 = 50

$$\text{Coil sides per slot} = \frac{\text{Number of coil sides}}{\text{Number of slots}} = \frac{50}{25} = 2$$

The coil sides are represented by parallel straight lines as shown in Fig. 1. Since two coil sides are accommodated in one slot, each slot will have one top coil side and one bottom coil side. The top coil sides are shown by solid (or continuous) lines and bottom coil sides are shown by broken (or discontinuous) lines. In Fig. 1 the coil sides are numbered such that the top coil sides are represented by odd numbers and bottom coil sides by even numbers.

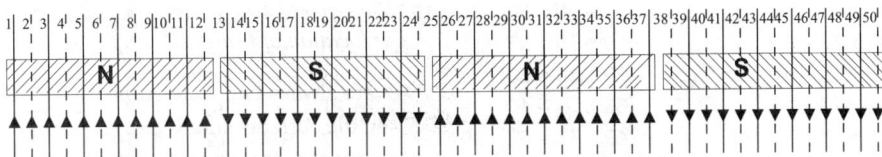

Fig. 1: *Representation of coil sides, poles and direction of current.*

$$\text{Coil sides per pole} = \frac{\text{Number of coil sides}}{\text{Number of poles}} = \frac{50}{4} = 12.5$$

Here, the coil sides per pole is not an integer. Hence we can assume that the first two poles enclose twelve coil sides each and the next two poles enclose thirteen coil sides each as shown in Fig. 1. The enclosure of coil sides by the poles are represented by shaded rectangles as shown in Fig. 1. The direction of current through conductors lying under north pole are marked upwards and that of south pole are marked downwards.

The winding pitches are calculated as shown below.

Let us choose progressive wave winding.

$$\text{Back pitch, } Y_b = \frac{2C}{p} + K = \frac{2 \times 25}{4} + K = 12.5 + K$$

Let, K = 0.5,

∴ Back pitch, $Y_b = 12.5 + 0.5 = 13$ coil sides

Winding pitch, $Y = \frac{2C+2}{p/2} = \frac{(2 \times 25) + 2}{4/2} = 26$ coil sides

∴ Front pitch, $Y_f = Y - Y_b = 26 - 13 = 13$ coil sides

Commutator pitch, $Y_c = \frac{Y}{2} = \frac{26}{2} = 13$ coil sides

Winding calculations		Back connections			Front connections		
$1 + 13 = 14$	$26 + 13 = 39$	1	←	14	14	→	27
$14 + 13 = 27$	$39 + 13 = 52 - 50 = 2$	27	←	40	40	→	3
$27 + 13 = 40$	$2 + 13 = 15$	3	←	16	16	→	29
$40 + 13 = 53 - 50 = 3$	$15 + 13 = 28$	29	←	42	42	→	5
$3 + 13 = 16$	$28 + 13 = 41$	5	←	18	18	→	31
$16 + 13 = 29$	$41 + 13 = 54 - 50 = 4$	31	←	44	44	→	7
$29 + 13 = 42$	$4 + 13 = 17$	7	←	20	20	→	33
$42 + 13 = 55 - 50 = 5$	$17 + 13 = 30$	33	←	46	46	→	9
$5 + 13 = 18$	$30 + 13 = 43$	9	←	22	22	→	35
$18 + 13 = 31$	$43 + 13 = 56 - 50 = 6$	35	←	48	48	→	11
$31 + 13 = 44$	$6 + 13 = 19$	11	←	24	24	→	37
$44 + 13 = 57 - 50 = 7$	$19 + 13 = 32$	37	←	50	50	→	13
$7 + 13 = 20$	$32 + 13 = 45$	13	←	26	26	→	39
$20 + 13 = 33$	$45 + 13 = 58 - 50 = 8$	39	←	2	2	→	15
$33 + 13 = 46$	$8 + 13 = 21$	15	←	28	28	→	41
$46 + 13 = 59 - 50 = 9$	$21 + 13 = 34$	41	←	4	4	→	17
$9 + 13 = 22$	$34 + 13 = 47$	17	←	30	30	→	43
$22 + 13 = 35$	$47 + 13 = 60 - 50 = 10$	43	←	6	6	→	19
$35 + 13 = 48$	$10 + 13 = 23$	19	←	32	32	→	45
$48 + 13 = 61 - 50 = 11$	$23 + 13 = 36$	45	←	8	8	→	21
$11 + 13 = 24$	$36 + 13 = 49$	21	←	34	34	→	47
$24 + 13 = 37$	$49 + 13 = 62 - 50 = 12$	47	←	10	10	→	23
$37 + 13 = 50$	$12 + 13 = 25$	23	←	36	36	→	49
$50 + 13 = 63 - 50 = 13$	$25 + 13 = 38$	49	←	12	12	→	25
$13 + 13 = 26$	$38 + 13 = 51 - 50 = 1$	25	←	38	38	→	1

Winding table

1 ← 14 → 27 ← 40 → 3 ← 16 → 29 ← 42 → 5 ← 18 → 31 ← 44
→ 7 ← 20 → 33 ← 46 → 9 ← 22 → 35 ← 48 → 11 ← 24 → 37 ← 50
→ 13 ← 26 → 39 ← 2 → 15 ← 28 → 41 ← 4 → 17 ← 30 → 43 ← 6
→ 19 ← 32 → 45 ← 8 → 21 ← 34 → 47 ← 10 → 23 ← 36 → 49 ← 12
→ 25 ← 38 → 1

The front and back connections of the coil sides are determined as shown below.

Let us start with top coil side marked 1.

Bottom coil side of the first coil = Top coil side of first coil + Back pitch
(Back connection)

$$= 1 + Y_b = 1 + 13 = 14$$

Top coil side of the coil lying one pole pitch from first coil (Front connection) $\Big\}$ = Bottom coil side of first coil + Front pitch

$$= 14 + Y_f = 14 + 13 = 27$$

Similarly, all the back and front connections are determined. The entire winding calculations, details of back and front connections and the winding table are shown above.

One back connection and one front connection are shown in Fig. 2. All the back connections are shown in Fig. 3 and all the front connections are shown in Fig. 4. The meeting points of coil ends formed by front connections are terminated on the commutator segments as shown in Fig. 4.

In simpelx wave winding the number of brushes are always two. One of the brush will be at positive potential and the other at negative potential. On observing the current entering at the commutator segments, we can place positive brush at segment-13. Because in segment-13, the current enter the commutator segment from the two conductors connected to them.

In segment-7, the current leaves the commutator segment through the conductors connected to them. Hence negative brush can be placed at segment-7. The complete winding diagram is shown in Fig. 5.

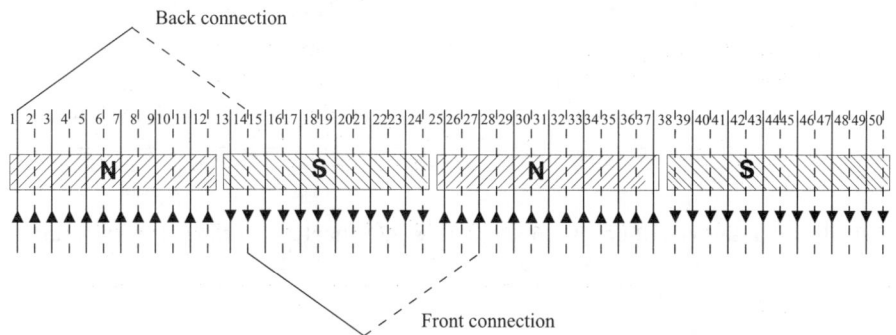

Fig. 2: The first back and front connection.

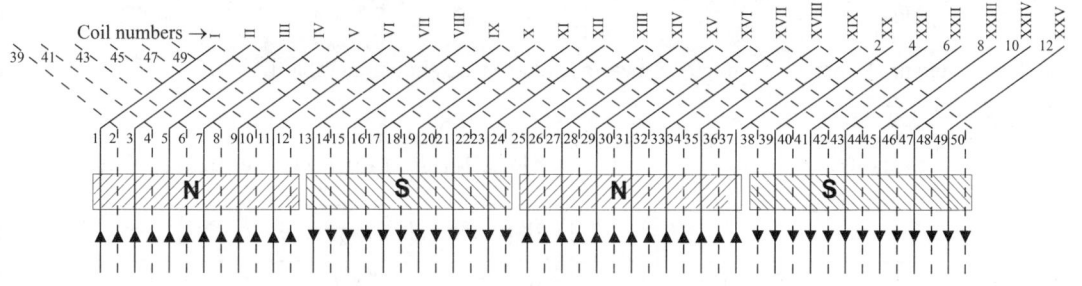

Fig. 3: Back connections of the winding.

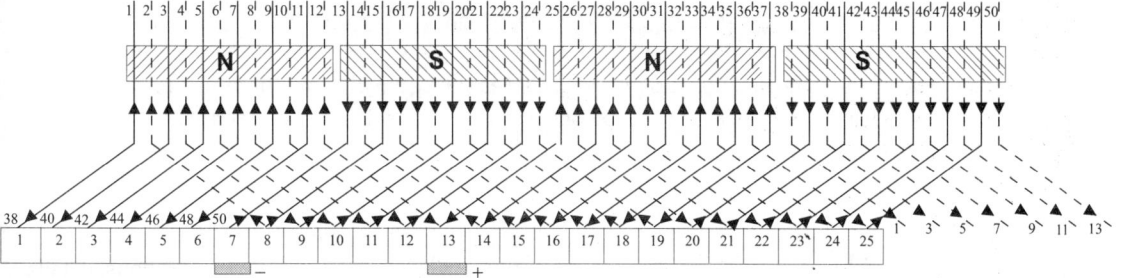

Fig. 4: Front connections and location of brushes.

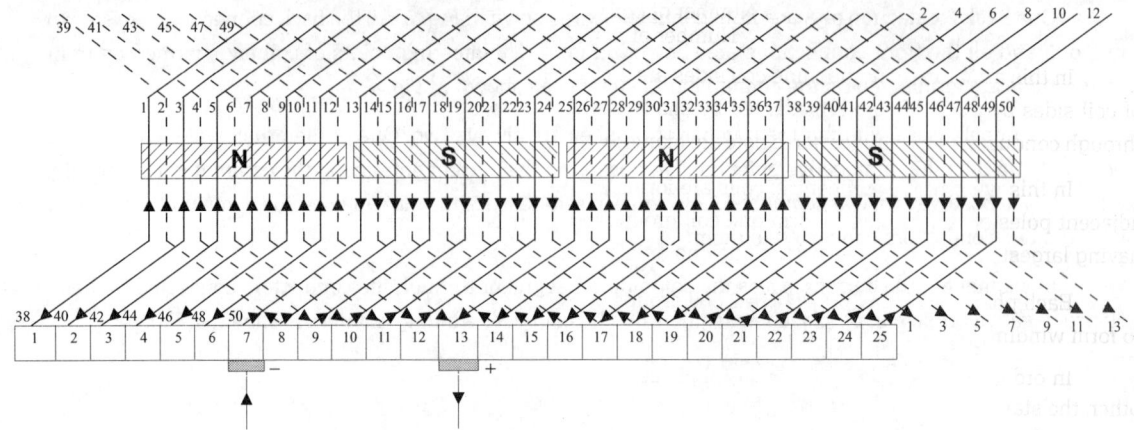

Fig. 5: Simplex wave winding of a 4 pole DC generator armature with 25 slots.

EXAMPLE 5.6

Draw a winding diagram of a 3-phase, 36 slot, 4 pole, AC armature for the single layer unbifurcated winding.

SOLUTION

Slots per pole per phase, $q = \dfrac{36}{3 \times 4} = 3$

Slots per pole $= \dfrac{36}{4} = 9$

The allotment of slots for conductors of various phases are shown in Table 1.

Table 1: Slot Allotment

Slots in a pole group	Slot allotment for phase sequence RYB		
	R-phase	B-phase	Y-phase
1, 2, 3, 4, 5, 6, 7, 8, 9	1, 2, 3	4, 5, 6	7, 8, 9
10, 11, 12, 13, 14, 15, 16, 17, 18	10, 11, 12	13, 14, 15	16, 17, 18
19, 20, 21, 22, 23, 24, 25, 26, 27	19, 20, 21	22, 23, 24	25, 26, 27
28, 29, 30, 31, 32, 33, 34, 35, 36	28, 29, 30	31, 32, 33	34, 35, 36

In single layer winding the number of slots and coil sides will be equal. Since the armature has 36 slots, there will be 36 coil sides, so that each slot occupy one coil side. The coil sides are represented by parallel straight lines as shown in Fig. 1.

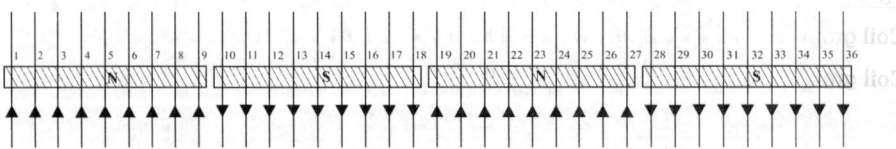

Fig. 1: Representation of coil sides, poles and direction of current.

$$\text{Coil sides per pole} = \frac{\text{Number of coil sides}}{\text{Number of poles}} = \frac{36}{4} = 9$$

In this armature winding, nine coil sides are enclosed by a pole at any one time instant. The enclosure of coil sides by the poles are represented by shaded rectangles as shown in Fig. 1. The direction of current through conductors lying under north pole are marked upwards and that of south pole are marked downwards.

In this winding the concentric coils are formed such that coil sides of similar phase group under two adjacent poles are connected to form one coil group. The span of coils will be different with outermost coil having largest span and inner most coil having smallest span.

Each phase will have, p/2 = 4/2 = 2 coil groups. The two coil groups of a phase are connected in series to form winding of a phase.

In order to provide 120° phase shift between emf induced in R, Y and B-phases with respect of each other, the start of R, Y and B-phases should have an angular spacing of 120 °e.

In this example,

$$\text{Slot angle, } \beta = \frac{180°}{\text{Slots per pole}} = \frac{180°}{9} = 20\,°e$$

Therefore, the number of slots between start of R-phase and Y-phase should be 6 slots (6 × 20° = 120 °e). Similarly, the start of B-phase should be 6 slots away from start of Y-phase for a phase sequence of RYB.

Let us start R-phase from slot 1 and so Y-phase start from 7th slot (1 + 6 = 7) and B-phase start from 13th slot (7 + 6 = 13).

R-Phase winding diagram

The R-phase coil sides forming pole group and coil group are listed in Table 2. The winding calculations, coil connections and winding table of R-phase are also shown below.

The R-phase winding has two coil groups which are connected in series to form R-phase winding. The series connection is made by following the current directions of the coil group.

Table 2: R-Phase slots and Coil groups

Coil sides in pole group	Coil group
1, 2, 3	1, 12
10, 11, 12	2, 11
	3, 10
19, 20, 21	19, 30
28, 29, 30	20, 29
	21, 28

Winding calculations

1 + 9 + 2 = 12
2 + 9 = 11
3 + 9 − 2 = 10
19 + 9 + 2 = 30
20 + 9 = 29
21 + 9 − 2 = 28

Coil connections

1   ← 12 → 1
2   ← 11 → 2
3   ← 10 → 3
19 ← 30 → 19
20 ← 28 → 20
21 ← 28 → 21

Winding table

Coil group-1: 1 ← 12 → 2 ← 11 → 3 ← 10

Coil group-2: 19 ← 30 → 20 ← 29 → 21 ← 28

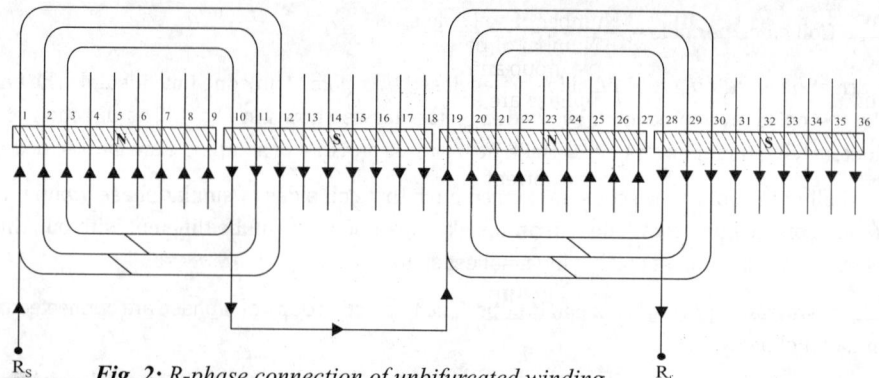

Fig. 2: *R-phase connection of unbifurcated winding.*

B-Phase winding diagram

The B-phase coil sides forming pole group and coil group are listed in Table 3. The winding calculations, coil connections and winding table of B-phase are also shown below.

The B-phase winding has two coil groups which are connected in series to form B-phase winding. The series connection is made by following the current directions of the coil group.

Table 3: B-Phase slots and Coil groups		Winding calculations	Coil connections

Coil sides in pole group	Coil group	Winding calculations	Coil connections
13, 14, 15	13, 24	$13 + 9 + 2 = 24$	$13 \leftarrow 24 \rightarrow 13$
22, 23, 24	14, 23	$14 + 9 = 23$	$14 \leftarrow 23 \rightarrow 14$
	15, 22	$15 + 9 - 2 = 22$	$15 \leftarrow 22 \rightarrow 15$
31, 32, 33	31, 6	$31 + 9 + 2 = 42 - 36 = 6$	$31 \leftarrow 6 \rightarrow 31$
4, 5, 6	32, 5	$32 + 9 = 41 - 36 = 5$	$32 \leftarrow 5 \rightarrow 32$
	33, 4	$33 + 9 - 2 = 40 - 36 = 4$	$33 \leftarrow 4 \rightarrow 33$

Winding table

   Coil group-1: $13 \leftarrow 24 \rightarrow 14 \leftarrow 23 \rightarrow 15 \leftarrow 22$

   Coil group-2: $31 \leftarrow 6 \rightarrow 32 \leftarrow 5 \rightarrow 33 \leftarrow 4$

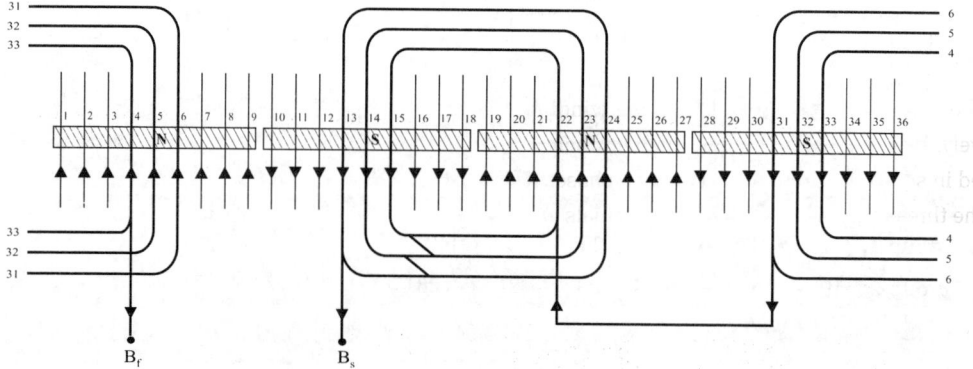

Fig. 3: *B-phase connection of unbifurcated winding.*

Y-Phase winding diagram

The Y-phase coil sides forming pole group and coil group are listed in Table 4. The winding calculations, coil connections and winding table of B-phase are also shown below.

The Y-phase winding has two coil groups which are connected in series to form Y-phase winding. The series connection is made by following the current directions of the coil group.

Table 4: Y-Phase slots and Coil groups

Coil sides in pole group	Coil group
7, 8, 9	7, 18
16, 17, 18	8, 17
	9, 16
25, 26, 27	25, 36
34, 35, 36	26, 35
	27, 34

Winding calculations

$7 + 9 + 2 = 18$

$8 + 9 = 17$

$9 + 9 - 2 = 16$

$25 + 9 + 2 = 36$

$26 + 9 = 35$

$27 + 9 - 2 = 34$

Coil connections

$7 \leftarrow 18 \rightarrow 7$

$8 \leftarrow 17 \rightarrow 8$

$9 \leftarrow 16 \rightarrow 9$

$25 \leftarrow 36 \rightarrow 25$

$26 \leftarrow 35 \rightarrow 26$

$27 \leftarrow 34 \rightarrow 27$

Winding table

Coil group-1: $7 \leftarrow 18 \rightarrow 8 \leftarrow 17 \rightarrow 9 \leftarrow 16$

Coil group-2: $25 \leftarrow 36 \rightarrow 26 \leftarrow 35 \rightarrow 27 \leftarrow 34$

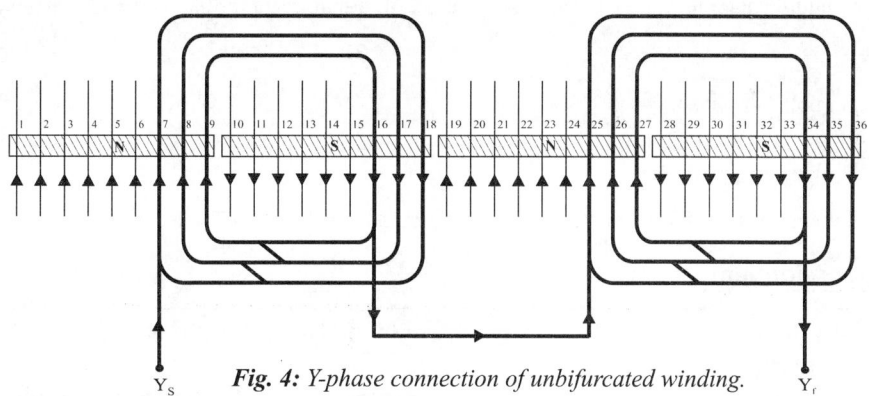

Fig. 4: *Y-phase connection of unbifurcated winding.*

3-Phase winding diagram

The R-phase, B-phase and Y-phase winding diagrams are shown separately in Figs. 2, 3 and 4 respectively. In each phase winding diagram there are two coil groups and the coil groups of a phase are connected in series to form the winding of a phase. The combined three-phase winding diagram is shown in Fig. 5. The three-phase winding has six terminals which can be connected in star or delta.

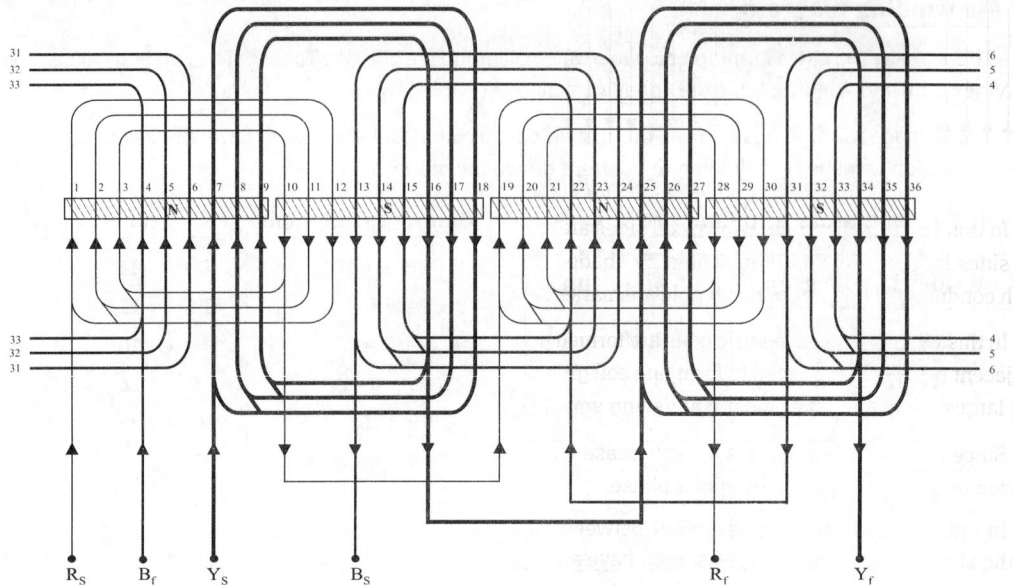

Fig. 5: 3-phase, 36 slot, 4 pole, single layer unbifurcated winding.

EXAMPLE 5.7

Draw a winding diagram of a 3-phase, 48 slot, 4 pole, AC armature for the single layer bifurcated winding.

SOLUTION

Slots per pole per phase, $q = \dfrac{48}{3 \times 4} = 4$

Slots per pole = $\dfrac{48}{4} = 12$

The allotment of slots for conductors of various phases are shown in Table 1.

Table 1: Slot Allotment

Slots in a pole group	Slot allotment for phase sequence RYB		
	R-phase	B-phase	Y-phase
1, 2, 3, 4, 5, 6, 7, 8, 9, 10, 11, 12	1, 2, 3, 4	5, 6, 7, 8	9, 10, 11, 12
13, 14, 15, 16, 17, 18, 19, 20, 21, 22, 23, 24	13, 14, 15, 16	17, 18, 19, 20	21, 22, 23, 24
25, 26, 27, 28, 29, 30, 31, 32, 33, 34, 35, 36	25, 26, 27, 28	29, 30, 31, 32	33, 34, 35, 36
37, 28, 39, 40, 41, 42, 43, 44, 45, 46, 47, 48	37, 38, 39, 40	41, 42, 43, 44	45, 46, 47, 48

In single layer winding number of slots and coil sides will be equal. Since the armature has 48 slots, there will be 48 coil sides. So that each slot occupy one coil side. The coil sides are represented by parallel straight lines as shown in Fig. 1.

$$\text{Coil sides per pole} = \frac{\text{Number of coil sides}}{\text{Number of poles}} = \frac{48}{4} = 12$$

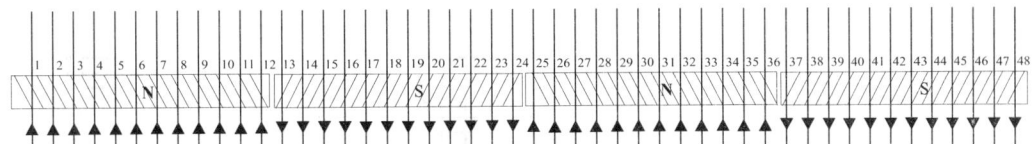

Fig.1: Representation of coil sides, poles and direction of current.

In this armature winding, twelve coil sides are enclosed by a pole at any one time instant. The enclosure of coil sides by the poles are represented by shaded rectangles as shown in Fig. 1. The direction of current through conductors lying under north pole are marked upwards and that of south pole are marked downwards.

In this winding the concentric coils are formed such that half the coil sides of similar phase group under two adjacent poles are connected to form one coil group. The span of coils will be different with outermost coil having largest span and inner most coil having smallest span.

Since the number of poles is 4, each phase will have 4 coil groups. The four coil groups of a phase are connected in series to form winding of a phase.

In order to provide 120° phase shift between emf induced in R, Y and B-phases with respect of each other, the start of R, Y and B-phases should have an angular spacing of 120 °e.

In this example,

$$\text{Slot angle, } \beta = \frac{180°}{\text{Slots per pole}} = \frac{180°}{12} = 15\,°e$$

Therefore, the number of slots between start of R-phase and Y-phase should be 8 slots (8 × 15° = 120 °e). Similarly, the start of B-phase should be 8 slots away from start of Y-phase for a phase sequence of RYB.

Let us start R-phase from slot 3 and so Y-phase start from 11th slot (3 + 8 = 11) and B-phase start from 19th slot (11 + 8 = 19).

R-Phase winding diagram

The R-phase coil sides forming pole group and coil group are listed in Table 2. The winding calculations, coil connections and winding table of R-phase are also shown below.

The R-phase winding has four coil groups which are connected in series to form R-phase winding. The series connection is made by following the current directions of the coil group.

Table 2: R-Phase slots and Coil groups

Coil sides in pole group	Coil group
1, 2, 3, 4	3, 14
13, 14, 15, 16	4, 13
25, 26, 27, 28	15, 26
37, 38, 39, 40	16, 25
	27, 38
	28, 37
	39, 2
	40, 1

Winding calculations

$3 + 12 - 1 = 14$

$4 + 12 - 3 = 13$

$15 + 12 - 1 = 26$

$16 + 12 - 3 = 25$

$27 + 12 - 1 = 38$

$28 + 12 - 3 = 37$

$39 + 12 - 1 = 50 - 48 = 2$

$40 + 12 - 3 = 49 - 48 = 1$

Coil connections

$3 \leftarrow 14 \rightarrow 3$

$4 \leftarrow 13 \rightarrow 4$

$15 \leftarrow 26 \rightarrow 15$

$16 \leftarrow 25 \rightarrow 16$

$27 \leftarrow 38 \rightarrow 27$

$28 \leftarrow 37 \rightarrow 28$

$39 \leftarrow 2 \rightarrow 39$

$40 \leftarrow 1 \rightarrow 40$

Winding table

Coil group-1: $3 \leftarrow 14 \rightarrow 4 \leftarrow 13$ | **Coil group-2:** $25 \leftarrow 16 \rightarrow 26 \leftarrow 15$

Coil group-3: $27 \leftarrow 38 \rightarrow 28 \leftarrow 37$ | **Coil group-4:** $1 \leftarrow 40 \rightarrow 2 \leftarrow 39$

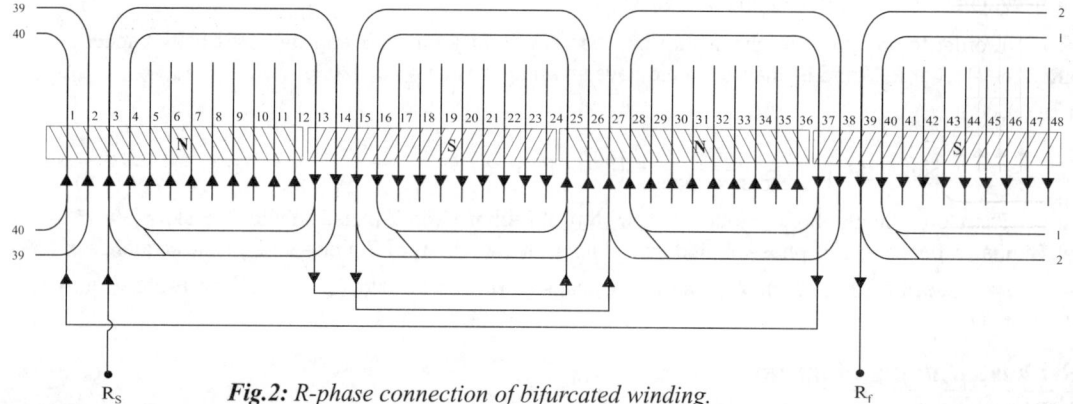

Fig.2: R-phase connection of bifurcated winding.

B-Phase winding diagram

The B-phase coil sides forming pole group and coil group are listed in Table 3. The winding calculations, coil connections and winding table of B-phase are also shown below.

The B-phase winding has four coil groups which are connected in series to form B-phase winding. The series connection is made by following the current directions of the coil group.

Table 3: B-Phase slots and Coil groups

Coil sides in pole group	Coil group
17, 18, 19, 20	19, 30
29, 30, 31, 32	20, 29
41, 42, 43, 44	31, 42
5, 6, 7, 8	32, 41
	43, 6
	44, 5
	7, 18
	8, 17

Winding calculations

$19 + 12 - 1 = 30$

$20 + 12 - 3 = 29$

$31 + 12 - 1 = 42$

$32 + 12 - 3 = 41$

$43 + 12 - 1 = 54 - 48 = 6$

$44 + 12 - 3 = 53 - 48 = 5$

$7 + 12 - 1 = 18$

$8 + 12 - 3 = 17$

Coil connections

$19 \leftarrow 30 \rightarrow 19$

$20 \leftarrow 29 \rightarrow 20$

$31 \leftarrow 42 \rightarrow 31$

$32 \leftarrow 41 \rightarrow 32$

$43 \leftarrow 6 \rightarrow 43$

$44 \leftarrow 5 \rightarrow 44$

$7 \leftarrow 18 \rightarrow 7$

$8 \leftarrow 17 \rightarrow 8$

Winding table

Coil group-1: $19 \leftarrow 30 \rightarrow 20 \leftarrow 29$ | **Coil group-2:** $41 \leftarrow 32 \rightarrow 42 \leftarrow 31 \rightarrow$

Coil group-3: $43 \leftarrow 6 \rightarrow 44 \leftarrow 5$ | **Coil group-4:** $17 \leftarrow 8 \rightarrow 18 \leftarrow 7 \rightarrow$

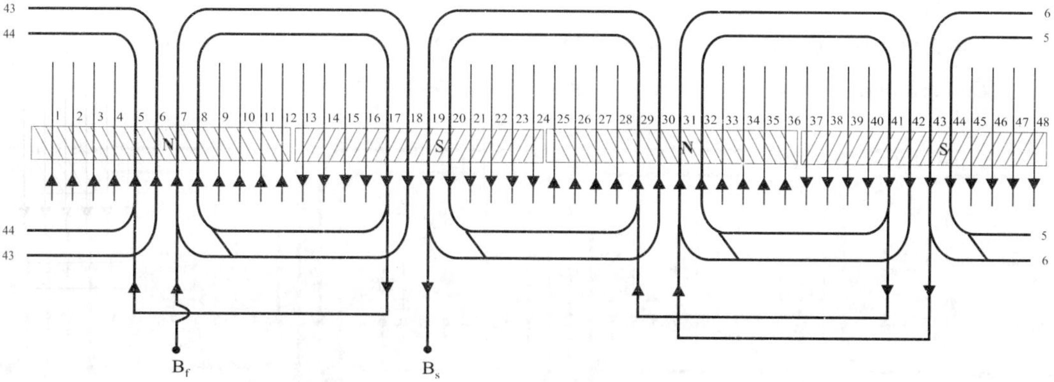

Fig.3: B-phase connection of bifurcated winding.

Y-Phase winding diagram

The Y-phase coil sides forming pole group and coil group are listed in Table 4. The winding calculations, coil connections and winding table of B-phase are also shown below.

The Y-phase winding has four coil groups which are connected in series to form Y-phase winding. The series connection is made by following the current directions of the coil group.

Table 4: Y-Phase slots and Coil groups

Coil sides in pole group	Coil group
9, 10, 11, 12	11, 22
21, 22, 23, 24	12, 21
33, 34, 35, 36	23, 34
45, 46, 47, 48	24, 33
	35, 46
	36, 45
	47, 10
	48, 9

Winding calculations

$$11 + 12 - 1 = 22$$
$$12 + 12 - 3 = 21$$
$$23 + 12 - 1 = 34$$
$$24 + 12 - 3 = 33$$
$$35 + 12 - 1 = 46$$
$$36 + 12 - 3 = 45$$
$$47 + 12 - 1 = 58 - 48 = 10$$
$$48 + 12 - 3 = 57 - 48 = 9$$

Coil connections

$$11 \leftarrow 22 \rightarrow 11$$
$$12 \leftarrow 21 \rightarrow 12$$
$$23 \leftarrow 34 \rightarrow 23$$
$$24 \leftarrow 33 \rightarrow 24$$
$$35 \leftarrow 46 \rightarrow 35$$
$$36 \leftarrow 45 \rightarrow 36$$
$$47 \leftarrow 10 \rightarrow 47$$
$$48 \leftarrow 9 \rightarrow 48$$

Winding table

Coil group-1: $11 \leftarrow 22 \rightarrow 12 \leftarrow 21$

Coil group-2: $23 \leftarrow 34 \rightarrow 24 \leftarrow 33$

Coil group-3: $35 \leftarrow 46 \rightarrow 36 \leftarrow 45$

Coil group-4: $47 \leftarrow 40 \rightarrow 48 \leftarrow 9$

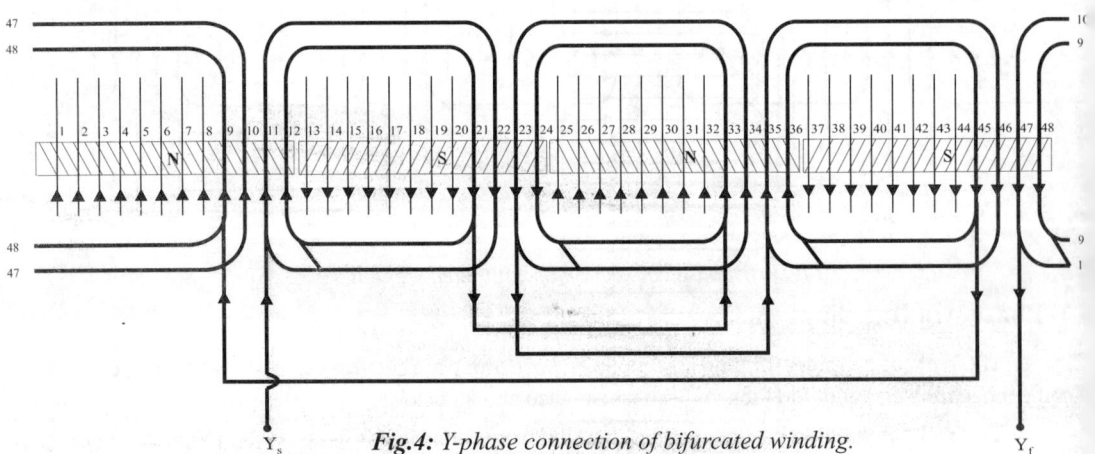

Fig.4: Y-phase connection of bifurcated winding.

3-phase winding diagram

The R-phase, B-phase and Y-phase winding diagrams are shown separately in Figs. 2, 3 and 4 respectively. In each phase winding diagram there are four coil groups and the coil groups of a phase are connected in series to form the winding of a phase. The combined three-phase winding diagram is shown in Fig. 5. The three-phase winding has six terminals which can be connected in star or delta.

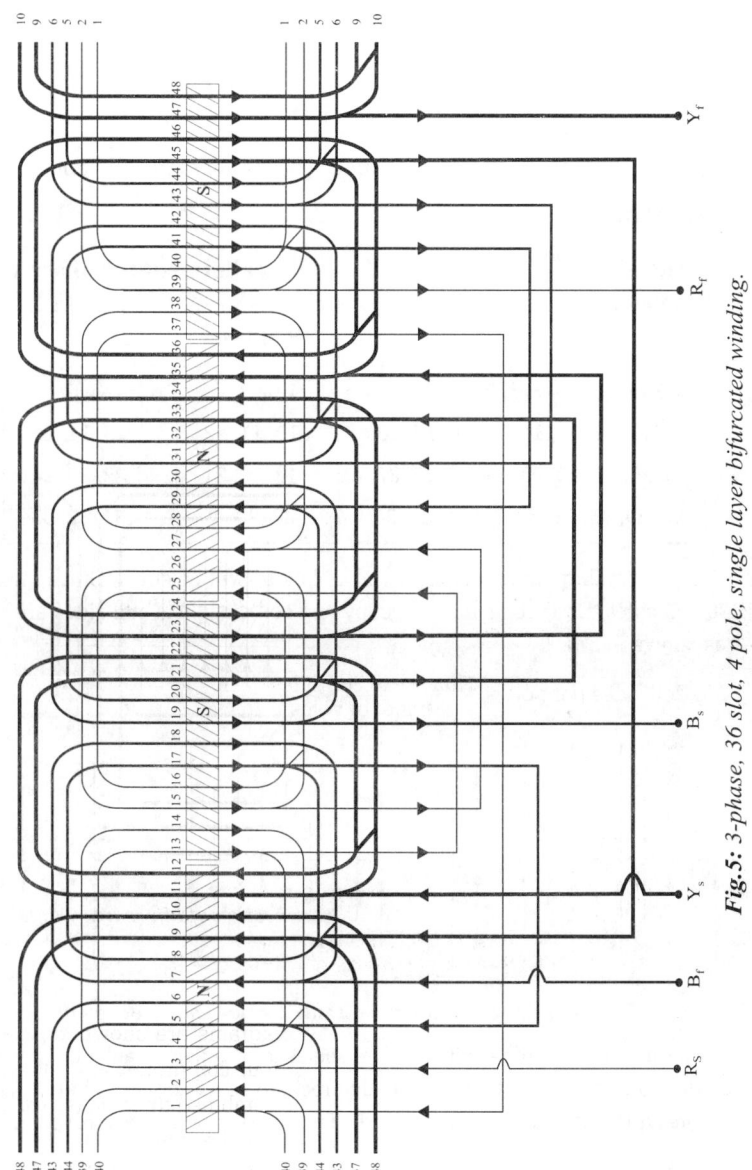

Fig.5: 3-phase, 36 slot, 4 pole, single layer bifurcated winding.

EXAMPLE 5.8

Draw a winding diagram for a 4 pole, 36 slot, 3-phase mush connected armature.

SOLUTION

Slots per pole per phase, $q = \dfrac{36}{3 \times 4} = 3$

Coil span $= \dfrac{36}{4} = 9$

Thus a phase group has 3 slots.

The allotment of slots for conductors of various phases are shown in Table 1.

Table 1: Slot Allotment

Slots in a pole group	Slot allotment for phase sequence RYB		
	R-phase	B-phase	Y-phase
1, 2, 3, 4, 5, 6, 7, 8, 9	1, 2, 3	4, 5, 6	7, 8, 9
10, 11, 12, 13, 14, 15, 16, 17, 18	10, 11, 12	13, 14, 15	16, 17, 18
19, 20, 21, 22, 23, 24, 25, 26, 27	19, 20, 21	22, 23, 24	25, 26, 27
28, 29, 30, 31, 32, 33, 34, 35, 36	28, 29, 30	31, 32, 33	34, 35, 36

In single layer winding number of slots and coil sides will be equal. Since the armature has 36 slots, there will be 36 coil sides. So that each slot occupy one coil side. The coil sides are represented by parallel straight lines as shown in Fig. 1.

$$\text{Coil sides per pole} = \frac{\text{Number of coil sides}}{\text{Number of poles}} = \frac{36}{4} = 9$$

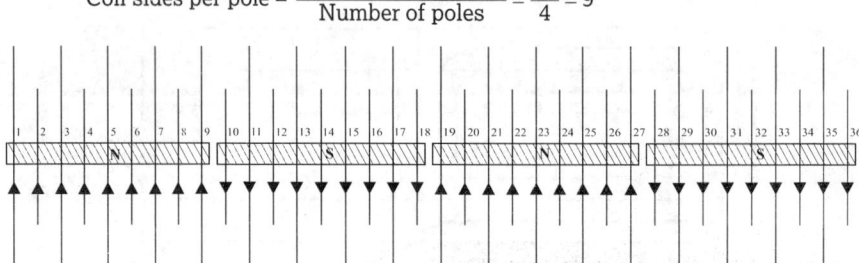

Fig. 1: Representation of coil sides, poles and direction of current.

In this armature winding, nine coil sides are enclosed by a pole at any one time instant. The enclosure of coil sides by the poles are represented by shaded rectangles as shown in Fig. 1. The direction of current through conductors lying under north pole are marked upwards and that of south pole are marked downwards.

In this winding the long coil side of a phase in a pole group is connected to short coil side of same phase in next pole group.

Since number of poles p is 4, each phase will have, p/2 = 4/2 = 2 coil groups. The two coil groups of a phase are connected in series to form winding of a phase.

In order to provide 120° phase shift between emf induced in R, Y and B-phases with respect of each other, the start of R, Y and B-phases should have an angular spacing of 120 °e.

In this example,

$$\text{Slot angle, } \beta = \frac{180°}{\text{Slots per pole}} = \frac{180°}{9} = 20 \text{ °e}$$

Therefore, the number of slots between start of R-phase and Y-phase should be 6 slots (6 × 20° = 120 °e). Similarly, the start of B-phase should be 6 slots away from start of Y-phase for a phase sequence of RYB.

Let us start R-phase from slot 1 and so Y-phase start from 7th slot (1 + 6 = 7) and B-phase start from 13th slot (7 + 6 = 13).

R-Phase winding diagram

The R-phase coil sides forming pole group and coil group are listed in Table 2. The winding calculations, coil connections and winding table of R-phase are also shown below.

The R-phase winding has two coil groups which are connected in series to form R-phase winding. The series connection is made by following the current directions of the coil group.

Table 2: R-Phase slots and Coil groups

Slots in pole group	Long coil side	Short coil side	Coil group
1, 2, 3	1	10	1, 10
10, 11, 12	3	12	3, 12
19, 20, 21	11	20	11, 20
28, 29, 30	19	28	19, 28
	21	30	21, 30
	29	2	29, 2

Winding calculations

1 + 9 = 10

3 + 9 = 12

11 + 9 = 20

19 + 9 = 28

21 + 9 = 30

29 + 9 = 38 − 36 = 2

Coil connections

1 ← 10 → 1

3 ← 12 → 3

11 ← 20 → 11

19 ← 28 → 19

21 ← 30 → 21

29 ← 2 → 29

Winding table

Coil group-1: 1 ← 10 → 3 ← 12 → 20 ← 11

Coil group-2: 19 ← 28 → 21 ← 30 → 2 ← 29

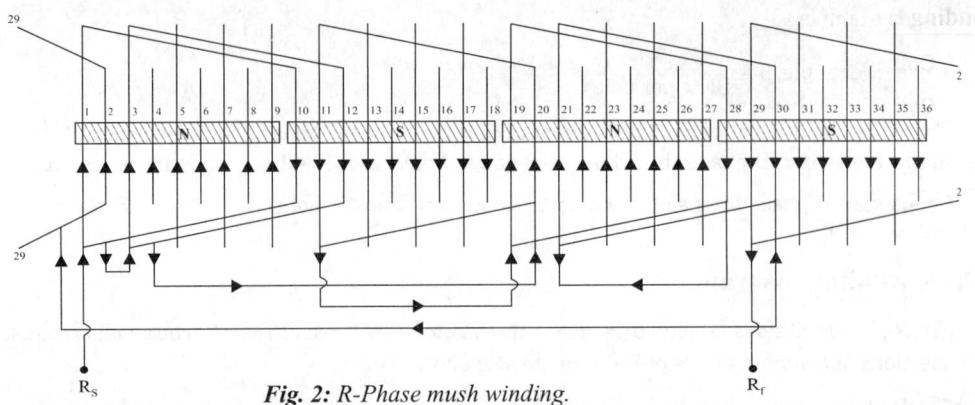

Fig. 2: R-Phase mush winding.

B-Phase winding diagram

The B-phase coil sides forming pole group and coil group are listed in Table 3. The winding calculations, coil connections and winding table of B-phase are also shown below.

The B-phase winding has two coil groups which are connected in series to form B-phase winding. The series connection is made by following the current directions of the coil group.

Table 3: B-Phase slots and Coil groups

Slots in pole group	Long coil side	Short coil side	Coil group
4, 5, 6	13	22	13, 22
13, 14, 15	15	24	15, 24
22, 23, 24	23	32	23, 32
31, 32, 33	31	4	31, 4
	33	6	33, 6
	5	14	5, 14

Winding calculations

$13 + 9 = 22$

$15 + 9 = 24$

$23 + 9 = 32$

$31 + 9 = 40 - 36 = 4$

$33 + 9 = 42 - 36 = 6$

$5 + 9 = 14$

Coil connections

$13 \leftarrow 22 \rightarrow 13$

$15 \leftarrow 24 \rightarrow 15$

$23 \leftarrow 32 \rightarrow 23$

$31 \leftarrow 4 \rightarrow 31$

$33 \leftarrow 6 \rightarrow 33$

$5 \leftarrow 14 \rightarrow 5$

Winding table

Coil group-1:   13 ← 22 → 15   ← 24 → 32   ← 23

Coil group-2:   31 ← 4   → 33   ← 6   → 14   ← 5

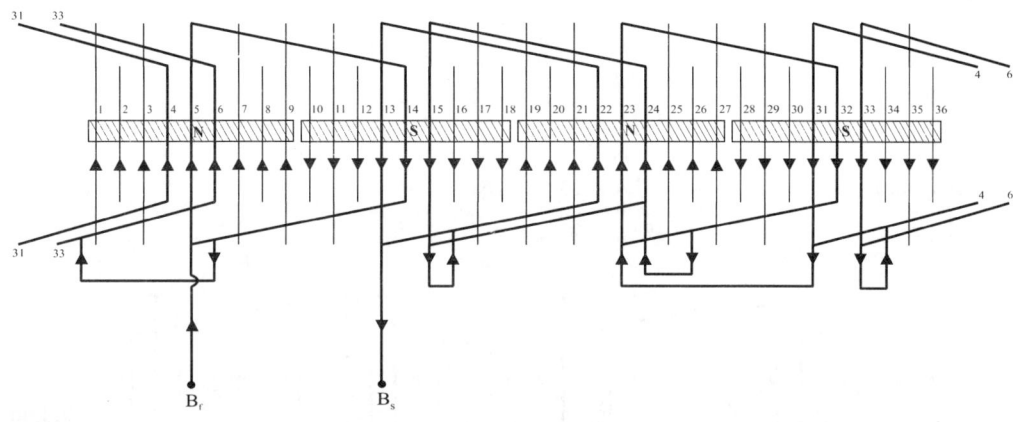

Fig. 3: B-Phase mush winding.

Y-Phase winding diagram

The Y-phase coil sides forming pole group and coil group are listed in Table 4. The winding calculations, coil connections and winding table of B-phase are also shown below.

The Y-phase winding has two coil groups which are connected in series to form Y-phase winding. The series connection is made by following the current directions of the coil group.

Table 4: Y-Phase slots and Coil groups

Slots in pole group	Long coil side	Short coil side	Coil group
7, 8, 9	7	16	7, 16
16, 17, 18	9	18	9, 18
25, 26, 27	17	26	17, 26
34, 35, 36	25	34	25, 34
	27	36	27, 36
	35	8	35, 8

Winding calculations	Coil connections
$7 + 9 = 16$	$7 \leftarrow 16 \rightarrow 7$
$9 + 9 = 18$	$9 \leftarrow 18 \rightarrow 9$
$17 + 9 = 26$	$17 \leftarrow 26 \rightarrow 17$
$25 + 9 = 34$	$25 \leftarrow 34 \rightarrow 25$
$27 + 9 = 36$	$27 \leftarrow 36 \rightarrow 27$
$35 + 9 = 44 - 36 = 8$	$35 \leftarrow 8 \rightarrow 35$

Winding table

Coil group-1: $\quad 7 \quad \leftarrow \quad 16 \quad \rightarrow \quad 9 \quad \leftarrow \quad 18 \quad \rightarrow \quad 26 \quad \leftarrow \quad 17$

Coil group-2: $\quad 25 \quad \leftarrow \quad 34 \quad \rightarrow \quad 27 \quad \leftarrow \quad 36 \quad \rightarrow \quad 8 \quad \leftarrow \quad 35$

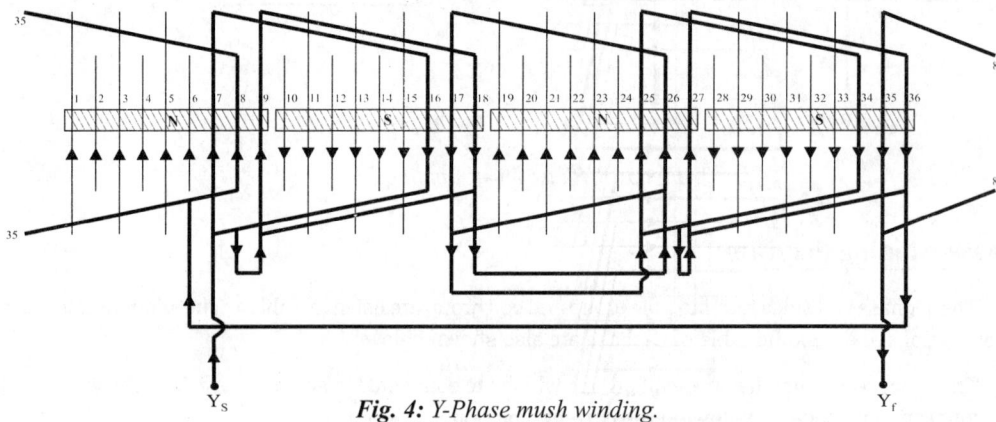

Fig. 4: *Y-Phase mush winding.*

3-phase winding diagram

The R-phase, B-phase and Y-phase winding diagrams are shown separately in Figs. 2, 3 and 4 respectively. In each phase winding diagram there are two coil groups and the coil groups of a phase are connected in series to form the winding of a phase. The combined three-phase winding diagram is shown in Fig. 5. The three-phase winding has six terminals which can be connected in star or delta.

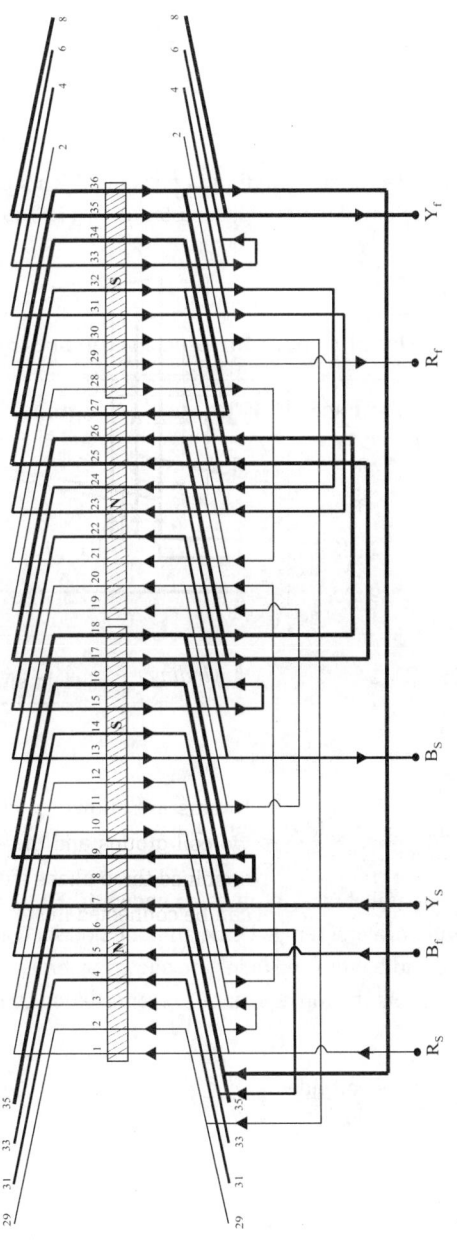

Fig. 5: 3-phase, 36 slot, 4 pole, single layer mush winding.

EXAMPLE 5.9

Draw the 3-phase winding diagram in the developed form for a lap wound 24 slots, 4 pole AC machine armature.

SOLUTION

$$\text{Slots per pole, n} = \frac{S}{p} = \frac{24}{4} = 6$$

$$\text{Slots per pole per phase, q} = \frac{n}{3} = \frac{6}{3} = 2$$

Here, n and q are integers are so integral slot winding is possible. Let us choose full pitch coils for winding. The allotments of slots for conductors of various phases are shown in Table 1.

Table 1: Slot Allotment

Slots in a pole group	Slot allotment for phase sequence RYB		
	R-phase	B-phase	Y-phase
1, 2, 3, 4, 5, 6	1, 2	3, 4	5, 6
7, 8, 9, 10, 11, 12	7, 8	9, 10	11, 12
13, 14, 15, 16, 17, 18	13, 14	15, 16	17, 18
19, 20, 21, 22, 23, 24	19, 20	21, 22	23, 24

Given that number of slots, S = 24

In simplex lap winding, S = C

∴ Number of coils, C = 24

Number of coil sides = 2C = 2 × 24 = 48

$$\text{Coil sides per slot} = \frac{\text{Number of coil sides}}{\text{Number of slots}} = \frac{48}{24} = 2$$

The coil sides are represented by parallel straight lines as shown in Fig. 1. Since two coil sides are accommodated in one slot, each slot will have one top coil side and one bottom coil side. The top coil sides are shown by solid (or continuous) lines and bottom coil sides are shown by broken (or discontinuous) lines. In Fig. 1 the coil sides are numbered such that the top coil sides are represented by odd numbers and bottom coil sides by even numbers.

$$\text{Coil sides per pole} = \frac{\text{Number of coil sides}}{\text{Number of poles}} = \frac{48}{4} = 12$$

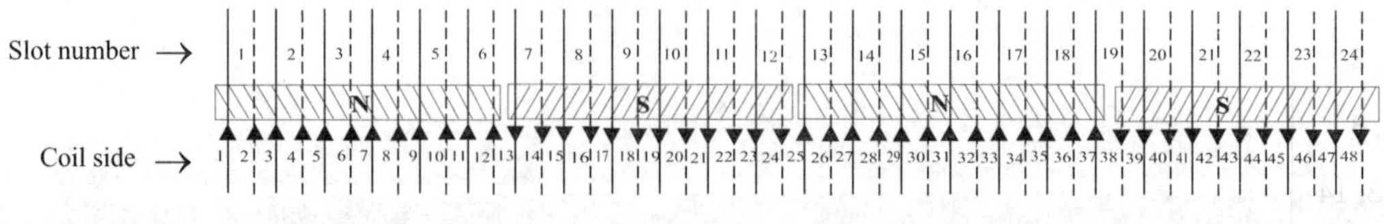

Fig. 1: Representation of coil sides, poles and direction of current.

In this armature winding, twelve coil sides are enclosed by a pole at any one time instant. The enclosure of coil sides by the poles are represented by shaded rectangles as shown in Fig. 1. The direction of current through conductors lying under north pole are marked upwards and that of south pole are marked downwards.

The winding pitches are calculated as shown below.

Let us choose progressive lap winding.

$$\text{Back pitch, } Y_b = \frac{2C}{p} + K = \frac{2 \times 24}{4} + K = 12 + K$$

In integral slot winding with full pitch coils,
$\text{Coil span} = \dfrac{2C}{p}$ and $K = 1$

Let, K = 1,

$$\therefore \text{ Back pitch, } \quad Y_b = 12 + 1 = 13 \text{ coil sides}$$

$$\text{Winding pitch, } Y = 2 \text{ (For progressive lap, Y is always +2)}$$

$$\text{Front pitch, } \quad Y_f = Y_b - Y = 13 - 2 = 11 \text{ coil sides}$$

The front and back connections of the coil sides are determined as shown below.

Let us start with top coil side marked 1.

Bottom coil side of the first coil = Top coil side of first coil + Back pitch
(Back connection)

$$= 1 + Y_b = 1 + 13 = 14$$

Top coil side of the second coil = Bottom coil side of first coil – Front pitch
(Front connection)

$$= 14 - Y_f = 14 - 11 = 3$$

Similarly, all the R-phase, B-phase and Y-phase winding connections are determined. The entire winding calculations, details of R-phase, B-phase and Y-phase winding connections and the winding table are shown below.

In order to provide 120° phase shift between emf induced in R, Y and B-phases with respect of each other, the start of R, Y and B-phases should have an angular spacing of 120 °e.

In this example,

$$\text{Slot angle, } \beta = \frac{180°}{\text{Slots per pole}} = \frac{180°}{6} = 30 \text{ °e}$$

Therefore, the number of slots between start of R-phase and Y-phase should be 4 slots (4 × 30° = 120 °e). Similarly, the start of B-phase should be 4 slots away from start of Y-phase for a phase sequence of RYB.

Let us start R-phase from slot 1 and so Y-phase start from 5th slot (1 + 4 = 5) and B-phase start from 9th slot (5 + 4 = 9).

R-Phase winding diagram

Table 2: R-Phase slots, Coil sides and Coil groups

Pole group		Coil sides of coil groups			
Slots	Coil sides	Coil group-1	Coil group-2	Coil group-3	Coil group-4
1, 2	1, 2, 3, 4	1, 3, 14, 16			
7, 8	13, 14, 15, 16		13, 15, 26, 28		
13, 14	25, 26, 27, 28			25, 7, 38, 40	
19, 20	37, 38, 39, 40				37, 39, 2, 4

Winding calculations

$1 + 13 = 14$	$13 + 13 = 26$	$25 + 13 = 38$	$37 + 13 = 50 - 48 = 2$
$14 - 11 = 3$	$26 - 11 = 15$	$38 - 11 = 27$	$50 - 11 = 39$
$3 + 13 = 16$	$15 + 13 = 28$	$27 + 13 = 40$	$39 + 13 = 52 - 48 = 4$

Back connections

$1 \leftarrow 14$	$13 \leftarrow 26$	$25 \leftarrow 38$	$37 \leftarrow 2$
$3 \leftarrow 16$	$15 \leftarrow 28$	$27 \leftarrow 40$	$39 \leftarrow 4$

Front connections

$14 \rightarrow 3$	$26 \rightarrow 15$	$38 \rightarrow 27$	$2 \rightarrow 39$

Winding table

Coil group-1: $1 \leftarrow 14 \rightarrow 3 \leftarrow 16$ **Coil group-3:** $25 \leftarrow 38 \rightarrow 27 \leftarrow 40$

Coil group-2: $13 \leftarrow 26 \rightarrow 15 \leftarrow 28$ **Coil group-4:** $37 \leftarrow 2 \rightarrow 39 \leftarrow 4$

The four coil groups are connected in series to form R-phase winding. The series connection is made by following the current direction of the coil groups.

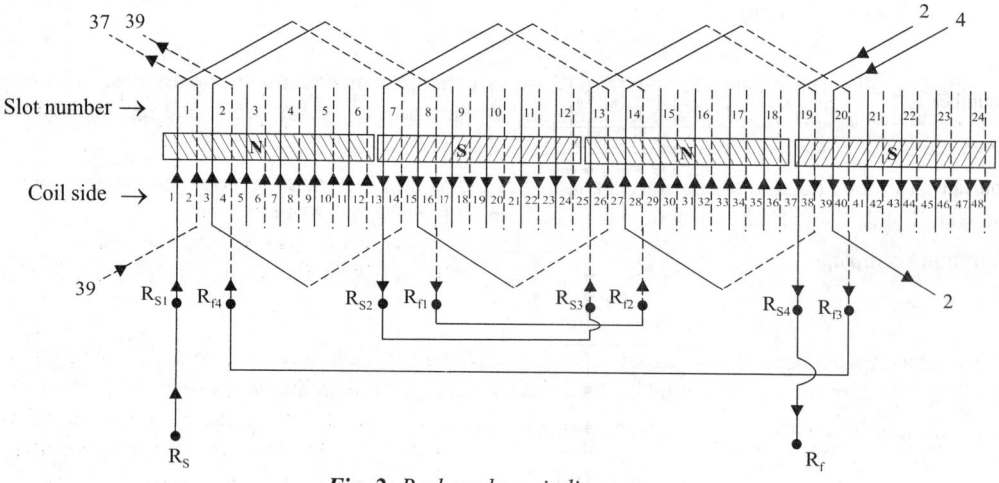

Fig. 2: R-phase lap winding.

B-Phase winding diagram

Table 3: B-Phase slots, Coil sides and Coil groups

Pole group		Coil sides of coil groups			
Slots	Coil sides	Coil group-1	Coil group-2	Coil group-3	Coil group-4
9, 10	17, 18, 19, 20	17, 19, 30, 32			
15, 16	29, 30, 31, 32		29, 31, 42, 44		
21, 22	41, 42, 43, 44			41, 43, 6, 8	
3, 4	5, 6, 7, 8				5, 7, 18, 20

Winding calculations

$17 + 13 = 30$	$29 + 13 = 42$	$41 + 13 = 54 - 48 = 6$	$5 + 13 = 18$
$30 - 11 = 19$	$42 - 11 = 31$	$54 - 11 = 43$	$18 - 11 = 7$
$19 + 13 = 32$	$31 + 13 = 44$	$43 + 13 = 56 - 48 = 8$	$7 + 13 = 20$

Back connections

$17 \leftarrow 30$	$29 \leftarrow 42$	$41 \leftarrow 6$	$5 \leftarrow 18$
$19 \leftarrow 32$	$31 \leftarrow 44$	$43 \leftarrow 8$	$7 \leftarrow 20$

Front connections

$30 \rightarrow 19$	$42 \rightarrow 31$	$6 \rightarrow 43$	$18 \rightarrow 7$

Winding table

Coil group-1: $17 \leftarrow 30 \rightarrow 19 \leftarrow 32$ **Coil group-3:** $41 \leftarrow 6 \rightarrow 43 \leftarrow 8$

Coil group-2: $29 \leftarrow 42 \rightarrow 31 \leftarrow 44$ **Coil group-4:** $5 \leftarrow 18 \rightarrow 7 \leftarrow 20$

The four coil groups are connected in series to form B-phase winding. The series connection is made by following the current direction of the coil groups.

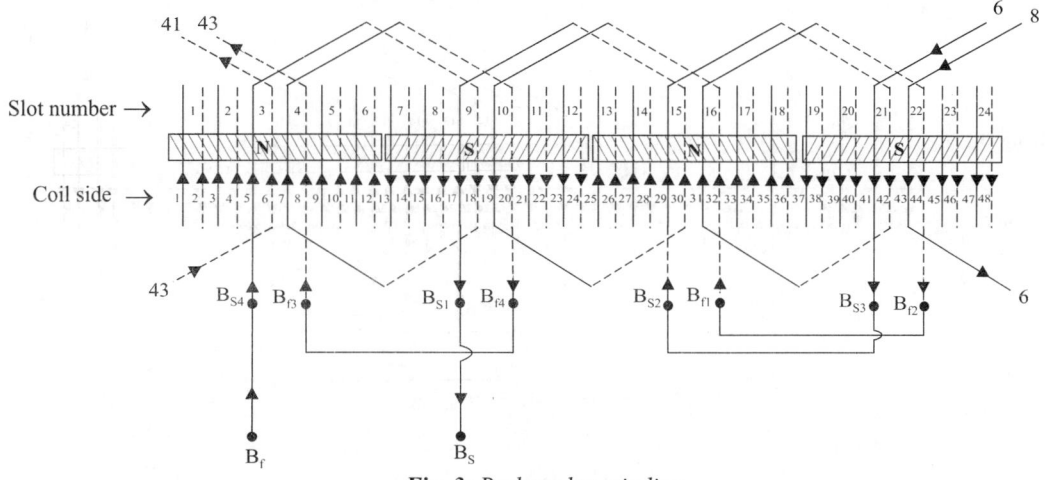

Fig. 3: B-phase lap winding.

Y-Phase winding diagram

Table 4: Y-Phase slots, Coil sides and Coil groups

Pole group		Coil sides of coil groups			
Slots	Coil sides	Coil group-1	Coil group-2	Coil group-3	Coil group-4
5, 6	9, 10, 11, 12	9, 11, 22, 24			
11, 12	21, 22, 23, 24		21, 23, 34, 36		
17, 18	33, 34, 35, 36			33, 35, 46, 48	
23, 24	45, 46, 47, 48				45, 47, 10, 12

Winding calculations

$9 + 13 = 22$	$21 + 13 = 34$	$33 + 13 = 46$	$45 + 13 = 58 - 48 = 10$
$22 - 11 = 11$	$34 - 11 = 23$	$46 - 11 = 35$	$58 - 11 = 47$
$11 + 13 = 24$	$23 + 13 = 36$	$35 + 13 = 48$	$47 + 13 = 60 - 48 = 12$

Back connections

$9 \leftarrow 22$	$21 \leftarrow 34$	$33 \leftarrow 46$	$45 \leftarrow 10$
$11 \leftarrow 24$	$23 \leftarrow 36$	$35 \leftarrow 48$	$47 \leftarrow 12$

Front connections

$22 \rightarrow 11$	$34 \rightarrow 23$	$46 \rightarrow 35$	$10 \rightarrow 47$

Winding table

Coil group-1: $9 \leftarrow 22 \rightarrow 11 \leftarrow 24$ **Coil group-3:** $33 \leftarrow 46 \rightarrow 35 \leftarrow 48$

Coil group-2: $21 \leftarrow 34 \rightarrow 23 \leftarrow 36$ **Coil group-4:** $45 \leftarrow 10 \rightarrow 47 \leftarrow 12$

The four coil groups are connected in series to form Y-phase winding. The series connection is made by following the current direction of the coil groups.

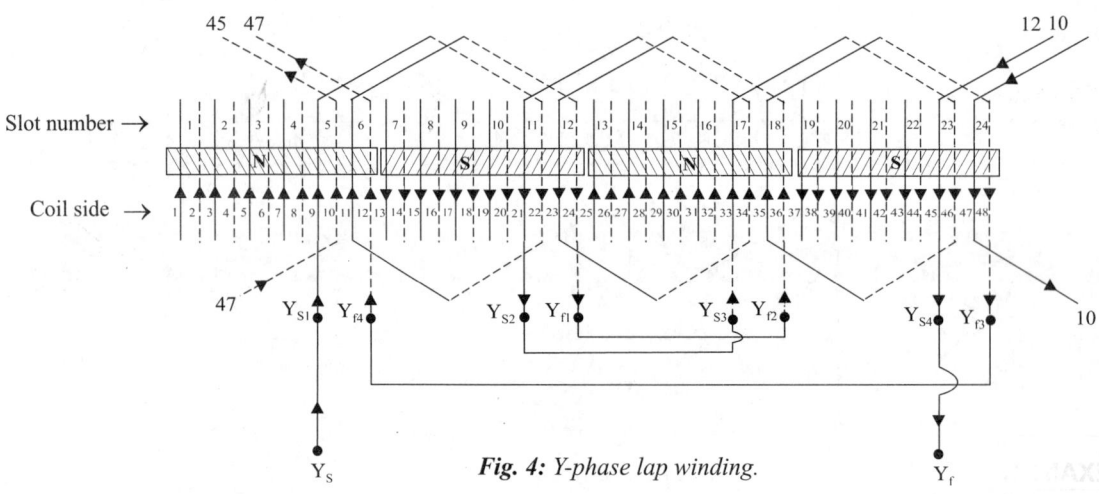

Fig. 4: *Y-phase lap winding.*

3-Phase winding diagram

The R-phase, B-phase and Y-phase winding diagrams are shown separately in Figs. 2, 3 and 4 respectively. In each phase winding diagram there are four coil groups and the coil groups of a phase are connected in series to form the winding of a phase. The combined three-phase winding diagram is shown in Fig. 5. The three-phase winding has six terminals which can be connected in star or delta.

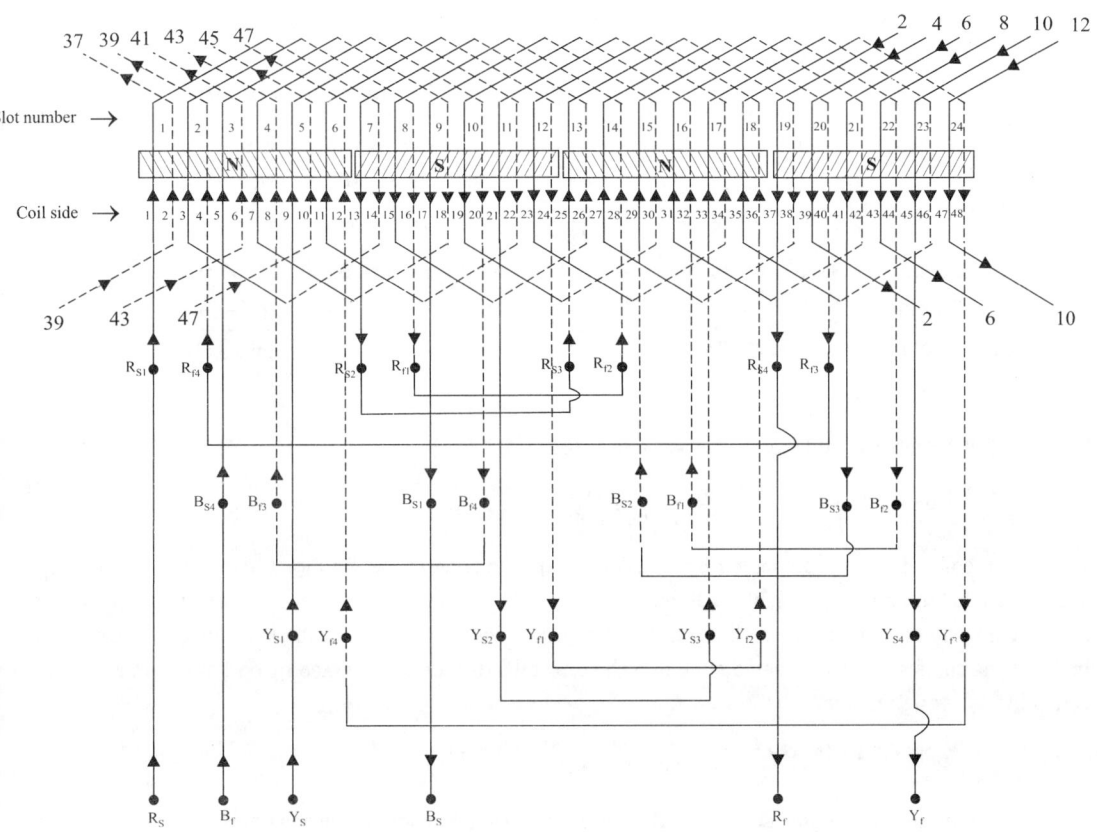

Fig. 5: *Simplex lap winding of a 3-phase, 4 pole AC machine armature with 12 slots.*

EXAMPLE 5.10

Draw the 3-phase winding diagram in the developed form for a lap wound 36 slots, 4 pole AC machine armature.

SOLUTION

$$\text{Slots per pole, n} = \frac{S}{p} = \frac{36}{4} = 9$$

$$\text{Slots per pole per phase, q} = \frac{n}{3} = \frac{9}{3} = 3$$

Here n and q are integers and so integral slot winding is possible. Let us choose full pitch coils for winding. The allotment of slots for conductors of various phases are shown in Table 1.

Table 1: Slot Allotment

Slots in a pole group	Slot allotment for phase sequence RYB		
	R-phase	B-phase	Y-phase
1, 2, 3, 4, 5, 6, 7, 8, 9	1, 2, 3	4, 5, 6	7, 8, 9
10, 11, 12, 13, 14, 15, 16, 17, 18	10, 11, 12	13, 14, 15	16, 17, 18
19, 20, 21, 22, 23, 24, 25, 26, 27	19, 20, 21	22, 23, 24	25, 26, 27
28, 29, 30, 31, 32, 33, 34, 35, 36	28, 29, 30	31, 32, 33	34, 35, 36

Given that number of slots = 36

$\boxed{\text{In simplex lap winding, S = C}}$

$\therefore$ Number of coils, C = 36

Number of coil sides = 2C = 2 × 36 = 72

Coil sides per slot = $\dfrac{\text{Number of coil sides}}{\text{Number of slots}} = \dfrac{72}{36} = 2$

The coil sides are represented by parallel straight lines as shown in Fig. 1. Since two coil sides are accommodated in one slot, each slot will have one top coil side and one bottom coil side. The top coil sides are shown by solid (or continuous) lines and bottom coil sides are shown by broken (or discontinuous) lines. In Fig. 1 the coil sides are numbered such that the top coil sides are represented by odd numbers and bottom coil sides by even numbers.

Coil sides per pole = $\dfrac{\text{Number of coil sides}}{\text{Number of poles}} = \dfrac{72}{4} = 18$

In this armature winding, eighteen coil sides are enclosed by a pole at any one time instant. The enclosure of coil sides by the poles are represented by shaded rectangles as shown in Fig. 1. The direction of current through conductors lying under north pole are marked upwards and that of south pole are marked downwards.

The winding pitches are calculated as shown below.

Let us choose progressive lap winding.

Back pitch, $Y_b = \dfrac{2C}{p} + K = \dfrac{2 \times 36}{4} + K = 18 + K$

Let, K = 1,

$\boxed{\begin{array}{l}\text{In integral slot winding} \\ \text{with full pitch coils,} \\[4pt] \text{Coil span} = \dfrac{2C}{p} \text{ and } K = 1\end{array}}$

$\therefore$ Back pitch, Y_b = 18 + 1 = 19 coil sides

Winding pitch, Y = 2 (For progressive lap, Y is always +2)

Front pitch, $Y_f = Y_b - Y = 19 - 2 = 17$ coil sides

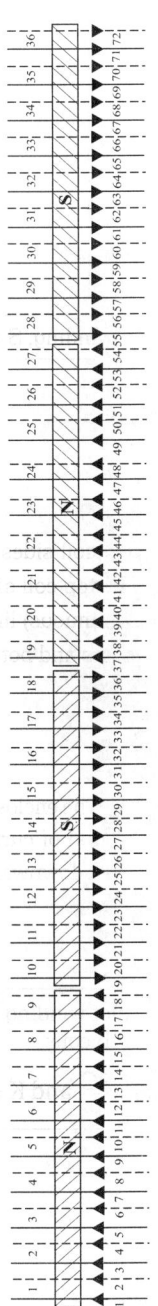

Fig. 1: Representation of coil sides, poles and direction of current.

The front and back connections of the coil sides are determined as shown below.

Let us start with top coil side marked 1.

Bottom coil side of the first coil = Top coil side of first coil + Back pitch

(Back connection)

$$= 1 + Y_b = 1 + 19 = 20$$

Top coil side of the second coil = Bottom coil side of first coil – Front pitch

(Front connection)

$$= 20 – Y_f = 20 – 17 = 3$$

Similarly, all the R-phase, B-phase and Y-phase winding connections are determined. The entire winding calculations, details of R-phase, B-phase and Y-phase winding connections and the winding table are shown ahead.

In order to provide 120° phase shift between emf induced in R, Y and B-phases with respect of each other, the start of R, Y and B-phases should have an angular spacing of 120 °e.

In this example,

Slot angle, $\beta = \dfrac{180°}{\text{Slots per pole}} = \dfrac{180°}{9} = 20 °e$

Therefore, the number of slots between start of R-phase and Y-phase should be 6 slots (6 × 20° = 120° e). Similarly, the start of B-phase should be 6 slots away from start of Y-phase for a phase sequence of RYB.

Let us start R-phase from slot 1 and so Y-phase start from 7th slot (1 + 6 = 7) and B-phase start from 13th slot (7 + 6 = 13).

R-Phase winding diagram

Table 2: R-Phase slots, Coil sides and Coil groups

Pole group		Coil sides of coil groups			
Slots	Coil sides	Coil group-1	Coil group-2	Coil group-3	Coil group-4
1, 2, 3	1, 2, 3, 4, 5, 6	1, 3, 5, 20, 22, 24	19, 21, 23, 38, 40, 42	37, 39, 41, 56, 58, 60	55, 57, 59, 2, 4, 6
10, 11, 12	19, 20, 21, 22, 23, 24				
19, 20, 21	37, 38, 39, 40, 41, 42				
28, 29, 30	55, 56, 57, 58, 59, 60				

Winding calculations

$1 + 19 = 20$	$19 + 19 = 38$
$20 - 17 = 3$	$38 - 17 = 21$
$3 + 19 = 22$	$21 + 19 = 40$
$22 - 17 = 5$	$40 - 17 = 23$
$5 + 19 = 24$	$23 + 19 = 42$

$37 + 19 = 56$	$55 + 19 = 74 - 72 = 2$
$56 - 17 = 39$	$74 - 17 = 57$
$39 + 19 = 58$	$57 + 19 = 76 - 72 = 4$
$58 - 17 = 41$	$76 - 17 = 59$
$41 + 19 = 60$	$59 + 19 = 78 - 72 = 6$

Back connections

1 ↓ 20	19 ↓ 38	37 ↓ 56	55 ↓ 2
3 ↓ 22	21 ↓ 40	39 ↓ 58	57 ↓ 4
5 ↓ 24	23 ↓ 42	41 ↓ 60	59 ↓ 6

Front connections

20 → 3	38 → 21	56 → 39	74 → 57
22 → 5	40 → 23	58 → 41	76 → 59

Winding table

Coil group-1: 1 ← 20 → 3 ← 22 → 5 ← 24

Coil group-2: 19 ← 38 → 21 ← 40 → 23 ← 42

Coil group-3: 37 ← 56 → 39 ← 58 → 41 ← 60

Coil group-4: 55 ← 2 → 57 ← 4 → 59 ← 6

The four coil groups are connected in series to form R-phase winding. The series connection is made by following the current direction of the coil groups.

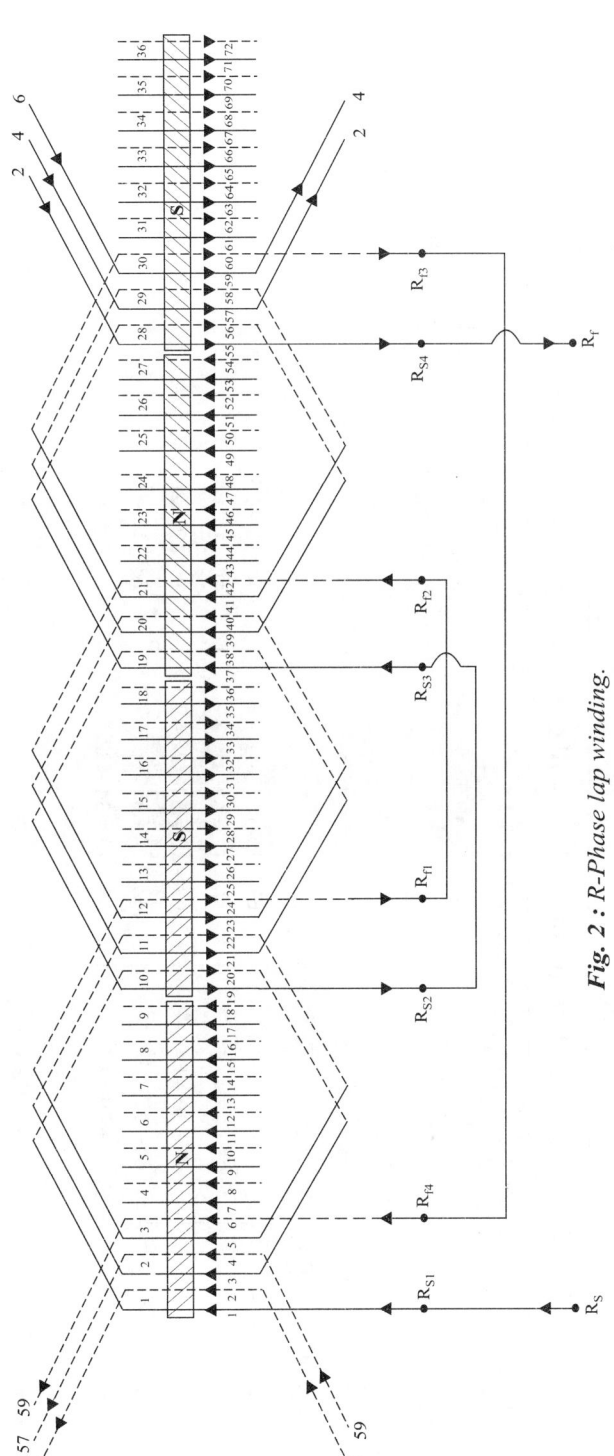

Fig. 2 : R-Phase lap winding.

B-Phase winding diagram

Table 3: B-Phase slots, Coil sides and Coil groups

Pole group		Coil sides of coil groups			
Slots	Coil sides	Coil group-1	Coil group-2	Coil group-3	Coil group-4
13, 14, 15	25, 26, 27, 28, 29, 30	25, 27, 29, 44, 46, 48			
22, 23, 24	43, 44, 45, 46, 47, 48		43, 45, 47, 62, 64, 66		
31, 32, 33	61, 62, 63, 64, 65, 66			61, 63, 65, 8, 10, 12	
4, 5, 6	7, 8, 9, 10, 11, 12				7, 9, 11, 26, 28, 30

Winding calculations

25 + 19 = 44	43 + 19 = 62	61 + 19 = 80 − 72 = 8	7 + 19 = 26
44 − 17 = 27	62 − 17 = 45	80 − 17 = 63	26 − 17 = 9
27 + 19 = 46	45 + 19 = 64	63 + 19 = 82 − 72 = 10	9 + 19 = 28
46 − 17 = 29	64 − 17 = 47	82 − 17 = 65	28 − 17 = 11
29 + 19 = 48	47 + 19 = 66	65 + 19 = 84 − 72 = 12	11 + 19 = 30

Back connections

25 ← 44	43 ← 62	61 ← 8	7 ← 26
27 ← 46	45 ← 64	63 ← 10	9 ← 28
29 ← 48	47 ← 66	65 ← 12	11 ← 30

Front connections

44 → 27	62 → 45	80 → 63	26 → 9
46 → 29	64 → 47	82 → 65	28 → 11

Winding table

Coil group-1: 25 ← 44 → 27 ← 46 → 29 ← 48

Coil group-2: 43 ← 62 → 45 ← 64 → 47 ← 66

Coil group-3: 61 ← 8 → 63 ← 10 → 65 ← 12

Coil group-4: 7 ← 26 → 9 ← 28 → 11 ← 30

The four coil groups are connected in series to form B-phase winding. The series connection is made by following the current direction of the coil groups.

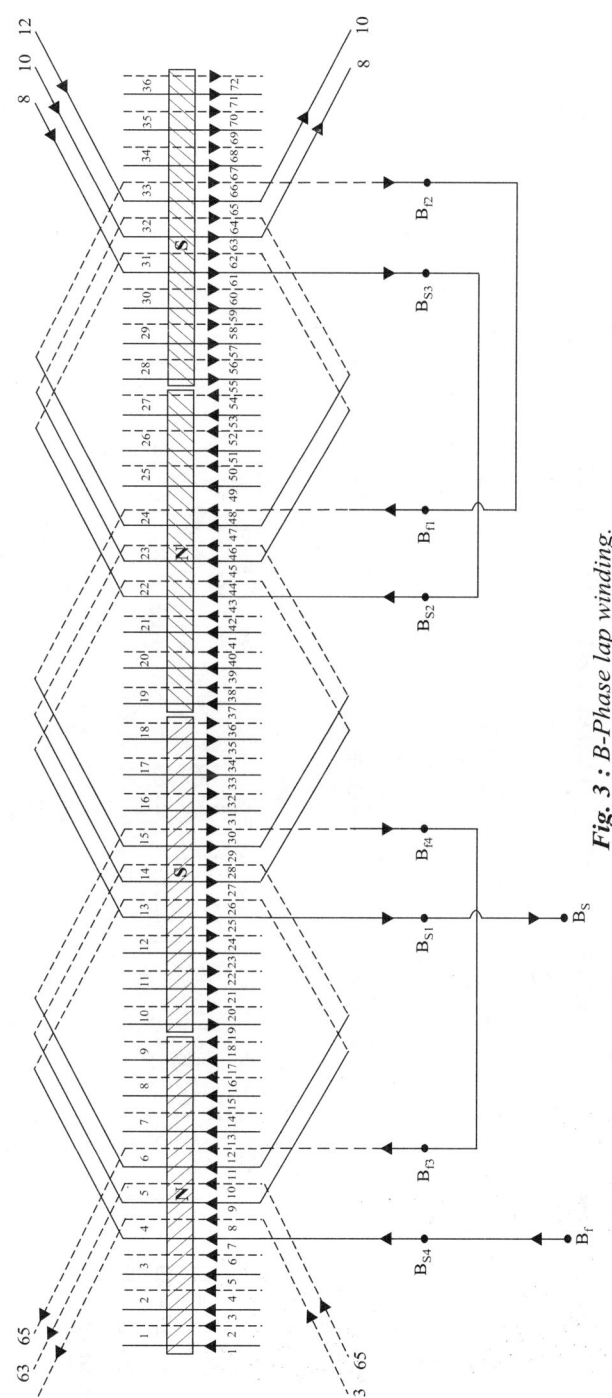

Fig. 3 : B-Phase lap winding.

Y-Phase winding diagram

Table 4: Y-Phase slots, Coil sides and Coil groups

Pole group		Coil sides of coil groups			
Slots	Coil sides	Coil group-1	Coil group-2	Coil group-3	Coil group-4
7, 8, 9	13, 14, 15, 16, 17, 18	13, 15, 17, 32, 34, 36			
16, 17, 18	31, 32, 33, 34, 35, 36		31, 33, 35, 50, 52, 54		
25, 26, 27	49, 50, 51, 52, 53, 54			49, 51, 53, 68, 70, 72	
34, 35, 36	67, 68, 69, 70, 71, 72				67, 69, 71, 14, 16, 18

Winding calculations

$13 + 19 = 32$	$31 + 19 = 50$	$49 + 19 = 68$	$67 + 19 = 86 - 72 = 14$
$32 - 17 = 15$	$50 - 17 = 33$	$68 - 17 = 51$	$86 - 17 = 69$
$15 + 19 = 34$	$33 + 19 = 52$	$51 + 19 = 70$	$69 + 19 = 88 - 72 = 16$
$34 - 17 = 17$	$52 - 17 = 35$	$70 - 17 = 53$	$88 - 17 = 71$
$17 + 19 = 36$	$35 + 19 = 54$	$53 + 19 = 72$	$71 + 19 = 90 - 72 = 18$

Back connections

$13 \leftarrow 32$	$31 \leftarrow 50$	$49 \leftarrow 68$	$67 \leftarrow 14$
$15 \leftarrow 34$	$33 \leftarrow 52$	$51 \leftarrow 70$	$69 \leftarrow 16$
$17 \leftarrow 36$	$35 \leftarrow 54$	$53 \leftarrow 72$	$71 \leftarrow 18$

Front connections

$32 \rightarrow 15$	$50 \rightarrow 33$	$68 \rightarrow 51$	$86 \rightarrow 69$
$37 \rightarrow 17$	$52 \rightarrow 35$	$70 \rightarrow 53$	$88 \rightarrow 71$

Winding table

Coil group-1: 13 ← 32 → 15 ← 34 → 17 ← 36

Coil group-2: 31 ← 50 → 33 ← 52 → 35 ← 54

Coil group-3: 49 ← 68 → 51 ← 70 → 53 ← 72

Coil group-4: 67 ← 14 → 69 ← 16 → 71 ← 18

The four coil groups are connected in series to form Y-phase winding. The series connection is made by following the current direction of the coil groups.

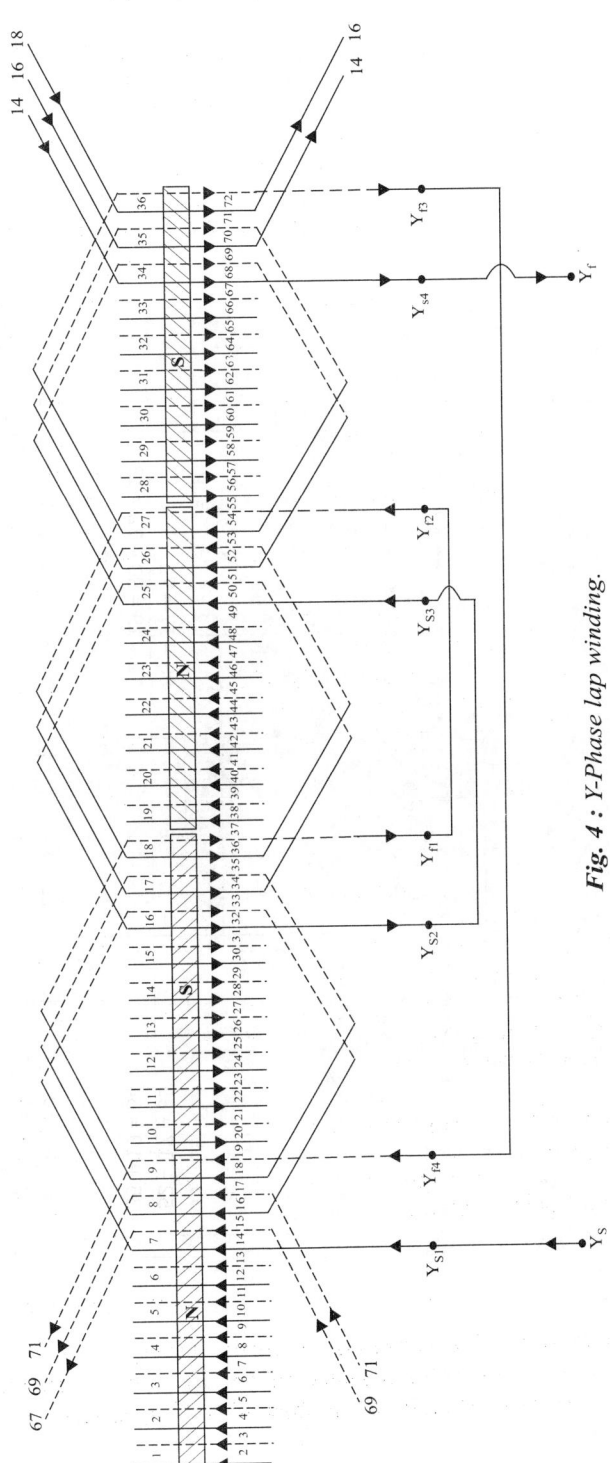

Fig. 4 : Y-Phase lap winding.

3-Phase winding diagram

The R-phase, B-phase and Y-phase winding diagrams are shown separately in Figs. 2, 3 and 4 respectively. In each phase winding diagram there are four coil groups and the coil groups of a phase are connected in series to form the winding of a phase. The combined three-phase winding diagram is shown in Fig. 5. The three-phase winding has six terminals which can be connected in star or delta.

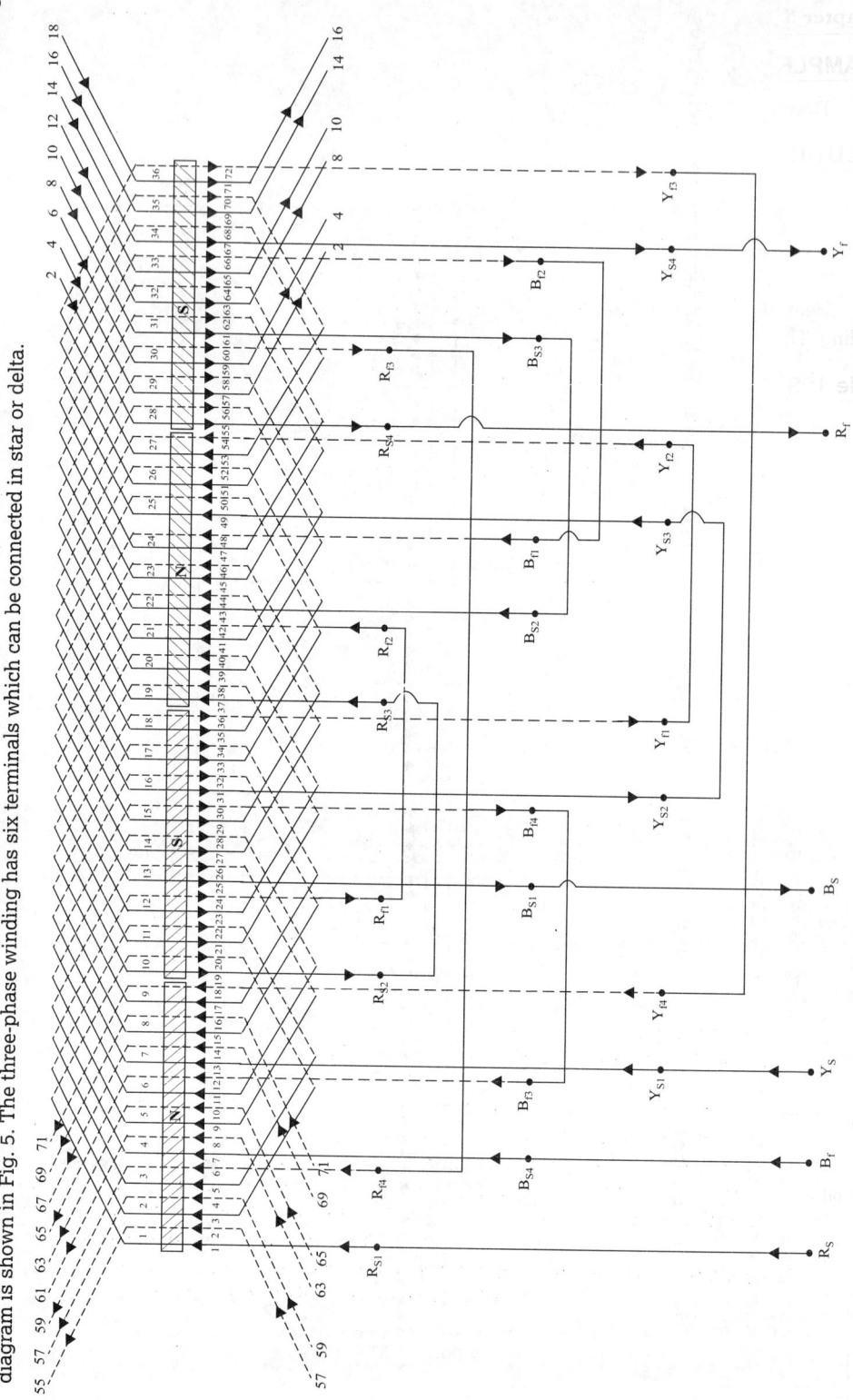

Fig. 5 : Lap winding of a 3-phase, 4 pole AC machine armature with 36 slots.

EXAMPLE 5.11

Draw a 3-phase winding diagram for a wave wound 24 slots, 4 pole AC machine armature.

SOLUTION

Slots per pole, $n = \dfrac{S}{p} = \dfrac{24}{4} = 6$

Slots per pole per phase, $q = \dfrac{n}{3} = \dfrac{6}{3} = 2$

Here n and q are integers and so integral slot winding is possible. Let us choose full pitch coils for winding. The allotment of slots for conductors of various phases are shown in Table 1.

Table 1: Slot Allotment

Slots in a pole group	Slot allotment for phase sequence RYB		
	R-phase	B-phase	Y-phase
1, 2, 3, 4, 5, 6	1, 2	3, 4	5, 6
7, 8, 9, 10, 11, 12	7, 8	9, 10	11, 12
13, 14, 15, 16, 17, 18	13, 14	15, 16	17, 18
19, 20, 21, 22, 23, 24	19, 20	21, 22	23, 24

Given that number of slots = 24

∴ Number of coils, C = 24

Number of coil sides = 2C = 2 × 24 = 48

Coil sides per slot = $\dfrac{\text{Number of coil sides}}{\text{Number of slots}} = \dfrac{48}{24} = 2$

In simplex wave winding, S = C

The coil sides are represented by parallel straight lines as shown in Fig. 1. Since two coil sides are accommodated in one slot, each slot will have one top coil side and one bottom coil side. The top coil sides are shown by solid (or continuous) lines and bottom coil sides are shown by broken (or discontinuous) lines. In Fig. 1 the coil sides are numbered such that the top coil sides are represented by odd numbers and bottom coil sides by even numbers.

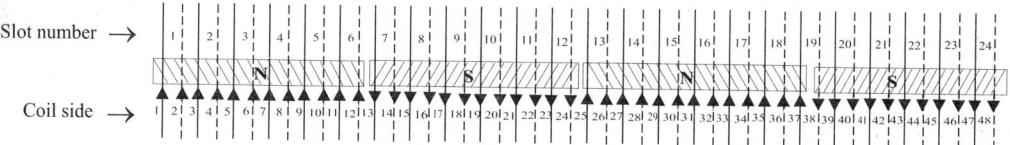

Fig. 1: Representation of coil sides, poles and direction of current.

Coil sides per pole = $\dfrac{\text{Number of coil sides}}{\text{Number of poles}} = \dfrac{48}{4} = 12$

In this armature winding 12 coil sides are enclosed by a pole at any one time instant. The enclosure of coil sides by the poles are represented by shaded rectangles as shown in Fig. 1. The direction of current through conductors lying under north pole are marked upwards and that of south pole are marked downwards.

The winding pitches are calculated as shown below.

Let us choose progressive wave winding.

$$\text{Back pitch, } Y_b = \frac{2C}{p} + K = \frac{2 \times 24}{4} + K = 12 + K$$

Let, $K = 1$,

$\therefore$ Back pitch, $Y_b = 12 + 1 = 13$ coil sides

$$\text{Winding pitch, } Y = \frac{4C}{p} = \frac{4 \times 24}{4} = 24 \text{ coil sides}$$

$\therefore$ Front pitch, $Y_f = Y - Y_b = 24 - 13 = 11$ coil sides

> In integral slot winding with full pitch coils,
>
> Coil span $= \dfrac{2C}{p}$ and $K = 1$

The front and back connections of the coil sides are determined as shown below.

Let us start with top coil side marked 1.

Bottom coil side of the first coil = Top coil side of first coil + Back pitch

(Back connection)

$$= 1 + Y_b = 1 + 13 = 14$$

Top coil side of the coil lying one pole $\Big\}$ = Bottom coil side of first coil + Front pitch
pitch from first coil (Front connection)$\Big\}$

$$= 14 + Y_f = 14 + 11 = 25$$

Similarly, all the R-phase, B-phase and Y-phase connections are determined. The entire winding calculations, details of R-phase, B-phase and Y-phase connections and the winding table are also shown below.

In wave winding there will be two coil groups in each phase. The direction of connection for second coil group is reverse that of first coil group and this is referred as clockwise and anticlockwise winding calculations.

In order to provide 120° phase shift between emf induced in R, Y and B-phases with respect to each other, the start of R, Y and B-phases should have an angular spacing of 120 °e.

In this example,

$$\text{Slot angle, } \beta = \frac{180°}{\text{Slots per pole}} = \frac{180°}{6} = 30 \text{ °e}$$

Therefore, the number of slots between start of R-phase and Y-phase should be 4 slots (4 × 30° = 120 °e). Similarly, the start of B-phase should be 4 slots away from start of Y-phase for a phase sequence of RYB.

Let us start R-phase from slot 1 and so Y-phase start from 5th slot (1 + 4 = 5) and B-phase start from 9th slot (5 + 4 = 9).

R-Phase winding diagram

Table 2: R-Phase slots, Coil sides and Coil groups

Pole group		Coil sides of coil groups	
Slots	Coil sides	Coil group-1	Coil group-2
1, 2	1, 2, 3, 4		
7, 8	13, 14, 15, 16	1, 14, 25, 38, 3, 16, 27, 40	
13, 14	25, 26, 27, 28		
19, 20	37, 38, 39, 40		4, 39, 28, 15, 2, 37, 26, 13

Clockwise winding calculations

$1 + 13 = 14$

$14 + 11 = 25$

$25 + 13 = 38$

$38 + 11 = 49 - 48 = 1 + 2 = 3$

$3 + 13 = 16$

$16 + 11 = 27$

$27 + 13 = 40$

$40 + 11 = 51 - 48 = 3 + 1 = 4$

Anticlockwise winding calculations

$4 + 48 - 13 = 39$

$39 - 11 = 28$

$28 - 13 = 15$

$15 - 11 = 4 - 2 = 2$

$2 + 48 - 13 = 37$

$37 - 11 = 26$

$26 - 13 = 13$

Back connections

$1 \leftarrow 14$

$25 \leftarrow 38$

$3 \leftarrow 16$

$27 \leftarrow 40$

$4 \leftarrow 39$

$28 \leftarrow 15$

$2 \leftarrow 37$

$26 \leftarrow 13$

Front connections

$14 \rightarrow 25$

$38 \rightarrow 3$

$16 \rightarrow 27$

$40 \rightarrow 4$

$39 \rightarrow 28$

$15 \rightarrow 2$

$37 \rightarrow 26$

Winding table

Coil group-1: $\quad 1 \leftarrow 14 \rightarrow 25 \leftarrow 38 \rightarrow 3 \leftarrow 16 \rightarrow 27 \leftarrow 40$

Coil group-2: $\quad 4 \leftarrow 39 \rightarrow 28 \leftarrow 15 \rightarrow 2 \leftarrow 37 \rightarrow 26 \leftarrow 13$

The two coil groups are connected in series to form R-phase winding. The series connection is made by following the current direction of the coil groups.

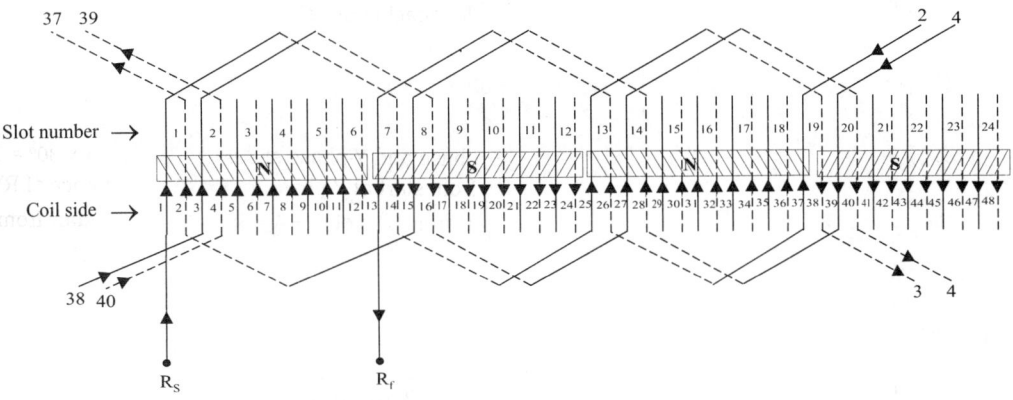

Fig. 2: *R-Phase wave winding.*

B-Phase winding diagram

Table 3: B-Phase slots, Coil sides and Coil groups

Pole group		Coil sides of coil groups	
Slots	Coil sides	Coil group-1	Coil group-2
9, 10	17, 18, 19, 20		
15, 16	29, 30, 31, 32	17, 30, 41, 6, 19, 32, 43, 8	
21, 22	41, 42, 43, 44		
3, 4	5, 6, 7, 8		20, 7, 44, 31, 18, 5, 42, 29

Clockwise winding calculations

$17 + 13 = 30$

$30 + 11 = 41$

$41 + 13 = 54 - 48 = 6$

$6 + 11 = 17 + 2 = 19$

$19 + 13 = 32$

$32 + 11 = 43$

$43 + 13 = 56 - 48 = 8$

$8 + 11 = 19 + 1 = 20$

Anticlockwise winding calculations

$20 - 13 = 7$

$7 + 48 - 11 = 44$

$44 - 13 = 31$

$31 - 11 = 20 - 2 = 18$

$18 - 13 = 5$

$5 + 48 - 11 = 42$

$42 - 13 = 29$

Back connections

$17 \leftarrow 30$

$41 \leftarrow 6$

$19 \leftarrow 32$

$43 \leftarrow 8$

$20 \leftarrow 7$

$44 \leftarrow 31$

$18 \leftarrow 5$

$42 \leftarrow 29$

Front connections

$30 \rightarrow 41$

$6 \rightarrow 19$

$32 \rightarrow 43$

$8 \rightarrow 20$

$7 \rightarrow 44$

$31 \rightarrow 18$

$5 \rightarrow 42$

Winding table

Coil group-1: $17 \leftarrow 30 \rightarrow 41 \leftarrow 6 \rightarrow 19 \leftarrow 32 \rightarrow 43 \leftarrow 8$

Coil group-2: $20 \leftarrow 7 \rightarrow 44 \leftarrow 31 \rightarrow 18 \leftarrow 5 \rightarrow 42 \leftarrow 29$

The two coil groups are connected in series to form B-phase winding. The series connection is made by following the current direction of the coil groups.

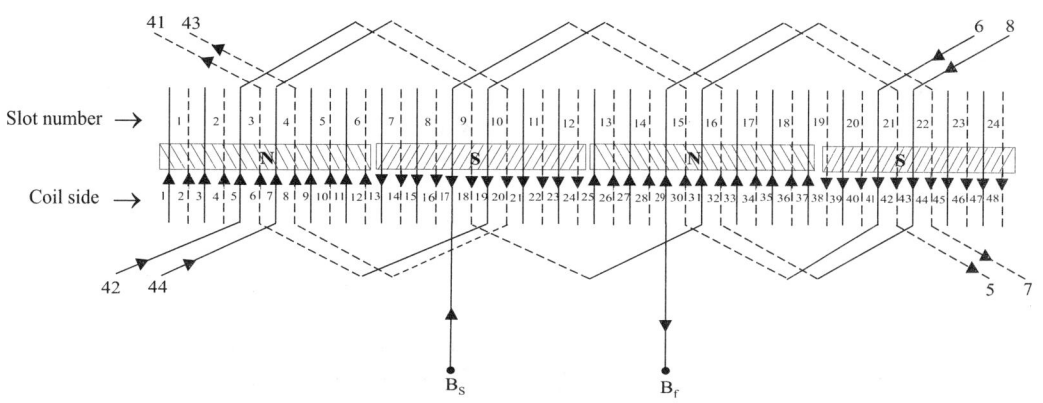

Fig. 3: B-phase wave winding.

Y-Phase winding diagram

Table 4: Y-Phase slots, Coil sides and Coil groups

Pole group		Coil sides of coil groups	
Slots	Coil sides	Coil group-1	Coil group-2
5, 6	9, 10, 11, 12	9, 22, 33, 46, 11, 24, 35, 48	
11, 12	21, 22, 23, 24		
17, 18	33, 34, 35, 36		
23, 24	45, 46, 47, 48		12, 47, 36, 23, 10, 45, 34, 21

Clockwise winding calculations

$9 + 13 = 22$

$22 + 11 = 33$

$33 + 13 = 46$

$46 + 11 = 57 - 48 = 9 + 2 = 11$

$11 + 13 = 24$

$24 + 11 = 35$

$35 + 13 = 48$

$48 + 11 = 59 - 48 = 11 + 1 = 12$

Anticlockwise winding calculations

$12 + 48 - 13 = 47$

$47 - 11 = 36$

$36 - 13 = 23$

$23 - 11 = 12 - 2 = 10$

$10 + 48 - 13 = 45$

$45 - 11 = 34$

$34 - 13 = 21$

Back connections

$9 \leftarrow 22$

$33 \leftarrow 46$

$11 \leftarrow 24$

$35 \leftarrow 48$

$12 \leftarrow 47$

$36 \leftarrow 23$

$10 \leftarrow 45$

$34 \leftarrow 21$

Front connections

$22 \rightarrow 33$

$46 \rightarrow 11$

$24 \rightarrow 35$

$48 \rightarrow 12$

$47 \rightarrow 36$

$23 \rightarrow 10$

$45 \rightarrow 34$

Winding table

Coil group-1:   9 ← 22 → 33 ← 46 → 11 ← 24 → 35 ← 48

Coil group-2: 12 ← 47 → 36 ← 23 → 10 ← 45 → 34 ← 21

The two coil groups are connected in series to form Y-phase winding. The series connection is made by following the current direction of the coil groups.

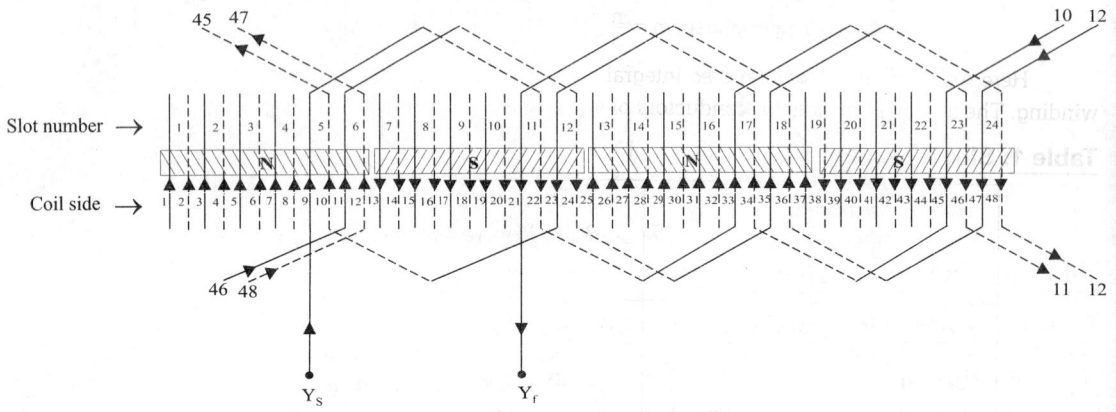

Fig.4: Y-Phase wave winding.

3-Phase winding diagram

The R-phase, B-phase and Y-phase winding diagrams are shown separately in Figs. 2, 3 and 4 respectively. In each phase winding diagram there are two coil groups and the two coil groups automatically connect in series to form the winding of a phase. The combined winding diagram is shown in Fig. 5. The three-phase winding diagram has six terminals which can be connected in star or delta.

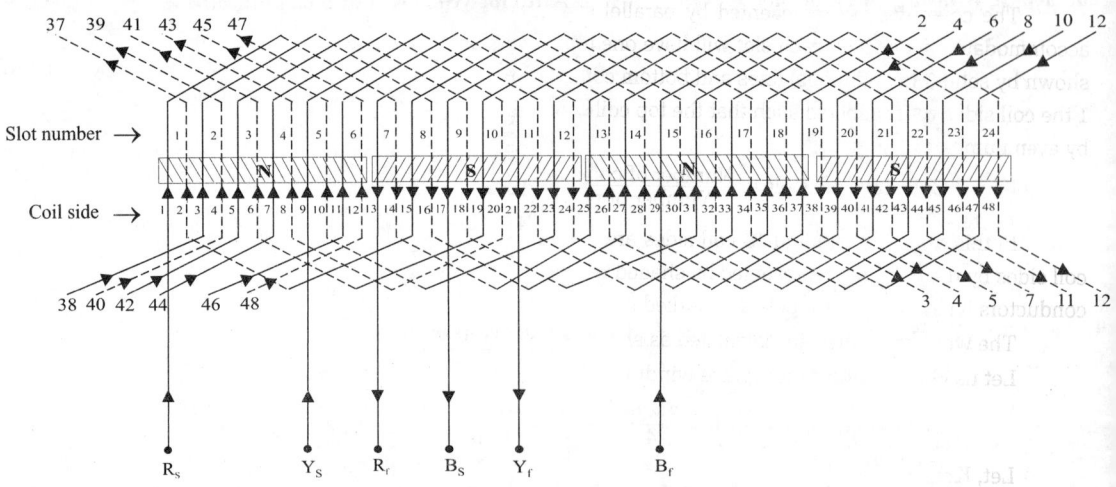

Fig. 5: Simplex wave winding of a 3-phase, 4 pole AC machine armature with 24 slots.

EXAMPLE 5.12

Draw the 3-phase winding diagram for a wave wound 36 slots, 4 pole AC armature machine.

SOLUTION

$$\text{Slots per pole, n} = \frac{S}{p} = \frac{36}{4} = 9$$

$$\text{Slots per pole per phase, m} = \frac{n}{3} = \frac{9}{3} = 3$$

Here n and m are integers and so integral slot winding is possible. Let us choose full pitch coils for winding. The allotment of slots for conductors of various phases are shown in Table 1.

Table 1: Slot Allotment

Slots in a pole group	Slot allotment for phase sequence RYB		
	R-phase	B-phase	Y-phase
1, 2, 3, 4, 5, 6, 7, 8, 9	1, 2, 3	4, 5, 6	7, 8, 9
10, 11, 12, 13, 14, 15, 16, 17, 18	10, 11, 12	13, 14, 15	16, 17, 18
19, 20, 21, 22, 23, 24, 25, 26, 27	19, 20, 21	22, 23, 24	25, 26, 27
28, 29, 30, 31, 32, 33, 34, 35, 36	28, 29, 30	31, 32, 33	34, 35, 36

Given that number of slots = 36

$$\therefore \text{ Number of coils, C} = 36$$

$$\boxed{\text{In simplex wave winding, S} = \text{C}}$$

Number of coil sides = 2C = 2 × 36 = 72

$$\text{Coil sides per slot} = \frac{\text{Number of coil sides}}{\text{Number of slots}} = \frac{72}{36} = 2$$

The coil sides are represented by parallel straight lines as shown in Fig. 1. Since two coil sides are accommodated in one slot, each slot will have one top coil side and one bottom coil side. The top coil sides are shown by solid (or continuous) lines and bottom coil sides are shown by broken (or discontinuous) lines. In Fig. 1 the coil sides are numbered such that the top coil sides are represented by odd numbers and bottom coil sides by even numbers.

$$\text{Coil sides per pole} = \frac{\text{Number of coil sides}}{\text{Number of poles}} = \frac{72}{4} = 18$$

In this armature winding 18 coil sides are enclosed by a pole at any one time instant. The enclosure of coil sides by the poles are represented by shaded rectangles as shown in Fig. 1. The direction of current through conductors lying under north pole are marked upwards and that of south pole are marked downwards.

The winding pitches are calculated as shown below.

Let us choose progressive wave winding.

$$\text{Back pitch, Y}_b = \frac{2C}{p} + K = \frac{2 \times 36}{4} + K = 18 + K$$

Let, K = 1,

$$\boxed{\begin{array}{l}\text{In integral slot winding} \\ \text{with full pitch coils,} \\ \\ \text{Coil span} = \frac{2C}{p} \text{ and } K = 1\end{array}}$$

$$\therefore \text{ Back pitch,} \quad \text{Y}_b = 18 + 1 = 19 \text{ coil sides}$$

$$\text{Windng pitch, Y} = \frac{4C}{p} = \frac{4 \times 36}{4} = 36 \text{ coil sides}$$

$$\therefore \text{ Front pitch,} \quad \text{Y}_f = \text{Y} - \text{Y}_b = 36 - 19 = 17 \text{ coil sides}$$

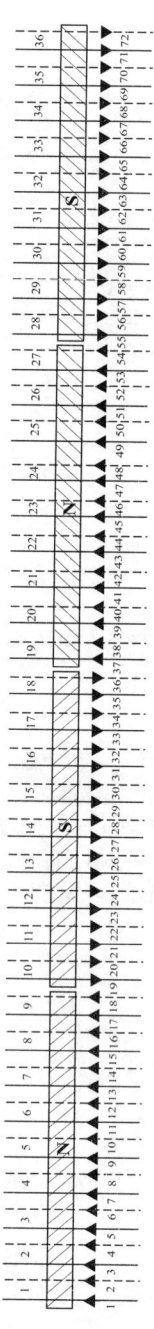

Fig. 1: Representation of coil sides, poles and direction of current.

The front and back connections of the coil sides are determined as shown below.

Let us start with top coil side marked 1.

Bottom coil side of the first coil = Top coil side of first coil + Back pitch
(Back connection)

$$= 1 + Y_b = 1 + 19 = 20$$

Top coil side of the coil lying one pole } = Bottom coil side of first coil + Front pitch
pitch from first coil (Front connection)

$$= 20 + Y_f = 20 + 17 = 37$$

Similarly, all the R-phase, B-phase and Y-phase connections are determined. The entire winding calculations, details of R-phase, B-phase and Y-phase connections and the winding table are also shown below.

In wave winding there will be two coil groups in each phase. The direction of connection for second coil group is reverse that of first coil group and this is referred as clockwise and anticlockwise winding calculations.

In order to provide 120° phase shift between emf induced in R, Y and B-phases with respect of each other, the start of R, Y and B-phases should have an angular spacing of 120 °e.

In this example,

$$\text{Slot angle, } \beta = \frac{180°}{\text{Slots per pole}} = \frac{180°}{9} = 20 \text{ °e}$$

Therefore, the number of slots between start of R-phase and Y-phase should be 6 slots (6 × 20° = 120 °e). Similarly, the start of B-phase should be 6 slots away from start of Y-phase for a phase sequence of RYB.

Let us start R-phase from slot 1 and so Y-phase start from 7th slot (1 + 6 = 7) and B-phase start from 13th slot (7 + 6 = 13).

R-Phase winding diagram

Table 2: R-Phase slots, Coil sides and Coil groups

Pole group		Coil sides of coil groups	
Slots	Coil sides	Coil group-1	Coil group-2
1, 2, 3	1, 2, 3, 4, 5, 6	1, 20, 37, 56, 3, 22, 39, 58, 5, 24, 41, 60	
10, 11, 12	19, 20, 21, 22, 23, 24		
19, 20, 21	37, 38, 39, 40, 41, 42		
28, 29, 30	55, 56, 57, 58, 59, 60		6, 59, 42, 23, 4, 57, 40, 21, 2, 55, 38, 19

Clockwise winding calculations

$1 + 19 = 20$

$20 + 17 = 37$

$37 + 19 = 56$

$56 + 17 = 73 - 72 = 1 + 2 = 3$

$3 + 19 = 22$

$22 + 17 = 39$

$39 + 19 = 58$

$58 + 17 = 75 - 72 = 3 + 2 = 5$

$5 + 19 = 24$

$24 + 17 = 41$

$41 + 19 = 60$

$60 + 17 = 77 - 72 = 5 + 1 = 6$

Anticlockwise winding calculations

$6 + 72 - 19 = 59$

$59 - 17 = 42$

$42 - 19 = 23$

$23 - 17 = 6 - 2 = 4$

$4 + 72 - 19 = 57$

$57 - 17 = 40$

$40 - 19 = 21$

$21 - 17 = 4 - 2 = 2$

$2 + 72 - 19 = 55$

$55 - 17 = 38$

$38 - 19 = 19$

Back connections

$1 \leftarrow 20$

$37 \leftarrow 56$

$3 \leftarrow 22$

$39 \leftarrow 58$

$5 \leftarrow 24$

$41 \leftarrow 60$

$6 \leftarrow 59$

$42 \leftarrow 23$

$4 \leftarrow 57$

$40 \leftarrow 21$

$2 \leftarrow 55$

$38 \leftarrow 19$

Front connections

$20 \rightarrow 37$

$56 \rightarrow 3$

$22 \rightarrow 39$

$58 \rightarrow 5$

$24 \rightarrow 41$

$60 \rightarrow 6$

$59 \rightarrow 42$

$23 \rightarrow 4$

$57 \rightarrow 40$

$21 \rightarrow 2$

$55 \rightarrow 38$

Winding table

Coil group-1: 1 ← 20 → 37 ← 56 → 3 ← 22 → 39 ← 58 → 5 ← 24 → 41 ← 60

Coil group-2: 6 ← 59 → 42 ← 23 → 4 ← 57 → 40 ← 21 → 2 ← 55 → 38 ← 19

The two coil groups are connected in series to form R-phase winding. The series connection is made by following the current direction of the coil groups.

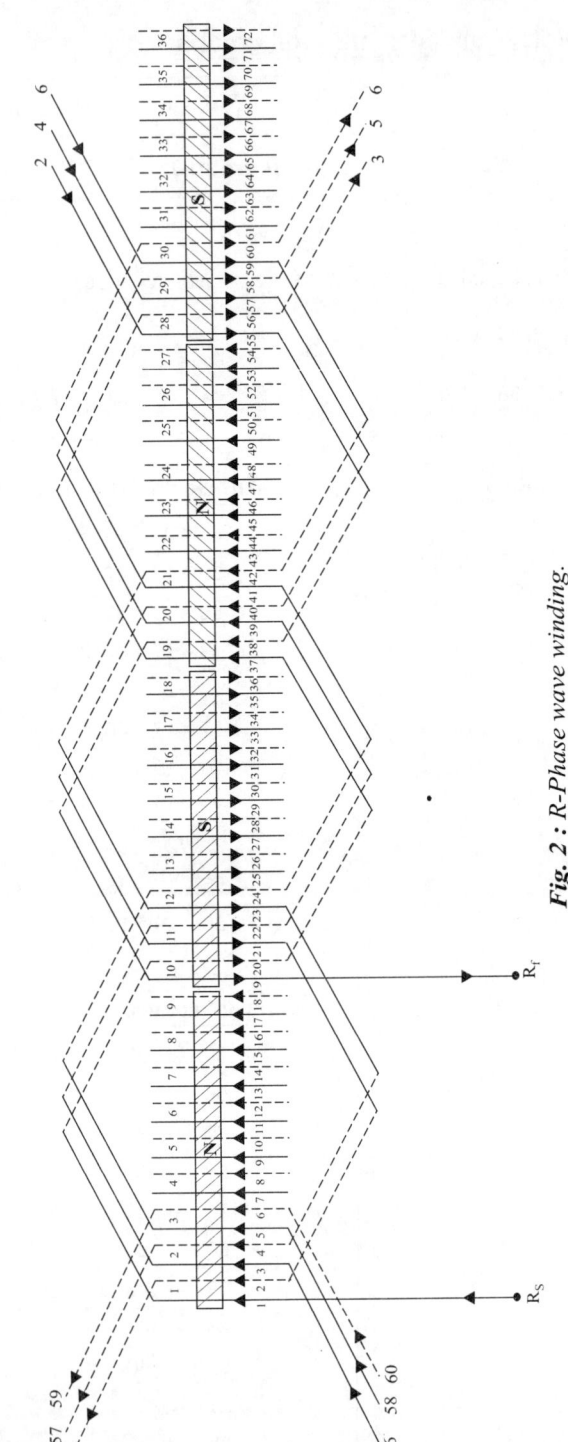

Fig. 2 : R-Phase wave winding.

B-Phase winding diagram

Table 3: B-Phase slots, Coil sides and Coil groups

Pole group		Coil sides of coil groups	
Slots	Coil sides	Coil group-1	Coil group-2
13, 14, 15	25, 26, 27, 28, 29, 30		
22, 23, 24	43, 44, 45, 46, 47, 48	25, 44, 61, 8, 27, 46, 63, 10, 29, 48, 65, 12	
31, 32, 33	61, 62, 63, 64, 65, 66		
4, 5, 6	7, 8, 9, 10, 11, 12		30, 11, 66, 47, 28, 9, 64, 45, 26, 7, 62, 43

Clockwise winding calculations

$25 + 19 = 44$

$44 + 17 = 61$

$61 + 19 = 80 - 72 = 8$

$8 + 17 = 25 + 2 = 27$

$27 + 19 = 46$

$46 + 17 = 63$

$63 + 19 = 82 - 72 = 10$

$10 + 17 = 27 + 2 = 29$

$29 + 19 = 48$

$48 + 17 = 65$

$65 + 19 = 84 - 72 = 12$

$12 + 17 = 29 + 1 = 30$

Anticlockwise winding calculations

$30 - 19 = 11$

$11 + 72 - 17 = 66$

$66 - 19 = 47$

$47 - 17 = 30 - 2 = 28$

$28 - 19 = 9$

$9 + 72 - 17 = 64$

$64 - 19 = 45$

$45 - 17 = 28 - 2 = 26$

$26 - 19 = 7$

$7 + 72 - 17 = 62$

$62 - 19 = 43$

Back connections

$25 \leftarrow 44$

$61 \leftarrow 8$

$27 \leftarrow 46$

$63 \leftarrow 10$

$29 \leftarrow 48$

$65 \leftarrow 12$

$30 \leftarrow 11$

$66 \leftarrow 47$

$28 \leftarrow 9$

$64 \leftarrow 45$

$26 \leftarrow 7$

$62 \leftarrow 43$

Front connections

$44 \rightarrow 61$

$8 \rightarrow 27$

$46 \rightarrow 63$

$10 \rightarrow 29$

$48 \rightarrow 65$

$12 \rightarrow 30$

$11 \rightarrow 66$

$47 \rightarrow 28$

$9 \rightarrow 64$

$45 \rightarrow 26$

$7 \rightarrow 62$

Winding table

Coil group-1: 25 ← 44 → 61 ← 8 → 27 ← 46 → 63 ← 10 → 29 ← 48 → 65 ← 12

Coil group-2: 30 ← 11 → 66 ← 47 → 28 ← 9 → 64 ← 45 → 26 ← 7 → 63 ← 43

The two coil groups are connected in series to form B-phase winding. The series connection is made by following the current direction of the coil groups.

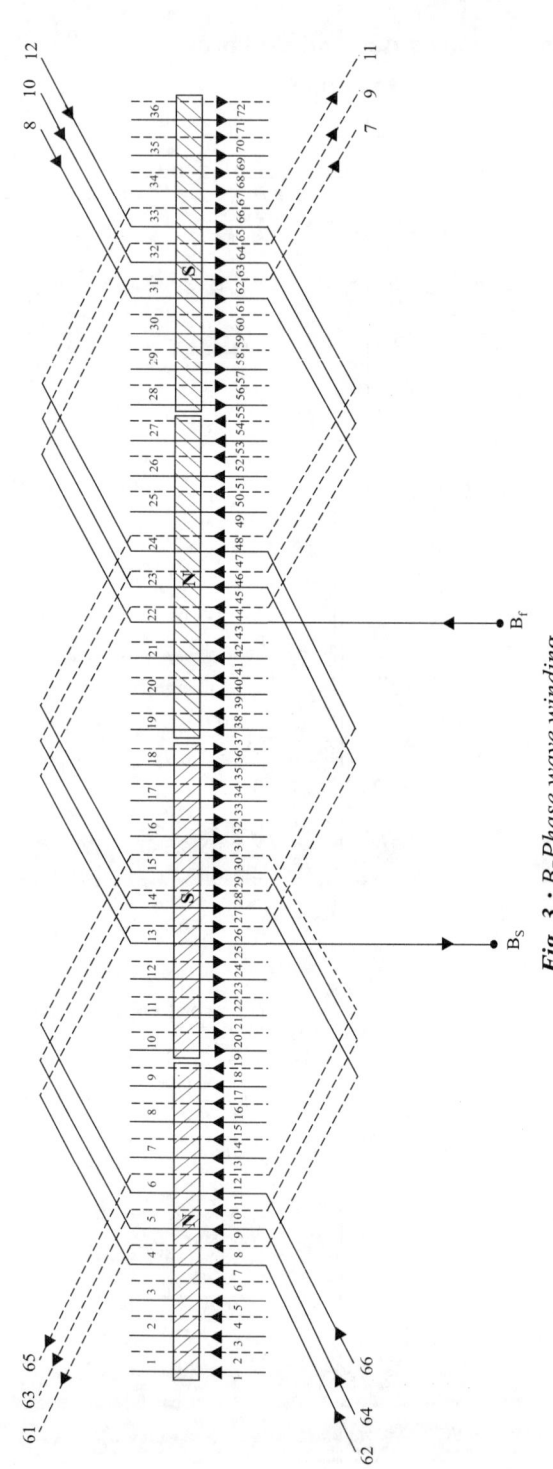

Fig. 3 : B-Phase wave winding.

Y-Phase winding diagram

Table 4: Y-Phase slots, Coil sides and Coil groups

Pole group		Coil sides of coil groups	
Slots	Coil sides	Coil group-1	Coil group-2
7, 8, 9	13, 14, 15, 16, 17, 18		
16, 17, 18	31, 32, 33, 34, 35, 36	13, 32, 49, 68, 15, 34, 51, 70, 17, 36, 53, 72	
25, 26, 27	49, 50, 51, 52, 53, 54		
34, 35, 36	67, 68, 69, 70, 71, 72		18, 71, 54, 35, 16, 69, 52, 33, 14, 67, 50, 31

Clockwise winding calculations

$$13 + 19 = 32$$
$$32 + 17 = 49$$
$$49 + 19 = 68$$
$$68 + 17 = 85 - 72 = 13 + 2 = 15$$
$$15 + 19 = 34$$
$$34 + 17 = 51$$
$$51 + 19 = 70$$
$$70 + 17 = 87 - 72 = 15 + 2 = 17$$
$$17 + 19 = 36$$
$$36 + 17 = 53$$
$$53 + 19 = 72$$
$$72 + 17 = 89 - 72 = 17 + 1 = 18$$

Anticlockwise winding calculations

$$18 + 72 - 19 = 71$$
$$71 - 17 = 54$$
$$54 - 19 = 35$$
$$35 - 17 = 18 - 2 = 16$$
$$16 + 72 - 19 = 69$$
$$69 - 17 = 52$$
$$52 - 19 = 33$$
$$33 - 17 = 16 - 2 = 14$$
$$14 + 72 - 19 = 67$$
$$67 - 17 = 50$$
$$50 - 19 = 31$$

Back connections

$$13 \leftarrow 32$$
$$49 \leftarrow 68$$
$$15 \leftarrow 34$$
$$51 \leftarrow 70$$
$$17 \leftarrow 36$$
$$53 \leftarrow 72$$
$$18 \leftarrow 71$$
$$54 \leftarrow 35$$
$$16 \leftarrow 69$$
$$52 \leftarrow 33$$
$$14 \leftarrow 67$$
$$50 \leftarrow 31$$

Front connections

$$32 \rightarrow 49$$
$$68 \rightarrow 15$$
$$34 \rightarrow 51$$
$$70 \rightarrow 17$$
$$36 \rightarrow 53$$
$$72 \rightarrow 18$$
$$71 \rightarrow 54$$
$$35 \rightarrow 16$$
$$69 \rightarrow 52$$
$$33 \rightarrow 14$$
$$67 \rightarrow 50$$

Winding table

Coil group-1: 13 ← 32 → 49 ← 68 → 15 ← 34 → 51 ← 70 → 17 ← 36 → 53 ← 72

Coil group-2: 18 ← 71 → 54 ← 35 → 16 ← 69 → 52 ← 33 → 14 ← 67 → 50 → 31

The two coil groups are connected in series to form Y-phase winding. The series connection is made by following the current direction of the coil groups.

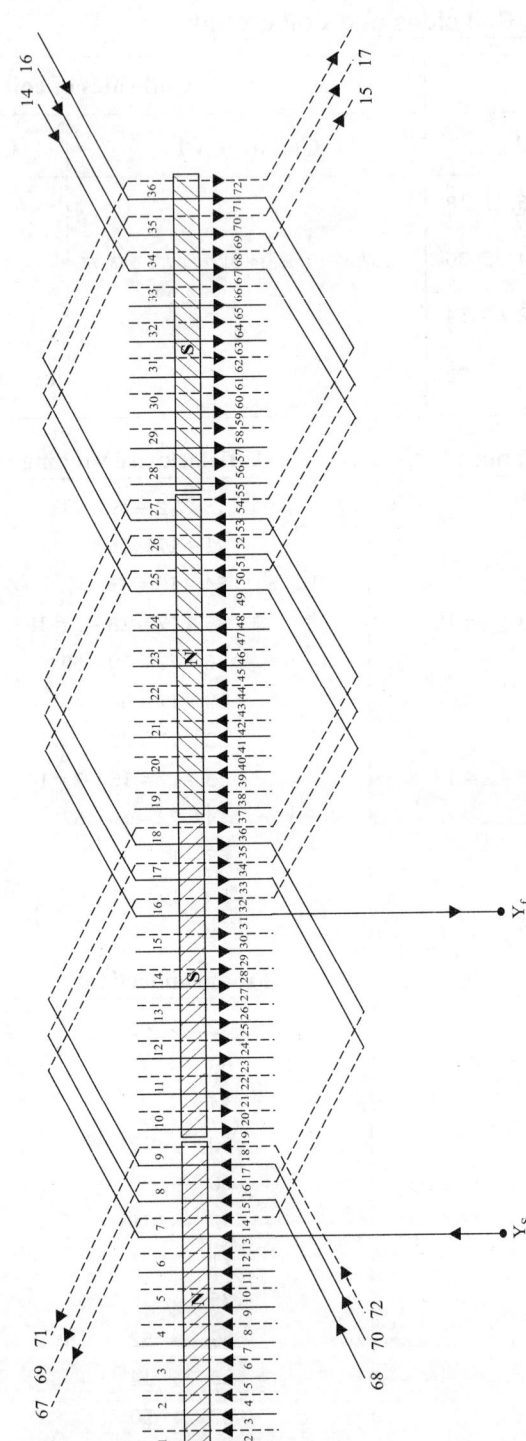

Fig. 4 : Y-Phase wave winding.

3-Phase winding diagram

The R-phase, B-phase and Y-phase winding diagrams are shown separately in Figs. 2, 3 and 4 respectively. In each phase winding diagram there are two coil groups and the two coil groups automatically connect in series to form the winding of a phase. The combined winding diagram is shown in Fig. 5. The three-phase winding diagram has six terminals which can be connected in star or delta.

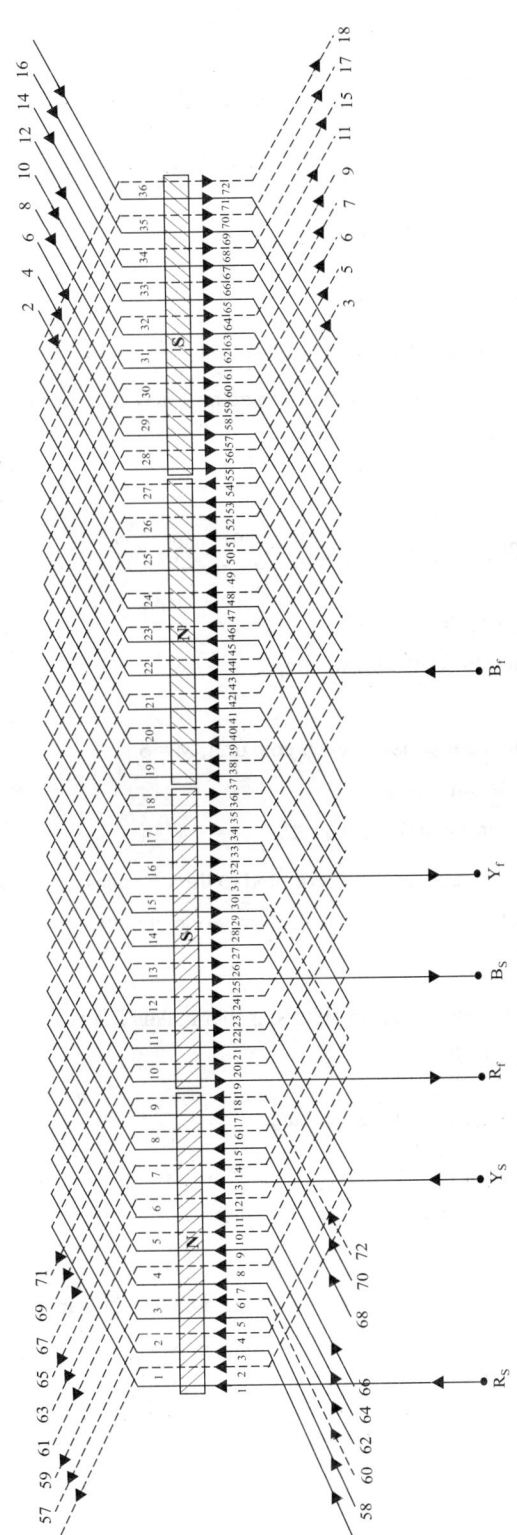

Fig. 5: Simplex wave winding of a 3-phase, 4 pole AC machine armature with 36 slots.

EXAMPLE 5.13

Draw a 3-phase winding diagram for a lap wound 30 slots, 4 pole AC machine armature.

SOLUTION

Given that,

Number of slots, $S = 30$

Number of phases, $m = 3$

Number of poles, $p = 4$

Slots per pole per phase, $q = \dfrac{S}{mp} = \dfrac{30}{3 \times 4} = 2.5$

Here, q is not an integer, therefore we have to go for fractional slot winding.

Let, us express q in the form shown below.

$$q = \frac{M}{d} = I + \frac{n}{d} \qquad \qquad \text{.....(1)}$$

$\therefore$ Slots per pole per phase, $q = \dfrac{S}{mp} = \dfrac{30}{3 \times 4} = \dfrac{30}{12} = \dfrac{30/6}{12/6}$

> Dividing numerator and denominator by 6.

$$\therefore q = \frac{5}{2} = 2\frac{1}{2} = 2 + \frac{1}{2} \qquad \qquad \text{.....(2)}$$

On comparing equations (1) and (2) we get,

I = 2, n = 2

Number of slots per phase in each pole unit, M = 5

Number of pole unit, d = 2

Slots per pole unit = mM = 3 $\times$ 5 = 15

The 15 slots of one pole unit can be distributed to three phases as shown below.

Let, D = Distance between coil sides of a coil in terms of slots.

Now, $D = \dfrac{1 + mMP}{d}$

where, P is a smallest possible value of integer that makes D as an integer.

Here, $D = \dfrac{1 + 3 \times 5P}{d} = \dfrac{1 + 15P}{2}$

On choosing, P = 1, we will get D as an integer.

Let, P = 1, $\therefore D = \dfrac{1 + 15 \times 1}{2} = 8$

The distribution of slots is by,

1, 1 + D, 1 + 2D, 1 + 3D, ...

In the above equation if the sum exceeds mM then subtract the sum by mM or integral multiple of mM.

The 15 slots of a pole unit can be distributed at the rate of 5 slots per phase. The first 15 slots with slot number 1 to 15 are slots of first pole unit.

Let us distribute first 15 slots as shown below.

Let us start from slot 1. The first five slots are allotted to R-phase.

$$1, \; 1 + D = 1 + 8 = 9, \; 1 + 2D = 1 + 2 \times 8 = 17 - 15 = 2$$

$$1 + 3D = 1 + 3 \times 8 = 25 - 15 = 10, \; 1 + 4D = 1 + 4 \times 8 = 33 - 30 = 3$$

$\boxed{mM = 15}$

$\boxed{2mM = 30}$

∴ Slots of R-phase: 1, 9, 2, 10, 3 ⟹ 1, 2, 3, 9, 10

The next 5 slots are allotted to B-phase.

$$1 + 5D = 1 + 5 \times 8 = 41 - 30 = 11$$

$\boxed{2mM = 30}$

$$1 + 6D = 1 + 6 \times 8 = 49 - 45 = 4$$

$\boxed{3mM = 45}$

$$1 + 7D = 1 + 7 \times 8 = 57 - 45 = 12$$

$$1 + 8D = 1 + 8 \times 8 = 65 - 60 = 5$$

$\boxed{4mM = 60}$

$$1 + 9D = 1 + 9 \times 9 = 73 - 60 = 13$$

∴ Slots of B-phase: 11, 4, 12, 5, 13 ⟹ 4, 5, 11, 12, 13

The remaining 5 slots in the pole unit are allotted to Y-phase.

∴ Slots of Y-phase: 6, 7, 8, 14, 15

Similarly the slots of next pole unit (slot number 16 to 30) are distributed as shown below.

Let us start from slot 16.

Let us allot first 5 slots to R-phase.

$$16, \; 16 + D = 16 + 8 = 24$$

$$16 + 2D = 16 + 2 \times 8 = 32 - 15 = 17$$

$\boxed{mM = 15}$

$$16 + 3D = 16 + 3 \times 8 = 40 - 15 = 25$$

$$16 + 4D = 16 + 4 \times 8 = 48 - 30 = 18$$

$\boxed{2mM = 30}$

∴ Slots of R-phase: 16, 24, 17, 25, 18 ⟹ 16, 17, 18, 24, 25

The next 5 slots are allotted to B-phase.

$$16 + 5D = 16 + 5 \times 8 = 56 - 30 = 26$$

$\boxed{2mM = 30}$

$$16 + 6D = 16 + 6 \times 8 = 64 - 45 = 19$$

$\boxed{3mM = 45}$

$$16 + 7D = 16 + 7 \times 8 = 72 - 45 = 27$$

$$16 + 8D = 16 + 8 \times 8 = 80 - 60 = 20$$

$\boxed{4mM = 60}$

$$16 + 9D = 16 + 9 \times 8 = 88 - 60 = 28$$

∴ Slots of B-phase: 26, 19, 27, 20, 28 ⟹ 19, 20, 26, 27, 28

The remaining 5 slots in the pole units are allotted to Y-phase.

∴ Slots of Y-phase: 21, 22, 23, 29, 30

The slot allotment for various phases are listed in Table 1. Here, slot per pole q is 2 ½ and so one slot in each phase is made common between two consecutive phases and coil sides in this slot are placed such that top coil side is one phase and bottom coil side is next phase.

Table 1: Slot Allotment

Slots in a pole pair	Slot allotment for phase sequence RYB					
	R-phase	R, B-phase	B-phase	B, Y-phase	Y-phase	Y, R-phase
1, 2, 3, 4, 5, 6, 7, 8	1, 2	3	4, 5	-	6, 7	8
9, 10, 11, 12, 13, 14, 15	9, 10	-	11, 12	13	14, 15	-
16, 17, 18, 19, 20, 21, 22, 23, 24	16, 17	18	19, 20	-	21, 22	23
24, 25, 26, 27, 28, 29, 30	24, 25	-	26, 27	28	29, 30	-

The coil sides are represented by parallel straight lines as shown in Fig. 1. Since two coil sides are accommodated in one slot, each slot will have one top coil side and one bottom coil side. The top coil sides are shown by solid (or continuous) lines and bottom coil sides are shown by broken (or discontinuous) lines. In Fig. 1 the coil sides are numbered such that the top coil sides are represented by odd numbers and bottom coil sides by even numbers.

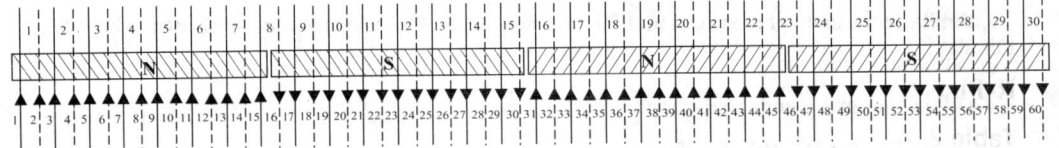

Fig. 1: *Representation of coil sides, poles and direction of current.*

Number of slots, S = 30

∴ Number of coils, C = 30

Number of coil sides = 2C = 2 × 30 = 60

$$\text{Coil sides per pole} = \frac{\text{Number of coil sides}}{\text{Number of poles}} = \frac{60}{4} = 15$$

In this armature winding, 15 coil sides are enclosed by a pole at any one time instant. The enclosure of coil sides by the poles are represented by shaded rectangles as shown in Fig. 1. The direction of current through conductors lying under north pole are marked upwards and that of south pole are marked downwards

The winding pitches are calculated as shown below.

Let us choose progressive lap winding.

$$\text{Back pitch, } Y_b = \text{Coil span} + K = \frac{2C}{p} + K = \frac{2 \times 30}{4} + K = 15 + K$$

Let, K = 0,

∴ Back pitch, $Y_b = 15 + 0 = 15$ coil sides

Winding pitch, Y = 2 (For progressive lap, Y is always +2)

Front pitch, $Y_f = Y_b - Y = 15 - 2 = 13$ coil sides

> In fractional slot winding,
>
> Coil span = $\frac{2C}{p}$ and K is any number that makes Y_b as an odd integer.

The front and back connections of the coil sides are determined as shown below.

Let us start with top coil side marked 1.

Bottom coil side of the first coil = Top coil side of first coil + Back pitch
(Back connection)

$$= 1 + Y_b = 1 + 15 = 16$$

Top coil side of the second coil = Bottom coil side of first coil − Front pitch
(Front connection)

$$= 16 - Y_f = 16 - 13 = 3$$

Similarly, all the R-phase, B-phase and Y-phase winding connections are determined. The entire winding calculations, details of R-phase, B-phase and Y-phase winding connections and the winding table are shown below.

In order to provide 120° phase shift between emf induced in R, Y and B-phases with respect of each other, the start of R, Y and B-phases should have an angular spacing of 120 °e.

In this example,

$$\text{Slot angle, } \beta = \frac{180°}{\text{Slots per pole}} = \frac{180°}{S/p} = \frac{180°}{30/4} = \frac{180°}{7.5} = 24 \, °e$$

Therefore, the number of slots between start of R-phase and Y-phase should be 5 slots (5 × 24° = 120 °e). Similarly, the start of B-phase should be 5 slots away from start of Y-phase for a phase sequence of RYB.

Let us start R-phase from slot 1 and so Y-phase start from 6th slot (1 + 5 = 6) and B-phase start from 11th slot (6 + 5 = 11).

R-Phase winding diagram

Table 2: R-Phase slots, Coil sides and Coil groups

Pole group		Coil sides of coil groups			
Slots	Coil sides	Coil group-1	Coil group-2	Coil group-3	Coil group-4
1, 2, 3	1, 2, 3, 4, 5	1, 3, 5, 16, 18, 20			
8, 9, 10	16, 17, 18, 19, 20		17, 19, 32, 34		
16, 17, 18	31, 32, 33, 34, 35			31, 33, 35, 46, 48, 50	
23, 24, 25	46, 47, 48, 49, 50				47, 49, 2, 4

Winding calculations

1 + 15 = 16	17 + 15 = 32	31 + 15 = 46	47 + 15 = 62 − 60 = 2
16 − 13 = 3	32 − 13 = 19	46 − 13 = 33	2 + 60 − 13 = 49
3 + 15 = 18	19 + 15 = 34	33 + 15 = 48	49 + 15 = 64 − 60 = 4
18 − 13 = 5		48 − 13 = 35	
5 + 15 = 20		35 + 15 = 50	

Back connections

1 ← 16	17 ← 32	31 ← 46	47 ← 2
3 ← 18	19 ← 34	33 ← 48	49 ← 4
5 ← 20		35 ← 50	

Front connections

16 → 3	32 → 19	46 → 33	2 → 49
18 → 5		48 → 35	

Winding table

Coil group-1: $1 \leftarrow 16 \rightarrow 3 \leftarrow 18 \rightarrow 5 \leftarrow 20$

Coil group-2: $17 \leftarrow 32 \rightarrow 19 \leftarrow 34$

Coil group-3: $31 \leftarrow 46 \rightarrow 33 \leftarrow 48 \rightarrow 35 \leftarrow 50$

Coil group-4: $47 \leftarrow 2 \rightarrow 49 \leftarrow 4$

The four coil groups are connected in series to form R-phase winding. The series connection is made by following the current direction of the coil groups.

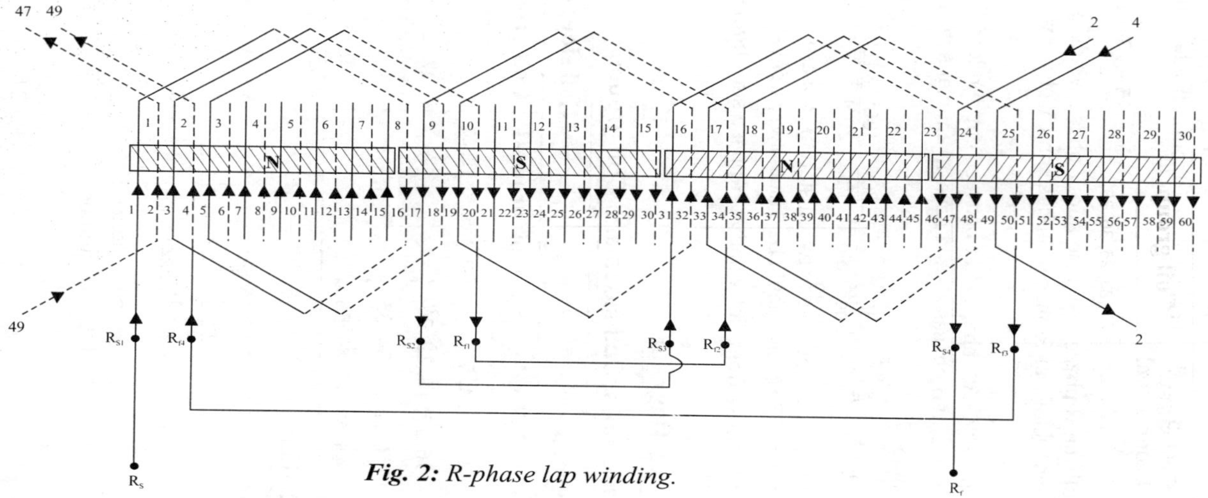

Fig. 2: R-phase lap winding.

B-Phase winding diagram

Table 3: B-Phase slots, Coil sides and Coil groups

Pole group		Coil sides of coil groups			
Slots	Coil sides	Coil group-1	Coil group-2	Coil group-3	Coil group-4
11, 12, 13	21, 22, 23, 24, 25	21, 23, 25, 36, 38, 40			
18, 19, 20	36, 37, 38, 39, 40		37, 39, 52, 54		
26, 27, 28	51, 52, 53, 54, 55			51, 53, 55, 6, 8, 10	
3, 4, 5	6, 7, 8, 9, 10				7, 9, 22, 24

Winding calculations

$21 + 15 = 36$	$37 + 15 = 52$	$51 + 15 = 66 - 60 = 6$	$7 + 15 = 22$
$36 - 13 = 23$	$52 - 13 = 39$	$6 + 60 - 13 = 53$	$22 - 13 = 9$
$23 + 15 = 38$	$39 + 15 = 54$	$53 + 15 = 68 - 60 = 8$	$9 + 15 = 24$
$38 - 13 = 25$		$8 + 60 - 13 = 55$	
$25 + 15 = 40$		$55 + 15 = 70 - 60 = 10$	

Back connections

21 ← 36	37 ← 52	51 ← 6	7 ← 22
23 ← 38	39 ← 54	53 ← 8	9 ← 24
25 ← 40		55 ← 10	

Front connections

36 → 23	52 → 39	6 → 53	22 → 9
38 → 25		8 → 55	

Winding table

Coil group-1: 21 ← 36 → 23 ← 38 → 25 ← 40

Coil group-2: 37 ← 52 → 39 ← 54

Coil group-3: 51 ← 6 → 53 ← 8 → 55 ← 10

Coil group-4: 7 ← 22 → 9 ← 24

The four coil groups are connected in series to form B-phase winding. The series connection is made by following the current direction of the coil groups.

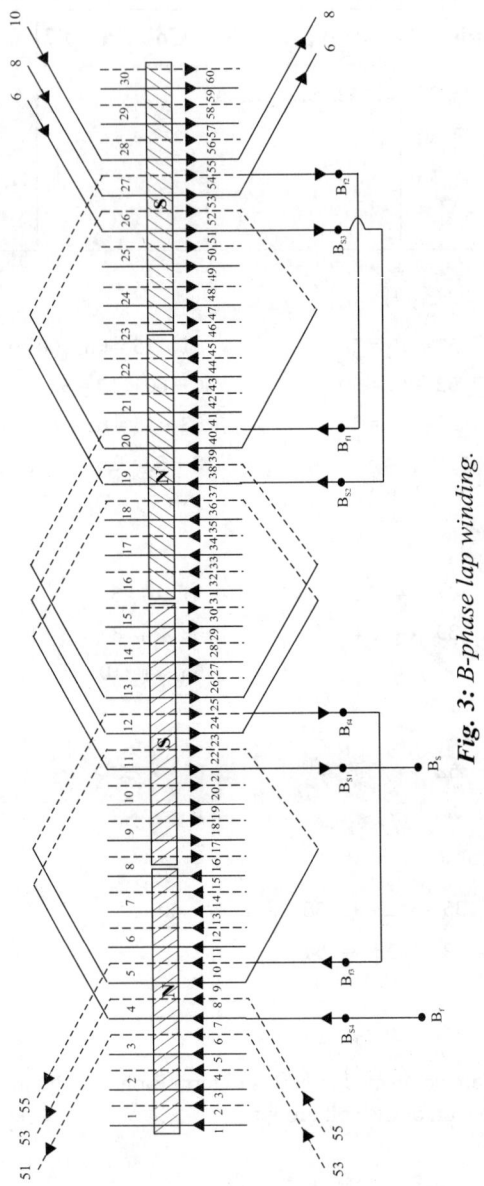

Fig. 3: B-phase lap winding.

Y-Phase winding diagram

Table 4: Y-Phase slots, Coil sides and Coil groups

Pole group		Coil sides of coil groups			
Slots	Coil sides	Coil group-1	Coil group-2	Coil group-3	Coil group-4
6, 7, 8	11, 12, 13, 14, 15	11, 13, 15, 26, 28, 30			
13, 14, 15	26, 27, 28, 29, 30		27, 29, 29, 44		
21, 22, 23	41, 42, 43, 44, 45			41, 43, 45, 56, 58, 60	
28, 29, 30	56, 57, 58, 59, 60				57, 59, 12, 14

Winding calculations

$11 + 15 = 26$	$27 + 15 = 42$	$41 + 15 = 56$	$57 + 15 = 72 - 60 = 12$
$26 - 13 = 13$	$42 - 13 = 29$	$56 - 13 = 43$	$12 + 60 - 13 = 59$
$13 + 15 = 28$	$29 + 15 = 44$	$43 + 15 = 58$	$59 + 15 = 74 - 60 = 14$
$28 - 13 = 15$		$58 - 13 = 45$	
$15 + 15 = 30$		$45 + 15 = 60$	

Back connections

$11 \leftarrow 26$	$27 \leftarrow 42$	$41 \leftarrow 56$	$57 \leftarrow 12$
$13 \leftarrow 28$	$29 \leftarrow 44$	$43 \leftarrow 58$	$59 \leftarrow 14$
$15 \leftarrow 30$		$45 \leftarrow 60$	

Front connections

$26 \rightarrow 13$	$42 \rightarrow 29$	$56 \rightarrow 43$	$12 \rightarrow 59$
$28 \rightarrow 15$		$58 \rightarrow 45$	

Winding table

Coil group-1: $11 \leftarrow 26 \rightarrow 13 \leftarrow 28 \rightarrow 15 \leftarrow 30$

Coil group-2: $27 \leftarrow 42 \rightarrow 29 \leftarrow 44$

Coil group-3: $41 \leftarrow 56 \rightarrow 43 \leftarrow 58 \rightarrow 45 \leftarrow 60$

Coil group-4: $57 \leftarrow 12 \rightarrow 59 \leftarrow 14$

The four coil groups are connected in series to form Y-phase winding. The series connection is made by following the current direction of the coil groups.

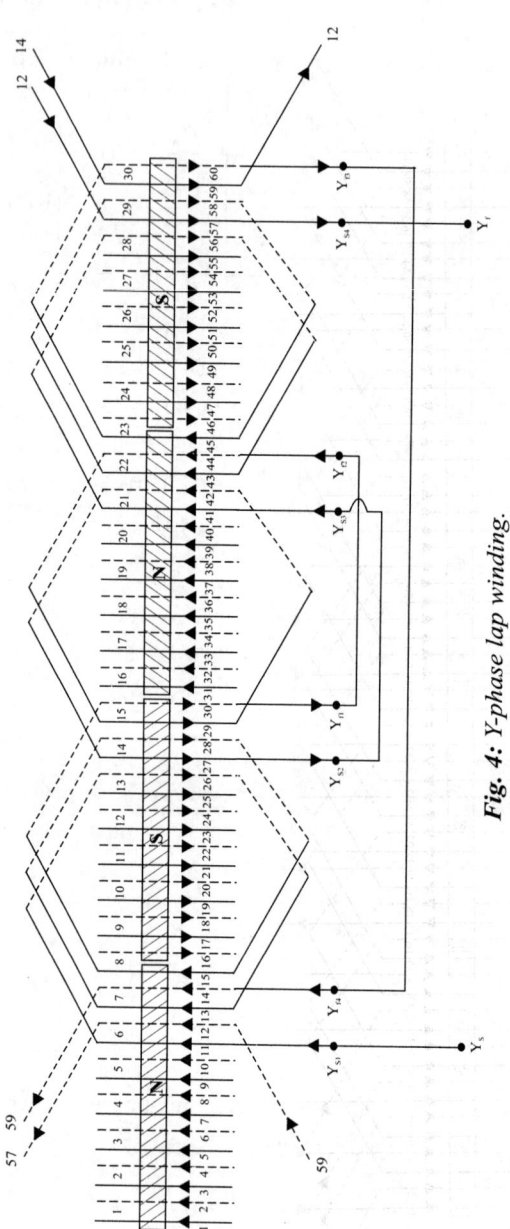

***Fig. 4:** Y-phase lap winding.*

3-phase winding diagram

The R-phase, B-phase and Y-phase winding diagrams are shown separately in Fig. 2, 3 and 4 respectively. In each phase winding diagram there are four coil groups and the coil groups of a phase are connected in series to form the winding of a phase. The combined winding diagram is shown in Fig. 5. The three-phase winding diagram has six terminals which can be connected in star or delta.

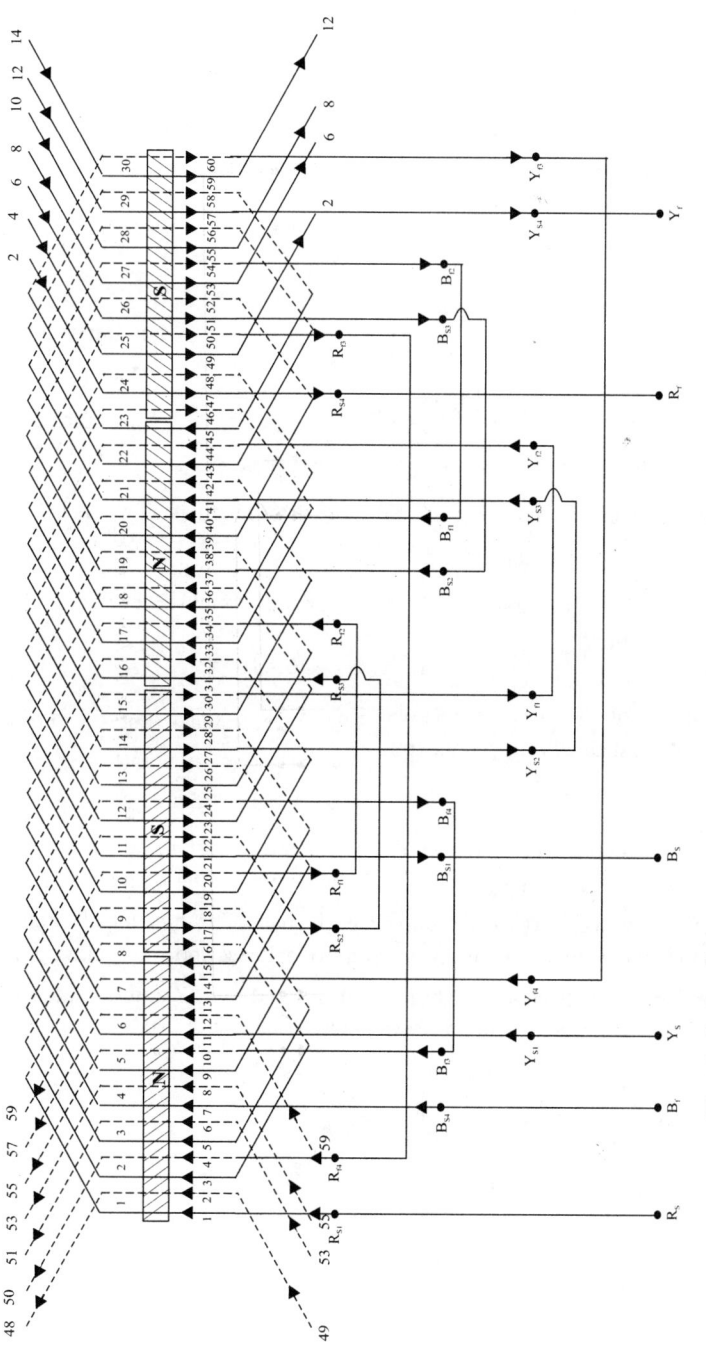

Fig. 5: Fractional slot lap winding 3-phase, 4 pole AC machine with 30 slots.

EXAMPLE 5.14

Give the layout of a double layer wave winding for a 3-phase, 4 pole, 30 slot armature.

SOLUTION

The number of slots and poles in this example are same is that of Example 5.13. Since the distribution of slots for lap and wave winding are same, refer to Example 5.13 for slot allotment.

The winding pitches are calculated as shown below.

Let us choose progressive wave winding.

$$\text{Back pitch, } Y_b = \text{Coil span} + K = \frac{2C}{p} + K = \frac{2 \times 30}{4} + K = 15 + K$$

Let, K = 0,

∴ Back pitch, $Y_b = 15 + 0 = 15$ coil sides

Winding pitch, $Y = \frac{4C}{p} = \frac{4 \times 30}{4} = 30$

Front pitch, $Y_f = Y_b - Y = 30 - 15 = 15$ coil sides

> In fractional slot winding,
> Coil span = $\frac{2C}{p}$ and K is any number that makes Y_b as odd integer.

The front and back connections of the coil sides are determined as shown below.

Let us start with top coil side marked 1.

Bottom coil side of the first coil = Top coil side of first coil + Back pitch
(Back connection)

$$= 1 + Y_b = 1 + 15 = 16$$

Top coil side of the coil lying one pole ⎫ = Bottom coil side of first coil + Front pitch
pitch from first coil (Front connection) ⎭

$$= 16 + Y_f = 16 + 15 = 31$$

Similarly, all the R-phase, B-phase and Y-phase connections are determined. The entire winding calculations, details of R-phase, B-phase and Y-phase connections and the winding table are also shown below.

In wave winding there will be two coil groups in each phase. The direction of connection for second coil group is reverse that of first coil group and this is referred as clockwise and anticlockwise winding calculations.

In order to provide 120° phase shift between emf induced in R, Y and B-phases with respect of each other, the start of R, Y and B-phases should have an angular spacing of 120 °e.

In this example,

$$\text{Slot angle, } \beta = \frac{180°}{\text{Slots per pole}} = \frac{180°}{S/p} = \frac{180°}{30/4} = \frac{180°}{7.5} = 24\text{ °e}$$

Therefore, the number of slots between start of R-phase and Y-phase should be 5 slots (5 × 24° = 120 °e). Similarly, the start of B-phase should be 5 slots away from start of Y-phase for a phase sequence of RYB.

Let us start R-phase from slot 1 and so Y-phase start from 6th slot (1 + 5 = 6) and B-phase start from 11th slot (6 + 5 = 11).

R-Phase winding diagram

Table 2: R-Phase slots, Coil sides and Coil groups

Pole group		Coil sides of coil groups	
Slots	Coil sides	Coil group-1	Coil group-2
1, 2, 3	1, 2, 3, 4, 5		
8, 9, 10	16, 17, 18, 19, 20	1, 16, 31, 46, 3, 18, 33, 48, 5, 20, 35, 50	
16, 17, 18	31, 32, 33, 34, 35		
23, 24, 25	46, 47, 48, 49, 50		4, 49, 34, 19, 2, 47, 32, 17

Winding calculations

$1 + 15 = 16$

$16 + 15 = 31$

$31 + 15 = 46$

$46 + 15 = 61 - 60 = 1 + 2 = 3$

$3 + 15 = 18$

$18 + 15 = 33$

$33 + 15 = 48$

$48 + 15 = 63 - 60 = 3 + 2 = 5$

$5 + 15 = 20$

$20 + 15 = 35$

$35 + 15 = 50$

$50 + 15 = 65 - 60 = 5 - 1 = 4$

$4 + 60 - 15 = 49$

$49 - 15 = 34$

$34 - 15 = 19$

$19 - 15 = 4 - 2 = 2$

$2 + 60 - 15 = 47$

$47 - 15 = 32$

$32 - 15 = 17$

Back connections

$1 \leftarrow 16$

$31 \leftarrow 46$

$3 \leftarrow 18$

$33 \leftarrow 48$

$5 \leftarrow 20$

$35 \leftarrow 50$

$4 \leftarrow 49$

$34 \leftarrow 19$

$2 \leftarrow 47$

$32 \leftarrow 17$

Front connections

$16 \rightarrow 31$

$46 \rightarrow 3$

$18 \rightarrow 33$

$48 \rightarrow 5$

$20 \rightarrow 35$

$50 \rightarrow 4$

$49 \rightarrow 34$

$19 \rightarrow 2$

$47 \rightarrow 32$

Winding table

Coil group-1: 1 $\leftarrow$ 16 $\rightarrow$ 31 $\leftarrow$ 46 $\rightarrow$ 3 $\leftarrow$ 18 $\rightarrow$ 33 $\leftarrow$ 48 $\rightarrow$ 5 $\leftarrow$ 20 $\rightarrow$ 35 $\leftarrow$ 50

Coil group-2: 4 $\leftarrow$ 49 $\rightarrow$ 34 $\leftarrow$ 19 $\rightarrow$ 2 $\leftarrow$ 47 $\rightarrow$ 32 $\leftarrow$ 17

The two coil groups are connected in series to form R-phase winding. The series connection is made by following the current direction of the coil groups.

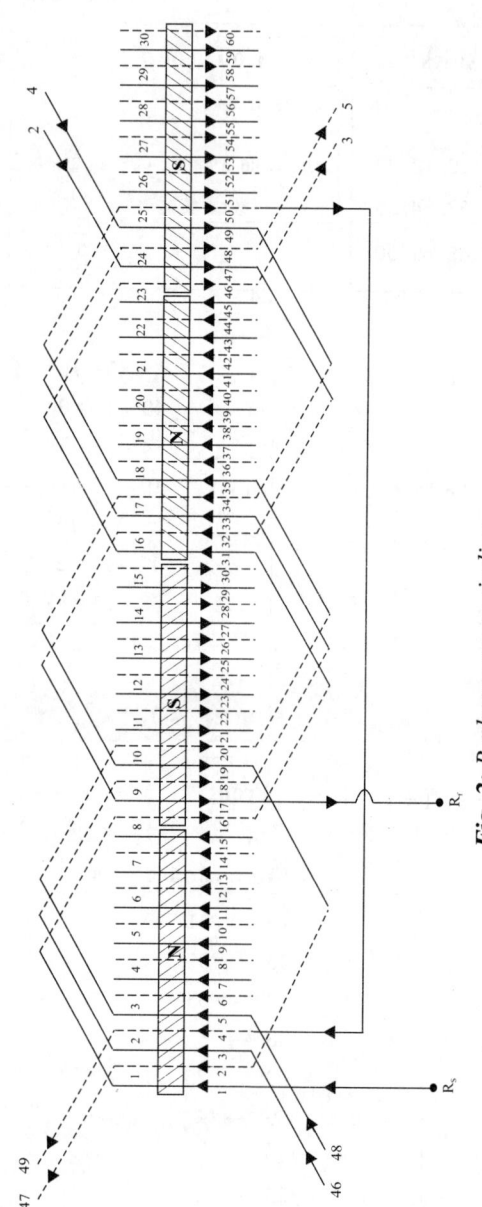

Fig. 2: R-phase wave winding.

B-Phase winding diagram

Table 3: B-Phase slots, Coil sides and Coil groups

Pole group		Coil sides of coil groups	
Slots	Coil sides	Coil group-1	Coil group-2
11, 12, 13	21, 22, 23, 24, 25		
18, 19, 20	36, 37, 38, 39, 40	21, 36, 51, 6, 23, 38, 53, 8, 25, 40, 15, 10	
26, 27, 28	51, 52, 53, 54, 55		
3, 4, 5	6, 7, 8, 9, 10		24, 9, 54, 39, 22, 7, 52, 37

Winding calculations

$21 + 15 = 36$

$36 + 15 = 51$

$51 + 15 = 66 - 60 = 6$

$6 + 15 = 21 + 2 = 23$

$23 + 15 = 38$

$38 + 15 = 53$

$53 + 15 = 68 - 60 = 8$

$8 + 15 = 23 + 2 = 25$

$25 + 15 = 40$

$40 + 15 = 55$

$55 + 15 = 70 - 60 = 10$

$10 + 15 = 25 - 1 = 24$

$24 - 15 = 9$

$9 + 60 - 15 = 54$

$54 - 15 = 39$

$39 - 15 = 24 - 2 = 22$

$22 - 15 = 7$

$7 + 60 - 15 = 52$

$52 - 15 = 37$

Back connections

$21 \leftarrow 36$

$51 \leftarrow 6$

$23 \leftarrow 38$

$53 \leftarrow 8$

$25 \leftarrow 40$

$55 \leftarrow 10$

$24 \leftarrow 9$

$54 \leftarrow 39$

$22 \leftarrow 7$

$52 \leftarrow 37$

Front connections

$36 \rightarrow 51$

$6 \rightarrow 23$

$38 \rightarrow 53$

$8 \rightarrow 25$

$40 \rightarrow 55$

$10 \rightarrow 24$

$9 \rightarrow 54$

$39 \rightarrow 22$

$7 \rightarrow 52$

Winding table

Coil group-1: 21 ← 36 → 51 ← 6 → 23 ← 38 → 53 ← 8 → 25 ← 40 → 55 ← 10

Coil group-2: 24 → 9 ← 54 ← 39 → 22 ← 7 → 52 ← 37

The two coil groups are connected in series to form B-phase winding. The series connection is made by following the current direction of the coil groups.

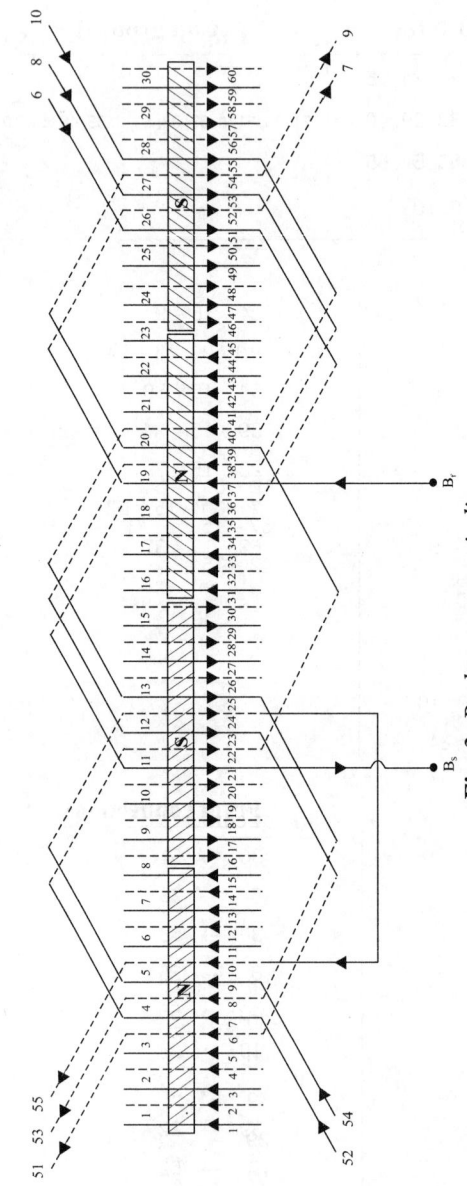

Fig. 3: B-phase wave winding.

Y-Phase winding diagram

Table 4: Y-Phase slots, Coil sides and Coil groups

Pole group		Coil sides of coil groups	
Slots	Coil sides	Coil group-1	Coil group-2
6, 7, 8	11, 12, 13, 14 , 15		
13, 14, 15	26, 27, 28, 29, 30	11, 26, 41, 56, 13, 28 , 43, 58, 15, 30, 45, 60	
21, 22, 23	41, 42, 43, 44, 45		
28, 29, 20	56, 57, 58, 59, 60		14, 59, 44, 29, 14, 57, 42, 27

Winding calculations

11 + 15 = 26	14 + 60 – 15 = 59
26 + 15 = 41	59 – 15 = 44
41 + 15 = 56	44 – 15 = 29
56 + 15 = 71 – 60 = 11 + 2 = 13	29 – 15 = 14 – 2 = 12
13 + 15 = 28	12 + 60 – 15 = 57
28 + 15 = 43	57 – 15 = 42
43 + 15 = 58	42 – 15 = 27
58 + 13 = 73 – 60 = 13 + 2 = 15	
15 + 15 = 30	
30 + 15 = 45	
45 + 15 = 60	
60 + 15 = 75 – 60 = 15 – 1 = 14	

Back connections

11 ← 26
41 ← 56
13 ← 28
43 ← 58
15 ← 30
45 ← 60
14 ← 59
44 ← 29
12 ← 57
42 ← 27

Front connections

26 → 41
56 → 13
28 → 43
58 → 15
30 → 45
60 → 14
59 → 44
29 → 12
57 → 42

Winding table

Coil group-1: $11 \leftarrow 26 \rightarrow 41 \leftarrow 56 \rightarrow 13 \leftarrow 28 \rightarrow 43 \leftarrow 58 \rightarrow 15 \leftarrow 30 \rightarrow 45 \leftarrow 60$

Coil group-2: $14 \leftarrow 59 \rightarrow 44 \leftarrow 29 \rightarrow 12 \leftarrow 57 \rightarrow 42 \leftarrow 27$

The two coil groups are connected in series to form Y-phase winding. The series connection is made by following the current direction of the coil groups.

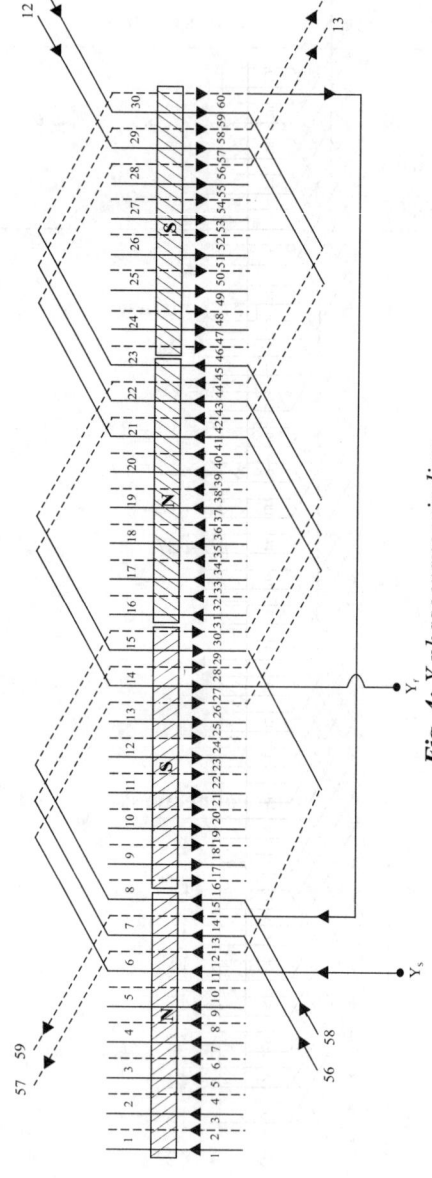

Fig. 4: Y-phase wave winding.

3-phase winding diagram

The R-phase, B-phase and Y-phase winding diagrams are shown separately in Fig. 2, 3 and 4 respectively. In each phase winding diagram there are two coil groups and the two coil groups automatically connect in series to form the winding of a phase. The combined winding diagram is shown in Fig. 5. The three-phase winding diagram has six terminals which can be connected in star or delta.

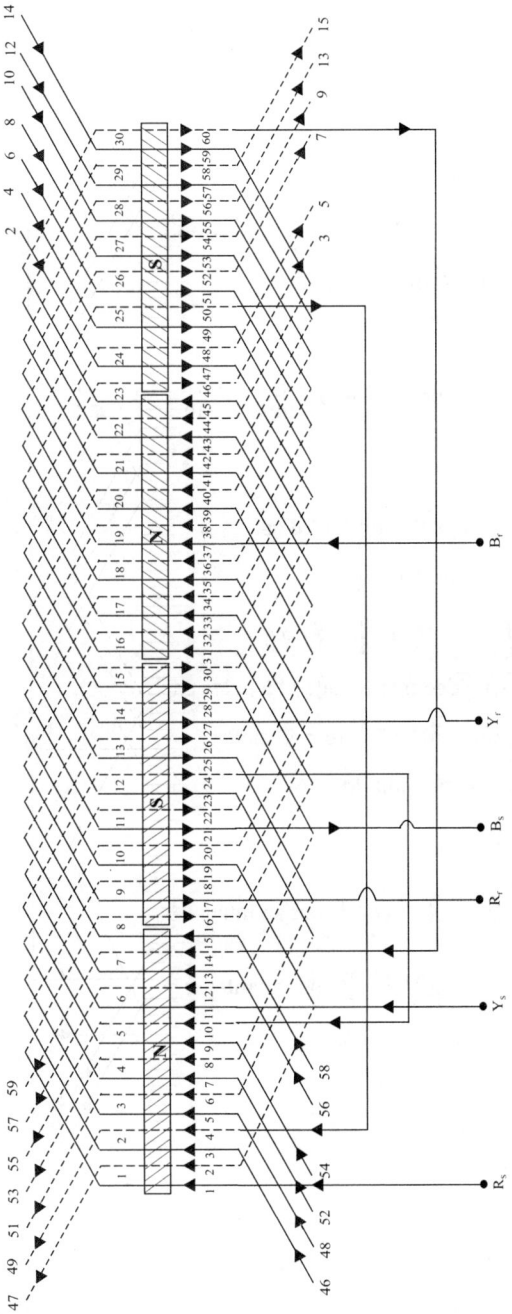

Fig. 5: Fractional slot wave winding of a 3-phase, 4 pole AC machine armature with 30 slots.

EXAMPLE 5.15

Work out the detailed slot distribution for a 96 slots, 10 pole, 3-phase AC machine armature.

SOLUTION

Given that,

Number of slots,   S = 96

Number of phases, m = 3

Number of poles,   p = 10

Slots per pole per phase, $q = \dfrac{S}{mp} = \dfrac{96}{3 \times 10} = \dfrac{96}{30}$

Let us express q in the form,

$$q = \dfrac{M}{d} = I + \dfrac{n}{d} \qquad \qquad \qquad .....(1)$$

$\therefore$ Slots per pole per phase, $q = \dfrac{S}{mp} = \dfrac{96}{3 \times 10} = \dfrac{96}{30} = \dfrac{96/6}{30/6}$

> Dividing numerator and denominator by 6.

$\therefore q = \dfrac{16}{5} = 3\dfrac{1}{5} = 3 + \dfrac{1}{2} \qquad \qquad .....(2)$

On comparing equations (1) and (2) we get,

M = 16, d = 5, I = 3, n = 1,

Number of pole units = $\dfrac{p}{d} = \dfrac{10}{2} = 5$

$\therefore$ 10 poles can be divided into 2 units of 5 poles each (d = 5).

Slots in a unit = mM = 3 × 16 = 48 slots

Slots per phase per unit, M = 16

> Also, $M = qd = \dfrac{16}{5} \times 5 = 16$

$$D = \dfrac{1 + mMP}{d}$$

Let, P = 1, $D = \dfrac{1 + 3 \times 16 \times 1}{5} = \dfrac{49}{5}$ not an integer

Let, P = 2, $D = \dfrac{1 + 3 \times 16 \times 2}{5} = \dfrac{97}{5}$ not an integer

Let, P = 3, $D = \dfrac{1 + 3 \times 16 \times 1}{5} = \dfrac{145}{5} = 29$

Let, P = 3 and so, D = 29.

The distribution of slots is given by,

1, 1 + D, 1 + 2D, 1 + 3D, ...

In the above equation of sum exceeds mM then subtract the sum by mM or integral multiple of mM.

The 48 slots of a pole unit can be distributed at the rate of 16 slots per phase. The first 48 slots with slot number 1 to 15 are slots of first pole unit.

Let us distribute first 48 slots as shown below.

Let us start from slot 1. The first 16 slots are allotted to R-phase.

$1, \; 1 + D = 1 + 29 = 30, \; 1 + 2D = 1 + 2 \times 29 = 59 - 48 = 11$

$\boxed{mM = 48}$

$1 + 3D = 1 + 3 \times 29 = 88 - 48 = 40$

$1 + 4D = 1 + 4 \times 29 = 117 - 96 = 21$

$\boxed{2mM = 96}$

$1 + 5D = 1 + 5 \times 29 = 146 - 144 = 2$

$\boxed{3mM = 144}$

$1 + 6D = 1 + 6 \times 29 = 175 - 144 = 31$

$1 + 7D = 1 + 7 \times 29 = 204 - 192 = 12$

$\boxed{4mM = 192}$

$1 + 8D = 1 + 8 \times 29 = 233 - 192 = 41$

$1 + 9D = 1 + 9 \times 29 = 262 - 240 = 22$

$\boxed{5mM = 240}$

$1 + 10D = 1 + 10 \times 29 = 291 - 288 = 3$

$\boxed{6mM = 288}$

$1 + 11D = 1 + 11 \times 29 = 320 - 288 = 32$

$1 + 12D = 1 + 12 \times 29 = 349 - 336 = 13$

$\boxed{7mM = 336}$

$1 + 13D = 1 + 13 \times 29 = 378 - 336 = 42$

$1 + 14D = 1 + 14 \times 29 = 407 - 384 = 23$

$\boxed{8mM = 384}$

$1 + 15D = 1 + 15 \times 29 = 436 - 432 = 4$

$\boxed{9mM = 432}$

$\therefore$ Slots of R-phase: 1, 30, 11, 40, 21, 2, 31, 12, 41, 22, 3, 32, 13, 42, 23, 4

On arranging R-phase slots in increasing order we get,

Slots of R-phase: 1, 2, 3, 4, 11, 12, 13, 21, 22, 23, 30, 31, 32, 40, 41, 42

The next 16 slots are allotted to B-phase.

$1 + 16D = 1 + 16 \times 29 = 465 - 432 = 33$

$\boxed{9mM = 432}$

$1 + 17D = 1 + 17 \times 29 = 494 - 480 = 14$

$\boxed{10mM = 432}$

$1 + 18D = 1 + 18 \times 29 = 523 - 480 = 43$

$1 + 19D = 1 + 19 \times 29 = 552 - 528 = 24$

$\boxed{11mM = 528}$

$1 + 20D = 1 + 20 \times 29 = 581 - 576 = 5$

$\boxed{12mM = 576}$

$1 + 21D = 1 + 21 \times 29 = 610 - 576 = 34$

$1 + 22D = 1 + 22 \times 29 = 639 - 624 = 15$

$\boxed{13mM = 624}$

$1 + 23D = 1 + 23 \times 29 = 668 - 624 = 44$

$1 + 24D = 1 + 24 \times 29 = 697 - 672 = 25$

$\boxed{14mM = 672}$

$1 + 25D = 1 + 25 \times 29 = 726 - 720 = 6$

$\boxed{15mM = 720}$

$1 + 26D = 1 + 26 \times 29 = 755 - 720 = 35$

$1 + 27D = 1 + 27 \times 29 = 784 - 768 = 16$

$\boxed{16mM = 768}$

$1 + 28D = 1 + 28 \times 29 = 813 - 768 = 45$

$1 + 29D = 1 + 29 \times 29 = 842 - 816 = 26$

$\boxed{17mM = 816}$

$1 + 30D = 1 + 30 \times 29 = 871 - 864 = 7$

$\boxed{18mM = 864}$

$1 + 31D = 1 + 31 \times 29 = 900 - 864 = 36$

∴ Slots of B-phase: 33, 14, 43, 24, 5, 34, 15, 44, 25, 6, 35, 16, 45, 26, 7, 36

On arranging B-phase slots in increasing order we get,

Slots of B-phase: 5, 6, 7, 14, 15, 16, 24, 25, 26, 33, 34, 35, 36, 43, 44, 45

The remaining 16 slots in the pole unit are allotted to Y-phase.

∴ Slots of Y-phase : 8, 9, 10, 17, 18, 19, 20, 27, 28, 29, 37, 38, 39, 46, 47, 48

Similarly the slots of next pole unit are distributed as shown below.

Let us start from slot 49.

Let us allot first 16 slots to R-phase.

$$49, 49 + D = 49 + 29 = 78$$

$$49 + 2D = 49 + 2 \times 29 = 107 - 48 = 59 \qquad \boxed{mM = 48}$$

$$49 + 3D = 49 + 3 \times 29 = 136 - 48 = 88$$

$$49 + 4D = 49 + 4 \times 29 = 165 - 96 = 69 \qquad \boxed{2mM = 96}$$

$$49 + 5D = 49 + 5 \times 29 = 194 - 144 = 50 \qquad \boxed{3mM = 144}$$

$$49 + 6D = 49 + 6 \times 29 = 223 - 144 = 79$$

$$49 + 7D = 49 + 7 \times 29 = 252 - 192 = 60 \qquad \boxed{4mM = 192}$$

$$49 + 8D = 49 + 8 \times 29 = 281 - 192 = 89$$

$$49 + 9D = 49 + 9 \times 29 = 310 - 240 = 70 \qquad \boxed{5mM = 240}$$

$$49 + 10D = 49 + 10 \times 29 = 339 - 288 = 51 \qquad \boxed{6mM = 288}$$

$$49 + 11D = 49 + 11 \times 29 = 368 - 288 = 80$$

$$49 + 12D = 49 + 12 \times 29 = 397 - 336 = 61 \qquad \boxed{7mM = 336}$$

$$49 + 13D = 49 + 13 \times 29 = 426 - 336 = 90$$

$$49 + 14D = 49 + 14 \times 29 = 455 - 384 = 71 \qquad \boxed{8mM = 384}$$

$$49 + 15D = 49 + 15 \times 29 = 484 - 432 = 52 \qquad \boxed{9mM = 432}$$

∴ Slots of R-phase: 49, 78, 59, 88, 69, 50, 79, 60, 89, 70, 51, 80, 61, 90, 75, 52

On arranging R-phase slots in increasing order we get,

Slots of R-phase: 49, 50, 51, 52, 59, 60, 61, 69, 70, 71, 78, 79, 80, 88, 89, 90

The next 16 slots are allotted to B-phase.

$$49 + 16D = 49 + 16 \times 29 = 513 - 432 = 81 \qquad \boxed{9mM = 432}$$

$$49 + 17D = 49 + 17 \times 29 = 542 - 480 = 62 \qquad \boxed{10mM = 432}$$

$$49 + 18D = 49 + 18 \times 29 = 571 - 480 = 91$$

$$49 + 19D = 49 + 19 \times 29 = 600 - 528 = 72 \qquad \boxed{11mM = 528}$$

$$49 + 20D = 49 + 20 \times 29 = 629 - 576 = 53 \qquad \boxed{12mM = 576}$$

$$49 + 21D = 49 + 21 \times 29 = 658 - 576 = 82$$

$$49 + 22D = 49 + 22 \times 29 = 687 - 624 = 63 \qquad \boxed{13mM = 624}$$

$$49 + 23D = 49 + 23 \times 29 = 716 - 624 = 92$$

$$49 + 24D = 49 + 24 \times 29 = 745 - 672 = 73 \qquad \boxed{14mM = 672}$$

$49 + 25D = 49 + 25 \times 29 = 774 - 720 = 54$

$49 + 26D = 49 + 26 \times 29 = 803 - 720 = 83$

$49 + 27D = 49 + 27 \times 29 = 832 - 768 = 64$

$49 + 28D = 49 + 28 \times 29 = 861 - 768 = 93$

$49 + 29D = 49 + 29 \times 29 = 890 - 816 = 74$

$49 + 30D = 49 + 30 \times 29 = 919 - 864 = 55$

$49 + 31D = 49 + 31 \times 29 = 948 - 864 = 84$

$\boxed{15mM = 720}$

$\boxed{16mM = 768}$

$\boxed{17mM = 816}$

$\boxed{18mM = 864}$

$\therefore$ Slots of B-phase: 81, 62, 91, 72, 53, 82, 63, 92, 73, 54, 83, 64, 93, 74, 55, 84

On arranging B-phase slots in increasing order we get,

Slots of B-phase: 53, 54, 55, 62, 63, 64, 72, 73, 74, 81, 82, 83, 84, 91, 92, 93

The remaining 16 slots in the pole unit are allotted to Y-phase.

$\therefore$ Slots of Y-phase : 56, 57, 58, 65, 66, 67, 68, 75, 76, 77, 85, 86, 87, 94, 95, 96

Table 1: Slot Allotment

Slots in a pole pair	Slot allotment for phase sequence RYB					
	R-phase	R, B-phase	B-phase	B, Y-phase	Y-phase	Y, R-phase
1, 2, 3, 4, 5, 6, 7, 8, 9, 10	1, 2, 3	4	5, 6, 7		8, 9, 10	
11, 12, 13, 14, 15, 16, 17, 18, 19, 20	11, 12, 13		14, 15, 16		17, 18, 19	20
21, 22, 23, 24, 25, 26, 27, 28, 29	21, 22, 23		24, 25, 26		27, 28, 29	
30, 31, 32, 33, 34,35, 36, 37, 38, 39	30, 31, 32		33, 34,35	36	37, 38, 39	
40, 41, 42, 43, 44, 45, 46, 47, 48	40, 41, 42		43, 44, 45		46, 47, 48	
49, 50, 51, 52, 53, 54, 55, 56, 57, 58	49, 50, 51	52	53, 54, 55		56, 57, 58	
59, 60, 61, 62, 63, 64, 65, 66, 67, 68	59, 60, 61		62, 63, 64		65, 66, 67	68
69, 70, 71, 72, 73, 74, 75, 76, 77	69, 70, 71		72, 73, 74		75, 76, 77	
78, 79, 80, 81, 82, 83, 84, 85, 86, 87	78, 79, 80		81, 82, 83	84	85, 86, 87	
88, 89, 91, 92, 93, 94, 95, 96	88, 89, 90		91, 92, 93		94, 95, 96	

EXAMPLE 5.16

Work out the detailed slot allotment for a 54 slots, 8 pole, 3-phase AC machine armature.

SOLUTION

Given that,

Number of slots,   S = 54

Number of phases, m = 3

Number of poles,   p = 8

Slots per pole per phase, $q = \dfrac{S}{mp} = \dfrac{54}{3 \times 8} = \dfrac{54}{24}$

Let us express q in the form,

$$q = \frac{M}{d} = I + \frac{n}{d} \qquad \qquad .....(1)$$

$\therefore$ Slots per pole per phase, $q = \dfrac{S}{mp} = \dfrac{54}{3 \times 8} = \dfrac{54}{24} = \dfrac{54/6}{24/6}$

> Dividing numerator and denominator by 6.

$$\therefore q = \frac{9}{4} = 2\frac{1}{4} = 2 + \frac{1}{4} \qquad \qquad .....(2)$$

On comparing equations (1) and (2) we get,

M = 9, d = 4, I = 2, n = 1,

Number of pole units = $\dfrac{P}{d} = \dfrac{8}{4} = 2$

$\therefore$ 8 poles can be divided into 2 units of 4 poles each (d = 4).

Slots in a unit = mM = 3 × 9 = 27 slots

Slots per phase per unit, M = 9

> Also, $M = qd = \frac{9}{4} \times 4 = 9$

$$D = \frac{1 + mMP}{d}$$

Let, P = 1, $D = \dfrac{1 + 3 \times 9 \times 1}{4} = 7$

The distribution of slots is given by,

1, 1 + D, 1 + 2D, 1 + 3D, ...

In the above equation of sum exceeds mM then subtract the sum by mM or integral multiple of mM.

The 27 slots of a pole unit can be distributed at the rate of 9 slots per phase. The first 27 slots with slot number 1 to 27 are slots of first pole unit.

Let us distribute first 27 slots as shown below.

Let us start from slot 1. The first 9 slots are allotted to R-phase.

1, 1 + D = 1 + 7 = 8

1 + 2D = 1 + 2 × 7 = 15

1 + 3D = 1 + 3 × 7 = 22

1 + 4D = 1 + 4 × 7 = 29 – 27 = 2

> mM = 27

1 + 5D = 1 + 5 × 7 = 36 – 27 = 9

1 + 6D = 1 + 6 × 7 = 43 – 27 = 16

1 + 7D = 1 + 7 × 7 = 50 – 27 = 23

1 + 8D = 1 + 8 × 7 = 57 – 54 = 3

> 2mM = 54

$\therefore$ Slots of R-phase: 1, 8, 15, 22, 2, 9, 16, 23, 3

On arranging R-phase slots in increasing order we get,

Slots of R-phase: 1, 2, 3, 8, 9, 15, 16, 22, 23

The next 9 slots are allotted to B-phase.

1 + 9D = 1 + 9 × 7 = 64 – 54 = 10

> 2mM = 54

1 + 10D = 1 + 10 × 7 = 71 – 54 = 17

1 + 11D = 1 + 11 × 7 = 78 – 54 = 24

$1 + 12D = 1 + 12 \times 7 = 85 - 81 = 4$ $\boxed{3mM = 81}$

$1 + 13D = 1 + 13 \times 7 = 92 - 81 = 11$

$1 + 14D = 1 + 14 \times 7 = 99 - 81 = 18$

$1 + 15D = 1 + 15 \times 7 = 106 - 81 = 25$

$1 + 16D = 1 + 16 \times 7 = 113 - 108 = 5$ $\boxed{4mM = 108}$

$1 + 17D = 1 + 17 \times 7 = 120 - 108 = 12$

$\therefore$ Slots of B-phase: 10, 17, 24, 4, 11, 18, 25, 5, 12

On arranging B-phase slots in increasing order we get,

Slots of B-phase: 4, 5, 10, 11, 17, 18, 24, 25

The remaining 16 slots in the pole unit are allotted to Y-phase.

$\therefore$ Slots of Y-phase : 6, 7, 12, 13, 14, 19, 20, 21, 26, 27

Similarly the slots of next pole unit are distributed as shown below.

Let us start from slot 28.

Let us allot first 9 slots to R-phase.

$28, 28 + D = 28 + 7 = 35$

$28 + 2D = 28 + 2 \times 7 = 42$

$28 + 3D = 28 + 3 \times 7 = 49$

$28 + 4D = 28 + 4 \times 7 = 56 - 27 = 29$ $\boxed{mM = 27}$

$28 + 5D = 28 + 5 \times 7 = 63 - 27 = 36$

$28 + 6D = 28 + 6 \times 7 = 70 - 27 = 43$

$28 + 7D = 28 + 7 \times 7 = 77 - 27 = 50$

$28 + 8D = 28 + 8 \times 7 = 84 - 54 = 30$ $\boxed{2mM = 54}$

$\therefore$ Slots of R-phase: 28, 35, 42, 49, 29, 36, 43, 50, 30

On arranging R-phase slots in increasing order we get,

Slots of R-phase: 28, 29, 30, 35, 36, 42, 43, 49, 50

The next 9 slots are allotted to B-phase.

$28 + 9D = 28 + 9 \times 7 = 91 - 54 = 37$

$28 + 10D = 28 + 10 \times 7 = 98 - 54 = 44$

$28 + 11D = 28 + 11 \times 7 = 105 - 54 = 51$

$28 + 12D = 28 + 12 \times 7 = 112 - 81 = 31$ $\boxed{3mM = 81}$

$28 + 13D = 28 + 13 \times 7 = 119 - 81 = 38$

$28 + 14D = 28 + 14 \times 7 = 126 - 81 = 45$

$28 + 15D = 28 + 15 \times 7 = 133 - 81 = 52$

$28 + 14D = 28 + 14 \times 7 = 140 - 108 = 32$ $\boxed{4mM = 108}$

$28 + 15D = 28 + 15 \times 7 = 147 - 108 = 39$

$\therefore$ Slots of B-phase: 37, 44, 51, 31, 38, 45, 52, 32, 39

On arranging B-phase slots in increasing order we get,

Slots of B-phase: 31, 32, 37, 38, 39, 44, 45, 51, 52

The remaining 16 slots in the pole unit are allotted to Y-phase.

∴ Slots of Y-phase : 33, 34, 40, 41, 46, 47, 48, 53, 54

Table 1: Slot Allotment

Slots in a pole pair	Slot allotment for phase sequence RYB					
	R-phase	R, B-phase	B-phase	B, Y-phase	Y-phase	Y, R-phase
1, 2, 3, 4, 5, 6, 7	1, 2	3	4, 5		6, 7	
8, 9, 10, 11, 12, 13, 14	8, 9		10, 11	12	13, 14	
15, 16, 17, 18, 19, 20, 21	15, 16		17, 18		19, 20	21
22, 23, 24, 25, 26, 27	22, 23		24, 25		26, 27	
28, 29, 30, 31, 32, 33, 34	28, 29	30	31, 32		33, 34	
35, 36, 37, 38, 39, 40, 41	35, 36		37, 38	39	40, 41	
42, 43, 44, 45, 47, 48	42, 43		44, 45		46, 47	48
49, 50, 51, 52, 53, 54	49, 50		51, 52		53, 54	

5.6 SHORT-ANSWER QUESTIONS

Q5.1 List the different types of slots that are used in rotating machines.

The different types of slots are,

1. Parallel sided slots with flat bottom
2. Tapered slots with flat bottom
3. Parallel sided slots with circular bottom
4. Tapered slots with circular bottom.
5. Circular slots.

Q5.2 State the difference between the armature winding of DC machine and the stator winding of AC machine.

The armature winding of DC machine has closed coils but the stator winding of AC machine has open coils.

Q5.3 Mention the two types of armature winding used in DC machine and on what factor the type of winding is decided?

(AU, Apr' 18, 2 M)

The two types of winding employed in dc machines are lap and wave winding. The choice between the two is based on voltage and current rating of the machine. For large voltage rating wave winding is preferred. For large current rating lap winding is preferred. Generally, when the current per parallel path is lessthan 200 A, wave winding is provided. If the current per parallel path is more than 200 A then the only choice is lap winding.

Q5.4 Define winding pitch.

The winding pitch is defined as the distance between the starts of two consecutive coils measured in terms of coil sides.

Q5.5   *What is back pitch?*

The distance between top and bottom coil sides of a coil measured around the back of the armature is called the back pitch. The back pitch is measured in terms of coil sides.

Q5.6   *What is front pitch in DC machine armature winding?*

The front pitch is the distance between two coil sides connected to the same commutator segment. It is measured in terms of coil sides.

Q5.7   *Define commutator pitch in DC machine armature winding.*

The commutator pitch is defined as the distance between the two commutator segments to which the two ends (start and finish) of a coil are connected. It is measured in terms of commutator segment.

Q5.8   *What is equalizer connection?*

The equalizer connections are low resistance copper conductors employed in lap winding to equalize the induced emfs in parallel paths. The difference in induced emf in parallel paths is due to slight unsymmetry in flux per pole in DC machines.

Q5.9   *Distinguish between lap and wave windings used in DC machine.*

The lap and wave windings primarily differ from each other in the following two factors.

i)   The number of circuits between the positive and negative brushes, i.e., number of parallel paths.

ii)   The manner in which the coil ends are connected to the commutator segments.

Q5.10   *How will you select the back and front pitch for a lap winding of DC machine armature?*

The back pitch should be nearly equal to coil sides per pole and it should be an odd integer.

$$\therefore \text{Back pitch, } Y_b = \frac{\text{Number of coil sides}}{\text{Number of poles}} + K = \frac{2C}{p} + K$$

Here K is a constant that will make 2C/p as an odd integer

For lap winding, Winding pitch, $Y = Y_b - Y_f$.

Also winding pitch is always two in lap winding.

$$\therefore \text{Front pitch, } Y_f = Y_b - Y = Y_b - 2$$

Q5.11   *How will you select the back and front pitch for a wave winding of a DC machine armature?*

The back pitch should be nearly equal to coil sides per pole and it should be an odd integer.

$$\therefore \text{Back pitch, } Y_b = \frac{\text{Number of coil sides}}{\text{Number of poles}} + K = \frac{2C}{p} + K$$

Here K is a constant that will make 2C/p as an odd integer

For wave winding, Winding pitch, $Y = \frac{2C + 2}{p/2}$

Also, Winding pitch, $Y = Y_b + Y_f$

$$\therefore \text{Front pitch, } Y_f = Y - Y_b$$

Q5.12   *List the characteristics of a lap winding.*

The characteristics of lap winding are

1.   The number of parallel paths is equal to number of poles.

2.   The current through a conductor is I_a/p where I_a is the armature current and p is number of poles.

3.   The winding will have large number of conductors with smaller area of cross-section.

4.   The voltage induced in various parallel paths may differ slightly due to asymmetry in the magnetic circuit.

Q5.13   *List the characteristics of wave winding.*

The characteristics of wave winding are,

1.   The number of parallel paths is two.

2.   The current through a conductor is $I_a/2$, where I_a is the armature current.

3.   The winding will have less number of conductors with larger area of cross-section.

4.   The emf induced in both the parallel paths will be always equal.

Q5.14   *What is simplex and multiplex winding?*

In simplex lap winding the number of parallel paths is equal to number of poles and in simplex wave winding the number of parallel paths is two. In multiplex windings the number of parallel paths will be multiples of simplex winding.

In duplex winding the number of parallel paths will be double that of simplex winding and in triplex winding the number of parallel paths is thrice that of simplex winding and so on.

Q5.15   *What are dummy coils?*

The coils which are placed in armature slot for mechanical balance but not connected electrically to the armature winding are called dummy coils.

Q5.16   *What is meant by equalizer connection?*

In lap winding, due to difference in the induced emf in various parallel paths, there may be circulating currents in the brushes and winding. The connections that are made to equalize the difference in induced emf and to avoid circulating currents through brushes are called equalizer connections.

Q5.17   *How equalizer connections will made?*

In armature winding the points at same potential under ideal conditions are connected to same equalizer rings. The connections are made using copper conductors usually in the form of rings and so they are called equalizer rings.

Q5.18   *Why equalizer connections are not needed in wave winding?*

In simplex wave winding there are only two parallel paths. The conductors forming a parallel path will be distributed equally under all the poles. Hence both the parallel paths are equally affected by the asymmetry in the magnetic circuit and so there is no circulating current. Therefore there is no necessity for equalizer connections.

5.7   EXERCISES

I.   Fill in the blanks

1.   In .... winding the finish of a coil is connected to start of next coil.

2.   In .... winding the finish of a coil is connected to start of a coil which is lying one pole pitch away from the finish.

3.   In .... winding the number of parallel paths is double that of .... winding.

4.   The active length of copper wire in the slot is called ....

5.   The active portions of the conductors in a coil are called ....

6.   The end portion of the coil connecting the two coil sides is called ....

7.   In .... winding the conductors will be arranged in two layers in a slot.

8.   The distance between two coil sides connected to same commutator segment is called ....

9.   The maximum voltage between adjacent segments at load should not exceed ....

10.   In .... winding the joining to the commutator progresses in the same direction as that of coil progress in the armature.

11.   The winding pitch for lap winding is always .... and the commutator pitch is always ....

12.   The .... are used for mechanical balancing of armature.

13.   When dummy coils are provided the winding is called ....

14.   In lap winding the points at same potential under ideal conditions are connected to ....

Answers			
1.	lap	8.	front pitch
2.	wave	9.	30 V
3.	duplex, simplex	10.	progressive
4.	conductor	11.	two, one
5.	coil sides	12.	dummy coil
6.	overhang	13.	forced winding
7.	double layer	14.	one equalizer ring

II.   State whether the following statements are True/False

1.   In wave winding the slots should be multiple of pole pair to avoid dummy coils.

2.   With increase in slot depth, the slot reactance increases.

3.   In multiplex winding the number of parallel paths are multiples of simplex winding.

4.   Equalizer connections are not needed for wave winding.

5.   For a specified power rating the volume of copper is same for lap and wave winding.

6.   The wave winding has to be employed when the armature current is more than 400 A.

7.   In full pitch coils the coil span will not be equal to pole pitch.

8.   The number of coils in armature is equal to number of commutator segments.

9.   With increase in number of armature coils the voltage between adjacent commutator segments increase.

10.   The DC machine armature winding are closed windings.

11.   The simplex lap winding has one closed electrical circuit.

12.   The winding pitch and commutator pitch of lap winding in DC machine are independent of number of slots and poles.

13.   The winding pitch and commutator pitch of wave winding winding in DC machine are independent of number of slots and poles.

14.   In the DC machine armature the direction of currents in all the conductors will be same at any one time instant.

15.   In wave winding the number of parallel paths is independent of poles.

Answers					
1.	False	6.	False	11.	True
2.	True	7.	False	12.	True
3.	True	8.	True	13.	False
4.	True	9.	False	14.	False
5.	True	10.	True	15.	True

III   Unsolved problems

E5.1   Draw the winding diagram in the developed from for a simplex lap wound 36 slots, 6 pole DC generator armature 36 commutator segment. Show the position of brushes.

E5.2   Draw the winding diagram for a 4 pole, 21 slots simplex wave wound DC generator with a commutator having 21 segment. Indicate the position of brushes.

E5.3   Draw a winding diagram of a 3-phase, 36 slots, 6 pole, AC armature for the single layer unbifurcated winding.

E5.4   Draw a winding diagram of a 3-phase, 36 slots, 6 pole, AC armature for the single layer bifurcated winding.

E5.5   Draw a winding diagram for a 4 pole, 24 slots, 3-phase mush connected armature.

E5.6   Draw the 3-phase integral slot winding diagram in the developed form for a lap wound 48 slots, 6 pole AC machine armature.

E5.7   Draw the 3-phase integral slot winding diagram for a wave wound 48 slots, 6 pole AC machine armature.

E5.8   Draw a 3-phase fractional slot winding diagram for a lap wound 45 slots, 4 pole AC machine armature.

E5.9   Give the layout of a double layer fractional slot wave winding for a 3-phase, 4 pole, 45 slot armature.

> **Chapter 6**

DESIGN OF DC MACHINES

List of symbols

Symbol	Meaning	Unit
AT_a	Armature mmf per pole	AT
AT_c	mmf required for core	AT
AT_g	mmf required for air-gap	AT
AT_i	mmf required for interpole	AT
AT_p	mmf required for pole	AT
AT_t	mmf required for teeth	AT
AT_y	mmf required for yoke	AT
AT_{fl}	mmf developed per pole at full load	AT
AT_{fo}	Total mmf per pole at no-load and normal voltage	AT
AT_{se}	Ampere-turns to be developed by series field coil	AT
A_b	Total brush area per spindle	mm^2
A_B	Total contact area of all brushes	mm^2
A_c	Area of cross-section of armature core	mm^2
A_p	Area of pole body (shank)	mm^2 or m^2
a	Number of parallel paths	-
a_a	Area of cross-section of armature conductor	mm^2
a_b	Contact area of each brush	mm^2
a_f	Area of cross-section of conductor in field winding	mm^2
a_i	Area of cross-section of interpole conductor	mm^2
a_{se}	Area of cross-section of series field conductor	mm^2
ac	Specific electric loading	amp.cond./m
at_c	mmf per metre corresponding to flux density in core	AT/m
at_p	mmf per metre corresponding to flux density in pole body	AT/m
at_t	mmf per metre corresponding to flux density of tooth at one third height from narrow end	AT/m

Symbol	Meaning	Unit
at_y	mmf per metre corresponding to flux density in yoke	AT/m
B_c	Flux density in armature core	Wb/m^2
B_g	Maximum gap density in the air-gap at no-load	Wb/m^2
B_p	Flux density in pole body	Wb/m^2
B_{av}	Specific magnetic loading (or) average gap density	Wb/m^2
B_{gi}	Flux density under interpole	Wb/m^2
B_{gm}	Maximum flux density in the air-gap on load condition	Wb/m^2
b	Pole arc	m
b_p	Width of the pole body	m
C	Total number of armature coils	-
C_{min}	Minimum number of armature coils	-
C_1	Clearence allowed for staggering the brushes	mm
C_2	Clearence for allowing end play	mm
C_b	Clearence between the brushes	mm
C_l	Leakage coefficient	-
C_o	Output coefficient	$kW/m^3\text{-}rps$
c	Cooling coefficient	$°CW\text{-}m^2$
D	Diameter of armature	m
D_c	Commutator diameter	m
D_i	Inner diameter of armature	m
D_m	Mean diameter of armature	m
D_{my}	Mean diameter of yoke	m
D_{max}	Maximum value of armature diameter	m
d_c	Depth of core	m
d_{if}	Inner diameter of field coil	m
d_{of}	Outer diameter of field coil	m
d_f	Depth of field coil	mm
d_{fc}	Diameter of bare field conductor	mm
d_{fci}	Diameter of field conductor including insulation thickness	mm
d_p	Diameter of pole body	mm
d_s	Depth of slot	mm
d_{se}	Depth of series field winding	mm
d_{mf}	Mean diameter of field coil	m
E	Generated emf or back emf (or) voltage rating of machine	V
E_c	Average voltage between adjacent commutator segments on no-load	V
E_f	Voltage across each shunt field coil	V
E_{cm}	Maximum voltage between adjacent commutator segments	V
E_{rav}	Average reactance voltage (in the coil)	V

Symbol	Meaning	Unit
e_z	Average voltage per conductor on no-load	V
e_{z_max}	Maximum voltage per conductor on load	V
h_f	Height of field coil	mm
h_p	Height of pole body	mm
h_s	Height of pole shoe	mm
h_{se}	Height of series field winding	mm
h_{pl}	Height of field pole	m
I	Full load current	A
I_a	Armature current	A
I_b	Current carried by each brush	A
I_f	Current in field winding	A
I_L	Load current	A
I_{se}	Current through series field winding at full load	A
I_z	Current through an armature conductor	A
K_f	Field form factor	-
K_g	Total gap contraction factor	-
K_{gi}	Interpole gap contraction factor	-
L	Length of armature	m
L_c	Length of commutator	m
L_i	Net iron length	m
L_o	Length of outer most turn	m
L_p	Length of pole core	m
L_{coil}	Inductance of a coil in armature	H
L_{ip}	Length of interpole	mm
L_{mtf}	Length of mean turn of field coil	m
L_{mta}	Length of mean turn of armature coil	m
L_{mtse}	Length of mean turn of series field coil	m
L_{max}	Maximum value of armature core length	m
L_{pi}	Net iron length of pole core	m
l_g	Length of air-gap	mm
l_{gi}	Length of air-gap under the interpole	mm
l_c	Length of flux path in core	m
N_c	Number of coils between commutator segments	-
N_{cs}	Number of commutator segments	-
n	Speed	rps
n_b	Number of brushes per spindle	-
P	Rated power output	kW
P_a	Power developed in armature	kW
P_b	Brush contact pressure on commutator	N/m^2

Symbol	Meaning	Unit
P_c	Total commutator loss	W
P_{bc}	Brush contact loss	W
P_{bf}	Brush friction loss	W
p	Number of poles	-
P_i	Power input	W
Q_f	Permissible copper loss in each field coil	W
q_f	Permissible loss per unit winding surface for normal temperature rise	W/m^2
R_f	Resistance of field coil	Ω
S	Cooling surface of field coil	m^2
S_a	Number of armature slots	-
S_f	Slot space factor	-
S_c	Commutator surface area	m^2
S_{fse}	Copper space factor for series field coil	m^2
T_c	Turns per coil (in armature)	-
T_f	Number of turns in each field coil	-
T_i	Number of turns in interpole	-
T_{se}	Number of turns in series field coil	-
t_b	Thickness of each brush	mm
u	Number of coil sides/slot	-
V_a	Peripheral velocity of armature	m/s
V_{am}	Maximum peripheral velocity of armature	m/s
V_c	Peripheral speed of commutator	m/s
w_b	Width of the brush	mm
w_d	Width of the duct	mm
Y	Winding pitch	coil sides
Y_b	Back pitch	coil sides
Y_f	Front pitch	coil sides
Y_c	Commutator pitch	coil sides
Z	Total number of armature conductors	-
Z_s	Number of conductors per slot	-
θ	Temperature rise	$°C$
ρ	Resistivity	$\Omega\text{-}m$
Ψ	Ratio of pole arc to pole pitch	-
η	Efficiency	-
τ	Pole pitch	m
δ_a	Current density in armature conductor	A/mm^2

Symbol	Meaning	Unit
δ_b	Current density in brushes	A/mm^2
δ_f	Current density in the field winding	A/mm^2
δ_i	Current density in interpole winding	A/mm^2
ϕ	Flux per pole	Wb
ϕ_c	Flux in armature core	Wb
ϕ_p	Flux in pole body	Wb
β_c	Commutator segment pitch	mm
λ	Specific permeance	H/m
μ	Coefficient of friction in commutator	-

6.1 CONSTRUCTION

The DC machines used for industrial electric drives have three major parts. They are field system, armature and commutator. The field system is located on the stationary part of the machine called *stator* and consists of main poles, interpoles and frame or yoke.

The main poles are designed to produce the magnetic flux. The interpoles are placed in between the main poles. They are employed to improve the commutation condition. The frame provides mechanical support to machine and also serve as a path for flux.

The armature is the rotating part (or rotor) of a DC machine and consists of armature core with slots and armature winding accommodated in slots. The conversion of energy from mechanical to electrical or vice-versa takes place in armature.

The commutator is mounted on the rotor of a DC machine. The commutator and brush arrangement works like a mechanical dual converter. In case of generator it rectifies the induced AC to DC. In case of motor it inverts the DC supply to AC. (In motor the commutator reverses the current through the armature conductors to get unidirectional torque).

Constructional elements of DC machine

Stator		Rotor	
-	Yoke or Frame	-	Armature core
-	Field pole	-	Armature winding
-	Pole shoe	-	Commutator
-	Field winding	**Others**	
-	Interpole	-	Brush
		-	Brush holder

6.2 OUTPUT EQUATION   (JNTKU, Apr' 19, 5 M) (AU, Apr' 19, 6 M)

The output of a machine can be expressed in terms of its main dimensions, specific magnetic and electric loadings and speed. The equation which relates the power output to D, L, B_{av}, **ac** and n of the machine is known as output equation.

Output equation and Output coefficient of DC machine

The following equations are used to derive the output equation.

Induced emf in armature, $E = \dfrac{\phi ZN}{60}\dfrac{p}{a}$ $\Rightarrow$ $E = \dfrac{\phi Znp}{a}$ $\boxed{n = \dfrac{N}{60}}$ .....(6.1)

Current through each conductor, $I_z = \dfrac{I_a}{a} \Rightarrow I_a = a\, I_z$ .....(6.2)

Specific magnetic loading, $B_{av} = \dfrac{p\phi}{\pi DL} \Rightarrow p\phi = \pi DL\, B_{av}$ .....(6.3)

Specific electric loading, $ac = \dfrac{I_z Z}{\pi D} \Rightarrow I_z Z = \pi D\, \textbf{ac}$ .....(6.4)

where,   n = Speed in rps   Z = Number of armature conductors
          I_a = Armature current   D = Diameter of rotor
          p = Number of poles   L = Length of rotor
          a = Number of parallel paths ϕ = Flux per pole

In DC generator the electrical power generated in the armature is given by the product of induced emf and armature current. In case of DC motor the mechanical equivalent of electrical power in armature is given by the product of induced emf (back emf) and armature current.

$\therefore$ Power developed in armature, $P_a = E\, I_a \times 10^{-3}$ in kW .....(6.5)

$$= \dfrac{\phi Znp}{a} \times a I_z \times 10^{-3}$$

> Using equations (6.1) and (6.2)

$$= p\phi \times I_z Z \times n \times 10^{-3}$$

$$= \pi DL\, B_{av} \times \pi D\, \textbf{ac} \times n \times 10^{-3}$$

> Using equations (6.3) and (6.4)

$$= \pi^2 B_{av}\, \textbf{ac} \times 10^{-3} \times D^2\, L\, n$$

$$= C_o\, D^2\, L\, n$$ .....(6.6)

where, $C_o = \pi^2 B_{av}\, \textbf{ac} \times 10^{-3}$ .....(6.7)

The equation, $P_a = C_o\, D^2\, Ln$ is called *output equation* of DC machine and the term C_o is called *output coefficient* of DC machine.

The output coefficient, C_o in terms of maximum gap density, B_g is given by,

     $C_o = \pi^2\, \psi B_g\, \textbf{ac} \times 10^{-3}$ .....(6.8)

where, $B_g = \dfrac{B_{av}}{\psi}$ .....(6.9)

       $\psi = \dfrac{b}{\tau}$ = Ratio of pole arc to pole pitch .....(6.10)

The term $D^2 L$ in the output equation is proportional to volume of active part. Therefore if C_o is constant then we can say the power output is directly proportional to the product of volume of active part and speed.

     i.e., $P_a \; \alpha \;$ Volume of active part $\times$ Speed

If C_o is varied then power output is directly proportional to the four quantities and they are B_{av}, **ac**, volume of active part and speed.

     i.e., $P_a \; \alpha \; B_{av} \times \textbf{ac} \times$ Volume of active part $\times$ Speed.

Power developed by the armature, P_a is different from the rated power output P, of the machine. The relationship between the two are,

     $P_a = \dfrac{P}{\eta}$ - For DC generator .....(6.11)

     $P_a = P$ - For DC motors .....(6.12)

where, η = Efficiency of DC generator

Armature torque in terms of main dimensions   *(AU, Apr' 17, 8 M)*

Let, T_a = Armature torque

Power developed in armature in terms of speed and torque is given by,

Power developed in armature, $P_a = \dfrac{2\pi N T_a}{60}$ $\boxed{n = \dfrac{N}{60}}$

$$= 2\pi n T_a$$

Power developed in armature in terms of main dimensions is given by,

Power developed in armature, $P_a = C_o D^2 L n$

On equating the above two equations of power developed in armature we get,

$$2\pi n T_a = C_o D^2 L n$$

$\therefore$ Armature torque, $T_a = \dfrac{C_o}{2\pi} D^2 L$

$$= \dfrac{\pi^2 \, B_{av} \, ac \times 10^{-3}}{2\pi} D^2 L \qquad \boxed{\text{Using equation (6.7)}}$$

$$= \dfrac{1}{2}\pi \, B_{av} \, \mathbf{ac} \, D^2 L \times 10^{-3} \qquad\qquad .....(6.13)$$

6.2.1 Choice of Armature Diameter

The factors to be considered for the choice of armature diameter are,

1. Peripheral speed
2. Pole pitch
3. Specific electric loading
4. Induced emf per conductor
5. Power output

In DC machines the peripheral speed lies in the range 15 to 50 m/s. Normally the peripheral speed should not exceed 30 m/s. If the speed exceeds 30 m/s then special rotor construction methods have to be employed to prevent the damage due to excessive centifugal force.

The diameter should be suitable for accommodating desired number of poles with normal values of pole pitch. The normal values of pole pitch are given in Table 6.1.

Table 6.1: Pole Pitch

Pole	Pole pitch *mm*
2	upto 240
4	240 to 400
6	350 to 450
above 6	450 to 500

An expression for maximum (or limiting) value of armature diameter can be derived as shown below.

Induced emf in armature, E = Emf per conductor $\times$ Conductors per parallel path

$$\therefore E = e_z \times \dfrac{Z}{a} \qquad\qquad .....(6.14)$$

Specific electric loading, $\mathbf{ac} = \dfrac{I_z Z}{\pi D} = \dfrac{I_a}{a} \dfrac{Z}{\pi D} \qquad \Rightarrow \qquad \therefore \dfrac{I_a Z}{a} = \pi D \, \mathbf{ac} \qquad .....(6.15)$

Power developed in armature, $P_a = E\,I_a \times 10^{-3}$ in kW

$$= e_z \frac{Z}{a} I_a \times 10^{-3} = e_z\,\pi D\,\mathbf{ac} \times 10^{-3} \qquad \boxed{\text{Using equations (6.14) and (6.15)}}$$

$$\therefore D = \frac{P_a}{\pi\,\mathbf{ac}\,e_z \times 10^{-3}}$$

$\therefore$ Maximum value of armature diameter, $D_{max} = \dfrac{P_a \times 10^3}{\pi\,\mathbf{ac}\,e_z}$ .....(6.16)

From equation (6.16) it can be observed that the maximum value of armature diameter depends on specific electric loading and induced emf per conductor.

6.2.2 Choice of Armature Length

The factors to be considered for the choice of armature length are,

1. Cost
2. Ventilation
3. Voltage between adjacent commutator segments
4. Specific magnetic loading

When the length of the core is large, the ratio of inactive copper to active copper will be small. Hence the machine may cost less. But when the core length is very large then ventilation of the core will be difficult. The centre portion of the core tends to attain a high temperature rise and so the core must be ventilated (or cooled) by special methods.

An expression for maximum (or limiting) value of core length can be derived as shown below.

Using equations (6.40), (6.44) and (6.46) of section 6.6.5 we get,

Maximum voltage between adjacent segments, $E_{cm} = 4\,N_c\,T_c\,e_z$

Let, $E_{cm} = 30\ V$

$\therefore$ Induced emf in a conductor, $e_z = \dfrac{7.5}{T_c\,N_c}$ .....(6.17)

where, T_c = Turns per coil

N_c = Number coils between adjacent segments = 1   -   For simplex lap winding

= p/2   -   For simplex wave winding

We know that,

Induced emf in a conductor = $B_{av}\,L\,V_a$ .....(6.18)

where, V_a = Peripheral speed

On equating equations (6.18) and (6.19), we get an expression for maximum value of core length L.

$\therefore B_{av}\,L_{max}\,V_a = \dfrac{7.5}{T_c\,N_c}$

Maximum value of armature core length, $L_{max} = \dfrac{7.5}{T_c\,N_c\,B_{av}\,V_a}$ .....(6.19)

From equation (6.19) it can be observed that the maximum value of armature core length depends on specific magnetic loading and peripheral speed.

6.3 MAIN DIMENSIONS OF DC MACHINES

In rotating machines the active part is cylindrical in shape. The volume of the cylinder is given by the product of area of cross section and length. If D is the diameter and L is the length of cylinder, then the volume is given by $\pi D^2 L/4$. Therefore D and L are specified as main dimensions.

In case of DC machine, D represent the diameter of armature and L represent the length of armature. The Fig. 6.1 shows the main dimensions of DC machine.

Here,   D = Diameter of armature

l_g = Length of air-gap

L = Length of armature

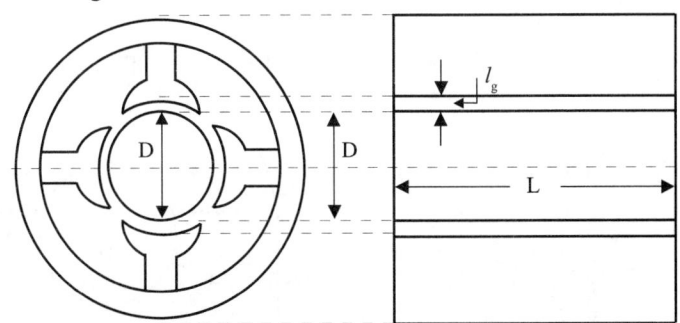

Fig. 6.1: *Main dimensions of DC machines.*

6.3.1 Magnetic Circuit   *(AU, Apr' 18, 5 M)*

The path of magnetic flux is called ***magnetic circuit***. The magnetic circuit of DC machine comprises yoke (or) frame, poles, air-gap, armature teeth and armature core.

The flux produced by field coils emerges from north pole and cross the air-gap to enter the armature teeth. Then it flows through armature core and again cross the air-gap to enter the south pole. The circuit close through the yoke of the machine. The magnetic circuit of a 4 pole DC machine is shown in Fig. 6.2.

In a DC machine the number of magnetic circuits is equal to number of poles. The magnetic circuit of a 4 pole DC machine is shown in Fig. 6.2 consists of four magnetic circuits. The various magnetic circuits are interlinked and form a symmetrical arrangement. Each complete magnetic path is linked with the coils on two adjacent poles, so that the total mmf acting around each path is sum of the ampere-turns developed by these two coils.

The duty of each field pole or coil is to develop the mmf required to establish the following flux in the magnetic circuit

1. The full value of the flux for the length of one pole, once across the air-gap and in one set of armature teeth.

2. One half the full value of the flux along one half the length of the single path in the armature core and yoke.

Therefore each field coil has to establish the flux over one half of the complete circuit as shown by thick or bold line in Fig. 6.2.

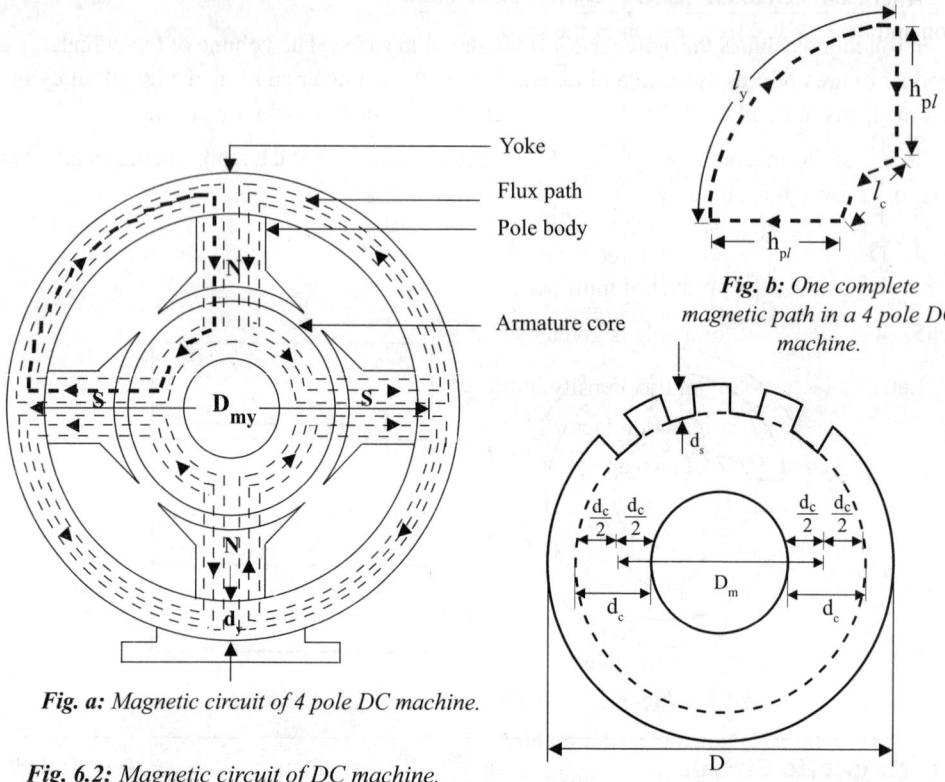

Yoke

Flux path

Pole body

Armature core

Fig. b: *One complete magnetic path in a 4 pole DC machine.*

Fig. a: *Magnetic circuit of 4 pole DC machine.*

Fig. 6.2: *Magnetic circuit of DC machine.*

Fig. c: *Cross-section of armature core.*

When the leakage flux is neglected, the magnetic circuit of a DC machine consists of five main sections.

1. Yoke   4. Armature teeth
2. Pole and Pole shoe   5. Armature core
3. Air-gap

The total mmf to be developed by each pole is given by the sum of mmf required for the above five sections. The mmf for each section can be estimated as explained below.

1. The working flux density for each section is determined so as to avoid saturation. The normal range of the working flux density in these sections are given in Table 6.2.

Table 6.2: Working Flux Density in Various Parts of DC Machine

Part	Working flux density in *Wb/m²*
Yoke	1.3 - 1.6
Pole	1.2 - 1.7
Air-gap	0.4 - 1.6
Armature teeth	1.5 - 2.2
Armature core	1.0 - 1.5

Alternatively the flux density can be calculated from the knowledge of flux passing through the section and the area of cross-section of the section.

$$\text{Flux density, B} = \frac{\text{Flux}}{\text{Area of cross-section}}$$

2. Determine the mmf per metre (**at**) required for each section from the B-H curve, using the estimated value of flux density.

3. Estimate the length of each section of magnetic path.

4. Determine the mmf required for each section of magnetic path. The mmf required for a section is given by the product of mmf per metre for that section and length of the section.

5. The total mmf for a pole is given by the sum of mmf required for all the five sections.

Let, B_g = Maximum flux density in the core

K_g = Gap contraction factor

l_g = Length of air-gap

l_c = Length of a magnetic path in core

l_y = Length of a magnetic path in yoke

d_s = Depth of slot

d_c = Depth of core

h_{pl} = Height of field pole

p = Number of poles

D_m = Mean diameter of armature

D_{my} = Mean diameter of yoke

at_t = mmf per metre corresponding to flux density of tooth at one-third height from narrow end.

at_c = mmf per metre corresponding to flux density in core

at_p = mmf per metre corresponding to flux density in pole body

at_y = mmf per metre corresponding to flux density in yoke

mmf for air-gap, $AT_g = 800{,}000\, B_g\, K_g\, l_g$

mmf for teeth, $AT_t = at_t \times d_s$

mmf for core, $AT_c = at_c \times l_c / 2$

mmf for pole, $AT_p = at_p \times h_{pl}$

mmf for yoke, $AT_y = at_y \times l_y / 2$

The values of at_t, at_c, at_p and at_y are determined from B-H curves.

$$\text{Length of flux path in core, } l_c = \frac{\pi\, D_m}{p} = \frac{\pi\, (D - 2d_s - d_c)}{p}$$

$$\text{Length of flux path in yoke, } l_y = \frac{\pi\, D_{my}}{p} = \frac{\pi\, (D + 2l_g + 2h_{pl} + d_y)}{p}$$

$$\left.\begin{array}{l}\text{Total mmf per pole at}\\\text{no-load and normal voltage}\end{array}\right\} AT_{fo} = AT_g + AT_t + AT_c + AT_p + AT_y \qquad \qquad .....(6.20)$$

6.3.2 Length of Air-gap

In rotating electrical machines a small gap is provided between the rotor and stator to avoid the friction between the stationary and rotating parts.

A larger value of air-gap results in lesser noise, better cooling, reduced pole face losses, reduced circulating currents and less distortion of field form. Also larger air-gap results in higher field mmf which reduces armature reaction.

In general, mmf required for air-gap, $AT_g = 800,000 \, B_g \, K_g \, l_g$ .....(6.21)

where, $K_g = 1.15 =$ Gap contraction factor.

In DC machines the mmf required for air-gap is normally taken as 0.5 to 0.7 times the armature mmf per pole.

$$\text{Armature mmf per pole} = \frac{I_z (Z/2)}{p} = \frac{I_z Z}{2p}$$

$$= \frac{ac\pi D}{2p}$$

$$= \frac{ac \, \tau}{2} \qquad\qquad .....(6.22)$$

$\left.\begin{array}{l}\therefore \text{ mmf required for} \\ \text{air-gap in DC machine}\end{array}\right\} AT_g = (0.5 \text{ to } 0.7) \times \dfrac{ac \, \tau}{2}$ .....(6.23)

On equating the equation (6.21) and (6.24) we get,

$$800,000 \, B_g \, K_g \, l_g = (0.5 \text{ to } 0.7) \times \frac{ac \, \tau}{2}$$

$$\therefore \text{ Air-gap length, } l_g = \frac{(0.5 \text{ to } 0.7) \times ac \, \tau}{1,600,000 \, B_g \, K_g} = \frac{(0.5 \text{ to } 0.7) \times ac \, \tau}{1.6 \times 10^6 \, B_g \, K_g} \qquad .....(6.24)$$

The usual values of air-gap lies between 0.01 to 0.015 times of pole pitch.

6.4 CHOICE OF SPECIFIC LOADINGS   *(AU, Nov' 17, 16 M) (AU, Apr' 19, 13 M) (AU, Nov' 18, 13 M)*

6.4.1 Specific Magnetic Loading

Each unit area of armature surface is capable of receiving a certain magnetic flux. Hence the flux per unit area is an important parameter to estimate the intensity of magnetic loading and it is also a criterion to decide the volume of active material. This flux per unit area is expressed as the average value of the flux density at the armature surface or specific magnetic loading (by assuming that the armature is smooth). It is denoted by B_{av} .

The *specific magnetic loading* or average flux density, B_{av} is given by the ratio of flux per pole and area under a pole.

$$\therefore \text{ Specific magnetic loading, } B_{av} = \frac{\text{Flux per pole}}{\text{Area under a pole}} = \frac{\text{Flux per pole}}{\text{Pole pitch} \times \text{Length of armature}}$$

$$= \frac{\phi}{\dfrac{\pi D}{p} \times L} = \frac{p\phi}{\pi DL} \qquad \boxed{\text{Pole pitch, } \tau \ = \ \pi D/p} \; .....(6.25)$$

From equation (6.25) we can say that the specific magnetic loading is also given by the ratio of total flux around the air-gap and the area of flux path at the air-gap.

$$\therefore \text{ Specific magnetic loading, } B_{av} = \frac{\text{Total flux around the air-gap}}{\text{Area of flux path at the air-gap}} = \frac{p\phi}{\pi DL} \quad .....(6.26)$$

The value of specific magnetic loading of DC machine will be in the range of 0.4 to 0.8 Wb/m^2.

6.4.2 Specific Electric Loading

Every section of armature is capable of carrying certain amount of current. Hence ampere-turn per unit section of armature periphery (circumference) is an important parameter to estimate the intensity of electric loading and it is also a criterion to decide the volume of active material. This ampere-turn per unit section of armature periphery is expressed as the specific electric loading. It is denoted by **ac**.

The *specific electric loading* is given by the ratio of total armature ampere conductors and armature periphery (circumference) at air-gap.

$$\therefore \text{ Specific electric loading, } \mathbf{ac} = \frac{\text{Total armature conductors}}{\text{Armature periphery at air-gap}} = \frac{I_z Z}{\pi D} \quad .....(6.27)$$

The value of specific electric loading of DC machine will be in the range of 15000 to 50000 *amp. cond./m*.

6.4.3 Choice of Specific Magnetic Loading

The choice of average gap density or specific magnetic loading depends on the following,

1. Flux density in teeth
2. Frequency of flux reversal
3. Size of machine

Large values of flux density in teeth results in increased field mmf. Higher values of field mmf increases the iron loss, copper loss and cost of copper. The B_{av} is chosen such that the flux density at the root of the teeth does not exceed 2.2 Wb/m^2.

If the frequency of flux reversals is high then iron losses in armature core and teeth would be high. Therefore we should not use a high value of flux density in the air-gap of machines which have a high frequency.

It is possible to use increased values of flux density as the size of the machine increases. As the diameter, D of the machine increases, the width of the tooth also increases, permitting an increased value of gap flux density without causing saturation in the machine. The value of air-gap flux density, B_g varies between 0.55 to 1.15 Wb/m^2 and the corresponding values of B_{av} are 0.4 to 0.8 Wb/m^2.

6.4.4 Choice of Specific Electric Loading

The choice of specific electric loading depends on the following,

1. Temperature rise
2. Speed of machine
3. Voltage
4. Size of machine
5. Armature reaction
6. Commutation

A higher value of **ac** results in a high temperature rise of windings. The temperature rise depends on the type of enclosure and cooling techniques employed in the machine. If the speed of machine is high, the ventilation of the machine is better and therefore, greater losses can be dissipated. Thus a higher value of **ac** can be used for machine having high speed.

In high voltage machines, large space is required for insulation and therefore there is less space for conductors. This means that in high voltage machines, the space left for conductors is less and therefore we should use a small value of **ac.** In large size machines it is easier to find space for accommodating conductors. Hence specific electric loading can be increased with increase in linear dimensions.

With high values of **ac**, armature reaction will be severe. To counter this the field mmf is increased and so the cost of the machine goes high. High values of **ac** worsens the commutation condition in machines. From the point of view of commutation a small value of **ac** is desirable. The value of **ac** usually lies between 15000 to 50000 *amp.cond/m.*

6.5 SELECTION OF NUMBER OF POLES   *(KTU, Feb' 18, 10 M) (VTU, Dec' 19, 6 M) (AU, Apr' 18, 5 M)*

The number of poles used in DC machines has an important bearing upon the magnetic and electric circuits. In case of AC machines, number of poles is fixed by the supply frequency and the speed of the machine. But in the case of DC machine, any number of poles can be used. However there is always a very small range of number of poles that gives a design which is sound from the commercial point of view. The selection of number of poles depends on,

1. Frequency	4. Length of commutator
2. Weight of iron parts	5. Labour charges
3. Weight of copper	6. Flash over and distortion of field form.

The number of poles are chosen such that the frequency lies between 25 to 50 *Hz*. With large number of poles the flux carried by the yoke reduces. Hence for a given flux, with large number of poles, area of cross-section of yoke can be reduced, which results in reduction of iron parts. Also by increasing the number of poles, the weight of iron in the armature core can be decreased. The overall diameter of the machine decreases as the number of poles is increased. Therefore from commercial point of view a large number of poles results in reduced cost.

The weight of copper in armature and field windings decreases with increase in number of poles. With increase in number of poles the length of the commutator reduces and so the overall length of the machine also reduces. With the increase in number of poles labour charges will increase.

The use of large number of poles results in increased danger of flash over between adjacent brush arms. With increase in number of poles there is reduction in distortion of field form under load conditions.

Advantages of large number of poles   *(AKTU, Dec' 19, 5 M)*

The large number of poles results in reduction of the following,

1. Weight of armature core and yoke

2. Cost of armature and field conductors

3. Overall length and diameter of machine

4. Length of commutator

5. Distortion of field form under load conditions.

Disadvantages of large number of poles

The large number of poles results in increase of the following,

1. Frequency of flux reversals
2. Labour charges
3. Possibility of flash over between brush arms.

Guiding factor for choice of number of poles *(KTU, Feb' 18, 8 M)*

1. The frequency should lie between 25 to 50 Hz.

2. The value of current per parallel path is limited to 200 A, thus the current per brush arm should not be more than 400 A. If p is number of poles then,

   Current per parallel path $= I_a / p$ - For lap winding .....(6.28)
   $= I_a / 2$ - For wave winding

   Current per brush arm $= 2I_a / p$ - For lap winding .....(6.29)
   $= I_a$ - For wave winding

3. The armature mmf should not be too large. The normal values of armature mmf per pole are listed in Table 6.3.

Table 6.3: Armature mmf per pole

Output in kW	Armature mmf per pole in AT
upto 100	5000 or less
100 to 500	5000 to 7500
500 to 1500	7500 to 10,000
over 1500	upto 12,500

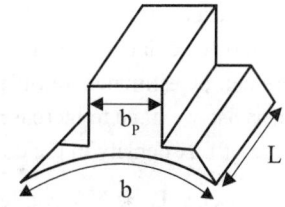

Fig. 6.3: Field Pole.

4. If there are more than one choice for number of poles which satisfies the above three conditions, then choose the largest value for poles. This results in reduction in iron and copper.

Pole proportions

The cross-section of the poles should be circular in order that the length of mean turn of the field winding is minimum (since a circle gives minimum periphery for a given area). But circular poles cannot be laminated, hence the next best alternative is square pole section.

In a square section the width of the pole body is equal to the length of the machine. Some manufacturers prefer a square pole face. For square pole face, the pole arc (b) is equal to the length of the machine. Some manufacturers prefer rectangular pole sections.

$\therefore$ L = b_p - For square pole section

L = b - For square pole face

Usually the ratio of pole arc to pole pitch or the ratio L/τ is specified.

$\psi = b/\tau = 0.64$ to 0.72

$L/\tau = 0.45$ to 1.1

6.6 DESIGN OF ARMATURE   *(AU, Apr' 19, 15 M)*

6.6.1 Armature Core Design

The armature of a DC machine consists of core and winding. The armature core is cylindrical in shape with slots on the outer periphery of the armature. The core is formed with circular laminations of thickness 0.5 *mm*. The winding is placed on the slots in the armature core. The design of armature core involves the design of main dimensions D and L, number of slots, slot dimensions and depth of core. The design of main dimensions D and L has been discussed in sections 3.2.2 to 3.2.4.

6.6.2 Depth Of Armature Core

The depth of armature core cannot be independently designed, because it depends on the diameter of armature (D), inner diameter of armature (D_i) and the depth of slot (d_s). The Fig. 6.4 shows the cross-section of armature.

From Fig. 6.4,

$$D = D_i + 2d_c + 2d_s$$

$\therefore$ Depth of core, $d_c = \dfrac{1}{2}(D - D_i - 2d_s)$ .....(6.30)

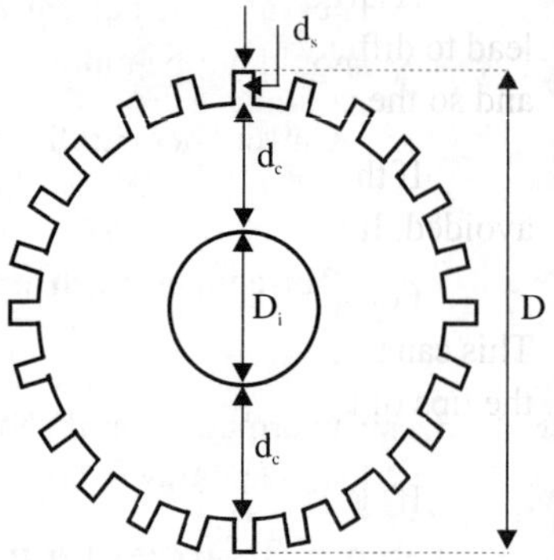

Fig. 6.4: Cross-section of armature.

After estimating D, D_i and d_s the available depth of core, d_c can be calculated. With this value of d_c, the flux density in the core can be estimated and if it does not exceed 1.5 Wb/m^2 then the available depth of core is sufficient, otherwise we have to increase the diameter of the armature D to give sufficient depth for core. The usual value of flux density in the core is 1.0 to 1.5 Wb/m^2.

Let, L_i = Net iron length of the armature

ϕ = Flux per pole

Area of the armature core, $A_c = L_i\, d_c$ .....(6.31)

Flux in the armature core, $\phi_c = \phi/2$ .....(6.32)

$\therefore$ Flux density in the core, $B_c = \phi_c/A_c$ .....(6.33)

For a specified flux density in the core, the depth of core can be estimated as shown below.

From equation (6.33) we get,

Area of armature core, $A_c = \phi_c/B_c$

On equating the above equation for area of armature core with equation (6.31) we get,

$$L_i\, d_c = \dfrac{\phi_c}{B_c}$$

$\therefore$ Depth of armature core, $d_c = \dfrac{\phi_c}{L_i\, B_c}$ Using equation (6.32) .....(6.34)

$$= \dfrac{1}{2}\dfrac{\phi}{L_i\, B_c}$$ .....(6.35)

6.6.3 Number Of Armature Slots

The factors to be considered for selection of number of armature slots are,

1. Slot width (or pitch)
2. Cooling of armature conductors
3. Flux pulsations
4. Commutation
5. Cost.

A large number of slots results in smaller slot pitch and so the width of tooth is also small. This may lead to difficulty in construction. But large number of slots will lead to less number of conductors per slot and so the cooling of armature conductors is better.

If the air-gap reluctance per pair of pole is constant then the flux pulsations and oscillations can be avoided. It can be proved that the air-gap reluctance is constant if the slots per pole is an integer plus 1/2.

For sparkless commutation the flux pulsations and oscillations under the interpole must be avoided. This can be achieved with large number of slots per pole. In fact, the number of slots in the region between the tips of two adjacent poles should be at least 3.

If, $\psi = \dfrac{\text{Pole arc}}{\text{Pole pitch}}$, then

$\left.\begin{array}{l}\text{Slots in the region between the} \\ \text{tips of two adjacent poles}\end{array}\right\} = (1 - \psi) \times \text{Slots per pole}$

$$= (1 - \psi) \times \frac{S_a}{p} \qquad\qquad .....(6.36)$$

Here, $(1 - \psi) \times \dfrac{S_a}{p} \geq 3$

Let, $(1 - \psi) \times \dfrac{S_a}{p} = 3$

$\therefore$ Slot per pole, $\dfrac{S_a}{p} = \dfrac{3}{1 - \psi}$

The typical value of $\Psi = 0.67$. For this typical value of Ψ,

$$\text{Slot per pole} = \frac{3}{1 - \psi} = \frac{3}{1 - 0.67} = 9 \qquad\qquad .....(6.37)$$

From the above calculations we can say that the slots per pole should be greater than or equal to 9, for better commutation. When large number of slots are used the cost of lamination and the cost of insulation will be high.

Guiding factors for number of armature slots

1. The slot pitch should lie between 25 to 35 mm. For small machines it can be 20 mm or even less than 20 mm.

2. The slot loading should not exceed 1500 ampere conductors.

   Slot loading = Number of conductors in the slot × Current per conductor.

3.   To reduce flux pulsation losses the slots per pole should be an integer plus 1/2 for lap winding and slots per pole arc should be an integer plus 1/2 for wave winding.

4.   To avoid sparking the number of slots per pole should have a minimum value of 9. The slots per pole varies from 9 to 16. In case of small machines it can be 8.

5.   The number of slots selected should be suitable for the type of winding. In case of simplex lap winding the number of slots should be a multiple of pole pair. In case of wave winding the number of slots should not be a multiple of pole pair to avoid dummy coils.

6.6.4 Slot Dimensions

The dimensions of the slot are slot width and depth. Usually the slot area is estimated from the knowledge of conductor area and slot space factor. The slot space factor lies in the range of 0.25 to 0.4 and the value depends on the thickness of insulation.

$$\text{Slot area} = \frac{\text{Conductor area}}{\text{Slot space factor}} \qquad .....(6.38)$$

After deciding the slot area, the depth of slot is assumed based on the diameter of the armature. The Table 6.4 can be used as a guideline for choosing the slot depth. Once the depth is finalised the width can be estimated from the slot area and depth.

The following factors can be considered before finalising the slot dimensions.

1. Flux density in tooth
2. Flux pulsations
3. Eddy current loss in conductors
4. Reactance voltage
5. Fabrication difficulties

Table 6.4: Slot Depth

Diameter of armature	Slot depth
m	*mm*
0.15	22
0.20	27
0.25	32
0.30	37
0.40	42
0.50	45

The dimensions of the slot and the number of slots will decide the dimensions of the tooth. The dimensions of the tooth should be chosen such that the flux density in any part of tooth does not exceed 2.1 Wb/m^2.

The slot opening should be as small as possible in order to reduce flux pulsation losses. With increase in depth of the slot the eddy current loss in conductors increases, specific permeance of slot increases, reactance voltage increases and it becomes difficult to fabricate the lamination with narrow width at the roots of teeth.

6.6.5 Number Of Armature Coils

The number of turns per coil and the number of coils are so chosen that the voltage between adjacent commutator segments is limited to a value where there is no possibility of a flashover. Normally, the maximum voltage between adjacent segments at load should not exceed 30 V.

The following relations are used to derive an expression for the number of armature coils, C

i) $e_z = B_{av} L V_a$

ii) $E = \phi Z n \, (p/a)$

iii) $E_{cm} \le 30 \, V$

iv) $B_{gm} = 1.3 \, B_g$

v) $B_g = B_{av} / K_f$

vi) $K_f = 0.66$

Average emf induced in a conductor, $e_z = \dfrac{\text{Emf per parallel path}}{\text{Conductor per parallel path}}$

$$= \dfrac{E}{Z/a} = \dfrac{E\,a}{Z} \qquad \qquad .....(6.39)$$

Also, Average emf induced in a conductor, $e_z = B_{av}\,L\,V_a$ $\qquad .....(6.40)$

where, $\quad B_{av} \;=\;$ Specific magnetic loading

$\qquad\quad L \;\;=\;$ Length of armature

$\qquad\quad V_a \;\;=\;$ Peripheral speed

In equation (6.40) if we replace B_{av} by B_{gm} then we get the maximum emf induced in a conductor on load.

$\therefore$ Maximum emf induced in a conductor on load, $e_{z\_max} = B_{gm}\,L\,V_a$ $\qquad .....(6.41)$

$$= 1.3\,B_g\,L\,V_a \qquad\qquad \boxed{B_{gm} = 1.3\,B_g}$$

$$= 1.3\,\dfrac{B_{av}}{K_f}\,L\,V_a \qquad\qquad \boxed{B_g = \dfrac{B_{av}}{K_f}}$$

$$= \dfrac{1.3}{0.66}\,B_{av}\,L\,V_a \qquad\qquad \boxed{K_f = 0.66}$$

$$\approx 2\,B_{av}\,L\,V_a = 2\,e_z \qquad \boxed{\text{Using equation (6.39)}}$$

$$= 2\,\dfrac{E\,a}{Z} \qquad\qquad .....(6.42)$$

Turns per coil, $T_c = \dfrac{\text{Total number of turns}}{\text{Number of coils}} = \dfrac{Z/2}{C} = \dfrac{Z}{2C}$ $\qquad .....(6.43)$

$\left.\begin{array}{l}\text{Maximum voltage between} \\ \text{adjacent segments on load}\end{array}\right\}$ $E_{cm} = \left(\begin{array}{c}\text{Number of conductors} \\ \text{between adjacent segment}\end{array}\right) \times \left(\begin{array}{c}\text{Maximum emf per} \\ \text{conductor on load}\end{array}\right)$

$$= \left(\begin{array}{c}\text{Number of coils} \\ \text{between segments}\end{array}\right) \times \left(\begin{array}{c}\text{Turns} \\ \text{per coil}\end{array}\right) \times 2 \times \left(\begin{array}{c}\text{Maximum emf per} \\ \text{conductor on load}\end{array}\right)$$

$$= N_c\,T_c\,2\,e_{z\_max} \qquad\qquad .....(6.44)$$

$$= N_c \times \dfrac{Z}{2C} \times 2 \times 2\,\dfrac{E\,a}{Z} \qquad\qquad .....(6.45)$$

$$\boxed{\text{Using equations (6.42) and (6.43)}}$$

$$= 2\,\dfrac{N_c\,E\,a}{C} \qquad\qquad .....(6.46)$$

When, $E_{cm} = 30$ V, $C = C_{min}$

$$\therefore \; 30 = 2\,\dfrac{N_c\,E\,a}{C_{min}}$$

$$\therefore \; C_{min} = \dfrac{N_c\,E\,a}{15} \qquad\qquad .....(6.47)$$

For lap winding, a = p and $N_c = 1$.

$$\therefore C_{min} = \frac{E\,p}{15} \qquad\qquad .....(6.48)$$

For wave winding a = 2 and $N_c = \frac{p}{2}$

$$\therefore C_{min} = \frac{(p/2) \times E \times 2}{15} = \frac{E\,p}{15} \qquad\qquad .....(6.49)$$

From equations (6.48) and (6.49) we can say that the minimum number of coils required for both lap and wave winding are same.

When C = Ep/15 the voltage between adjacent segments is maximum at 30 V, hence this value of C gives the minimum number of coils. With increase in the number of coils above Ep/15, the voltage between adjacent segments can be reduced. In the DC machines the number of coils is equal to the number of commutator segments. Hence Ep/15 is also equal to minimum number of commutator segments.

6.6.6 Area of Cross-section of Armature Conductor

The area of cross-section of the armature conductor can be estimated from the knowledge of current through a conductor and current density. The current through a conductor is estimated from the knowledge of power developed in armature, induced emf and number of parallel paths. The current density is assumed based on the efficiency, cost and allowable temperature rise.

A large value of current density results in smaller size of conductors, low cost, higher temperature rise, high copper loss and lesser area of slot. The range of values of current density for copper conductors is 4 to 7 A/mm^2. The typical values of current density for certain type of machines are given below.

Let, δ_a = Current density in armature conductor in A/mm^2

For large machine with strap wound armature, δ_a = 4.5 A/mm^2

For small machine with wire wound armature, δ_a = 5 A/mm^2

For high speed fan ventilated machine, δ_a = 6 to 7 A/mm^2

The current through a conductor and its area of cross section are estimated as shown below.

Power developed in armature, $P_a = E\,I_a \times 10^{-3}$

$$\therefore \text{Armature current, } I_a = \frac{P_a}{E \times 10^{-3}}$$

Current through an armature conductor, $I_z = \frac{I_a}{a}$

where, a = Number of parallel paths.

Area of cross-section of armature conductor, $a_a = \frac{I_a}{\delta_a}$

6.6.7 Armature Resistance

The length of mean turn of armature winding can be estimated from,

Length of mean turn of armature winding, $L_{mta} = 2L + 2.3\tau + 5d_s$

Resistance of copper wire can be estimated using, $R = \rho l/a$, where, ρ is resistivity, l is length and a is area of cross-section of wire.

$$\therefore \text{ Resistance of each conductor} = \frac{1}{2} \times \frac{\rho\, L_{mta}}{a_a}$$

$$\boxed{\text{Lenght of a conductor} = \frac{1}{2} L_{mta}}$$

In DC machine armature $\dfrac{Z}{a}$ conductors are connected in series in one parallel path.

Resistance of each parallel path $= \dfrac{Z}{a} \times \dfrac{\rho\, L_{mta}}{2\, a_a}$

The number of parallel paths in DC machine armature is a. The resistance of each parallel is same. In parallel combination of n equal resistances of value R, the equivalent resistance of parallel combination is R/n. Therefore, resistance of armature is obtained by dividing the resistance of one parallel path by a.

$\therefore$ Resistance of armature, $r_a = \dfrac{1}{a}\left(\dfrac{Z}{a}\,\dfrac{\rho\, L_{mta}}{2\, a_a}\right) = \dfrac{Z\,\rho\, L_{mta}}{2\, a^2\, a_a}$

Length of all conductors of the armature winding $\left.\right\}$ $L_a =$ Length of a conductor $\times$ Number of conductor $= \dfrac{1}{2}\, L_{mta}\, Z$

$\therefore$ Total volume of the conductors $= L_a \times$ Area of cross-section area of conductor $= L_a\, a_a$

Weight of armature winding $=$ Specific gravity of copper $\times$ Volume of conductor

$$= 8.9 \times 10^{-3} \times L_a\, a_a = L_a\, a_a \times 8.9 \times 10^{-3}$$

6.6.8 Design of Lap and Wave winding for DC machine

Step 1:   Find the range of slots from the range of slot pitch.

Armature slot pitch, $y_{sa} = 25$ to 35 *mm*.

Slots, $S_a = \pi D / y_{sa}$

where, D = Diameter armature.

Step 2:   In the above range of slots, list the values of slots which are not multiples of pole pairs

Step 3:   In order to reduce flux pulsations, the slots per pole should be integer $\pm\tfrac{1}{2}$. In this case dummy coils should be provided. To avoid dummy coils, take slots per pole arc as an integer $\pm\tfrac{1}{2}$. The integer can be in the range of ° to 16. List all the multiples of integer $\pm\tfrac{1}{2}$ from the list obtained in step 2.

Step 4:   Calculate slots per pole arc for all slots listed in step 3.

$$\text{Slots per pole arc} = \dfrac{\text{Pole arc}}{\text{Pole pitch}} \times \text{Slots per pole} = \psi \times \dfrac{S_a}{p}$$

Choose the value of slot for which slots per pole arc is nearly equal to integer $\pm\tfrac{1}{2}$.

Step 5:   Estimate the total number of armature conductors, using the equation of induced emf, E.

$$E = \dfrac{\phi Z n p}{a}$$

Find the conductors per slot and choose it to the nearest even number.

$$\text{Conductors per slot} = \dfrac{Z}{S_a}$$

Step 6:   Find the minimum number of coils using the equation, $C_{min} = \dfrac{Ep}{15}$.

Step 7:   Assume, u = 2,4,6,8,...... etc., where u = coil sides per slot.

Step 8: For each value of u, calculate number of coils, $C = \dfrac{1}{2}\,uS_a$. Choose the number of coils such that, it is greater than minimum number of coils. Also the value of u (corresponding to chosen value of C) should be a divisor of conductors per slot. If a suitable value of C is not obtained to satisfy the above condition, then make another choice of slots from the list obtained in step 3.

Step 9: Once the number of coils and slots are finalised. Estimate the new value of total number of conductors and number of turns per coil.

Total armature conductors, Z = Slots × Conductors per slot

Number of turns per coil $\quad = \dfrac{Z}{2C}$

6.7 DESIGN OF COMMUTATOR AND BRUSHES

The commutator and brush arrangement are used to convert the bidirectional internal armature current to unidirectional external load current or viceversa. The current flows through the brushes mounted on the commutator surface. The brushes are located at the magnetic neutral axis which is midway between two adjacent poles.

When a armature conductor pass through the magnetic neutral axis, the current in the conductor reverses from one direction to the other. Since the brushes are mounted on magnetic neutral axis, the coil undergoing current reversal is short circuited by carbon brush. During this short circuit period, the current must be reduced from its original value to zero and then built up to an equal value in the opposite direction. This process is called *commutation* and the time during which the current reversal takes place is called the time of commutation.

The phenomena of commutation is affected by resistance of the brush, reactance emf induced by leakage flux and rotational emf induced by armature flux. Based on the factors affecting the commutation, the process of commutation can be classified into following types.

1. Resistance commutation
2. Retarded commutation
3. Accelerated commutation
4. Sinusoidal commutation

When the reactance and rotational emfs are eliminated by the compoles, the commutation is assisted only by the resistance of the brush. This type of commutation is called *resistance commutation*. By employing high resistance carbon brushes, a straight line or linear resistance commutation can be achieved.

When the brushes are placed at geometrical neutral axis, a reactance voltage and rotational emf are induced in the coil under going commutation. This is due to the shift in magnetic neutral axis during load conditions. These voltages delays the process of current reversal and this type of commutation is called *delayed commutation*. In this case the commutation is completed before the current can reach its final value and so the current has to jump through the air in the form of a spark at the trailing edge of brush.

When the brushes are slightly shifted from magnetic neutral axis in the direction of rotation, the emf induced in the coil undergoing commutation will accelerate the process of current reversal. This type of commutation is called *accelerated commutation*. This type of commutation may give rise to burning of brushes at the leading edge.

When the process of current reversal is delayed at the leading edge and accelerated at the trailing edge, the commutation is called *sinusoidal commutation*.

The commutator is cylindrical in shape and placed at one end of the armature. It consists of a number of copper bars or segments separated from one another by a suitable insulating material of thickness 0.5 to 1mm. The number of commutator segments is equal to number of coils in the armature. Each segment is connected to an armature coil and the connections are made with the help of risers. In large machines the risers are made of copper strips and in small machines lugs are provided instead of risers.

The materials used for commutator segments are hard drawn copper or silver copper (0.05% silver). The silvered copper can withstand high temperature (350°C) used for soldering the risers. The insulators used between commutator segments are made of mica or micanite or resin bonded asbestos. The different types of materials used for brushes are natural graphite, hard carbon, electrographite and metal graphite.

A carbon brush placed on the commutator surface and the cross-section of commutator are shown in Fig. 6.5. The commutator segments are firmly held by V-shaped clamping rings secured by bolts.

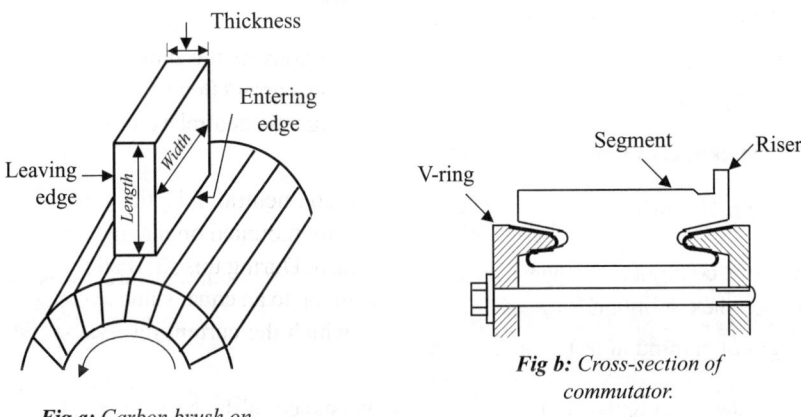

Fig a: Carbon brush on commutator surface.

Fig b: Cross-section of commutator.

Fig. 6.5: Commutator.

Commutator Design

In DC machines the number of commutator segments is same as number of coils in armature. Also the equation for minimum number of coils give the minimum number of segments.

$$\text{Number of coils, } C = \frac{1}{2} \times uS_a \qquad \qquad .....(6.50)$$

$$\text{Number of commutator segments} = C$$

$$\text{Minimum number of segment} = \frac{Ep}{15} \qquad \qquad .....(6.51)$$

$$\text{Commutator segment pitch, } \beta_c = \frac{\pi D_c}{C} \qquad \qquad .....(6.52)$$

Choose the diameter of commutator, D_c as 60 to 80% of the diameter of armature. The commutator diameter is also chosen such that the commutator peripheral speed is limited to 15 *m/s* to 30 *m/s*. The width of commutator segment, β_c is greater than or equal to 4 *mm*. Thickness of brush is selected such that it covers 1 to 3 commutator segments.

Current carried by each brush, $I_b = \dfrac{2\,I_a}{p}$ - For lap winding $\boxed{\text{Using equation (6.29)}}$

$\qquad\qquad\qquad\qquad\qquad\qquad = I_a$ - For wave winding   .....(6.53)

Total brush contact area per spindle, $A_b = I_b / \delta_b$ .....(6.54)

The number of brush locations are decided by the type of winding. In lap winding the number of brush locations is equal to number of poles and in wave winding it is always two. In each location there may be more than one brush mounted on a spindle.

The area of each individual brush should be chosen such that it does not carry more than 70 A. Hence the number of brushes in a spindle, n_b is selected such that each brush does not carry more than 70 A.

Let, a_b = Contact area of each brush

$\qquad n_b$ = Number of brushes per spindle

Now, contact area of brushes in a spindle, $A_b = n_b\, a_b$ .....(6.55)

$\qquad$ Also, $a_b = w_b\, t_b$ .....(6.56)

$\qquad \therefore A_b = n_b\, w_b\, t_b$

Usually, the thickness of brush, $t_b = (1 \text{ to } 3) \times \beta_c$ .....(6.57)

$\qquad \therefore$ Width of brush, $w_b = \dfrac{A_b}{n_b\, t_b} = \dfrac{a_b}{t_b}$ .....(6.58)

The length of the commutator depends on the space required for mounting the brushes and to dissipate the heat generated by the commutator losses.

$\qquad$ Length of commutator, $L_c = n_b\,(w_b + C_b) + C_1 + C_2$ .....(6.59)

C_b = Clearance between the brushes (usually 5 mm)

C_1 = Clearance allowed for staggering the brushes (10 mm for small machine and 30 mm for large machine)

C_2 = Clearance for allowing end play (usually 10 to 25 mm).

Estimation of power loss in commutator

The losses at the commutator are the brush contact losses and the brush friction losses. The brush contact loss depends on material, condition and quality of commutation obtained. Hence it is difficult to predetermine accurately the brush contact losses. The brush contact drop, V_b is independent of load current.

For a specified type of brush material, the brush drop can be estimated using the values listed in Table 6.5.

$\qquad \therefore$ Brush contact loss, $P_{bc} = V_b I_a$ .....(6.60)

where, I_a = Armature current

The brush friction loss can be calculated from the formula,

$\qquad P_{bf} = \mu\, P_b\, A_B\, V_c$ .....(6.61)

$\therefore$ Total commutator loss, $P_c = P_{bc} + P_{bf}$

where, P_b = Brush contact pressure on commutator

A_B = Total contact area of all brushes

= $p\,A_b$ -   For lap winding

= $2A_b$ -   For wave winding

μ = Coefficient of friction

V_c = Peripheral speed of commutator

Temperatur Rise in Commutator

Temperature rise, $\theta = \dfrac{P_c\,c}{S_c}$

where,   c   = Cooling coefficient

S_c = Commutator surface area

Commutator surface area, $S_c = \pi\,D_c\,L_c$

Cooling coefficient, $c = \dfrac{0.012}{1 + 0.1\,V_c}$

Table 6.5: Properties of Brush Materials

Type of material	Brush contact drop in V	Current density in A/mm^2	Pressure in kN/m^2	Commutator speed in m/s	Co-efficient of friction
Natural graphite	0.7 - 1.2	0.1	14	50 - 60	0.1 - 0.2
Hard carbon	0.7 - 1.8	0.065 - 0.085	14 - 20	20 - 30	0.15 - 0.25
Electro graphite	0.7 - 1.8	0.085 - 0.11	18 - 21	30 - 60	0.1 - 0.2
Metal graphite	0.4 - 0.7	0.1 - 0.2	18 - 21	20 - 30	0.1 - 0.2

6.8 DESIGN OF FIELD   (RGPV, Jun' 20, 10 M)

The field system consist of poles, pole shoe and field winding. The two types of field windings are shunt and series field winding. The shunt field winding have large number of turns made of thin conductors, because the current carried by them is very low. The series field winding is designed to carry heavy current and so it is made of thick conductors or strips.

The field coils are former wound, insulated and fixed over the field poles. In shunt machines, the full winding space along the height of the pole is used to accommodate shunt field winding. In compound machines, 80% of the winding space is taken by shunt field and the remaining 20% by series field. The Fig. 6.6 shows a pole carrying shunt field coil.

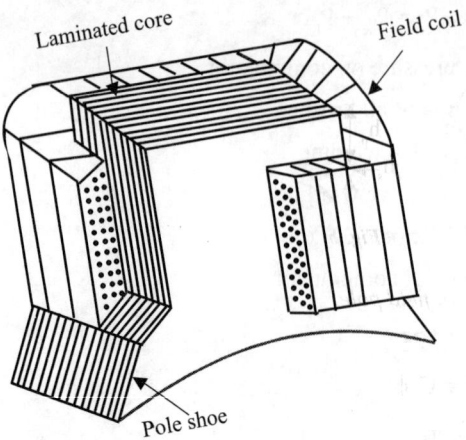

Fig. 6.6: *Pole and field coil of a DC machine.*

The factors to be considered for the design of field winding are,

     1. mmf per pole and flux density

     2. Loss dissipated from the surface of field coil

     3. Resistance of the field coil

     4. Current density in the field conductors.

Typical values of various terms used in field design

$$B_p = 1.2 \text{ to } 1.7 \; Wb/m^2 \qquad\qquad S_f = 0.4 \text{ to } 0.75$$

$$AT_{fl} = (1.1 \text{ to } 1.25) \, AT_a \qquad\qquad d_f = 30 \text{ to } 55 \; mm$$

$$q_f = 700 \; W/m^2 \qquad\qquad\qquad \rho = 2 \times 10^{-8} \; \Omega\text{-}m$$

$$\delta_f = 1.2 \text{ to } 3.5 \; A/mm^2$$

6.8.1 Design of Shunt Field Winding

The design of shunt field winding involves the determination of the following informations regarding the pole and shunt field winding. A brief discussion about the estimation of various quantities are presented below.

     1. Dimensions of the main field pole

     2. Dimensions of the field coil

     3. Current in the shunt field winding

     4. Resistance of the field coil

     5. Dimensions of the field conductor

     6. Number of turns in the field coil

     7. Losses in the field coil

Dimensions of the main field pole

The dimensions of the rectangular field pole are area of cross-section, length, width and height of pole body. For cylindrical poles the diameter has to be estimated instead of length and width. Fig. 6.7 shows the dimensions of field pole. The area of the pole body can be estimated from the knowledge of flux per pole, leakage coefficient and flux density in the pole. The leakage coefficient depends on the power output of the DC machine. A suitable value of leakage coefficient can be assumed from Table 6.6. The usual range of flux density in the pole is 1.2 to 1.7 Wb/m^2.

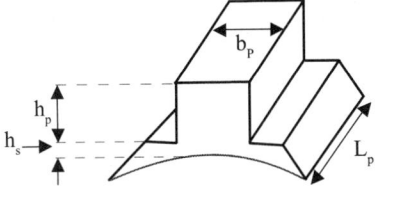

Fig. a: Rectangular pole.

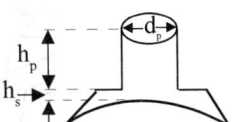

Fig. b: Cylindrical pole.

Fig. 6.7: Dimension of field pole.

Table 6.6: Leakage Coefficient

Output in kW	Leakage co-efficient, C_l
50	1.12 to 1.25
100	1.11 to 1.22
200	1.10 to 1.20
500	1.09 to 1.18
1000	1.08 to 1.16

Flux in the pole body, $\phi_p = C_l \phi$ .....(6.62)

Area of pole body, $A_p = \phi_p / B_p$ .....(6.63)

When circular poles are employed the area of cross-section will be a circle and so the diameter of the pole can be estimated from the equation of circle.

Area of pole body, $A_p = \dfrac{\pi d_p^2}{4}$

$\therefore$ Diameter of pole body, $d_p = \sqrt{\dfrac{4 A_p}{\pi}}$ .....(6.64)

When rectangular poles are employed the length of the pole is chosen as 10 to 15 mm less than the length of armature. The reduction in length of pole is to permit end play and to avoid magnetic centering.

Length of pole, $L_p = L - (0.001 \text{ to } 0.015)$ .....(6.65)

Net iron length of pole, $L_{pi} = 0.9 L_p$ .....(6.66

Width of the pole, $b_p = \dfrac{A_p}{L_{pi}}$ .....(6.67)

The height of the pole body is given by the sum of height of field coil, thickness of insulation and clearance.

Height of pole body, $h_p = h_f +$ Thickness of insulation and clearance .....(6.68)

Thickness of insulation and clearance $= 0.1 \lambda$ to 0.15λ

Total height of pole, $h_{pl} = h_p + h_s$ .....(6.69)

where, $h_s =$ Height of pole shoe $= 0.1$ to 0.2 times h_p.

Dimensions of field coil

The field coils are former wound and placed on the poles. The field coils may have rectangular or circular cross-section, depending on the type of poles. The field coils and their cross-sections are shown in Fig. 6.8. The dimensions of the field coil are depth, height and length of mean turn of field coil.

Usually the depth of field coil is assumed and the value depends on the diameter of armature. A suitable value of depth of field winding can be selected from Table 6.7.

The height of field coil depends on the surface required for cooling the coil and number of turns in the field coil. Here the height of field coil (h_f) and number of turns (T_f) cannot be independently designed. Therefore two equations are formed with h_f and T_f as variables and they are solved to estimate h_f and T_f. The formation of the two equations are explained in the section "Power Loss in the field coil".

The length of mean turn (L_{mt}) of field coil can be calculated using the dimensions of pole and depth of field coil. It is actually the length of the turn in centre of the field coil.

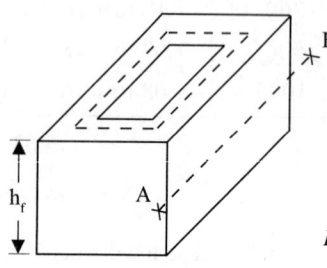

Fig a: Rectangular field coil.

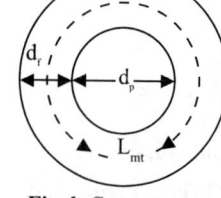

Fig b: Cross-section of rectangular field coil at AB.

Fig c: Cylindrical field coil.

Fig d: Cross-section of circular field coil at CD.

Fig. 6.8: *Dimensions of field coil.*

Table 6.7: Depth of Field Winding

Armature diameter in *m*	Depth of field winding in *mm*
0.2	30
0.35	35
0.5	40
0.65	45
1.00	50
above 1*m*	55

For rectangular field coil, with reference to Fig. 6.8,

Length of mean turn, $L_{mtf} = 2 (L_p + b_p + 2d_f)$ .....(6.70)

(or) Length of mean turn, $L_{mtf} = \dfrac{L_o + L_i}{2}$ .....(6.71)

where, L_o = Length of outer most turn

L_i = Length of inner most turn

For cylindrical field coil, with reference to Fig. 6.8,

Length of mean turn, $L_{mtf} = \pi (d_p + d_f)$ .....(6.72)

Current in the shunt field winding

The shunt field current can be estimated from the knowledge of voltage across field coil and the resistance of field coil. Each pole of a DC machine carries one field coil and all the field coils are connected in series to form shunt field winding. Hence the voltage across each field coil is given by,

Voltage across each shunt field coil, $E_f = \dfrac{\text{Voltage across shunt field winding}}{\text{Number of poles}}$

Normally the voltage across the shunt field winding is equal to 80 to 85% of rated machine voltage. Because in generators 15 to 20% of rated voltage is absorbed by the field rheostat provided for voltage regulation. In case of motors this allowance depends on the range of speed control.

Voltage across each shunt field coil, $E_f = \dfrac{(0.8 \text{ to } 0.85) \times V}{p}$ .....(6.73)

Field current, $I_f = \dfrac{E_f}{R_f}$ .....(6.74)

where R_f is the resistance of each field coil.

Resistance of field coil

The resistance of the field coil can be estimated from the knowledge of resistivity of copper, length of mean turn of field coil, number of turns in field coil and area of cross-section of field conductor.

We know that, Resistance $= \dfrac{\text{Resistivity} \times \text{Length}}{\text{Area}}$ ....(6.75)

$\therefore$ Resistance of field coil, $R_f = \dfrac{\text{Resistivity of copper} \times \text{Length of field coil}}{\text{Area of cross-section of field conductor}}$ .....(6.76)

Here, Length of field coil = Length of mean turn $\times$ Number of turns in field coil

$\therefore$ Resistance of field coil, $R_f = \dfrac{\rho \, L_{mtf} \, T_f}{a_f}$ .....(6.77)

Dimensions of field conductor

The dimensions of field conductor are area of cross-section and diameter. The area of cross-section of the field conductor can be estimated from the knowledge of field current (I_f) and current density (δ_f). The usual range of current density in the field conductor is 1.2 to 3.5 A/mm^2.

Area of cross-section of field conductor, $a_f = \dfrac{I_f}{\delta_f}$ .....(6.78)

Alternatively the area of cross-section of field conductor can be estimated from the knowledge of ampere-turns, resistivity, length of mean turn and voltage across each field coil.

From equation (6.77) we get,

Area of cross-section of field conductor, $a_f = \dfrac{\rho \, L_{mtf} \, T_f}{R_f}$ .....(6.79)

$= \dfrac{\rho \, L_{mtf} \, T_f \, I_f}{E_f}$

$\boxed{\text{Using equation (6.74)}}$

$\boxed{AT_{fl} = T_f \, I_f}$

$= \dfrac{\rho \, L_{mtf} \, AT_{fl}}{E_f}$ .....(6.80)

Usually the conductors with circular cross-section is used for field winding. Therefore the area of cross-section is also given by the equation for area of circle.

Area of cross-section, $a_f = \dfrac{\pi}{4} \, d_{fc}^2$ .....(6.81)

$\therefore$ Diameter of field conductor, $d_{fc} = \sqrt{\dfrac{4a_f}{\pi}}$ .....(6.82)

The conductors used for winding are coated with winding varnish in order to insulate one turn from the other. The diameter of the conductor including the thickness of insulation is necessary to estimate the copper space factor.

$\left.\begin{array}{l}\text{Diameter of field conductor}\\ \text{including insulation thickness}\end{array}\right\}$ $d_{fci} = d_{fc} + \text{Thickness of insulation}$ .....(6.83)

Copper space factor, $S_f = 0.75 \left(\dfrac{d_{fc}}{d_{fci}}\right)^2$ .....(6.84)

Number of turns in field coil

When the ampere-turns to be developed by the field coil is known the turns can be estimated from the knowledge of field current.

Field ampere-turns on load, $AT_{fl} = I_f T_f$

$\therefore$ Turns in field coil, $T_f = \dfrac{AT_{fl}}{I_f}$ .....(6.85)

The field ampere-turns is also related to resistance of field coil and dimensions of field coil, which in turn related to height of field coil. The height of field coil depends on the cooling surface required for dissipating the heat developed in the field coil. Hence two equations are formed with T_f and h_f as variables and they are solved to estimate T_f and h_f. (The formation of the two equations are explained in the section "Power Loss in the field coil").

Power Loss in the field coil

The power loss in the field coil is copper loss which depends on resistance and current. Heat is developed in the field coil due to this loss and the heat is dissipated through the surface of the coil. If there is no sufficient surface for heat dissipation then heat accumulates, which may lead to damage (or burning) of the coil.

In field coil design the loss dissipated per unit surface area is specified and from which the required surface area can be estimated. The surface area of field coil depends on length of mean turn, depth and height of field coil. Usually the length of mean turn is estimated from the dimensions of the pole and depth of winding is assumed and then the height of field winding is estimated in order to provide the required surface area.

The heat can be dissipated from all the four sides of a coil i.e., inner, outer, top and bottom surface of the coil.

Inner surface area of the field coil $= L_{mtf} (h_f - d_f)$

Outer surface area of the field coil $= L_{mtf} (h_f + d_f)$

Top surface area of the field coil $= L_{mtf} d_f$

Bottom surface area of the field coil $= L_{mtf} d_f$

$\left.\begin{array}{l}\text{Total surface area}\\ \text{of field coil}\end{array}\right\}$ $S = L_{mtf} (h_f - d_f) + L_{mtf} (h_f + d_f) + L_{mtf} d_f + L_{mtf} d_f$

$= 2 L_{mtf} h_f + 2 L_{mtf} d_f = 2 L_{mtf} (h_f + d_f)$ .....(6.86)

Permissible copper loss in the field coil $\left.\rule{0pt}{16pt}\right\}$ $Q_f = S\, q_f$ .....(6.87)

$\boxed{\text{Using equation (6.86)}}$

$$= 2\, L_{mtf}\, q_f\, (h_f + d_f)$$ .....(6.88)

where q_f is loss dissipated per unit area.

Actual copper loss in field coil $= I_f^2\, R_f = \dfrac{E_f^2}{R_f}$ .....(6.89)

$\boxed{\text{Using equation (6.77)}}$

$$= \frac{E_f^2}{\rho\, L_{mtf}\, T_f / a_f} = \frac{E_f^2\, a_f}{\rho\, L_{mtf}\, T_f}$$ .....(6.90)

On equating the permissible copper loss to actual copper loss, we can form an equation with T_f and h_f as variables.

$$\therefore\ 2\, L_{mtf}\, q_f\, (h_f + d_f) = \frac{E_f^2\, a_f}{\rho\, L_{mtf}\, T_f}$$ .....(6.91)

Another equation with T_f and h_f can be formed from the knowledge of conductor area in the field coil as shown below.

Conductor area in field coil $\left.\rule{0pt}{16pt}\right\}$ $=$ Number of turns $\times \left(\begin{array}{c} \text{Area of cross-section} \\ \text{of field conductor} \end{array} \right)$

$$= T_f\, a_f$$ .....(6.92)

Also, conductor area in field coil $\left.\rule{0pt}{16pt}\right\}$ $=$ Copper space factor $\times \left(\begin{array}{c} \text{Area of cross-section} \\ \text{of field coil} \end{array} \right)$

$$= S_f\, h_f\, d_f$$ .....(6.93)

On equating the equations (6.92) and (6.93) we get,

$$T_f\, a_f = S_f\, h_f\, d_f$$ .....(6.94)

The equations (6.91) and (6.94) can be solved to estimate h_f and T_f.

Procedure for shunt field design

Step 1:   Determine the dimensions of the pole. Assume a suitable value of leakage coefficient from Table 6.6 and flux density in the range 1.2 to 1.7 Wb/m^2.

Flux in the pole body, $\phi_p = C_l\, \phi$

Area of pole body, $A_p = \dfrac{\phi_p}{B_p}$

For cylindrical pole,

Diameter of pole body, $d_p = \sqrt{\dfrac{4\, A_p}{\pi}}$

For rectangular pole,

Length of pole, $L_p = L - (0.001\ \text{to}\ 0.015)$

Net iron length of pole, $L_{pi} = 0.9\, L_p$

Width of the pole, $b_p = \dfrac{A_p}{L_{pi}}$

Step 2:   Determine the length of mean turn of field coil. Assume a suitable depth of field winding from Table 6.7.

For rectangular field coils,

Length of mean turn, $L_{mtf} = 2 (L_p + b_p + 2d_f)$

For cylindrical field coils,

Length of mean turn, $L_{mtf} = \pi (d_p + d_f)$

Step 3:   Calculate the voltage across each shunt field coil.

Voltage across field coil, $E_f = \dfrac{(0.8 \text{ to } 0.85) \times V}{p}$

Step 4:   Calculate the area of cross-section of field conductor.

Area of cross-section of field conductor, $a_f = \dfrac{\rho L_{mtf} AT_{fl}}{E_f}$

Step 5:   Calculate the diameter of field conductor and copper space factor.

Diameter of field conductor, $d_{fc} = \sqrt{\dfrac{4a_f}{\pi}}$

$\left. \begin{array}{l} \text{Diameter of field conductor} \\ \text{including insulation thickness} \end{array} \right\}$ $d_{fci} = d_{fc} + \text{Thickness of insulation}$

Copper space factor, $S_f = 0.75 \left(\dfrac{d_{fc}}{d_{fci}}\right)^2$

Step 6:   Determine the number of turns (T_f) and height of field coil (h_f). They can be determined by solving the following two equations (equations (6.91) and (6.94)).

$$2 L_{mtf} q_f (h_f + d_f) = \dfrac{E_f^2 a_f}{\rho L_{mtf} T_f}$$

$$T_f a_f = S_f h_f d_f$$

Step 7:   Calculate the resistance of the field coil and field current.

Resistance of field coil, $R_f = \dfrac{\rho L_{mtf} T_f}{a_f}$

Field current, $I_f = \dfrac{E_f}{R_f}$

Step 8:   Check for current density in field coil.

Current density in field coil, $\delta_f = \dfrac{I_f}{a_f}$

The current density should not exceed 3.5 A/mm^2. If it exceeds 3.5 A/mm^2, then increase a_f by 5% and repeat steps 5 to 8 until δ_f is less than 3.5 A/mm^2.

Step 9:   Check for desired value of mmf.

Actual value of mmf, $AT_{actual} = I_f T_f$

The desired field mmf may be either specified in the problem or it may be taken as 1.1 to 1.25 times the armature mmf at full load. The actual value of mmf should be equal to or higher than desired value.

If the actual mmf is less than the desired value then increase the depth of field winding by 5% and repeat step 2 to step 9 until the desired mmf is achieved.

Step 10:   Check for temperature rise.

Actual copper loss $= I_f^2 R_f$

Surface area of field coil, $S = 2 L_{mtf} (h_f + d_f)$

Cooling coefficient, $c = \dfrac{0.14 \text{ to } 0.16}{1 + 0.1 V_a}$

where, V_a = Peripheral velocity of armature.

Temperature rise, $\theta = \dfrac{\text{Actual copper loss} \times c}{S}$

If termperature rise is within limits then the design is accepted. The allowable temperature rise depends on the class of insulation. If temperature rise exceeds the limit then repeat the design by increasing the depth of field winding by 5%.

6.8.2 Design of Series Field Winding   (AKTU, Dec' 19, 5 M)

In compound machines, each pole carries a shunt field coil and a series field coil. In each pole approximately 80% of the height is occupied by shunt field coil and 20% by series field coil. The design of shunt field coil for a compound machine is same as that for a shunt machine. The series field coil is provided to compensate for the reduction in field mmf due to armature reaction. The ampere-turns to be developed by the series field will be 15 to 25% of full load armature ampere-turns.

In case of series machines each pole carries a series field coil and all the field coils are connected in series to form series field winding. The ampere-turns to be developed by the series field coil in DC series machine is 1.15 to 1.25 times the full load armature mmf.

Design of series field coil

The step-by-step procedure for design of series field coil is given below.

Step 1:   Estimate the ampere-turns to be developed by series field coil.

$$\left.\begin{array}{l}\text{Armature ampere-turns}\\\text{at full load (per pole)}\end{array}\right\} = \dfrac{\text{Current through} \times \text{Number of armature}}{\text{a turn} \qquad \text{turns}} \Big/ \text{Number of poles}$$

$$= \dfrac{I_z \times Z / 2}{p} = \dfrac{I_z Z}{2p} \qquad\qquad .....(6.95)$$

For compound machines,

$$\left.\begin{array}{l}\text{Ampere-turns of}\\\text{series field coil}\end{array}\right\} AT_{se} = 0.15 \text{ to } 0.25 \times \dfrac{I_z Z}{2p} \qquad\qquad .....(6.96)$$

For series machines,

$$\left.\begin{array}{l}\text{Ampere-turns of}\\\text{series field coil}\end{array}\right\} AT_{se} = 1.15 \text{ to } 1.25 \times \dfrac{I_z Z}{2p} \qquad\qquad .....(6.97)$$

Step 2:   Calculate the number of turns in the series field coil

$$\text{Number of turns in series field coil}\Big\} \; T_{se} = \frac{AT_{se}}{I_{se}} \quad \text{(Rounded to nearest integer)} \qquad .....(6.98)$$

where, $I_{se} = I_a = $ Current through series field winding at full load.

Step 3:   Determine the area of cross-section of the series field conductor

$$\text{Area of cross-section of series field conductor}\Big\} \; a_{se} = \frac{I_{se}}{\delta_{se}} \qquad .....(6.99)$$

where, $\delta_{se} = $ Current density in series field conductor.

The current density in chosen as 2 to 2.3 A/mm^2. For low capacity machines circular conductors are used and for higher capacity machines rectangular conductors are used.

Step 4:   Estimate the dimensions of the field coil.

Conductor area in field coil $= T_{se} \, a_{se}$ $\qquad\qquad .....(6.100)$

$$\text{Also, conductor area in field coil}\Big\} = \text{Copper space factor} \times \text{Height of coil} \times \text{Depth of coil}$$

$$= S_{fse} \, h_{se} \, d_{se}$$

where, $S_{fse} = $ Copper space factor for series field coil.

$d_{sc} = $ Depth of series field coil.

When circular conductors are employed the value of copper space factor is 0.6 to 0.7. When rectangular conductors are employed, the space factor depends on thickness and type of insulation.

On equating the two equations for conductor area of field coil we get,

$$S_{fse} \, h_{se} \, d_{se} = T_{se} \, a_{se}$$

$$\therefore \text{ Height of field coil, } h_{se} = \frac{T_{se} \, a_{se}}{S_{fse} \, d_{se}} \qquad .....(6.101)$$

Choose a suitable depth and calculate the height of series field coil using the equation (6.101)

In case of compound machine, the total height of pole required to accommodate field winding will be sum of the height of shunt field coil and series field coil.

Step 5:   Estimate the resistance of series field coil.

$$\text{Resistance of series field coil}\Big\} = \frac{\rho \, L_{mtse} \, T_{se}}{a_{se}} \qquad .....(6.102)$$

where, $L_{mtse} = $ Length of mean turn of series field coil.

The length of mean turn of series field coil can also be estimated from the dimensions of the pole.

With reference to Fig. 6.8,

$$\text{Length of mean turn of series field coil}\Big\} \; L_{mtse} = 2 \, (L_p + b_p + 2d_{se}) \qquad .....(6.103)$$

6.8.3 Design of Interpoles *(KTU, Feb' 18, 4 M)*

The interpoles or commutating poles are small poles placed between main poles. The polarity of the interpole must be that of the main pole just ahead for a generator and just behind it for a motor, in the direction of rotation.

The interpoles are made of cast steel or punched from sheet steel with no special pole shoe. Normally the length of interpole is made equal to length of main pole.

The winding of the interpole must produce an mmf which is sufficient to neutralize the cross magnetizing armature mmf of interpolar axis and enough more to produce the flux density required to generate rotational voltage in the coil undergoing commutation to cancel the reactance voltage.

Since both the armature reaction and reactance voltage are proportional to armature current, the interpole winding should be connected in series with the armature for production of neutralizing effect at all conditions of load.

The reactance voltage can be estimated from pitchelmayer's equation,

Average Reactance voltage in the coil, $E_{rav} = 2T_c \, \mathbf{ac} \, V_a \, \lambda \, L$ .....(6.104)

(Pitchelmayer's equation)

where, T_c = Turns per coil

$\mathbf{ac}$ = Specific electric loading

V_a = Peripheral speed of armature

L = Length of armature

λ = Specific permeance

Inductance of a coil in armature, $L_{coil} = 2 \, T_c^2 \, L \, \lambda$ .....(6.105)

When each brush spans only one commutator segment, the flux density under interpole B_{gi} is calculated from the following formula which is derived from pitchelmayer's equation.

$$B_{gi} = \mathbf{ac} \, \lambda \, \frac{L}{L_{ip}}, \quad \text{where, } L_{ip} = \text{Length of interpole}$$

In general,

Flux density under interpole, $B_{gi} = 2 \, I_z \, Z_s \, \dfrac{L}{L_{ip}} \, \dfrac{1}{V_a \, T_c} \, \lambda$ .....(6.106)

where, Z_s = Conductors per slot

Let, l_{gi} = Length of air-gap under the interpole

K_{gi} = Inerpole gap contraction factor

$\left. \begin{array}{l} \text{mmf required} \\ \text{for interpole} \end{array} \right\}$ $AT_i = \left(\begin{array}{c} \text{mmf required to} \\ \text{establish } B_{gi} \end{array} \right) + \left(\begin{array}{c} \text{mmf required to overcome} \\ \text{armature reaction} \end{array} \right)$

mmf required to establish $B_{gi} = 800000 \, B_{gi} \, K_{gi} \, l_{gi}$

$\left. \begin{array}{l} \text{mmf required to overcome} \\ \text{armature reaction} \end{array} \right\} = \dfrac{I_z Z}{2p}$ - Without compensating winding   .....(6.107)

$= \dfrac{(1 - \psi) \, I_z \, Z}{2p}$ - With compensating winding   .....(6.108)

Number of turns in interpole, $T_i = \dfrac{AT_i}{I_a}$ .....(6.109)

Current density in interpole winding, $\delta_i = 2.5$ to 4 A/mm^2

Area of cross-section of interpole conductor, $a_i = \dfrac{I_a}{\delta_i}$ .....(6.110)

6.9 SUMMARY OF DESIGN EQUATIONS

1.   Power developed in armature, $P_a = C_o D^2 L n$

2.   Output coefficient, $C_o = \pi^2 B_{av}\, \mathbf{ac} \times 10^{-3}$ (or) $C_o = \pi^2 \Psi B_g\, \mathbf{ac} \times 10^{-3}$

3.   Specific magnetic loading, $B_{av} = \dfrac{p\phi}{\pi DL}$

4.   Specific electric loading, $\mathbf{ac} = \dfrac{I_z Z}{\pi D}$

5.   Total magnetic loading $= p\phi$

6.   Total electric loading $= I_z Z$

7.   Pole pitch, $\tau = \pi D/p$

8.   Choice of L/τ for DC machines

    i)   In general, $L/\tau = 0.7$ to 0.9

    ii)   For square pole criterion, $L/\tau = 0.7$

9.   Maximum flux density at air-gap, $B_g = B_{av} / \psi$

10. Maximum values of D and L in DC machines

    i)   Maximum permissible diameter, $D_{max} = \dfrac{P_a \times 10^3}{\pi\, \mathbf{ac}\, e_z}$

    ii)   Maximum permissible length, $L_{max} = \dfrac{7.5}{T_c\, N_c\, B_{av}\, V_a}$

11. Current per parallel path $= I_a/p$ -   For lap winding

                                 $= I_a/2$ -   For wave winding

12. mmf required for air-gap, $AT_g = (0.5$ to $0.7) \times \dfrac{\mathbf{ac}\, \tau}{2}$

13. Air-gap length, $l_g = \dfrac{(0.5 \text{ to } 0.7) \times \mathbf{ac}\, \tau}{1.6 \times 10^6\, B_g\, K_g}$

Design of Armature Core

14. Depth of armature core, $d_c = \dfrac{1}{2}(D - D_i - 2d_s)$ or $d_c = \dfrac{1}{2}\dfrac{\phi}{L_i\, B_c}$

15. Average emf induced in a conductor, $e_z = B_{av}\, L\, V_a$

16. Maximum emf induced in a conductor, $e_{z\,max} = B_{gm} L V_a$

17. Turns per Coil, $T_c = \dfrac{Z}{2C}$

18. Maximum flux density in the $\left.\begin{array}{l}\text{air-gap on load condition}\end{array}\right\}$ $B_{gm} = 1.3 B_g = 1.3 B_{av} / K_f \approx 2 B_{av}$ $\boxed{K_f = 0.66}$

19. Maximum voltage between $\left.\begin{array}{l}\text{adjacent segments on load}\end{array}\right\}$ $E_{cm} = N_c T_c 2 e_{z\,max}$ or $E_{cm} = \dfrac{N_c E a}{C}$

20. Minimum number of armature coils, $C_{min} = \dfrac{Ep}{15}$

21. Number of armature slots, $S_a = \dfrac{\pi D}{y_{sa}}$

22. Conductors per slot, $Z_s = \dfrac{Z}{S_a}$

23. Induced emf in armature, $E = \dfrac{\phi Z n p}{a}$

24. Slots per pole arc $= \psi \times \dfrac{S_a}{p}$

25. Armature current, $I_a = \dfrac{P_a}{E \times 10^{-3}}$

26. Current through an armature conductor, $I_z = \dfrac{I_a}{a}$

27. Area of cross-section of armature conductor, $a_a = \dfrac{I_z}{\delta_a}$

Magnetic circuit of DC machine

28. mmf for air-gap, $At_g = 800,000 \, B_g K_g l_g$

29. mmf for teeth, $At_t = \mathbf{at}_t \times d_s$

30. mmf for core, $At_c = \mathbf{at}_c \times l_c/2$

31. mmf for pole, $At_p = \mathbf{at}_p \times h_{pl}$

32. mmf for yoke, $At_y = \mathbf{at}_y \times l_y/2$

33. Length of flux path in core, $l_c = \dfrac{\pi D_m}{p} = \dfrac{\pi (D - 2d_s - d_c)}{p}$

34. Length of flux path in yoke, $l_y = \dfrac{\pi D_{my}}{p} = \dfrac{\pi (D + 2l_g + 2h_{pl} + d_y)}{p}$

35. Total mmf per pole at no-load and normal voltage, $AT_{fo} = AT_g + AT_t + AT_c + AT_p + AT_y$

Shunt field design

36.   Flux in pole body, $\phi_p = C_l\,\phi$

37.   Area of pole body, $A_p = \phi_p / B_p$

38.   For circular pole,

      i) Diameter of pole body, $d_p = \sqrt{\dfrac{4\,A_p}{\pi}}$

      ii) Length of mean turn, $L_{mtf} = \pi\,(\,d_p + d_f)$

39.   For rectangular pole,

      i)   Length of pole, $L_p = L - (0.001\ \text{to}\ 0.015)$ and $L_{pi} = 0.9L_p$

      ii) Width of pole, $b_p = \dfrac{A_p}{L_{pi}}$

      iii) Length of mean turn, $L_{mtf} = 2\,(L_p + b_p + 2d_f)$ or $L_{mtf} = \dfrac{L_o + L_i}{2}$

40.   Voltage across each field coil, $E_f = \dfrac{(0.8\ \text{to}\ 0.85) \times V}{p}$

41.   Resistance of each field coil, $R_f = \rho\,L_{mtf}\,T_f / a_f$

42.   Field current, $I_f = E_f / R_f$

43. $\left.\begin{array}{l}\text{Area of cross-section}\\ \text{of field conductor}\end{array}\right\}\ a_f = \dfrac{I_f}{\delta_f}$ or $a_f = \dfrac{\rho\,L_{mtf}\,T_f}{R_f}$ or $a_f = \dfrac{\rho\,L_{mtf}\,AT_{fl}}{E_f}$

44.   Diameter of field conductor, $d_{fc} = \sqrt{\dfrac{4\,a_f}{\pi}}$

45.   Diameter of field conductor including insulation thickness, $d_{fci} = d_{fc} +$ Insulation thickness

46.   Number of turns in field coil, $T_f = AT_{fl} / I_f$

47.   Permissible copper loss, $Q_f = S\,q_f$ or $Q_f = 2\,L_{mtf}\,q_f\,(h_f + d_f)$

48.   Conductor area in field coil, $A_{cf} = T_f\,a_f$ or $A_{cf} = S_f\,h_f\,d_f$

49.   Space factor, $S_f = 0.75\left(\dfrac{d_{fc}}{d_{fci}}\right)^2$

50.   Actual field copper loss $= \dfrac{E_f^2\,a_f}{\rho\,L_{mtf}\,T_f}$

51.   Cooling surface of field coil, $S = 2\,L_{mtf}\,(h_f + d_f)$

52.   Cooling coefficient, $c = \dfrac{0.14\ \text{to}\ 0.16}{1 + 0.1\,V_a}$

53.   Temperature rise, $\theta = \dfrac{\text{Actual field copper loss} \times c}{S}$

Series field design

54. Armature ampere-turn at full load (per pole) $= I_z\, Z\, /\, 2p$

55. Series field ampere-turn,

    i) Ampere-turns of series field coil $\Big\}\ AT_{se} = 0.15 \text{ to } 0.25 \times \dfrac{I_z\, Z}{2p}$ - For compound machine

    ii) Ampere-turns of series field coil $\Big\}\ AT_{se} = 1.15 \text{ to } 1.25 \times \dfrac{I_z\, Z}{2p}$ - For series machine

56. Number to turns in series field coil, $T_{se} = AT_{se}\, /\, I_{se}$

57. Area of cross-section of series field conductor, $a_{se} = I_{se}\, /\, \delta_{se}$

58. Conductor area in series field coil $= T_{se}\, a_{se}$ or $S_{fse}\, h_{se}\, d_{se}$

59. Resistance of series field coil $= \dfrac{\rho\, L_{mtse}\, T_{se}}{a_{se}}$

60. Length of mean turn, $L_{mtse} = 2\,(L_p + b_p + 2\, d_{se})$

Design of commutator and interpole

61. Number of coils, $C = \dfrac{1}{2} \times u\, S_a$

62. Number of commutator segments $= C$

63. Diameter of commutator, $D_c = (0.6 \text{ to } 0.8) \times D$

64. Commutator segment pitch, $\beta_c = \dfrac{\pi\, D_c}{C} \geq 4\ mm$

65. Current carried by each brush, $I_b = \dfrac{2\, I_a}{p}$ - For lap winding

    $= I_a$ - For wave winding

66. Total brush contact area per spindle, $A_b = \dfrac{I_b}{\delta_b}$

67. Width of brush, $w_b = \dfrac{A_b}{n_b\, t_b} = \dfrac{a_b}{t_b}$

68. Length of commutator, $L_c = n_b\,(w_b + C_b) + C_1 + C_2$

69. Brush friction loss, $P_{bf} = \mu\, P_b\, A_B\, V_c$

70. Average reactance voltage in coil, $E_{rav} = 2\, T_c\ \mathbf{ac}\, V_a \lambda\, L$ (Pitchelmayer's equation)

71. Flux density under interpole, $B_{gi} = 2\, I_z\, Z_s\, \dfrac{L}{L_{ip}}\ \dfrac{1}{V_a\, T_c}\, \lambda$

72.   Inductance of a coil in armature $= 2\,T_c^2\,L\,\lambda$

73. $\left.\begin{array}{l}\text{mmf required to overcome} \\ \text{armature reaction}\end{array}\right\} = \dfrac{I_z\,Z}{2p}$ -   Without compensating winding

$\qquad\qquad\qquad = \dfrac{(1-\psi)\,I_z\,Z}{2p}$ -   With compensating winding

74.   Number of turns in interpole, $T_i = AT_i\,/\,I_a$

75.   Area of cross-section of interpole conductor, $a_i = I_a/\delta_i$

76. $\left.\begin{array}{l}\text{mmf required} \\ \text{for interpole}\end{array}\right\}\ AT_i = \left(\begin{array}{c}\text{mmf required to} \\ \text{establish } B_{gi}\end{array}\right) + \left(\begin{array}{c}\text{mmf required to overcome} \\ \text{armature reaction}\end{array}\right)$

6.10 SOLVED PROBLEMS

EXAMPLE 6.1   *(JNTKU, Apr' 19, 7 M) (RGPV, Jun' 20, 10 M)*

Find the main dimensions of a 200 *kW*, 250 *V*, 6 pole, 1000 *rpm* DC generator. The maximum value of flux density in the gap is 0.87 *Wb/m²* and the ampere conductors per metre of armature periphery are 31000. The ratio of pole arc to pole pitch is 0.67 and the efficiency is 91 percent. Assume the ratio of length of core to pole pitch = 0.75.

GIVEN DATA

200 *kW*	N = 1000 *rpm*	$\Psi = 0.67$
250 *V*	$B_g = 0.87$ *Wb/m²*	$L/\tau = 0.75$
6 pole	**ac** = 31000 *amp.cond./m*	$\eta = 0.91$

SOLUTION

Given that, $\dfrac{L}{\tau} = 0.75$

$\therefore L = 0.75\,\tau = 0.75 \times \dfrac{\pi D}{p} = \dfrac{0.75 \times \pi}{6} D = 0.3927\,D$ $\boxed{\tau = \dfrac{\pi D}{p}}$ .....(1)

$\therefore$ Length of armature, L = 0.3927 D

Power developed in armature, $P_a = \dfrac{P}{\eta} = \dfrac{200}{0.91} = 219.78\ kW$

Output coefficient, $C_o = \pi^2\,B_{av}\,\text{ac} \times 10^{-3} = \pi^2\,\psi\,B_g\,\text{ac} \times 10^{-3}$

$\qquad\qquad\qquad = \pi^2 \times 0.67 \times 0.87 \times 31000 \times 10^{-3}$

$\qquad\qquad\qquad = 178.34\ kW/m^3\text{-}rps$

Also, the power developed in armature, $P_a = C_o\,D^2\,Ln$

$\therefore D^2 L = \dfrac{P_a}{C_o\,n} = \dfrac{219.78}{178.34 \times (1000\,/\,60)}$

$\therefore D^2 L = 0.0739$

$\therefore D^2 (0.3927D) = 0.0739$

$\therefore$ Diameter of armature, $D = \left(\dfrac{0.0739}{0.3927}\right)^{1/3} = 0.5731\ m$ $\boxed{\text{Using equation (1)}}$

$\therefore$ Length of armature, $L = 0.3927\ D = 0.3927 \times 0.5731 = 0.2251\ m$ $\boxed{\text{Using equation (1)}}$

RESULT

Diameter of the armature, $D\ =\ 0.5731\ m$

Length of the armature, $\ \ L\ =\ 0.2251\ m$

EXAMPLE 6.2

A 40 *HP*, 1000 *rpm* DC motor has ac = 30000 *amp.cond./m* and B_{av} = 0.44 *Wb/m²*. Estimate the HP of an 800 *rpm* DC motor which has B_{av} = 0.5 *Wb/m²*. The second machine has a current density 10% greater than that of 40 *HP* machine and the linear dimensions including those of slots are 20% greater than 40 *HP* machine. Assume that both the motors have same efficiency.

GIVEN DATA

Machine-I	**Machine-II**
$HP_1 = 40$	$N_2 = 800\ rpm$
$N_1 = 1000\ rpm$	$B_{av2} = 0.5\ Wb/m^2.$
$ac_1 = 30000\ amp.cond./m$	δ_2 = 10% greater than δ_1.
$B_{av1} = 0.44\ Wb/m^2$	Linear dimensions including slots are 20% greater.

SOLUTION

Machine-I

$\left.\begin{array}{l}\text{Power developed in}\\ \text{armature of machine}-\text{I}\end{array}\right\} = HP_1 \times 0.746$ $\boxed{\text{Neglecting friction losses}}$

Also, power developed in armature = $C_{o1}\ D^2_1\ L_1\ n_1$ $\boxed{C_{o1} = \pi^2\ B_{av1}\ ac_1 \times 10^{-3}}$

$$= \pi^2\ B_{av1}\ ac_1\ D^2_1\ L_1\ n_1 \times 10^{-3}$$

On equating the above two equations of power developed in armature, we get,

$$\pi^2\ B_{av1}\ ac_1 \times 10^{-3} \times D^2_1\ L_1\ n_1 = HP_1 \times 0.746$$

$$\therefore D^2_1\ L_1 = \frac{HP_1 \times 0.746}{\pi^2\ B_{av1}\ ac_1 \times 10^{-3} \times n_1}$$

$$= \frac{40 \times 0.746}{\pi^2 \times 0.44 \times 30000 \times 10^{-3} \times \dfrac{1000}{60}}$$

$$= 0.0137 \qquad\qquad\qquad\qquad\qquad .....(1)$$

Machine-II

For machine-II, δ increases by 10% and linear dimensions increases by 20%

Hence, $\delta_2 = 1.1\,\delta_1$, $a_{z2} = (1.2)^2\,a_{z1}$, $D_2 = 1.2\,D_1$ .....(2)

Let, ac_2 = Specific electric loading of machine-II

$$\therefore ac_2 = \frac{I_z Z}{\pi D_2}$$

$$= \frac{\delta_2\,a_{z2}\,Z}{\pi D_2} \qquad\qquad \boxed{\delta_2 = \frac{I_z}{a_{z2}} \;\Rightarrow\; I_z = \delta_2\,a_{z2}}$$

$$= \frac{\delta_2\,a_{z2}\,Z}{\pi D_2} = \frac{(1.1\,\delta_1)\,(1.2)^2\,a_{z1}\,Z}{\pi\,(1.2\,D_1)}$$

$$= 1.1\times1.2\times\frac{\delta_1\,a_{z1}\,Z}{\pi D_1} \qquad \boxed{\frac{\delta_1\,a_{z1}\,Z}{\pi D_1} = ac_1,\ \text{Specific electric loading of machine-I}}$$

$$= 1.1\times1.2\times30000$$

$$= 39600\ amp.cond./m$$

$$\left.\begin{array}{l}\therefore\text{Power developed in}\\[2pt]\text{armature of machine – II}\end{array}\right\} P_{a2} = C_{o2}\,D_2^2\,L_2\,n_2 \qquad \boxed{C_{o2} = \pi^2\,B_{av1}\,ac_1 \times 10^{-3}}$$

$$= \pi^2\,B_{av}\,ac_2 \times 10^{-3}\times D_2^2\,L_2\,n_2$$

$$= \pi^2 \times 0.5\times39600\times10^{-3}\times(1.2\,D_1)^2\,(1.2\,L_1)\times\frac{800}{60}$$

$$= \pi^2 \times 0.5\times39600\times10^{-3}\times1.2^3\,(D_1^2\,L_1)\times\frac{800}{60} \qquad \boxed{\begin{array}{l}\text{Using}\\\text{equation (2)}\end{array}}$$

$$= \pi^2 \times 0.5\times39600\times10^{-3}\times1.2^3\times0.0137\times\frac{800}{60} \qquad \boxed{\begin{array}{l}\text{Using}\\\text{equation (1)}\end{array}}$$

$$= 61.68\ kW = \frac{61.68}{0.746} = 82.68\ HP \approx 83\ HP$$

RESULT

The HP rating of Machine-II = 83 HP

EXAMPLE 6.3   (KTU, Feb' 18, 10 M)

The core diameter and length of an armature of a 250 kW, 500 V, 600 rpm, 6 pole, DC generator are 75 cm and 30 cm respectively. The lap connected armature has 720 conductors. Using the data obtained from this machine, obtain the main dimensions and the number of poles for a 350 kW, 500 V, 720 rpm, DC generator. Assume the full load efficiency as 0.85.

GIVEN DATA

Machine-I		Machine-II
$P_1 = 250\ kW$	$L_1 = 30\ cm = 0.3\ m$	$P_2 = 350\ kW$
$V_1 = 500\ V$	$Z_1 = 720$	$V_2 = 500\ V$
$N_1 = 600\ rpm$	Lap connected	$N_2 = 720\ rpm$
$D_1 = 75\ cm = 0.75\ m$	$p = 6$	$\eta = 0.85$

SOLUTION

Machine-I

$$\text{Induced emf, } E_1 = \frac{\phi Z_1 N_1}{60} \frac{p}{a} = \frac{\phi Z_1 N_1}{60}$$

| p = a, for lap connected armature. |

$$\therefore \text{ Flux per pole, } \phi = \frac{60 E_1}{Z_1 N_1} = \frac{60 V_1}{Z N_1} = \frac{60 \times 500}{720 \times 600} = 0.0694 \; Wb$$

| $E \approx V$ |

$$\therefore \text{ Full load current, } I_1 = \frac{P_1}{V_1 \times 10^{-3}} = \frac{250}{500 \times 10^{-3}} = 500 \; A$$

| $P = V_1 I_1 \times 10^{-3}$ |

$$\left. \begin{array}{l} \text{Current through each} \\ \text{armature conductor} \end{array} \right\} I_{z1} = \frac{\text{Armature current}}{\text{Number of parallel paths}} = \frac{I_a}{a} = \frac{I_1}{p}$$

| $I_a = I_1$ |

| a = p, for lap connected. |

$$= \frac{500}{6} = 83.3333 \; A$$

$$\text{Specific electric loading, } ac_1 = \frac{I_z Z_1}{\pi D} = \frac{83.3333 \times 720}{\pi \times 0.75} = 25464.7807 = 25465 \; amp.cond/m$$

$$\text{Specific magnetic loading, } B_{av1} = \frac{p\phi}{\pi D_1 L_1} = \frac{6 \times 0.0694}{\pi \times 0.75 \times 0.3} = 0.5891 \; Wb/m^2$$

Machine-II

$$\text{Power developed in armature, } P_{a2} = \frac{P_2}{\eta} = \frac{350}{0.85} = 411.7647 \; kW$$

$$\text{Armature current, } I_{a2} = \frac{P_{a2} \times 10^3}{V_2} = \frac{411.7647 \times 10^3}{500} = 823.5294 \; A$$

Let, the armature winding be lap connected and number of poles be 6, same as machine-I.

$$\therefore \text{ Current per brush arm, } I_{z2} = \frac{I_{a2}}{a} = \frac{I_{a2}}{p} = \frac{823.5294}{6} = 137.2549 \; A$$

$$\text{Frequency, } f = \frac{p N_2}{120} = \frac{6 \times 720}{120} = 36 \; Hz$$

When p = 6, the current per brush arm do not exceed 200 A and frequency lies in the range of 25 to 50 Hz. Therefore, the choice of p = 6 is acceptable.

From machine-I,

$$\frac{L_1}{D_1} = \frac{0.75}{0.3} \; \Rightarrow \; L_1 = \frac{0.75}{0.3} D_1 \; \Rightarrow \; L_1 = 2.5 D_1$$

$$\therefore L_2 = 2.5 D_2 \hspace{6cm} .....(1)$$

$$\text{Power developed in armature, } P_{a2} = C_{o2} D_2{}^2 L_2 n_2$$

| $C_o = \pi^2 B_{av} \, ac \times 10^{-3}$ |

$$= \pi^2 B_{av2} \, ac_2 \times 10^{-3} \times D_2{}^2 L_2 n_2$$

| $n = \dfrac{N}{60}$ | $P_{a2} = \dfrac{P_{a2}}{\eta}$ |

$$= \pi^2 B_{av2} \, ac_2 \times 10^{-3} D_2{}^2 L_2 \left(\frac{N}{60} \right)$$

| $ac_2 = ac_1$ |

$$\therefore D_2{}^2 L_2 = \frac{60 \, P_{a2}}{\pi^2 B_{av2} \, ac_2 \times 10^{-3} N_2}$$

| $B_{av2} = B_{av1}$ |

$$= \frac{60 \times 411.7647}{\pi^2 \times 25465 \times 0.5891 \times 10^{-3} \times 720} = 0.2318 \; m^3$$

$$\therefore D_2{}^2 L_2 = 0.2318 \; \Rightarrow \; D_2{}^2 (2.5 \, D_2) = 0.2318 \; \Rightarrow \; D_2{}^3 = \frac{0.2318}{2.5}$$

| Using equation (1) |

$\therefore$ Diameter of armature, $D_2 = \left(\dfrac{0.2318}{2.5}\right)^{\frac{1}{3}} = 0.4526\ m$

$\therefore$ Length of armature, $L_2 = 2.5\ D_2 = 2.5 \times 0.4526 = 1.1315\ m$

RESULT

Number of poles, $p = 6$

Diameter of armature, $D_2 = 0.4526\ m$

Length of armature, $L_2 = 1.1315\ m$

EXAMPLE 6.4
(AU, Apr' 17, 8 M)

Determine suitable values for the number of poles, D and L for a 1000 kW, 500 V, 300 rpm DC shunt generator. Assume average gap density = 1 Tesla and specific electric loading as 400 amp.cond./cm.

GIVEN DATA

$P = 1000\ kW$ $B_{av} = 1\ Tesla = 1\ Wb/m^2$

$V = 500\ V$ $ac = 400\ amp.cond./cm = 40000\ amp.cond./m.$

$N = 300\ rpm$

SOLUTION

The choice of poles depends on the following factors.

1. Frequency of flux reversals should lie between 25 Hz to 50 Hz.
2. The current per brush arm should not exceed 400 A.
3. For reduced cost the highest possible choice of poles should be chosen.

(Refer section 6.5 for the guiding factors for choice of number of poles)

Speed in rps, n = 300 / 60 = 5 rps

If p = 10, $f = \dfrac{pn}{2} = \dfrac{10 \times 5}{2} = 25\ Hz$

If p = 20, $f = \dfrac{pn}{2} = \dfrac{20 \times 5}{2} = 50\ Hz$

Hence choice of poles can be 10, 12, 14, 16, 18 or 20.

Armature current, $I_a = \dfrac{1000 \times 10^3}{500} = 2000\ A$

For p = 10, current per brush arm = $I_a / 10 = 200\ A$

For p = 20, current per brush arm = $I_a / 20 = 100\ A$

For all choice of poles the current limit is not violated. (But for minimum cost we can choose the maximum number of poles.)

Let number of poles,   p = 10

For square pole face, $L/\tau = 0.7$

$\dfrac{L}{\tau} = 0.7$ and $\tau = \pi D / p$ $L = \dfrac{\pi D}{p} 0.7$

$\therefore \dfrac{L}{\pi D / p} = 0.7$ $L = \dfrac{\pi \times 0.7}{10} D$

$\therefore$ Length of armature, L = 0.2199 D   .....(1)

The power developed in armature, $P_a = C_o D^2 L n$

Output coefficient, $C_o = \pi^2 B_{av} \, \textbf{ac} \times 10^{-3}$

Let the power output, $P \approx P_a$, $\therefore P = \pi^2 B_{av} \, \textbf{ac} \times 10^{-3} D^2 L n$

$$\therefore D^2 L = \frac{P}{\pi^2 B_{av} \, \textbf{ac} \times 10^{-3} n}$$

$$= \frac{1000}{\pi^2 \times 1 \times 40000 \times 10^{-3} \times 5} = 0.5066 \, m^3$$

$\therefore D^2 L = 0.5066$ $\boxed{\text{Using equation (1)}}$

$D^2 (0.2199 \, D) = 0.5066 \, m^3$

$\therefore$ Diameter of armature, $D = (0.5066 / 0.2199)^{1/3} = 1.3207 \, m$

$\therefore$ Length of armature, $L = 0.2199 \, D = 0.2199 \times 1.3207 = 0.2904 \, m$ $\boxed{\text{Using equation (1)}}$

RESULT

Number of poles, $p = 10$

Diameter of armature, $D = 1.3207 \, m$

Length of armature, $L = 0.2904 \, m$

EXAMPLE 6.5

Prove that the output of a DC machine with single turn coil is given by, $\dfrac{3a \, E \, v \, \textbf{ac}}{p \, N} \, kW$ where a, E, v, **ac**, p and N denote respectively the pair of armature paths, average voltage between adjacent commutator segment, peripheral speed of the armature in *m/s*, armature conductor per cm of periphery, pair of poles and speed in *rpm*.

SOLUTION

a = Pair of parallel paths

E = Voltage between adjacent commutator segment

v = Peripheral speed in *m/s*

ac = Specific electric loading in *amp.cond./cm*

p = Pair of poles

N = Speed in *rpm*

The output equation is, $P_a = C_o D^2 L n$

where, $C_o = \pi^2 B_{av} \, \textbf{ac}_1 \times 10^{-3}$

      $\textbf{ac}_1$ = Specific electric loading in *amp.cond./m*

In the output equation we have to eliminate the main dimensions D, L and B_{av}. The $\textbf{ac}_1$ should be changed to *amp.cond./cm* and speed should be expressed in *rpm*.

$$\text{Speed in rps, } n = \frac{\text{Speed in rpm}}{60} = \frac{N}{60} \qquad\qquad .....(1)$$

$$\left.\begin{array}{l}\text{Specific electric loading} \\ \text{in } amp.cond./m \end{array}\right\} \textbf{ac}_1 = \frac{\textbf{ac}_1}{100} \times 100 = \left(\begin{array}{c}\text{Specific electric loading} \\ \text{in } amp.cond./cm\end{array}\right) \times 100$$

$$= \textbf{ac} \times 100$$

$$\therefore \textbf{ac}_1 = \textbf{ac} \times 100 \qquad\qquad .....(2)$$

Let, N_c = Number of coils between adjacent commutator segment

$$\therefore N_c = \frac{\text{Number of poles}}{\text{Number of parallel paths}} = \frac{2 \times \text{Pair of poles}}{2 \times \text{Pair of parallel path}} = \frac{p}{a} \qquad .....(3)$$

Let, E = Average voltage between adjacent commutator segments.

$$\therefore E = 2\,T_c\,N_c\,B_{av}\,Lv$$

$$= 2 \times T_c \times \frac{p}{a}\,B_{av}\,Lv \qquad \boxed{\text{Using equation (3)}}$$

$$= 2 \times 1 \times \frac{p}{a}\,B_{av}\,Lv \qquad \boxed{T_c = 1}$$

$$\therefore \text{Specific magnetic loading, } B_{av} = \frac{aE}{2pvL} \qquad .....(4)$$

Peripheral speed, $v = \pi Dn$ in *m/s* (neglecting air-gap)

$$\therefore v = \pi D\,\frac{N}{60} \qquad \boxed{\text{Using equation (1)}}$$

$$\therefore D = \frac{60v}{\pi N} \qquad .....(5)$$

The output equation is, $P_a = C_o\,D^2\,L\,n$ $\qquad \boxed{C_o = \pi^2\,B_{avl}\,\mathbf{ac_1} \times 10^{-3}}$

$$= \pi^2\,B_{av}\,\mathbf{ac_1} \times 10^{-3} \times D^2\,L\,n$$

$$= \pi^2\left(\frac{aE}{2pvL}\right)(\mathbf{ac} \times 100) \times 10^{-3} \times \left(\frac{60v}{\pi N}\right)^2 \times L \times \frac{N}{60} \qquad \boxed{\begin{array}{l}\text{Using equations (1),}\\ \text{(2), (4) and (5)}\end{array}}$$

$$= \frac{100 \times 10^{-3} \times 60}{2} \times \frac{aE\ \mathbf{ac}\ v}{p\,N}$$

$$\therefore P_a = \frac{3\ aE\ \mathbf{ac}\ v}{pN} \text{ in } kW$$

Hence proved.

EXAMPLE 6.6   *(VTU, Dec' 19, 8 M) (AU, Apr' 19, 7 M) (AU, Apr' 18, 8 M)*

Find the main dimensions and the number of poles of a 37 *kW*, 230*V*, 1400 *rpm* shunt motor so that a square pole face is obtained. The average gap density is 0.5 *Wb/m²* and the ampere conductors per metre are 22000. The ratio of pole arc to pole pitch is 0.7 and the full load efficiency is 90%.

GIVEN DATA

37 *kW*	Square pole face	$\Psi = 0.7$
230 *V*	$B_{av} = 0.5$ *Wb/m²*	$\eta = 90\ \%$
1400 *rpm*	ac = 22000 *amp.cond./m*	

SOLUTION

If poles, p = 2, then $f = \dfrac{pN}{120} = \dfrac{2 \times 1400}{120} = 23.33$ *Hz*

If poles, p = 4, then $f = \dfrac{pN}{120} = \dfrac{4 \times 1400}{120} = 46.67$ *Hz*

Power input, $P_i = VI \times 10^{-3}$ *kW*

Also, Power input, $P_i = \dfrac{\text{Power output}}{\text{Efficiency}} = \dfrac{P}{\eta}$

∴ Load current, $I = \dfrac{P}{\eta \times V \times 10^{-3}} = \dfrac{37}{0.9 \times 230 \times 10^{-3}} = 178.74\ A$

Let, armature current, $I_a \approx I = 178.74\ A$

The armature current is less than 200 A. Hence the current per parallel path will not exceed the upper limit of 200 A. When poles, p = 4, the frequency, f = 46 Hz, which lies in the range of 25 to 50 Hz.

Hence poles, p = 4 is the best choice.

Given that, $\dfrac{\text{Pole arc}}{\text{Pole pitch}} = 0.7$

For square pole face, length of armature is equal to pole arc

∴ $\dfrac{\text{Pole arc}}{\text{Pole pitch}} = \dfrac{\text{Length}}{\text{Pole pitch}} = \dfrac{L}{\tau} = 0.7$

∴ $L = 0.7\ \tau = 0.7 \times \dfrac{\pi D}{p} = \dfrac{0.7 \times \pi}{4} \times D = 0.5498\ D$

$\boxed{\tau = \dfrac{\pi D}{p}}$

∴ Length of armature, L = 0.5498 D .....(1)

Output coefficient, $C_o = \pi^2\ B_{av}\ ac \times 10^{-3}$

$= \pi^2 \times 0.5 \times 22000 \times 10^{-3}$

$= 108.57\ kW/m^3\text{-}rps$

Given that, power output, P = 37 kW.

For DC motors power developed in armature, $P_a \approx P = 37\ kW$

Also, power developed in armature, $P_a = C_o\ D^2\ Ln$

∴ $D^2 L = \dfrac{P_a}{C_o\ n} = \dfrac{37}{108.57 \times (1400/60)} = 0.0146\ m^3$

∴ $D^2 L = 0.0146\ m^3$

$D^2 (0.5498D) = 0.0146$

∴ Diameter of armature, $D = \left(\dfrac{0.0146}{0.5498}\right)^{1/3} = 0.2983\ m \approx 0.3\ m$ Using equation (1)

∴ Length of armature, L = 0.5498 D = 0.5498 × 0.3 = 0.165 m   Using equation (1)

RESULT

Number of poles,   p = 4
Diameter of armature, D = 0.3 m
Length of armature,   L = 0.165 m

EXAMPLE 6.7

A 4 pole, 25 HP, 440 V, 600 rpm, series motor. Assuming diameter of armature = 0.3373 m, lenght of armature = 0.1775, B_{av} = 0.66 Wb/m², a = 3. Calculate flux per pole, total armature conductor, induced emf.

GIVEN DATA

4 pole	B_{av} = 0.66 Wb/m²
25 HP	a = 3
440 V	D = 0.3373 m
b/τ = 0.67	L = 0.1775 m

SOLUTION

Flux per pole, $\phi = B_{av} \dfrac{\pi DL}{p}$

$$= \frac{0.66 \times \pi \times 0.3373 \times 0.1775}{4} = 0.0310 \, Wb$$

Total armature conductor, $Z = \dfrac{Ea \times 60}{\phi N p}$

$$= \frac{440 \times 3 \times 60}{0.0310 \times 600 \times 4} = 1064.5 \approx 1065$$

Induced emf, $E = \dfrac{\phi ZN}{60} \dfrac{p}{a}$

$$= \frac{0.0310 \times 1065 \times 600 \times 4}{60 \times 3} = 440 \, V$$

RESULT

Flux per pole, ϕ = 0.0310 Wb

Total armature conductor, Z = 1065

Induced emf, E = 440 V

EXAMPLE 6.8
(AKTU, Dec' 19, 7 M)

A 4 pole, 25 HP, 500V, 600 rpm series motor has an efficiency of 82%. The pole faces are square and the ratio of pole arc to pole pitch is 0.67. Take B_{av} = 0.55 Wb/m^2 and ac = 17000 $amp.cond./m$. Obtain the main dimensions of the core and particulars of a suitable armature winding.

GIVEN DATA

4 pole	B_{av} = 0.55 Wb/m^2	600 rpm
25 HP	η = 82%	Series motor
500 V	ac = 17000 $amp.cond./m$	Square pole face
b/τ = 0.67		

SOLUTION

Given that, b/τ = 0.67

For square pole face, L = b   ; $\therefore$ L/τ = 0.67

$$\therefore L = 0.67\tau = 0.67 \times \frac{\pi D}{p} = \frac{0.67 \times \pi}{4} D = 0.5262 \, D$$

$\boxed{\tau = \dfrac{\pi D}{p}}$

$\therefore$ Length of armature, L = 0.5262 D

$\qquad\qquad\qquad\qquad\qquad$ ..... (1)

Output coefficient, $C_o = \pi^2 B_{av} \, ac \times 10^{-3}$

$$= \pi^2 \times 0.55 \times 17000 \times 10^{-3}$$

$$= 92.2808 \, kW/m^3\text{-}rps$$

For DC motor, power developed in armature, $P_a \approx P = 25 \, HP$

$\boxed{1 \, HP = 0.746 \, kW}$

$$= 25 \times 0.746 \, kW = 18.65 \, kW$$

Also power developed in armature, $P_a = C_o D^2 L n$

$$\therefore D^2 L = \frac{P_a}{C_o n} = \frac{18.65}{92.2808 \times (600/60)} = 0.0202 \ m^3$$

$$\therefore D^2 L = 0.0202 \ m^3 \qquad \qquad \qquad ..... (2)$$

$$\therefore D^2 (0.5262 \ D) = 0.0202 \qquad \boxed{\text{Using equation (1)}}$$

$$\therefore \text{Diameter of armature, } D = \left(\frac{0.0202}{0.5262}\right)^{1/3} = 0.3373 \ m$$

$$\therefore \text{Length of armature, } L = 0.5262 \ D = 0.5262 \times 0.3373 = 0.1775 \ m \qquad \boxed{\text{Using equation (1)}}$$

Power input, $P_i = \dfrac{P}{\eta} = \dfrac{18.65}{0.82} = 22.7439 \ kW$

Load current, $I = \dfrac{P_i}{V \times 10^{-3}} = \dfrac{22.7439}{500 \times 10^{-3}} = 45.4878 \ A \qquad \boxed{P_i = V I \times 10^{-3}}$

In series motor, $I_a = I = 45.4878 \ A$

Since armature current is less than 200 A, wave winding is preferred.

Specific magnetic loading, $B_{av} = \dfrac{\phi p}{\pi D L}$

$$\therefore \text{Flux per pole, } \phi = B_{av} \frac{\pi D L}{p} = \frac{0.55 \times \pi \times 0.3373 \times 0.1775}{4} = 0.0259 \ Wb$$

Induced emf, $E = \dfrac{\phi Z N}{60} \dfrac{p}{a} \qquad \boxed{E \approx V}$

Total armature conductor, $Z = \dfrac{Ea \times 60}{\phi N p} = \dfrac{500 \times 2 \times 60}{0.0259 \times 600 \times 4} = 965 \qquad \boxed{\begin{array}{l}\text{For wave winding}\\ a = 2\end{array}}$

The slot pitch, y_{sa} lie in the range of 25 to 35 mm.

$$\therefore \text{Number of slots, } \quad S_a = \frac{\pi D}{y_{sa}}$$

When $y_{sa} = 25 \ mm, \quad S_a = \dfrac{\pi \times 0.3373}{25 \times 10^{-3}} = 42$

When $y_{sa} = 35 \ mm, \quad S_a = \dfrac{\pi \times 0.3373}{35 \times 10^{-3}} = 30$

The number of slots should lie in the range 30 to 42. For wave winding the slots per pole should not be a multiple of pole pair. Here pole pair = 2. Hence number of slots should be a odd number. The allowable choice of slots are 31, 33, 35, 37, 39 and 41.

To reduce flux pulsations and to avoid dummy coils, the slots per pole arc should be an integer ± ½. The slots per pole arc is given by number of slots under pole arc.

$$\text{Slots per pole arc} = \frac{\text{Pole arc}}{\text{Pole pitch}} \times \text{Slots per pole}$$

$$= \psi \times \frac{S_a}{p} = 0.67 \times \frac{S_a}{4} = 0.1675\, S_a$$

When $S_a = 31$; $0.1675\, S_a = 5.19$ $\qquad$ When $S_a = 37$; $0.1675\, S_a = 6.19$

When $S_a = 33$; $0.1675\, S_a = 5.52$ $\qquad$ When $S_a = 39$; $0.1675\, S_a = 6.53$

When $S_a = 35$; $0.1675\, S_a = 5.86$ $\qquad$ When $S_a = 41$; $0.1675\, S_a = 6.86$

Let us choose an integer 6.

$$\text{integer} \pm \tfrac{1}{2} = 6 \pm \tfrac{1}{2} = 5.5 \text{ or } 6.5$$

When $S_a = 33$ or 39, slots per pole arc is nearly equal to the chosen integer $\pm \tfrac{1}{2}$.

Let number of slots, $S_a = 33$.

Minimum number of coils, $C_{min} = \dfrac{E\,p}{15} = \dfrac{500 \times 4}{15} = 133$

Conductors per slot, $Z_s = \dfrac{Z}{S_a} = \dfrac{965}{33} = 29.24 \approx 30$

Number of coils, $C = \dfrac{1}{2} u\, S_a$

Let us assume different values of coil sides per slot, u

$\qquad$ $u = 6$, $C = (1/2) \times 6 \times 33 = 99$ $\qquad$ $u = 10$, $C = (1/2) \times 10 \times 33 = 165$

$\qquad$ $u = 8$, $C = (1/2) \times 8 \times 33 = 132$

Choose the number of coils such that conductors per slot is divisible by coil sides per slot. Here conductors per slot 30, is divisible by coil sides per slot 10. Also when $u = 10$, the number coils is more than the required minimum number of coils. Hence the best choice for number of coils is 165.

The following are the final values.

Number of slots, $S_a = 33$
Number of coils, $C = 165$
Conductors per slot, $Z_s = 30$

New value of total armature conductors, $Z = $ Slots $\times$ Conductors per slot

$$= 33 \times 30 = 990$$

Number of turns per coil, $T_c = \dfrac{Z}{2C} = \dfrac{990}{2 \times 165} = 3$

RESULT

Diameter of armature, $D = 0.3373\ m$
Length of armature, $L = 0.1775\ m$
Type of armature winding $=$ Wave winding
Number of slots, $S_a = 33$
Number of coils, $C = 165$
Conductors per slot, $Z_s = 30$
Number of turns per coil, $T_c = 3$

EXAMPLE 6.9
(AU, Nov' 18, 13 M) (AU, Nov' 17, 16 M)

Determine the diameter and length of armature core for a 55 *kW*, 110 V, 1000 *rpm*, 4 pole shunt generator, assuming specific electric and magnetic loadings of 26000 *amp.cond./m* and 0.5 *Wb/m²* respectively. The pole arc should be about 70% of pole pitch and length of core about 1.1 times the pole arc. Allow 10 *A* for the field current and assume a voltage drop of 4 *V* for the armature circuit. Specify the winding used and also determine suitable values for the number of armature conductors and number of slots.

GIVEN DATA

55 *kW*	B_{av} = 0.5 *Wb/m²*	N = 1000 *rpm*
110 V	**ac** = 26000 *amp.cond./m*	I_f = 10 *A*
4 pole	b = 0.7 τ	$I_a R_a$ = 4 *V*
Shunt generator	L = 1.1 b	

SOLUTION

Given that, L = 1.1 b and b = 0.7 τ

$$\therefore L = 1.1b = 1.1(0.7\tau) = 0.77\tau = 0.77 \times \frac{\pi D}{p}$$

$$\boxed{\tau = \frac{\pi D}{p}}$$

$$= \frac{0.77 \times \pi}{4} \times D = 0.6048\,D$$

$\therefore$ Length of armature, L = 6048 D   ..... (1)

Induced emf, $E = V + I_a R_a = 110 + 4 = 114\ V$

Load current, $I = \dfrac{P}{V \times 10^{-3}} = \dfrac{55}{110 \times 10^{-3}} = 500\ A$

$$\boxed{P = VI \times 10^{-3}}$$

Armature current, $I_a = I + I_f = 500 + 10 = 510\ A$

Power developed in armature, $P_a = EI_a \times 10^{-3}$

$$= 114 \times 510 \times 10^{-3} = 58.14\ kW$$

Output coefficient, $C_o = \pi^2\,B_{av}\,\mathbf{ac} \times 10^{-3}$

$$= \pi^2 \times 0.5 \times 26000 \times 10^{-3}$$

$$= 128.3\ kW/m^3\text{-}rps$$

Also, power developed in armature, $P_a = C_o\,D^2 L\,n$

$$\therefore D^2 L = \frac{P_a}{C_o\,n} = \frac{58.14}{128.3 \times (1000/60)} = 0.0272\ m^3$$

$\therefore D^2 L = 0.0272\ m^3$

$\therefore D^2 (0.6048\,D) = 0.0272$
Using equation (1)

$\therefore$ Diameter of armature, $D = \left(\dfrac{0.0272}{0.6048}\right)^{1/3} = 0.3556\ m$

$\therefore$ Length of armature, L = 0.6048 D = 0.6048 × 0.3556 = 0.2151 *m*
Using equation (1)

The armature current, $I_a = 510 \ A$

If wave winding is used then, current per parallel path $= \dfrac{I_a}{2} = \dfrac{510}{2} = 255 \ A$

If lap winding is used then, current per parallel path $= \dfrac{I_a}{4} = \dfrac{510}{4} = 127.5 \ A$

In wave winding the current per parallel path is greater than the upper limit of 200 A. **Hence wave winding is not suitable.**

In lap winding the current per parallel path is less than the upper limit of 200 A. **Hence lap winding is preferred.**

Specific magnetic loading, $B_{av} = \dfrac{p\phi}{\pi DL}$

$\therefore$ Flux per pole, $\phi = \dfrac{B_{av}\,\pi DL}{p} = \dfrac{0.5 \times \pi \times 0.3556 \times 0.2151}{4} = 0.03 \ Wb$

For lap winding, induced emf, $E = \dfrac{\phi ZN}{60}$

$\therefore$ Number of armature conductor, $Z = \dfrac{60 \times E}{\phi N} = \dfrac{60 \times 114}{0.03 \times 1000} = 228$

The armature slot pitch, y_{sa} lie in the range of 25 to 35 mm.

Number of armature slots, $S_a = \dfrac{\pi D}{y_{sa}}$

When $y_{sa} = 25 \ mm$, $S_a = \dfrac{\pi \times 0.3556}{25 \times 10^{-3}} = 45$

When $y_{sa} = 35 \ mm$, $S_a = \dfrac{\pi \times 0.3556}{35 \times 10^{-3}} = 32$

The number of slots should lie in the range 32 to 45. For lap winding slots per pole should be a multiple of pole pair. Here pole pair = 2. Hence number of slots should be even number. The allowable choice of slots are 32, 34, 36, 38, 40, 42 and 44. To reduce flux pulsations, the slots per pole should be integer $\pm\frac{1}{2}$.

Let, slots per pole $= 9 \pm \frac{1}{2} = 8.5$ or 9.5

Let us choose 9.5 slots per pole.

$\therefore$ Number of slots, S_a = Slots per pole $\times$ Number of poles

$$= 9.5 \times 4 = 38 \text{ slots}$$

The designed value of 38 slots is one of the allowed choice of slots. Hence the choice of 38 slots is acceptable.

Conductors per slot, $Z_s = \dfrac{Z}{S_a} = \dfrac{228}{38} = 6$ | For double layer winding conductors per slot should be even. |

Minimum number of coils, $C_{min} = \dfrac{Ep}{15} = \dfrac{114 \times 4}{15} = 30$

Number of coils, $C = \frac{1}{2} \times u \ S_a$

Let us assume different values of coil sides per slot, u

When $u = 2$, $C = \dfrac{1}{2} \times 2 \times 38 = 38$

When $u = 4$, $C = \dfrac{1}{2} \times 4 \times 38 = 76$

Choose the number of coils such that the conductors per slot is divisible by coil sides per slot. Here conductors per slot 6 is divisible by coil sides per slot 2. Hence the best choice for number of coils is 38.

∴ The following are final values,

Number of slots, S_a = 38

Number of coils,   C = 38

Conductors per slot, Z_s = 6

New value of total armature conductors, Z = Slots × Conductors per slot = 38 × 6 = 228

Number of turns per coil, $T_c = \dfrac{Z}{2C} \dfrac{228}{2 \times 38} = 3$

RESULT

Diameter of armature,   D = 0.3556 m
Length of armature,   L = 0.2151 m
Number of slots, S_a = 38
Number of coils,   C = 38
Total armature conductors, Z = 228
Conductors per slot, Z_s = 6
Turns per coil, T_s = 3

EXAMPLE 6.10

Design the shunt field winding of a 6 pole, 440 V, DC generator allowing a voltage drop of 15 % in the regulator. The following design data are available. mmf per pole = 7200 AT; mean length of turn = 1.2 m; winding depth = 3.5 cm; watts per sq.m. of cooling surface = 650.

Calculate the inner, outer and end cooling surfaces of the cylindrical field coil. Take diameter of the insulated wire to be 0.4 mm greater than the bare wire. Assume 2 micro-ohm-cm as the resistivity of copper at the working temperature.

GIVEN DATA

AT_{fl} = 7200 AT $d_{fci} = d_{fc} + 0.4$ mm $q_f = 650$ W/m²

L_{mt} = 1.2 m $\rho = 2\,\mu\,\Omega\text{-}cm = 2 \times 10^{-8}$ Ω-m   p = 6

d_f = 3.5 cm = 0.035 m   Voltage drop in field regulator = 15 % V_g = 440 V

SOLUTION

Given that,

Voltage drop in field regulator = 15% of V_g

∴ Voltage across field winding = 0.85 V_g

Voltage across each field coil, $E_f = \dfrac{0.85\,V_g}{p} = \dfrac{0.85 \times 440}{6} = 62.33\ V$

$\left.\begin{array}{l}\text{Area of cross-section}\\ \text{of field conductor}\end{array}\right\}a_f = \dfrac{AT_{fl}\,\rho\,L_{mt}}{E_f} = \dfrac{7200 \times 2 \times 10^{-8} \times 1.2}{62.33}$

$= 2.772 \times 10^{-6}\ m^2 = 2.772\ mm^2$

Let the conductors have circular cross-section.

$\therefore$ Diameter of bare conductor, $d_{fc} = \sqrt{\dfrac{4a_f}{\pi}} = \sqrt{\dfrac{4 \times 2.772}{\pi}} = 1.88 \; mm$

Diameter of insulated conductor, $d_{fci} = 1.88 + 0.4 = 2.28 \; mm$

Space factor for winding, $S_f = 0.75 \left(\dfrac{d_{fc}}{d_{fci}}\right)^2 = 0.75 \times \left(\dfrac{1.88}{2.28}\right)^2 = 0.51$

Cooling surface of the coil, $S = 2 \, L_{mt} \, (h_f + d_f) = 2 \times 1.2 \, (h_f + 0.035)$

$$= 2.4 \, h_f + 0.084$$

Permissible loss, $Q_f = q_f \, S = 650 \, [2.4 \, h_f + 0.084] = 1560 \, h_f + 54.6$ .....(1)

$\left.\begin{array}{l} \text{Actual copper loss} \\ \text{in field coil} \end{array}\right\}$ $Q_f = \dfrac{E_f^2}{R_f} = \dfrac{E_f^2}{\dfrac{T_f \, \rho \, L_{mt}}{a_f}} = \dfrac{E_f^2 \, a_f}{T_f \, \rho \, L_{mt}}$

$$= \dfrac{62.33^2 \times 2.772 \times 10^{-6}}{T_f \times 2 \times 10^{-8} \times 1.2} = \dfrac{0.45 \times 10^6}{T_f}$$ .....(2)

On equating equations (1) and (2) we get

$1560 \, h_f + 54.6 = \dfrac{0.45 \times 10^6}{T_f}$

$\therefore T_f = \dfrac{0.45 \times 10^6}{1540 \, h_f + 54.6}$ .....(3)

Conductor area in field coil $= S_f \, h_f \, d_f = 0.51 \times h_f \times 0.035 = 0.01785 \, h_f$ .....(4)

Also, conductor area $= T_f \, a_f = T_f \times 2.772 \times 10^{-6}$ .....(5)

On equating equations (4) and (5) we get,

$0.01785 \, h_f = T_f \times 2.772 \times 10^{-6}$

$\therefore h_f = \dfrac{2.772 \times 10^{-6}}{0.01785} \, T_f$

$$= 1.553 \times 10^{-4} \, T_f$$ .....(6)

Using equation (6), the equation (3) can be written as,

$T_f = \dfrac{0.45 \times 10^6}{1560 \times 1.553 \times 10^{-4} \, T_f + 54.6}$

$\therefore T_f = \dfrac{0.45 \times 10^6}{0.242 \, T_f + 54.6}$

On cross-multiplying the above equation we get,

$0.242 \, T_f^2 + 54.6 \, T_f - 0.45 \times 10^{-6} = 0$

The roots above quadratic equation is T_f.

$\therefore T_f = \dfrac{-54.6 \pm \sqrt{54.6^2 + 4 \times 0.242 \times 0.45 \times 10^6}}{2 \times 0.242} = 1255$

Taking only positive value.

From equation (6),

$$h_f = 1.553 \times 10^{-4} \, T_f$$
$$= 1.553 \times 10^{-4} \times 1255 = 0.195 \ m$$

Here the field coil is cylindrical and the dimensions of the field coil are shown in Fig. 1.

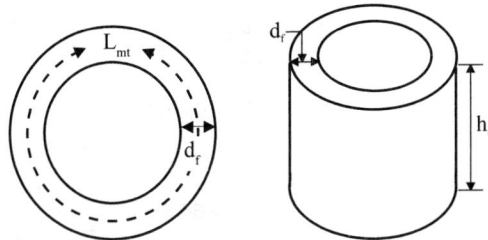

Fig. 1: Dimensions of field coil.

Here, $L_{mt} = \pi \, d_{mf}$

where, d_{mf} = Mean diameter of field coil

Let, d_{if} = Inner diameter of field coil

        d_{of} = Outer diameter of field coil

$\therefore d_{if} = d_{mf} - d_f$

$d_{of} = d_{mf} + d_f$

$\therefore d_{mf} = \dfrac{L_{mt}}{\pi} = \dfrac{1.2}{\pi} = 0.382 \ m$

$d_{if} = 0.382 - 0.035 = 0.347 \ m$

$d_{of} = 0.382 + 0.035 = 0.417 \ m$

Length of inner most turn, $L_{if} = \pi \, d_{if} = \pi \times 0.347 = 1.09 \ m$

Length of outer most turn, $L_{of} = \pi \, d_{of} = \pi \times 0.416 = 1.307 \ m$

Inner cooling surface, $h_f \, L_{if} = 0.195 \times 1.09 = 0.213 \ m^2$

Outer cooling surface, $h_f \, L_{of} = 0.195 \times 1.307 = 0.255 \ m^2$

$$\text{Top cooling surface} = \left(\frac{\pi \, d_{of}^2}{4} - \frac{\pi \, d_{if}^2}{4} \right)$$

$$= \frac{\pi \, (0.416)^2}{4} - \frac{\pi \, (0.347)^2}{4} = 0.041 \ m^2$$

Bottom cooling surface = $0.041 \ m^2$

Note : *Top and bottom surfaces are end cooling surfaces.*

RESULT

Area of cross-section of field conductor, a_f = 2.772 mm^2

Diameter of bare conductor, d_{fc} = 1.88 mm

Diameter of insulated conductor, d_{fci} = 2.28 mm

Number turns in field coil, T_f = 1255

Height of field coil, h_f = 0.195 m

Inner cooling surface of a field coil, $h_f L_{if}$ = 0.213 m^2

Outer cooling surface of a field coil, $h_f L_{of}$ = 0.255 m^2

Top cooling surface of a field coil = 0.041 m^2

Bottom cooling surface of a field coil = 0.041 m^2

EXAMPLE 6.11

Calculate the size of the conductor and number of turns for the field coil of a 6 poles, 460 V, DC shunt motor. The coil is to supply 4000 AT at the working temperature, where ρ = 0.02 *micro-ohm-m*. The length of the inside turn is 0.74 *m*, the space factor of the winding is 0.52 and the permissible dissipation per sq.m. of external surface (excluding the two ends) is 1200 W. Solution should not be attempted by assuming a numerical value for the winding depth.

GIVEN DATA

AT_{fl} = 4000 AT $p = 6$ $L_i = 0.74\ m$

$h_f = 0.13\ m$ $S_f = 0.52$ $\rho = 0.02\ \mu\,\Omega\text{-}m = 2 \times 10^{-8}\ \Omega\text{-}m$

$q_f = 1200\ W/m^2$ $V_g = 460\ V$

SOLUTION

Let the voltage drop in field = 15%

∴ Voltage across each field coil, $E_f = \dfrac{(1-0.15)\ V_g}{p} = \dfrac{0.85 \times 460}{6}$

$$= 65.17\ V$$

Length of mean turn, $L_{mt} = 2\ (L_p + b_p + 2d_f) = 2\ (L_p + b_p) + 4d_f$

$$= L_i + 4d_f = 0.74 + 4d_f$$

Cooling surface area available, $S = 2\ L_{mt}\ h_f$

$$= 2\ (0.74 + 4d_f)\ 0.13 = 0.1924 + 1.04\ d_f$$

Permissible power loss per coil, $Q_f = Sq_f = (0.1924 + 1.04\ d_f)\ 1200$

$$= 230.88 + 1248\ d_f \qquad\qquad .....(1)$$

Actual copper loss in field coil, $Q_f = I_f^2\ R_f = \dfrac{I_f\ T_f}{T_f}\ I_f\ R_f$

$$= \dfrac{AT_{fl}}{T_f}\ E_f$$

$$= \dfrac{4000 \times 65.17}{T_f} = \dfrac{260680}{T_f} \qquad\qquad .....(2)$$

Multiply and divide by T_f

$\boxed{I_f\ T_f = AT_{fl}}$

$\boxed{I_f\ R_f = E_f}$

On equating equations (1) and (2) we get,

$$230.88 + 1248\, d_f = \frac{260680}{T_f}$$

$$\therefore T_f = \frac{260680}{230.88 + 1248\, d_f} \qquad\qquad .....(3)$$

Conductor area $= S_f\, h_f\, d_f = 0.52 \times 0.13\, d_f = 0.0676\, d_f \qquad .....(4)$

Also, Conductor Area $= T_f\, a_f = T_f\, \dfrac{AT_{fl}\, \rho\, L_{mt}}{E_f}$

$$= \frac{T_f \times 4000 \times 2 \times 10^{-8}\,(0.74 + 4d_f)}{65.17}$$

$$= T_f\,(9.08 \times 10^{-7} + 4.91 \times 10^{-6}\, d_f) \qquad .....(5)$$

On equating equations (4) and (5) we get

$$0.0676\, d_f = T_f\,(9.08 \times 10^{-7} + 4.91 \times 10^{-6}\, d_f)$$

$$\therefore T_f = \frac{0.0676\, d_f}{9.08 \times 10^{-7} + 4.91 \times 10^{-6}\, d_f} \qquad .....(6)$$

On equating equations (3) and (6) we get,

$$\frac{260680}{230.88 + 1248\, d_f} = \frac{0.0676\, d_f}{9.08 \times 10^{-7} + 4.91 \times 10^{-6}\, d_f}$$

On cross multiplying the above equation we get,

$$0.237 + 1.28\, d_f = 15.61\, d_f + 84.36\, d_f^2$$

$$\therefore 84.36\, d_f^2 + 14.33\, d_f - 0.237 = 0$$

The roots of above quadratic equation is d_f.

$$\therefore d_f = \frac{-14.33 \pm \sqrt{14.33^2 + 4 \times 84.36 \times 0.237}}{2 \times 84.36}$$

> Taking only positive value.

$$= 0.01518\ m$$

$$= 15.18\ mm$$

From equation (3),

$$T_f = \frac{260680}{230.88 + 1248\, d_f} = \frac{260680}{230.88 + 1248\,(0.01518)} = 1043$$

$$L_{mt} = L_i + 4d_f = 0.74 + 4 \times 0.01518 = 0.8\ m$$

$$a_f = \frac{AT_{fl}\, \rho\, L_{mt}}{E_f} = \frac{4000 \times 2 \times 10^{-8} \times 0.8}{65.17}$$

$$= 0.982 \times 10^{-6}\ m^2 = 0.982\ mm^2$$

RESULT

Depth of field coil,	d_f	$= 15.18\ mm$
Number of turns in field coil,	T_f	$= 1043$
Area of cross-section of field conductor,	a_f	$= 0.982\ mm^2$

EXAMPLE 6.12

Calculate the reactance voltage induced per coil for a single turn two layer winding with two conductors per slot, of a 250 kW, 525 V, 6 pole lap wound DC generator driven at 220 rpm. The number of armature conductors is 600. The inductance per coil is 0.0057 mH. The brush covers one commutator segment.

If the armature diameter is 1.6 m and core length is 0.3 m, determine the flux density under the interpole. The length of interpole is 0.18 m.

GIVEN DATA

250 kW	N = 220 rpm	Z = 600
525 V	L_{coil} = 0.0057 mH	L = 0.3 m
6 pole	$t_b = \beta_c$	D = 1.6 m
Lap wound	L_{ip} = 0.18 m	T_c = 1
Double layer winding	Two conductor per slot	

SOLUTION

$$\text{Armature current, } I_a = \frac{P \times 10^3}{V} = \frac{250 \times 10^3}{525} = 476.19 \ A \qquad \boxed{P = VI \times 10^{-3}}$$

$$\text{Current in each conductor, } I_z = \frac{I_a}{p} = \frac{476.19}{6} = 79.365 \ A$$

$$\text{Specific electric loading, ac} = \frac{I_z Z}{\pi D} = \frac{79.365 \times 600}{\pi \times 1.6} = 9473.5 \ AT/m$$

$$\text{Peripheral speed of armature, } V_a = \pi D n = \pi \times 1.6 \times \frac{220}{60} = 18.43 \ m/s$$

$$\text{Inductance of a coil, } L_{coil} = 2 \ T_c^2 \ L \ \lambda$$

$$\therefore \text{Specific permeance, } \lambda = \frac{L_{coil}}{2 \ T_c^2 \ L} = \frac{0.0057 \times 10^{-3}}{2 \times 1^2 \times 0.3}$$

$$= 9.5 \times 10^{-6} \ Wb/AT\text{-}m$$

The pitchelmayer's equation is applicable to machines in which the brushes spans over one commutator segment.

$$\text{Average reactance voltage, } E_{rav} = 2 \ T_c \ \textbf{ac} \ V_a \ \lambda \ L$$

$$= 2 \times 1 \times 9473.5 \times 18.43 \times 9.5 \times 10^{-6} \times 0.3$$

$$= 0.9952 \ V$$

$$\text{Flux density under interpole, } B_{gi} = \textbf{ac} \ \lambda \left(\frac{L}{L_{ip}}\right)$$

$$= 9473.5 \times 9.5 \times 10^{-6} \times \frac{0.3}{0.18}$$

$$= 0.15 \ Wb/m^2$$

RESULT

Average reactance voltage, E_{rav} = 0.9952 V

Flux density under interpole, B_{gi} = 0.15 Wb/m²

EXAMPLE 6.13
(RGPV, Jun' 20, 10 M)

In a 500 kW, 440 V, 375 rpm, 8 pole, compound generator the external diameter of armature is 1.1 m, gross armature length is 0.3 m, armature slot pitch is 2.5 cm and flux per pole is 0.0875 Wb. Determine, number of conductors in armature winding, number of slots in armature, resistance of the armature winding.

GIVEN DATA

500 kW	$D = 1.1\ m$
440 V	$L = 0.3\ m$
375 rpm	$y_{ss} = 2.5\ cm = 0.025\ m$
8 pole	$\phi = 0.0875\ Wb$

SOLUTION

Let us assume lap winding. Therefore, $p = a$

$$\text{Induced emf, } E = \frac{\phi Z N}{60}\frac{p}{a} = \frac{\phi Z N}{60}$$

$$\therefore \text{ Number of armature conductor, } Z = \frac{60\,E}{\phi N} = \frac{60\,V}{\phi N} \qquad \boxed{E \approx V}$$

$$= \frac{60 \times 440}{0.0875 \times 375} = 804.57$$

Let, Z be divisible by 8.

$$\therefore \text{ Number of armature conductors, } Z = 808$$

$$\text{Number of armature slots, } S_a = = \frac{\pi D}{y_{ss}} = \frac{\pi \times 1.1}{0.025} = 138.23 \approx 138$$

$$\text{Armature current, } I_a = = \frac{P}{V} = \frac{500 \times 10^3}{440} = 1136.3636\ A$$

Let, current density, $\delta = 4.5\ A/mm^2$

$$\therefore \text{ Area of cross-section, } a_a = \frac{I_a}{\delta_a} = \frac{1136.3636}{4.5} = 252.5252\ mm^2 = 252.5252 \times 10^{-6}\ m^2$$

Let, depth of slot, $d_s = 37\ mm = 0.37\ m$

$$\text{Pole pitch, } \tau = \frac{\pi D}{p} = \frac{\pi \times 1.1}{8} = 0.432\ m$$

$$\text{Length of mean turn, } L_{mt} = 2L + 2.3\tau + 5d_s$$

$$= (2 \times 0.3) + (2.3 \times 0.432) + (5 \times 0.037) = 1.7786\ m$$

$$\text{Resistance of armature, } r_a = \frac{Z \rho L_{mt}}{2a^2 a_a} \qquad \boxed{\text{Resistivity of copper, } \rho = 0.0172 \times 10^{-6}\ \Omega\text{-}m}$$

$$= \frac{808 \times 0.0172 \times 10^{-6} \times 1.7786}{2 \times 8^2 \times 252.5252 \times 10^{-6}} = 0.7647 \times 10^{-3}\ \Omega$$

RESULT

Number of armature conductor,	Z	$= 808$
Number of armature slots,	S_a	$= 138$
Resistance of armature,	r_a	$= 0.7647 \times 10^{-3}\ \Omega$

EXAMPLE 6.14

Determine the total commutator losses for a 1000 kW, 500 V, 800 rpm, 10 pole generator. Given that commutator diameter = 1.0 m, current density at brush contact = 75 × 10^{-3} A/mm^2, brush pressure = 14.7 kN/m^2, coefficient of friction = 0.28, brush contact drop = 2.2 V.

GIVEN DATA

P = 1000 kW	D_c = 1.0 m	V = 500 V
δ_b = 75 × 10^{-3} A/mm^2	N = 800 rpm	P_b = 14.7 kN/m^2
p = 10	μ = 0.28	V_b = 2.2 V

SOLUTION

Armature current, $I_a = \dfrac{P}{V} = \dfrac{1000 \times 10^3}{500} = 2000\ A$ $\boxed{P = VI \times 10^{-3}}$

Brush contact loss, $P_{bc} = V_b\,I_a = 2.2 \times 2000 = 4400\ W$

Commutator pheripheral speed, $V_c = \pi\,D_c\,n$

$$= \pi \times 1 \times (800\,/\,60) = 41.89\ m/s$$

Current per brush area = $\dfrac{2I_a}{p} = \dfrac{2 \times 2000}{10} = 400\ A$

Current per brush is limited to 70 A

$\therefore$ Number of brushes per spindle, $n_b = \dfrac{400}{70} \approx 6$

Let number of brushes per spindle, $n_b = 8$

$\therefore$ Current carried by each brush, $I_b = \dfrac{400}{8} = 50\ A$

Area of each brush, $a_b = \dfrac{I_b}{\delta_b} = \dfrac{50}{75 \times 10^{-3}} = 666.67\ mm^2$ $\boxed{\delta_b = 75 \times 10^{-3}\ A/mm^2}$

Contact area of each brush arm, $A_b = n_b\,a_b = 8 \times 666.67 = 5333.3\ mm^2$ $\boxed{\begin{array}{l} A_B = pA_b\ \text{for lap} \\ \text{winding.} \end{array}}$

Total brush contact area, $A_B = pA_b = 10 \times 5333.3 = 53{,}333\ mm^2 = 53{,}333 \times 10^{-6}\ m^2$

Brush friction loss, $P_{bf} = \mu\,P_b\,A_B\,V_c = 0.28 \times 14.7 \times 10^3 \times 53{,}333 \times 10^{-6} \times 41.89$

$$= 9196\ W$$

$\therefore$ Total commutator loss, $P_c = P_{bc} + P_{bf} = 4400 + 9196 = 13596\ W = 13.596\ kW$

RESULT

Total commutator loss, $P_c = 13.596\ kW$

EXAMPLE 6.15

Design a suitable commutator for a 350 *kW*, 600 *rpm*, 440 *V*, 6 pole DC generator having an armature diameter of 0.75 *m*, brush pressure is 15 *kN/m²*, coefficient of friction is 0.28 and number of coils 288. Assume suitable values wherever necessary. Also estimate commutator loss and temperature rise.

GIVEN DATA

P = 350 *kW*	D = 0.75 *m*	p = 6	P_b = 15 *kN/m²*
V = 440 *V*	C = 288	N = 600 *rpm*	μ = 0.28

SOLUTION

Let the diameter of commutator, D_c = 0.75 D = 0.75 × 0.75 = 0.56 *m*

Peripheral speed of commutator, $V_c = \pi\, D_c\, n = \pi \times 0.56 \times \dfrac{600}{60} = 17.59\ m/s$

V_c > 15 *m/s.*   Hence reduce D_c

Let, D_c = 0.64 D = 0.64 × 0.75 = 0.48 *m*

Now, $V_c = \pi \times 0.48 \times (600 / 60) = 15.08\ m/s$

Now, $V_c \approx 15$ *m/s.* Hence, D_c = 0.48 *m* is acceptable.

The minimum width of commutator segment = 4 *mm*

Let, N_{cs} = Number of commutator segments

In DC machine, number of commutator segments is equal to number of coils.

$\therefore\ N_{cs}$ = C = 288

Commutator pitch, $\beta_c = \dfrac{\pi\, D_c}{C} = \dfrac{\pi \times 0.48}{288} = 5.2 \times 10^{-3}\ m = 5.2\ mm$

The commutator pitch is above the minimum value of 4 *mm*

Armature current, $I_a = \dfrac{P \times 10^3}{V} = \dfrac{350 \times 10^3}{440} = 795.45\ A$ $\boxed{P = VI \times 10^{-3}}$

Current per brush area = $\dfrac{2\,I_a}{p} = \dfrac{2 \times 795.45}{6} = 265.15\ A$

Current per brush is limited to 70 *A*

$\therefore$ Number of brushes, $n_b = \dfrac{265.15}{70} \approx 4$

Let, number of brushes, n_b = 6

Current carried by each brush, $I_b = \dfrac{265.15}{6} = 44.19\ A$

Let, current density in brush, δ_b = 75 × 10⁻³ *A/mm²*

$\therefore$ Area of each brush, $a_b = \dfrac{I_b}{\delta_b} = \dfrac{44.19}{75 \times 10^{-3}} = 589.2\ mm^2$

Let thickness of brush = 3 × Width of commutator segment

$\therefore\ t_b = 3\,\beta_c$

$\therefore\ t_b = 3 \times 5.2 = 15.6\ mm$

Width of brush, $w_b = \dfrac{a_b}{t_b} = \dfrac{589.2}{15.6} = 37.77 \; mm \approx 38 \; mm$

Let, $C_b = 5 \; mm$ (clearance between brushes)

$C_1 = 20 \; mm$ (allowance for staggering)

$C_2 = 20 \; mm$ (allowance for end play)

Length of commutator, $L_c = n_b \, (w_b + C_b) + C_1 + C_2$

$$= 6 \times (38 + 5) + 20 + 20 = 298 \; mm$$

$$= 0.298 \; m \approx 0.3 \; m$$

Brush contact loss, $P_{bc} = V_b \, I_a$

$$= 2.2 \times 795.45$$

$$= 1750 \; W$$

Assume	$A_B = p \, A_b$
$V_b = 2.2 \; V$	$A_b = n_b \, a_b$

Brush friction loss, $P_{bf} = \mu \, P_b \, A_B \, V_c = \mu \, P_b \, P \, n_b \, a_b \, V_c$

$$= 0.28 \times 15 \times 10^3 \times 6 \times 6 \times 589.2 \times 10^{-6} \times 15$$

$$= 1336.3 \; W$$

Total commutator loss, $P_C = P_{bc} + P_{bf}$

$$= 1750 + 1336.3$$

$$= 3086.3 \; W$$

$$= 3.0863 \; kW$$

Cooling coefficient, $c = \dfrac{0.012}{1 + 0.1 \, V_c}$

$$= \dfrac{0.012}{1 + 0.1 \times 15} = 0.0048 \; m^2 \, {}^\circ C/W$$

Surface area of commutator, $S_c = \pi \, D_c \, L_c$

$$= \pi \times 0.48 \times 0.3$$

$$= 0.4524 \; m^2$$

Temperature rise, $\theta = \dfrac{P_c \, c}{S_c}$

$$= \dfrac{3086.3 \times 0.0048}{0.4524}$$

$$= 32.7^\circ \, C$$

RESULT

Number of commutator segments, N_{cs} = 288

Diameter of commutator, D_c = 0.48 m

Width of commutator segment, β_c = 5.2 mm

Number of brushes,	n_b	=	6
Thickness of brush,	t_b	=	15.6 mm
Width of brush,	w_b	=	38 mm
Length of commutator,	L_c	=	0.3 m
Total commutator loss,	P_c	=	3.0863 kW
Temperature rise,	θ	=	32.7 $^\circ C$

EXAMPLE 6.16

A 4 pole, 400 V, 960 rpm, shunt motor has an armature of 0.3 m in diameter and 0.2 m in length. The commutator diameter is 0.22 m. Give full details of a suitable winding including the number of slots, number of commutator segments and number of conductors in each slot for an average flux density of approximately 0.55 Wb/m^2 in the air-gap.

GIVEN DATA

4 pole	$D = 0.3\ m$	$L = 0.2\ m$
400 V	960 rpm	$D_c = 0.22\ m$
Shunt motor	$B_{av} = 0.55\ Wb/m^2$	

SOLUTION

Specific magnetic loading, $B_{av} = \dfrac{\phi p}{\pi DL}$

$\therefore$ Flux per pole, $\phi = \dfrac{B_{av}\pi DL}{p} = \dfrac{0.55 \times \pi \times 0.3 \times 0.2}{4} = 0.026\ Wb$

Induced emf, $E = \dfrac{\phi ZN}{60}\dfrac{p}{a}$

Let us choose lap winding, $\therefore$ p = a.

Number of armature conductors, $Z = \dfrac{60 \times E}{\phi N} = \dfrac{60 \times 400}{0.026 \times 960} = 962$

The slot pitch, y_{sa} lies between 25 mm to 35 mm.

Number of slots, $S_a = \dfrac{\pi D}{y_{sa}}$

When $y_{sa} = 25\ mm$, $S_a = \dfrac{\pi \times 0.3}{25 \times 10^{-3}} = 38$

When $y_{sa} = 35\ mm$, $S_a = \dfrac{\pi \times 0.3}{35 \times 10^{-3}} = 27$

The slots should lie in the range of 27 to 38. For lap winding slots per pole should be a multiple of pole pair. Here pole pair = 2. Hence the number of slots should be even. The allowable choice are 28, 30, 32, 34, 36 and 38.

To reduce flux pulsations slots per pole should be an integer $\pm$ ½.

Let slots per pole = 8 $\pm$ ½ = 8.5 or 7.5

Let us choose 7.5 slots per pole.

| For double layer winding conductors per slot should be even. |

$\therefore$ Number of slots, S_a = Slots per pole × Number of poles = 7.5 × 4 = 30 slots

The designed value of 30 slots is one of the allowed choice of slots.

Conductors per slot, $Z_s = \dfrac{Z}{S_a} = \dfrac{962}{30} = 32$

Minimum number of coils, $C_{min} = \dfrac{Ep}{15} = \dfrac{400 \times 4}{15} = 107$

Number of coils, $C = \frac{1}{2} \times u \, S_a$

Let us assume different values of coil sides per slot, u.

When $u = 2$, $C = (1/2) \times 2 \times 30 = 30$	When $u = 6$, $C = (1/2) \times 6 \times 30 = 90$
When $u = 4$, $C = (1/2) \times 4 \times 30 = 60$	When $u = 8$, $C = (1/2) \times 8 \times 30 = 120$

Choose the number of coils such that conductors per slot is divisible by coil sides per slot. Here the conductors per slot 32 is divisible by coil sides per slot 8. Also when u = 8, the number of coils is more than the required minimum number of coils. Hence the best choice for number of coils is 120.

The following are the final values.

$$\text{Number of slots,} \quad S_a = 30$$
$$\text{Number of coils,} \quad C = 120$$
$$\text{Conductors per slot,} \quad Z_s = 32$$

New value of total armature conductors, $Z = \text{Slots} \times \text{Conductors per slot}$

$$= 30 \times 32 = 960$$

Number of turns per coil, $T_c = \dfrac{Z}{2C} = \dfrac{960}{2 \times 120} = 4$

CHECK

Number of commutator segments is equal to number of coils.

$\therefore$ Number of commutator segments, $N_{cs} = 120$

Commutator segment pitch, $\beta_c = \dfrac{\pi D_c}{C} = \dfrac{\pi \times 0.22}{120} = 5.76 \times 10^{-3} \, m = 5.76 \, mm$

The commutator segment pitch is greater than the minimum value of 4 *mm*. Hence the design is acceptable.

RESULT

Number of slots,	S_a	= 30
Total armature conductors,	Z	= 960
Number of coils,	C	= 120
Turns per coil,	T_c	= 4
Conductors per slot,	Z_s	= 32
Number of commutator segment,	N_{cs}	= 120
Commutator segment pitch,	β_c	= 5.76 *mm*

6.11 COMPUTER PROGRAMS

PROGRAM 6.1

Write a C program to design the diameter and length of armature of a DC generator for specifications in Example 6.1.

Method-1

```
/* Program to design diameter and length of armature*/
/* VARIABLE DECLARATION:
. . . . . . . . . . . . . . . . . . . . . . .
p =  Number of poles.
P = Rated power output.
ac = Specific electric loading.
N = Speed in rpm.
Eta = efficiency.
Psi = Ratio of pole arc to pole pitch.
Pa = Armature power.
Bg = Maximum gap density in the air gap at no load.
Co = Output coefficient.
LDow = Ratio of length of core to polepitch.
D = Diameter of armature.
L = Length of armature.
D2L = Product of square of diameter and length.
Dow = Pole pitch.
. . . . . . . . . . . . . . . . . . . . . . . . . . . . . .*/

#include<stdio.h>
#include<conio.h>
#include<math.h>
#include<iostream.h>
void main()
{
int P,p,N;
float Eta,Psi,Pa,ac,Bg,Co,D,L,LDow,D2L,Dow;
clrscr();
printf("INPUT DATA\n");
printf("Enter power output, P =");
scanf("%d", &P);
printf("Enter efficiency, Eta = ");
scanf("%f", &Eta);
printf("Enter ratio of pole arc to pole pitch, Psi = ");
scanf("%f", &Psi);
printf("Enter flux density in gap,  Bg = ");
```

```
scanf("%f", &Bg);
printf("Enter specific electric loading, ac = ");
scanf("%f", &ac);
printf("Enter speed in rpm, N = ");
scanf("%d", &N);
printf("Enter number of poles, p = ");
scanf("%d", &p);
printf("Enter ratio of length of core to pole pitch, LDow =");
scanf("%f", &LDow);
Pa = P/(Eta);                              /* Pa = Armature power */
Co = (3.14*3.14*Psi*Bg*ac*10e-4);         /* Co = Output coefficient */
D2L = Pa/(Co*N/(60));
Dow = LDow*3.14/(p);                       /* Dow = Pole pitch */
D = pow(D2L/(Dow),0.33);                   /* Diameter of armature */
L = Dow*D;                                 /* Length of armature */
printf("OUTPUT\n");

printf("Diameter of Armature, D = %8.4f m\n", D);
printf("Length of armature, L = %8.4f m\n", L);
getch();
}
```

Input data

```
Enter power output,                                    P    = 200
Enter efficiency,                                      Eta  = 0.91
Enter ratio of pole arc to pole pitch,                 Psi  = 0.67
Enter specific electric loading,                       ac   = 31000
Enter speed in rpm,                                    N    = 1000
Enter number of poles,                                 p    = 6
Enter ratio of length of core to pole pitch, LDOW = 0.75
```

Output

```
Diameter of Armature, D =  0.5766 m
Length of armature,   L =  0.2263 m
```

Method-2

```
/* Program to design diameter and length of armature*/

#include<stdio.h>
#include<conio.h>
#include<math.h>
#include<iostream.h>
void main()
```

```
{
clrscr();
int P,p,N;
float Eta,Psi,ac,Pa,Bg,Co,D,L,LDow,D2L,Dow;
P=200;          /* P = Rated power output */
Eta = 0.91;     /* Eta = efficiency */
Psi = 0.67;     /* Psi = Ratio of pole arc to pole pitch */
Bg = 0.87;      /* Bg = Maximum gap density in the air-gap at no load */
ac = 31000;     /* ac = Specific electric loading */
N = 1000;       /* N = Speed in rpm */
p = 6;          /* p =  Number of poles */
LDow = 0.75;    /* LDow = Ratio of length of core to pole pitch */

Pa = P/Eta;                         /* Pa = Armature power */
Co = (3.14*3.14*Psi*Bg*ac*10e-4);   /* Co = Output coefficient */
D2L = Pa/(Co*N/(60));
Dow = LDow*3.14/(p);                /* Dow = Pole pitch */
D = pow(D2L/(Dow),0.33);            /* D   = Diameter of armature */
L = Dow*D;                          /* L   = Length of armature */

printf("OUTPUT\n");
printf("Diameter of Armature, D = %8.4f m\n", D);
printf("Length of armature,   L = %8.4f m\n", L);
getch();
}
```

Output

```
Diameter of Armature, D =   0.5766 m
Length of armature,    L =   0.2263 m
```

PROGRAM 6.2

Write a C program to estimate the number of poles, diameter and length of armature of a shunt motor for specifications in Example 6.6.

/* Program to design the Number of poles, diameter and length of armature dimensions*/

```
#include<stdio.h>
#include<conio.h>
#include<math.h>
void main()
{
int P,p,N,a,b,f;
float Eta,Psi,Pa,Bav,Co,D,L,D2L,Dow,f1,f2,n,ac;
clrscr();
```

```
    Pa = 37;              /* Pa = Power developed in armature. */
    Eta = 0.9;            /* Eta = efficiency. */
    Psi = 0.7;            /* Psi = Ratio of pole arc to pole pitch. */
    Bav = 0.5;            /* Bav = Specific magnetic loading. */
    ac = 22000;           /* ac = Specific electric loading. */
    N = 1400;             /* N = Speed in rpm. */
    LDow = 0.7;           /* LDow = Ratio of length of core to pole pitch. */
    a = 2;
    b = 4;
    f1 = a*N/(120);
    f2 = b*N/(120);
    n = N/60;                         /* n = Speed in rps. */
    Co = (3.14*3.14*Bav*ac*10e-4);    /* Co = Output coefficient. */
    D2L = Pa/(Co*n);                  /* D2L = Product of square of diameter
                                              and length. */

    if(25<f2<=50)
    {
    p = b;
    }
    else if(25<f1<=50)
    {
    p = a;
    }
    Dow = Psi*3.14/(p);               /* Dow = Pole pitch. */
    D = pow(D2L/(Dow),0.33);          /* D = Diameter of armature. */
    L = Dow*D;                        /* L = Length of armature. */

    printf("OUTPUT\n");
    printf("Number of poles, p =%d\n",p);
    printf("Diameter of Armature, D =%8.2f m\n",D);
    printf("Length of armature, L =%8.2f m\n",L);
    getch();
    }
```

Output

```
    Number of poles,      p = 4
    Diameter of Armature, D = 0.30 m
    Length of armature,   L = 0.17 m
```

Program 6.3

Write a C program to estimate the flux per pole, induced emf and total armature conductor of a DC series motor for the specifications in Example 6.7.

```
/* Program to estimate flux, induced emf and armature conductor */
```

```
#include<stdio.h>
#include<conio.h>
#include<math.h>
#include<iostream.h>
void main()
{
clrscr();
int p,a,N,V,Z,E;
float D,L,Bav,PHI;
D = 0.3373;              /* D = Diameter of armature.*/
L = 0.1775;              /* L = Length of armature.*/
p = 4;                   /* p = Number of poles. */
Bav = 0.66;              /* Bav = Specific magnetic loading.*/
a = 3;                   /* a = Number of parallel path. */
N = 600;                 /* N = Speed. */
V = 440;                 /* V = Voltage. */

PHI = Bav*3.14*D*L/(p); /*PHI = Flux per pole.*/
Z = V*a*60/(PHI*N*p);    /*Z = Total armature conductor.*/
E = PHI*Z*N/(60)*(p/a); /*E = Induced emf.*/

printf("OUTPUT\n");
printf("Flux per pole, PHI = %8.4f Wb\n", PHI);
printf("Total armature conductor, Z = %8f\n", Z);
printf("Induced emf, E = %8.4f\n", E);
getch();
}
```

Output

```
Flux per pole,           PHI = 0.0310 Wb
Total armature conductor, Z   = 1065
Induced emf,             E    = 440 V
```

Program 6.4

Write a C program to estimate the average reactance voltage and flux density under interpole of a single turn two layer winding for specifications in Example 6.12.

```
/* Program to estimate the Average reactance voltage and Flux density
under interpole. */

#include<stdio.h>
#include<conio.h>
#include<math.h>
void main()
```

```
{
int Po,V,p,N,Z,Tc;
float Lcoil,L,D,Lip,Erav,Bgi,LAMDA,ac,Ia,Iz,Va;
clrscr();
Po = 250;                  /* Po = Power ouput. */
V = 525;                   /* V =  Voltage. */
p = 6;                     /* p = Number of poles. */
N = 220;                   /* N = Speed in rpm. */
Lcoil = 0.0057;            /* Lcoil = Inductance of coil in armature. */
Lip = 0.18;                /* Lip = Length of interpole. */
L = 0.3;                   /* L = Length of armature. */
D = 1.6;                   /* D = Diameter of armature. */
Tc = 1;                    /* Tc = Turns per coil. */
Z = 600;                   /* Z = Armature conductor. */


Ia = Po*10e+3/(V);               /* Ia = Armature current. */
Iz = Ia/(p);                     /* Iz = Current through an armature
                                       conductor. */
ac = Iz*Z/(3.141*D);             /* ac = Specific electric loading. */
Va = 3.141*D*N/(60);             /* Va = Peripheral velocity of armature. */
LAMDA = Lcoil*10e-3/(2*Tc*Tc*L); /* LAMDA = Specific permeance. */
Erav = 2*Tc*ac*Va*LAMDA*10e-6*L; /* Erav = Average reactance voltage. */
Bgi = ac*LAMDA*10e-6*L/(Lip);    /* Bgi = Flux density under interpole. */

printf("OUTPUT\n");
printf("Average reactance voltage, Erav = %8.4f V\n",Erav);
printf("Flux density under interpole, Bgi = %8.2f Wb/m3\n",Bgi);
getch();
}
```

Output

```
Average reactance voltage,    Erav  =  0.9952 V
Flux density under interpole, Bgi   =  0.15 Wb/m3
```

PROGRAM 6.5

Write a C program to estimate the number of armature conductor, number of armature slots and resistance of armature of a compund generator for specifications in Example 6.13.

```
/* Program to estimate armature conductor, armature slots and armature resistance */
```

```
#include<stdio.h>
#include<conio.h>
#include<math.h>
#include<iostream.h>
void main()
{
int N,E,P,V,a,Sa;
float D,Yss,PHI,Z,Rho,Ia,DELa,Aa,L,Dow,Ds,Lmt,Ra;
clrscr();
D = 1.1;                    /* D = Diameter of armature. */
Yss = 0.025;               /* Yss = Armature slot pitch. */
N = 375;                   /* N = Speed. */
PHI = 0.0875;              /* PHI = Flux per pole. */
E = 440;                   /* E = Voltage. */
Rho = 0.0172;              /* Rho = Resistivity of copper. */
P = 500;                   /* P = Output power.*/
DELa = 4.5;                /* DELa = Current density. */
a = 8;                     /* a = Number of parallel path. */
L = 0.3;                   /* L = Length of armature.*/
Dow = 0.432;               /* Dow = Pole pitch.*/
Ds = 0.037;                /* Ds = Depth of slot. */


Sa = 3.14*D/Yss;                        /* Sa = Number of armature
                                               slots.*/

Z = 60*E/(PHI*N);                       /* Z = Number of armature
                                               conductor.*/

Ia = P*10e+2/(E);                       /* Ia = Armature current.*/
Aa = Ia/(DELa*10e-1);                   /* Aa = Area of crossection.*/
Lmt = 2*L+2.3*Dow+5*Ds;                 /* Lmt = Length of mean turn.*/
Ra = Z*Rho*10e-6*Lmt/(2*a*a*Aa*10e-6);/* Ra = Resistance of armature.*/

printf("OUTPUT\n");
printf(" Number of armature slots, Sa =%8.4f\n", Sa);
printf("Number of armature conductor, Z = %8.4f\n", Z);
printf("Resistance of armature, Ra = %8.4f ohm\n", Ra);
getch();
}
```

Output

```
Number of armature slots,      Sa = 138
Number of armature conductor,  Z = 808
Resistance of armature,        Ra = 0.0008 ohm
```

Program 6.6

Write a C program to estimate the number of a slots, armature conductor, number of coils, turns per coils, conductor per slot, number of commutator segment and commutator segment pitch of a shunt motor for specifications in Example 6.16.

```c
/* Program to estimate
      1. Number of slots.
      2. Total armature conductors.
      3. Number of coils.
      4. Turns per coil.
      5. Conductors per slot.
      6. Number of commutator segment.
      7. Commutator segment pitch.   */

#include<stdio.h>
#include<conio.h>
#include<math.h>
void main()
{
int p,E,Sa,Z,C,Tc,Zs,Ncs,N;
float D,L,Bav,Dc,PHI,BETAC;
clrscr();
p = 4;                 /* p = Number of poles. */
D = 0.3;               /* D = Diameter of armature. */
L =0.2;                /* L = Length of armature. */
E = 400;               /* E = Voltage rating of machine. */
N = 960;               /* N = Speed in rpm. */
Bav = 0.55;            /* Bav = Specific magnetic loading. */
Dc = 0.22;             /* Dc = Diameter of commutator. */

Sa = 7.5*p;                  /* Sa = Number of armature slots. */
PHI =  Bav*3.141*D*L/(p);    /* PHI = Flux per pole. */
Z = 60*E/(PHI*N);            /* Z = Total number of armature conductor. */
C = 0.5*8*Sa;                /* C = Number of coils. */
Zs = Z/(Sa);                 /* ZS = Number of conductors per slots. */
Tc = Z/(2*C);                /* Tc = Turns per coil. */
Ncs = C;                     /* Ncs = Number of commutator segments. */
BETAC = 3.141*Dc/C;          /* BETAC = Commutator segment pitch. */
```

```
    printf("OUTPUT\n");
    printf("Number of slots, Sa = %d\n", Sa);
    printf("Total armature conductors, Z = %d\n", Z);
    printf("Number of coils, C =%d\n", C);
    printf("Turns per coil, Tc =%d\n", Tc);
    printf("Conductors per slot, Zs = %d\n", Zs);
    printf("Number of commutator segment, Ncs = %d\n", Ncs);
    printf("Commutator segment pitch, BETAC = %8.2f mm\n", BETAC);
    getch();
    }
```

Output

Number of slots,	Sa	=	30
Total armature conductors,	Z	=	964
Number of coils,	C	=	120
Turns per coil,	Tc	=	4
Conductors per slot,	Zs	=	32
Number of commutator segment,	Ncs	=	120
Commutator segment pitch,	BETAC	=	5.77 mm

6.12 SHORT-ANSWER QUESTIONS

Q6.1 List the constructional elements of a DC machine.

The major constructional elements of a DC machine are stator, rotor, brushes and brush holders. The various parts of stator and rotor are listed below.

Stator - Yoke (or) Frame   **Rotor** - Armature core
- Field pole   - Armature winding
- Pole shoe   - Commutator
- Field winding
- Interpole

Q6.2 What is output equation?

The equation which relates the power output to the main dimensions (D and L), Specific loadings (B_{av} and **ac**) and speed (n) of a machine is known as output equation.

The output equation of DC machine is, $P_a = C_0 D^2 L n$, in kW

where, P_a = Power developed in armature of DC machine.
    C_0 = Output coefficient.

Q6.3 Write the expression for output coefficient of DC machine.

Output coefficient, $C_0 = \pi^2 B_{av} \, \mathbf{ac} \times 10^{-3}$, in $kW/m^3\text{-}rps$

Q6.4 **What is the relation between the power developed in armature and the power output in DC machine?**

For generators, $P_a = P/\eta$

For motors, $P_a = P$

where, P_a = Power developed in armature

P = Rated power output

η = Efficiency.

Q6.5 **Write the expression for power developed in the armature of DC machine in terms of maximum gap density.**

Power developed in armature, $P_a = C_o D^2 L n$

$$C_o = \pi^2 \Psi B_g \, \mathbf{ac} \times 10^{-3}$$

where, B_g = Maximum gap density.

Q6.6 **Mention the factors governing the length of armature core in DC machines.**

The factors to be considered for the choice of armature length are,

1. Cost
2. Ventilation
3. Voltage between adjacent commutator segments
4. Specific magnetic loading.

Q6.7 **State the factors which limit the armature diameter of a DC machine.**

The factors which limit the armature diameter of a DC machine are,

1. Peripheral speed   4. Induced emf per conductor
2. Pole pitch   5. Power output
3. Specific electric loading

Q6.8 **Write the equation for maximum value of the main dimensions of a DC machine.**

Maximum value of armature length, $L_{max} = \dfrac{7.5}{T_c \, N_c \, B_{av} \, V_a}$

Maximum value of armature diameter, $D_{max} = \dfrac{P_a \times 10^3}{\pi \, \mathbf{ac} \, e_z}$

Q6.9 **What is active part?**

In electrical machines the core and winding of the machine are together called active part. Because, the energy conversion takes place only in the active part of the machine.

Q6.10 **What are the main dimensions of a rotating machine?**

The main dimensions of a rotating machine are the armature diameter or stator bore, D and armature or stator core length, L.

Q6.11 **What are the constituents of magnetic circuits in a DC machine.**   *(AU, Nov' 18, 2 M)*

The magnetic circuit of DC machine comprises yoke (or) frame, poles, air-gap, armature teeth and armature core.

Q6.12 *Define specific elelctric and magnetic loading.* *(AU, Nov' 17, 2 M) (AU, Nov' 18, 2 M)*

Specific Magnetic Loading

The specific magnetic loading is given by the ratio of total flux around the air-gap and the area of flux path at the air-gap.

$$\therefore B_{av} = \frac{\text{Total flux around the air-gap}}{\text{Area of flux path at the air-gap}} = \frac{p\phi}{\pi DL}$$

Specific Electric Loading

The specific electric loading is given by the ratio of total armature ampere conductors and armature periphery (circumference) at air-gap.

$$\therefore ac = \frac{\text{Total armature conductors}}{\text{Armature periphery at air-gap}} = \frac{I_z Z}{\pi D}$$

Q6.13 *Write down the expression for brush friction losses.* *(AU, Apr' 17, 2 M)*

The brush friction loss can be calculated from the formula,

$$P_{bf} = \mu\, P_b\, A_B\, V_c$$

where, P_b = Brush contact pressure on commutator

  A_B = Total contact area of all brushes = $p\, A_b$ - For lap winding

    = $2A_b$ - For wave winding

  A_b = Total brush area per spindle

  μ = Coefficient of friction

  V_c = Peripheral speed of commutator

Q6.14 *What are the factors that can be varied to vary the power output of a rotating electrical machine?*

The power output of a rotating electrical machine depends of specific electric loading, specific magnetic loading, volume of active part and speed. Hence by varying these four quantities the power output of a machine can be varied.

Q6.15 *In a DC machine, What are the limiting values of armature peripheral speed and voltage between adjacent commutator segments?*

Maximum armature peripheral speed, V_{am} = 30 *m/s*.

Maximum voltage between commutator segments, E_{cm} = 30 *V*.

Q6.16 *Give the expression for the torque developed by a DC motor in terms of main dimensions of armature.*

Let, T_a = Torque developed in the armature

We know that,

  Power = Torque × Angular velocity $\Rightarrow$ $P_a = T_a \times 2\pi n$

$$\boxed{\begin{aligned} P_a &= C_o\, D^2\, L\, n \\ &= \pi^2\, B_{av}\, \mathbf{ac} \times 10^{-3}\, D^2 \times L\, n \end{aligned}}$$

$$\therefore \text{Torque, } T_a = \frac{1}{2\pi n}\, P_a$$

$$= \frac{1}{2\pi n}\, \pi^2\, B_{av}\, \mathbf{ac} \times 10^{-3}\, D^2 L n$$

$$= \frac{\pi}{2}\, B_{av}\, \mathbf{ac}\, D^2\, L \times 10^{-3}$$

Q6.17 *Calculate the main dimensions for a 500 kW, 1kV, 600 rpm, 6 pole DC machine. Take L/τ = 1 and C$_o$ = 220 kW/m^3-rps.*

Solution

Given that, $L / \tau = 1$,

$$\therefore L = \tau = \frac{\pi D}{p} = \frac{\pi}{6} D = 0.5236 \, D \qquad .....(1)$$

We know that, $P_a = C_o \, D^2 \, L \, n$

$$\therefore D^2 L = \frac{P_a}{C_o n} = \frac{500}{220 \times 600 / 60} = 0.2273 \, m^3$$

$\therefore D^2 L = 0.2273$

$\therefore D^2 (0.5236 \, D) = 0.2273$　　　　　$\boxed{\text{Using equation (1)}}$

$$\therefore D = \left(\frac{0.2273}{0.5236}\right)^{1/3} = 0.76 \, m$$

$\therefore L = 0.5236 \, D = 0.5236 \times 0.76 = 0.4 \, m$

Q6.18 *Discuss the factors which affect the proportions of the armature core in DC machine. (or size of DC machine)*

The size of armature core and so the DC machine is decided by the main dimensions D and L.
The factors which affect the D and L of the armature core are the following,

　　1. Dimensions of the pole

　　2. Moment of inertia

　　3. Peripheral speed

　　4. Voltage between adjacent commutator segments.

Q6.19 *List out the design considerations for DC motors operating on solid state circuits.*

The DC machines operating on solid state circuits should have following features.

　　1. Low value of inductance of the armature which helps faster switching of thyristors.

　　2. Low moment of inertia of the armature which helps is quick reversal of direction and regenerative braking.

　　3. Small air-gap in order to reduce the starting and no-load current.

Q6.20 *What are the applications of DC special motors?*

The DC special motors are used in closed loop control system as power actuators and to provide linear motions. They are also used as clutches, couplings, eddy current brakes, very high speed drives, etc.

Q6.21 *Define the terms "active copper" and "inactive copper" in the design of DC machines.*

The copper wires or strips are used for armature and field windings. The armature winding consists of number of coils accommodated in the slots. The portions of coils lying on the slots are called active copper because they decide the electric loading of the machine. The overhang portions of the coils are called inactive copper because they simply connect the conductors to form coils.

Q6.22 *Calculate the output coefficient of a DC shunt generator from the given data.*

　　　　$B_g = 0.89 \, Wb/m^2$, **ac** = 32000 *amp.cond./m*, $\Psi = 0.66$.

Solution

Output coefficient, $C_o = \pi^2 \, \Psi \, B_g \, \textbf{ac} \times 10^{-3}$

$$= \pi^2 \times 0.66 \times 0.89 \times 32000 \times 10^{-3}$$

$$= 185.5 \, kW/m^3\text{-}rps.$$

Q6.23 What is the range of specific magnetic loading in DC machine?

The usual range of specific magnetic loading in DC machine is 0.4 to 0.8 Wb/m^2.

Q6.24 What are the factors to be considered for the choice of specific magnetic loading?

The choice of specific magnetic loading depends on the following,

1. Flux density in teeth

2. Frequency of flux reversals

3. Size of machine

Q6.25 What is the range of specific electric loading in DC machine?

The usual range of specific electric loading in DC machine is 15000 to 50000 *amp.cond./m.*

Q6.26 What are the factors that influence the choice of specific electric loading?

The choice of specific electric loading depends on the following,

1. Temperature rise   4. Size of machine

2. Speed of machine   5. Armature reaction

3. Voltage   6. Commutation.

Q6.27 What is the purpose of constructing the pole body by laminated sheets?

The laminated poles offers homogeneous construction, (because while casting internal blow holes may develop and while forging internal cracks may develop).

Also the laminated poles offers the flexibility of increasing the length by keeping the diameter fixed, in order to increase the power output (or capacity) of the machine.

Q6.28 What are the factors to be considered for the selection of number of poles in DC machine?

The factors to be considered for the selection of number of poles in DC machine are,

1. Frequency   4. Length of commutator

2. Weight of iron parts   5. Labour charges

3. Weight of copper   6. Flash over and distortion of field form

Q6.29 What are the quantities that are affected by the number of poles?

Weight of iron and copper, length of commutator and dimension of brushes are the quantities affected by number of poles.

Q6.30 List the advantages and disadvantages of large number of poles.   *(AU, Nov' 17, 2 M)*

Advantages of large number of poles

The large number of poles results in reduction of the following,

1. Weight of armature core and yoke

2. Cost of armature and field conductors

3. Overall length and diameter of machine

4. Length of commutator

5. Distortion of field form under load condition.

Disadvantages of large number of poles

The large number of poles results in increase of the following,

1. Frequency of flux reversals
2. Labour charges
3. Possibility of flash over between brush arms.

Q6.31 **Why square pole is preferred?**

If the cross-section of the pole body is square then the length of the mean turn of field winding is minimum. Hence to reduce the copper requirement a square cross-section is preferred for the poles of DC machine.

Q6.32 **What is square pole and square pole face?**

In square pole, the width of the pole body is made equal to the length of the armature. In square pole face, the pole arc is made equal to the length of the armature.

Q6.33 **Mention guiding factors for the selection of number of poles.**

The guiding factors for choice of number of poles are,

1. The frequency should lie between 25 to 50 Hz.
2. The value of current per parallel path is limited to 200 A, thus the current per brush arm should not be more than 400 A.
3. The armature mmf should not be too large. The mmf per pole should be in the range 5000 to 12,500 AT.
4. Choose the largest value of poles which satisfies the above three conditions.

Q6.34 **What are the advantages of large length of air-gap in DC machines?**

In DC machines a larger value of air-gap results in lesser noise, better cooling, reduced pole face losses, reduced circulating currents, less distortion of field form and lesser armature reaction.

Q6.35 **What are the factors to be considered for estimating the length of air-gap in DC machine?**

The factors to be considered for estimating the length of air-gap are armature reaction, cooling, iron losses, distortion of field form and noise.

Q6.36 **Mention the factors that governing the choice of number of armature slots in a DC machine.**

The factors governing the choice of number of armature slots are,

1. Slot pitch   4. Commutation
2. Slot loading   5. Suitability for winding
3. Flux pulsations.

Q6.37 **What are the factors to be considered for deciding the slot dimensions?**

The factors to be considered for deciding the slot dimensions are,

1. Flux density in tooth
2. Flux pulsations
3. Eddy current loss in conductors
4. Reactance voltage
5. Fabrication difficulties.

Q6.38 *What are the effects caused by excessive deep slot?*

In DC machines with increase in depth of slot the eddy current loss in conductors increases, specific permeance of slot increases, reactance voltage increases and it becomes difficult to fabricate the lamination with narrow width at the roots of teeth.

Q6.39 *State any three conditions in deciding the choice of number of slots for a large DC machine.*

1. The slot loading should be less than 1500 *amp.cond.*
2. The number of slots per pole should be greater than or equal to 9 to avoid sparking
3. The slot pitch should lie between 25 to 35 *mm.*

Q6.40 *Why the armature core of DC machine is laminated?*

The armature of DC machine is laminated to reduce hysteresis and eddy current losses.

Q6.41 *Explain how depth of armature core for a DC machine is determined.*

Let, ϕ = Flux per pole L_i = Net iron length of armature

ϕ_c = Flux in armature core d_c = Depth of armature core

B_c = Flux density in armature core A_c = Area of cross-section of armature core

Now, $\phi_c = \dfrac{\phi}{2}$ and $A_c = \dfrac{\phi_c}{B_c} = \dfrac{\phi}{2\,B_c}$ .....(1)

Also, $A_c = L_i\, d_c$ .....(2)

On equating equations (1) and (2) we get,

$$L_i\, d_c = \frac{\phi}{2\,B_c} \quad \Rightarrow \quad d_c = \frac{\phi}{2\,L_i\, B_c}$$

Q6.42 *What is split coil?*

The split coils will have more than two coil sides. When all the top coil sides of a coil are lying in one slot and their corresponding bottom coil sides are accommodated in two different slots then the coil is called split coil.

Q6.43 *What are the components of the magnetic circuit of a DC machine?*

In DC machine the magnetic circuit consists of yoke, pole, pole shoe, teeth and armature core.

Q6.44 *State the uses of yoke in a DC machine.*

The yoke serve as a path for flux in DC machine and it also serve as an enclosure for the machine.

Q6.45 *What factor decides the number of turns in a winding?*

The number of turns in a winding is decided by the emf per turn and flux density. The emf per turn depends on the type of insulation employed.

Q6.46 *How the area of cross-section of a conductor is estimated?*

The area of cross-section of a conductor is estimated based on temperature rise, resistivity and cooling methods. Assume a current density based on the above factors. For copper conductors the normal range of current density is 2.5 to 5 *A/mm²*.

Now, area of cross-section of a conductor = $\dfrac{\text{Current per conductor}}{\text{Current density}}$

Q6.47 Discuss the parameters governing the selection of conductor dimensions.

The following are the parameters governing the selection of conductor dimensions

    1. Allowable temperature rise   4. Resisitvity of the material of the conductor

    2. Maximum current through a conductor   5. Cooling method

    3. Current density

Q6.48 What is current density? Where it is used in the design of a machine ?

The current density is the ratio of current through a conductor to its area of cross-section, usually expressed in A/mm^2.

$$\text{Current density, } \delta = \frac{\text{Current, I}}{\text{Area of cross-section, } a_z} \text{ in } A/mm^2$$

The current density is used to estimate the area of cross-section used in field winding, armature winding, stator winding, rotor bars and end rings.

Q6.49 State the parameters governing slot utilization factor or slot space factor.

The following factors decides the slot utilization factor.

    1. Voltage rating

    2. Thickness of insulation

    3. Number of conductors per slot

    4. Area of cross-section of the conductor

    5. Dimensions of the conductor (or type of conductor like circular, square, strip, etc.)

Q6.50 What do you understand by slot pitch ?

The slot pitch is defined as the distance between centres of two adjacent slots measured in linear scale.

$$\text{Slot pitch, } Y_s = \frac{\pi D}{S_a}$$

where, D = Diameter of armature

          S_a = Number of slots in armature

Q6.51 Discuss the parameters governing selection of slot dimension.

The dimensions of the slot decides most of the electrical characteristics of the machine. The dimensions of the slots are slot area, slot depth and slot width at various sections. The area of the slot is decided based on the area required for conductors and the area required for insulation.

Generally 25 to 40% of area is occupied by the conductors and the remaining area is occupied by insulations. The depth and width is decided based on the arrangement of conductors in a slot.

Q6.52 Define slot space factor or slot insulation factor.   *(AU, Apr' 19, 2 M)*

The slot space factor is defined as the ratio of conductor area to slot area.

$$\text{Slot space factor} = \frac{\text{Conductor area}}{\text{Slot area}}$$

Q6.53 How to determine the number of slots for lap and wave winding in DC machine?

For lap winding the number of slots should be a multiple of pole pair to avoid dummy coils and slots per pole should be integer plus $^1/_2$ to reduce flux pulsation losses.

For wave winding the number slots should not be a multiple of pole pair to avoid dummy coil and slots per pole arc should be integer plus $^1/_2$ to reduce flux pulsation losses.

Q6.54 *What is the purpose of slot insulation?*

The conductors are placed on the slots in the armature. When the armature rotates the insulation of the conductors may damage due to vibrations. This may lead to a short circuit with armature core if the slots are not insulated.

Q6.55 *What factor decides the minimum number of armature coils in DC machine?* *(AU, Apr' 18, 2 M)*

The maximum voltage between adjacent commutator segments decides the minimum number of coils.

Q6.56 *What is the duty of each field pole in DC machine?*

The duty of each field pole or coil is to develop the mmf required to establish the following flux in the magnetic circuit.

1. The full value of the flux for the length of one pole, once across the air-gap and in one set of armature teeth.

2. One half the full value of the flux along one half the length of the single path in the armature core and yoke.

Q6.57 *How does the series and shunt field windings differ from each other?*

The shunt field winding have large number of turns made of thin conductors, because the current carried by them is very low. The series field winding is designed to carry heavy current and so it is made of thick conductors or strips.

Q6.58 *What is length of mean turn of field coil?*

The length of turn in the centre of the coil is called length of mean turn. It can be calculated from the dimensions of the pole and the depth of field coil.

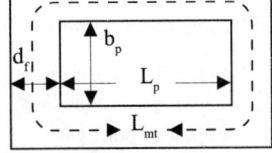

Length of mean turn, $L_{mt} = 2 (L_p + b_p + 2d_f)$

Fig. Q6.57: Cross-section of rectangular field coil.

Q6.59 *Mention the factors to be considered for the design of shunt field coil.*

The factors to be considered for the design of shunt field coil are,

1. mmf per pole and flux density.
2. Loss dissipated from the surface of field coil.
3. Resistance of the field coil.
4. Current density in the field conductors.

Q6.60 *Define copper space factor of a coil.*

The copper space factor of a coil is defined as the ratio of conductor area and the area of cross-section of the coil.

$$\text{Copper space factor} = \frac{\text{Conductor area}}{\text{Area of cross-section of the coil}}$$

where, Conductor area = Number of turns × Area of cross-section of conductor.

Q6.61 *How the ampere-turns of the series field coil is estimated?*

In compound machines the ampere-turns to be developed by the series field coil is estimated as 15 to 25% of full load armature mmf.

In series machines the ampere-turns to be developed by the series field coil is estimated as 1.15 to 1.25 times the full load armature mmf.

Q6.62 What is the height occupied by series field coil in a field pole?

In a field pole of compound machine, approximately 80% of the height is occupied by shunt field coil and 20% by series field coil.

Q6.63 What is meant by commutation?

The process of current reversal in a coil is called commutation.

Q6.64 What is resistance commutation?

When the current reversal in the coils are assisted only by the resistance of the brushes, the commutation is called resistance commutation.

Q6.65 What are the different types of commutation?

The different types of commutation are,

1. Resistance commutation
2. Retarded commutation
3. Accelerated commutation
4. Sinusoidal commutation.

Q6.66 What is sinusoidal commutation?

When the process of current reversal is delayed at the leading edge and accelerated at the trailing edge, the commutation is called sinusoidal commutation.

Q6.67 State the conditions of electrical symmetry of commutator windings (DC machine armature windings)?

In DC machine armature windings electrical symmetry is achieved if the conductors are placed symmetrically with regard to the field systems. For this the number of slots and commutator segments should be multiples of pair of poles.

Q6.68 Discuss the parameters governing the length of commutator.

The length of the commutator depends upon the number of brushes and clearance between the brushes. The surface area required to dissipate the heat generated by the commutator losses is provided by keeping sufficient length of commutator.

Q6.69 What are the factors that influence the choice of commutator diameter?

The factors that influence the choice of commutator diameter are,

1. The peripheral speed
2. The peripheral voltage gradient should be limited to 3 V/mm
3. Number of coils in the armature.

Q6.70 What is the purpose of the mica strip between two adjacent commutator segments?

Mica is placed in between two commutator segments in order to insulate the segments from each other.

Q6.71 What are the factors to be considered in the design of commutator?   *(AU, Apr' 19, 2 M)*

The following are the factors to be considered in the design of commutator.

1. Peripheral speed
2. Voltage between adjacent commutator segments
3. Number of coils in the armature
4. The number of brushes
5. Commutator losses.

Q6.72 What type of copper is used for commutator segments?

The commutator segments are made of hard drawn copper or silver copper (0.05% silver).

Q6.73 What is the need for brushes in DC machine?

The brushes are used in DC machines to collect or draw current from the rotating armature.

Q6.74 What are the materials used for brushes in DC machines?

The various materials used for brushes in DC machines are,

1. Natural graphite
2. Hard carbon
3. Electrographite
4. Metal graphite

Q6.75 How do design the number of brushes for a DC machine?

The number of brush locations are decided by the type of winding. In lap-winding the number of brush locations is equal to number of poles and in wave winding it is always two.

In each location there may be more than one brush mounted on a spindle, whenever the current per brush location is more than 70 A. Hence the number of brushes in a spindle are selected such that each brush does not carry more than 70 A.

Q6.76 What are the effects of armature reaction?

The various effects of armature reaction are reduction in induced emf, increase in iron losses, delayed commutation, sparking and ring firing.

Q6.77 How the effects of armature reaction can be reduced?

The effects of armature reaction can be reduced by,

1. Increasing the length of air-gap at pole tips
2. Increasing the reluctance at pole tips
3. Providing compensating winding and interpoles.

Q6.78 How the polarities of interpole are decided?

The polarity of the interpole must be that of the main pole just ahead (in the direction of rotation) for a generator and just behind (in the direction of rotation) for a motor.

Q6.79 What is the effect of interpole on main pole?

In case of generator the interpole will magnetize the leading edge and demagnetize the trailing edge of main pole. In case of motor the interpoles will demagnetize the leading edge and magnetize the trailing edge of main pole.

Q6.80 State the use of interpoles.

The interpole are used in DC machines to neutralize the cross magnetizing armature mmf at the interpolar axis and to neutralize the reactance voltage in the coil undergoing commutation.

Q6.81 State the relation between the armature diameter and commutator diameter for various ratings of DC machines?

The diameter of the commutator is chosen as 60 to 80% of the armature diameter. The limiting factor is the peripheral speed. Typical choice of commutator diameter for various voltage ratings are listed here.

For 350 to 700 V machine, commutator diameter = 0.62 D

For 200 to 250 V machine, commutator diameter = 0.68 D

For 100 to 125 V machine, commutator diameter = 0.75 D

6.13 EXERCISES

I.   Fill in the blanks

1.   The .... are placed in between the main poles.
2.   The number of poles in DC machine is selected such that the frequency lies between........
3.   The basic constructional elements of a rotating electrical machine are .... and ....
4.   In electrical machines the core and the winding of the machine are together called ....
5.   The .... is defined as the total flux around the armature periphery.
6.   The .... is defined as the total number of ampere conductors around the armature periphery.
7.   The .... and .... of armature are called main dimensions.
8.   The unit of output coefficient is ....
9.   In small DC machines, the friction, windage and iron losses are approximately equal to .... of the ....
10.   In large DC machines, if P is the kW rating of the machine and η is efficiency then the power developed in the armature of generator is .... and that of motor is ....
11.   The main dimensions of a machine .... with .... in speed and output coefficient.
12.   When large number of poles are provided the flux carried by the yoke ....
13.   The weight of copper .... with increase in number of poles.
14.   For square pole face the .... is equal to the length of the armature.
15.   The armature core is made of laminations of thickness ....
16.   A large number of slots results in .... slot pitch.
17.   For better commutation the slots per pole should be ....
18.   The brushes are placed at ....
19.   The process of current reversal in a coil is called ....
20.   For a .... the polarity of interpole must be same as that of main pole just ahead in the direction of rotation.

Answers

1. interpoles	8. kW/m^3-rps or kVA/m^3-rps	15. 0.5 *mm*
2. 25 to 50 *Hz*	9. one-third, total losses	16. smaller
3. stator, rotor	10. P/η, P	17. ≥ 9
4. active part	11. decreases, increases	18. magnetic neutral axis
5. total magnetic loading	12. reduces	19. commutation
6. total electric loading	13. decreases	20. generator
7. diameter, length	14. pole arc	

II.   State whether the following statements are True/False

1.   The maximum value of armature length depends on specific magnetic loading and peripheral speed.
2.   The maximum value of armature diameter depends on specific electric loading and induced emf per conductor.
3.   High values of specific electric loading helps the commutation.
4.   The overall diameter of a DC machine increases with increase in number of poles.
5.   When the poles have circular cross-section the length of mean turn is minimum.
6.   A larger length of air-gap reduces armature reaction.
7.   When large number of slots per pole are provided the sparking during commutation increases.
8.   In DC machine each field coil has to establish the flux over one of the complete magnetic circuit.
9.   The shunt field coil will have thick conductors and series field coil will have thin conductors.
10.   In compound machines, the series field compensates for reduction in flux due to armature reaction.
11.   During commutation the coil is short circuited by carbon brush.
12.   The diameter of commutator should be higher than the diameter of armature.
13.   The interpole winding is connected parallel to armature.
14.   The interpole will not reduce sparking at commutator segments.

Answers

1.	True	8.	False
2.	True	9.	False
3.	False	10.	True
4.	False	11.	True
5.	True	12.	False
6.	True	13.	False
7.	False	14.	False

III.   Unsolved problems

E6.1 Estimate the main dimensions of a 4 pole, 100 kW, 1500 rpm DC generator assuming specific electric and magnetic loadings as 19000 $amp.cond./m$ and 0.4 Wb/m^2 respectively. Assume that the length of armature is equal to the pole pitch.

$$(D = 0.41 \; m \; ; \; L = 0.32 \; m)$$

E6.2 Find the suitable values for number of poles, diameter and length of armature core of a 400 kW, 500 V, 180 rpm DC generator. Assume B_{av} = 0.6 Wb/m^2 and ac = 35000 $amp.cond./m$. Choose L/τ = 1.2.

$$(p = 20 \; ; D = 1.5 \; m \; ; L = 0.28 \; m)$$

E6.3 Determine the main dimensions of a DC shunt generator with the following specifications:
5 kW, 220 V, 1500 rpm, 4 pole, ac = 200 $amp.cond./cm$, B_{av} = 0.6 Wb/m^2, η = 90% and pole arc = 0.7τ. Choose $L = \tau$.

$$(D = 0.15 \; m \; ; L = 0.08 \; m)$$

E6.4 Calculate the main dimensions of a 50 kW, 4 pole, 600 rpm DC generator from the following

data : B_{av} = 0.8 Wb/m^2, ac = 200 $amp\ cond./cm$, $\dfrac{\text{Pole arc}}{\text{Pole pitch}} = 0.6$, $\dfrac{\text{Core length}}{\text{Pole arc}} = 0.75$, efficiency = 0.85.

(D = 0.4722 m ; L = 0.1667 m)

E6.5 Find the main dimensions and number of poles of a 100 kW, 230 V, 1000 rpm shunt motor so that a square pole face is obtained. The average gap density is 0.85 Wb/m^2 and ampere conductors per metre are 22000. The ratio of pole arc to pole pitch = 0.67. The full load efficiency is 91%.

(p = 6, D = 0.4525 m, L = 0.1587 m)

E6.6 Determine the main dimensions of a 80 kW, 4 pole, 600 rpm DC shunt generator, the full load terminal voltage being 220 V. The maximum gap density is 0.75 Wb/m^2 and the armature ampere conductors per metre are 27,000. Assume a square pole face.

(D = 0.4703 m, L = 0.2586 m)

E6.7 Design shunt field winding of a 4 pole, 240 V DC generator which has the following design data. mmf per pole = 5060 AT, mean length of turn = 1.21 m, winding depth = 45 mm, cooling surface per coil = 0.24 m^2. Calculate the inner, outer diameter of the cylindrical coil. Assume 2 $micro$-ohm-cm as the resistivity of copper at working temperature.

> **Ans :** $d_{fci} = d_{fc} + 0.2 = 1.948$ mm $d_{if} = 0.3402$ m
>
> $a_f = 2.401$ mm^2 $d_{of} = 0.4302$ m
>
> $T_f = 614$ $h_f = 0.0542$ m

E6.8 Determine the total commutator losses for a 600 kW, 440 V, 500 rpm, 6 pole generator. Given that commutator diameter = 2.1 m, current density at brush contact = 75 × 10^{-3} A/mm^2, brush pressure = 15.5 kN/m^2, coefficient of friction = 0.32, brush contact drop = 2.8 V.

(P_c = 9.840 kW)

E6.9 Calculate the reactance voltage induced per coil for a single turn two layer winding with two conductors per slot, of a 300 kW, 500 V, 6 pole lap wound DC generator driven at 2650 rpm. The number of armature conductors is 600. The inductance per coil is 0.0065 mH. The brush covers one commutator segment.

If the armature diameter is 2.2 m and core length is 0.35 m, determine the flux density under the interpole. The length of interpole is 0.2 m.

(E_{rav} = 1.6926 V, B_{gi} = 0.14 Wb/m^2)

E6.10 Calculate the size of the conductor and number of turns for the field coil of a 6 poles, 500 V, DC shunt motor. The coil is to supply 4500 AT at the working temperature, where ρ = 0.04 $micro$-ohm-m. The length of the inside turn is 0.82 m, the space factor of the winding is 0.56 and the permissible dissipation per sq.m. of external surface (excluding the two ends) is 1400 W. Solution should not be attempted by assuming a numerical value for the winding depth.

(d_f = 29.91 mm, T_f = 913, a_f = 2.287 mm^2)

> # Chapter 7

DESIGN OF TRANSFORMERS

List of symbols

Symbol	Meaning	Unit
A_c	Area of conductor (or copper) in the window	m^2
A_{cc}	Area of circumscribing circle	m^2
A_{gi}	Gross core area	m^2
A_i	Net core area	m^2
A_w	Window area	m^2
A_Y	Area of yoke	m^2
AT	Ampere turns or Total mmf	AT
AT_o	Total magnetizing mmf	AT
a	Width of the largest stamping	m
a_p	Area of cross-section of primary winding conductor	mm^2
a_s	Area of cross-section of secondary winding conductor	mm^2
at_c	mmf per meter for flux densities in core	AT/m
at_y	mmf per meter for flux densities in yoke	AT/m
B_m	Maximum flux density	Wb/m^2
b	Breadth of the rectangle	m
b_p	Radial width of primary winding	mm
b_s	Radial width of secondary winding	mm
C_1	Clearance between transformer winding and tank along width	mm
C_2	Clearance between transformer winding and tank along length	mm
C_3	Clearance between transformer frame and tank at bottom	mm
C_4	Clearance between transformer frame and tank at top	mm
D	Distance between core centres	m
D_c	Depth of core	m
D_y	Depth of yoke	m
D_{oc}	Outer diameter of coil	m

Symbol	Meaning	Unit
d	Diameter of circumscribing circle	m
d_t	Diameter of cooling tube	m
E	Induced emf	V
E_p	Primary induced emf	V
E_s	Secondary induced emf	V
E_t	Emf per turn	V
f	Frequency	Hz
f_a	Instantaneous axial force	kN
f_{r_av}	Average radial force	kN
f_{r_max}	Instantaneous maximum value of radial force	kN
f_{a_asy}	Instantaneous axial force in assymmetric winding	kN
H	Overall height of the transformer frame	m
H_T	Height of tank	m
H_w	Height of the window or Length of the limb	m
H_Y	Height of the yoke	m
i_{m_sc}	Maximum instantaneous value of fault current	A
I_l	Loss component of no-load current	A
I_m	Magnetizing component of no-load current	A
I_o	No-load current	A
I_p	Current in primary winding	A
I_s	Current in secondary winding	A
I_{LP}	Line current on primary	A
I_{LS}	Line current on secondary	A
I_{LV}	Rated current of low voltage winding	A
K_c	Core area factor	-
K_w	Window space factor	-
K_{pk}	Peak factor	-
L_c	Axial height of winding	m
L_o	Mean circumference of duct	m
L_{mt}	Length of mean turn of transformer winding	m
L_{mtp}	Length of primary winding	m
L_{mts}	Length of secondary winding	m
L_T	Length of tank	m
l_c	Length of flux path through core	m
l_i	Mean length of flux path in iron	m
l_t	Length of cooling tube	m
n_t	Total number of cooling tubes	-
P_c	Copper loss	W

Symbol	Meaning	Unit
P_i	Iron loss	W
R	Reluctance of iron path	-
S_f	Stacking factor	-
S_T	Heat dissipating surface of the tank	m^2
S_{dt}	Total dissipating surface area of cooling tubes	m^2
S_t	Surface area of cooling tubes	m^2
T_p	Number of turns in primary winding	-
T_s	Number of turns in secondary winding	-
T_{HV}	Number of turns in high voltage winding	-
T_{LV}	Number of turns in low voltage winding	-
V_p	Terminal voltage of primary winding	V
V_s	Terminal voltage of secondary winding	V
V_{LP}	Line voltage on primary side	V
V_{LS}	Line voltage on secondary side	V
V_{LV}	Rated voltage of low voltage winding	V
V_{HV}	Rated voltage of high voltage winding	V
W	Length of yoke or Overall width of transformer frame	m
W_c	Width of the core	m
W_T	Width of the tank	m
W_w	Width of the window	m
θ	Allowable temperature rise	m
μ	Permeability	H/m
μ_r	Relative permeability	-
μ_o	Absolute permeability	H/m
δ	Current density	A/mm^2
λ_{conv}	Specific heat dissipation due to convection	$W/m^2\text{-}°C$
λ_{rad}	Specific heat dissipation due to radiation	$W/m^2\text{-}°C$
ϕ_m	Maximum flux in core	Wb
ϕ_o	Flux in duct	Wb
ϕ_p	Leakage flux in primary winding	Wb
ϕ_s	Leakage flux in secondary winding	Wb

7.1 CONSTRUCTION

A transformer consists of two windings coupled through a magnetic medium. The two windings work at different voltage level. The two windings of the transformer are called *High voltage winding* and *Low voltage winding*. Both the windings are wound on a common core. One of the winding is connected to AC supply and it is called *primary*. The other winding is connected to load and it is called *secondary*. The transformer is used to transfer electrical energy from high voltage winding to low voltage winding or vice-versa through magnetic field.

The construction of transformers varies greatly, depending on their applications, winding voltage and current ratings and operating frequencies. The two major types of construction of transformers (used in transmission and distribution of electrical energy) are core type and shell type. Depending on the application, these transformers can be classified as distribution transformers and power transformers.

The transformer is extremely important as a component in many different types of electric circuits, from small-signal electronic circuits to high voltage power transmission systems.

The most important function performed by transformers are,

1. Changing voltage and current level in an electric system.
2. Matching source and load impedances for maximum power transfer in electronic and control circuitry.
3. Electrical isolation between one or more electrical circuits linked by magnetic flux.

7.1.1 Core Type Transformer

In single-phase core type transformer, the magnetic core is built of laminations to form a rectangular frame and the windings are arranged concentrically with each other around the legs or limbs. The top and bottom horizontal portion of the core are called yoke. The yokes connect the two limbs and have a cross sectional area equal to or greater than that of limbs.

Each limb carries one half of primary and secondary. The two windings are closely coupled together to reduce the leakage reactance. The low voltage winding is wound near the core and high voltage winding is wound over low voltage winding away from core in order to reduce the amount of insulating materials required.

7.1.2 Shell Type Transformer

In single-phase shell type transformer the windings are put around the central limb and the flux path is completed through two side limbs. The central limb carries total mutual flux while the side limbs forming a part of a parallel magnetic circuit carry half the total flux. The cross-sectional area of the central limb is twice that of each side limbs.

Table 7.1: Comparison of Core and Shell Type Transformer *(PTU, May' 19, 2 M)*

Core type	Shell type
1. Easy in design and construction.	1. Comparatively complex.
2. Has low mechanical strength due to non-bracing of windings.	2. High mechanical strength.
3. Reduction of leakage reactance is not easily possible.	3. Reduction of leakage reactance is highly possible.
4. The assembly can be easily dismantled for repair work.	4. It cannot be easily dismantled for repair work.
5. Better heat dissipation from windings.	5. Heat is not easily dissipated from windings since it is surrounded by core.
6. Has longer mean length of core and shorter mean length of coil turn. Hence best suited for EHV (Extra High Voltage) requirements.	6. It is not suitable for EHV (Extra High Voltage) requirements.

7.1.3   Distribution Transformer

Transformer upto 200 kVA (or 500 kVA) are used to step down distribution voltage to a standard service voltage or from transmission voltage to distribution voltage are known as ***distribution transformers***. They are kept in operation all the 24 hours a day whether they are carrying any load or not.

The load on the distribution transformer varies from time to time and the transformer will be on no-load most of the time. Hence in distribution transformer the copper loss (which depends on load) will be more when compared to core loss (which occurs as long as transformer is in operation). Hence distribution transformers are designed with less iron loss and designed to have the maximum efficiency at a load much lesser than full load. Also it should have good regulation to maintain the variation of supply voltage with in limits and so it is designed with small value of leakage reactance.

7.1.4   Power Transformer

The transformers used in sub-stations and generating stations are called ***power transformers***. They have ratings above 200 kVA. Usually a sub-station will have number of transformers working in parallel. During heavy load periods all the transformers are put in operation and during light load periods some transformers are disconnected. Therefore the power transformers should be designed to have maximum efficiency at or near full load. Power transformers are designed to have considerably greater leakage reactance that is permissible in distribution transformers in order to limit the fault current. In the case of power transformers inherent voltage regulation is less important than the current limiting effect of higher leakage reactance.

Table 7.2: Comparison of Distribution and Power Transformer *(AKTU, Dec' 19, 5 M) (AU, Apr' 17, 6 M)*

Distribution transformer	Power transformer
1. Power rating upto 200 kVA.	1. Power rating above 200 kVA.
2. Used for distribution of electrical energy so it is placed near load centres.	2. Used for transmission of electrical energy so it is installed in sub-stations and power stations.
3. Designed for high energy efficiency.	3. Designed for high power efficiency.
4. Designed for low value of regulation.	4. Designed for high value of regulation.
5. Designed with small value of leakage reactance.	5. Designed with higher value of leakage reactance.

7.2   kVA OUTPUT FOR SINGLE-PHASE TRANSFORMER   *(HTU, Dec' 18, 10 M)*
(OUTPUT EQUATION OF SINGLE-PHASE TRANSFORMER)   *(RGPV, Jun' 18, 10 M)*

The equation which relates the rated *kVA* output of a transformer to the area of core and window is called output equation. In transformers the output *kVA* depends on flux density and ampere-turns. The flux density is related to core area and the ampere-turns is related to window area.

The simplified cross-section of core type and shell type single-phase transformers are shown in Figs. 7.1 and 7.2. The low voltage winding is placed nearer to the core in order to reduce the insulation requirement. The space inside the core is called window and it is the space available for accommodating the primary and secondary winding. The window area is shared between the winding and their insulations.

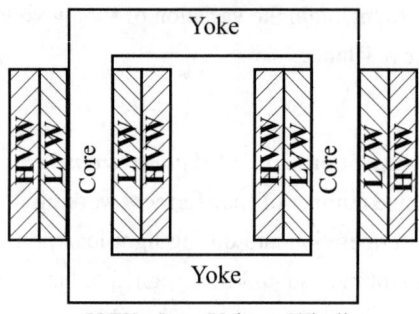

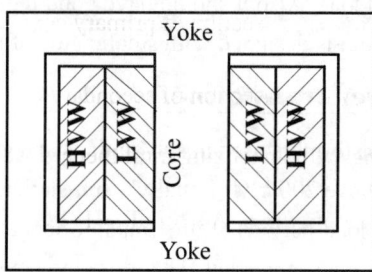

LVW - Low Voltage Winding   ;   HVW - High Voltage Winding

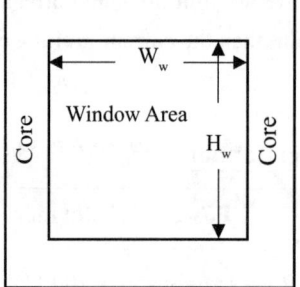

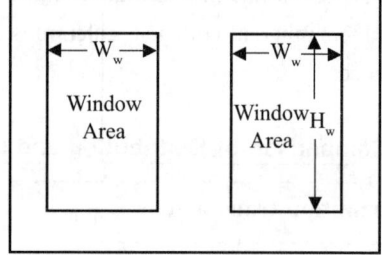

Fig. 7.1: Cross-section of core type single-phase transformer.

Fig. 7.2: Cross-section of shell type single-phase transformer.

We know that,

Induced emf in a winding of transformer, $E = 4.44 \, f \, \phi_m \, T$

∴ Emf per turn, $E_t = E \, / \, T = 4.4 \, f \, \phi_m$

The induced emf and emf per turn in primary and secondary windings are,

$$E_p = 4.44 \, f \, \phi_m \, T_p \qquad\qquad .....(7.1)$$

$$\therefore E_t = \frac{E_p}{T_p} = 4.44 \, f \, \phi_m \qquad\qquad .....(7.2)$$

$$E_s = 4.44 \, f \, \phi_m \, T_s$$

$$\therefore E_t = \frac{E_s}{T_s} = 4.44 \, f \, \phi_m$$

The window in single-phase transformer contains one primary and one secondary winding. The window space factor, K_w is the ratio of conductor area in window to total area of window.

$$\text{Window space factor, } K_w = \frac{\text{Conductor area in window}}{\text{Total area of window}} = \frac{A_c}{A_w} \qquad \qquad .....(7.3)$$

$$\therefore \text{ Conductor area in window, } A_c = K_w A_w \qquad \qquad .....(7.4)$$

The current density, δ is same in both the windings.

$$\therefore \text{ Current density, } \delta = \frac{I_p}{a_p} = \frac{I_s}{a_s} \qquad \qquad .....(7.5)$$

$$\text{Area of cross-section of primary conductor, } a_p = \frac{I_p}{\delta} \qquad \qquad .....(7.6)$$

$$\text{Area of cross-section of secondary conductor, } a_s = \frac{I_s}{\delta} \qquad \qquad .....(7.7)$$

If we neglect magnetizing mmf then primary ampere-turns is equal to secondary ampere-turns.

$$\therefore \text{ Ampere-turns, } AT = I_p T_p = I_s T_s \qquad \qquad .....(7.8)$$

Conductor area in window, $A_c = \begin{array}{c}\text{Conductor area of} \\ \text{primary winding}\end{array} + \begin{array}{c}\text{Conductor area of} \\ \text{secondary winding}\end{array}$

$$= \begin{bmatrix}\text{Number of primary} \\ \text{turns} \times \text{area of} \\ \text{cross-section of} \\ \text{primary conductor}\end{bmatrix} + \begin{bmatrix}\text{Number of secondary} \\ \text{turns} \times \text{area of} \\ \text{cross-section of} \\ \text{secondary conductor}\end{bmatrix}$$

$$= T_p a_p + T_s a_s = T_p \frac{I_p}{\delta} + T_s \frac{I_s}{\delta} \qquad \boxed{\begin{array}{c}\text{Using equations} \\ (7.6) \text{ and } (7.7)\end{array}}$$

$$= \frac{1}{\delta}(T_p I_p + T_s I_s) = \frac{1}{\delta}(AT + AT) \qquad \boxed{\text{Using equation (7.8)}}$$

$$= \frac{2AT}{\delta} \qquad \qquad .....(7.9)$$

On equating equations (7.4) and (7.9) we get,

$$K_w A_w = 2\frac{AT}{\delta}$$

$$\therefore \text{ Ampere-turns, } AT = \frac{1}{2} K_w A_w \delta \qquad \qquad .....(7.10)$$

The kVA rating of single-phase transformer is given by,

$$\text{kVA rating, } Q = V_p I_p \times 10^{-3} \approx E_p I_p \times 10^{-3} \qquad \boxed{E_p \approx V_p}$$

$$= 4.44 \text{ f } \phi_m \, T_p \, I_p \times 10^{-3} \qquad \boxed{\text{Using equation (7.1)}}$$

$$\therefore Q = 4.44 \, f \, \phi_m \, AT \times 10^{-3}$$

$AT = I_p \, T_p$
Substituting for AT from equation (7.10).

$$= 4.44 \, f \, \phi_m \, \frac{K_w A_w \delta}{2} \times 10^{-3}$$

$B_m = \dfrac{\phi_m}{A_i}$

$$= 2.22 \, f \, \phi_m \, K_w \, A_w \, \delta \times 10^{-3}$$

$$= 2.22 \, f \, B_m \, A_i \, K_w \, A_w \, \delta \times 10^{-3} \qquad \qquad .....(7.11)$$

The equation (7.11) is the output equation of single-phase transformer.

7.3 kVA OUTPUT FOR THREE-PHASE TRANSFORMER (OUTPUT EQUATION OF THREE-PHASE TRANSFORMER)

(VTU, Dec' 19, 6 M) (AU, Nov' 18, 6 M)
(PTU, May' 19, 5 M)
(UTU, Dec' 13, 10 M) (PU, Nov' 19, 6 M) (KTU, Feb' 18, 4 M) (MU, May' 19, 10 M) (JNKTU, Apr' 19, 7 M)

The simplified cross-section of core type three-phase transformer is shown in Fig. 7.3. The cross-section has three limbs and two windows. Each limb carries the low voltage and high voltage winding of a phase.

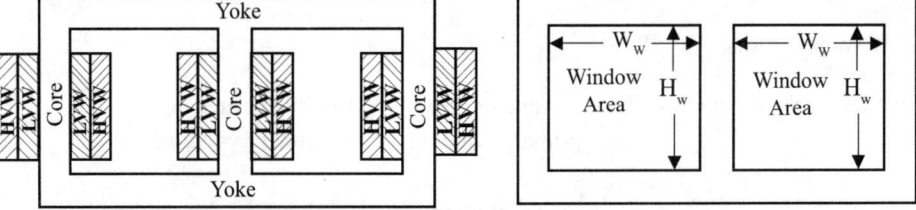

LVW - Low Voltage Winding   ;   HVW - High Voltage Winding

Fig. 7.3: Cross-section of core type three-phase transformer.

We know that,

Induced emf per phase in a winding, $E = 4.44 \, f \, \phi_m \, T$ .....(7.12)

$$\therefore \text{Emf per turn, } E_t = \frac{E}{T} = 4.44 \, f \, \phi_m$$

The induced emf and emf per turn in primary and secondary windings per phase are,

$$E_p = 4.44 \, f \, \phi_m \, T_p \qquad \qquad .....(7.13)$$

$$\therefore E_t = \frac{E_p}{T_p} = 4.44 \, f \, \phi_m \qquad \qquad .....(7.14)$$

$$E_s = 4.44 \, f \, \phi_m \, T_s$$

$$\therefore E_t = \frac{E_s}{T_s} = 4.44 \, f \, \phi_m$$

In case of three-phase transformer, each window has two primary and two secondary windings. The window space factor, K_w is the ratio of conductor area in window to total area of window,

$$K_w = \frac{\text{Conductor area in window}}{\text{Total area of window}} = \frac{A_c}{A_w} \qquad \qquad .....(7.15)$$

$\therefore$ Conductor area in window, $A_c = K_w A_w$ .....(7.16)

> **Note :** *In three-phase transformers, one window is considered.*

The current density, δ is same in both the windings.

$\therefore$ Current density, $\delta = \dfrac{I_p}{a_p} = \dfrac{I_s}{a_s}$ .....(7.17)

where, I_p = Primary current per phase

I_s = Secondary current per phase

Area of cross-section of primary conductor, $a_p = \dfrac{I_p}{\delta}$ .....(7.18)

Area of cross-section of secondary conductor, $a_s = \dfrac{I_s}{\delta}$ .....(7.19)

If we neglect magnetizing mmf then primary ampere-turns per phase is equal to secondary ampere-turns per phase.

$\therefore$ Ampere-turns, $AT = I_p T_p = I_s T_s$ .....(7.20)

Conductor area in window, $A_c = \left(\begin{array}{l} 2 \times \text{Number of} \\ \text{primary turns} \times \text{area} \\ \text{of cross-section of} \\ \text{primary conductor} \end{array} \right) + \left(\begin{array}{l} 2 \times \text{Number of} \\ \text{secondary turns} \times \text{area} \\ \text{of cross-section of} \\ \text{secondary conductor} \end{array} \right)$

$= 2\, T_p\, a_p + 2\, T_s\, a_s$

$= 2\, T_p \dfrac{I_p}{\delta} + 2\, T_s \dfrac{I_s}{\delta}$ | Using equations (7.18) and (7.19) |

$= \dfrac{2}{\delta}(T_p I_p + T_s I_s)$

$= \dfrac{2}{\delta}(AT + AT)$ | Using equation (7.20) |

$= \dfrac{4\, AT}{\delta}$ ....(7.21)

On equating equations (7.16) and (7.21) we get,

$\dfrac{4AT}{\delta} = K_w A_w$

$\therefore$ Ampere-turn, $AT = \dfrac{K_w A_w \delta}{4}$ .....(7.22)

The kVA rating of three-phase transformer is given by,

$$\text{kVA rating, } Q = 3 \times \text{Volt–ampere per phase} \times 10^{-3} = 3\, V_p\, I_p \times 10^{-3}$$

$$= 3\, E_p\, I_p \times 10^{-3}$$

$$\boxed{E_p \approx V_p}$$

$$= 3 \times 4.44\, f\, \phi_m\, T_p\, I_p \times 10^{-3}$$

$$\boxed{\text{Using equation (7.13)}}$$

$$= 3 \times 4.44\, f\, \phi_m\, AT \times 10^{-3}$$

$$\boxed{AT = I_p\, T_p}$$

$$\boxed{\begin{array}{l}\text{Substituting for AT from}\\ \text{equation (7.22).}\end{array}}$$

$$= 3 \times 4.44\, f\, \phi_m \times \frac{K_w\, A_w\, \delta}{4} \times 10^{-3}$$

$$= 3.33\, f\, \phi_m\, K_w\, A_w\, \delta \times 10^{-3}$$

$$\boxed{B_m = \frac{\phi_m}{A_i}}$$

$$= 3.33\, f\, B_m\, A_i\, K_w\, A_w\, \delta \times 10^{-3} \qquad \dots(7.23)$$

The equation (7.23) is the output equation of three-phase transformer.

7.3.1 Emf Per Turn   *(VTU, Dec' 19, 6 M)*

The transformer design starts with the selection of an appropriate value for emf per turn. Hence an equation for emf per turn can be developed by relating output kVA, magnetic and electric loading. In transformers the ratio of specific magnetic and electric loading is specified rather than actual value of specific loadings.

$$\text{Ratio of specific magnetic and electric loading, } r = \frac{\phi_m}{AT} \qquad \dots(7.24)$$

The volt-ampere per phase of a transformer is given by the product of voltage and current per phase. Considering the primary voltage and current per phase we can write,

$$\text{kVA per phase, } Q = V_p\, I_p \times 10^{-3} = E_p\, I_p \times 10^{-3}$$

$$\boxed{V_p \approx E_p}$$

$$= 4.44\, f\, \phi_m\, T_p\, I_p \times 10^{-3}$$

$$\boxed{\text{Using equation (7.13)}}$$

$$= 4.44\, f\, \phi_m\, AT \times 10^{-3}$$

$$\boxed{\text{Using equation (7.20)}}$$

$$= 4.44\, f\, \phi_m \times \frac{\phi_m}{r} \times 10^{-3}$$

$$\boxed{\text{Using equation (7.24)}}$$

On rewriting the above equation we get,

$$\phi_m^2 = \frac{Q\, r}{4.44\, f \times 10^{-3}}$$

$$\therefore \phi_m = \sqrt{\frac{Q\, r \times 10^3}{4.44\, f}} \qquad \dots(7.25)$$

We know that,

Emf per turn, $E_t = 4.44\ f\ \phi_m$ Using equation (7.14)

$$= 4.44\ f\ \sqrt{\dfrac{Q\,r \times 10^3}{4.44\ f}}$$ Using equation (7.25)

$$= \sqrt{4.44\ f\ r \times 10^3}\ \sqrt{Q} \qquad \qquad .....(7.26)$$

$$= K\ \sqrt{Q} \qquad \qquad .....(7.27)$$

where, $K = \sqrt{4.44\ f\ r \times 10^3} \qquad \qquad .....(7.28)$

     Using equation (7.24)

$$= \sqrt{4.44\ f \times \dfrac{\phi_m}{AT} \times 10^3} \qquad \qquad .....(7.29)$$

From equation (7.27) we can say that the emf per turn is directly proportional to K. The value of K depends on the type, service condition and method of construction of transformer. The value of K for different types of transformers are listed in Table 7.3.

Table 7.3: Values of K for different types of Transformer

Transformer type	K
Single-phase shell type	1.0 to 1.2
Single-phase core type	0.75 to 0.85
Three-phase shell type	1.3
Three-phase core type, distribution transformer	0.45
Three-phase core type, power transformer	0.6 to 0.7

Note : In equation (7.27) the Q is kVA rating for single-phase transformer and Q is kVA per phase for three-phase transformer.

7.4 OVERALL DIMENSIONS

The overall dimensions can be expressed in terms of following main dimensions of the transformer.

     H_w = Height of window

     W_w = Width of window

     a   = Width of largest stamping

     d   = Diameter of circumscribing circle

     D = Distance between core centres

     H_y = Height of yoke

     D_y = Depth of yoke

The overall dimesnsions of transformer are overall height and width of transformer frame.

Let,   H = Overall height of transformer frame

     W = Overall width of transformer frame

These dimensions for various types of transformers are shown in Figs. 7.4 to 7.6.

Figure 7.4 shows a vertical and horizontal cross-section of the core and winding assembly of a core type single-phase transformer. The Fig. 7.5 shows a vertical and horizontal cross-section of the core and winding assembly of a core type three-phase transformer. Figure 7.6 shows a vertical and horizontal cross-section of a shell type single-phase transformer.

For single-phase core type transformer,

$\qquad$ Overall height of transformer frame, $H = H_w + 2H_y$

$\qquad$ Overall width of transformer frame, $W = W_w + 2a$

For single-phase shell type transformer,

$\qquad$ Overall height of transformer frame, $H = H_w + 2a$

$\qquad$ Overall width of transformer frame, $W = 2(W_w + a)$

For three-phase core type transformer,

$\qquad$ Overall height of transformer frame, $H = H_w + 2H_y$

$\qquad$ Overall width of transformer frame, $W = 2W_w + 3a$

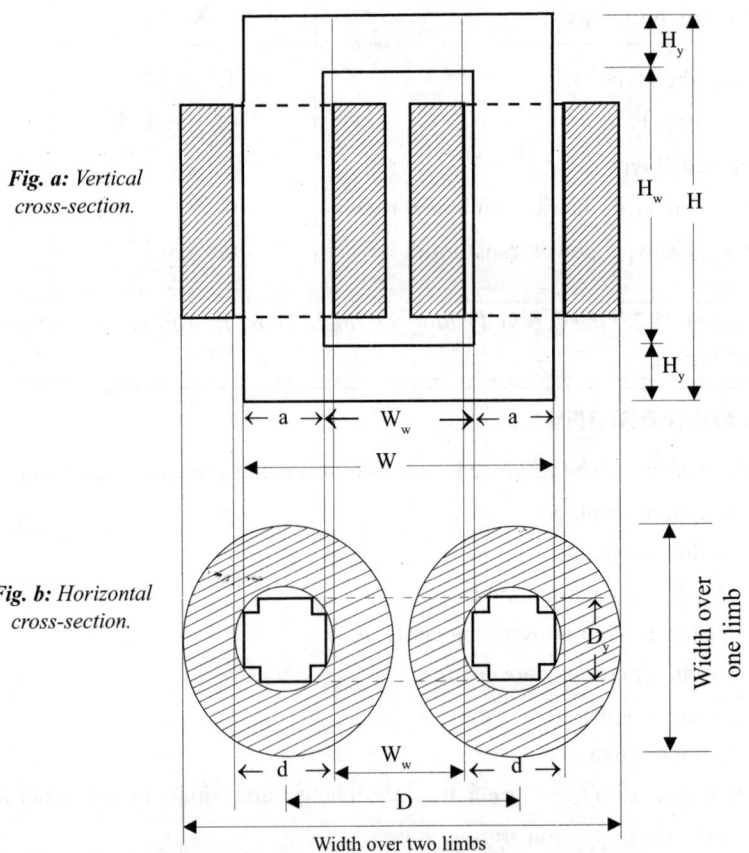

Fig. a: Vertical cross-section.

Fig. b: Horizontal cross-section.

Fig. 7.4: Single-phase core type transformer.

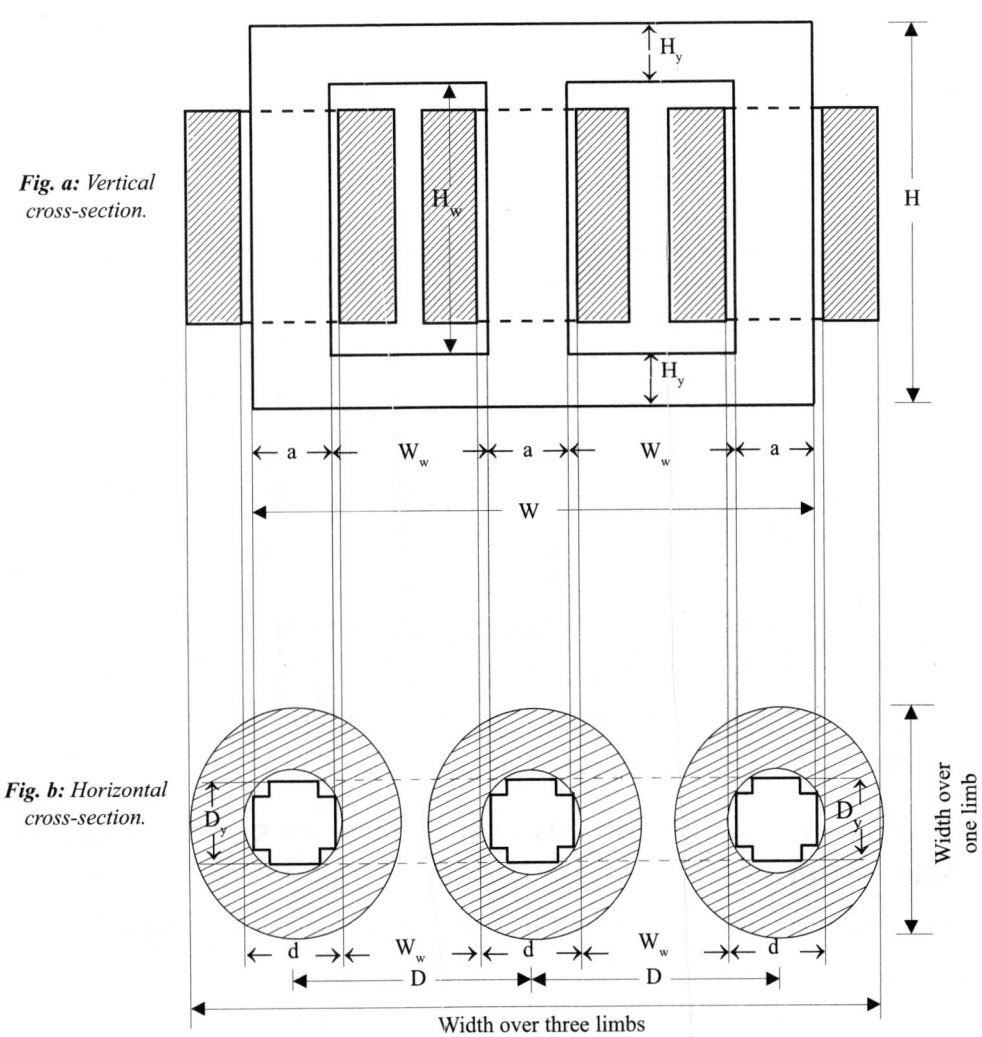

Fig. a: Vertical cross-section.

Fig. b: Horizontal cross-section.

Fig. 7.5: *Three-phase core type transformer.*

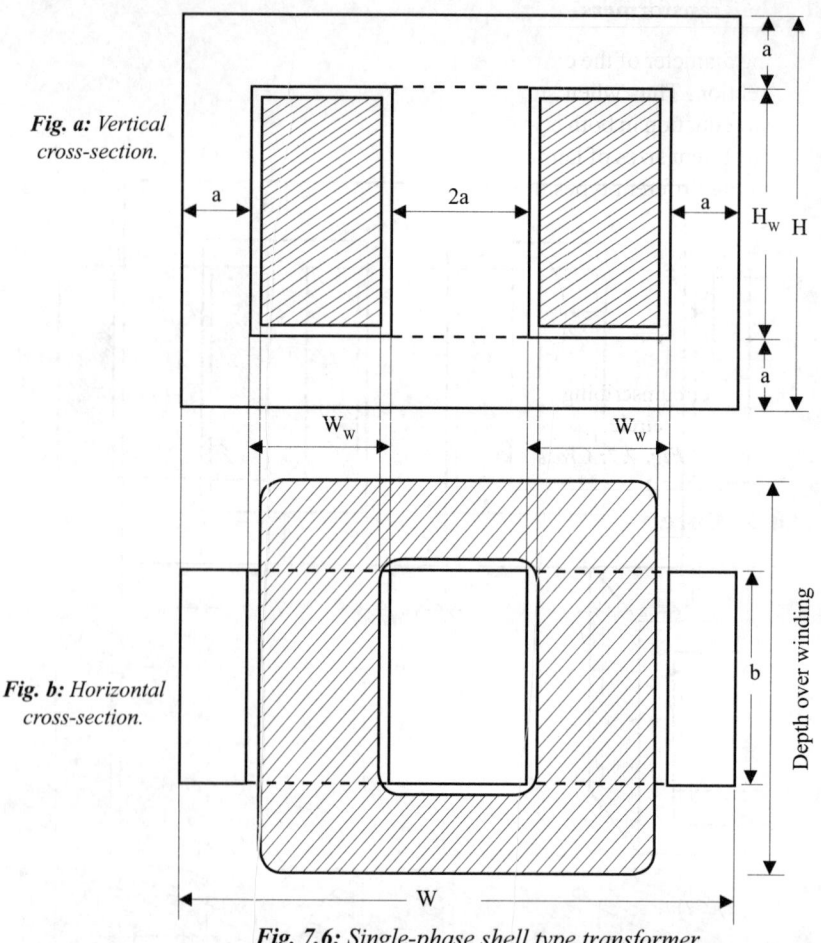

Fig. a: Vertical cross-section.

Fig. b: Horizontal cross-section.

Fig. 7.6: Single-phase shell type transformer.

7.5 DESIGN OF YOKE, CORE AND WINDING FOR CORE AND SHELL TYPE TRANSFORMERS

Note : *Design procedures for yoke, core and winding of core and shell type transformer are same.*

7.5.1 Design of Cores

For core type transformer the cross-section may be rectangular, square or stepped. When circular coils are required for distribution and power transformers, the square and stepped cores are used.

For shell type transformer the cross-section may be rectangular. When rectangular cores are used the coils are also rectangular in shape. The rectangular core is suitable for small and low voltage transformers. In core type transformer with rectangular cores, the ratio of depth to width of the core is 1.4 to 2. In shell type transformers with rectangular cores the width of the central limb is 2 to 3 times the depth of the core. Figure 7.7 shows the cross-section of transformer cores.

The excessive leakage fluxes produced during short circuit and over loads, develops severe mechanical stresses on the coils. On circular coils these forces are radial and there is no tendency for the coil to change its shape. But on rectangular and square coils the forces are perpendicular to the conductors and tends to deform the shape of coil. Hence circular coils are employed in high voltage and high capacity transformers.

In square cores the diameter of the circumscribing circle is larger than the diameter of stepped cores of same area of cross-section. Thus when stepped cores are used the length of mean turn of winding is reduced with consequent reduction in both cost of copper and copper loss. However with larger number of steps a large number of different sizes of laminations have to be used. This results in higher labour charges for shearing and assembling different types of laminations.

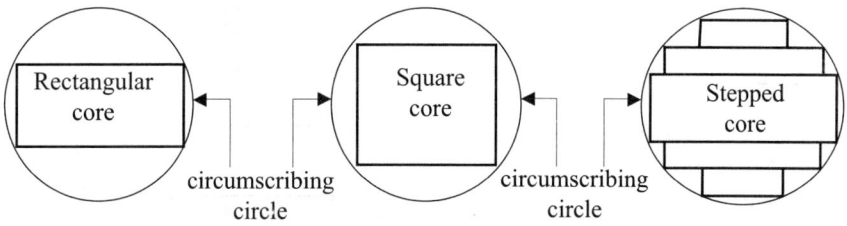

Fig. 7.7: Cross-section of transformer cores.

7.5.2 Rectangular Core

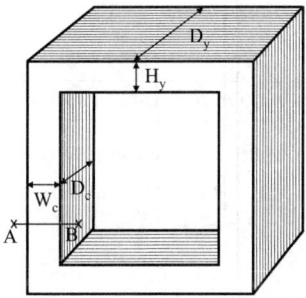

Fig. a: Rectangular core.

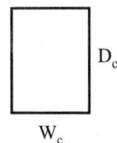

Fig. b: Cross-section at A-B.

Fig. 7.8: *Rectangular core of a single-phase core type transformer.*

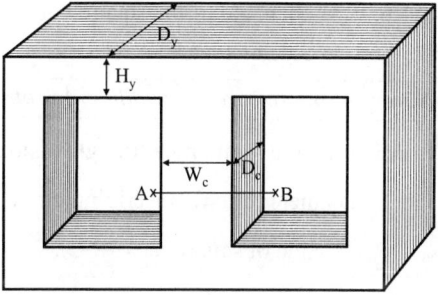

Fig. a: Rectangular core.

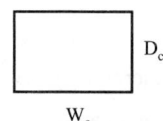

Fig. b: Cross-section at A-B.

Fig. 7.9: *Rectangular core of a single-phase shell type transformer.*

Let, D_c = Depth of core

W_c = Width of core

For core type transformers,

D_c = 1.4 to 2 times W_c

For shell type transformers,

W_c = 2 to 3 times D_c

In transformers with rectangular cores, the coils are made in rectangular shape. Therefore, the design of core is straight forward by calculating the area of core to maintain the required flux density. Then, the width and depth of core can be estimated from the core area to satisfy the given ratio of depth to width of core.

7.5.3 Square Core

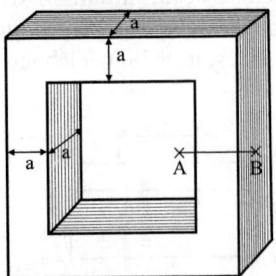

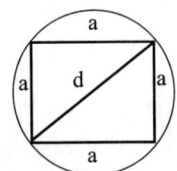

Fig. a: Square core.   *Fig. b: Cross-section at AB.*

Fig. 7.10: *Square core of a single-phase core type transformer.*

Let, d = Diameter of circumscribing circle

Also, d = Diagonal of the square core

   a = Side of square

Note : *In square core the width and depth of core are equal and equal to side of square.*

Considering the right angle triangle inside the square shown in Fig. 7.10 we can write,

Diameter of circumscribing circle, $d = \sqrt{a^2 + a^2} = \sqrt{2\,a^2} = \sqrt{2}\,a$

$\therefore$ Side of square, $a = \dfrac{d}{\sqrt{2}}$ .....(7.30)

Gross core area, A_{gi} = Area of square = a^2

$\qquad\qquad = \left(\dfrac{d}{\sqrt{2}}\right)^2$ | Using equation (7.30) |

$\qquad\qquad = 0.5\ d^2$ .....(7.31)

Let stacking factor, $S_f = 0.9$

Net core area, A_i = Stacking factor × Gross core area

$\qquad\qquad = 0.9 \times 0.5\ d^2$ | Using equation (7.31) |

$\qquad\qquad = 0.45\ d^2$ .....(7.32)

Note : *The gross core area is the area including insulation area and net core area is the area of iron alone excluding insulation area.*

Area of circumscribing circle, $A_{cc} = \dfrac{\pi}{4}\,d^2$

$\therefore \dfrac{\text{Net core area}}{\text{Area of circumscribing circle}} = \dfrac{A_i}{A_{cc}} = \dfrac{0.45\ d^2}{(\pi/4)\,d^2} = 0.57$

$\therefore \dfrac{\text{Gross core area}}{\text{Area of circumscribing circle}} = \dfrac{A_{gi}}{A_{cc}} = \dfrac{0.5\ d^2}{(\pi/4)d^2} = 0.64$

$\therefore$ Core area factor, $K_c = \dfrac{\text{Net core area}}{\text{Square of circumscribing circle}} = \dfrac{A_i}{d^2} = \dfrac{0.45\ d^2}{d^2} = 0.45$

7.5.4 Two Stepped Core or Cruciform Core

In stepped cores the dimensions of the steps should be chosen, such as to occupy the maximum area within a circle. The dimensions of the two step to give maximum area for the core in the given area of circle (or for a given diameter of circle) are determined as follows.

Let,   a = Length of the rectangle

        b = Breadth of the rectangle

        d = Diameter of the circumscribing circle

Also, d = Diagonal of the rectangle

        θ = Angle between the diagonal and length of the rectangle.

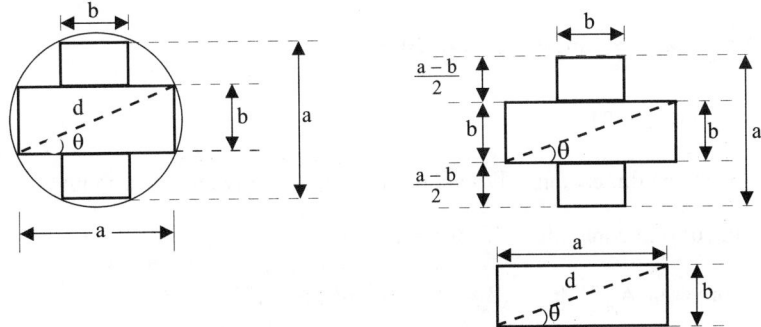

Fig. 7.11: Cross-section of two stepped core.

The cross-section of two stepped core is shown in Fig. 7.11. The maximum core area for a given diameter, d is obtained when θ is maximum value. Hence differentiate gross core area, A_{gi} with respect to θ and equate to zero to solve for maximum value of θ.

From Fig. 7.11 we get,

$$\cos \theta = \frac{a}{d} \quad \Rightarrow \quad a = d \cos \theta \qquad \qquad .....(7.33)$$

$$\sin \theta = \frac{b}{d} \quad \Rightarrow \quad b = d \sin \theta \qquad \qquad .....(7.34)$$

The two stepped core can be divided into three rectangles. The area of three rectangles gives the gross core area. With reference to Fig. 7.11 we can write,

$$\text{Gross core area, } A_{gi} = ab + \left(\frac{a-b}{2}\right)b + \left(\frac{a-b}{2}\right)b = ab + \frac{2(a-b)}{2}b$$

$$= ab + ab - b^2 = 2\,ab - b^2$$

$$= 2\,(d\cos\theta)(d\sin\theta) - (d\sin\theta)^2 \qquad \boxed{\text{Using equations (7.33) and (7.34)}}$$

$$= d^2\, 2\sin\theta\cos\theta - d^2\sin^2\theta \qquad \boxed{\sin2\theta = 2\sin\theta\cos\theta}$$

$$= d^2\sin2\theta - d^2\sin^2\theta \qquad \qquad .....(7.35)$$

To get maximum value of θ, differentiate A_{gi} with respect to θ, and equate to zero,

$$\therefore \quad \frac{d}{d\theta} A_{gi} = 0 \qquad\qquad\qquad\qquad\qquad\qquad\qquad .....(7.36)$$

On differentiating equation (7.35) with respect to θ we get,

$$\frac{d}{d\theta} A_{gi} = d^2 \cos 2\theta \times 2 - d^2 2 \sin\theta \cos\theta$$

> Using equation (7.36)

Let, $d^2 \cos 2\theta \times 2 - d^2\, 2\sin\theta \cos\theta = 0$

$$\therefore \quad d^2\, 2 \sin\theta \cos\theta = d^2 \cos 2\theta \times 2 \quad\Rightarrow\quad d^2 \sin 2\theta = d^2 \cos 2\theta \times 2$$

$$\therefore \quad \frac{\sin 2\theta}{\cos 2\theta} = 2 \quad\Rightarrow\quad \tan 2\theta = 2 \quad\Rightarrow\quad 2\theta = \tan^{-1} 2$$

$$\therefore \quad \theta = \frac{1}{2}\tan^{-1} 2 = 31.72°$$

When $\theta = 31.72°$, the dimensions of the core (a and b) will give the maximum area for core.

Using the value of θ in equation (7.35) we get,

$$\text{Gross core area, } A_{gi} = d^2 \sin(2 \times 31.72°) - d^2 \sin 31.72°$$

$$= d^2 (\sin(2 \times 31.72°) - \sin 31.72°) = 0.618\, d^2$$

Let,

Stacking factor, $S_f = 0.9$

Net core-area , $A_i = $ Stacking factor $\times$ Gross core area

$$= 0.9 \times 0.618\, d^2 = 0.56\, d^2$$

Note :

When $\theta = 31.72°$,	$a = d \cos\theta$	$b = d \sin\theta$	
	$= d \cos 31.72°$	$= d \sin\ 31.72°$	Using equations (7.33) and (7.34)
	$= 0.85\, d$	$= 0.53\, d$	

$$\therefore \quad \frac{\text{Net core area}}{\text{Area of circumscribing circle}} = \frac{0.56\, d^2}{(\pi/4)\, d^2} = 0.71 \qquad\qquad .....(7.37)$$

$$\therefore \quad \frac{\text{Gross core area}}{\text{Area of circumscribing circle}} = \frac{0.618\, d^2}{(\pi/4)\, d^2} = 0.79 \qquad\qquad .....(7.38)$$

$$\therefore \text{ Core area factor, } K_c = \frac{\text{Net core area}}{\text{Square of circumscribing circle}}$$

$$= \frac{A_i}{d^2} = \frac{0.56\, d^2}{d^2} = 0.56 \qquad\qquad .....(7.39)$$

7.5.5 Multi-Stepped Cores

We can prove that the area of circumscribing circle is more effectively utilised by increasing the number of steps. The most economical dimensions of various steps for a multistepped core can be calculated as shown for cruciform (or two stepped) core. The results are tabulated in Table 7.4.

Table 7.4: Ratios of Area of Core and Circumscribing Circle

Ratio	Square core	Cruciform core	3-stepped core	4-stepped core
$\dfrac{\text{Gross core area}}{\text{Area of circumscribing circle}} = \dfrac{A_{gi}}{A_{cc}}$	0.64	0.79	0.84	0.87
$\dfrac{\text{Net core area}}{\text{Area of circumscribing circle}} = \dfrac{A_i}{A_{cc}}$	0.58	0.71	0.75	0.78
$K_c = \dfrac{\text{Net core area}}{\text{Square of circumscribing circle}} = \dfrac{A_i}{d^2}$	0.45	0.56	0.6	0.62

Choice of flux density in the core
(AKTU, Dec' 19, 3 M) (MU, May' 19, 5 M)

The flux density decides the area of cross-section of core and core loss. Higher values of flux density results in smaller core area, lesser cost, reduction in length of mean turn of winding, higher iron loss and large magnetizing current.

The choice of flux density depends on the service condition (i.e., distribution or transmission) and the material used for laminations of the core. The laminations made with cold rolled silicon steel can work with higher flux densities than the laminations made with hot rolled silicon steel. Usually the distribution transformers will have low flux density to achieve lesser iron loss.

When hot rolled silicon steel is used for laminations the following values can be used for maximum flux density (B_m).

B_m = 1.1 to 1.4 Wb/m^2 - For distribution transformers

B_m = 1.2 to 1.5 Wb/m^2 - For power transformers

When cold rolled silicon steel is used for laminations the following values can be used for maximum flux density (B_m).

B_m = 1.55 Wb/m^2 - For transformers with voltage rating upto 132 kV

B_m = 1.6 Wb/m^2 - For transformers with voltage rating 132 kV to 275 kV

B_m = 1.7 Wb/m^2 - For transformers with voltage rating 275 kV to 400 kV

7.5.6 Design of Yoke

The core of the transformer can be divided into vertical limbs that carry the windings and the top and horizontal sections that simply provides closed path for flux. The vertical limbs are considered as actual core and the design of core is the design of vertical limbs.

The top and bottom horizontal portion of the core are called yoke (Refer Figs. 7.1 to 7.3). The flux density in the yoke is made low by increasing the area of yoke by 10 to 15% higher than that of core or vertical limbs. The reduction of flux density will lead to reduction in magnetizing current and iron loss.

In small capacity transformers the cross-section of core and yoke will be of same shape. i.e., if cross-section of core is cruciform then cross-section of yoke is also cruciform. In large capacity transformers, multi-stepped core is employed to reduce the length of mean turn. But the yoke is made of simpler cross-section in order to reduce cost. Four types of cross-section of yoke are used in practice: rectangle, cruciform, rectangular section with a step upward and rectangular section with a step downward.

The area of yoke is chosen either same as gross core area or slightly higher.

$$\text{Area of yoke, } A_y = (1.15 \text{ to } 1.25) \times A_{gi} \qquad - \text{ For hot rolled steel}$$

$$\text{Area of yoke, } A_y = A_{gi} \qquad - \text{ For grain oriented steel}$$

Let, D_y = Depth of yoke

H_y = Height of yoke

Now, the area of yoke in terms of depth and height is,

$$\text{Area of yoke, } A_y = D_y \, H_y \qquad \qquad .....(7.40)$$

For rectangular cores,

$$D_y = \text{Depth of core} = D_c$$

$$\therefore \text{ Height of yoke, } H_y = \frac{A_y}{D_y} = \frac{A_y}{D_c} \qquad \boxed{\text{Using equation (7.40)}}$$

For square or stepped cores,

$$D_y = \text{Width of largest stamping} = a$$

$$\therefore \text{ Height of yoke, } H_y = \frac{A_y}{D_y} = \frac{A_y}{a} \qquad \boxed{\text{Using equation (7.40)}}$$

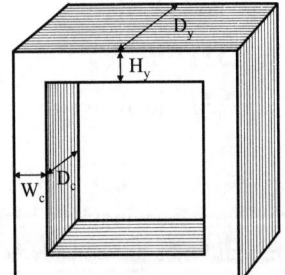

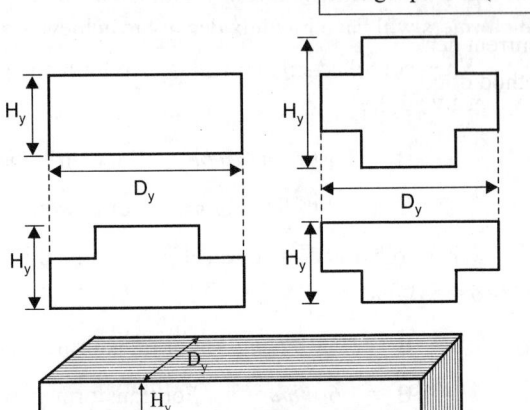

Fig. 7.12: Cross-section of yoke.

7.6 DESIGN OF WINDING

The transformer has one high voltage winding and one low voltage winding. The design of winding involves the determination of number of turns and area of cross-section of the conductor used for winding. The number of turns is estimated using voltage rating and emf per turn (or by using ampere-turns and rated current). The area of cross-section is estimated using rated current and current density.

Usually the number of turns of low voltage winding is estimated first using the given data and it is corrected to nearest integer. Then the number of turns of high voltage winding are chosen to satisfy the voltage rating of the transformer.

$$\text{Number of turns in low voltage winding, } T_{LV} = \frac{V_{LV}}{E_t} \text{ or } \frac{AT}{I_{LV}} \qquad \qquad .....(7.41)$$

where, V_{LV} = Rated voltage of low voltage winding

I_{LV} = Rated current of low voltage winding

$$\text{Number of turns in high voltage winding, } T_{HV} = T_{LV} \times \frac{V_{HV}}{V_{LV}} \qquad \qquad ....(7.42)$$

where, V_{HV} = Rated voltage of high voltage winding

Note : In step-up transformer, $T_{LV} = T_P,\ V_{LV} = V_P,\ T_{HV} = T_S$ *and* $V_{HV} = V_S$

In step-down transformer, $T_{HV} = T_p,\ V_{HV} = V_P,\ T_{LV} = T_S$ *and* $V_{LV} = V_S$

$$\text{Rated current in a winding} = \frac{\text{kVA per phase} \times 10^3}{\text{Voltage rating of the winding}} \qquad \qquad ....(7.43)$$

The area of cross-section of primary and secondary winding conductors are estimated by assuming a current density. The choice of current density depends on the allowable temperatue rise, copper loss and method of cooling. The range of current density for various types of transformers are given below.

δ = 1.1 to 2.2 *A/mm²* - For distribution transformers.

δ = 1.1 to 2.2 *A/mm²* - For small power transformers with self oil cooling.

δ = 2.2 to 3.2 *A/mm²* - For large power transformers with self oil cooling or air-blast.

δ = 5.4 to 6.2 *A/mm²* - For large power transformers with forced circulation of oil or with water cooling coils.

$$\text{Area of cross-section of primary winding conductor, } a_p = \frac{I_p}{\delta} \qquad \qquad ....(7.44)$$

$$\text{Area of cross-section of secondary winding conductor,} a_s = \frac{I_s}{\delta} \qquad \qquad ....(7.45)$$

Note : In transformers same current density is assumed for primary and secondary.

7.6.1   Resistance of Transformer Winding

Let, L_{mtp}, L_{mts} = Length of primary and secondary windings respectively

T_p, T_s = Number of turns in primary and secondary windings respectively

a_p, a_s = Area of cross-section of primary and secondary conductor respectively.

The resistance can be estimated using the basic equation for resistance of a metallic wire, $R = \rho l/a$, where, ρ is resistivity, l is length and a is area of cross-section.

Total length of winding conductor is given by product of number of turns in the winding and length of mean turn of the winding.

$$\therefore \text{ Resistance of primary winding, } r_p = \rho \times \frac{T_p \, L_{mtp}}{a_p}$$

$$\therefore \text{ Resistance of secondary winding, } r_s = \rho \times \frac{T_s \, L_{mts}}{a_s}$$

Total I²R loss in windings, $P_c = I_p^{\,2} \, r_p + I_s^{\,2} \, r_s$
(Total copper loss)

$$\therefore \text{ Total resistance of transformer referred to primary, } R_p = \frac{P_c}{I_p^{\,2}} = \frac{I_p^{\,2} \, r_p + I_s^{\,2} \, r_s}{I_p^{\,2}}$$

$$= r_p + \left(\frac{I_s}{I_p}\right)^2 r_s = r_p + \frac{1}{k^2} \, r_s$$

$$\therefore \text{ Total resistance of transformer referred to secondary, } R_s = \frac{P_c}{I_s^{\,2}} = \frac{I_p^{\,2} \, r_p + I_s^{\,2} \, r_s}{I_s^{\,2}}$$

$$= \left(\frac{I_p}{I_s}\right)^2 r_p + r_s = k^2 \, r_p + r_s$$

where, k = Transformer ratio

$$= \frac{V_s}{V_p} = \frac{T_s}{T_p} = \frac{I_p}{I_s}$$

$$\text{Per unit resistance, } \varepsilon_r = \frac{I_p \, R_p}{V_p} = \frac{I_s \, R_s}{V_s} \qquad\qquad\qquad .....(7.46)$$

7.6.2 Leakage Reactance of Transformer Winding

The flux developed in transformer can be divided into useful flux and leakage flux. The useful flux will link both primary and secondary and help to transfer of power between windings. The leakage flux will link only one of the winding and so will not help to transfer of power or energy. Due to this leakage flux in transformer, there will be a self reactance in each winding. This self reactance of transformer winding is alternatively known as leakage reactance of transformer.

7.6.3 Leakage Flux in Transformer Winding

The estimation of leakage reactance requires knowledge of the distribution of leakage flux in the primary and secondary windings. The distribution of the leakage flux depends upon the geometrical configuration of the coils and the width of duct detween coils.

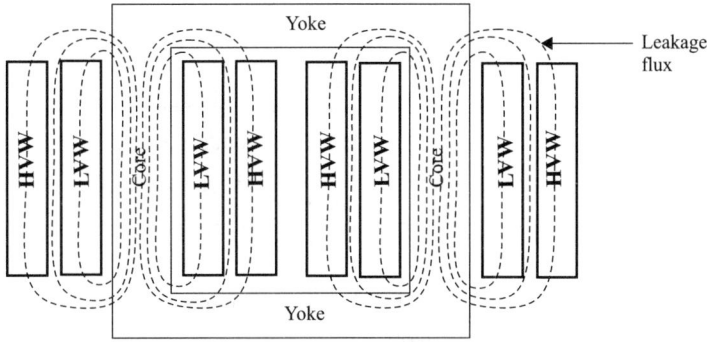

Fig. a: *Core type concentric winding..*

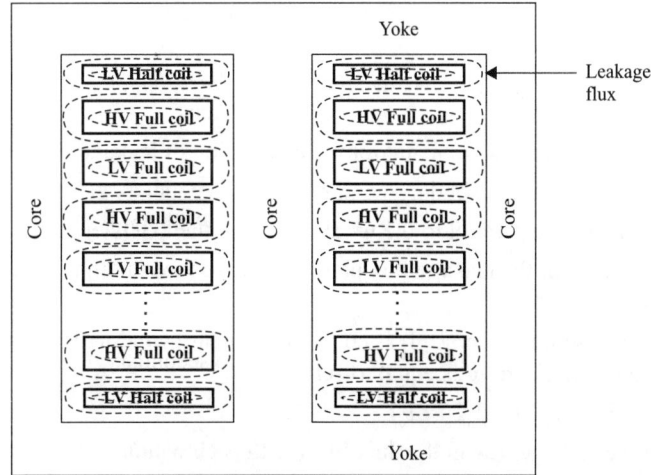

Fig. b: *Sandwich coils in shell type transformer.*

Fig. 7.13: *Leakage flux in sandwich winding.*

The popular arrangement of the core and the windings of core type transformer is shown in Fig. 7.13a. This concentric arrangement has cylindrical windings of equal axial length. The leakage field is mainly flow into the space or duct between the windings and runs parallel with the core for nearly the full length of the coils.

The distribution of the leakage flux in the case of shell type transformer having sandwich coil as shown in Fig. 7.13b. This field will mainly flow into the ducts between the windings and runs parallel along the width of the sandwich coils.

7.6.4 Estimation of Leakage Reactance of Core Type Transformer Winding

The arrangement of windings of a core type transformer is shown in Fig. 7.14.   *(MU, May' 19, 10 M)*

Let, ϕ_p =   Leakage flux in primary winding

ϕ_s =   Leakage flux in secondary winding

ϕ_o =   Flux in duct

L_o =   Mean circumference of duct

L_c =   Axial height of windings

b_p, b_s =   Radial width of primary and secondary windings

a   =   Width of radial duct

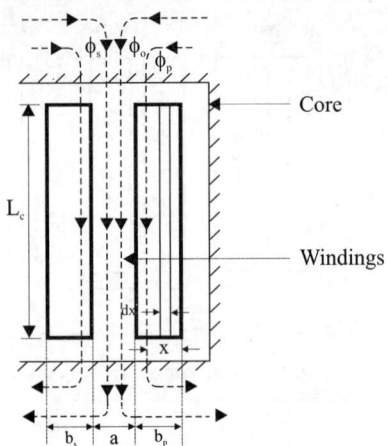

Fig. 7.14: *Leakage flux and mmf distribution*
in core type transformer.

The following assumptions are made for the calculation of leakage flux and reactance.

1. The primary and the secondary windings have an equal axial length.

2. The flux paths are parallel to the windings along the axial height.

3. The reluctance of the leakage flux path external to the winding is very high and so no leakage flux external to winding.

4. The primary winiding mmf $I_p T_p$ is equal to secondary winding $I_s T_s$. Therefore, the magnetizing mmf and hence magnetizing current is equal to zero.

$$\text{Total mmf, } AT = I_p T_p = I_s T_s$$

5. Half of the leakage flux in the duct links with each winding.

6. The length of the mean turn of the primary and secondary windings are equal.

7. The leakage flux path through yokes are negligible.

8. The windings are uniformly distributed and therefore the winding mmf varies linearly from zero from one end to AT to the other end.

The flux leakages can be estimated separately for conductor and duct portion and then added to get total leakage flux.

Leakage flux in Conductor portion

Consider an small strip of width dx at a distance x from the edge of primary winding along its width as shown in Fig. 7.14. The total width of primary winding is b_p.

Now,   mmf acting across the strip = Ampere-turn $\times \dfrac{x}{b_p} = I_p T_p \dfrac{x}{b_p}$ .....(7.47)

Permeance of strip = $\mu_o \dfrac{\text{Area of flux path}}{\text{Length of flux path}}$

$$= \mu_o \dfrac{L_{mtp} \, dx}{L_c}$$ .....(7.48)

Leakage flux in the strip = mmf × Permeance

$$= I_p T_p \frac{x}{b_p} \times \mu_o \frac{L_{mtp} dx}{L_c} \qquad \boxed{\text{Using equations (7.47) and (7.48)}}$$

$$= \mu_o \frac{L_{mtp}}{L_c} I_p T_p \frac{x}{b_p} dx \qquad .....(7.49)$$

The leakage flux given by equation (7.49) links with $\dfrac{x}{b_p} T_p$ turns of primary winding.

Let, $d\phi_{cp}$ = Flux linkages of the strip of width x

Now, $d\phi_{cp}$ = Leakage flux in the strip of width x × Total turn linked by flux

$$= \mu_o \frac{L_{mtp}}{L_c} I_p T_p \frac{x}{b_p} dx \times \frac{x}{b_p} T_p$$

$$= \mu_o \frac{L_{mtp}}{L_c} I_p T_p^2 \left(\frac{x}{b_p}\right)^2 dx \qquad .....(7.50)$$

Now, the leakage flux in conductor portion of primary winding can be obtained by integrating $d\phi_{cp}$ between limits o to b_p.

$$\therefore \phi_{cp} = \int_0^{b_p} d\phi_{cp} = \int_0^{b_p} \mu_o \frac{L_{mtp}}{L_c} I_p T_p^2 \left(\frac{x}{b_p}\right) dx$$

$$= \mu_o \frac{L_{mtp}}{L_c} \frac{I_p T_p^2}{b_p^2} \int_0^{b_p} x^2 dx$$

$$= \mu_o \frac{L_{mtp}}{L_c} \frac{I_p T_p^2}{b_p^3} \left[\frac{x^3}{3}\right]_0^{b_p}$$

$$= \mu_o \frac{L_{mtp}}{L_c} \frac{I_p T_p^2}{b_p^2} \frac{b_p^3}{3} = \mu_o \frac{L_{mtp}}{L_c} I_p T_p^2 \frac{b_p}{3}$$

Similarly,

Leakage flux in secondary, $\phi_{cs} = \mu_o \dfrac{L_{mts}}{L_c} I_s T_s^2 \dfrac{b_s}{3}$

Leakage flux in duct portion

mmf acting across duct = $I_p T_p$

Permeance of duct = $\mu_o \dfrac{\text{Area of flux path}}{\text{Length of flux path}} = \mu_o \dfrac{L_o a}{L_c}$

Flux in duct, ϕ_o = mmf × Permeance

$$= I_p T_p \times \frac{L_o a}{L_c} = \mu_o I_p T_p \frac{L_o a}{L_c}$$

Half of the duct flux links with primary winding and another half link with secondary winding and so half of duct flux links with the entire primary winding with turns T_p.

$\left.\begin{array}{l}\therefore \text{ Flux linkages of primary} \\ \text{winding due to duct flux}\end{array}\right\} \phi_{op} = \dfrac{1}{2}\, \phi_o\, T_p = \dfrac{1}{2}\, \mu_o\, I_p T_p\, \dfrac{L_o}{L_c}\, a\, T_p$

$$= \dfrac{1}{2}\, \mu_o\, I_p\, T_p^2\, \dfrac{L_o}{L_c}\, a$$

Similarly,

$\left.\begin{array}{l}\text{Flux linkages of secondary} \\ \text{winding due to duct flux}\end{array}\right\} \phi_{os} = \dfrac{1}{2}\, \mu_o\, I_s T_s^2\, \dfrac{L_o}{L_c}\, a$

Total Leakage flux

Total leakage flux is sum of leakage flux in conductor portion and duct.

Hence, total flux linkages of primary winding, $\phi_p = \phi_{cp} + \phi_{op}$

$$= \mu_o\, \dfrac{L_{mtp}}{L_c}\, I_p\, T_p^2\, \dfrac{b_p}{3} + \dfrac{1}{2}\, \mu_o\, I_p\, T_p^2\, \dfrac{L_o}{L_c}\, a$$

$$= \mu_o\, I_p T_p^2\, \dfrac{L_{mt}}{L_c}\left(\dfrac{b_p}{3} + \dfrac{a}{2}\right) \qquad \boxed{\text{Let, } L_o = L_{mtp} = L_{mt}}$$

Similary,

Total flux linkages of secondary winding, $\phi_s = \phi_{cs} + \phi_{os}$

$$= \mu_o\, I_s T_s^2\, \dfrac{L_{mt}}{L_c}\left(\dfrac{b_s}{3} + \dfrac{a}{2}\right)$$

Leakage Reactance

Leakage inductance of primary winding, $L_p = \dfrac{\text{Flux linkages}}{\text{Current}} = \dfrac{\phi_p}{I_p} = \dfrac{1}{I_p} \times \phi_p$

$$= \dfrac{1}{I_p} \times \mu_o\, I_p\, T_p^2\, \dfrac{L_{mt}}{L_c}\left(\dfrac{b_p}{3} + \dfrac{a}{2}\right)$$

$$= \mu_o\, T_p^2\, \dfrac{L_{mt}}{L_c}\left(\dfrac{b_p}{3} + \dfrac{a}{2}\right)$$

Leakage reactance of primary winding, $x_p = 2\pi\, f\, L_p = 2\pi f \mu_o\, T_p^2\, \dfrac{L_{mt}}{L_c}\left(\dfrac{b_p}{3} + \dfrac{a}{2}\right)$

Similarly,

Leakage reactance of secondary winding, $x_s = 2\pi f \mu_o\, T_s^2\, \dfrac{L_{mt}}{L_c}\left(\dfrac{b_s}{3} + \dfrac{a}{2}\right)$

$\left.\begin{array}{l}\text{Leakage reactance of secondary} \\ \text{winding referred to primary side}\end{array}\right\} x_s' = x_s\, \dfrac{1}{k^2} = x_s\left(\dfrac{T_p}{T_s}\right)^2 \qquad \boxed{\text{Transformer ratio, } k = \dfrac{T_s}{T_p}}$

$$= 2\pi f \mu_o\, T_s^2\, \dfrac{L_{mt}}{L_c}\left(\dfrac{b_s}{3} + \dfrac{a}{2}\right)\left(\dfrac{T_p}{T_s}\right)^2$$

$$= 2\pi f \mu_o\, T_p^2\, \dfrac{L_{mt}}{L_c}\left(a + \dfrac{b_p + b_s}{3}\right)$$

$\therefore$ Total reactance of transformer referred to primary side $\left.\right\} \quad X_p = x_p + x_s'$

$$= 2\pi f \mu_o\, T_p^2\, \frac{L_{mt}}{L_c}\left(\frac{b_p}{3} + \frac{a}{2}\right) + 2\pi f \mu_o\, T_p^2\left(\frac{b_s}{3} + \frac{a}{2}\right)$$

$$= 2\pi f \mu_o T_p^2\, \frac{L_{mt}}{L_c}\left(a + \frac{b_p + b_s}{3}\right) \qquad \qquad .....(7.51)$$

Per unit reactance, $\varepsilon_x = \dfrac{I_p X_p}{V_p}$ \qquad \boxed{\text{Using equation (7.51)}}

$$= \frac{I_p}{V_p}\, 2\pi f \mu_o\, T_p^2\, \frac{L_{mt}}{L_c}\left(a + \frac{b_p + b_s}{3}\right) \qquad \boxed{T_p / V_p = E_t}$$

$$= 2\pi f \mu_o\, \frac{AT}{E_t}\, \frac{L_{mt}}{L_c}\left(a + \frac{b_p + b_s}{3}\right) \qquad \boxed{I_p\, T_p = AT} \quad .....(7.52)$$

In certain arrangement, each winding is constructed with two coils connected in series with each coil having half the number of turns. In this case, the equation (7.51) is modified as shown in equation (7.53).

$$X_{p2} = 2\pi f \mu_o\, T_p^2\, \frac{L_{mt}}{L_c}\left(a + \frac{b_p + b_s}{3} \times \frac{1}{2}\right)$$

$$= 2\pi f \mu_o\, T_p^2\, \frac{L_{mt}}{L_c}\left(a + \frac{b_p + b_s}{6}\right) \qquad \qquad .....(7.53)$$

Per unit reactance, $\varepsilon_{x2} = \dfrac{I_p X_{p2}}{V_p}$

$$= \frac{I_p}{V_p}\, 2\pi f \mu_o\, T_p^2\, \frac{L_{mt}}{L_c}\left(a + \frac{b_p + b_s}{6}\right)$$

$$= 2\pi f \mu_o\, \frac{AT}{E_t}\, \frac{L_{mt}}{L_c}\left(a + \frac{b_p + b_s}{6}\right) \qquad \qquad .....(7.54)$$

7.6.5 Leakage Reactance of Shell Type Transformer Winding with Sandwich Coils

The leakage flux distribution in shell type transformer using sandwich coil is shown in Fig. 7.15. In this arrangement, alternate coils will be low voltage winding coil and high voltage winding coil.

Let the high voltage winding have n coils with each high voltage coil sandwiched between two coils of low voltage winding. This requires the low voltage winding to have two half coils, one at each end of the winding and n − 1 full coils. Each half coil of low voltage winding contains half the number of turns that of full low voltage coil. Thus if there are n coils in the high voltage winding, there will be n − 1 full coils and 2 half coils in low voltage winding.

The n coils of a winding can be considered as 2n half coils and equation (7.53) can be used to estimate the leakage reactance of primary and secondary.

Here each half coil of primary will have $T_p/2n$ turns and that of secondary will have $T_s/2n$ turns.

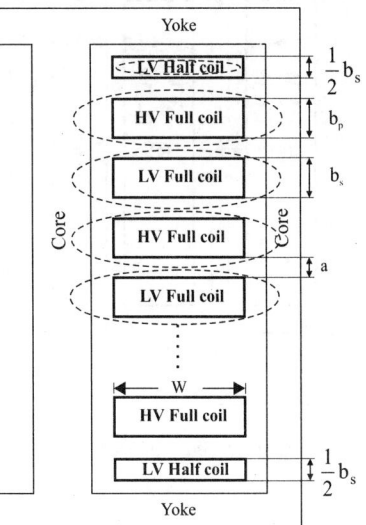

Fig. 7.15: Sandwich winding.

Leakage reactance of each unit referred to primary side $\Big\} X_u = 2\pi f \mu_o \left(\dfrac{T_p}{2n}\right)^2 \dfrac{L_{mt}}{W} \left(a + \dfrac{b_p + b_s}{6}\right)$

∴ Total reactance of transformer referred on the primary side $\Big\} X_p = 2n \times 2\pi f \mu_o \left(\dfrac{T_p}{2n}\right)^2 \dfrac{L_{mt}}{W} \left(a + \dfrac{b_p + b_s}{6}\right)$

$$= \pi f \mu_o \dfrac{L_{mt}}{W} \dfrac{T_p^2}{n} \left(a + \dfrac{b_p + b_s}{6}\right) \qquad .....(7.55)$$

Similarly,

Total reactance of transformer referred to the secondary side $\Big\} X_s = \pi f \mu_o \dfrac{L_{mt}}{W} \dfrac{T_s^2}{n} \left(a + \dfrac{b_p + b_s}{6}\right)$

Per unit reactance, $\varepsilon_x = \dfrac{\pi f \mu_o}{n} \dfrac{AT}{E_t} \dfrac{L_{mt}}{W} \left(a + \dfrac{b_p + b_s}{6}\right) \qquad .....(7.56)$

7.6.6 Mechanical Forces *(UTU, Dec' 13, 10 M) (HTU, Dec' 18, 10 M) (PU, Nov' 19, 8 M) (MU, May' 19, 10 M)*

Mechanical forces are developed due to interaction of leakage fluxes with current carrying primary and secondary conductors in a transformer. This is similar to motoring action and so the direction of force can be obtained by applying Flemings left hand rule. The leakage fluxes can be divided into two components: Radial leakage flux and Axial leakage flux.

The reaction of primary and secondary current with radial leakage flux will develop a radial mechanical force which will try to expand the outer winding and compress the inner winding. The reaction of primary and secondary current with axial leakage flux will develop an axial mechanical force which will try to compress the windings. The forces acting on windings are shown in Fig. 7.16.

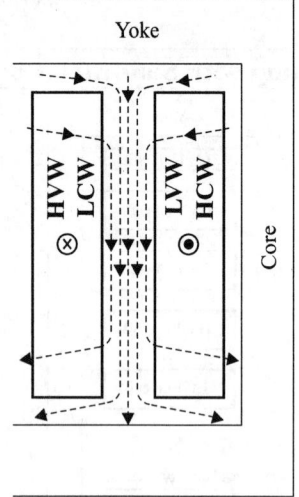

Fig. (a): Leakage field.

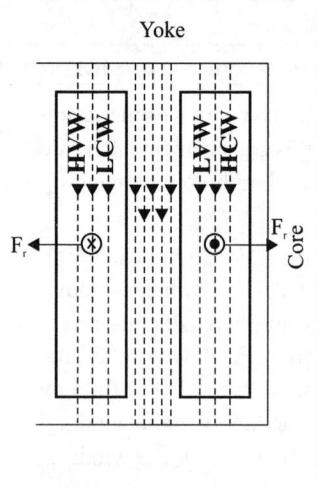

Fig. (b): Axial leakage field.

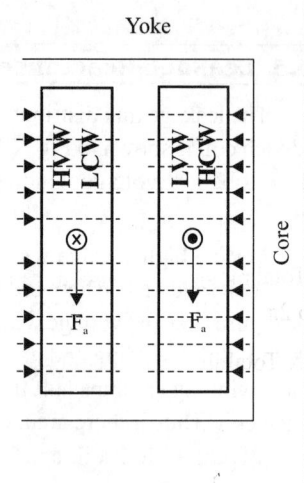

Fig. (c): Radial leakage field.

Fig. 7.16: Leakage fields and mechanical forces.

The mechanical forces produced in a transformer may be classified as,

1. Radial forces

2. Axial forces

3. Unbalanced axial forces due to magnetic assymetry.

Radial forces

Consider a core type transformer shown in Fig. 7.17.

Let, T = Number of turns

i = Instantaneous current through winding

B_{av} = Average flux density in duct

L_c = Length of coil

The flux density varies from 0 to B.

where, $B = \mu H = \mu_o \dfrac{\text{Instantaneous mmf}}{\text{Length}} = \mu_o \dfrac{iT}{L_c}$

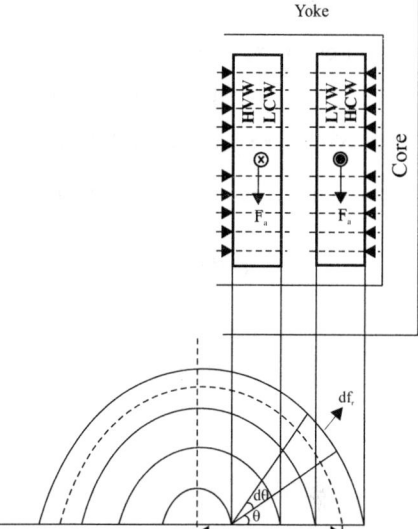

Fig. 7.17: Radial forces on windings.

$\therefore$ Average flux density, $B_{av} = \dfrac{0+B}{2} = \dfrac{\mu_o}{2} \dfrac{iT}{L_c}$ .....(7.57)

Let, R = Mean radius of the circular winding

Consider a strip of angular width $d\theta$ in a total angular distance of $\theta = 2\pi$.

Now the instantaneous radial force, df_r acting on the strip is given by,

df_r = Turns × Flux density × Current × Length of strip

$= T\, B_{av}\, i\, R d\theta$

$= T \dfrac{\mu_o}{2} \dfrac{iT}{L_c} i R\, d\theta$

$\boxed{\text{Using equation (7.57)}}$

$= \dfrac{\mu_o\, i^2\, T^2\, R}{2\, L_c} d\theta$

Total radial instantaneous radial force can be obtained by integrating the above equation with limits $\theta = 0$ to 2π.

$\therefore$ Total instantaneous force, $f_r = \displaystyle\int_0^{2\pi} df_r = \int_0^{2\pi} \dfrac{\mu_o\, i^2\, T^2\, R}{2\, L_c}\, d\theta$

$= \dfrac{\mu_o\, i^2\, T^2\, R}{2\, L_c} \displaystyle\int_0^{2\pi} d\theta$

$= \dfrac{\mu_o\, i^2\, T^2\, R}{2\, L_c}\, 2\pi$

$= \dfrac{\mu_o\, i^2\, T^2\, L_{mt}}{2\, L_c}$

where, $L_{mt} = 2\pi R =$ Length of mean turn of the coil.

Let, $I =$ rms value of current

$I_m =$ Maximum value of current

Now, $i = I_m \sin \omega t = \sqrt{2}\, I \sin \omega t$

The average value of radial force, F_{r_av} is obtained as shown below.

Average radial force, $F_{r\_av} = \dfrac{1}{2\pi} \displaystyle\int_0^{2\pi} \dfrac{\mu_o\, i^2\, T^2\, L_{mt}}{2\, L_c}\, d(\omega t)$

$\qquad = \dfrac{1}{2\pi} \displaystyle\int_0^{2\pi} \dfrac{\mu_o\, (\sqrt{2}\, I \sin \omega t)^2\, T^2\, L_{mt}}{2\, L_c}\, d(\omega t)$

$\qquad = \dfrac{\mu_o\, I^2\, T^2\, L_{mt}}{2\pi\, L_c} \displaystyle\int_0^{2\pi} \sin^2 \omega t\, d(\omega t)$

$\qquad = \dfrac{\mu_o\, I^2\, T^2\, L_{mt}}{2\pi\, L_c} \displaystyle\int_0^{2\pi} \dfrac{(1 - \cos 2\omega t)}{2}\, d(\omega t)$

$\qquad = \dfrac{\mu_o\, I^2\, T^2\, L_{mt}}{2\pi\, L_c} \dfrac{1}{2} \left[\omega t - \dfrac{\sin 2\omega t}{2} \right]_0^{2\pi}$

$\qquad = \dfrac{\mu_o\, I^2\, T^2\, L_{mt}}{2\pi\, L_c}\, 2\pi$

$\qquad = \dfrac{\mu_o\, I^2\, T^2\, L_{mt}}{2\, L_c}$

Let, $\varepsilon_x =$ Per unit leakage reactance

Now, short circuit fault current $= \dfrac{I}{\varepsilon_x}$

Maximum value of fault current $= \sqrt{2}\, \dfrac{I}{\varepsilon_x}$

Under worst case fault condition the instantaneous value of short circuit current may be double that of maximum value of fault current.

Maximum instantaneous value of fault current, $i_{m\_sc} = 2\sqrt{2}\, \dfrac{I}{\varepsilon_x}$

Therefore, instantaneous maximum value of radial force under worst case fault condition can be obtained from equation of radial force by replacing I by i_{m_sc}.

$\therefore$ Instantaneous maximum value of radial force, $f_{r\_max} = \dfrac{\mu_o}{2} (i_{m\_sc}\, T)^2\, \dfrac{L_{mt}}{L_c}$

$\qquad = \dfrac{\mu_o}{2} \left(\dfrac{2\sqrt{2}\, I}{\varepsilon_x}\, T \right)^2 \dfrac{L_{mt}}{L_c}$

$\qquad = \dfrac{4\, \mu_o\, I^2\, T^2\, L_{mt}}{\varepsilon_x\, L_c}$

Axial forces

The axial forces does not exist in symmetrical windings. In sandwich coils with 2n sections axial forces exist in half coils at two ends. The number of turns in half coils are T/2n.

Let, W be the width of half coil which is also same as length of flux path. Therefore, the equation for axial forces can be obtained from the equation of radial force by replacing T by T/2n and L_c by W.

$$\therefore \text{ Instantaneous axial force, } f_a = \frac{\mu_o}{2} \left(i \frac{T}{2n} \right)^2 \frac{L_{mt}}{W}$$

$$= \frac{\mu_o}{8n^2} (i\,T)^2 \frac{L_{mt}}{W}$$

Forces due to assymmetry

The mechanical forces will be different when the construction of coils are not symmetric. When the length of high voltage and low voltage windings are unequal the axial forces can be obtained by multiplying the equation for radial force by k, where k is lengthwise unbalance factor and replacing the length of flux path L_c by $a + b_p + b_s$.

Here, a = Width duct between primary and secondary winding

b_p = Width of primary winding

b_s = Width of secondary winding

$$\left. \begin{array}{l} \therefore \text{ Instantaneous axial force} \\ \text{in asymmetric winding} \end{array} \right\} f_{a\_asy} = k \times \frac{\mu_o}{2} (iT)^2 \frac{L_{mt}}{(a + b_p + b_s)}$$

$$= \frac{\mu_o}{2} k \frac{(iT)^2 L_{mt}}{(a + b_p + b_s)}$$

7.7 ESTIMATION OF NO-LOAD CURRENT OF TRANSFORMER *(HTU, Dec' 18, 10 M)*

The no-load current of a transformer has two components. They are magnetizing component and loss component. The magnetizing current depends on the mmf required to establish the desired flux. The loss component of no-load current depends on the iron loss.

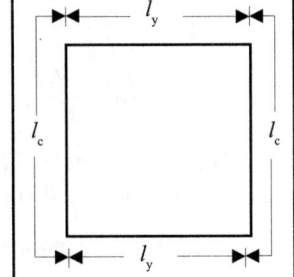

7.7.1 No-Load Current of Single-Phase Transformer

Total length of core = $2\,l_c$

Total length of yoke = $2\,l_y$

Here, $l_c = H_w$ = Height of window

and $l_y = W_w$ = Width of window

Fig. 7.18: Core type transformer.

mmf for core = mmf per meter for maximum flux density in core × Total length of core

$$= \mathbf{at_c} \times 2\,l_c = 2\,\mathbf{at_c}\,l_c$$

mmf for yoke = mmf per meter for maximum flux density in yoke × Total length of yoke

$$= at_y \times 2 \, l_y = 2 \, at_y \, l_y$$

Total magnetizing mmf, AT_o = mmf for core + mmf for yoke + mmf for joints

$$= 2 \, at_c \, l_c + 2 \, at_y \, l_y + \text{mmf for joints} \qquad \qquad .....(7.58)$$

Alternatively, AT_o can be estimated from flux density and total length of iron path.

Let, l_i = Total length of iron path

R = Reluctance of iron path

$\mu = \mu_r \, \mu_o$ = Permeability

μ_r = Relative permeability

μ_o = Absolute permeability

Now, $AT_o = R \, \phi_m$

$$= \frac{l_i}{\mu \, A_i} \times \phi_m$$

$$= \frac{l_i}{\mu_r \, \mu_o} \, B_m$$

$$\boxed{R = \frac{l_i}{\mu \, A_i}}$$

$$\boxed{\mu = \mu_r \, \mu_o}$$

$$\boxed{B_m = \frac{\phi_m}{A_i}}$$

Maximum value of magnetizing current $= \dfrac{AT_o}{T_p}$ $\qquad \qquad .....(7.59)$

If the magnetizing current is sinusoidal then,

rms value of magnetizing current, $I_m = \dfrac{AT_o}{\sqrt{2} \, T_p}$ $\qquad \qquad .....(7.60)$

Note : *Since maximum value of flux is used in the design of transformer, the current given by equation (7.59) will be maximum value.*

When the magnetizing current is non-sinusoidal, the peak factor, K_{pk} should be used in place of $\sqrt{2}$.

$$\therefore \, I_m = \frac{AT_o}{K_{pk} \, T_p} \qquad \qquad .....(7.61)$$

The values of at_c and at_y are taken from B-H curves for transformer steel. The joints in a magnetic circuit may be taken as short air-gaps in parallel with iron paths.

Loss component of no-load current, $I_l = \dfrac{P_i}{V_p}$ $\qquad \qquad .....(7.62)$

where, P_i = Iron loss in W

V_p = Terminal voltage of primary winding

The iron losses are calculated by finding the weight of cores and yokes. The loss per kg of iron is taken from the loss curves given by the manufacture of transformer laminations.

No-load current, $I_o = \sqrt{I_m^2 + I_l^2}$ $\qquad \qquad .....(7.63)$

7.7.2 No-load Current of Three-Phase Transformer *(PU, Nov' 19, 6 M)*

Total length of core = $3\,l_c$

Total length of yoke = $2\,l_y$

Here, $l_c = H_w$ = Height of window

$l_y = 2W_w + d$

where, W_w = Width of window

d = Diameter of circumscribing circle

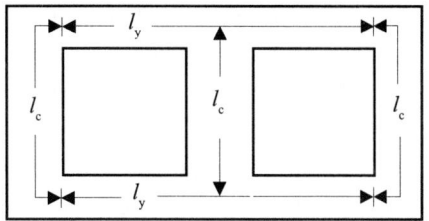

Fig. 7.19: Three-phase core type transformer.

mmf for core = mmf per meter for maximum flux density in core × Total length of core

$$= \mathbf{at}_c \times 3\,l_c = 3\,\mathbf{at}_c\,l_c$$

mmf for yoke = mmf per meter for maximum flux density in yoke × Total length of yoke

$$= \mathbf{at}_y \times 2\,l_y = 2\,\mathbf{at}_y\,l_y$$

Total magnetizing mmf required for the transformer $\Big\}$ = mmf for core + mmf for yoke + mmf for joints

$$= 3\,\mathbf{at}_c\,l_c + 2\,\mathbf{at}_y\,l_y + \text{mmf for joints}$$

Total magnetizing mmf required per phase $\Big\}$ $AT_o = \dfrac{3\,\mathbf{at}_c\,l_c + 2\,\mathbf{at}_y\,l_y + \text{mmf for joints}}{3}$ .....(7.64)

Maximum value of magnetizing current per phase $\Big\} = \dfrac{AT_o}{T_p}$

If magnetizing current is sinusoidal then,

rms value of magnetizing current per phase, $I_m = \dfrac{AT_o}{\sqrt{2}\,T_p}$

If the magnetizing current is non-sinusoidal then K_{pk} should be used in the place of $\sqrt{2}$.

$$\therefore I_m = \frac{AT_o}{\sqrt{2}\,T_p} \quad \text{(or)} \quad I_m = \frac{AT_o}{K_{pk}\,T_p} \qquad .....(7.65)$$

The values of $\mathbf{at}_c$ and $\mathbf{at}_y$ are taken from B-H curves.

Let, P_i = Total iron loss for the three phases

$$\therefore P_i = 3\,V_p\,I_l$$

$\therefore$ Loss component of no-load current, $I_l = \dfrac{P_i}{3\,V_p}$

$\therefore$ No-load current per phase, $I_o = \sqrt{I_m^2 + I_l^2}$ .....(7.66)

7.8 TEMPERATURE RISE IN TRANSFORMERS   *(HTU, Dec' 18, 10 M) (PTU, May' 19, 5 M)*

The losses developed in the transformer cores and windings are converted into thermal energy and cause heating of corresponding transformer parts. The heat dissipation in transformer occurs by Conduction, Convection and Radiation. The paths of heat flow in transformer are the following,

1. From the internal most heated spots of a given part (of core or winding) to their outer surface in contact with the oil.

2. From the outer surface of a transformer part to the oil that cools it.

3. From the oil to the walls of a cooler, eg. wall of tank.

4. From the walls of the cooler to the cooling medium air or water.

In the path 1 mentioned above heat is transferred by conduction. In the path 2 and 3 mentioned above heat is transferred by convection of the oil. In path 4 the heat is dissipated by both convection and radiation.

7.8.1 Cooling of Transformers   *(AU, Apr' 19, 13 M) (AU, Apr' 17, 6 M) (AU, Apr' 18, 5 M)*

The various methods of cooling transformers are, *(MU, May' 19, 5 M) (UTU, Dec' 13, 5 M) (KTU, Feb' 18, 8 M)*

1. Air natural	6. Forced circulation of oil
2. Air blast	7. Oil forced-air natural
3. Oil natural	8. Oil forced-air forced
4. Oil natural-air forced	9. Oil forced-water forced
5. Oil natural-water forced	

The choice of cooling method depends upon the size, type of application and type of conditions obtaining at the site where the transformer in installed.

Air natural is used for transformers upto 1.5 *MVA*. Since cooling by air is not so effective and proves insufficient for transformers of medium sizes, oil is used as a coolant. Oil is used for almost all transformers except for the transformers used for special applications. Both plain walled and corrugated walled tanks are used in oil cooled transformer.

In oil natural-air forced method the oil circulating under natural head transfers heat to tank walls. The air is blown through the hollow space to cool the transformer. In oil natural-water forced method, copper cooling coils are mounted above the transformer core but below the surface of oil. Water is circulated through the cooling coils to cool the transformer.

In oil forced-air natural method of cooling, oil is circulated through the transformer with the help of a pump and cooled in a heat exchanger by natural circulation of air. In oil forced-air forced method, oil is cooled in external heat exchanger using air blast produced by fans. In oil forced-water forced method, heated oil is cooled in a water heat exchanger. In this method pressure of oil is kept higher than that of water to avoid leakage of oil.

Natural cooling is suitable upto 10 *MVA*. The forced oil and air circulation are employed for transformers of capacities above 10 *MVA*. The forced oil and water is used for transformers designed for power plants.

7.8.2 Transformer oil as a Cooling Medium

For the transformer oil, the specific heat dissipation due to convection of oil is given by,

$$\lambda_{conv} = 40.3 \left(\frac{\theta}{H}\right)^{1/4} W/m^2\text{-}°C \qquad\qquad .....(7.67)$$

where, θ = Temperature difference of the surface relative to oil

H = Height of dissipating surface

The average working temperature of oil is 50 °C to 60 °C.

For θ = 20 °C and H = 0.5 to 1 m, the λ_{conv} = 80 to 100 $W/m^2\text{-}°C$.

The corresponding figure for convection of air is 8 $W/m^2\text{-}°C$. Thus the convection due to oil is more than 10 times than that of air.

7.8.3 Temperature Rise in Plain Walled Tanks

The transformer core and winding assembly is placed inside a container called tank. The walls of the tank dissipate heat by both radiation and convection. For a temperature rise of 40 °C above the ambient temperature of 20°C, the specific heat dissipation are as follows,

Specific heat dissipating due to radiation, λ_{rad} = 6 $W/m^2\text{-}°C$

Specific heat dissipation due to convection, λ_{con} = 6.5 $W/m^2\text{-}°C$

Let, λ_{spec} = Total specific heat dissipation in plain walled tanks.

Now, $\lambda_{spec} = \lambda_{rad} + \lambda_{conv} = 6 + 6.5 = 12.5\ W/m^2\text{-}°C$

$$\text{Temperature rise, } \theta = \frac{\text{Total loss}}{\text{Specific heat} \times \text{Heat dissipating}} = \frac{P_i + P_c}{\lambda_{spec}\ S_T} \qquad .....(7.68)$$
$$\quad\ \text{dissipation} \quad\ \text{surface of the tank}$$

where, P_i = Iron loss

P_c = Copper loss

S_T = Heat dissipating surface of the tank

The heat dissipating surface of the tank is given by total area of vertical sides plus one half area of the top cover. If the oil is in contact with the cover then the total heat dissipating surface of the tank is given by total area of vertical sides plus full area of the top cover. The area of bottom of the tank should be neglected as it has very little cooling effect.

For transformers of low capacity the plain walled tanks have sufficient surface to keep the temperature rise within the limits. But for transformers of large output, the plain walled tanks are not sufficient to dissipate the losses. This is because, the volume and hence losses increase as cube of linear dimensions, while the dissipating surface increases as the square of linear dimensions. Thus an increase in rating results in an increase in loss to be dissipated per unit area giving a higher temperature rise. Modern oil immersed power transformers with natural oil cooling and a plain tank may be provided for output not exceeding 20 to 30 kVA.

Transformer rated for larger outputs must be provided with means to improve the conditions of heat dissipation. This may be achieved by providing corrugations, cooling tubes and radiators.

7.9 DESIGN OF TANK AND COOLING TUBES OF TRANSFORMERS

7.9.1 Design of Tank

(HTU, Dec' 18, 10 M) (PU, Nov' 19, 8 M) (MU, May' 19, 10 M)

The dimensions of the tank are decided by the dimensions of the transformer frame and clearance required on all the sides. The dimensions of the tank are shown in Fig. 7.20.

Let, C_1 = Clearance between winding and tank along the width.

C_2 = Clearance between the winding and tank along the length.

C_3 = Clearance between the transformer frame and the tank at the bottom.

C_4 = Clearance between the transformer frame and the tank at the top.

D_{oc} = Outer diameter of coil.

With reference to Fig. 7.20 we can write,

$$\text{Width of the tank, } W_T = 2D + D_{oc} + 2C_1 \quad - \text{ For three-phase transformer}$$

$$= D + D_{oc} + 2C_1 \quad - \text{ For single-phase transformer} \quad .....(7.69)$$

$$\text{Length of the tank, } L_T = D_{oc} + 2\,C_2 \quad\quad\quad\quad\quad\quad\quad .....(7.70)$$

$$\text{Height of the tank, } H_T = H + C_3 + C_4 \quad\quad\quad\quad\quad\quad .....(7.71)$$

The clearance on the sides depends on voltage and power rating of the winding. The clearance at the top depends on the oil height above the assembled transformer and the space for mounting the terminals and tap changing gear. The clearance at the bottom depends on the space required for mounting the transformer frame inside the tank. The typical values of the clearances are listed in Table 7.5.

Table 7.5: Clearances between Transformer Frame and Tank

Voltage	kVA rating	Clearance in *mm*			
		C_1	C_2	C_3	C_4
Upto 11 *kV*	< 1000 *kVA*	40	50	75	375
Upto 11 *kV*	1000 to 5000 *kVA*	70	90	100	400
11 *kV* to 33 *kV*	< 1000 *kVA*	75	100	75	450
11 *kV* to 33 *kV*	1000 to 5000 *kVA*	85	125	100	475

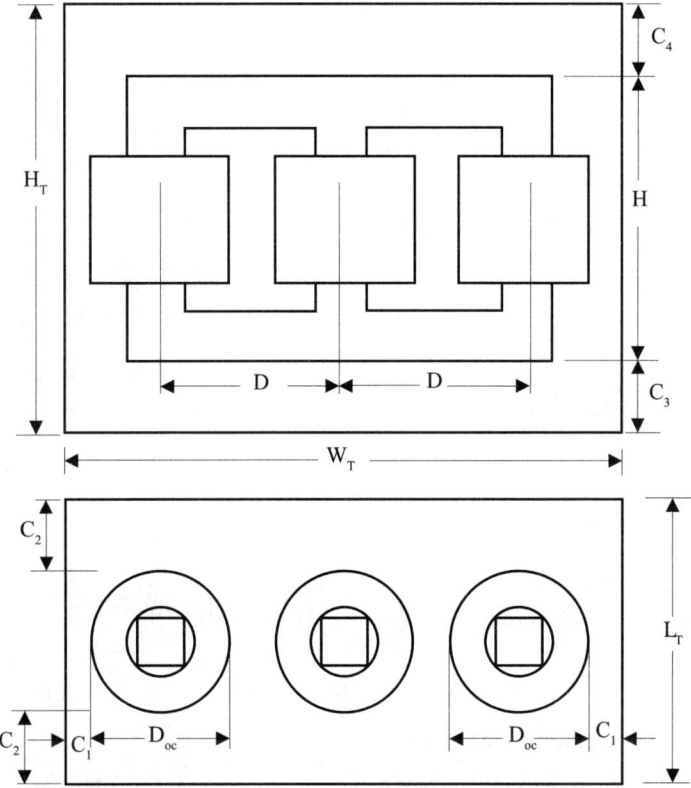

Fig. 7.20: *Dimensions of transformer tank.*

7.9.2 Design of Cooling Tubes

The transformers are provided with cooling tubes to increase the heat dissipating area. The tubes are mounted on the vertical sides of the transformer tank. But the increase in dissipation of heat is not proportional to increase in area, because the tubes would screen (conceals) some of the tank surface preventing radiations from the screened surface. On the other-hand the tubes will improve the circulation of oil. This improves the dissipation of loss by convection. The circulation of oil is due to more effective pressure heads produced by columns of oil in tubes.

The improvement in loss dissipation by convection is equivalent to loss dissipated by 35% of tube surface area. Hence to account for this improvement in dissipation of loss by convection an additional 35% tube area is added to actual tube surface area or the specific heat dissipation due to convection is taken as 35% more than that without tubes.

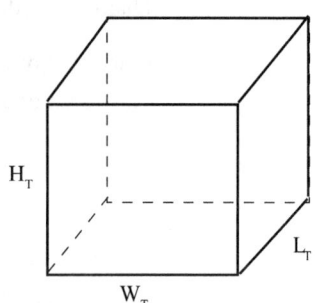

Fig. 7.21: *Dimension of tank.*

Let, S_T = Dissipating surface of the tank

$$= 2 (L_T H_T + W_T H_T)$$

$$= 2 H_T (L_T + W_T)$$

Neglecting top and bottom
Refer Fig. 7.21

Let, S_{dt} = Total dissipating surface of cooling tubes

Now, $S_{dt} = X\, S_T$

X_i = Percentage increase in tube surface area

∴ Loss dissipated by tank walls, $P_{LDTW} = (\lambda_{rad} + \lambda_{conv})\, S_T$

> Tank walls dissipate loss by radiation and convection

∴ Loss dissipated by cooling tubes, $P_{LDC} = \lambda_{conv}\, X_i\, XS_T$

> Cooling tubes dissipate loss by convection

. ∴ Total loss dissipated, P_{LD} = Loss dissipated $+$ Loss dissipated by
by tank walls $\quad$ cooling tubes

$$= P_{LDTW} + P_{LDC} = (\lambda_{rad} + \lambda_{conv})\, S_T + \lambda_{conv}\, X_i\, XS_T \qquad .....(7.72)$$

Let, P_{TL} = Total transformer loss

P_i = Iron loss

P_c = Copper loss

Now, $P_{TL} = P_i + P_c$

$$\left.\begin{array}{l}\text{Temperature rise in}\\ \text{transformer with cooling tubes}\end{array}\right\}\theta = \frac{\text{Total transformer loss}}{\text{Total Loss dissipated}} = \frac{P_{TL}}{P_{LD}}$$

$$= \frac{P_i + P_c}{S_T\left((\lambda_{rad} + \lambda_{conv}) + \lambda_{conv}\, X_i\, X\right)} \qquad \boxed{\text{Using equation (7.72)}}$$

From the above equation we can write,

$$(\lambda_{rad} + \lambda_{conv}) + \lambda_{conv}\, X_i\, X = \frac{P_i + P_c}{\theta\, S_T} \quad \Rightarrow \quad X = \frac{1}{\lambda_{conv}\, X_i}\left(\frac{P_i + P_c}{\theta\, S_T} - (\lambda_{rad} + \lambda_{conv})\right)$$

Total dissipating surface area of cooling tubes, $S_{dt} = X\, S_T$

$$= \frac{1}{\lambda_{conv}\, X_i}\left(\frac{P_i + P_c}{\theta\, S_T} - (\lambda_{rad} + \lambda_{conv})\right) S_T$$

$$= \frac{1}{\lambda_{conv}\, X_i}\left(\frac{P_i + P_c}{\theta} - (\lambda_{rad} + \lambda_{conv})\, S_T\right)$$

Let, l_t = Length of the tube

d_t = Diameter of the tube

∴ Surface area of each cooling tube, $S_t = \pi\, d_t\, l_t$
(Surface area of a cylinder)

Total number of tubes, $n_t = \dfrac{\text{Total dissipating surface area of cooling tubes}}{\text{Surface area of each cooling tube}}$

$$= \frac{S_{dt}}{S_t}$$

The standard diameter of the cooling tubes is 50 *mm* and the length of the tube depends on the height of the tank. The tubes are arranged with a centre to centre spacing of 75 *mm*.

7.10 EFFECT OF FREQUENCY ON IRON LOSSES

The specific iron loss, $P_i = K_h f B_m^2 + K_e f^2 B_m^2$ in W/kg .....(7.73)

where K_h and K_e are constants.

Induced emf, $E = 4.44 f B_m A_i T$

Hence if E is constant then the product $f B_m$ will be constant.

The eddy current loss $= K_e f^2 B_m^2$ .....(7.74)

In equation (7.74) $f B_m$ = constant as long as E is constant and so eddy current loss remain constant, even though the frequency is changed.

The hysteresis loss $= K_h f B_m^2$ .....(7.75)

Let, $K_1 = f B_m$

$\therefore$ Hysteresis loss $= K_h K_1 B_m$

$$= K_h K_1 \frac{K_1}{f} = K_h \frac{K_1^2}{f}$$ .....(7.76)

From equation (7.76) it is clear that the hysteresis loss will decrease with increase in frequency if the voltage is kept constant.

Therefore, the total iron losses decreases if the frequency is increased and the applied voltage is kept constant. If the flux density is also changed, the new induced emf can be determined from the emf equation. For the new voltage, total iron loss and the magnetising current and the no load current can be calculated.

The change in frequency will not have much effect on leakage inductance and therefore leakage reactance will increase linearly with frequency. Due to skin effect, the effective resistance increases with increase in frequency. This effect is not normally important with small changes in frequency.

7.11 SUMMARY OF DESIGN EQUATIONS

1. i) Emf per turn, $E_t = 4.44 f \phi_m$

   ii) Emf per turn, $E_t = K\sqrt{Q}$, where $K = \sqrt{4.44 f \dfrac{\phi_m}{AT} \times 10^3}$

2. Window space factor, $K_w = \dfrac{\text{Conductor area in window}}{\text{Total area of window}} = \dfrac{A_c}{A_w}$

3. kVA rating, $Q = 2.22 f B_m A_i K_w A_w \delta \times 10^{-3}$ (Single-phase transformer)

4. kVA rating, $Q = 3.33 f B_m A_i K_w A_w \delta \times 10^{-3}$ (Three-phase transformer)

5. Ampere-turns, $AT = I_p T_p = I_s T_s$

6. Total conductor area in window of single-phase transformer, $A_c = \dfrac{2 AT}{\delta}$

7. Total conductor area in window of 3-phase transformer, $A_c = \dfrac{4 AT}{\delta}$

Overall Dimensions

8.   For single-phase core type transformer,

        i)   Overall height of transformer frame, $H = H_w + 2H_y$

        ii)   Overall width of transformer frame, $W = W_w + 2a$

9.   For single-phase shell type transformer,

        i)   Overall height of transformer frame, $H = H_w + 2a$

        ii)   Overall width of transformer frame, $W = 2(W_w + a)$

10.   For three-phase core type transformer,

        i)   Overall height of transformer frame, $H = H_w + 2H_y$

        ii)   Overall width of transformer frame, $W = 2W_w + 3a$

Square Core

11.   Side of square, $a = \dfrac{d}{\sqrt{2}}$

12.   Gross core area, $A_{gi} = 0.5\ d^2$

13.   Net core area, $A_i = 0.45\ d^2$

14. $\dfrac{\text{Net core area}}{\text{Area of circumscribing circle}} = 0.58$

15. $\dfrac{\text{Gross core area}}{\text{Area of circumscribing circle}} = 0.64$

16.   Core area factor, $K_c = 0.45$

Cruciform Core

17.   Dimensions of cruciform core for maximum core area

        (i)   Length of rectangle, $a = 0.85\ d$

        (ii)   Breadth of rectangle, $b = 0.53\ d$

18.   Gross core area, $A_{gi} = 0.618\ d^2$

19.   Net core area, $A_i = 0.56\ d^2$

20. $\dfrac{\text{Net core area}}{\text{Area of circumscribing circle}} = 0.71$

21. $\dfrac{\text{Gross core area}}{\text{Area of circumscribing circle}} = 0.79$

22.   Core area factor, $K_c = 0.56$

Design of Yoke

23.   Choice of area of yoke

        i) Area of yoke, $A_Y = (1.15 \text{ to } 1.25) \times A_{gi}$ -   For hot rolled steel

        ii) Area of yoke, $A_Y = A_{gi}$ -   For grain oriented steel

24. Area of yoke, $A_Y = D_Y H_Y$

25. Choice of depth of yoke

    i) Depth of yoke, $D_Y = D_C$ -   For rectangular core

    ii) Depth of yoke, $D_Y = a$ -   For square or stepped core

26. i) Height of yoke, $H_Y = \dfrac{A_Y}{D_Y} = \dfrac{A_Y}{D_C}$ -   For rectangular core

    ii) Height of yoke, $H_Y = \dfrac{A_Y}{D_Y} = \dfrac{A_Y}{a}$ -   For square or stepped core

Design of winding

27. i) Number of turns in low voltage winding, $T_{LV} = \dfrac{V_{LV}}{E_t}$

    ii) Number of turns in low voltage winding, $T_{LV} = \dfrac{AT}{I_{LV}}$

28. Number of turns in high voltage winding, $T_{HV} = T_{LV} \times \dfrac{V_{HV}}{V_{LV}}$

29. Rated current in a winding $= \dfrac{\text{kVA per phase} \times 10^3}{\text{Voltage rating of the winding}}$

30. Area of cross-section of primary conductor, $a_p = \dfrac{I_p}{\delta}$

31. Area of cross-section of secondary conductor, $a_s = \dfrac{I_s}{\delta}$

Single-phase transformer

32. mmf for core $= 2 \, \mathbf{at}_c \, l_c$

33. mmf for yoke $= 2 \, \mathbf{at}_y \, l_y$

34. Total magnetizing mmf, AT_o = mmf for core + mmf for yoke + mmf for joints

$$= 2 \, \mathbf{at}_c \, l_c + 2 \, \mathbf{at}_y \, l_y + \text{mmf for joints}$$

35. Alternate formulae for AT_o

    i) $AT_o = R \, \phi_m$

    ii) $AT_o = \dfrac{l_i}{\mu_r \, \mu_o} \, B_m$

36. Maximum value of magnetizing current $= \dfrac{AT_o}{T_p}$

37. rms value of magnetizing current, $I_m = \dfrac{AT_o}{K_{pk} \, T_p}$ $\boxed{\text{For sinusoidal flux, } K_{pk} = \sqrt{2}}$

38. Loss component of no-load current, $I_l = \dfrac{P_i}{V_p}$

39. No-load current, $I_o = \sqrt{I_m^2 + I_l^2}$

Three-phase transformer

40.   mmf for core = 3 **at**$_c$ l_c

41.   mmf for yoke = 2 **at**$_y$ l_y

42.   Total magnetizing mmf per phase, AT_o = mmf for core + mmf for yoke + mmf for joints

$$= \frac{3 \text{ at}_c \, l_c + 2 \text{ at}_y \, l_y + \text{mmf for joints}}{3}$$

43.   rms value of magnetizing current per phase, $I_m = \dfrac{AT_o}{K_{pk} \, T_p}$ $\boxed{\text{For sinusoidal flux, } K_{pk} = \sqrt{2}}$

44.   Loss component of no-load current per phase, $I_l = \dfrac{P_i}{3 \, V_p}$

45.   No-load current per phase, $I_o = \sqrt{I_m^2 + I_l^2}$

Design of Winding

46.   Resistance of primary winding, $r_p = \rho \times \dfrac{T_p \, L_{mtp}}{a_p}$

47.   Resistance of secondary winding, $r_s = \rho \times \dfrac{T_s \, L_{mts}}{a_s}$

48.   Total instantaneous force, $f_r = \dfrac{\mu_o \, i^2 \, T^2 \, L_{mt}}{2 \, L_c}$

49.   Average radial force, $F_{r\_av} = \dfrac{\mu_o \, I^2 \, T^2 \, L_{mt}}{2 \, L_c}$

50.   Maximum instantaneous value of fault current, $i_{m\_sc} = 2\sqrt{2}\dfrac{I}{\varepsilon_x}$

51.   Instantaneous maximum value of radial force, $f_{r\_max} = \dfrac{4 \, \mu_o \, I^2 \, T^2 \, L_{mt}}{\varepsilon_x \, L_c}$

52.   Instantaneous axial force, $f_a = \dfrac{\mu_o}{8n^2} (i\,T)^2 \dfrac{L_{mt}}{W}$

53.   Instantaneous axial force in asymmetric winding $\left.\begin{array}{c}\\\\\end{array}\right\} f_{a\_asy} = \dfrac{\mu_o}{2} \, k \, \dfrac{(iT)^2 \, L_{mt}}{(a + b_p + b_s)}$

Cooling of transformer

54.   Temperature rise in plain walled tanks, $\theta = \dfrac{P_i + P_c}{(\lambda_{rad} + \lambda_{conv}) \, S_t}$

55.   Temperature rise in transformer with cooling tubes, $\theta = \dfrac{P_i + P_c}{S_T((\lambda_{rad} + \lambda_{conv}) + \lambda_{conv} \, X_i \, X)}$

56.   Loss dissipated by tank walls, $P_{LDTW} = (\lambda_{rad} + \lambda_{conv}) \, S_T$

57.   Loss dissipated by cooling tubes, $P_{LDC} = \lambda_{conv} \, X_i \, XS_T$

58. Total loss dissipated, $P_{LD} = P_{LDTW} + P_{LDC}$

59. Total dissipating surface area of cooling tubes, $S_{dt} = \dfrac{1}{\lambda_{conv}\, X_i} \left(\dfrac{P_i + P_c}{\theta} - (\lambda_{rad} + \lambda_{conv})\, S_T \right)$

60. Surface area of each cooling tube, $S_t = \pi\, d_t\, l_t$

61. Total number of tubes, $n_t = \dfrac{S_{dt}}{S_t}$

62. Width of the tank, $W_T = 2D + D_{oc} + 2C_1$ - Three-phase transformer

$\qquad\qquad\qquad\qquad\quad = D + D_{oc} + 2C_1$ - Single-phase transformer

63. Length of the tank, $L_T = D_{oc} + 2C_2$

64. Height of the tank, $H_T = H + C_3 + C_4$

7.12 SOLVED PROBLEMS

EXAMPLE 7.1

Calculate the core and window areas required for a 1000 kVA, 6600/400 V, 50 Hz, single-phase core type transformer. Assume a maximum flux density of 1.25 Wb/m^2 and a current density of 2.5 A/mm^2. Voltage per turn = 30.5 V. Window space factor = 0.32.

GIVEN DATA

kVA = 1000   f $= 50\ Hz$ $B_m = 1.25\ Wb/m^2$

V_p = 6600 V $V_s = 400\ V$ $\delta = 2.5\ A/mm^2$

E_t = 30.5 V $K_w = 0.32$ 1-phase, Core type

SOLUTION

Emf per turn, $E_t = 4.44\ f\ \phi_m$

$$\therefore\ \phi_m = \frac{E_t}{4.44\ f} = \frac{30.5}{4.44 \times 50}$$

$$= 0.1374\ Wb$$

Flux density, $B_m = \dfrac{\phi_m}{A_i}$

$$\therefore \text{Net core area, } A_i = \frac{\phi_m}{B_m} = \frac{0.1374}{1.25}$$

$$= 0.1099\ m^2 = 0.1099 \times 10^6\ mm^2$$

kVA rating of transformer, $Q = 2.22\ f\ B_m\ A_i\ K_w\ A_w \delta \times 10^{-3}$

$\therefore$ Window area, $A_w = \dfrac{Q}{2.22\, f\, B_m\, A_i\, K_w\, \delta \times 10^{-3}}$

$$= \dfrac{1000}{2.22 \times 50 \times 1.25 \times 0.1099 \times 10^6 \times 0.32 \times 2.5 \times 10^{-3}}$$

$$= 0.0820 \ m^2$$

$$= 0.0820 \times 10^6 \ mm^2$$

RESULT

Net core area, A_i = 0.1099 m^2 = 0.1099 $\times 10^6$ mm^2

Window area, A_w = 0.0820 m^2 = 0.0820 $\times 10^6$ mm^2

EXAMPLE 7.2   *(MU, May' 19, 10 M) (PTU, May' 19, 10 M) (AU, Apr' 17, 10 M) (AU, Nov' 17, 16 M) (AU, Apr' 18, 8 M)*

Estimate the main dimensions including winding conductor area of a 3-phase, Δ-Y core type transformer rated at 300 *kVA*, 6600/440 V, 50 *Hz*. A suitable core with 3-steps having a circumscribing circle of 0.25 *m* diameter and a leg spacing of 0.4 *m* is available. Emf per turn = 8.5 *V*, δ = 2.5 *A/mm²*, K_w= 0.28, S_f = 0.9 (stacking factor).

GIVEN DATA

3-phase, Δ-Y	50 *Hz*	E_t = 8.5 *V*
3-stepped core	δ = 2.5 *A/mm²*	Core type
300 *kVA*	d = 0.25	K_w = 0.28
leg spacing = 0.4 *m*	S_f = 0.9	6600/440 *V*

SOLUTION

Let 440 *V* side be secondary and 6600 *V* be primary. Here the secondary is star connected and primary is delta connected.

$\therefore$ Secondary voltage per phase, $V_s = \dfrac{440}{\sqrt{3}} = 254\ V$ | Star connected secondary. |

Also, $E_s \approx V_s$

Emf per turn, $E_t = \dfrac{E_s}{T_s}$

$\therefore$ Number of secondary turns per phase, $T_s = \dfrac{E_s}{E_t} = \dfrac{254}{8.5} = 29.88 \approx 30$

Phase voltage ratio of transformer, $\dfrac{V_s}{V_p} = \dfrac{254}{6600}$ | Delta connected primary. |

Number of primary turns per phase $T_p = T_s \times \dfrac{V_p}{V_s}$

$$= 30 \times \dfrac{6600}{254} = 779.5 \approx 780$$

kVA rating of transformer, $Q = \sqrt{3}\ V_{LP}\ I_{LP} \times 10^{-3} = \sqrt{3}\ V_{LS}\ I_{LS} \times 10^{-3}$

where, $\quad V_{LP}$ = Line voltage on primary side

$\quad\quad I_{LP}$ = Line current on primary side

$\quad\quad V_{LS}$ = Line voltage on secondary side

$\quad\quad I_{LS}$ = Line current on secondary side

Line current on primary side, $I_{LP} = \dfrac{Q}{\sqrt{3}\ V_{LP} \times 10^{-3}}$

$$= \frac{300}{\sqrt{3} \times 6600 \times 10^{-3}} = 26.24\ A$$

Since primary is delta connected,

Phase current on primary, $I_p = \dfrac{I_{LP}}{\sqrt{3}} = \dfrac{26.24}{\sqrt{3}} = 15.15\ A$

$\left.\begin{array}{l}\text{Area of cross-section}\\ \text{of primary conductor}\end{array}\right\}\ a_p = \dfrac{I_p}{\delta} = \dfrac{15.15}{2.5} = 6.06\ mm^2$

Line current on secondary sides, $I_{LS} = \dfrac{Q}{\sqrt{3} \times V_{LS} \times 10^{-3}}$

$$= \frac{300}{\sqrt{3} \times 440 \times 10^{-3}} = 393.65\ A$$

Since secondary is star connected,

Phase current on secondary, $I_S = I_{LS} = 393.65\ A$

$\left.\begin{array}{l}\text{Area of cross-section}\\ \text{of secondary conductor}\end{array}\right\}\ a_s = \dfrac{I_s}{\delta} = \dfrac{393.65}{2.5} = 157.46\ mm^2$

Conductor area in window, $A_c = 2\ (a_p\ T_p + a_s\ T_s)$

$$= 2\ (6.06 \times 780 + 157.46 \times 30)$$

$$= 18901.2\ mm^2 = 18901.2 \times 10^{-6}\ m^2$$

$$= 0.0189\ m^2$$

Window Area, $A_w = \dfrac{A_c}{K_w}$

$$= \frac{0.0189}{0.28} = 0.0675\ m^2$$

Area of circumscribing circle $= \dfrac{\pi d^2}{4} = \dfrac{\pi\ (0.25)^2}{4} = 0.0491\ m^2$

For 3-stepped core, $\dfrac{\text{Gross core area}}{\text{Area of circumscribing circle}} = 0.84$

Gross core area, $A_{gi} = 0.84 \times$ Area of circumscribing circle

$$= 0.84 \times 0.0491 = 0.0412\ m^2$$

Net core area, $A_i = S_f \times A_{gi}$

$$= 0.9 \times 0.0412 = 0.0371\ m^2$$

Given that, leg spacing = 0.4 m

Width of window, W_w = Leg spacing = 0.4 m

Height of window, $H_w = \dfrac{A_w}{W_w} = \dfrac{0.0675}{0.4} = 0.17\ m$

RESULT

Number of primary turns per phase,	T_p	= 780
Number of secondary turns per phase,	T_s	= 30
Area of cross-section of primary conductor,	a_p	= 6.06 mm^2
Area of cross-section of secondary conductor,	a_s	= 157.46 mm^2
Net core area,	A_i	= 0.0371 m^2
Window area,	A_w	= 0.0675 m^2
Height of window,	H_w	= 0.17 m
Width of window,	W_w	= 0.4 m

EXAMPLE 7.3

A 3-phase, 50 Hz, oil cooled core type transformer has the following dimensions: Distance between core centres = 0.2 m, height of window = 0.24 m, diameter of circumscribing circle = 0.14 m, flux density in the core = 1.25 Wb/m^2, current density in the conductor = 2.5 A/mm^2. Assume a window space factor of 0.2 and core area factor = 0.56. The core is 2-stepped. Estimate kVA rating of the transformer.

GIVEN DATA

3-phase	D = 0.2 m	δ = 2.5 A/mm^2
50 Hz	H_w = 0.24 m	K_w = 0.2
Core type	d = 0.14 m	K_c = 0.56
2-stepped core	B_m = 1.25 Wb/m^2	

SOLUTION

kVA rating of three-phase transformer, $Q = 3.33\ f\ B_m\ A_i\ K_w\ A_w\ \delta \times 10^{-3}$

To estimate the kVA, we have to calculate A_w and A_i

Width of window, $W_w = D - d = 0.2 - 0.14 = 0.06\ m$

Window Area, $A_w = H_w \times W_w$

$\qquad\qquad = 0.24 \times 0.06 = 0.0144\ m^2$

Core area factor, $K_c = \dfrac{A_i}{d^2}$

Net core area, $A_i = K_c\ d^2$

$\qquad\qquad = 0.56 \times 0.14^2 = 0.01098\ m^2 = 0.011\ m^2$

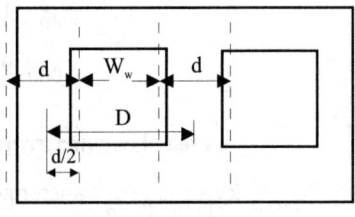

Fig. 1

kVA rating, $Q = 3.33 \, f \, B_m \, A_i \, K_w \, A_w \, \delta \times 10^{-3}$

$\qquad = 3.33 \times 50 \times 1.25 \times 0.011 \times 0.2 \times 0.0144 \times 2.5 \times 10^6 \times 10^{-3}$

$\qquad = 16.4835 \, kVA = 16.5 \, kVA$

RESULT

kVA rating of the transformer, $Q = 16.5 \, kVA$

EXAMPLE 7.4
(AKTU, Dec' 19, 7 M) (HTU, Dec' 18, 10 M)

Determine the dimensions of core and window for a 5 kVA, 50 Hz, 1-phase, core type transformer. A rectangular core is used with long side twice as long as short side. The window height is 3 times the width. Voltage per turn = 1.8 V, space factor = 0.2, $\delta = 1.8 \, A/mm^2$, $B_m = 1 \, Wb/m^2$.

GIVEN DATA

Q = 5 kVA	Core-type	$\delta = 1.8 \, A/mm^2$
f = 50 Hz	Rectangular core	$B_m = 1 \, Wb/m^2$
1-phase	$E_t = 1.8 \, V$	Long side = 2 × Short side
$H_w = 3W_w$	$K_w = 0.2$	

SOLUTION

Emf per turn, $E_t = 4.44 \, f \, \phi_m$

$\therefore \phi_m = \dfrac{E_t}{4.44 \, f}$

$\qquad = \dfrac{1.8}{4.44 \times 50} = 0.0081 \, Wb$

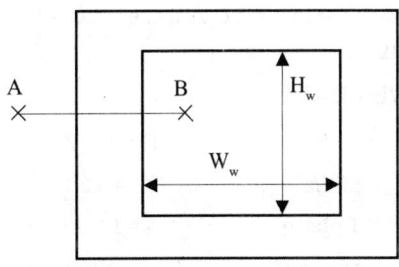

Fig. 1

Net core area, $A_i = \dfrac{\phi_m}{B_m} = \dfrac{0.0081}{1} = 0.0081 \, m^2$

Gross core area, $A_{gi} = \dfrac{A_i}{S_f} = \dfrac{0.0081}{0.9} = 0.009 \, m^2$

Cross-section of the core is rectangle.

Gross core area, A_{gi} = Length × Breadth = a × b

$\qquad = 2b \times b = 2b^2$

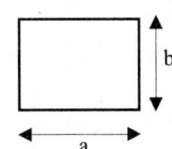

Fig. 2: Cross-section of core at A-B.

Given that, a = 2b

$\therefore b = \sqrt{\dfrac{A_{gi}}{2}} = \sqrt{\dfrac{0.009}{2}} = 0.067 \, m$

$\qquad a = 2b = 2 \times 0.067 = 0.134 \, m$

kVA rating of transformer, $Q = 2.22 \, f \, B_m \, A_i \, K_w \, A_w \, \delta \times 10^{-3}$

$\therefore$ Window area, $A_w = \dfrac{Q}{2.22 \, f \, B_m A_i K_w \delta \times 10^{-3}}$

$\qquad = \dfrac{5}{2.22 \times 50 \times 1 \times 0.0081 \times 0.2 \times 1.8 \times 10^6 \times 10^{-3}}$

$\qquad = 0.0154 \, m^2$

Window area, $A_w = H_w W_w$

$$= 3W_w \times W_w = 3 W_w^2$$

$$\therefore W_w = \sqrt{\frac{A_w}{3}} = \sqrt{\frac{0.0154}{3}} = 0.0716 \, m$$

$$H_w = 3 W_w = 3 \times 0.0716 = 0.2148 \, m$$

RESULT

Net core area,	A_i	=	$0.0081 \, m^2$
Dimensions of the core,	$a \times b$	=	$0.134 \times 0.067 \, m$
Window area,	A_w	=	$0.0154 \, m^2$
Dimensions of window, $H_w \times W_w$		=	$0.2148 \times 0.0716 \, m$

EXAMPLE 7.5

Determine the dimension of the core and window for a 10 kVA, 50 Hz, 1-phase, core type transformer. A rectangular core is used with long side twice as long as short side. The window height is 3 times the width. Voltage per turn = 1.6 V, space factor = 0.2, δ = 1.8 A/mm², B_m = 1 Wb/m², stacking factor = 0.9, window area = 0.0195 m².

GIVEN DATA

10 kVA	$E_t = 1.6 \, V$	$S_f = 0.9$
50 Hz	$K_w = 0.2$	$A_w = 0.0195 \, m^2$
1-phase	$\delta = 1.8 \, A/mm^2$	Long side = 2 × Short side
$H_w = 3W_w$	$B_m = 1 \, Wb/m^2$	

SOLUTION

Emf per turn, $E_t = 4.44 \, f \, \phi_m$

$$\therefore \phi_m = \frac{E_t}{4.44 \, f}$$

$$= \frac{1.6}{4.44 \times 50} = 0.0072 \, Wb$$

Net core area, $A_i = \dfrac{\phi_m}{B_m} = \dfrac{0.0072}{1} = 0.0072 \, m^2$

Gross core area, $A_{gi} = \dfrac{A_i}{S_f} = \dfrac{0.0072}{0.9} = 0.008 \, m^2$

Cross-section of the core is rectangle.

Gross core area, A_{gi} = Length × Breadth = a × b

$$= 2b \times b = 2b^2$$

$$\therefore b = \sqrt{\frac{A_{gi}}{2}} = \sqrt{\frac{0.008}{2}} = 0.0632 \, m$$

$$a = 2b = 2 \times 0.0632 = 0.1264 \, m$$

Window area, $A_w = H_w W_w$

$\boxed{\text{Given that, } H_w = 3W_w}$

$= 3W_w \times W_w = 3\,W_w^{\,2}$

$\therefore W_w = \sqrt{\dfrac{A_w}{3}} = \sqrt{\dfrac{0.0195}{3}} = 0.081\, m$

$H_w = 3\,W_w = 3 \times 0.081 = 0.243\, m$

RESULT

Dimensions of the core,   a × b   = 0.1264 × 0.0632 m

Dimensions of window, $H_w \times W_w$ = 0.243 × 0.081 m

EXAMPLE 7.6
(VTU, Dec' 19, 10 M)

Determine the dimensions of the core, the number of turns, the cross-section area of conductors in primary and secondary windings of a 100 kVA, 2200/480 V, 1-phase, core type transformer, to operate at a frequency of 50 Hz, by assuming the following data.

Volt per turn = 7.5 V. Maximum flux density =1.2 Wb/m². Ratio of effective cross-sectional area of core to square of diameter of circumscribing circle is 0.6. Ratio of height to width of window is 2. Window space factor = 0.28. Current density = 2.5 A/mm².

GIVEN DATA

100 kVA	$K_w = 0.28$	$H_w/W_w = 2$
2200/480 V	$E_t = 7.5\ V$	1-phase
50 Hz	$A_i/d^2 = 0.6$	$\delta = 2.5\ A/mm^2$
Core type	$B_m = 1.2\ Wb/m^2$	

SOLUTION

Emf per turn, $E_t = 4.44\,f\,\phi_m$

$\therefore \phi_m = \dfrac{E_t}{4.44\,f}$

$= \dfrac{7.5}{4.44 \times 50} = 0.0338\, Wb$

$\therefore$ Net core area, $A_i = \dfrac{\phi_m}{B_m} = \dfrac{0.0338}{1.2} = 0.0282\, m^2$

$\boxed{B_m = \dfrac{\phi_m}{A_i}}$

Given that, $\dfrac{A_i}{d^2} = 0.6$.   Hence the core is 3-stepped core

$\left.\begin{array}{l}\therefore \text{Diameter of} \\ \text{circumscribing circle}\end{array}\right\}\, d = \sqrt{\dfrac{A_i}{0.6}} = \sqrt{\dfrac{0.0282}{0.6}} = 0.2168\, m$

kVA rating of 1-phase transformer, $Q = 2.22\,f\,B_m\,A_i\,K_w\,A_w\,\delta \times 10^{-3}$

$\therefore$ Window area, $A_w = \dfrac{Q}{2.22\,f\,B_m A_i K_w \delta \times 10^{-3}}$

$= \dfrac{100}{2.22 \times 50 \times 1.2 \times 0.0282 \times 0.28 \times 2.5 \times 10^6 \times 10^{-3}}$

$= 0.038 \ m^2$

We know that,

Window area, $A_w = H_w W_w = 2\,W_w \times W_w = 2\,W_w^{\,2}$

$\boxed{\dfrac{H_w}{W_w} = 2, \ \therefore \ H_w = 2\,W_w}$

$\therefore$ Width of window, $W_w = \sqrt{\dfrac{A_w}{2}} = \sqrt{\dfrac{0.038}{2}} = 0.1378 \ m$

Heigth of window, $H_w = 2\,W_w = 2 \times 0.1378 = 0.2756 \ m$

Let, 480 V side be secondary and 2200 V side be primary.

$\therefore$ Secondary voltage, $V_s = 480 \ V$

Also, $E_s \approx V_s$

Emf per turn, $E_t = \dfrac{E_s}{T_s}$

$\therefore$ Number of turns in secondary, $T_s = \dfrac{E_s}{E_t} = \dfrac{V_s}{E_t} = \dfrac{480}{7.5} = 64$

The voltage ratio of transformer, $\dfrac{V_s}{V_p} = \dfrac{480}{2200}$

$\therefore$ Number of turns in primary, $T_p = T_s \times \dfrac{V_p}{V_s} = 64 \times \dfrac{2200}{480}$

$= 293$

Considering primary voltage and current,

kVA rating of single-phase transformer, $Q = V_p I_p \times 10^{-3}$

$\therefore$ Current in primary, $I_p = \dfrac{Q}{V_p \times 10^{-3}} = \dfrac{100}{2200 \times 10^{-3}} = 45.45 \ A$

Area of cross–section of primary conductor $\left.\right\}\, a_p = \dfrac{I_p}{\delta} = \dfrac{45.45}{2.5} = 18.18 \ mm^2$

Considering secondary voltage and current,

kVA rating of single-phase transformer, $Q = V_s I_s \times 10^{-3}$

$\therefore$ Current in secondary, $I_s = \dfrac{Q}{V_s \times 10^{-3}} = \dfrac{100}{480 \times 10^{-3}} = 208.33 \ A$

Area of cross–section of secondary conductor $\left.\right\}\, a_s = \dfrac{I_s}{\delta} = \dfrac{208.33}{2.5} = 83.33 \ mm^2$

RESULT

Net core area,	A_i	=	$0.0282\ m^2$
Diameter of circumscribing circle,	d	=	$0.2168\ m$
Window area,	A_w	=	$0.038\ m^2$
Window dimension,	$H_w \times W_w$	=	$0.2756 \times 0.1378\ m$
Number of turns in primary,	T_P	=	293
Number of turns in secondary,	T_S	=	64
Area of cross-section of primary conductor,	a_p	=	$18.18\ mm^2$
Area of cross-section of secondary conductor,	a_s	=	$83.33\ mm^2$

EXAMPLE 7.7

Calculate the dimension of the core, the number of turns and cross-sectional area of conductors in the primary and secondary windings of a 200 kVA, 2500 / 440 V, 50 Hz, 1-phase, shell type transformer. Maximum flux density, = 1.1 Wb/m^2, current density = 2.4 A/mm^2, AT = 9687.5 $amp\text{-}turn$.

GIVEN DATA

200 kVA $\delta = 2.4\ A/mm^2$

2500/440 V $B_m = 1.1\ Wb/m^2$

1-phase   · $AT = 9687.5\ amp\text{-}turn$

Shell type

SOLUTION

Primary current, $I_p = \dfrac{Q}{V_p \times 10^{-3}} = \dfrac{200}{2500 \times 10^{-3}} = 80\ A$ $\boxed{Q = V_p\,I_p \times 10^{-3}}$

Secondary current, $I_s = \dfrac{Q}{V_s \times 10^{-3}} = \dfrac{200}{440 \times 10^{-3}} = 455\ A$ $\boxed{Q = V_s\,I_s \times 10^{-3}}$

Secondary turns, $T_s = \dfrac{AT}{I_s} = \dfrac{9687.5}{455} = 21.29 \approx 22$ $\boxed{AT = T_s\,I_s}$

Primary turns, $T_p = T_s \times \dfrac{V_p}{V_s} = 22 \times \dfrac{2500}{440} = 125$

Area of cross-section of primary conductor $\left.\right\}\ a_p = \dfrac{I_p}{\delta} = \dfrac{80}{2.4} = 33.33\ mm^2$

Area of cross-section of secondary conductor $\left.\right\}\ a_s = \dfrac{I_s}{\delta} = \dfrac{455}{2.4} = 189.58\ mm^2$

RESULT

Secondary turns,	T_s =	22
Primary turns,	T_p =	125
Area of cross-section of primary conductor,	A_p =	33.33 mm^2
Area of cross-section of secondary conductor,	A_s =	189.58 mm^2

EXAMPLE 7.8   (KTU, Feb' 18, 12 M)

Calculate the dimension of the core, the number of turns and cross-sectional area of conductors in the primary and secondary windings of a 100 kVA, 2300 / 400 V, 50 Hz, 1-phase, shell type transformer. Ratio of magnetic and electric loadings equal to 480 × 10⁻⁸ (i.e., ratio of flux and secondary mmf at full load). B_m = 1.1 Wb/m², δ = 2.2 A/mm², K_w = 0.3, Stacking factor = 0.9.

$$\frac{\text{Depth of stacked core}}{\text{Width of central limb}} = 2.6 \qquad \frac{\text{Height of window}}{\text{Width of window}} = 2.5$$

GIVEN DATA

100 kVA	ϕ_m/AT = 480 × 10⁻⁸	D_c / W_c = 2.6
2300/400 V	B_m = 1.1 Wb/m²	50 Hz
1-phase	δ = 2.2 A/mm²	H_w/W_w = 2.5
Shell type	K_w = 0.3	S_f = 0.9

SOLUTION

The kVA rating of 1-phase transformer $\Big\}$
$$Q = 2.22 f B_m A_i K_w A_w \delta \times 10^{-3}$$

$$\boxed{\phi_m = B_m A_i}$$

$$= 2.22 f (B_m A_i)(K_w A_w) \delta \times 10^{-3}$$

$$\boxed{AT = \frac{K_w A_w \delta}{2} \Rightarrow K_w A_w \delta = 2AT}$$

$$= 2.22 f \phi_m 2AT \times 10^{-3}$$

$$\boxed{AT = \frac{\phi_m}{480 \times 10^{-8}}}$$

$$= 2.22 f \phi_m \times 2 \times \frac{\phi_m}{480 \times 10^{-8}} \times 10^{-3}$$

$$\therefore \phi_m^2 = \frac{Q \times 480 \times 10^{-8}}{4.44 \times f \times 10^{-3}} = \frac{100 \times 480 \times 10^{-8}}{4.44 \times 50 \times 10^{-3}}$$

$$\phi_m = \sqrt{\frac{100 \times 480 \times 10^{-8}}{4.44 \times 50 \times 10^{-3}}} = 0.0465 \, Wb$$

Net core area, $A_i = \dfrac{\phi_m}{B_m} = \dfrac{0.0465}{1.1} = 0.0423 \, m^2$

Gross-core area, $A_{gi} = \dfrac{A_i}{S_f} = \dfrac{0.0423}{0.9} = 0.047 \, m^2$

Given that, $\dfrac{D_c}{W_c} = 2.6 \Rightarrow D_c = 2.6 \times W_c$ .....(1)

We know that,

$$A_{gi} = D_c \times W_c = 2.6 \times W_c \times W_c = 2.6 \times W_c^2 \quad \boxed{\text{Using equation (1)}}$$

$\therefore$ Width of core, $W_c = \sqrt{\dfrac{A_{gi}}{2.6}} = \sqrt{\dfrac{0.047}{2.6}} = 0.1345 \ m$

Depth of core, $D_c = 2.6 \times W_c$

$$= 2.6 \times 0.1345 = 0.3497 \ m$$

Given that, $\dfrac{\phi_m}{AT} = 480 \times 10^{-8}$

$\therefore AT = \dfrac{\phi_m}{480 \times 10^{-8}} = \dfrac{0.0465}{480 \times 10^{-8}} = 9687.5 \ amp\text{-}turn$

$\therefore$ Primary current, $I_p = \dfrac{Q}{V_p \times 10^{-3}} = \dfrac{100}{2300 \times 10^{-3}} = 43.48 \ A$

Secondary current, $I_s = \dfrac{Q}{V_s \times 10^{-3}} = \dfrac{100}{400 \times 10^{-3}} = 250 \ A$

Secondary turns, $T_s = \dfrac{AT}{I_s} = \dfrac{9687.5}{250} = 38.75 \approx 40$

Transformer voltage ratio, $\dfrac{V_s}{V_p} = \dfrac{400}{2300}$

Primary turns, $T_p = T_s \times \dfrac{V_p}{V_s} = 40 \times \dfrac{2300}{400} = 230$

Area of cross-section of primary conductor $\Big\} \ a_p = \dfrac{I_p}{\delta} = \dfrac{43.48}{2.2} = 19.76 \ mm^2$

Area of cross-section of secondary conductor $\Big\} \ a_s = \dfrac{I_s}{\delta} = \dfrac{250}{2.2} = 113.64 \ mm^2$

Window dimensions

The kVA rating of single-phase transformer $\Big\} \ Q = 2.22 \ f \ B_m \ A_i \ K_w \ A_w \ \delta \times 10^{-3}$

$\therefore$ Window area, $A_w = \dfrac{Q}{2.22 \ f \ B_m A_i K_w \delta \times 10^{-3}}$

$$= \dfrac{100}{2.22 \times 50 \times 1.1 \times 0.0423 \times 0.3 \times 2.2 \times 10^6 \times 10^{-3}}$$

$$= 0.0293 \ m^2$$

We know that,

Area of window, $A_w = H_w W_w = 2.5 \ W_w \ W_w = 2.5 \ W_w^2$

$\therefore$ Width of window, $W_w = \sqrt{\dfrac{A_w}{2.5}} = \sqrt{\dfrac{0.0293}{2.5}} = 0.1083 \ m$

Height of window, $H_w = 2.5 \ W_w = 2.5 \times 0.1083 = 0.2708 \ m$

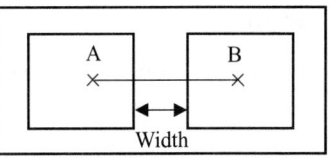

Width

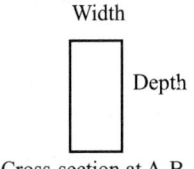

Cross-section at A-B

Fig. 1

$\boxed{Q = V_p \ I_p \times 10^{-3}}$

$\boxed{Q = V_s \ I_s \times 10^{-3}}$

$\boxed{AT = T_s \ I_s}$

$\boxed{\dfrac{H_w}{W_w} = 2.5 \ \Rightarrow \ H_w = 2.5 \ W_w}$

RESULT

Area of cross-section of core,	A_i	$= 0.0423\ m^2$
Core cross-section,	$W_c \times D_c$	$= 0.1345 \times 0.3497\ m$
Area of window,	A_w	$= 0.0293\ m^2$
Window dimensions,	$H_w \times W_w$	$= 0.2708 \times 0.1083\ m$
Number of primary turns,	T_p	$= 230$
Number of secondary turns,	T_s	$= 40$
Area of cross-section of primary conductor,	a_p	$= 19.76\ mm^2$
Area of cross-section of secondary conductor,	a_s	$= 113.64\ mm^2$

EXAMPLE 7.9   *(AU, Nov' 18, 7 M)*

Determine the dimensions of core and yoke for a *200 kVA, 50 Hz*, single-phase core type transformer. A cruciform core is used with distance between centers of adjacent limbs equal to 1.6 times the width of core laminations. Assume voltage per turn is 14 V, maximum flux density, B_m = 1.1 *Wb/m²*, window space factor = 0.32, current density = 3 *A/mm²*, stacking factor = 0.9. The net iron area is 0.56 d² for cruciform core, width of largest stamping = 0.85 d.

GIVEN DATA

200 *kVA*	$K_w = 0.32$	$a = 0.85\ d$	$E_t = 14\ V$
50 *Hz*	$B_m = 1.1\ Wb/m^2$	$S_f = 0.9$	$D = 1.6\ a$
$\delta = 3\ A/mm^2$	$A_i = 0.56\ d^2$		

SOLUTION

Emf per turn, $E_t = 4.44\ f\ \phi_m = 4.44\ f\ B_m\ A_i$ $\boxed{\phi_m = B_m\ A_i}$

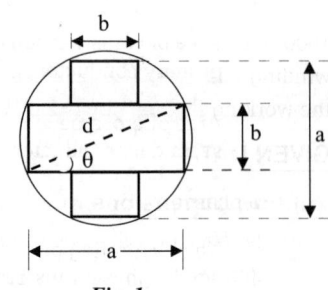

∴ Net core area, $A_i = \dfrac{E_t}{4.44\ f\ B_m}$

$= \dfrac{14}{4.44 \times 50 \times 1.1} = 0.0573\ m^2$

Given that,

Net iron area, $A_i = 0.56\ d^2$

∴ Diameter of circumscribing circle, $d = \sqrt{\dfrac{A_i}{0.56}}$

$= \sqrt{\dfrac{0.0573}{0.56}} = 0.32\ m$

Width of largest stamping, a = 0.85 d

$= 0.85 \times 0.32 = 0.272\ m$

Distance between core centers, D = 1.6 a

$= 1.6 \times 0.272 = 0.435\ m$

Width of window, $W_w = D - d$

$= 0.435 - 0.32 = 0.115\ m$

Fig. 1

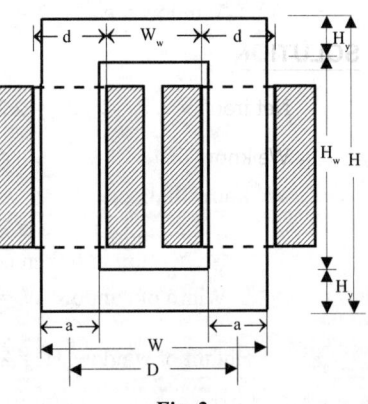

Fig. 2

kVA rating, $Q = 2.22 \, f \, B_m \, K_w \, \delta \, A_w \, A_i \times 10^{-3}$

$\therefore$ Window area, $A_w = \dfrac{Q}{2.22 \, f \, B_m \, K_w \, \delta \, A_i \times 10^{-3}}$

$= \dfrac{200}{2.22 \times 50 \times 1.1 \times 0.32 \times 3 \times 10^6 \times 0.0573 \times 10^{-3}} = 0.0297 \, m^2$

Height of window, $H_w = \dfrac{A_w}{W_w} = \dfrac{0.0297}{0.115} = 0.258 \, m$

Let us use the same stepped section for the yoke and core.

$\therefore$ Depth of yoke, $D_y = a = 0.272 \, m$

Height of yoke, $H_y = a = 0.272 \, m$

Overall height of frame, $H = H_w + 2H_y = 0.258 + 2 \times 0.272 = 0.802 \, m$

Overall length of frame, $W = D + a = 0.435 + 0.272 = 0.707 \, m$

RESULT

Height of window, $H_w = 0.258 \, m$

Width of window, $W_w = 0.115 \, m$

Overall height of frame, $H = 0.802 \, m$

Overall length of frame, $W = 0.707 \, m$

EXAMPLE 7.10 (AU, Apr' 19, 13 M)

A single-phase, 400 V, 50 Hz, transformer is built from stampings having a relative permeability of 1000. The length of the flux path is 2.5 m. The area of cross-section of the core is $2.5 \times 10^{-3} \, m^2$ and the primary winding has 800 turns. Estimate the maximum flux and no-load current of the transformer. The iron loss at the working flux density is 2.6 W/kg. Iron weight $7.8 \times 10^3 \, kg/m^3$, stacking factor is 0.9.

GIVEN DATA

1-phase	$B_m = 2.6 \, W/kg$	Length of flux path, $l_i = 2.5 \, m$
50 Hz	$T_p = 800$ turns	Iron weight $= 7.8 \times 10^3 \, kg/m^3$
400 V	$A_{gi} = 2.5 \times 10^{-3} \, m^2$	
$S_f = 0.9$	$\mu_r = 1000$	

SOLUTION

Net iron area, $A_i = S_f \, A_{gi} = 0.9 \times 2.5 \times 10^{-3} = 2.25 \times 10^{-3} \, m^2$

We know that,

$E_p = 4.44 \, f \, \phi_m \, T_p$

$\therefore$ Maximum flux in core, $\phi_m = \dfrac{E_p}{4.44 \, f \, T_p} = \dfrac{400}{4.44 \times 50 \times 800}$

$= 2.25 \times 10^{-3} \, Wb$

Reluctance, $R = \dfrac{l_i}{\mu \, A_i}$

Magnetizing mmf, $AT_o = R \phi_m$

$$= \frac{l_i}{\mu A_i} \phi_m = \frac{l_i \ \phi_m}{\mu_r \mu_o \ A_i}$$

$$\boxed{\mu = \mu_r \ \mu_o}$$

$$\boxed{\mu_o = 4\pi \times 10^{-7} \ H/m}$$

$$= \frac{2.5 \times 2.25 \times 10^{-3}}{1000 \times 4\pi \times 10^{-7} \times 2.25 \times 10^{-3}} = 1989 \ AT$$

Magnetizing current, $I_m = \dfrac{AT_o}{\sqrt{2} \ T_p}$

$$= \frac{1989}{\sqrt{2} \times 800} = 1.758 \ A$$

Volume of core = Net core area × Length of flux path = $A_i \times l_i$

$$= 2.25 \times 10^{-3} \times 2.5$$

$$= 5.625 \times 10^{-3} \ m^3$$

∴ Weight of core = Volume × Density

$$= 5.625 \times 10^{-3} \times 7.8 \times 10^3 = 43.875 \ kg$$

Iron loss, P_i = Loss/kg × Weight

$$= 2.6 \times 43.875 = 114 \ W$$

Loss component of no-load current, $I_l = \dfrac{P_i}{E_p} = \dfrac{114}{400} = 0.285 \ A$

∴ No-load current, $I_o = \sqrt{I_m^2 + I_l^2}$

$$= \sqrt{1.758^2 + 0.285^2} = 1.781 \ A$$

RESULT

Maximum flux in core, ϕ_m = 2.25 × 10⁻³ *Wb*

No-load current, I_o = 1.781 *A*

EXAMPLE 7.11

A single-phase, 500 *V*, 50 *Hz*, transformer is built from stampings having a relative permeability of 1500. The length of the flux path is 2.9 *m*. The area of cross-section of the core is 2.9 × 10⁻³ *m²* and the primary winding has 900 turns. Estimate the reluctance, maximum flux and magnetizing current of the transformer. Stacking factor is 0.9, AT_o = 2200.

GIVEN DATA

1-phase	T_p = 900 turns	AT_o = 2200
50 *Hz*	μ_r = 1500	
500 *V*	A_{gi} = 2.9 × 10⁻³ *m²*	
S_f = 0.9	l_i = 2.9 *m*	

SOLUTION

Net iron area, $A_i = S_f A_{gi} = 0.9 \times 2.9 \times 10^{-3} = 2.61 \times 10^{-3} \ m^2$

We know that,

$$E_p = 4.44 \ f \ \phi_m \ T_p$$

$\therefore$ Maximum flux in core, $\phi_m = \dfrac{E_p}{4.44 \ f \ T_p} = \dfrac{500}{4.44 \times 50 \times 900}$

$$= 2.5 \times 10^{-3} \ Wb$$

Reluctance, $R = \dfrac{l_i}{\mu \ A_i}$

$$= \dfrac{2.9}{1500 \times 4\pi \times 10^{-7} \times 2.61 \times 10^{-3}} = 58976.17 \ AT$$

Magnetizing current, $I_m = \dfrac{AT_0}{\sqrt{2} \ T_p}$

$$= \dfrac{2200}{\sqrt{2} \times 900} = 1.7285 \ A$$

RESULT

Reluctance,	R =	$58976.17 \ AT/Wb$
Maximum flux in core,	ϕ =	$0.0025 \ Wb$
Magnetizing current,	I_m =	$1.7285 \ A$

EXAMPLE 7.12 *(UTU, Dec' 13, 10 M)*

A 100 kVA, 2000/400 V, 50 Hz, single-phase, shell type transformer has the following particulars: B_{max} = 1.1 Wb/m², current density, δ = 2.2 A/mm², K_w = 0.33, volt per turn = 11 V, core is rectangular and stamping are 7 cm wide. Length of window = 2(width of window). Obtain net iron area, area of window, dimensions and weight of core. Specific gravity of iron = 7.8 gm/cm³. Sketch the core and show how LV and HV windings are arranged.

GIVEN DATA

100 kVA	$K_w = 0.33$	Specific gravity of iron = 7.8 gm/cm³ = 7.8 × 10³ kg/m³
2000/400 V	$B_m = 1.1 \ Wb/m^2$	$H_w = 2W_w$
50 Hz	$\delta = 2.2 \ A/mm^2$	a = 7 cm = 0.07 m
1-phase	$E_t = 11$	

SOLUTION

Emf per turn, $E_t = 4.44 \ f \ \phi_m = 4.44 \ f \ B_m \ A_i$ $\boxed{\phi_m = B_m \ A_i}$

$\therefore$ Net iron area, $A_i = \dfrac{E_t}{4.44 \ f \ B_m} = \dfrac{11}{4.44 \times 50 \times 1.1} = 0.045 \ m^2$

Gross core area, $A_{gi} = \dfrac{A_i}{S_f} = \dfrac{0.045}{0.9} = 0.05 \ m^2$

Cross-section of the core is rectangle. With reference to Fig. 1b.

Gross core area, $A_{gi} = 2a \times b = 2 \times 0.07 \times b$

$\therefore b = \dfrac{A_{gi}}{2 \times 0.07} = \dfrac{0.05}{2 \times 0.07} = 0.3571\,m = 0.36\,m = 36\,cm$

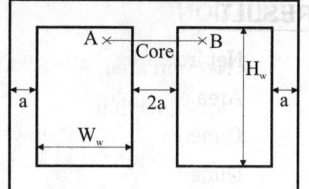

Fig. a: *Dimensions of core.*

kVA rating of transformer, $Q = 2.22\,f\,B_m\,A_i\,K_w\,A_w\,\delta \times 10^{-3}$

$\therefore$ Area of window, $A_w = \dfrac{Q}{2.22\,f\,B_m\,A_i\,K_w\,\delta \times 10^{-3}}$

$= \dfrac{100}{2.22 \times 50 \times 1.1 \times 0.045 \times 0.33 \times 2.2 \times 10^6 \times 10^{-3}}$

$= 0.0251\,m^2$

Also, Area of window, $A_w = H_w \times W_w = 2\,W_w \times W_w$

$\therefore$ Width of window, $W_w = \sqrt{\dfrac{A_w}{2}} = \sqrt{\dfrac{0.0251}{2}} = 0.112\,m$

Height of window, $H_w = 2\,W_w = 2 \times 0.112 = 0.224\,m$

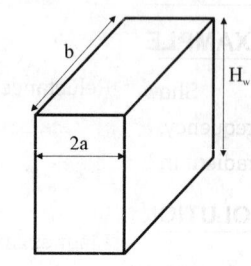

Fig. b: *Cross-section of AB.*

Fig. 1: *Shell type transformer.*

With reference to Fig. 1b,

Volume of core = Width × Depth × Height

$= 2a \times b \times H_w = 2a\,b\,H_w$

Weight of core = Volume of core × Specific gravity of iron

$= 2a\,b\,H_w \times 7.8 \times 10^3$

$= 2 \times 0.07 \times 0.36 \times 0.224 \times 7.8 \times 10^3 = 88\,kg$

Primary turns, $T_p = \dfrac{V_p}{E_t} = \dfrac{2000}{11} = 181.8 \approx 182$ turns

$\boxed{\text{Transformer ratio, } k = \dfrac{V_s}{V_p} = \dfrac{T_s}{T_p} = \dfrac{I_p}{I_s}}$

Secondary turns, $T_s = \dfrac{V_s}{V_p} \times T_p = \dfrac{400}{2000} \times 182 = 36.4 \approx 37$ turns

Primary current, $I_p = \dfrac{Q}{V_p \times 10^{-3}} = \dfrac{100}{2000 \times 10^{-3}} = 50\,A$

Area of cross-section of primary conductor, $a_p = \dfrac{I_p}{\delta} = \dfrac{50}{2.2} = 22.7272\,mm^2$

Secondary current, $I_s = \dfrac{Q}{V_s \times 10^{-3}} = \dfrac{100}{400 \times 10^{-3}} = 250\,A$

Area of cross-section of secondary conductor, $a_s = \dfrac{I_s}{\delta} = \dfrac{250}{2.2} = 113.6363\,mm^2$

Refer Fig. 7.2 for arrangement of low voltage and high voltage winding.

RESULT

Net iron area,	A_i	=	$0.045\ m^2$
Area of window,	A_w	=	$0.0251\ m^2$
Dimensions of window,	$H_w \times W_w$	=	$0.224 \times 0.112\ m$
Dimensions of core,	$2a \times b$	=	$0.14 \times 0.36\ m$
Weight of core		=	$88\ kg$

EXAMPLE 7.13

(JNTKU, Apr' 19, 7 M)

Show that the output of a 3-phase core type transformer is, $Q = 5.23\ f\ B_m\ H\ d^2\ H_w \times 10^{-3}\ kVA$, where, $f =$ Frequency, Hz; $B_m =$ Maximum flux density, Wb/m^2; $d =$ Effective diameter of the core, m; $H =$ Magnetic potential gradient in limb, A/m; $H_w =$ Height of window, m.

SOLUTION

Let us assume, $A_i =$ Area of circumscribing circle, A_{cc}

$$\therefore A_i = A_{cc} = \frac{\pi}{4}\ d^2 \qquad\qquad .....(1)$$

In a three-phase core type transformer each limb has one primary and one secondary winding.

$$\therefore \text{Total mmf over one limb} = T_p\ I_p + T_s\ I_s = 2\ T_p\ I_p \qquad\boxed{T_p\ I_p = T_s\ I_s}$$

$$\therefore \text{Magnetic potential gradient, } H = \frac{\text{mmf}}{\text{Lenght of field}} = \frac{\text{mmf}}{\text{Height of limb}}$$

$$= \frac{2\ T_p\ I_p}{H_w}$$

$$\therefore T_p\ I_p = \frac{H\ H_w}{2} \qquad\qquad .....(2)$$

Considering primary voltage and current,

$$\left.\begin{array}{l}\text{kVA output of a}\\\text{three-phase transformer}\end{array}\right\} Q = 3\ E_p\ I_p \times 10^{-3} \qquad\boxed{E_p = 4.44\ f\ \phi_m\ T_p}$$

$$\boxed{\phi_m = B_m\ A_i}$$

$$= 3 \times 4.44\ f\ \phi_m\ T_p\ I_p \times 10^{-3}$$

$$= 3 \times 4.44\ f\ B_m\ A_i\ T_p\ I_p \times 10^{-3}$$

$$= 3 \times 4.44\ f\ B_m\ \frac{\pi}{4}\ d^2\ \frac{H\ H_w}{2} \times 10^{-3} \qquad\boxed{\begin{array}{l}\text{Using equations (1)}\\\text{and (2)}\end{array}}$$

$$= \frac{3 \times 4.44\ \pi}{4 \times 2}\ f\ B_m\ H\ d^2\ H_w \times 10^{-3}$$

$$= 5.23\ f\ B_m\ H\ d^2\ H_w \times 10^{-3}\ kVA$$

Note : Magnetic field gradient is a variation of magnetic field in a particular direction. In transformer core, the flux is uniform and so the magnetic field gradient is same as magnetizing force, H which is ampere-turn per unit length.

EXAMPLE 7.14

(JNTKU, Apr' 19, 6 M)

The current densities in the primary and secondary windings of a transformer are 2.2 and 2.1 A/mm^2 respectively. The ratio of transformation is 10: 1 and the length of mean turn of the primary is 10% greater than that of the secondary. Calculate the resistance of the secondary winding given that the primary winding resistance is 10 Ω.

GIVEN DATA

$\delta_p = 2.2 \ A/mm^2$ $r_p = 10 \ \Omega$

$\delta_s = 2.1 \ A/mm^2$ $L_{mtp} = \dfrac{110}{100} L_{mts} \Rightarrow \dfrac{L_{mts}}{L_{mtp}} = \dfrac{1}{1.1}$

$k = \dfrac{1}{10} = 0.1$

SOLUTION

Resistance of primary winding, $r_p = \rho \ \dfrac{T_p \ L_{mtp}}{a_p}$

$$= \rho \ \frac{T_p \ L_{mtp}}{\dfrac{I_p}{\delta_p}}$$

$$= \rho \ \frac{T_p \ L_{mtp} \ \delta_p}{I_p}$$

Similarly,

Resistance of secondary winding, $r_s = \rho \ \dfrac{T_s \ L_{mts} \ \delta_s}{I_s}$

$\therefore$ Ratio of secondary and $\left.\begin{matrix}\\\\\end{matrix}\right\} = \dfrac{r_p}{r_s} = r_s \times \dfrac{1}{r_p} = \rho \ \dfrac{T_s \ L_{mts} \ \delta_s}{I_s} \times \dfrac{1}{\rho} \ \dfrac{I_p}{T_p \ L_{mtp} \ \delta_p}$
primary resistance

$$= \left(\frac{T_s}{T_p}\right)\left(\frac{L_{mts}}{L_{mtp}}\right)\left(\frac{\delta_s}{\delta_p}\right)\left(\frac{I_p}{I_s}\right)$$

$$= k \left(\frac{L_{mts}}{L_{mtp}}\right)\left(\frac{\delta_s}{\delta_p}\right) k$$

$\therefore$ Secondary resistance, $r_s = k^2 \left(\dfrac{L_{mts}}{L_{mtp}}\right)\left(\dfrac{\delta_s}{\delta_p}\right) r_p$

$$= \left(\frac{1}{10}\right)^2 \times \left(\frac{1}{1.1}\right) \times \left(\frac{2.1}{2.2}\right) \times 10$$

$$= 0.01 \times 0.9091 \times 0.9545 \times 10$$

$$= 0.0868 \ \Omega$$

$$a_p = \frac{I_p}{\delta_p}$$

$$a_s = \frac{I_s}{\delta_s}$$

Transformer ratio,

$$k = \frac{I_p}{I_s} = \frac{T_s}{T_p}$$

RESULT

Resistance of secondary winding, $r_s = 0.0868 \ \Omega$

EXAMPLE 7.15

(MU, May' 19, 10 M)

A 100 kVA, 2000/400 V, 50 Hz, single-phase, shell type transformer has sandwich coils. There are two full high voltage coils, one full low voltage coil and 2 half low voltage coils. Calculate the value of leakage reactance referred to high voltage side. The given data is: depth of half voltage coil = 40 mm, depth of low volatage coils = 35 mm, depth of duct between high voltage and low voltage = 16 mm, width of winding = 0.12 m, length of mean turn = 1.5 m, the number of turns in high voltage winding are 200.

GIVEN DATA

100 kVA	b_p = 40 mm = 0.04 m	W = 0.12 m
2000/400 V	b_s = 36 mm = 0.036 m	L_{mt} = 1.5 m
50 Hz	a = 16 mm = 0.016 m	n = 2
1-phase	T_p = 200	

SOLUTION

Leakage reactance referred to high voltage side $\left.\right\}$ $X_p = \pi f \mu_0 \dfrac{L_{wt}}{W} \dfrac{T_p^2}{n}\left(a + \dfrac{b_p + b_s}{6}\right)$ $\qquad$ $\boxed{\mu_0 = 4\pi \times 10^{-7}\ H/m}$

$$= \pi \times 50 \times 4\pi \times 10^{-7} \times \frac{1.5}{0.12} \times \frac{200^2}{2} \times \left(0.016 + \frac{0.04 + 0.036}{6}\right)$$

$$= 49.348 \times (0.016 + 0.0127) = 1.4163\ \Omega$$

High voltage winding current at full load, $I_p = \dfrac{Q}{V_p}$

$$= \frac{100 \times 1000}{2000} = 50\ A$$

$\therefore$ Per unit leakage reactance, $\varepsilon_x = \dfrac{I_p X_p}{V_p}$

$$= \frac{50 \times 1.4163}{2000} = 0.0354$$

RESULT

Leakage reactance referred to high voltage side $\left.\right\}$ X_p = 1.4163 Ω

EXAMPLE 7.16

A 500 kVA, 6600/440 V, 50 Hz, 3-phase, delta/ star, core type transformer has 400 turns on the high voltage winding. The height of winding is 0.5 m and the length of mean turn 1.2 m. Calculate the instantaneous radial force on the high voltage if a short circuit occurs at the terminals of the low voltage winding with high voltage energised. The leakage impedance is 4 percent. Take the double effect multiplier as 1.6. Also calculate the force at full load.

GIVEN DATA

500 kVA	T = 400
6600/440 V	L_c = 0.5
50 Hz	L_{mts} = 1.2
3-phase	ε_x = 4% = 0.04

SOLUTION

Full load current per phase on high voltage side $\left.\right\}$ $I = \dfrac{kVA \times 10^3}{3\ V_p}$ $\qquad$ $\boxed{kVA = 3\ V_p\ I_p \times 10^{-3}}$

$\boxed{I_p = I}$

$$= \frac{500 \times 10^3}{3 \times 6600} = 25.2525\ A$$

$\therefore$ Instantaneous peak value of $\left.\vphantom{\begin{matrix}a\\b\end{matrix}}\right\}$ $i_{m\_sc} = 1.6 \times \dfrac{\sqrt{2}\,I}{\varepsilon_x}$
short circuit current

$$= \frac{1.6 \times \sqrt{2} \times 25.2525}{0.04} = 1428.5\ A$$

Instantaneous maximum radial
force on the high voltage coil at $\left.\vphantom{\begin{matrix}a\\b\\c\end{matrix}}\right\}$ $f_{r\_max} = \dfrac{\mu_0}{2}(i_{m\_sc}\,T)^2 \dfrac{L_{mt}}{L_c}$
short circuit condition

$$= \frac{4\pi \times 10^{-7}}{2} \times (1428.5 \times 400)^2 \times \frac{1.2}{0.5}$$

$$= 492347.3\ N = 492.3473 \times 10^3\ N$$

$$= 492.3473\ kN$$

Force at full load, $F_r = \dfrac{\mu_0}{2}(I\,T)^2 \dfrac{L_{mt}}{L_c}$

$$= \frac{4\pi \times 10^{-7}}{2} \times (25.2525 \times 400)^2 \times \frac{1.2}{0.5}$$

$$= 153.8579\ N$$

RESULT

Instantaneous maximum radial force, f_{r_max} =   492.3473 kN

Force at full load, F_r =   153.8579 N

EXAMPLE 7.17

A 500 *kVA*, 6600/400 *V*, 50 *Hz*, single-phase, core type transformer has the following data:

Width of high voltage winding = 28 *mm*, width of low voltage winding = 24 *mm*, width of duct = 16 *mm*, height of coils = 0.36 *m*, length of mean turn = 1.26 *m*, number of turn in high voltage winding = 195, per unit impedence = 0.037, doubling effect multiplier = 1.9.

(a) Find the instantaneous radial force on high voltage winding under short circuit conditions if the height of high voltage and low voltage winding is equal. (b) Find the instantaneous axial force on high voltage winding under short circuit conditions if the high voltage winding is 5 percent shorter than the low voltage winding at one end.

GIVEN DATA

500 *kVA*	a = 16 *mm*	T = 195
6600/400 *V*	b_p = 28 *mm*	ε_x = 0.037
50 *Hz*	b_s = 24 *mm*	k = 5% = 0.05
1-phase	L_{mts} = 1.26	L_c = 0.36 *m*

SOLUTION

RMS value of full load current, $I_p = \dfrac{kVA \times 10^3}{V_p}$

$\boxed{kVA = 3\,V_p\,I_p \times 10^{-3}}$

$\boxed{I_p = I}$

$$= \frac{500 \times 10^3}{6600} = 75.7575\ A$$

$$\left.\begin{array}{l}\text{Instantaneous peak current}\\ \text{under short circuit condition}\end{array}\right\} i_{m\_sc} = 1.9 \times \frac{\sqrt{2}\, I_p}{\varepsilon_x}$$

$$= 1.9 \times \frac{\sqrt{2} \times 75.7575}{0.037} = 5501.6 \ A$$

Instantaneous radial force, $F_r = \frac{\mu_o}{2} (i_{m\_sc}\, T)^2 \frac{L_{mt}}{L_c}$

$$= \frac{4\pi \times 10^{-7}}{2} \times (5501.6 \times 195)^2 \times \frac{1.26}{0.36}$$

$$= 2531017.6 \ N = 2531.0176 \times 10^3 \ N$$

$$= 2531.0176 \ kN$$

Instantaneous axial force, $F_a = \frac{\mu_o}{2} k\, (i_{m\_sc}\, T)^2 \frac{L_{mt}}{2(a + b_p + b_s)}$

$$= \frac{4\pi \times 10^{-7}}{2} \times 0.05 \times (5501.6 \times 195)^2 \times \frac{1.26}{(0.016 + 0.028 + 0.024)}$$

$$= 669975.2 \ N = 669.9752 \times 10^3 \ N = 669.9752 \ kN$$

RESULT

Instantaneous radial force, F_r = 2531.0176 kN

Instantaneous axial force, F_a = 669.9752 kN

EXAMPLE 7.18

(VTU, Dec' 19, 10 M) (AU, Apr' 18, 8 M) (AU, Apr' 17, 10 M)

The tank of 1250 kVA, natural oil cooled transformer has the dimensions length, width and height as 0.65 × 1.55 × 1.85 m respectively. The full load loss =13.1 kW. The loss dissipation due to radiations = 6 W/m²-°C, loss dissipation due to convection = 6.5 W/m²-°C, improvement in convection due to provision of tubes = 40%, temperature rise = 40 °C, length of each tube = 1 m, diameter of tube = 50 mm. Find the number of tubes for this transformer. Neglect the top and bottom surface of the tank as regards the cooling.

GIVEN DATA

kVA = 1250	λ_{conv} = 6.5 W/m²-°C	$L_T \times W_T \times H_T$ = 0.65 × 1.55 × 1.85 m
l_t = 1 m	λ_{rad} = 6 W/m²-°C	Improvement in cooling, X_i = 40%
d_t = 50 mm	θ = 40°C	Total load, $P_i + P_c$ = 13.1 kW

SOLUTION

L_T = Length of tank = 0.65 m

W_T = Width of tank = 1.55 m

H_T = Height of tank = 1.85 m

$\left.\begin{array}{l}\text{Heat dissipating}\\ \text{surface of tank}\end{array}\right\}$ S_T = Total area of vertical sides

$$= 2\,(L_T H_T + W_T H_T)$$

$$= 2\,H_T(L_T + W_T)$$

$$= 2 \times 1.85 \times (0.65 + 1.55) = 8.14 \ m^2$$

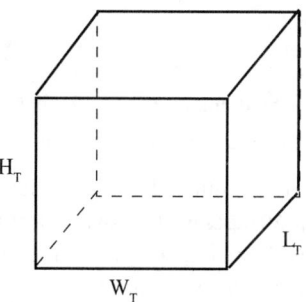

Fig. 1: *Tank dimension.*

Loss dissipated by tank walls, $P_{LDTW} = (\lambda_{rad} + \lambda_{conv}) S_T$

$\boxed{\text{Loss dissipated by radiation and convection}}$

$$= (6 + 6.5) S_T = 12.5 S_T$$

Heat dissipating area of tubes $= X S_T$

Loss dissipated by cooling tubes, $P_{LDC} = \lambda_{conv} X_i X S_T$

$\boxed{\text{Loss dissipated by convection}}$

$$= 6.5 \times \frac{140}{100} \times X S_T = 9.1 X S_T$$

∴ Total loss dissipated = Loss dissipated   Loss dissipated by
                         by tank walls   +cooling tubes

$$= 12.5 S_T + 9.1 X S_T$$

$$= S_T (12.5 + 9.1 X)$$

Temperature rise in
transformer with cooling tubes $\left.\right\} \theta = \dfrac{\text{Total loss}}{\text{Total loss dissipated}}$

$\boxed{\begin{array}{l}\text{Total loss,} = 13.1 \; kW \\ \qquad\quad = 13.1 \times 10^3 \; W\end{array}}$

$$= \frac{13.1 \times 10^3}{S_T (12.5 + 9.1 X)}$$

On rearranging the above equation we get,

$$12.5 + 9.1 X = \frac{13.1 \times 10^3}{\theta S_T}$$

$$\therefore X = \frac{1}{9.1} \left(\frac{13.1 \times 10^3}{\theta S_T} - 12.5 \right)$$

$$= \frac{1}{9.1} \left(\frac{13.1 \times 10^3}{40 \times 8.14} - 12.5 \right) = 3.0476$$

Total dissipating surface area of cooling tubes, $S_{dt} = X S_T$

$$= 3.0476 \times 8.14 = 24.8075 \; m^2$$

Surface area of each cooling tube, $S_t = \pi \, d_t \, l_t$

$$= \pi \times 50 \times 10^{-3} \times 1 = 0.157 \; m^2$$

Total number of cooling tubes $= \dfrac{\text{Total dissipating surface area of cooling tubes}}{\text{Surface area of each cooling tube}} = \dfrac{S_{dt}}{S_t}$

$$= \frac{24.8075}{0.157} = 158$$

The diameter of the tube is 50 mm and the standard distance between the tubes is half of the diameter and so, let distance between tubes = 25 mm.

The width of the tank is 1550 mm. If we leave an edge spacing of 62.5 mm on either sides then we can arrange 20 tubes widthwise with a spacing of 75 mm between centres of tubes. On lengthwise we can arrange 8 tubes with same spacing as that of widthwise tubes. But one row is not sufficient to accommodate the required 158 cooling tubes. Hence three rows of cooling tubes are provided on both lengthwise and widthwise. The plan of the cooling tubes is shown in Fig. 2. For symmetry of construction two additional tubes are provided.

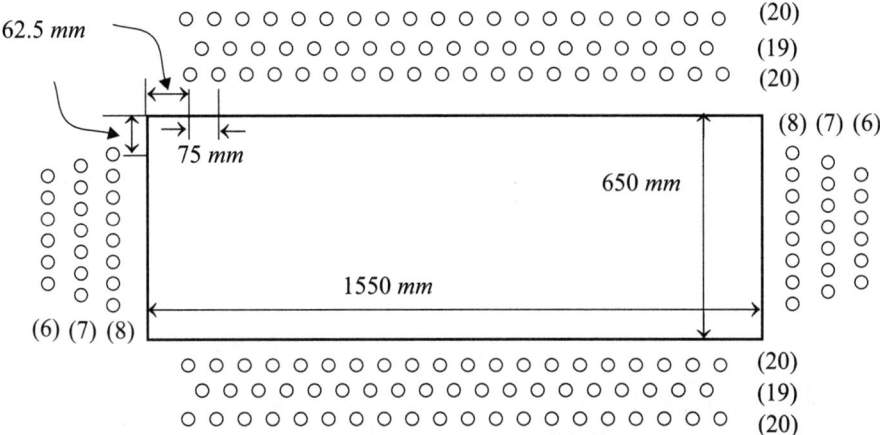

62.5 mm

75 mm

650 mm

1550 mm

(20)
(19)
(20)

(8) (7) (6)

(6) (7) (8)

(20)
(19)
(20)

Fig. 2: *Plan showing the arrangement of cooling tubes.*

RESULT

The total number of tubes = 160

Cooling tubes are arranged as 3 rows on widthwise with each row consisting of 20, 19 and 20 tubes and 3 rows on lengthwise each row consisting of 8, 7 and 6 tubes.

EXAMPLE 7.19   *(AU, Nov' 18, 13 M)*

A 250 kVA, 3-phase core type transformer has a total loss of 4800 W on full load. The transformer tank is 1.25 m in height and $1m \times 0.5$ m in plan. Design a suitable scheme for cooling tubes if the average temperature rise is to be limited to 35 °C. The diameter of the tube is 50 mm and are spaced 75 mm from each other. The average height of the tube is 1.05 m.

GIVEN DATA

kVA = 250	$L_T \times W_T \times H_T = 0.5 \times 1 \times 1.25$ m	6600/400 V
$\theta = 35°C$	Total power loss = 4800 W	3-phase
$d_t = 50$ mm	Distance between tube centres = 75 mm	Core type
$l_t = 1.05$ m		

SOLUTION

L_T = Length of tank = 0.5 m

H_T = Height of tank = 1.25 m

W_T = Width of tank = 1 m

Heat dissipating surface of tank $\}$ S_T = Total Area of vertical sides

$= 2 (L_T H_T + W_T H_T) = 2 H_T(L_T + W_T)$

$= 2 \times 1.25 \times (0.5 + 1) = 3.75$ m^2

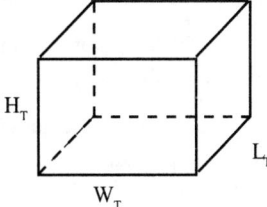

Fig. 1: *Tank dimensions.*

Let, specific loss dissiption due to radiation, λ_{rad} = 6 W/m^2-$°C$

Specific loss dissipation due to convection, λ_{conv} = 6.5 W/m^2-$°C$

Loss dissipated by tank walls, $P_{LDTW} = (\lambda_{rad} + \lambda_{conv}) S_T$

> Loss dissipated by radiation and convection

$$= (6 + 6.5) S_T = 12.5 S_T$$

Heat dissipating area of tubes = $X S_T$

Loss dissipated by cooling tubes, $P_{LDC} = \lambda_{conv} X_i X S_T$

> Loss dissipated by convection

$$= 6.5 \times \frac{135}{100} \times X S_T = 8.8 X S_T$$

∴ Total loss dissipated = Loss dissipated by tank walls + Loss dissipated by cooling tubes

$$= 12.5 S_T + 8.8 X S_T = S_T (12.5 + 8.8 X)$$

Temperature rise in transformer with cooling tubes $\Big\} \theta = \dfrac{\text{Total loss}}{\text{Total loss dissipated}}$

> Total loss = 4800 W

$$= \frac{4800}{S_T (12.5 + 8.8 X)}$$

On rearranging the above equation we get,

$$X = \frac{1}{8.8} \left[\frac{4800}{\theta\, S_T} - 12.5 \right] \qquad \Rightarrow \qquad X = \frac{1}{8.8} \left[\frac{4800}{35 \times 3.75} - 12.5 \right] = 2.7354$$

Total dissipating surface area of cooling tubes, $S_{dt} = X S_T = 2.7354 \times 3.75 = 10.2578 \ m^2$

Surface area of each cooling tube, $S_t = \pi\, d_t\, l_t = \pi \times 50 \times 10^{-3} \times 1.05 = 0.1649 \ m^2$

Number of cooling tubes, $n_t = \dfrac{\text{Total dissipating surface area of cooling tubes}}{\text{Surface area of each cooling tube}} = \dfrac{S_{dt}}{S_t}$

$$= \frac{10.2578}{0.1649} = 62.206 \approx 62$$

The width of the tank is 1000 mm. If we leave an edge spacing of 87.5 mm on either sides, then we can arrange 12 tubes widthwise with a spacing of 75 mm between the centres of tubes.

The length of the tank is 500 mm. If we leave an edge spacing of 100 mm on either sides, then we can arrange 5 tubes lengthwise with a spacing of 75 mm between the centres of tubes.

But one row is not sufficient to accommodate the required 62 cooling tubes. Hence 2 rows of cooling tubes are provided on both lengthwise and widthwise. The plan of the cooling tubes is shown in Fig. 2. For symmetry of construction two additional tubes are provided.

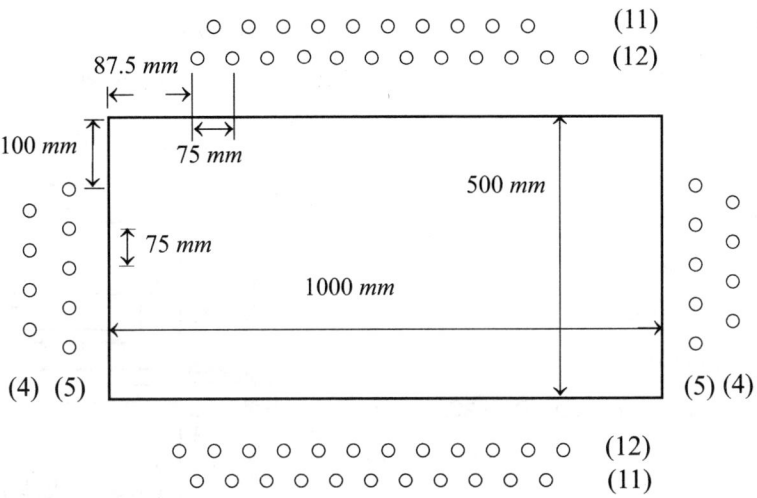

Fig. 2: *Plan showing the arrangement of cooling tubes.*

RESULT

Total number of cooling tubes = 64

Cooling tubes are arranged as 2 rows on widthwise with each row consisting of 12 and 11 tubes and 2 rows on lengthwise with each row consisting of 5 and 4 tubes.

EXAMPLE 7.20   *(AU, Nov' 17, 16 M)*

A 1000 *kVA*, 6600/440 *V*, 50 *Hz*, 3-phase, delta-star, core type, oil immersed natural cooled transformer has the following design data: Distance between centers of adjacent limbs = 0.47 *m*, outer diameter of high voltage winding = 0.44 *m*, height of the frame = 1.24 *m*, core loss = 3.7 *kW* and I²R loss = 10.5 *kW*. Design a suitable tank for the transformer. The average temperature rise of the oil should not exceed 35°C. The specific heat dissipation for the tank walls is 6 *W/m²-°C* and 6.5 *W/m²-°C* due to radiation and convection respectively. Assume that the convection is improved by 35% due to the provision to tubes.

GIVEN DATA

1000 *kVA*	D = 0.47 *m*	λ_{rad} = 6 *W/m²-°C*
6600/440 *V*	D_{oc} = 0.44 *m*	λ_{conv} = 6.5 *W/m²-°C*
50 *Hz*	H = 1.24 *m*	Core loss = 3.7 *kW*
3-phase	θ = 35°C	X_i = 35 %
Delta-star connected	I²R loss = 10.5 *kW*	

SOLUTION

Design of tank

Let us assume the clearance required between transformer core-winding assembly and tank as follow. (Refer Table 7.4)

$C_1 = 70\ mm = 0.07\ m$

$C_2 = 90\ mm = 0.09\ m$

$C_3 = 100\ mm = 0.1\ m$

$C_4 = 400\ mm = 0.4\ m$

With reference to Fig. 1.

Width of tank, $W_T = 2D + D_{oc} + 2C_1$

$\qquad = (2 \times 0.47) + 0.44 + (2 \times 0.07)$

$\qquad = 1.52\ m$

Lengh of tank, $L_T = D_{oc} + 2C_2$

$\qquad = 0.44 + (2 \times 0.09)$

$\qquad = 0.62\ m$

Height of tank, $H_T = H + C_3 + C_4$

$\qquad = 1.24 + 0.1 + 0.4$

$\qquad = 1.74\ m$

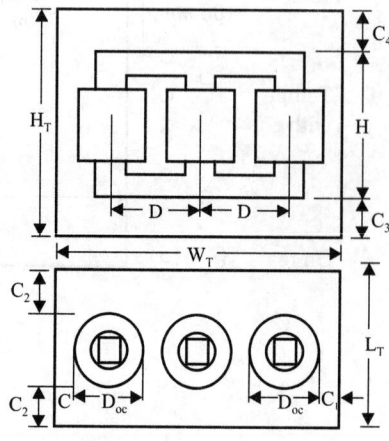

Fig. 1: *Dimensions of transformer core and tank.*

Design of cooling tubes

With reference to Fig. 2.

Heat dissipating surface of tank $\}$ S_T = Total area of vertical sides

$\qquad = 2\ (L_T H_T + W_T H_T)$

$\qquad = 2\ H_T(L_T + W_T)$

$\qquad = 2 \times 1.74 \times (0.62 + 1.52) = 7.45\ m^2$

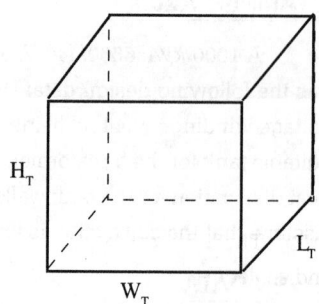

Fig. 2: *Tank dimension.*

Loss dissipated by tank walls, $P_{LDTW} = (\lambda_{rad} + \lambda_{conv})\ S_T = (6 + 6.5)\ S_T = 12.5\ S_T$

Heat dissipating area of tubes = $X\ S_T$

Loss dissipated by cooling tubes, $P_{LDC} = \lambda_{conv}\ X_i\ X\ S_T = 6.5 \times \dfrac{135}{100} \times X\ S_T = 8.8\ X\ S_T$

∴ Total loss dissipated = Loss dissipated   Loss dissipated by
                          by tank walls + cooling tubes

$$= 12.5\,S_T + 8.8\,XS_T = S_T\,(12.5 + 8.8\,X)$$

Total loss = Core loss + I^2R loss

$$= 3.7 + 10.5 = 14.2\ kW$$

Temperature rise in transformer $\Big\}$ $\theta = \dfrac{\text{Total loss}}{\text{Total loss dissipated}}$
with cooling tubes

$$= \frac{14.2 \times 10^3}{S_T\,(12.5 + 8.8\,X)}$$

On rearranging the above equation we get,

$$12.5 + 8.8\,X = \frac{14.2 \times 10^3}{\theta S_T}$$

$$\therefore X = \frac{1}{8.8}\left[\frac{14.2 \times 10^3}{\theta\,S_T} - 12.5\right]$$

$$= \frac{1}{8.8}\left[\frac{14.2 \times 10^3}{35 \times 7.45} - 12.5\right] = 4.768$$

Total dissipating surface area of cooling tubes, $S_{dt} = X\,S_T = 4.768 \times 7.45 = 35.5216\ m^2$

The heigth of transformer tank is 1.74 m and so the average length of tubes can be chosen as 1.5 m and the diameter of each tube is 50 mm.

Surface area of each cooling tube, $S_t = \pi\,d_t\,l_t = \pi \times 50 \times 10^{-3} \times 1.5 = 0.2356\ m^2$

Number of cooling tubes, $n_t = \dfrac{\text{Total dissipating surface area of cooling tubes}}{\text{Surface area of each cooling tube}}$

$$= \frac{35.5216}{0.2356} = 150.77 \approx 150$$

The diameter of the tube is 50 mm and the standard distance between the tubes is half of the diameter and so, let distance between tubes = 25 mm.

The width of the tank is 1520 mm. If we leave an edge spacing of 47.5 mm on either sides then we can arrange 19 tubes widthwise with a spacing of 75 mm between centres of tubes.

The length of the tank is 620 mm. If the leave an edge spacing of 47.5 mm on either sides then we can arrange 7 tubes lengthwise with a spacing of 75 mm between centres of tubes.

One row each on lengthwise and widthwise are not sufficient to accommodate 150 tubes, hence additional two rows on lengthwise with 18 and 17 tubes and additional two rows on widthwise with 7 tubes and 6 tubes are provided. The plan of cooling tubes is shown in Fig.3.

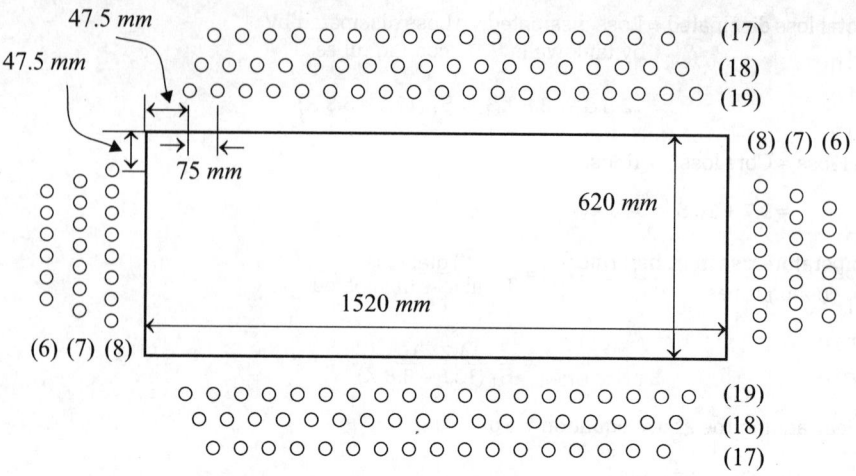

Fig. 3: *Plan showing the arrangement of cooling tubes.*

RESULT

Width of tank, W_T = 1.52 m

Lengh of tank, L_T = 0.62 m

Height of tank, H_T = 1.74 m

Total number of tubes, n_t = 150

7.13   COMPUTER PROGRAMS

PROGRAM 7.1

Write a C program to design the core and window dimensions of a single-phase core type transformer for the specifications in Example 7.1.

Method-1

```
/* Program to design core and window dimensions*/
/* VARIABLE DECLARATION
.............................................
Et = Emf per turn.
f = frequency.
Bm = Maximum flux density.
Q = kVA rating.
DEL = Current density.
Kw = Window space factor.
PHIm = Maximum flux in core.
Ai = Net core area.
Aw = Window area.
..............................................*/
```

```
#include<stdio.h>
#include<conio.h>
#include<math.h>
#include<iostream.h>
void main()
{
int f,Q;
float PHIm,Bm,Ai,DEL,Kw,Aw,Et;
clrscr();
printf(ÏNPUT DATA\n");
printf(""Enter Emf per turn, Et = ");
scanf("%f", &Et);
printf(Ënter Frequency, f = ");
scanf("%d", &f);
printf("Enter Maximum flux density, Bm = ");
scanf("%f", &Bm);
printf("Enter kVA rating, Q = ");
scanf("%d", &Q);
printf("Enter Current density, DEL = ");
scanf("%f", &DEL);
printf(Enter window space factor, Kw = ");
scanf("%f", &Kw);

PHIm = Et/(4.44*f);                        /* PHIm = Maximum flux in
                                                   core. */
Ai = PHIm/Bm;                              /* Ai   = core area. */
Aw = Q/(2.22*f*Bm*Ai*10e+6*Kw*DEL*10e-5);  /* Aw   = Window area. */
printf("OUTPUT\n");
printf("Core area, Ai = %8.4f m2\n", Ai);
printf("Window area, Aw = %8.4f m2\n", Aw);
getch();
}
```

Input data

```
Enter Emf per turn,         Et  = 30.5
Enter Frequency,            f   = 50
Enter Maximum flux density, Bm  = 1.25
Enter kVA rating,           Q   = 1000
Enter Current density,      DEL = 2.5
Enter Window space factor,  Kw  = 0.32
```

Output

```
Core area,    Ai = 0.1099 m2
Window area,  Aw = 0.0820 m2
```

Method-2

```
/* Program to design core and window dimensions*/

#include<stdio.h>
#include<conio.h>
#include<math.h>
#include<iostream.h>
void main()
{
clrscr();
int f,Q;
float PHIm,Bm,Ai,DEL,Kw,Aw,Et;

Et = 30.5;                      /* Et = Emf per turn. */
f = 50;                         /* f = Frequency. */
Bm = 1.25;                      /* Bm = Maximum flux density. */
Q = 1000;                       /* Q = kVA rating. */
DEL = 2.5;                      /* DEL = Current density. */
Kw = 0.32;                      /* Kw = window space factor.*/

PHIm = Et/(4.44*f);                              /* PHIm = Maximum flux
                                                          in core.*/
Ai = PHIm/Bm;                                    /* Ai = core area.*/
Aw = Q/(2.22*f*Bm*Ai*10e+6*Kw*DEL*10e-5); /* Aw = Window area.*/

printf("OUTPUT\n");
printf("Core area, Ai = %8.4f m2\n", Ai);
printf("Window area, Aw = %8.4f m2\n", Aw);
getch();
}
```

Output

```
Core area,   Ai = 0.1099 m2
Window area, Aw = 0.0820 m2
```

PROGRAM 7.2 •

Write a C program to design core and window dimensions of a single-phase core transformer for the specifications in Example 7.5.

```
/* Program to design core and window dimensions. */

#include<stdio.h>
#include<conio.h>
#include<math.h>
```

```
#include<iostream.h>
void main()
{
clrscr();
int f;
float Et,PHIm,Ai,Sf,Agi,b,a,Ww,Hw,Aw,Bm;
Et = 1.6;                  /* Et = Emf per turn.*/
f = 50;                    /* f = Frequency.*/
Bm = 1;                    /* Bm = Maximum flux density.*/
Sf = 0.9;                  /* Sf = Stacking factor.*/
Aw = 0.0195;               /* Aw = Window area */

PHIm = Et/(4.44*f); /* PHIm = Maximum flux in core.*/
Ai = PHIm/Bm;       /* Ai = Net core area.*/
Agi = Ai/Sf;        /* Agi = Gross core area.*/
b = sqrt((Agi/2));  /* b = Breadthof rectangle.*/
a = 2*b;            /* a = Width of largest stamping.*/

Ww = sqrt((Aw/3));  /* Ww = Width of window.*/
Hw = 3*Ww;          /* Hw = Height of window.*/

printf("OUTPUT\n");
printf("Dimension of core are a and b\n");
printf("Breadth of rectangle, b = %8.4f m\n", b);
printf("Width of largest stamping, a = %8.4f m\n", a);
printf("Dimension of window are Hw and Ww\n");
printf("Width of window, Ww = %8.4f m\n");
printf("Height of window, Hw = %8.4f m\n");
getch();
}
```

Output

```
Dimension of core are a and b
Breadth of rectangle,       b = 0.0632 m
Width of largest stamping, a = 0.1264 m
Dimension of window are Hw and Ww
Width of window,     Ww = 0.0806 m
Height of window,    Hw = 0.2418 m
```

PROGRAM 7.3

Write a C program to estimate the number of turns and area of cross-section in primary and secondary conductor of a single-phase shell type transformer for specifications in Example 7.7.

```c
/* Program to estimate Number of turns and area of cross-
sections of primary and secondary conductors. */

#include<Stdio.h>
#include<conio.h>
#include<math.h>
#include<iostream.h>
void main()
{
clrscr();
int Q,Vp,Vs,Ts,Tp;
float Ip,Is,AT,DEL,Ap,As;
Q = 200;                    /* Q = Kva rating.*/
Vp = 2500;                  /* Vp = Primary voltage.*/
Vs = 440;                   /* Vs = secondary voltage.*/
AT = 9687.5;                /* AT = Ampere-turns.*/
DEL = 2.4;                  /* DEL = Current density.*/

Ip = Q/(Vp*10e-4);          /* Ip = Primary current.*/
Is = Q/(Vs*10e-4);          /* Is = Secondary current.*/
Ts = AT/Is;                 /* Ts = Secondary turns.*/
Tp = Ts*Vp/(Vs);            /* Tp = Primary turns.*/
Ap = Ip/DEL;                /* Ap = Area of cross-section of
                                  primary conductor.*/

As = Is/DEL;                /* As = Area of cross-section of
                                  secondary conductor.*/
printf("OUTPUT\n");
printf("Secondary turns, Ts = %f\n", Ts);
printf("Primary turns, Tp = %f\n", Tp);
printf("Area of cross-section of primary conductor, Ap =  %8.4f
        mm2\n", Ap);

printf("Area of cross-section of secondary conductor, As =  %8.4f
        mm2\n", As);
getch();
}
```

Output

```
Secondary turns,                                    Ts = 22
Primary turns,                                      Tp = 125
Area of cross-section of primary conductor,    Ap = 33.3333 mm2
Area of cross-section of secondary conductor, As = 189.3939 mm2
```

PROGRAM 7.4

Write a C program to estimate the reluctance, maximum flux and magnetizing current of a single-phase transformer for specifications in Example 7.11.

```c
/* Program to estimate reluctance, maximum flux and magnetizing current.*/
#include<stdio.h>
#include<conio.h>
#include<math.h>
#include<iostream.h>
void main()
{
clrscr();
int Ep,f,Tp;
float li,R,PHIm,ATo,Im,Ai,Agi,Sf,Newr;
Newr = 1500;             /* Newr = Relative permeability.*/
li = 2.9;                /* li = Length of fluxpath.*/
f = 50;                  /* f = Frequency.*/
Tp = 900;                /* Tp = Number of turns in primary winding.*/
Ep = 500;                /* Ep = Primary induced emf.*/
ATo = 2200;              /* ATo = Magnetizing mmf.*/
Sf = 0.9;                /* Sf = Stacking factor. */
Agi = 2.9;               /* Agi = Gross core area. */

Ai = Sf*Agi*10e-4;              /* Ai = Net iron area.*/
PHIm = Ep/(4.44*f*Tp);          /* PHIm = maximum flux in core.*/
R = li/(4*3.14*10e-7*Newr*Ai);  /* R = Reluctance.*/
Im = ATo/(1.4142*Tp);           /* Im = Magnetizing current.*/

printf("OUTPUT\n");
printf("Reluctance, R = %8.4f AT/Wb\n", R);
printf("Maximum flux in core, PHIm = %8.4f Wb\n", PHIm);
printf("Magnetizing current, Im = %8.4f A\n", Im);
getch();
}
```

Output

```
Reluctance,            R    = 58976.1719 AT/Wb
Maximum flux in core, PHIm = 0.0025 Wb
Magnetizing current, Im    = 1.7285 A
```

PROGRAM 7.5

Write a C program to estimate the resistance of secondary winding of a transformer for specifications in Example 7.14.

```
/* Program to estimate Resistance of secondary winding */

#include<stdio.h>
#include<conio.h>
#include<math.h>
void main()
{
float DELp,DELs,k,LmtpLmts,rs,rp;
clrscr();
rp = 10;                 /* rp = Primary winding resistance. */
DELp = 2.2;              /* DELp = Current density in primary winding. */
DELs = 2.1;              /* DELs = Current density in secondary winding. */
k = 0.1;                 /* k = Ratio transformation. */
LmtpLmts = 0.9091;       /* LmtpLmts = Fraction of length of primary winding and
                                        secondary winding. */

rs = k*k*(LmtpLmts)*(DELs/DELp)*rp;   /* rs = Secondary winding resistance. */

printf("OUTPUT\n");
printf("Resistance of secondary winding, rs = %8.4f ohm\n", rs);
getch();
}
```

Output

Resistance of secondary winding, rs = 0.0868 ohm

PROGRAM 7.6

Write a C program to estimate the leakage reactance referred to high voltage side of a single-phase shell type transformer for specifications in Example 7.15.

```
/* Program to estimate Leakage reactance referred to high voltage side. */

#include<stdio.h>
#include<conio.h>
#include<math.h>
void main()
```

```
{
int f,n,Tp ;
float W,Lmt,a,bs,bp,Xp ;
clrscr();
f = 50;          /* f = Frequency */
n = 2;           /* n = full high voltage coils. */
W = 0.12;        /* W = Width of winding. */
Lmt = 1.5;       /* Lmt = Length of mean turn. */
Tp = 200;        /* Tp = Number of turns in primary winding.*/
a = 0.016;       /* a = depth of duct between high voltage and low voltage. */
bs = 0.036;      /*bs = Depth of low voltage coil.*/
bp = 0.04;       /* bp = Depth of half voltage coil.*/


Xp = 3.141*f*4*3.141*10e-8*(Lmt/(W))*Tp*Tp/(n)*(a+bp+bs/(6));

printf("OUTPUT\n");
printf("Leakage reactance referred to high voltage, Xp = %8.4f ohm\n", Xp);
getch();
}
```

Output

```
Leakage reactance referred to high voltage, Xp = 1.4203 ohm
```

PROGRAM 7.7

Write a C program to estimate the force at full load of a three-phase core type transformer for specifications in Example 7.16.

```
/* Program to estimate instantaneous maximum radial force and force at full load */

#include<stdio.h>
#include<conio.h>
#include<math.h>
void main()
{
int f,kVA,T,Vp;
float Lmt,Lc,I,Fr,EPSILONx,im_sc,fr_max;
clrscr();
f = 50;          /* f = Frequency */
kVA = 500;       /* kVA = kVA rating*/
T = 400;         /* T = Turns on high voltage winding. */
```

```
Lmt = 1.2;          /* Lmt = Length of mean turn. */
Lc = 0.5;           /* Lc = Height of winding.*/
Vp =  6600;         /* Vp = Terminal voltage for primary winding.*/
EPSILONx = 0.04;    /* EPSILONx = Leakage impedence. */


I = kVA*1000/(3*Vp);                        /* I = Full load current per
                                               phase on high voltage
                                               side.*/
im_sc = 1.6*1.4142*I/(EPSILONx);            /* im_sc = Instantaneous peak
                                               value of short circuit
                                               current. */
fr_max = 4*3.14*10e-8/(2)*im_sc*T*Lmt/(Lc); /* fr_max = Instantaneous
                                               maximum radial
                                               force. */
Fr = (4*3.141*10e-7/2)*(I*T*I*T)*(Lmt/Lc);  /* Fr = Force at full load. */
printf("OUTPUT\n");
printf(Instantaneous maximum radial force, fr_max = %8.4 kN/n", fr_max);
printf("Force at full load, Fr = %8.4f N/n", Fr);
getch();
}
```

Output

```
Instantaneous maximum radial force, fr_max = 492.0873 kN
Force at full load,                         Fr = 153.8608 N
```

PROGRAM 7.8

Write a C program to estimate the radial force and axial force of a single-phase core type transformer for specifications in Example 7.17.

```
/* Program to estimate Instataneous radial and axial force on high
voltage winding. */
#include<stdio.h>
#include<conio.h>
#include<math.h>
void main()
{
int Vp,T,kVA;
float EPSILONx,Lmts,Lc,Ip,im_sc,Fr,Fa,k,bp,bs,a;
clrscr();
kVA = 500;              /* kVA = kVA rating. */
Vp = 6600;              /* Vp = Primary winding voltage. */
EPSILONx = 0.037;       /* epsilon = Per unit impedence. */
```

```
Lmts = 1.26;              /* Lmts = Length of mean turn. */
Lc = 0.36;                /* Lc = Axial height of winding. */
T = 195;                  /* T = number of turn. */
a = 0.016;                /* a = Width of duct. */
bp = 0.028;               /*bp = Radial width of primary winding. */
bs = 0.024;               /*bs = Radial width of secondary winding. */
k = 0.05;                 /*k =  High voltage winding. */

Ip  = kVA*1000/(Vp);            /* Ip = RMS value of full load
                                         current.*/

im_sc = 1.9*(1.41*Ip/(EPSILONx));  /* im_sc = Instantaneous peak current
                                            under short circuit. */

Fr = 6.2831*10e-7*(im_sc*T)*(im_sc*T)*(Lmts/(Lc));      /* Fr = Radial force. */

Fa = 6.2831*10e-7*k*(im_sc*T)*(im_sc*T)*(Lmts/(a+bp+bs)); /* Fa = Axial
                                                             force.*/
printf("OUTPUT\n");
printf(" Instantaneous radial force, Fr =%8.4f kN\n", Fr);
printf("Instantaneous axial force, Fa = %8.4f kN\n", Fa);
getch();
}
```

Output

```
Instantaneous radial force,  Fr = 2531.0216 kN
Instantaneous axial force,   Fa = 669.9740 kN
```

7.14 SHORT-ANSWER QUESTIONS

Q7.1 *What are the various types of transformers?*

The transformers can be classified based on construction, applications, frequency range, number of windings and type of connection.

Based on construction the transformers are classified as,

    1. Core type   2. Shell type

Based on applications the transformers are classified as,

    1. Distribution transformers   4. Instrument transformers

    2. Power transformers   5. Electronics transformers

    3. Special transformers

Based on frequency range the transformers are classified as,

    1. Power frequency transformer   5. Wide band transformer

    2. Audio frequency transformer   6. Narrow band transformer

    3. UHF transformers   7. Pulse transformer

Based on the number of windings the transformers are classified as,

    1. Auto transformer   2. Two winding transformer

Based on the type of connection the transformers are classified as

    1. Single-phase transformer   3. Three-phase transformer

Q7.2 What are the different types of heat transfer methods found in electrical machines? *(AU, Apr' 18, 2 M)*

There are three types heat transfer methods are given by,

  1. Conduction (Solid and liquids)
  2. Convection (Liquids and gases)
  3. Radiation (Electro-magnetic waves).

Q7.3 What are the constructional elements of a transformer?

The constructional elements of a transformer are core, high and low voltage windings, cooling tubes or radiators and tank.

Q7.4 What is the range of efficiency of a transformer?

The efficiency of a commercial transformer will be in the range of 94% to 99%. Among the available electrical machines the transformers have the highest efficiency.

Q7.5 What is transformer bank?

A transformer bank consists of three independent single-phase transformers with their primary and secondary windings either in star or delta connected to form a 3-phase transformer.

Q7.6 What are the salient features of distribution transformer? *(AU, Apr' 18, 5 M)*

  1. The distribution transformers will have low iron loss and higher value of copper loss.
  2. The capacity of transformers will be upto 500 *kVA*.
  3. The transformers will have plain walled tanks or provided with cooling tubes or radiators.
  4. The leakage reactance and regulation will be low.
  5. Designed for high energy efficiency.

Q7.7 What is yoke section of a transformer? What is its use?

The sections of the core which connect the limbs are called yoke. The yoke is used to provide a closed path for flux.

Q7.8 Mention the uses of distribution transformer.

The distribution transformers are used at load centres to step down the transmission line voltage to a standard service voltage required for consumers.

Q7.9 What are distribution transformers?

The transformers used at load centres to step down distribution voltage to a standard service voltage required for consumers are called distribution transformers.

Q7.10 What are power transformers?

The transformers used in sub-stations and generating stations for step-down or step-up the voltage are called power transformers.

Q7.11 State the uses of power transformers.

  1. In generating stations the power transformers are used to step-up the voltage to a higher level required for primary transmission.
  2. In sub-stations the power transformers are used to step-down the voltage from to a lower level required for secondary transmission.

Q7.12 How the design of distribution transformer differs from that of a power transformer? *(AU, Nov' 17, 2 M)*

  1. The distribution transformers are designed to have low iron loss and higher copper loss, whereas in power transformers the copper loss will be lesser than iron loss.
  2. The distribution transformers are designed to have the maximum efficiency at a load much lesser than full load, whereas the power transformers are designed to have maximum efficiency at or near full load.
  3. In distribution transformer the leakage reactance is kept low to have better regulation, whereas in power transformers the leakage reactance is kept high to limit the short circuit current.

Q7.13 Distinguish between core and shell type transformers.

In core type transformer the coil surrounds the core, whereas in shell type transformer the core surrounds the coil.

Q7.14 What is the advantage in shell type transformer over core type transformer.

In shell type transformers the coils are well supported on all the sides and so they can withstand higher mechanical stresses developed during short circuit conditions. Also the leakage reactance will be less in shell type transformers.

Q7.15 In transformers, why the low voltage winding is placed near the core?

The winding and core are both made of metals and so an insulation have to be placed inbetween them. The thickness of insulation depends on the voltage rating of the winding. In order to reduce the insulation requirement the low voltage winding is placed near the core.

Q7.16 What is window space factor?   *(KTU, Feb' 18, 4 M) (AU, Apr' 19, 2 M) (AU, Apr' 17, 2 M)*

The window space factor is defined as the ratio of conductor area in window to total area of window.

Q7.17 Write down the output equation of 1-phase and 3-phase transformer.

The equation which relates the rated kVA output of the transformer to core and window area is called output equation.

Output kVA of single-phase transformer, $Q = 2.22 \, f \, B_m \, A_i \, K_w \, A_w \, \delta \times 10^{-3}$

Output kVA of three-phase transformer, $Q = 3.33 \, f \, B_m \, A_i \, K_w \, A_w \, \delta \times 10^{-3}$

Q7.18 How will you select the emf per turn of a transformer?

The equation of emf per turn in terms of kVA rating, flux, frequency and ampere-turn is given by,

Emf per turn, $E_t = K\sqrt{Q}$

where, $K = \sqrt{4.44 \, f \, \dfrac{\phi_m}{AT} \times 10^3}$

For a given kVA rating the emf per turn is directly proportional to K. The value of K depends on the type, service condition and method of construction of transformer. The value of K for different types of transformers are listed in the following Table.

Transformer type	K
Single-phase shell type	1.0 to 1.2
Single-phase core type	0.75 to 0.85
Three-phase shell type	1.3
Three-phase core type, distribution transformer	0.45
Three-phase core type, power transformer	0.6 to 0.7

Q7.19 Why circular coils are preferred in transformers?

The excessive leakage fluxes produced during short circuit and over loads, develop severe mechanical stresses on the coils. On circular coils these forces are radial and there is no tendency for the coil to change its shape. But on rectangular coils the forces are perpendicular to the conductors and tends to deform the coil in circular form.

Q7.20 What are the advantages and disadvantages of stepped cores?   *(AU, Nov' 18, 2 M)*

Advantages

For same area of cross-section the stepped cores will have lesser diameter of circumscribing circle than square cores. This results in reduction in length of mean turn of the winding with consequent reduction in both cost of copper and copper loss.

Disadvantages

With large number of steps a large number of different sizes of laminations have to be used. This results in higher labour charges for shearing and assembling different types of laminations.

Q7.21 Define conductor space factor (or) copper space factor.

The conductor space factor is the ratio of conductor area and window area in case of transformers. (In case of rotating machines it is the ratio of conductor area and slot area.)

(AU, Nov' 18, 2 M)

Q7.22 What is the objective behind using sheet steel stampings in the construction of electrical machines ? (or) how is iron loss is reduced in transformer ?

The stampings are used to reduce the eddy current losses. The stampings are stacked such that they offer small area of cross-section for the eddy current path so resistance will be high which in-turn reduces eddy current. Moreover, the stampings are insulated by a thin coating of varnish, hence when the stampings are stacked to form a core, the resistance for the eddy current is very high.

Q7.23 What are the various types of cross-sections used for the core of the transformers?

The various types of cross-section used are square, rectangle, cruciform and multistepped cores.

Q7.24 Give typical values of K_C for various types of transformers.

The various values of core area factor, K_C are given below:

    Square core, K_C = 0.45
    Cruciform core, K_C = 0.56
    Three stepped core, K_C = 0.6
    Four stepped core, K_C = 0.62

Q7.25 The area of yoke in a transformer is taken as 15 to 20% larger than that of core why? Why not increase the size of core also?

The dimensions of the yoke is kept larger to reduce the flux density in the yoke which in-turn reduces iron losses. If core area is increased then working flux density will be lesser and to compensate for reduction in flux we have to increase the number of turns. Therefore core area is not increased.

Q7.26 What type of steel is commonly used for the core of transformer?

The hot rolled and cold rolled silicon steel with 3 to 5% silicon are used for the laminations of the core of transformers. The hot rolled silicon stell allows a maximum flux density of 1.45 Wb/m^2 and the cold rolled silicon steel permits a maximum flux density of 1.8 Wb/m^2.

Q7.27 How the laminations of the core are insulated?

In laminations made of hot rolled silicon steel a thin coating of kaolin or varnish is applied to insulate them. In laminations of cold rolled silicon steel a phosphate based coating is applied to insulate them. In high capacity transformers above 10 MVA rating in addition to phosphate coating, a coating of kaolin or varnish is applied.

(HTU, Dec' 18, 2 M)

Q7.28 What do you meant by stacking factor (or iron space factor)? What is its usual value?

In transformers, the core is made of laminations and the laminations are insulated from each other by a thin coating of varnish. Hence when the laminations are stacked to form the core, the actual iron area will be less than the core area. The ratio of iron area and total core area is called stacking factor.

$$\text{Stacking factor, } S_f = \frac{\text{Area of cross-section of iron in the core}}{\substack{\text{Area of cross-section of the core} \\ \text{including the insulation area}}}$$

The usual value of stacking factor is 0.9.

Q7.29 *Draw the cruciform section of the transformer core or cross-section of two stepped core and give the optimum dimensions in terms of circumscribing circle diameter, d.*

The optimum dimensions of a and b in terms of d are,

a = 0.85 d, b = 0.53 d

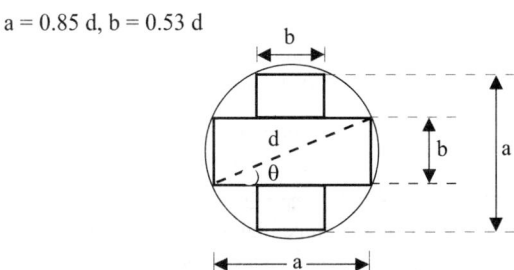

Fig. Q7.29: Cross-section of two stepped core.

Q7.30 *Draw the cross-section of three stepped core. How many different size of laminations are used in it?*

The cross-section of three stepped core is shown in figure below. It is constructed by using three different sizes of laminations.

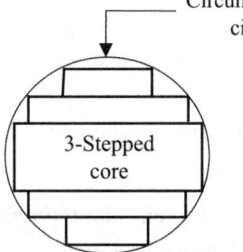

Fig. Q7.30: Cross-section of three stepped transformer core.

Q7.31 *Why stepped cores are used in transformers?*   *(HTU, Dec' 18, 2 M) (AU, Nov' 17, 2 M) (AU, Apr' 18, 2 M)*

When stepped cores are used the diameter of the circumscribing circle is minimum for a given area of the core. This helps in reducing the length of mean turn of the winding with consequent reduction in both cost of copper and copper loss.

Q7.32 *What is the range of flux densities used in the design of transformer?*

When hot rolled silicon steel is used the range of flux density is,

B_m = 1.1 to 1.4 Wb/m^2 -   For distribution transformer

B_m = 1.2 to 1.5 Wb/m^2 -   For power transformer

When cold rolled silicon steel is used the typical choice of flux density is,

B_m = 1.5 Wb/m^2 -   For transformers with voltage rating upto 132 kV

B_m = 1.6 Wb/m^2 -   For transformers with voltage rating 132 kV to 275 kV

B_m = 1.7 Wb/m^2 -   For transformers with voltage rating 275 kV to 400 kV

Q7.33 *List the different types of windings used in core type transformers.*

The different types of windings employed in core type transformers are,

1. Cylindrical winding
2. Helical winding
3. Double helical winding
4. Multi-layer helical winding
5. Cross-over winding
6. Disc and continuous disc winding
7. Aluminium foil winding

Q7.34 What are the factors to be considered to choose the type of winding for a core type transformer?

The type of winding used for core type transformers depends on the following factors.   *(HTU, Dec' 18, 2 M)*

1. Current density   4. Surge voltage
2. Short circuit current   5. Impedance
3. Temperature rise   6. Transport facilities

Q7.35 What is tertiary winding?

Some three-phase transformers may have a third winding called tertiary winding apart from primary and secondary. It is also called auxiliary winding (or) stabilizing winding.

The tertiary winding is provided in a transformer for any one of the following reasons,

1. To supply small additional load at a different voltage.
2. To give supply to phase compensating devices such as capacitors which work at different voltage.
3. To limit short circuit current.
4. To indicate voltage in high voltage testing transformer.

Q7.36 How the tertiary winding is connected? Why?

The tertiary winding is normally connected in delta. When the teritary is connected in delta, the unbalance in phase voltage during unsymmetrical faults in primary or secondary is compensated by the circulating currents flowing in the closed delta.

Q7.37 What are the main classification of transformer winding?

The various types of transformer windings are,

1. Cylindrical winding   3. Continuous disc type winding
2. Crossover winding   4. Helical winding

Q7.38 How to design the windings of a transformer?

The winding design consists of estimation of number of turns and area of cross-section of winding conductors. Usually the number of turns in low voltage winding is estimated by assuming emf per turn and the turns of high voltage winding are estimated from the voltage ratio.

$$\left.\begin{array}{l}\text{Number of turns in}\\\text{low voltage winding}\end{array}\right\} = \frac{\text{Rated voltage of low voltage winding}}{\text{Emf per turn}}$$

$$\left.\begin{array}{l}\text{Number of turns in}\\\text{high voltage winding}\end{array}\right\} = \left(\begin{array}{l}\text{Number of turns in}\\\text{low voltage winding}\end{array}\right) \times \frac{\text{Rated voltage of high voltage winding}}{\text{Rated voltage of low voltage winding}}$$

The area of cross-section of the winding conductors are determined by assuming a suitable current density.

$$\left.\begin{array}{l}\text{Area of cross-section}\\\text{of winding conductor}\end{array}\right\} = \frac{\text{Rated current of the winding}}{\text{Current density}}$$

Q7.39 The voltage per turn of a 500 kVA, 11 kV/415 V, Δ /Y, 3-phase transformer is 8.7 V. Calculate the number of turns per phase of LV and HV windings.   *(AU, Apr' 18, 2 M)*

Solution

Phase voltage of LV winding $= \dfrac{415}{\sqrt{3}} = 239.6\ V$ $\boxed{\text{Star connected}}$

Phase voltage of HV winding $= 11{,}000\ V$ $\boxed{\text{Delta connected}}$

$$\left.\begin{array}{l}\text{Number of turns}\\\text{in LV winding}\end{array}\right\} = \frac{\text{Phase voltage of LV winding}}{\text{Emf per turn}} = \frac{239.6}{8.7} = 27.54 \approx 28$$

$$\left.\begin{array}{l}\text{Number of turns}\\\text{in HV winding}\end{array}\right\} = \text{Number of turns of LV winding} \times \text{Phase voltage ratio}$$

$$= 28 \times \frac{11,000}{239.6} = 1285.5 \approx 1286$$

Q7.40 Name the various types of cross-section used for core type of transformer?

1. Square core
2. Rectangle core
3. Cruciform core
4. Multi stepped cores.

Q7.41 How is leakage reactance of winding estimated?

It is estimated by primarily estimating the distribution of leakage flux and the resulting flux leakages of the primary and the secondary windings. The distribution of the leakage flux depends upon the geometrical configuration of the coils and the neighboring iron masses and also on the permeability of the iron.

Q7.42 What types of forces acts on the coils of a transformer in the event of a short circuit on a transformer?

During short circuit conditions the radial forces will be acting on the coil, which is due to short circuit currents.

Q7.43 What is the range of current densities used in the design of transformer winding?

The choice of current density depends on the allowable temperature rise, copper loss and method of cooling. The range of current density for various types of transformers are given below:

$\delta = 1.1$ to 2.2 A/mm^2 - For distribution transformers

$\delta = 1.1$ to 2.2 A/mm^2 - For small power transformers with self oil cooling

$\delta = 2.2$ to 3.2 A/mm^2 - For large power transformers with self oil cooling or air-blast

$\delta = 5.4$ to 6.2 A/mm^2 - For large power transformers with forced
circulation of oil or with water cooling coils

Q7.44 List the different methods of cooling of transformers.   *(AU, Apr' 19, 2 M)*

The different methods of cooling of transformers are,

1. Air Natural (AN)
2. Air Blast (AB)
3. Oil Natural (ON)
4. Oil Natural Air Forced (ONAF)
5. Oil Natural Water Forced (ONWF)
6. Oil Forced (OF)
7. Oil Forced Air Natural (OFAN)
8. Oil Forced Air Forced (OFAF)
9. Oil Forced Water Forced (OFWF)

Q7.45 What are the factors to be considered for selecting the cooling method of a transformer?

The choice of cooling method depends on *kVA* rating of transformer, size, application and the site condition where it will be installed.

Q7.46 How the heat dissipates in a transformer?

The heat dissipation in a transformer occurs by Conduction, Convection and Radiation.

Q7.47 Why transformer oil is used as a cooling medium?

When transformer oil is used as a coolant the heat dissipation by convection is 10 times more than the convection due to air. Hence the transformer oil is used as coolant.

Specific heat dissipation by convection due to air = 8 W/m^2-$°C$

Specific heat dissipation by convection due to oil = 80 to 100 W/m^2-$°C$

Q7.48 Why cooling tubes are provided?

Cooling tubes are provided to increase heat dissipating area of the tank.

Q7.49 How the heat dissipation is improved by the provision of cooling tubes?

The cooling tubes will improve the circulation of oil. The circulation of oil is due to more effective pressure heads produced by columns of oil in tubes. The improvement in cooling is accounted by taking the specific heat dissipation due to convection as 35% more than that without tubes.

Q7.50 Classify the transformer according to cooling methods?

According to cooling methods the transformers can be classified as follows,

1. Air cooled transformers
2. Oil cooled transformers
3. Water cooled transformers.

Q7.51 Mention the various methods of cooling for larger power transformers.

The various methods of cooling large power transformers are,

1. Oil natural air forced
2. Oil natural water forced
3. Oil forced air natural
4. Oil forced air forced
5. Oil forced water forced.

Q7.52 What are the properties of transformer oil?

The various properties of transformer oil are,

1. High dielectric strength
2. High resistivity and density
3. Low viscosity
4. Low impurity
5. Reasonable cost and flash point.

Q7.53 Explain simple cooling and mixed cooling of transformer.

In simple cooling the transformer is cooled by a single method of cooling. In mixed cooling the transformer is equipped with two or more cooling methods and at any one time one method is employed depending on the amount of heat to be removed.

Q7.54 What are the merits and demerits of using water for forced cooling of transformers?

The advantage in forced water cooling is that large amount of heat can be removed quickly from the transformer. The disadvantage in forced water cooling is that the water may leak into oil and the oil may be contaminated.

Q7.55 In mines applications transformers with oil cooling should not be used why?

The oil used for transformer cooling is inflammable (i.e., can easily set on fire), hence leakage of cooling oil may create fire accidents in mines. Therefore oil cooled transformers are not used in mines.

Q7.56 What is breather?

The breather is a device fitted in transformer for breathing. In small oil cooled transformers some air-gap is provided between the oil level and tank top surface. When the oil is heated, it expands and air is expelled out of transformer to the atmosphere through breather. When the oil is cooled, it shrinks and air is drawn from the atmosphere into the transformer through breather. This action of transformer is called breathing.

Q7.57 Why silica gel is used in breather?

The silica gel is used to absorb the moisture when the air is drawn from atmosphere into the transformer.

Q7.58 What is conservator?

A conservator is a small cylindrical drum fitted just above the transformer main tank. It is used to allow the expansion and contraction of oil without contact with surrounding atmosphere.

When conservator is fitted in a transformer, the tank is fully filled with oil and the conservator is half filled with oil.

Q7.59 How will you fix up the tank dimensions based on overall dimensions of transformer frame?

The dimensions of the tank are decided by the dimensions of the transformer frame and clearance required on all the sides. The clearance on the sides depends on voltage and power rating of the winding. The clearance at the top depends on the oil height above the assembled transformer and the space for mounting the terminals and tap changing gear. The clearance at the bottom depends on the space required for mounting the transformer frame inside the tank.

(AU, Apr' 17, 2 M)

Q7.60 How the magnetic curves are used for calculating the no-load current of a transformer?

The **B-H** curve can be used to find the mmf per metre for the flux densities in yoke and core. The loss curve can be used to estimate the iron loss per *kg* for the flux densities in yoke and core.

Q7.61 How the leakage reactance of the transformer is reduced?

In transformers the leakage reactance is reduced by interleaving the high voltage and low voltage winding.

Q7.62 Define transformation ratio.

It is defind as the ratio of secondary terminal voltage to primary terminal voltage.
It is denoted by k.

$$\text{Transformer ratio, } k = \frac{V_s}{V_p} = \frac{T_s}{T_p} = \frac{I_p}{I_s}$$

Q7.63 Write the expression for temperature rise in plain walled tanks.

$$\text{Temperature rise, } \theta = \frac{\text{Total loss}}{\text{Specific heat dissipation} \times \text{Heat dissipating surface of the tank}}$$

$$= \frac{P_i + P_c}{\lambda_{spec} \, S_T}$$

where, P_i = Iron loss

P_c = Copper loss

S_T = Heat dissipating surface of the tank

7.15 EXERCISES

I.   Fill in the blanks

1.   The thickness of laminations used for transformer core is ....

2.   A transformer consists of two windings coupled through ....

3.   Depending on the applications in power system the transformers are classified as .... and ....
      transformers.

4.   Based on construction the transformers can be classified into .... and .... transformers.

5.   In distribution transformers .... loss will be greater than .... loss.

6.   The power transformers will have maximum efficiency at ....

7.   The ratio of net core area and square of circumscribing circle is called ....

8.   The main dimensions of the transformer are .... and ....

9.   The walls of transformer tank dissipate heat by .... and ....

10.   The .... houses the transformer core and winding assembly.

11.   The provision of cooling tubes improves heat dissipation due to ....

12.   The efficiency of a commercial transformer will be in the range of ....

13.   The .... is defined as the ratio of copper area in window to total area of window.

14.   The cold rolled grain oriented silicon steel permits a maximum flux density of ......

15.   In transformers the ...... is reduced by interleaving the high and low voltage winding.

Answers

1. 0.3 to 0.5 *mm*   6. full load   11. convection

2. magnetic field   7. core area factor   12. 94% to 99%

3. distribution, power   8. height of window,   13. window space factor
                                    width of window

4. shell, core   9. radiation, convection   14. 1.8 *Wb/m²*

5. copper, iron   10. tank   15. leakage reactance

II.   State whether the following statements are True/False

1.   In transformers the functions of primary and secondary winding are interchangable.

2.   The kVA rating and ampere-turns of both the winding of a transformer are different.

3.   In shell type transformer the cross-section of the central limb is twice that of side limb.

4.   The cross-section of the yoke and core can be different.

5.   Power transformers will have lesser leakage reactance than distribution transformer.

6.   In transformers the output equation relates the kVA rating to area of core and window.

7.   The diameter of circumscribing circle decreases with increase in number of steps.

8.   The laminations made of cold rolled steel can work with higher flux densities.

9.   The primary and secondary winding will have different current densities.

10.   The bottom area of tank has very little cooling effect.

11.   The convection of heat by oil is 10 times more than that of air.

12.   In transformers with increase in kVA rating, the losses increases as cube of linear dimensions but the heat dissipating surface increases as the square of linear dimensions.

13.   The tertiary winding of a three-phase transformer is normally connected in star.

14.   The dimensions of the yoke is kept larger to reduce iron losses.

15.   In transformer the low voltage winding is placed near the core.

Answers

1. True	4. True	7. True	10. True	13. False
2. False	5. False	8. True	11. True	14. True
3. True	6. True	9. False	12. True	15. True

III.   Unsolved problems

E7.1   Determine the core and yoke dimensions for a 250 kVA, 50 Hz, single-phase, core type transformer, Emf per turn = 15 V, the window space factor = 0.33, current density = 3 A/mm^2 and B_{max} = 1.1 Wb/m^2. The distance between the centres of the square section core is twice the width of the core.

$$(D_y = H_y = a = 0.26 \ m \ ; D = 0.52 \ m).$$

E7.2   Calculate the dimensions of the core, the number of turns and cross sectional area of conductors in the primary and secondary windings of a 250 kVA, 6600 / 400 V, 50 Hz, single-phase shell type transformer. Ratio of magnetic to electric loadings = 560 × 10^{-8}, B_m = 1.1 Wb/m^2, δ = 2.5 A/mm^2, K_w = 0.32, Stacking factor = 0.9, Depth of stacked core / Width of core = 2.6, Height of window / Width of window = 2.0

$$(A_i = 0.0722 \ m^2 \ ; A_w = 0.0354 \ m^2 \ ; H_w \times W_w = 0.2548 \times 0.1247 \ ;$$
$$T_p = 379 \ ; T_s = 23 \ ; a_p = 15.15 \ mm^2 \ ; a_s = 250 \ mm^2)$$

E7.3   The tank of a 500 kVA, 1ϕ, 50 Hz, 6600/400 V transformer is 110 cm × 65 cm × 155 cm. If the load loss is 6.2 kW, find the suitable arrangements for the cooling tubes to limit the temperature rise to 35 °C. Take the diameter of the cooling tubes as 5 cm and average length of the tube as 110 cm.

(Number of tubes = 72, on 65 cm side two rows with 7 tubes and 6 tubes and on 110 cm side two rows with 12 tubes and 11 tubes)

E7.4   The tank of a 500 kVA, 50 Hz, 1ϕ, core type transformer is 1.05 × 0.62 × 1.6 m high. The mean temperature rise is limited to 35 °C. The loss disspating surface of tank is 5.34 m². Total loss is 5325 W. Find area of tubes and number of tubes needed. Take diameter and length of tubes as 50 cm and 120 cm respectively.

(Number of tubes = 52; on 0.62 m side two rows with 7 tubes and 6 tubes and on 1.05 m side one row with 13 tubes.)

E7.5   A single-phase, 500 V, 50 Hz, transformer is built from stampings having a relative permeability of 1000. The length of the flux path is 2.9 m. The area of cross-section of the core is 3.2 × 10^{-3} m^2 and the primary winding has 700 turns. Estimate the maximum flux and no-load current of the transformer. The iron loss at the working flux density is 2.8 W/kg. Iron weight 7.2 × 10^3 kg/m^3, stacking factor is 0.9.

$$(\phi_m = 3.22 \times 10^{-3} \ Wb, \ I_o = 2.4112 \ A)$$

E7.6   A 200 *kVA*, 2000/440 *V*, 50 *Hz*, single-phase, shell type transformer has sandwich coils. There are two full high voltage coils, one full low voltage coil and 2 half low voltage coils. Calculate the value of leakage reactance referred to high voltage side. The given data is: depth of half voltage coil = 42 *mm*, depth of low volatage coils = 38 *mm*, depth of duct between high voltage and low voltage = 18 *mm*, width of winding = 0.15 *m*, length of mean turn = 1.9 *m*, the number of turns in high voltage winding are 200.

$$(X_p = 1.5652 \ \Omega)$$

E7.7   A 350 *kVA*, 6000/400 *V*, 50 *Hz*, 3-phase, delta/ star, core type transformer has 200 turns on the high voltage winding. The height of winding is 0.3 *m* and the length of mean turn 1.2 *m*. Calculate the instantaneous radial force on the high voltage if a short circuit occurs at the terminals of the low voltage winding with high voltage energised. The leakage impedence is 3 percent. Take the double effect multiplier as 1.6. Also calculate the force at full load.

$$(F_r = 37.992 \ N)$$

Chapter 8

DESIGN OF THREE-PHASE INDUCTION MOTOR

List of symbols

Symbol	Meaning	Unit
AT_{cr}	mmf for rotor core	AT
AT_{cs}	mmf for stator core	AT
AT_g	mmf for air-gap	AT
AT_{tr}	mmf for rotor teeth	AT
AT_{ts}	mmf for stator teeth	AT
AT_{60}	mmf for I_m using flux density at 60° from interpolar axis	AT
A_p	Area per pole	m^2
a	Number of parallel path in armature winding	-
a_b	Area of cross-section of rotor bar	mm^2
a_e	Area of cross-section of end ring	mm^2
a_s	Area of cross-section of stator conductor	mm^2
a_r	Area of cross-section of rotor conductor	mm^2
at_{cr}	Specific magnetic loading for rotor core	Wb/m^2
at_{cs}	Specific magnetic loading for stator core	Wb/m^2
at_{tr}	Specific magnetic loading for rotor teeth	Wb/m^2
at_{ts}	Specific magnetic loading for stator teeth	Wb/m^2
ac	Specific electric loading	$amp.cond./m$
B_{g60}	Air-gap flux density at 60° from interpolar axis	Wb/m^2
B_{av}	Specific magnetic loading	Wb/m^2
B_{cr}	Flux density in rotor core	Wb/m^2
B_{cs}	Flux density in stator core	Wb/m^2
B_{tr60}	Rotor teeth flux density at 60° from interpolar axis	Wb/m^2
B_{ts60}	Stator teeth flux density at 60° from interpolar axis	Wb/m^2
$B_{tr1/3}$	Flux density at one-third height of rotor tooth from narrow end	Wb/m^2
$B_{ts1/3}$	Flux density at one-third height of stator tooth from narrow end	Wb/m^2
C_o	Output coefficient	$kVA/m^3\text{-}rps$

Symbol	Meaning	Unit
D	Diameter of stator bore	m
D_i	Inner diameter of rotor	m
D_o	Outer diameter of stator core	m
D_r	Diameter of rotor	m
d_{cr}	Depth of rotor core	m
d_{sc}	Depth of stator core	m
d_{ss}	Depth of stator slot	m
d_{sr}	Depth of rotor slot	m
d_e	Depth of end ring	mm
d_{rm}	Mean diameter of rotor core	m
d_s	Diameter of stator conductor	mm
d_{sm}	Mean diameter of stator core	m
d_{sr}	Depth of rotor slot	m
d_{ss}	Depth of stator slot	m
E_r	Rotor phase voltage at stand still	V
E_s	Stator phase voltage	V
E_{sc}	Stator voltage to circulate full load current at standstill	V
HP	Power output in horse power rating	HP
I_b	Rotor bar current	A
I_e	End ring current	A
I_{e_max}	Maximum value of end ring current	A
I_m	Magnetizing current	A
I_L	Line current	A
I_l	Loss component of no-load current	A
I_o	No-load current	A
I_r	Rotor current	A
I_{sc}	Short circuit current per phase	A
I_s	Stator current per phase	A
I_z	Current through a stator conductor	A
I_{sci}	Ideal short circuit current	A
K_{wr}	Rotor winding factor	-
K_{ws}	Stator winding factor	-
L	Length of stator core	m
L_i	Net iron length of stator core	m
L_o	Length of conductor in overhang	m
L_{mtr}	Length of mean turn rotor	m
L_{mts}	Length of mean turn of stator	m
l_{cs}	Length of mean flux path in stator core	m

Symbol	Meaning	Unit
l_{cr}	Length of mean flux path in rotor core	m
l_g	Length of air-gap	mm
m_s	Number of phases in stator winding	-
n_s	Synchronous speed	rps
p	Number of poles	-
P_o	Power output of induction motor	kW
q_s	Stator slots per poles per phase	-
q_r	Rotor slots per pole per phase	-
Q	kVA input	kVA
R_s	Motor resistance referred to stator	Ω
r_b	Resistance of each bar	Ω
r_e	Resistance of each ring	Ω
r_r	Rotor resistance per phase	Ω
r_r'	Rotor resistance per phase referred to stator	Ω
r_s	Stator resistance per phase	Ω
S_r	Number of rotor slots	-
S_s	Number of stator slots	-
T_r	Rotor turns per phase	-
T_s	Stator turns per phase	-
t_e	Thickness of end ring	mm
V_a	Peripheral speed	m/s
V_L	Line voltage	V
W_{tr}	Width of rotor tooth	mm
W_{ts}	Width of stator tooth	mm
$W_{ts1/3}$	Width of stator teeth at one-third height from narrow end	mm
$W_{tr1/3}$	Width of rotor teeth at one-third height from narrow end	mm
W_{ts_min}	Minimum width of stator tooth	mm
W_{tr_min}	Minimum width of rotor tooth	mm
X_h	Harmonic leakage reactance	Ω
X_m	Magnetic leakage reactance	Ω
X_s	Total leakage reactance of the motor referred to stator	Ω
X_{ss}	Stator slot leakage reactance	Ω
X_{sr}'	Rotor slot leakage reactance referred to stator	Ω
X_o	Overhang leakage reactance	Ω
X_z	Zigzag leakage reactance	Ω
y_{ss}	Stator slot pitch or width	mm
y_{sr}	Rotor slot pitch or width	mm
Z	Number of stator conductors	-
Z_s	Standstill impedence per phase	Ω

Symbol	Meaning	Unit
Z_{ss}	Conductors per stator slot	-
Z_{rs}	Conductor per rotor slot	-
λ	Specific slot permeance for stator	$Wb/AT\text{-}m$
λ_h	Specific harmonic permeance	$Wb/AT\text{-}m$
λ_o	Specific overhang permeance	$Wb/AT\text{-}m$
λ_{sr}'	Specific permeance of rotor slots referred to stator	$Wb/AT\text{-}m$
λ_{ss}	Specific slot permeance for stator slot	$Wb/AT\text{-}m$
λ_z	Specific zigzag permeance	$Wb/AT\text{-}m$
η	Efficiency	-
σ	Dispersion coefficient	-
δ_a	Current density in armature conductor	A/mm^2
δ_b	Current density in rotor bar	A/mm^2
δ_e	Current density in the end ring	A/mm^2
δ_r	Current density in rotor	A/mm^2
δ_s	Current density in stator conductors	A/mm^2
ϕ_m	Maximum flux in the core	Wb
ϕ_{sc}	Phase angle of short circuit current	$deg.$
τ	Pole pitch	m
τ_{rm}	Pole pitch at mean rotor diameter	m
τ_{sm}	Pole pitch at mean stator diameter	m
θ	Angle of skew	$deg.mech.$
θ_{sk}	Electrical angle of skew	$deg.or\ rad.elect.$

8.1 CONSTRUCTION

The two major parts of three-phase induction motor are stator and rotor. The stator consists of core and winding. The stator core is made of laminated sheet steel of thickness 0.5 *mm*. The stator core internal diameter and the length are the main dimensions of induction motor.

The two different types of rotor of three-phase induction motors are squirrel cage rotor and wound rotor (or slip ring rotor). In squirrel cage construction the rotor consists of core, copper or aluminium bars and end rings. The end rings are provided to short circuit the rotor bars at both the ends. The rotor core is made of lamintated sheet steel with thickness 0.5 *mm*. The aluminium bars and end rings are casted directly over the rotor core. When copper is employed, the rotor bars are inserted on the slots from the end of rotor and the end rings are joined to them by bracing.

The wound rotor consists of a core, winding, slip rings and brushes. The rotor core is made of laminations and it carries a three-phase winding. One end of each phase are connected to from a star point. The other end of each phase are connected to three slip rings. The slip rings are mounted on the rotor shaft and they are insulated from the rotor and from each other. Carbon brushes are mounted over the slip rings which offers the facility of connecting the external resistances to rotor winding.

Constructional elements of squirrel cage induction motor

Stator   - Frame   **Rotor**   - Rotor core

         - Stator core   - Rotor bars

         - Stator winding   - End ring

Constructional elements of slip ring induction motor

Stator   - Frame   **Rotor**   - Rotor core

         - Stator core   - Rotor winding

         - Stator winding   - Slip rings

8.2   OUTPUT EQUATION OF INDUCTION MOTOR   *(VTU, Dec' 19, 6 M) (AU, Apr' 17, 6 M)*

(PU, Nov' 19, 8 M) (AKTU, Dec' 19, 5 M) (KTU, Feb' 18, 8 M) (MU, May' 19, 10 M)

In case of induction motor the equation for kVA input is considered as output equation.

$$\text{kVA input, } Q = C_o D^2 L n_s \qquad\qquad\qquad .....(8.1)$$

$$\text{Output coefficient, } C_o = 11 K_{ws} B_{av} \text{ ac} \times 10^{-3} \qquad\qquad .....(8.2)$$

where, K_{ws} = Stator winding factor

         B_{av} = Specific magnetic loading

         ac = Specific electric loading

The rating of an induction motor is sometimes given in horse power. This rating refers to the power output at the shaft of the motor. The kVA input for the motor can be calculated from the following formula.

$$\text{kVA input, } Q = \frac{HP \times 0.746}{\eta \times \cos \phi} \qquad\qquad\qquad .....(8.3)$$

where,   HP   = Power output in horse power rating

         η = Efficiency

         $\cos \phi$ = Power factor

Also, kVA input can be calculated from line and phase voltage and current.

$$\text{kVA input, } Q = \sqrt{3} \; V_L I_L \times 10^{-3} = 3 \, E_s I_s \times 10^{-3}$$

Output equation and output coefficient of induction motor

The equations of induced emf, frequency, current through each conductor and total number of stator conductors of an induction motor are given below. These equations are obtained from the knowledge of machine theory.

Stator induced emf per phase, $E_s = 4.44 f \phi T_s K_{ws}$ .....(8.4)

Frequency of induced emf, $f = \dfrac{pn_s}{2}$ .....(8.5)

Current through each conductor, $I_z = \dfrac{I_s}{a}$ .....(8.6)

where, I_s = Stator current per phase

       a   = Number of parallel circuits or paths per phase in stator winding.

Number of stator conductors, $Z = $ Number of phases $\times 2\, T_s$

$$= 3 \times 2\, T_s = 6\, T_s \qquad\qquad .....(8.7)$$

Specific magnetic loading, $\quad B_{av} = \dfrac{p\phi}{\pi DL} \quad$ (or) $\quad p\phi = \pi DL\, B_{av} \qquad\qquad .....(8.8)$

Specific electric loading, $\qquad ac = \dfrac{I_z Z}{\pi D} \quad$ (or) $\quad I_z Z = \pi D\, ac \qquad\qquad .....(8.9)$

Consider a 3-phase machine having one circuit (one parallel path) per phase. The volt-ampere rating of one phase is given by the product of voltage per phase and current per phase. Hence the kVA input or kVA rating of 3-phase AC machine can be written as shown in equation (8.10).

kVA input, $\quad Q = 3\, E_s\, I_s \times 10^{-3} \qquad\qquad .....(8.10)$

$$= 3 \times 4.44\, f\, \phi\, T_s\, K_{ws} \times I_z \times 10^{-3} \qquad \boxed{a=1} \quad \begin{array}{l}\text{Using equations}\\ \text{(8.4) and (8.6)}\end{array}$$

$$= 3 \times 4.44 \times \frac{p\, n_s}{2} \times \phi\, T_s\, K_{ws}\, I_z \times 10^{-3} \qquad \boxed{\text{Using equations (8.5)}}$$

$$= 6.66 \times p\, n_s\, \phi\, T_s\, K_{ws}\, I_z \times 10^{-3}$$

$$= 1.11 \times p\phi \times I_z \times 6\, T_s \times n_s \times K_{ws} \times 10^{-3}$$

$$= 1.11 \times p\phi \times I_z Z \times n_s \times K_{ws} \times 10^{-3} \qquad \boxed{\text{Using equations (8.7)}}$$

$$= 1.11 \times \pi\, D\, L\, B_{av} \times \pi D\, ac \times n_s \times K_{ws} \times 10^{-3} \qquad \begin{array}{l}\boxed{\text{Using equations}}\\ \boxed{\text{(8.8) and (8.9)}}\end{array}$$

$$= 1.11\, \pi^2\, B_{av}\, ac\, K_{ws} \times 10^{-3} \times D^2\, L\, n_s$$

$$= 11\, B_{av}\, ac\, K_{ws} \times 10^{-3} \times D^2\, L\, n_s$$

$$= C_o\, D^2\, L\, n_s \qquad\qquad .....(8.11)$$

where, $C_o = 11\, B_{av}\, ac\, K_{ws} \times 10^{-3} \qquad\qquad .....(8.12)$

$\therefore$ kVA input, $\qquad Q = C_o\, D^2\, L\, n_s \qquad\qquad .....(8.13)$

Output coefficient, $\qquad C_o = 11\, B_{av}\, ac\, K_{ws} \times 10^{-3} \qquad\qquad .....(8.14)$

The equation, $Q = C_o\, D^2\, L\, n_s$ is called **output equation** and C_o is called **output coefficient**. The term $D^2\, L$ in the output equation is proportional to volume of active part. Therefore, if C_o is constant then we can say that the kVA rating is directly proportional to the product of volume of active part and speed.

$\therefore$ Q $\quad \alpha \quad$ Volume of active part $\times$ Speed $\hfill$ *(PTU, May' 19, 2 M)*

If C_o is varied then power output is directly proportional to the four quantities: B_{av}, ac, volume of active part and speed.

$\therefore$ Q $\quad \alpha \quad$ $B_{av} \times ac \times$ Volume of active part $\times$ Speed

Losses, efficiency and power factor

The various losses in induction motors are stator copper loss, rotor copper loss, stator iron losses and friction and windage losses.

For squirrel cage induction motors the efficiency varies from 0.72 to 0.91 and power factor varies from 0.66 to 0.9. For slip ring induction motors the efficiency varies from 0.84 to 0.91 and power factor varies from 0.7 to 0.92.

8.3 MAIN DIMENSIONS   *(JNTKU, Apr' 19, 9 M)   (MU, May' 19, 10 M)*

The main dimensions of induction motor are the diameter of stator bore, D and the length of stator core, L.

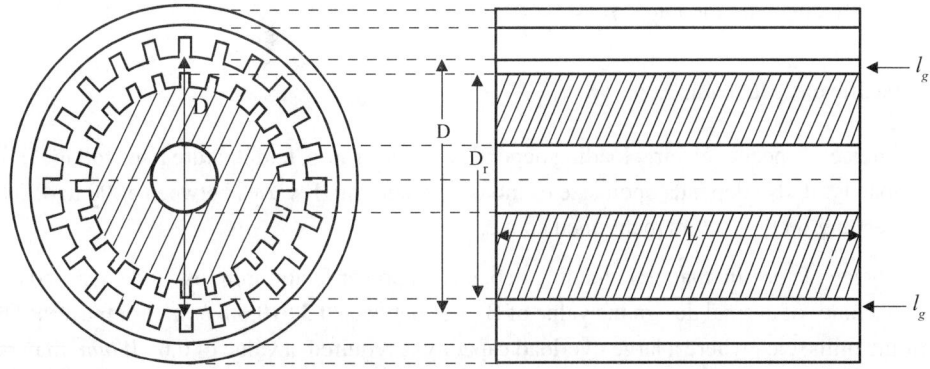

Fig. 8.1: *Main dimensions of induction motor.*

The product $D^2 L$ is determined from kVA input, specific electric and magnetic loadings. The seperation of D and L, from the product D^2L, depends on the ratio L/τ where τ $(= \pi D/p)$, is the pole pitch. In induction motors most of the operating characteristics are decided by L/τ ratio of the motor. The ratio of core length to pole pitch (L/τ) for various design features are listed below.

For minimum cost, L/τ = 1.5 to 2

For good power factor, L/τ = 1.0 to 1.25

For good efficiency, L/τ = 1.5

For good overall design, L/τ = 1

Generally L/τ lies between 0.6 to 2. It can be shown that, for best power factor the pole pitch τ is given by the equation,

$$\tau = \sqrt{0.18\,L} \qquad\qquad .....(8.15)$$

The diameter of the stator bore and hence the diameter of rotor is also limited by peripheral speed. Standard constructions are employed for peripheral speeds upto to 60 *m/s*. For higher peripheral speeds upto 75 *m/s*, special construction methods should be employed for rotor which results in higher cost. For a normal design, the diameter should be so chosen that the peripheral speed does not exceed about 30 *m/s*.

The stator is provided with radial ventilating ducts if the core length exceeds 125 *mm*. The width of each duct is about 8 to 10 *mm*.

8.4 CHOICE OF SPECIFIC LOADINGS

For the design of induction motor, the horse power rating, speed, power factor and efficiency are specified. Therefore, in order to calculate the value of D^2L, we must evaluate the output coefficient. The value of output coefficient depends upon the choice of specific electric loading (**ac**) and specific magnetic loading (B_{av}).

The equations for total and specific loadings of induction motor are same as that of DC machines discussed in Chapter-6, Section 6.4.

$$\therefore \text{ Total magnetic loading} \qquad = p\,\phi$$

$$\text{Total electric loading} \qquad = I_z Z$$

$$\text{Specific magnetic loading,} \quad B_{av} = \frac{p\phi}{\pi DL}$$

$$\text{Specific electric loading,} \qquad \mathbf{ac} = \frac{I_z Z}{\pi D}$$

The choice of specific electric loading depends on copper loss, temperature rise, voltage rating and overload capacity. It also depends upon size of motor and cooling. It varies between 5000 to 45000 *amp. cond./m* depending on the factors mentioned above.

The choice of specific magnetic loading depends on power factor, iron loss and over load capacity. For 50 *Hz* machines of normal design the value of B_{av} lies between 0.3 to 0.6 *Wb/m²*. For machines used in cranes, rolling mills, etc., where a large overload capacity is required, a value of 0.65 *Wb/m²* may be used.

8.4.1 Choice of Specific Electric Loading

A large value of **ac** results in higher copper losses and higher temperature rise. For machines with high voltage rating smaller values of **ac** should be preferred. Since for high voltage machines the space required for insulation is large.

For high overload capacity, lower values of **ac** should be selected. Since large values of **ac** results in large number of turns per phase, the leakage reactance will be high. Large values of leakage reactances results in reduced over load capacity.

8.4.2 Choice of Specific Magnetic Loading

With large values of B_{av}, the magnetizing current will be high, which results in poor power factor. However, in induction motors the flux density in the air gap should be such that there is no saturation in any part of the magnetic circuit.

A large value of B_{av} results in increased iron loss and decreased efficiency. With higher values of B_{av}, higher values of overload capacity can be obtained. Since the higher B_{av} provides, large values of flux per pole, the turns per phase will be less and so the leakage reactance will be less. Lower value of leakage reactance results in higher over load capacity.

8.5 DESIGN OF STATOR   *(PU, Nov' 19, 6 M) (RGPV, Jun' 20, 5 M)*

The stator design is common for both squirrel cage and slip ring induction motor.

8.5.1 Stator Winding

For small motors upto 5 *HP*, single layer windings like mush winding, whole coil concentric winding and bifurcated concentric winding are employed. For large capacity machines, double layer windings (either lap or wave winding) are employed with diamond shaped coils.

The stator winding can be designed for either star or delta depending on the running condition.

Stator turns per phase

The turns per phase, T_s can be estimated from stator phase voltage and maximum flux in the core. The maximum flux in the core can be estimated from B_{av}, D, L and p.

$$\text{Specific magnetic loading,} \quad B_{av} = \frac{p\phi}{\pi DL} \qquad \qquad .....(8.16)$$

$$\text{Here, } \phi = \phi_m \qquad \qquad \therefore \phi_m = \frac{B_{av}\, \pi\, DL}{p} \qquad \qquad ..... (8.17)$$

$$\text{Stator phase voltage,} \quad E_s = 4.44\, K_{ws}\, f\, \phi_m\, T_s \qquad \qquad ..... (8.18)$$

$$\text{Stator turns per phase,} \quad T_s = \frac{E_s}{4.44\, K_{ws}\, f\, \phi_m} \qquad \qquad ..... (8.19)$$

Length of mean turn

The approximate length of mean turn of the winding on induction motor stators for use on voltage upto 650 *V* may be calculated from the following empirical relationship.

$$\text{Length of mean turn of stator, } L_{mts} = 2L + 2.3\tau + 0.24 \qquad \qquad .....(8.20)$$

Here the values of L and τ are expressed in *m*.

Stator conductors

The area of cross-section of stator conductors can be estimated from the knowledge of current density, kVA rating of the machine and stator phase voltage. The current density in the stator winding is usually lies between 3 to 5 *A/mm²*.

$$\text{kVA rating of 3-phase machine, } Q = 3\, E_s\, I_s \times 10^{-3} \qquad \qquad .....(8.21)$$

where, E_s = Stator phase voltage

I_s = Stator phase current

$$\therefore I_s = \frac{Q}{3\, E_s \times 10^{-3}} \qquad \qquad .....(8.22)$$

Let, δ_s = Current density in stator conductors

Now, Area of cross-section of stator conductors, $a_s = \dfrac{I_s}{\delta_s}$

Diameter of stator conductor, $d_s = \sqrt{\dfrac{4\, a_s}{\pi}}$ $\boxed{a_s = \pi\, d_s^2 / 4}$

Round conductors are used for small diameters. If the diameter is more than 2 or 3 *mm* then bar or strip conductors are used.

8.5.2 Stator Core

The stator core is made of laminations of thickness 0.5 *mm*. The design of stator core involves selection of number of slots, estimation of dimensions of teeth and depth of stator core.

Stator slots

The different types of slots used in induction motor are open slots and semienclosed slots. The shape of the slots have an important effect upon the operating performance of the motor as well as the problem of installing the winding.

When open slots are used the winding coils can be formed and fully insulated before installing and also it is easier to replace the individual coils. Another advantage of open slots is that their use avoids excessive slot leakage thereby reducing the leakage reactance.

When semienclosed slots are used, the coils must be taped and insulated after they are placed in the slots. The advantages of semienclosed slots are less air-gap contraction factor giving a small value of magnetizing current, low tooth pulsation loss and much quieter operation (less noise). Semi enclosed slots are mostly preferred for induction motors.

In small motors where round conductors are used, the tapered slot with parallel sided tooth arrangement is useful as it gives the maximum slot area for a particular tooth flux density. In large and medium size machines where strip conductors are preferred, parallel sided slots with tapered teeth are used.

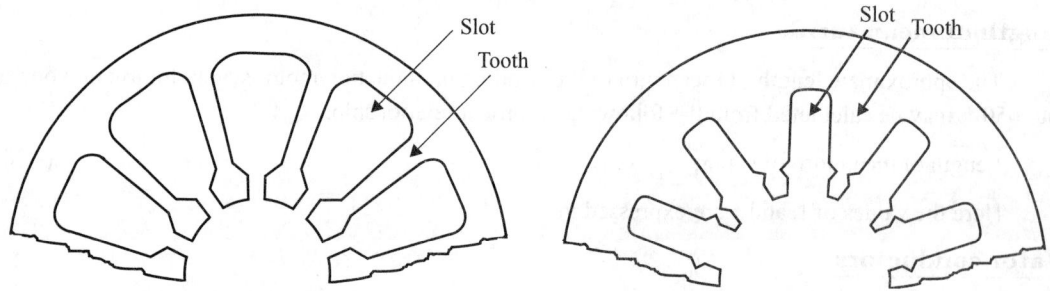

Fig. 8.2: Semienclosed slots with tapered slots and paralled sided teeth.   *Fig. 8.3: Open slots with parallel sided slots and tapered teeth.*

Choice of stator slot

The number of stator slots depends on tooth pulsation loss, leakage reactance, ventilation, magnetizing current, iron loss and cost. In general the number of slots should be selected to give an integral number of slot per pole per phase. The slot pitch at the air-gap for open type of slots should be between 15 to 25 *mm*. For semienclosed slots the slot pitch may be less than 15 *mm*.

$$\text{Stator slot pitch, } y_{ss} = \frac{\text{Gap surface}}{\text{Total number of stator slots}}$$

Let, S_s = Total stator slots

πD = Gap surface

$$\text{Number of stator slots, } S_s = \frac{\text{Gap surface}}{\text{Stator slot pitch}} = \frac{\pi D}{y_{ss}} \qquad .....(8.23)$$

Number of stator conductors, Z = Number of phase × Conductors per phase

$$= 3 \times 2\, T_s = 6\, T_s$$

where, T_s = Stator turns per phase

Conductors per slot, $Z_{ss} = \dfrac{\text{Number of stator conductors}}{\text{Number of stator slots}} = \dfrac{6T_s}{S_s}$ .....(8.24)

> *Note : Z_{ss} must be even for double layer winding.*

Guidelines for selecting stator slots   *(RGPV, Jun' 20, 5 M)*

Step 1:   The stator slot pitch various from 15 *mm* to 25 *mm*. Calculate the range of stator slots using the equation.

Number of stator slots, $S_s = \pi D / y_{ss}$

Minimum number of slots are obtained when $y_{ss} = 25$ *mm*

Maximum number of slots are obtained when $y_{ss} = 15$ *mm*.

Step 2:   Stator slots should be a multiple of q where q is slots per pole per phase.

Number of stator slots, S_s = Number of phase × poles × q

For q = 2, 3, 4, etc., calculate stator slots using the above equation.

Step 3:   Select the choices of stator slots which are common between the values obtained in step 1 and 2.

Step 4:   The best choice of stator slot is given by the value of slot in the list obtained from step 3 and statisfying the slot loading.

Slot loading = $I_z\, Z_{ss}$

where, I_z = Current through a conductor

   Z_{ss} = Conductors per slot

> *Note : Try to choose lesser number of slots which results in reduced labour and less amount of copper.*

Area of stator slot

Approximate area of each slot = $\dfrac{\text{Copper area per slot}}{\text{Space factor}}$

$$= \dfrac{Z_{ss} \times a_s}{\text{Space factor}}$$

The space factors vary from 0.25 to 0.4. High voltage machines have lower space factors due to large thickness of insulation. After obtaining the area of the slot, the dimensions of the slot should be adjusted. The slot should not be too wide to give a thin tooth. The width of the slot should be so adjusted such that the mean flux density in the tooth lies between 1.3 to 1.7 *Wb/m²*. The width of tooth should not be too large as it results in narrow and deep slots. The deeper slots give a large value of leakage reactance. In general the ratio of slot depth to slot width should be between 3 and 6.

Stator teeth

The dimensions of the slot determine the value of flux density in the teeth. A high value of flux density in the teeth is not desirable, as it leads to a higher iron loss and a greater magnetizing mmf. The maximum value of B_{ts} (the mean flux density in stator tooth) should not exceed 1.7 Wb/m^2.

$$\therefore \text{ Minimum teeth area per pole } = \frac{\phi_m}{1.7} \qquad \qquad .....(8.25)$$

Also, Teeth area per pole = Number of slots per pole × Net iron length × Width of tooth.

$$= \frac{S_s}{p} \times L_i \times W_{ts} \qquad \qquad .....(8.26)$$

When teeth area per pole is minimum, the width of tooth will be minimum. Therefore, replace W_{ts} in equation (8.26) by W_{ts_min} and equate to equation (8.25).

$$\therefore \frac{\phi_m}{1.7} = \frac{S_s}{p} \times L_i \times W_{ts\_min}$$

$$\therefore \text{ Minimum width of stator tooth, } W_{ts\_min} = \frac{\phi_m \, p}{1.7 \, S_s \, L_i} \qquad \qquad .....(8.27)$$

The minimum width of stator tooth is either near the gap surface or at one-third height of tooth from slot opening. A check for minimum tooth width using equation (8.27) should be applied before finally deciding the dimensions of stator slot.

Depth of stator core

The cross-section of stator core is shown in Fig. 8.4. The depth of stator core depends on the flux density in the core. The flux density in the stator core lies between 1.2 to 1.5 Wb/m^2.

The flux passing through the stator core is half of the flux per pole.

$$\therefore \text{ Flux in the stator core } = \frac{\phi_m}{2}$$

Let, B_{cs} = Flux density in stator core

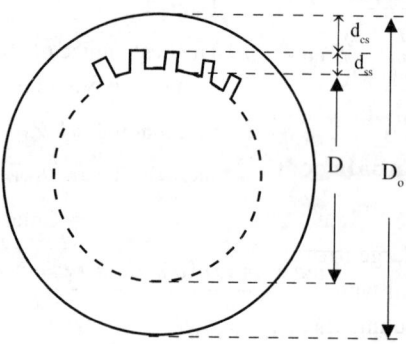

$$\text{Area of stator core} = \frac{\text{Flux through core}}{\text{Flux density in stator core}} = \frac{\phi_m}{2 \, B_{cs}}$$

Fig. 8.4: Cross-section of stator core.

Also, Area of stator core = Length × Depth = $L_i \times d_{cs}$

On equating the above two equations for area of stator core we get,

$$L_i \, d_{cs} = \frac{\phi_m}{2 \, B_{cs}}$$

$$\therefore \text{ Depth of core, } d_{cs} = \frac{\phi_m}{2 \, B_{cs} \, L_i} \qquad \qquad .....(8.28)$$

Outer diameter of stator core, D_o = D + 2 (Depth of stator slots + Depth of core)

$$= D + 2 \, (d_{ss} + d_{cs}) \qquad \qquad .....(8.29)$$

8.6 LENGTH OF AIR-GAP   *(HTU, Dec' 18, 10 M) (PU, Nov' 19, 8 M) (VTU, Dec' 19, 6 M) (AU, Apr' 17, 6 M)*

The length of air-gap in induction motor is decided by considering the following factors.

1. Power factor   4. Over-load capacity
2. Pulsation loss   5. Unbalanced magnetic pull
3. Cooling   6. Noise

Power factor

The mmf required to send the flux through air-gap is proportional to the product of flux density and the length of air-gap. Even with very small densities, the mmf required for air-gap is much more than that for the rest of the magnetic circuit. Therefore, it is the length of air-gap that primarily determines the magnetizing current drawn by the machine. Hence with larger air-gap length magnetizing current will be larger which results in increasing power factor angle, which in turn result in smaller value of power factor, cos ϕ.

Overload capacity

The length of air-gap affects the value of zigzag leakage reactance which forms a large part of total leakage reactance. If the length of air-gap is large then the zigzag leakage flux will be less and so the leakage reactance will be less. With lesser value of leakage reactance the over load capacity increases. Hence, greater is the length of air-gap, greater is the overload capacity.

Pulsation loss

With larger length of air-gap, the variation of reluctance due to slotting is small. The tooth pulsation loss, which is produced due to variation in reluctance of the air-gap, is reduced accordingly. Therefore, the pulsation loss is less with large air-gaps.

Unbalanced magnetic pull

If the length of air-gap is small, then even a small deflection or eccentricity of the shaft would produce a large irregularity in the length of air-gap. It is responsible for production of large unbalanced magnetic pull which has the tendency to bend the shaft still more at a place where it is already bent resulting in fouling of rotor with stator. If the length of air-gap of a machine is large, a small eccentricity would not be able to produce noticeable unbalanced magnetic pull.

Cooling

If the length of air-gap is large, the cylindrical surfaces of rotor and stator are separated by a large distance. This would afford better facilities for cooling at the gap surfaces especially when a fan is fitted for circulation of air.

Noise

The principal cause of noise in induction motors is the variation of reluctance of the path of the zig zag leakage flux. To ensure that the noise produced will not be objectionable, it is necessary to make the zigzag leakage as small as possible. This can be done by increasing the length of the air-gap.

From the above, we conclude that the length of air-gap in an induction machine should be as small as mechanically possible in order to reduce the magnetizing current which inturn increase the power factor. This is a major consideration. But if a higher overload capacity, better cooling, reduction in noise or reduction in unbalanced magnetic pull is important, then large air-gap lengths should be used.

Relations for calculation of length of air-gap

The following empirical formulae can be used to calculate the length of air-gap, l_g in mm.

Table 8.1: Length of air-gap

D in *m*	l_g in *mm*
0.15	0.35
0.20	0.50
0.25	0.60
0.30	0.70
0.45	1.3
0.55	1.8
0.65	2.5
0.80	4.0

1.   For small induction motor,

$$l_g = 0.2 + 2\sqrt{DL} \qquad .....(8.30)$$

2.   Alternate formula for small induction motors,

$$l_g = 0.125 + 0.35\,D + L + 0.015\,V_a \qquad .....(8.31)$$

3.   Another formula for general use,

$$l_g = 0.2 + D \qquad .....(8.32)$$

4.   For machines with journal bearings,

$$l_g = 1.6\sqrt{D} - 0.25 \qquad .....(8.33)$$

> **Note :** *In all the above formulae the values of D and L are expressed in metre and V$_a$ in m/s. But the result, l$_g$ in mm.*

Typical values of length of air-gap for 4 pole induction motors in relation to the main dimension D are listed in Table 8.1.

8.7 DESIGN OF SQUIRREL CAGE ROTOR

The squirrel cage rotor consists of a laminated core, rotor bars and end-rings. The rotor bars and end rings are made of aluminium or copper. The length of the rotor is same as that of stator. Some manufacturers, keep the length of rotor slightly higher than that of stator, in order to utilize the end fluxes. The diameter of the rotor is slightly lesser than the stator to avoid mechanical friction between the stationary stator and rotating rotor.

Diameter of rotor, $D_r = D - 2l_g$ .....(8.34)

where,   D = Diameter of stator bore

           l_g = Length of air-gap

8.7.1 Choice of Rotor Slots

With certain combinations of stator and rotor slots, the following problems may develop in the induction motor.

1.   The motor may refuse to start.

2.   The motor may crawl at some subsynchronous speed.

3.   Severe vibrations are developed and so the noise will be excessive.

The above effects are due to harmonic magnetic fields developed in the machine. The harmonic fields are due to winding, slotting, saturation and irregularities in air-gap.

The harmonic fields are superposed upon the fundamental sine wave field and induce emfs in the rotor windings and thus circulate harmonic currents. These harmonic currents, in turn, interact with the harmonic fields to produce harmonic torques.

In fact the harmonic fields may be thought of as separate low power motors that are direct coupled to the same shaft as the fundamental. Therefore, the net motor torque is equal to the sum of the torque due to the fundamental and the torques produced by a group of harmonic fields.

The space harmonic fields have more poles than the fundamental and therefore have lower synchronous speeds. Some of these fields revolve in the forward direction and some in the backward direction. As motor speeds above their respective synchronous values, the forward rotating harmonic fields produce braking torques. The backward rotating harmonic fields produce braking torque at all speeds. In addition the harmonic fields are responsible for increase in stray load losses and increased motor heating.

The squirrel cage rotor will circulate currents due to any harmonic emf produced by the gap flux except that has a wavelength equal to the pitch of the bars. The effects of space harmonic fields produced by windings are greatly intensified by slotting. The slots introduces steps in the mmf wave and produces further harmonics and also modulates the gap flux. Hence the choice of rotor slots is particularly important in the case of squirrel cage machines. Any bad combination of stator and rotor slots may result in awkward behaviour.

Harmonic induction torques $\qquad$ *(HTU, Dec' 18, 10 M) (RGPV, Jun' 20, 5 M) (UTU, Dec' 13, 10 M)*

Harmonic induction torques are torques produced by harmonic fields due to stator winding and slots.

Harmonic induction torques due to stator winding

The stator of induction motor has a three-phase winding which produces harmonics when sinusoidal current pass through the winding. The order of winding harmonics can be estimated from the following equation.

$$\text{Order of winding harmonics, } n_{wh} = 6\,N \pm 1$$

where, N is an interger

$+$ sign for forward rotating torque

$-$ sign for reverse rotating torque.

When N=1, $n_{wh} = 6\,N \pm 1 = 6 \times 1 \pm 1 = 7, 5$

When N=2, $n_{wh} = 6\,N \pm 1 = 6 \times 2 \pm 1 = 13, 11$

$\therefore$ When N = 1, winding produces forward rotating 7th harmonic and backward rotating 5th harmonic.

When N = 2, winding produces forward rotating 13th harmonic and backward rotating 11th harmonic.

In induction motors only 5th and 7th harmonics are more pronounced and produces dips in the torque-speed characteristics. With certain combinations of rotor and stator slots the dip due to 7th harmonic may become very pronounced.

The number of poles for the nth harmonic is n times the number of poles of the fundamental. Hence the synchronous speed of nth harmonic is $1/n^{th}$ of the synchronous speed of fundamental. The 7th harmonic produces dip in speed-torque characteristics at $1/7^{th}$ synchronous speed and the 5th harmonic produces dip at $-1/5^{th}$ synchronous speed.

If 7th harmonic is more pronounced, the machine may crawl at a speed little lesser than $1/7^{th}$ synchronous speed.

If the mechanical load on the shaft requires a constant load torque and if the torque developed by the rotor is below this load torque then the motor cannot accelerate upto its full speed but continues to run at a speed little lower than $1/7^{th}$ synchronous speed. This condition of the motor is called *crawling*.

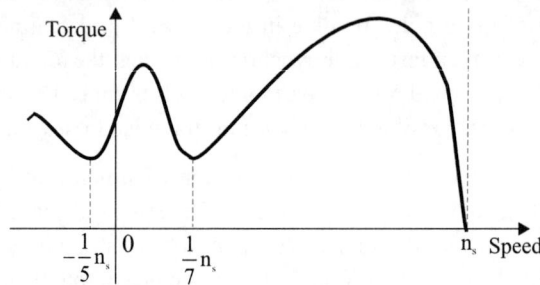

Fig 8.5: Dips caused by 5th and 7th harmonic in the torque-speed characteristics.

Harmonic induction torques due to stator and rotor slots

The slots in stator and rotor develop harmonics and the order of slot harmonics can be estimated from the following equation.

Order of slot harmonics, $n_{sh} = 6\,Aq \pm 1$

where, A = Interger

q_s = Stator slots per pole phase

+ sign for forward rotating torque

− sign for reverse rotating torque

Consider a motor with 4 poles and 36 slots stator. For this motor,

Slots per pole per phase, $q = \dfrac{36}{3 \times 4} = 3$

Let, A=1

Now, Order of stator slot harmonics, $n_{sh} = 6\,Aq \pm 1 = 6 \times 1 \times 3 \pm 1 = 19,\ 17$

Therefore, when A = 1, forward rotating 19th harmonic and backward rotating 17th harmonic are produced. This causes dips in the speed-torque characteristics at $1/19^{th}$ and $-1/17^{th}$ synchronous speeds respectively.

The effect of production of dips may be augmented by rotor slotting. Corresponding to the above 4 pole, 36 slot stator if we choose 76 rotor slots, there would be one rotor bar corresponding to every 19th harmonic pole (Slots per pole = 76/4 = 19). Thus the 19th harmonic torques would be very large and the rotor would vibrate considerably. Therefore it is necessary to avoid values of rotor slots exceeding stator slots by about 15%.

Harmonic synchronous torques

When the stator and rotor harmonics are of the same order then the torque will be alternately in opposite directions as they move past each other. But if their speeds happen to coincide they will lock together and if sufficiently powerful giving rise to a synchronous torque. In such a case the motor would crawl at constant subsynchronous speed.

When A = 1, the stator produces harmonics of the order,

$$n_{sh} = 6\, Aq_s \pm 1 = 6 \times 1 \times \frac{S_s}{3 \times p} \pm 1 = \frac{2\,S_s}{p} \pm 1$$

These stator slot harmonics revolve at a speed 1/n of synchronous speed with respect to stator.

When A = 1, the rotor slotting produces harmonic fields of the order,

$$n_{sh}' = 6\, Aq_s \pm 1 = 6 \times 1 \times \frac{S_r}{3 \times p} \pm 1 = \frac{2\,S_r}{p} \pm 1$$

These rotor slot harmonics revolve at a speed 1/n' that of the fundamental.

The speeds would be equal if, $2\,(S_s / p) \pm 1 = 2\,(S_r / p) \pm 1$

One of the possibilities for this to happen, is when $S_s = S_r$

When the number of rotor slots is equal to the number of stator slots, the speeds of all the harmonics produced by stator slotting coincide with the speed of corresponding rotor harmonics. Thus harmonics of every order would try to exert synchronous torques at their corresponding synchronous speeds and the machine would refuse to start. This is known as *cogging*.

The other possibility for equal stator and rotor harmonic speeds are, when $S_s - S_r = \pm p$.

Thus in order to avoid synchronous cusps the difference of stator and rotor slots should not be equal to $\pm p$ or a multiple of p. (Synchronous cusps are the synchronous torques produced due to harmonic synchronous speeds. Due to synchronous cusps the machine will crawl).

The difference in harmonic induction and synchronous torques is that the operating speed changes slightly for the former but is constant for the latter, for a small variation of the shaft load.

Vibration and noise

The rotation of rotor results in rapid variations in flux density in the air-gap thereby producing rapid changes in forces of attraction between stator and rotor teeth. The teeth, being cantilevers, respond to varying forces and are set into vibrations. Thus noise is produced. Some analysis shows that the vibration torques are produced if $(S_s - S_r) = \pm (p \pm 1)$ or $\pm (p \pm 2)$.

Voltage ripples

The harmonic fields produced by the stator current induce harmonic currents in the rotor which in turn reflects back additional harmonic fields into the stator. This produce ripples in the terminal voltage and additional iron losses. The voltage ripples produce high frequency currents in the supply lines, which in turn may produce inductive interference with communication circuits.

Rules for selecting rotor slots   *(PU, Nov' 19, 8 M) (RGPV, Jun' 20, 5 M) (AU, Apr' 18, 5 M)*

The following general rules should be followed concerning the choice of rotor slots for squirrel cage machines.

1. Number of stator slots should never be equal to rotor slots. Satisfactory results are obtained when S_r is 15 to 30% larger or smaller than the S_s.

2. The difference $(S_s - S_r)$ should not be equal to $\pm p$, $\pm 2p$ or $\pm 5p$ to avoid synchronous cusps.

3. The difference $(S_s - S_r)$ should not be equal to $\pm 3p$ for 3-phase machine to avoid magnetic locking.

4. The difference $(S_s - S_r)$ should not be equal to ± 1, ± 2, $\pm (p \pm 1)$ or $\pm (p \pm 2)$ to avoid noise and vibrations.

    Summarising, $(S_s - S_r)$ should not be equal to 0, $\pm p$, $\pm 2p$, $\pm 3p$, $\pm 5p$, ± 1, ± 2, $\pm (p \pm 1)$, $\pm (p \pm 2)$.

8.7.2 Design of Rotor Bars and Slots   *(AU, Apr' 19, 7 M)*

For a 3-phase machine, the rotor bar current is given by the equation,

$$\text{Rotor bar current, } I_b = \frac{6 I_s T_s}{S_r} K_{ws} \cos \phi \qquad \qquad .....(8.35)$$

$$\approx 0.85 \frac{6 I_s T_s}{S_r} \qquad \qquad .....(8.36)$$

where, I_s = Stator current per phase

T_s = Stator turns per phase

S_r = Number of rotor slots

The performance of an induction motor is greatly influenced by the resistance of rotor. Higher rotor resistance has higher starting torque but lesser efficiency. The rotor resistance is the sum of the resistance of the bars and the end rings. The cross-section of the bars and end rings are selected to meet both the requirements of starting torque as well as efficiency. The current density in the rotor bar, δ_b may be taken between 4 to 7 *A/mm²*.

$$\therefore \text{ Area of each rotor bar, } a_b = \frac{I_b}{\delta_b} \qquad \qquad .....(8.37)$$

In case of squirrel cage motor the cross-section of bars will take the shape of the slot and insulation is not used between bars and rotor core. The rotor slots for squirrel cage rotor may be either closed or semienclosed types. The semienclosed slots provides better overload capacity.

Advantages of closed slots

1. Low reluctance
2. Less magnetizing current
3. Quieter operation
4. Large leakage reactance and so starting current is limited

Disadvantage of closed slots

    1. Reduced over load capacity

Generally, the rotor slots and so the rotor bars are rectangular in shape. In rectangular bars, during starting most of current flows through top portion of the bar and so the effective rotor resistance is increased. This improves the starting torque.

8.7.3 Design of End Rings

The distribution of current in the bars and end rings of a squirrel cage motor is complicated. It can be shown that if flux distribution is sinusoidal then the bar current and end ring current will also be sinusoidal.

Consider a group of rotor bars under one pole pitch. Let one half send current to an end ring in one direction and the other half in the other direction. If the maximum value of the current in each bar is I_{b_max} and if the current is maximum in all the bars at the same time, then maximum value of the current in the end ring is given by,

$$\left.\begin{array}{c}\text{Maximum value of}\\\text{end ring current}\end{array}\right\} I_{e\_max} = \frac{\text{Bars per pole}}{2} \times \text{Current per bar}$$

$$= \frac{S_r}{2p} I_{b\_max}$$

However, current is not maximum in all the bars under one pole at the same time but varies according to sine law, hence the maximum value of the current in the end ring is the average of the current of half the bars under one pole.

$$\left.\begin{array}{c}\therefore \text{ Maximum value of}\\\text{end ring current}\end{array}\right\} I_{e\_max} = \frac{\text{Bars per pole}}{2} \times I_{b\_ave} = \frac{S_r/p}{2} \times \frac{2}{\pi} I_{b\_max}$$

$$= \frac{S_r/p}{2} \times \frac{2}{\pi} \times \sqrt{2} I_b = \frac{\sqrt{2}\, S_r\, I_b}{\pi p}$$

where, I_b = rms value of bar current

> **Note :** *Here the bar current is sinusoidal. Therefore,* $I_{b\_ave} = \dfrac{2}{\pi} I_{b\_max}$ *and* $I_{b\_max} = \sqrt{2}\, I_b.$

The end ring current also varies sinusoidally,

$$\left.\begin{array}{c}\therefore \text{ rms value of end}\\\text{ring current}\end{array}\right\} I_e = \frac{1}{\sqrt{2}} I_{e\_max} = \frac{1}{\sqrt{2}} \times \frac{\sqrt{2}\, S_r\, I_b}{\pi p} = \frac{S_r I_b}{\pi p} \qquad \qquad .....(8.38)$$

Let current density in the end ring δ_e be 4 to 7 A/mm^2

$$\therefore \text{ Area of cross-section of end ring, } a_e = \frac{I_e}{\delta_e} \qquad \qquad .....(8.39)$$

Also, Area of end ring, a_e = Depth of end ring × Thickness of end ring.

$$= d_e \times t_e \qquad \qquad .....(8.40)$$

The depth of end ring can be assumed depending on the inner and outer diameter of the rotor. Using the area of end ring and depth of end ring the thickness of end ring can be estimated.

8.7.4 Reduction of Harmonic Torques

The methods used for reduction or elimination of harmonic torques are chording, integral slot winding, skewing and increasing the length of air-gap.

Chording : The chorded windings with integral number of slots per pole per phase weakens the stator winding mmf harmonics.

Integral slot winding : Windings with fractional number of slots per pole per phase create asymmetrical mmf distribution around the air-gap and favour the creation of noise in the motor. Therefore, fractional slot windings are not used for induction motor stator and only integral slot windings are used.

Skewing : The motor noise, vibrations, cogging and synchronous cusps can be reduced or even entirely eliminated by skewing either the stator or the rotor. The practice generally followed in India is to skew the rotor.

In order to eliminate the effect of any harmonic, the rotor bars should be skewed through an angle so that the bars lie under alternate harmonic poles of the same polarity or in other words the bars must be skewed through two harmonic pole pitches.

Suppose it is desired to eliminate a harmonic of the order n in a machine with p poles. The number of nth order harmonic poles is np.

$$\therefore \text{ Mechanical angle between two adjacent harmonic poles} = \frac{360}{np} \qquad .....(8.41)$$

Therefore, for elimination of nth harmonic by skewing the rotor bars are skewed through two harmonic pole pitches,

$$\text{Mechanical angle of skew, } \theta = 2 \times \frac{360}{np} = \frac{720}{np} \,^{\circ}m \qquad .....(8.42)$$

$$\text{Electrical angle of skew, } \theta_{sk} = \frac{720}{np} \times \frac{p}{2}$$

$$\boxed{1^{\circ}m = \frac{p}{2} \,^{\circ}e} \quad \boxed{180^{\circ} = \pi \, rad}$$

$$= \frac{360}{n} \,^{\circ}e = \frac{2\pi}{n} \, rad.elect \qquad .....(8.43)$$

For infinetly distributed winding the distribution factor for n^{th} order harmonic is given by the following equation.

$$\text{Distribution factor of } n^{th} \text{ order harmonic, } K_{dn} = \frac{\sin \dfrac{n\,\theta_{sk}}{2}}{\dfrac{n\,\theta_{sk}}{2}}$$

$$\text{When, } \theta_{sk} = \frac{2\pi}{n}, \text{ then } K_{dn} = \frac{\sin \dfrac{n}{2}\dfrac{2\pi}{n}}{\dfrac{n}{2}\dfrac{2\pi}{n}} = \frac{\sin \pi}{\pi} = 0$$

From the above calculation it is clear that the distribution factor of nth order harmonic is zero when angle of skew is $2\pi/n$ and so the nth order harmonic emf is zero and so nth order harmonic is eliminated.

Increasing air-gap length : The increase in air-gap length reduces the harmonic torques but increases the no-load current and results in poor power factor. Hence only for mechanical reasons the air-gap is made larger.

8.8 Design of Wound Rotor (Slip Ring Rotor)

The wound rotor has the facility of adding external resistance to rotor circuit in order to improve the torque developed by the motor. The rotor consists of laminated core with semi-enclosed slots and carries a three-phase winding.

8.8.1 Rotor Windings

For small motors mush windings are employed for the rotor. It is usual to use several wires in parallel per turn, to keep the conductor small enough to go through the narrow slot opening.

For large motors, a double layer bar type wave winding is used. The winding has generally two bars per slot. The bars are pushed through partially closed slots and are bent to shape at the other end.

In motors of output more than $750\ kW$, large number of bars per slot are used to reduce the current handled by slip rings. This type of winding is called barrel winding and is usually wave wound.

8.8.2 Number of Rotor Turns

The rotor is equivalent to secondary of a transformer and the voltage between slip rings is maximum when the rotor is at rest. The rotor voltage on open circuit between slip rings should not exceed $500\ V$ for small machines where hand operated starters and switchgear are employed. For large size machines the voltage between slip rings can be upto $2000\ V$.

For the induction motor the turns ratio is given by, $\dfrac{E_r}{E_s} = \dfrac{K_{wr}\, T_r}{K_{ws}\, T_s}$ .....(8.44)

$\therefore$ Rotor turns per phase, $T_r = \dfrac{K_{ws}\, T_s}{K_{wr}} \times \dfrac{E_r}{E_s}$ .....(8.45)

The rotor ampere-turn is assumed as 85% of stator ampere-turn.

$\therefore$ Rotor ampere-turn = $0.85 \times$ Stator ampere-turn

$\therefore I_r T_r = 0.85\, I_s T_s$

Hence, rotor current, $I_r = \dfrac{0.85\, I_s T_s}{T_r}$ .....(8.46)

The current density for rotor conductors is assumed same as that of stator conductors.

The range of current density in rotor is 3 to 5 A/mm^2

Let, δ_r = Current density in rotor

$\therefore$ Area of rotor conductor, $a_r = \dfrac{I_r}{\delta_r}$ .....(8.47)

8.8.3 Number of Rotor Slots

The discussions made on choice of squirrel cage rotor slots are also applicable to the choice of wound rotor slots. For wound rotors the windings are always three-phase winding and they are star connected at one end and the other three end are terminated on three slip rings mounted on the shaft.

Since the windings are three-phase windings, the number of slots should be such that a balanced winding is obtained. Generally windings with an integral number of slots per pole per phase are used for the rotor. When fractional slot windings are used, it is preferable to have the number of slots as multiples of phases and pair of poles.

8.8.4 Rotor Teeth

The width of rotor slot should be such that the flux density in the rotor tooth does not exceed about 1.7 Wb/m^2. The maximum flux density for rotor tooth occurs at its root since its section is minimum there.

Let, W_{tr} = Width of rotor tooth

$$\text{Minimum teeth area per pole} = \frac{\text{Flux per pole}}{\text{Maximum flux density}} = \frac{\phi_m}{1.7}$$

Total teeth area per pole = Number of rotor slots per pole × Net iron length × Width of tooth

$$= \frac{S_r}{p} \times L_i \times W_{tr}$$

The minimum width of rotor tooth can be obtained by equating the minimum teeth area per pole to total teeth area per pole after replacing W_{tr} by W_{tr_min}.

$$\therefore \frac{S_r}{p} \times L_i \times W_{tr\_min} = \frac{\phi_m}{1.7}$$

$$\therefore W_{tr\_min} = \frac{\phi_m / 1.7}{\dfrac{S_r}{p} \times L_i} = \frac{\phi_m \, p}{1.7 \times S_r \times L_i} \qquad\qquad .....(8.48)$$

A check has to made, so that the actual minimum width of the tooth is not more than $W_{tr\ min}$ given by equation (8.48).

Actual minimum width of rotor tooth = Rotor slot pitch at the root – Rotor slot width

$$= \frac{\pi \, (D_r - 2d_{sr})}{S_r} - y_{sr} \qquad\qquad .....(8.49)$$

where, d_{sr} = Depth of rotor slot

y_{sr} = Rotor slot pitch or width

8.8.5 Rotor Core

The flux density in the rotor core is generally equal to stator core density.

$$\therefore \text{Depth of rotor core, } d_{cr} = \frac{\phi_m}{2 \times B_{cr} \times L_i} \qquad\qquad .....(8.50)$$

where, B_{cr} = Flux density in the rotor core

Inner diameter of rotor lamination, $D_i = D_r - 2(d_{sr} + d_{cr})$ $\qquad\qquad .....(8.51)$

where, d_{cr} = Depth of rotor core

Note : Refer Fig. 8.15 for dimensions of rotor core.

8.8.6 Slip Rings and Brushes

The wound rotor consists of three slip rings mounted on the shaft but insulated from it. The rings are made of either brass or phosphor bronze. The area of cross-section of slip rings are decided from the knowledge of rotor current and assuming a current density of 4 to 7 A/mm^2. The cross-section of the slip ring will be rectangle. The length and breadth of the rectangle are decided based on mechanical stability and constraints.

The brushes used in wound rotor machines are made of metal graphite. The metal graphite is an alloy of copper and carbon, with very low resistance and high mechanical stength. The dimensions of the brushes are decided by assuming a current density of 0.1 to 0.2 A/mm^2.

8.9 MAGNETIC LEAKAGE CALCULATIONS

In inductance machine the following leakage flux exists.

1. Stator slot leakage flux

2. Rotor slot leakage flux

3. Overhang leakage flux

4. Zigzag leakage flux

5. Harmonic leakage flux

The details of above leakage flux are discussed in Section 3.12 of Chapter-3. These leakage fluxes will induce an opposing emf to applied voltage and so create a voltage drop like a reactance. Therefore, the effect of leakage flux can be accounted by estimating a total leakage reactance.

8.9.1 Leakage Reactance

The leakage reactance of induction motor has the following five components.

1. Stator slot leakage reactance

2. Rotor slot leakage reactance

3. Overhang leakage reactance

4. Zigzag leakage reactance

5. Harmonic leakage reactance

The calculation of leakage reactance has been explained in Section 1.5.1 of Chapter-1.

The following equation for stator slot leakage reactance is obtained from equation (1.78).

Stator slot leakage reactance, $X_{ss} = 8 \pi f T_s^2 L \left(\dfrac{\lambda_{ss}}{pq_s} \right)$ .....(8.52)

where, λ_{ss} = Specific slot permeance for stator slot

q_s = Stator slots/ pole/ phase

The equation for other leakage reactance can be obtained from stator slot leakage reactance by replacing the specific slot permeance for stator, λ_{ss} by the respective permeance.

$\therefore$ Rotor slot leakage reactance referred to stator $\left. \right\}$ $X_{sr}' = 8 \pi f T_s^2 L \left(\dfrac{\lambda_{sr}'}{pq_s} \right)$ .....(8.53)

Overhang leakage reactance, $X_o = 8 \pi f \, T_s^2 \, L_o \left(\dfrac{\lambda_o}{pq_s} \right)$ .....(8.54)

Zigzag leakage reactance, $X_z = 8 \pi f \, T_s^2 \, L \left(\dfrac{\lambda_z}{pq_s} \right)$ .....(8.55)

Harmonic leakage reactance, $X_h = 8 \pi f \, T_s^2 \, L \left(\dfrac{\lambda_h}{pq_s} \right)$ .....(8.56)

where, $\lambda_{sr}' =$ Specific permeance of rotor slots referred to stator

$\qquad \lambda_o =$ Specific overhang permeance

$\qquad \lambda_z =$ Specific zigzag permeance

$\qquad \lambda_h =$ Specific harmonic permeance

$\qquad L_o =$ Length of conductor in overhang

Let, $X_s =$ Total leakage reactance referred to stator

$\therefore X_s = X_{ss} + X_{sr}' + X_o + X_z + X_h$ .....(8.57)

Using equations (8.52) to (8.56) the equation (8.57) can be expressed as shown below:

$X_s = \dfrac{8 \pi f \, T_s^2}{p \, q_s} \left(L \, (\lambda_{ss} + \lambda_{sr}' + \lambda_z + \lambda_h) + L_o \lambda_o \right)$

8.9.2 Specific Permeance

The specific permeance for stator and rotor slots can be estimated from slot dimensions as explained in Chapter-3 Section 3.14 for parallel sided, tapered and circular slots.

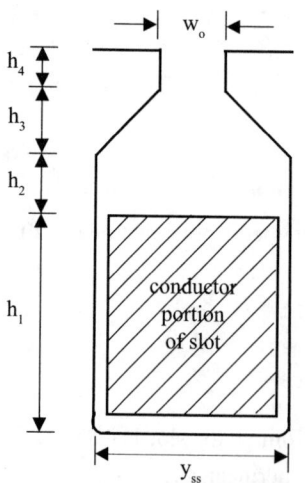

Fig. 8.6: *Parallel sided slot.*

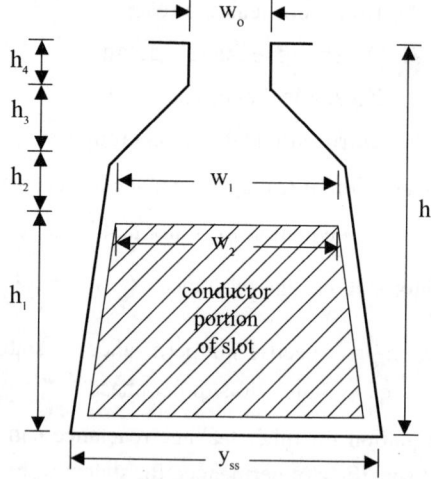

Fig. 8.7: *Tapered slot.*

For parallel sided slot, (Refer equation 3.65)

$$\lambda_{ss} \text{ or } \lambda_{sr} = \mu_o \left[\frac{h_1}{3W_o} + \frac{h_2}{y_{ss}} + \frac{2h_3}{y_{ss} + W_o} + \frac{h_4}{W_o} \right] \qquad .....(8.58)$$

For tapered slot, (Refer equation 3.70)

$$\lambda_{ss} \text{ or } \lambda_{sr} = \mu_o \left[\frac{2h_1}{3(y_{ss} + W_2)} + \frac{2h_2}{W_2 + W_1} + \frac{2h_3}{W_1 + W_o} + \frac{h_4}{W_o} \right] \qquad .....(8.59)$$

For circular slot, (Refer equation 3.71)

$$\lambda_{ss} \text{ or } \lambda_{sr} = \mu_o \left[0.66 + \frac{h}{W_o} \right] \qquad .....(8.60)$$

The specific permeance of rotor referred to stator can be estimated from following equation.

$$\lambda_{sr}' = \left(\frac{K_{ws}}{K_{wr}} \right)^2 \frac{S_s}{S_r} \lambda_{sr} \qquad .....(8.61)$$

The product of $L_o \lambda_o$ required to estimate overhang leakage reactance can be calculated from the following equation.

$$L_o \lambda_o = \mu_o \frac{K_s \tau^2}{\pi y_{ss}} \qquad .....(8.62)$$

where, K_s can be estimated from Fig. 8.9.

Alternatively, the specific permeance for overhang leakage can be calculated from the following equation.

$$\lambda_o = \mu_o \frac{k L_o}{2\sqrt{2} \, y_{ss}} \qquad .....(8.63)$$

where, L_o = Length of overhang

y_{ss} = Stator slot pitch

The constant k depends on type of winding and can be selected from Table 8.2.

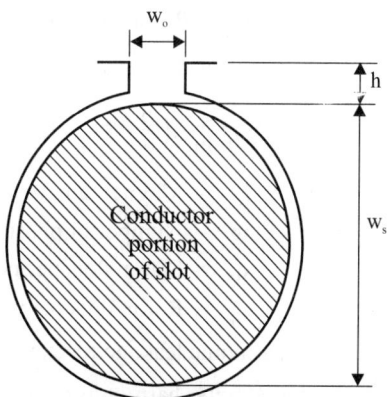

Fig. 8.8: Circular slot.

Table 8.2: Values of constant k

Rotor type	Stator winding type	k
Slip-ring	Barrel winding	0.55×10^{-6}
Slip-ring	Concentric winding	0.35×10^{-6}
Cage rotor	Barrel winding	0.37×10^{-6}
Cage rotor	Concentric winding	0.27×10^{-6}

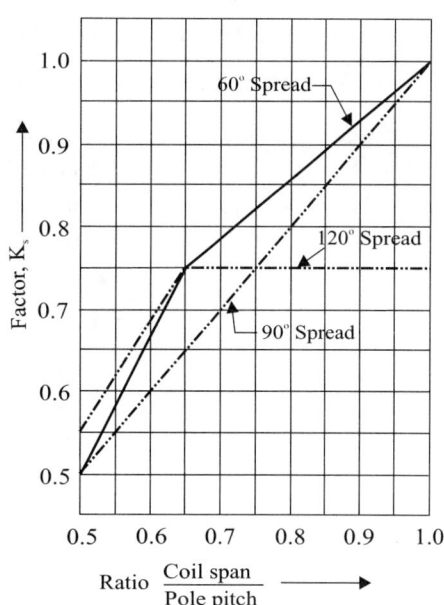

Fig. 8.9: Slot leakage factor.

The specific permeance for zigzag leakage can be estimated from the following equation.

$$\lambda_z = \mu_o \frac{W_{ts} \, W_{tr} \, (W_{ts}^2 + W_{tr}^2)}{12 \, l_g \, y_{ss}^2 \, y_{sr}} \qquad .....(8.64)$$

$$W_{ts} = y_{ss} - W_{os} \quad ; \quad W_{tr} = y_{sr} - W_{or}$$

where, W_{ts} = Width of stator tooth

W_{tr} = Width of rotor tooth

l_g = Length of air-gap

y_{ss} = Stator slot pitch

y_{sr} = Rotor slot pitch

W_{os} = Stator slot opening

W_{or} = Rotor slot opening

Alternatively, the zigzag leakage reactance can be calculated from the following equation.

$$X_z = \frac{5}{6} \frac{X_m}{m_s^2} \left(\frac{1}{q_s^2} + \frac{1}{q_r^2} \right) \qquad .....(8.65)$$

where, X_m = Magnetizing reactance

m_s = Number of phases in stator winding

q_s = Stator slots per pole per phase

q_r = Rotor slots per pole per phase

The harmonic leakage reactance is negligible in induction motor and usually neglected.

Alternatively, harmonic or differential leakage reactance can be estimated from the following equation.

$$X_h = X_m \, (K_{hs} + K_{hr}) \qquad .....(8.66)$$

where, K_{hs} and K_{hr} are constants (for stator and rotor respectively) and are taken from Fig. 8.10.

X_m = Magnetic leakage reactance

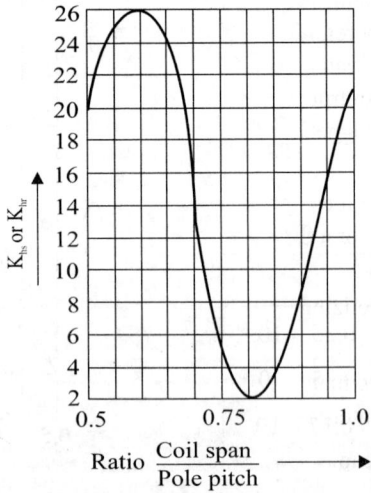

Fig. 8.10: *Values of factor K_{hs} and K_{hr}.*

8.10 OPERATING CHARACTERISTICS

(PU, Nov' 19, 6 M)

8.10.1 No-Load Current

The current drawn by induction motor when there is no-load on rotor shaft is called *no-load current*. The induction motor draws current on no-load to establish flux in the core and to supply for power loss on no-load. Therefore, the no-load current has two components: magnetizing component and loss component.

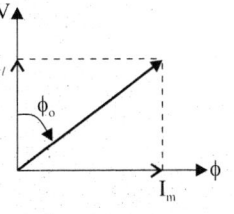

Fig. 8.11: *Vector diagram of no-load current.*

Let, I_m = Magnetizing component

I_l = Loss component

I_o = No-load current

The vector diagram of no-load current is shown in Fig. 8.11. Here, I_m and I_l are in perpendicular to each other. With reference to Fig. 8.11 we can write,

No-load current, $I_o = \sqrt{I_m^2 + I_l^2}$ .....(8.67)

The equations for I_m and I_l are given by equations (8.64) and (8.77) respectively.

8.10.2 Magnetizing Current

The magnetic circuit of a four pole induction motor is shown in Fig. 8.12. The flux produced by stator mmf passes through the following parts:

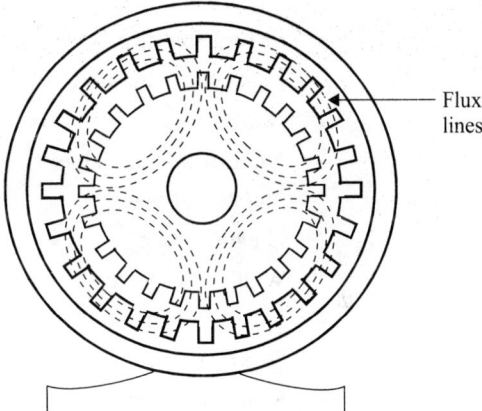

Flux lines

1. Air-gap
2. Rotor teeth
3. Rotor core
4. Stator teeth
5. Stator core

In induction motor, the permeability of iron in the stator and rotor teeth are not same. Therefore, if the value of flux density is calculated for the mean mmf, then the total flux obtained will be larger than true value which in turn gives a smaller value of magnetizing current smaller than the true value.

Fig. 8.12: *Magnetic circuit of 4 pole induction motor.*

Alternatively, if maximum values are considered then mmf is too small and so the magnetizing current is too large. In induction motor a flux tube crossing the air-gap at 60° from the interpolar axis will always give a good approximation.

Thus the calculation of magnetizing mmf should be based upon the value of flux density at 60° from the interpolar axis.

Let, AT_{60} = mmf to create magnetizing current considering flux density at 60° from interpolar axis.

AT_g = mmf for air-gap

AT_{ts} = mmf for stator teeth

AT_{tr} = mmf for rotor teeth

AT_{cs} = mmf for stator core

AT_{cr} = mmf for rotor core

Now the total mmf is given by,

Total mmf, $AT_{60} = AT_g + AT_{ts} + AT_{tr} + AT_{cs} + AT_{cr}$ .....(8.68)

Magnetizing current per phase, $I_m = \dfrac{0.427 \, p \, AT_{60}}{K_{ws} \, T_s}$ .....(8.69)

mmf for air-gap

Let, B_{av} = Average flux density in air-gap

B_{g60} = Air-gap flux density at 60° from interpolar axis

K_g = Gap contraction factor

l_g = Length of air-gap

We know that, $B_{g60} = 1.36 \, B_{av}$

∴ mmf for air-gap, $AT_g = 800{,}000 \, B_{g60} \, K_g \, l_g$ .....(8.70)

mmf for stator teeth

Let, B_{ts60} = Stator teeth flux density at 60° from interpolar axis

$B_{ts1/3}$ = Flux density at one-third height of stator tooth from narrow end

ϕ_m = Maximum flux in the core

S_s = Number of stator slots

p = Number of poles

L_i = Length of iron

$W_{t1/3}$ = Width of stator teeth at one-third height from narrow end

at_{ts} = Specific magnetic loading for stator teeth

d_{ss} = Depth of stator slot

y_{ss} = Stator slot width

The value of mmf for stator teeth is found out by finding flux density at a section one-third height of tooth from narrow end.

The equations for calculating mmf for stator teeth are given below:

$$B_{ts1/3} = \dfrac{\phi_m}{(S_s/p) \times L_i \times W_{ts1/3}} \qquad \qquad .....(8.71)$$

$$W_{ts1/3} = \dfrac{\pi(D + 2d_{ss}/3)}{S_s} - y_{ss} \qquad \qquad .....(8.72)$$

$B_{ts60} = 1.36 \, B_{ts1/3}$

mmf required for stator teeth, $AT_{ts} = at_{ts} \times d_{ss}$ .....(8.73)

Note: The mmf per metre at_{ts} for stator teeth is found from Fig. 8.13 corresponding to B_{ts60}.

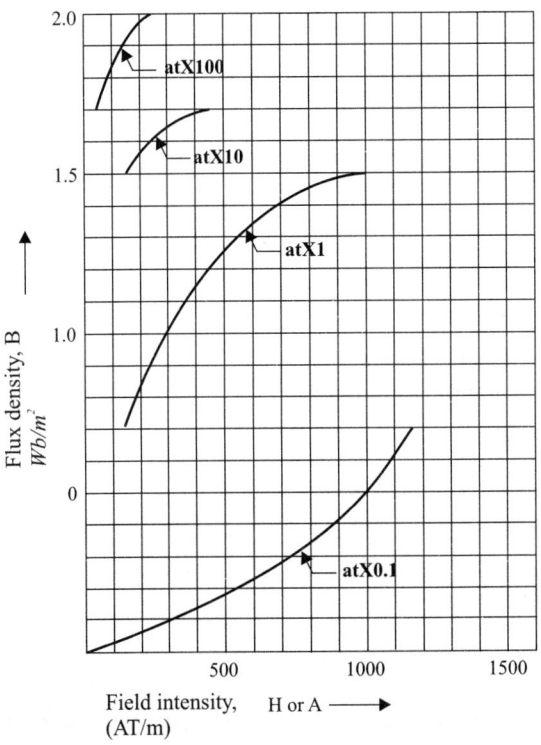

Fig. 8.13: *B-H curve for electrical steel (non-oriented).*

mmf for rotor teeth

Let, B_{tr60} = Rotor teeth flux density at 60° from interpolar axis

$B_{tr1/3}$ = Flux density at one-third height of rotor tooth from narrow end

S_r = Number of rotor slots

$W_{tr1/3}$ = Width of rotor teeth at one-third height from narrow end

D_r = Diameter of rotor

d_{sr} = Depth of rotor slot

y_{sr} = Rotor slot width

at_{tr} = Specific magnetic loading for rotor teeth

The value of flux density in rotor teeth is estimated at one-third height from narrow end. The following are design equations to estimate mmf for rotor teeth.

$$B_{ts1/3} = \frac{\phi_m}{(S_r/p) \times L_i \times W_{tr1/3}} \qquad \qquad .....(8.74)$$

$$W_{tr1/3} = \frac{\pi(D_r - 4d_{sr}/3)}{S_r} - y_{sr} \qquad \qquad .....(8.75)$$

$$B_{tr60} = 1.36\, B_{tr1/3}$$

∴ mmf required for rotor teeth, $AT_{tr} = \mathbf{at}_{tr} \times d_{sr}$ .....(8.76)

*Note : The mmf per meter, **at**$_{tr}$, for rotor teeth is found from Fig. 8.13 corresponding to B$_{tr60}$*

mmf for stator core

Let, d_{sm} = Mean diameter of stator core

  D = Diameter of stator bore

  d_{sc} = Depth of stator core

  d_{ss} = Depth of stator slot

  l_{cs} = Length of mean flux path in stator core

  τ_{sm} = Pole pitch at mean stator diameter

  AT_{cs} = mmf for stator core

  $\mathbf{at}_{cs}$ = Specific magnetic loading for stator core

With reference to Fig. 4.14.

$$d_{sm} = \frac{d_{sc}}{2} + d_{ss} + D + d_{ss} + \frac{d_{sc}}{2}$$

$$= D + 2d_{ss} + d_{sc}$$

Length of mean flux path in stator core is one-third of pole pitch at mean stator diameter.

$$\therefore l_{cs} = \frac{1}{3}\, \tau_{sm} = \frac{1}{3}\, \frac{\pi\, d_{sm}}{p} = \frac{\pi\,(D + 2\, d_{ss} + d_{sc})}{3p}$$

mmf for stator core, $AT_{cs} = \mathbf{at}_{cs} \times l_{cs}$

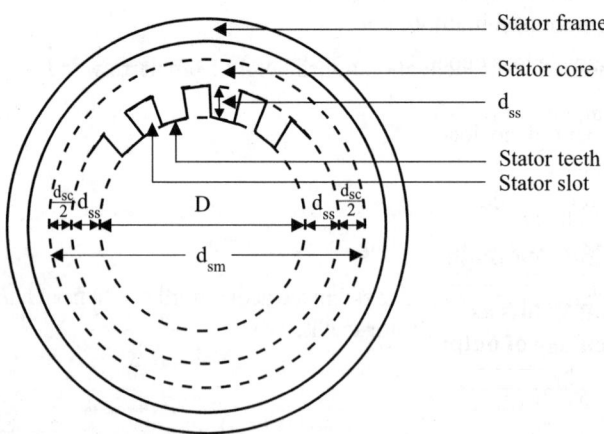

Fig. 8.14: Dimensions of stator core.

mmf for rotor core

Let, d_{rm} = Mean diameter of rotor core

D_r = Diameter of rotor

d_{sr} = Depth of rotor core

d_{ss} = Depth of rotor slot

l_{cr} = Length of mean flux path in rotor core

τ_{rm} = Pole pitch at mean rotor diameter

AT_{cr} = mmf for rotor core

at_{cr} = Specific magnetic loading for rotor core

With reference to Fig. 8.15.

$$d_{rm} = D_r - d_{sr} - \frac{d_{cr}}{2} - \frac{d_{cr}}{2} - d_{sr}$$

$$= D_r - 2d_{sr} - d_{cr}$$

Length of mean flux path in rotor core is one-third of pole pitch at mean rotor diameter.

$$l_{cr} = \frac{1}{3} \tau_{rm} = \frac{1}{3} \frac{\pi d_{rm}}{p} = \frac{\pi (D_r - 2 d_{sr} - d_{cr})}{3p}$$

$\therefore$ mmf for rotor core, $AT_{cr} = at_{cr} \times l_{cr}$

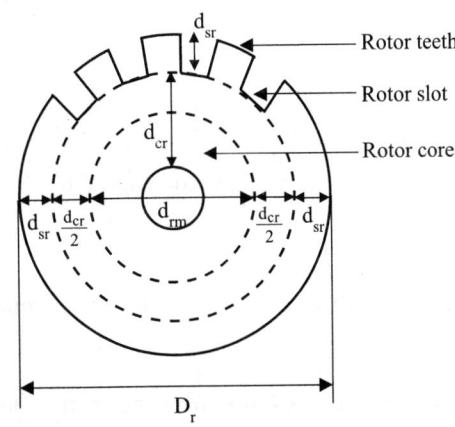

Fig. 8.15: Dimensions of rotor core.

8.10.3 Loss Component of No-Load Current

The loss component of no-load current is due to core loss, friction and windage and no-load copper loss. Usually no-load copper loss is neglected because the no-load current is low and copper loss is propotional to square of current and so no-load copper loss will have very low value.

Since rotor frequency is very low, the core loss in rotor is negligible. The core loss is considerable only in stator core and it is due to hysteresis and eddy current. The core loss can be estimated from the design data, loss/kg of the iron used in the core.

The friction and windage loss depends on capacity and hence size of motor, and it can be estimated using the Table 8.3.

$$\left. \begin{array}{l} \text{Loss component of no-load} \\ \text{current per phase} \end{array} \right\} I_l = \frac{\text{Total no-load loss}}{3 V_p} \qquad \qquad .....(8.77)$$

where, V_p = Phase voltage

TABLE 8.3: Friction and Windage loss

Output kW	F and W loss as percentage of output
0.75	5.5 %
3.70	3.5 %
7.50	2.7 %
37	1.5 %
75	1.2 %
150	1.0 %

8.10.4 Short Circuit Current

The short circuit current is given by the ratio of stator voltage required to circulate full load current at standstill to stator impedance at standstill.

∴ Short circuit current per phase, $I_{sc} = \dfrac{E_{sc}}{Z_s}$ .....(8.78)

where, E_{sc} = Stator voltage to circulate full load current at standstill

$\quad\quad Z_s$ = Standstill impedence per phase

The standstill impedence is given by vector sum of resistance of motor referred to stator and leakage reactance.

∴ Complex value of standstill impedence, $\overline{Z_s} = R_s + jX_s$

∴ Standstill impedence, $Z_s = \sqrt{R_s^2 + X_s^2}$ .....(8.79)

where, R_s = Motor resistance referred to stator

$\quad\quad X_s$ = Leakage reactance of motor referred to stator

Resistance of motor referred to stator

Let, r_s = Stator resistance per phase

$\quad T_s$ = Number of stator turns

$\quad \rho$ = Resistivity of copper = 0.021 Ω/m of 75^oC

$\quad L_{mts}$ = Length of mean turn of stator

$\quad r_r$ = Rotor resistance per phase

$\quad r_r'$ = Rotor resistance per phase referred to stator

The stator and rotor resistance can be estimated using following equations.

Stator resistance per phase, $r_s = \dfrac{\rho\, T_s\, L_{mts}}{a_s}$ .....(8.80)

Rotor resistance referred to stator, $r_r' = \left(\dfrac{K_{ws}\, T_s}{K_{wr}\, T_r}\right)^2 r_r$ .....(8.81)

Total motor resistance referred to stator, $R_s = r_s + r_r'$ .....(8.82)

Rotor resistance of slip ring rotor

Let, r_r = Rotor resistance of wound rotor

$\quad T_r$ = Number of turns in rotor winding

$\quad L_{mtr}$ = Length of mean turn of rotor

$\quad a_r$ = Area of cross-section of rotor winding conductor

∴ Rotor resistance per phase $\left.\begin{array}{l} \\ \text{in slip-ring motor} \end{array}\right\}\, r_r = \dfrac{\rho\, T_r\, L_{mtr}}{a_r}$ .....(8.83)

Rotor resistance of cage rotor

Let, L_b = Length of rotor bar

a_b = Area of cross-section of each bar

D_e = Diameter of end ring

a_e = Area of cross-section of end ring

∴ Resistance of each bar, $r_b = \dfrac{\rho L_b}{a_b}$ .....(8.84)

Resistance of each end ring, $r_e = \dfrac{\rho \pi D_e}{a_e}$ .....(8.85)

Copper loss in rotor = Loss due to rotor bars + Loss due to two end rings

$$= \dfrac{S_r \, r_b \, I_b^2 + 2 \, r_e \, I_e^2}{3}$$ .....(8.86)

where, S_r = Number of rotor bars

$\left.\begin{array}{l}∴ \text{Rotor resistance} \\ \text{per phase in cage motor}\end{array}\right\} r_r = \dfrac{\text{Copper loss in rotor}}{I_b^2}$ .....(8.87)

8.10.5 Circle Diagram

Circle diagram is a graphical method to predict or determine the performance characteristics of an induction motor. From the circle diagram we can estimate full load current, power factor, maximum power output, pull-out torque, full-load efficiency and slip. The circle diagram is constructed from the following design data:

I_m = Magnetizing current per phase

I_l = Loss component of no-load current per phase

X_s = Total standstill leakage reactance per phase referred to stator

R_s = Total resistance per phase referred to stator

Z_s = Standstill impedence per phase referred to stator

E_s = Stator voltage per phase

The procedure for drawing circle diagram

The procedure for drawing circle diagram is given below (Refer Fig. 8.16).

1. Draw two lines OX and OY perpendicular to each other.

2. Choose a current scale, for example 1 cm = 2 A.

3. Draw line OO' whose length is equal to the no-load current per phase, I_o at an angle ϕ_o with line OY.

where, $\phi_o = \tan^{-1} I_m / I_l$

$$I_o = \sqrt{I_m^2 + I_l^2}$$

4. Draw a line O'D passing through O' and parallel to line OX.

5. Draw line OB whose length is equal to short circuit current per phase, I_{sc} at an angle, ϕ_{sc} with line OY.

where, $\phi_{sc} = \tan^{-1} X_s/R_s$

$I_{sc} = E_s/Z_s$

6. Join the point O' and B, now the line O'B is the output line.

7. Choose mid-point, R in line O'B and construct the perpendicular bisector of O'B passing through point R and intersecting the line O'D at point C, where point C is the centre of circle having radius O'C.

8. Draw the circle O'BD, with centre at point C.

9. Draw a perpendicular to O'D such that it passes through point B and meeting the line O'D at point F. Then divide the line BF in such a way that

$$\frac{BG}{GF} = \frac{\text{Rotor resistance referred to stator}}{\text{Stator resistance}} = \frac{r_r'}{r_s}$$

10. Join O' to G. Now the line O'G is known as torque line.

Note : *The circle diagram is drawn by considering phase current and phase voltage therefore the results are obtained in terms of power per phase. In order to get total power for three-phase machine we have to multiply the results of power and torque by 3.*

---

11. Determine the power scale from current scale.

Power scale = Current scale $\times$ E_s

where, E_s = Stator voltage per phase

12. Extend line FB and mark point S such that line BS is rated power output per phase.

13. Draw a line parallel to output line O'B passing through point S and cutting the circle at point A. Now point A is the operating point for rated power output and line OA is full load current per phase.

14. Draw a perpendicular to line OX passing through point A. Let the crossing point of this perpendicular with line OX, line OD, torque line and output line be respectively H,J,K and L.

Now the line segments AH, AK, AL, LK, KJ and JH represent the following.

AH = Power input	LK = Rotor copper loss
AK = Rotor input	KJ = Stator copper loss
AL = Power output	JH = Constant loss

15. The location of point M on circle for maximum power output is obtained by drawing a perpendicular on the output line passing through point C. Draw a line parallel to line FS passing through point N on output line and cutting the circle at point M. Now the line MN represents maximum output.

16. The location of point P on circle for maximum torque is obtained by drawing a perpendicular on torque line passing through point T on torque line and point C on line O'D. Draw a line parallel to line BS passing through point Q on torque line and cutting the circle at point P. Now the line PQ represents maximum torque.

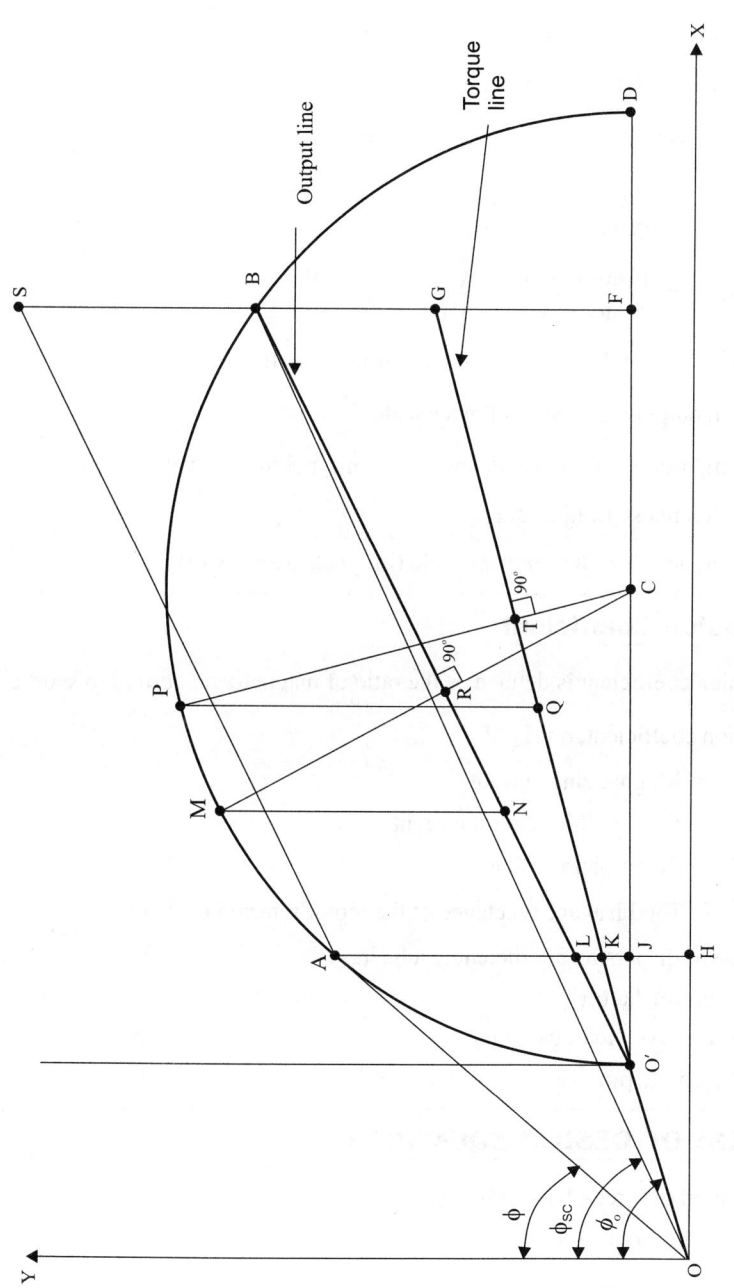

Fig. 8.16: Circle Diagram.

Performance characteristics from circle diagram

Stator current per phase at full load, I_s = OA × Current scale

Stator power factor at full load, cos ϕ = AH/OA

Constant loss = 3 × JH × Power scale

Rotor copper loss at full load = 3 × LK × Power scale

$$\text{Slip, s} = \frac{\text{Rotor copper loss}}{\text{Rotor input}} = \frac{LK}{AK}$$

$$\text{Efficiency, } \eta = \frac{\text{Rotor output}}{\text{Stator input}} = \frac{AL}{AH}$$

Torque = 3 × AK × Power scale (in synchronous watt)

Maximum output = 3 × MN × Power scale

Maximum torque = 3 × PQ × Power scale (in synchronous watt)

Line BG represents starting torque,

∴ Starting torque = 3 × BG × Power scale (in synchronous watt).

8.10.6 Dispersion Coefficient

The dispersion coefficient is defined as the ratio of magnetizing current to short circuit current.

$$\text{Dispersion coefficient, } \sigma = I_m / I_{sc} \qquad\qquad .....(8.88)$$

where, I_m = Magnetizing current

I_{sc} = E_s/X_s = Short circuit current

E_s = Stator phase voltage

X_s = Total leakage reactance of the motor referred to stator

Higher value of dispersion coefficient results in,

1. Poor power factor
2. Reduced over-load capacity
3. Reduced output

8.11 SUMMARY OF DESIGN EQUATIONS

1.   Output equation of three-phase induction motor

i) kVA input, $Q = C_o D^2 L\, n_s$

ii) kVA input, $Q = \dfrac{HP \times 0.746}{\eta \times \cos \phi}$

iii) kVA input, $Q = \sqrt{3}\ V_L\, I_L \times 10^{-3}$

iv) kVA input, $Q = 3\ E_s\, I_s \times 10^{-3}$

2. Output coefficient, $C_o = 11 \, K_{ws} \, B_{av} \, ac \times 10^{-3}$

3. Maximum flux, $\phi_m = \dfrac{B_{av} \, \pi DL}{p}$

4. Values of L/τ for various design features

   i) For minimum cost, L/τ =   1.5 to 2

   ii) For good power factor, L/τ =   1.0 to 1.25

   iii) For good efficiency, L/τ =   1.5

   iv) For good overall design, L/τ =   1

   v) For good power factor, τ = $\sqrt{0.18 \, L}$

Design of stator

5. Stator phase voltage, $E_s = 4.44 \, K_{ws} \, f \, \phi_m \, T_s$

6. Stator turns per phase, $T_s = \dfrac{E_s}{4.44 \, K_{ws} \, f \, \phi_m}$

7. Area of stator conductors, $a_s = \dfrac{I_s}{\delta_s}$

8. Stator slot pitch, $y_{ss} = \dfrac{\pi D}{S_s}$

9. Total number of stator conductors, $Z = 6 \, T_s$

10. Conductors per slot, $Z_{ss} = \dfrac{6 \, T_s}{S_s}$

11. Area of each slot $= \dfrac{Z_{ss} \, a_s}{\text{Space factor}}$

12. Length of mean turn, $L_{mts} = 2L + 2.3 \, \tau + 0.24$

13. Minimum teeth area per pole $= \dfrac{\phi_m}{1.7}$

14. Minimum width of stator tooth, $W_{ts\_min} = \dfrac{\phi_m}{1.7 \, (S_s / p) \, L_i}$

15. Area of stator core $= \dfrac{\phi_m}{2 \, B_{cs}}$

16. Outer diameter for stator core, $D_o = D + 2 \, (d_{ss} + d_{cs})$

17. i) Length of air-gap, $l_g = 0.2 + 2\sqrt{DL}$

   ii) Length of air-gap, $l_g = 0.125 + 0.35 \, D + L + 0.015 \, V_a$

   iii) Length of air-gap, $l_g = 0.2 + D$

   iv) Length of air-gap, $l_g = 1.6\sqrt{D} - 0.25$

> **Note :** *D and L are expressed in m and V$_a$ in m/s, but l$_g$ in mm.*

Squirrel cage rotor

18.   Diameter of rotor, $D_r = D - 2l_g$

19.   Angle between two adjacent harmonic poles $= \dfrac{360}{np}$

20.   Electrical angle of skew, $\theta_{sk} = \dfrac{2\pi}{n}$ rad.elect

21.   Rotor bar current, $I_b = 0.85 \dfrac{6 I_s T_s}{S_r}$

22.   Area of each rotor bar, $a_b = \dfrac{I_b}{\delta_b}$

23.   End ring current, $I_e = = \dfrac{S_r I_b}{\pi p}$

24.   Area of cross-section of end-ring, $a_e = \dfrac{I_e}{\delta_e}$ or $a_e = d_e \times t_e$

Wound rotor (Slip ring rotor)

25.   Rotor turns per phase, $T_r = \dfrac{K_{ws} T_s}{K_{wr}} \dfrac{E_r}{E_s}$

26.   Area of rotor conductor, $a_r = \dfrac{I_r}{\delta_r}$

27.   Minimum width of rotor tooth, $W_{tr\_min} = \dfrac{\pi (D_r - 2d_{sr})}{S_r} - y_{sr}$

28.   Depth of rotor core, $d_{cr} = \dfrac{\phi_m}{2 B_{cr} L_i}$

29.   Inner diameter of rotor lamination, $D_i = D_r - 2(d_{sr} + d_{cr})$

Leakage reactance

30.   Stator slot leakage reactance, $X_{ss} = 8 \pi f T_s^2 L \left(\dfrac{\lambda_{ss}}{pq_s}\right)$

31.   Rotor slot leakage reactance referred to stator $\left.\vphantom{\int}\right\} X_{sr}' = 8 \pi f T_s^2 L \left(\dfrac{\lambda_{sr}'}{pq_s}\right)$

32.   Overhang leakage reactance, $X_o = 8 \pi f T_s^2 L_o \left(\dfrac{\lambda_o}{pq_s}\right)$

$$\boxed{L_o \lambda_o = \mu_o \dfrac{k_s \lambda^2}{\pi y_{ss}}}$$

33.   Zigzag leakage reactance, $X_z = 8 \pi f T_s^2 L \left(\dfrac{\lambda_z}{pq_s}\right)$

34.   Harmonic leakage reactance, $X_h = 8 \pi f T_s^2 L \left(\dfrac{\lambda_h}{pq_s}\right)$

35.   Specific permeance of rotor referred to stator, $\lambda_{sr}' = \left(\dfrac{K_{ws}}{K_{wr}}\right)^2 \dfrac{S_s}{S_r} \lambda_{sr}$

36. Specific permeance for overhang leakage, $\lambda_o = \mu_o \dfrac{k\, L_o}{2\sqrt{2}\ y_{ss}}$

37. Specific permeance for zigzag leakage, $\lambda_z = \mu_o \dfrac{W_{ts}\, W_{tr}\, (W_{ts}^2 + W_{tr}^2)}{12\, l_g\, y_{ss}^2\, y_{sr}}$

Magnetizing current

38. No-load current, $I_o = \sqrt{I_m^2 + I_l^2}$

39. mmf for magnetizing current, $AT_{60} = AT_g + AT_{ts} + AT_{tr} + AT_{cs} + AT_{cr}$

40. mmf for air-gap, $AT_g = 800{,}000\ B_{g60}\, K_g\, l_g$

41. mmf required for stator teeth, $AT_{ts} = at_{ts} \times d_{ss}$

42. mmf required for rotor teeth, $AT_{tr} = at_{tr} \times d_{sr}$

43. mmf for stator core, $AT_{cs} = at_{cs} \times l_{cs}$

44. mmf for rotor core, $AT_{cr} = at_{cr} \times l_{cr}$

Short circuit current

45. Short circuit current per phase, $I_{sc} = \dfrac{E_s}{Z_s}$

46. Standstill impedence, $Z_s = \sqrt{R_s^2 + X_s^2}$

47. Stator resistance per phase, $r_s = \dfrac{\rho\, T_s\, L_{mts}}{a_s}$

48. Rotor resistance referred to stator, $r_r' = \left(\dfrac{K_{ws}\, T_s}{K_{wr}\, T_r}\right)^2 r_r$

49. Total motor resistance referred to stator, $R_s = r_s + r_r'$

50. Rotor resistance per phase in slip-ring motor $\Big\}\ r_r = \dfrac{\rho\, T_r\, L_{mtr}}{a_r}$

51. Resistance of each bar, $r_b = \dfrac{\rho\, L_b}{a_b}$

52. Resistance of each end ring, $r_e = \dfrac{\rho\, \pi\, D_e}{a_e}$

53. Rotor resistance per phase in cage motor $\Big\}\ r_r = \dfrac{\text{Copper loss in rotor}}{I_b^2}$

54. Dispersion coefficient, $\sigma = I_m\, /\, I_{sc}$

8.12 SOLVED PROBLEMS

EXAMPLE 8.1

Determine the approximate diameter and length of stator core, the number of stator slots and number of stator conductors for a 11 kW, 400 V, 3-phase, 4 pole, 1425 rpm, delta connected induction motor. B_{av} = 0.45 Wb/m^2, ac = 23000 $amp.cond./m$, full load efficiency = 0.85, pf = 0.88, L / τ = 1. The stator employes a double layer winding.

GIVEN DATA

11 kW	Delta connected	1425 rpm
3-phase	Double layer winding	B_{av} = 0.45 Wb/m^2
4 pole	ac = 23000 $amp.cond./m$	pf = 0.88
400 V	η = 0.85	L/τ =1

SOLUTION

Given that, $\dfrac{L}{\tau} = 1$

$$\therefore L = \tau = \frac{\pi D}{p} = \frac{\pi D}{4} = 0.7854\,D$$

$\therefore$ Length of stator core, L = 0.7854 D $\hspace{3cm}$ .....(1)

kVA input, $Q = \dfrac{\text{Output}}{\eta \times \text{pf}} = \dfrac{11}{0.85 \times 0.88} = 14.7\ kVA$

Synchronous speed, $n_s = \dfrac{2f}{p} = \dfrac{2 \times 50}{4} = 25\ rps$

Output coefficient, $C_o = 11\,K_{ws}\,B_{av}\,\text{ac} \times 10^{-3}$ $\hspace{1cm}$ $\boxed{\text{Let, } K_{ws} = 0.955}$

$\hspace{3cm} = 11 \times 0.955 \times 0.45 \times 23000 \times 10^{-3} = 108.7268\ kVA/m^3\text{-}rps$

Also, kVA input, $Q = C_o\,D^2\,L\,n_s$

$$\therefore D^2 L = \frac{Q}{C_o\,n_s} = \frac{14.7}{108.7268 \times 25} = 0.0054\ m^3$$

$\therefore$ $D^2 L$ = 0.0054

D^2 (0.7854 D) = 0.0054 $\hspace{3cm}$ $\boxed{\text{Using equation (1)}}$

$$\therefore \text{Diameter of stator bore, D} = \left(\frac{0.0054}{0.7854}\right)^{1/3} = 0.1902\ m$$

$\therefore$ Length of stator core, L = 0.7854 D = 0.7854 × 0.1902

$\hspace{4cm}$ = 0.1494 m

Let, Diameter of stator bore, D = 0.19 m

$\hspace{1.5cm}$ Length of stator core, L = 0.15 m

$$\text{Maximum flux per pole, } \phi_m = \frac{B_{av}\,\pi DL}{p} = \frac{0.45 \times \pi \times 0.19 \times 0.15}{4} = 0.01\ Wb$$

Stator turns per phase, $T_s = \dfrac{E_s}{4.44\, f\phi_m\, K_{ws}}$

Delta connected stator.
$\therefore E_s = V_L$

$$= \frac{400}{4.44 \times 50 \times 0.01 \times 0.955} = 188$$

The stator slots should be multiple of q, where q is slots per pole per phase.

Stator slots, S_s = Number of phase × Poles × q

For q = 2, $S_s = 3 \times 4 \times 2 = 24$

For q = 3, $S_s = 3 \times 4 \times 3 = 36$

For q = 4, $S_s = 3 \times 4 \times 4 = 48$

The stator slot pitch should lie between 15 mm to 25 mm.

When, $S_s = 36$, $y_{ss} = \dfrac{\pi D}{S_s} = \dfrac{\pi \times 0.19 \times 10^3}{36} = 16.58$ mm

When $S_s = 36$, the slot pitch (y_{ss}) lies between 15 mm to 25 mm. Hence the stator slots can be 36.

Conductors per slot, $Z_{ss} = \dfrac{6\, T_s}{S_s} = \dfrac{6 \times 188}{36} = 31.333$

Z_{ss} should be even integer for double layer winding and so it is 30 or 32.

Let, $Z_{ss} = 32$

$\therefore$ Total stator conductors, $Z = S_s \times Z_{ss} = 36 \times 32 = 1152$

New value of turns per phase, $T_s = \dfrac{Z_{ss}\, S_s}{6} = \dfrac{32 \times 36}{6} = 192$

RESULT

Diameter of stator bore,	D =	0.19 m
Length of stator core,	L =	0.15 m
Number of stator slots,	S_s =	36
Total stator conductor,	Z =	1152
Turns per phase,	T_s =	192

EXAMPLE 8.2   (KTU, Feb' 18, 10 M) (AKTU, Dec' 19, 10 M) (VTU, Dec' 19, 10 M) (PTU, May' 19, 10 M) (AU, Apr' 17, 10 M)

Estimate the stator core dimensions, number of stator slots and number of stator conductors per slot for a 100 kW, 3300 V, 50 Hz, 12 pole, star connected slip ring induction motor. B_{av} = 0.4 Wb/m², ac = 25000 amp.cond./m, η = 0.9, pf = 0.9. Choose main dimensions to give best power factor. The slot loading should not exceed 500 amp.cond.

GIVEN DATA

100 kW	3300 V	B_{av} = 0.4 Wb/m²
50 Hz	12 pole	ac = 25000 amp.cond./m.
η = 0.9	pf = 0.9	Slot loading $\leq$ 500 amp.cond.
Star connected	3-phase	

SOLUTION

For best power factor, $\tau = \sqrt{0.18\,L}$

$$\therefore \frac{\pi D}{p} = \sqrt{0.18\,L}$$

$$\boxed{\tau = \frac{\pi D}{p}}$$

On squaring the above equation we get,

$$\frac{\pi^2 D^2}{p^2} = 0.18\,L$$

$$\therefore D^2 = \frac{0.18 \times p^2}{\pi^2}\,L = \frac{0.18 \times 12^2}{\pi^2}\,L = 2.6262\,L$$

$$\therefore D^2 = 2.6262\,L \qquad\qquad\qquad\qquad .....(1)$$

kVA input, $Q = \dfrac{\text{Output}}{\eta \times pf}$

$$= \frac{100}{0.9 \times 0.9} = 123.457\ kVA$$

Output coefficient, $C_o = 11\ B_{av}\ \mathbf{ac}\ K_{ws} \times 10^{-3}$

$$\boxed{\text{Let, } K_{ws} = 0.96}$$

$$= 11 \times 0.4 \times 25000 \times 0.96 \times 10^{-3}$$

$$= 105.6\ kVA/m^3\text{-}rps$$

Synchronous speed, $n_s = \dfrac{2f}{p} = \dfrac{2 \times 50}{12} = 8.33\ rps$

Also, kVA input, $Q = C_o\ D^2\ L\ n_s$

$$\therefore D^2 L = \frac{Q}{C_o\,n_s} = \frac{123.457}{105.6 \times 8.33} = 0.1403\ m^3$$

$$\therefore D^2 L = 0.1403$$

$$2.6262\,L \times L = 0.1403$$

$$\boxed{\text{Using equation (1)}}$$

$$\therefore \text{Length of stator core, } L = \sqrt{\frac{0.1403}{2.6262}} = 0.2311\ m \approx 0.23\ m$$

From equation (1), $D^2 = 2.6262\,L$

$$\therefore \text{Diameter of stator bore, } D = \sqrt{2.6262\,L} = \sqrt{2.6262 \times 0.23} = 0.7772\ m \approx 0.78\ m$$

$$\therefore L = 0.23\ m \quad \text{and} \quad D = 0.78\ m$$

Stator voltage per phase, $E_s = \dfrac{3300}{\sqrt{3}} = 1905.256\ V$

$$\boxed{\text{Stator is star connected}}$$

Flux per pole, $\phi_m = \dfrac{B_{av}\ \pi D L}{p} = \dfrac{0.4 \times \pi \times 0.78 \times 0.23}{12} = 0.0188\ Wb$

Stator turns per phase, $T_s = \dfrac{E_s}{4.44\ f \phi_m\ K_{ws}} = \dfrac{1905.256}{4.44 \times 50 \times 0.0188 \times 0.96} = 476$

The stator slot pitch should lie between 15 to 25 *mm*.

Stator slot, $S_s = \dfrac{\pi D}{y_{ss}}$

When, $y_{ss} = 15$ *mm*, $\qquad S_s = \dfrac{\pi \times 0.78}{15 \times 10^{-3}} = 163$

When, $y_{ss} = 25$ *mm*, $\qquad S_s = \dfrac{\pi \times 0.78}{25 \times 10^{-3}} = 98$

The stator slots, S_s should lie between 98 to 163.

The stator slots be multiple of q, where q is slots per pole per phase.

Stator slots, S_s = Number of phase × Poles × q

When, q = 2, $\quad S_s = 3 \times 12 \times 2 = 72$

When, q = 3, $\quad S_s = 3 \times 12 \times 3 = 108$

When, q = 4, $\quad S_s = 3 \times 12 \times 4 = 144$

When, q = 5, $\quad S_s = 3 \times 12 \times 5 = 180$

The S_s values of 108 and 144 lie in the range of 98 to 163.

$\therefore S_s$ can be either 108 or 144.

Check for slot loading

Stator current per phase, $I_s = \dfrac{kVA \times 10^3}{\sqrt{3} \times V_L} = \dfrac{123.457 \times 10^3}{\sqrt{3} \times 3300} = 21.6\ A$ $\qquad$ $\boxed{\text{Since star connected, } I_L = I_{ph}}$

When $S_s = 108$ $\qquad\qquad\qquad\qquad\qquad$ When $S_s = 144$,

$Z_{ss} = \dfrac{6\ T_s}{S_s} = \dfrac{6 \times 476}{108} = 26.44 \approx 26$ $\qquad$ $Z_{ss} = \dfrac{6\ T_s}{S_s} = \dfrac{6 \times 476}{144} = 19.83 \approx 20$

Slot loading $= Z_{ss}\ I_s = 26 \times 21.6$ $\qquad\qquad$ Slot loading $= Z_{ss}\ I_s = 20 \times 21.6$

$\qquad\qquad = 561.6$ *amp.cond.* $\qquad\qquad\qquad\qquad = 432$ *amp.cond.*

When $S_s = 144$, the slot loading does not exceeds 500 *amp.cond.* Hence 144 slots is suitable for the machine.

Total stator conductors, $Z = S_s \times Z_{ss} = 144 \times 20 = 2880$

New value of turns per phase, $T_s = \dfrac{Z_{ss}\ S_s}{6} = \dfrac{20 \times 144}{6} = 480$

RESULT

Diameter of stator bore, $\quad$ D $\ =\ $ 0.78 *m*

Length of stator core, $\qquad$ L $\ =\ $ 0.23 *m*

Number of stator slots, $\qquad S_s\ =\ $ 144

Total stator conductors, $\quad$ Z $\ =\ $ 2880

Turns per phase, $\qquad\qquad T_s\ =\ $ 480

EXAMPLE 8.3

(AU, Nov' 17, 16 M) (AU, Apr' 18, 8 M)

Determine the D and L of a 70 *HP*, 415 *V*, 3-phase, 50 *Hz*, star connected, 6 pole induction motor for which **ac** = 30000 *amp.cond./m* and B_{av} = 0.51 *Wb/m²*. Take η = 90 % and pf = 0.91. Assume τ = L. Estimate the number of stator conductors required for a winding in which the conductors are connected in two parallel paths. Choose a suitable number of conductors per slots, so that the slot loading does not exceed 750 *amp.cond.*

GIVEN DATA

70 *HP*	415 *V*	B_{av} = 0.51 *Wb/m²*
3-phase	50 *Hz*	**ac** = 30000 *amp.cond./m*
η = 0.9	pf = 0.91	Slot loading ≤ 750 *amp.cond.*
6 pole	τ = L	Star connected

Conductors are connected in two parallel paths.

SOLUTION

Given that, τ = L

$$\therefore L = \frac{\pi D}{p} = \frac{\pi D}{6} = 0.5236\ D$$

$$\boxed{\tau = \frac{\pi D}{p}}$$

$$L = 0.5236\ D \qquad\qquad\qquad .....(1)$$

$$\text{kVA input, } Q = \frac{HP \times 0.746}{\eta \times pf}$$

$$= \frac{70 \times 0.746}{0.9 \times 0.91} = 63.76\ kVA$$

Output coefficient, C_o = 11 B_{av} **ac** K_{ws} × 10⁻³

$$\boxed{K_{ws} = 0.955}$$

$$= 11 \times 0.51 \times 30000 \times 0.955 \times 10^{-3}$$

$$= 160.7265\ kVA/m^3\text{-}rps$$

Synchronous speed, $n_s = \frac{2f}{p} = \frac{2 \times 50}{6} = 16.667\ rps$

Also, kVA input, Q = C_o D² L n_s

$$\therefore D^2 L = \frac{Q}{C_o\, n_s} = \frac{63.76}{160.7265 \times 16.667} = 0.0238\ m^3$$

$$\therefore D^2 L = 0.0238$$

$$\boxed{\text{Using equation (1)}}$$

$$D^2 (0.5236\ D) = 0.0238$$

$$D = \left(\frac{0.0238}{0.5236}\right)^{1/3} = 0.35688 \approx 0.36\ m$$

∴ Diameter of stator bore, D = 0.36 *m*

L = 0.5236 D = 0.5236 × 0.36 = 0.1885 ≈ 0.19 *m*

$$\boxed{\text{Using equation (1)}}$$

∴ Length of stator core, L = 0.19 *m*

Flux per pole, $\phi_m = \dfrac{B_{av}\, \pi DL}{p}$

$$= \dfrac{0.51 \times \pi \times 0.36 \times 0.19}{6} = 0.0183 \; Wb$$

> **Star connected stator**
> $\therefore E_s = V_L / \sqrt{3}$

Turns per phase, $T_s = \dfrac{E_s}{4.44\, f\phi_m\, K_{ws}}$

$$= \dfrac{415 / \sqrt{3}}{4.44 \times 50 \times 0.0183 \times 0.955} = 61.756 \approx 62$$

Since the conductor are placed in two parallel paths,

Total stator conductor, $Z = 6\, T_s \times 2 = 12\, T_s = 12 \times 62 = 744$

The slot pitch, y_{ss} should lie between 15 to 25 mm.

When y_{ss} = 15 mm, $\quad S_s = \dfrac{\pi D}{y_{ss}} = \dfrac{\pi \times 0.36}{15 \times 10^{-3}} = 75$

When y_{ss} = 25 mm, $\quad S_s = \dfrac{\pi D}{y_{ss}} = \dfrac{\pi \times 0.36}{25 \times 10^{-3}} = 45$

The number of stator slots lie in the range of 45 to 75.

The stator slots should be multiple of q, where q is slots per pole per phase.

$\quad$ Stator slots, S_s = Number of phase × Poles × q

$\quad$ When $\quad$ q = 2, $\quad$ $S_s = 3 \times 6 \times 2 = 36$
$\quad$ When $\quad$ q = 3, $\quad$ $S_s = 3 \times 6 \times 3 = 54$
$\quad$ When $\quad$ q = 4, $\quad$ $S_s = 3 \times 6 \times 4 = 72$

The values of S_s which lies between 45 to 75 are S_s = 54 and S_s = 72.

Stator current per phase, $I_s = \dfrac{kVA \times 10^3}{\sqrt{3}\, V_L}$

> Stator is star connected, $\therefore I_L = I_s$

$$= \dfrac{63.76 \times 10^3}{\sqrt{3} \times 415} = 88.7 \; A$$

Current through conductor, $I_z = \dfrac{I_s}{a}$

> Conductors are connected in two parallel paths.

$$= \dfrac{88.7}{2} = 44.35 \; A$$

Check for slot loading

When S_s = 54	When S_s = 72
Conductors per slot $\Big\}$ $Z_{ss} = \dfrac{744}{54} = 13.77 \approx 14$	Conductors per slot $\Big\}$ $Z_{ss} = \dfrac{744}{72} = 10.33 \approx 10$
Slot loading $= Z_{ss}\, I_z = 14 \times 44.35$	Slot loading $= Z_{ss}\, I_z = 10 \times 44.35$
$= 620.9 \; amp.cond.$	$= 443.5 \; amp.cond.$

In both the cases the limiting value of slot loading 750 *amp.cond.* is not exceeded.

For lower fabrication cost, $S_s = 54$

For low temperature rise, $S_s = 72$

Let, $S_s = 54$

$\therefore Z_{ss} = 14$

Total stator conductors, $Z = Z_{ss} \times S_s = 14 \times 54 = 756$

New value of turns per phase, $T_s = \dfrac{Z_{ss} S_s}{6 \times 2} = \dfrac{756}{6 \times 2} = 63$

RESULT

Diameter of stator bore,	D	= 0.36 m
Length of stator core,	L	= 0.19 m
Turns per phase,	T_s	= 63
Number of stator slots,	S_s	= 54
Conductors per slot,	Z_{ss}	= 14

EXAMPLE 8.4

(AU, Nov' 17, 16 M)

Estimate the main dimensions, air-gap length, stator slots, stator turns per phase and cross-sectional area of stator and rotor conductors for a 3-phase, 15 *HP*, 400 *V*, 6 pole, 50 *Hz*, 975 *rpm*, induction motor. The motor is suitable for star delta starting. $B_{av} = 0.45\ Wb/m^2$, **ac** = 20000 *amp.cond./m*, L/τ = 0.85, η = 0.9, pf = 0.85.

GIVEN DATA

3-phase	400 V	$B_{av} = 0.45\ Wb/m^2$
15 HP	L/τ = 0.85	**ac** = 20000 *amp.cond./m.*
50 Hz	6 pole	η = 0.9
pf = 0.85	975 rpm	Star-delta starting

SOLUTION

Given that, L/τ = 0.85

$\therefore L = 0.85\ \tau = 0.85 \times \dfrac{\pi D}{p} = 0.445\ D$

$\boxed{\tau = \dfrac{\pi D}{p}}$

$\therefore$ Length of stator bore, L = 0.445 D   .....(1)

kVA input, $Q = \dfrac{HP \times 0.746}{\eta \times pf} = \dfrac{15 \times 0.746}{0.9 \times 0.85} = 14.63\ kVA$

Output coefficient, $C_o = 11\ B_{av}\ \textbf{ac}\ K_{ws} \times 10^{-3}$

$= 11 \times 0.45 \times 20000 \times 0.955 \times 10^{-3}$

$= 94.545\ kVA/m^3\text{-}rps$

Synchronous speed, $n_s = \dfrac{2f}{p} = \dfrac{2 \times 50}{6} = 16.667\ rps$

Also, kVA input, $Q = C_o\, D^2\, L\, n_s$

$$\therefore D^2L = \frac{Q}{C_o\, n_s} = \frac{14.63}{94.545 \times 16.667} = 9.284 \times 10^{-3}\ m^3$$

$$\therefore D^2L = 9.284 \times 10^{-3}$$

$$\therefore D^2(0.445\,D) = 9.284 \times 10^{-3} \qquad \boxed{\text{Using equation (1)}}$$

$$\therefore \text{Diameter of stator bore, } D = \left(\frac{9.284 \times 10^{-3}}{0.445}\right)^{1/3} = 0.2753\ m \approx 0.275\ m$$

$$\therefore \text{Length of stator core, } L = 0.445 \times 0.275 = 0.1224\ m \approx 0.12\ m$$

$$\phi_m = \frac{B_{av}\,\pi DL}{p} = \frac{0.45 \times \pi \times 0.275 \times 0.12}{6} = 7.775 \times 10^{-3}\ Wb$$

For star delta starting, the motor should be designed for delta connection. In delta connection, phase voltage is equal to line voltage.

$$\text{Turns per phase, } T_s = \frac{E_s}{4.44\, f\phi_m\, K_{ws}} = \frac{400}{4.44 \times 50 \times 7.775 \times 10^{-3} \times 0.955} \qquad \boxed{E_s = V_L}$$

$$= 242.66 \approx 242$$

Total stator conductors, $Z = 6\, T_s = 6 \times 242 = 1452$

Slot pitch lies between 15 mm to 25 mm.

When $y_{ss} = 15\ mm$

$$S_s = \frac{\pi D}{y_{ss}} = \frac{\pi \times 0.275}{15 \times 10^{-3}} = 57.6 \approx 58$$

When $y_{ss} = 25\ mm$

$$S_s' = \frac{\pi D}{y_{ss}} = \frac{\pi \times 0.275}{25 \times 10^{-3}} = 34.55 \approx 34$$

The number of slots lies between 34 and 58.

The stator slots should be multiple of q, where q is slots per pole per phase.

Stator slots, $S_s =$ Number of phase × Poles × q

When q = 2, $S_s = 3 \times 6 \times 2 = 36$
When q = 3, $S_s = 3 \times 6 \times 3 = 54$
When q = 4, $S_s = 3 \times 6 \times 4 = 72$

Let, $S_s = 36$

$$\therefore Z_{ss} = \frac{6T_s}{S_s} = \frac{1452}{36} = 40.33 \approx 40$$

New value of total stator conductors, $Z = S_s \times Z_{ss} = 36 \times 40 = 1440$

New value of turns per phase, $T_s = \frac{Z_{ss}\, S_s}{6} = \frac{40 \times 36}{6} = 240$

kVA input, $Q = \sqrt{3}\, V_L \times I_L \times 10^{-3} = 3\, E_s\, I_s\, 10^{-3}$

$$\therefore I_s = \frac{Q \times 10^3}{3E_s} = \frac{14.62 \times 10^3}{3 \times 400} = 12.183 \ A$$

Let, current density in stator conductor, $\delta_s = 3 \ A/mm^2$

$\therefore$ Area of cross-section of stator conductor, $a_s = \dfrac{I_s}{\delta_s} = \dfrac{12.183}{3} = 4.061 \ mm^2$

Length of air-gap, $l_g = 0.2 + 2\sqrt{DL} = 0.2 + 2 \times \sqrt{0.275 \times 0.12} = 0.5633 \ mm$

Let, $l_g = 0.6 \ mm$

Rotor slots

Let, S_r = Number of rotor slots

        S_s = Number of stator slots

We know that, $(S_s - S_r)$ cannot be, 0, $\pm p$, $\pm 2p$, $\pm 3p$, $\pm 5p$, ± 1, ± 2, $\pm (p \pm 1)$, $\pm (p \pm 2)$

Here, $p = 6$

     $\therefore (S_s - S_r)$ cannot be, 0, ± 6, ± 12, ± 18, ± 30, ± 1, ± 2, ± 5, ± 7, ± 8, ± 4

Here, $(S_s - S_r)$ can be, $\pm 3, \pm 9, \pm 10, \pm 11$, etc.

Let, $S_s - S_r = +3$

$\therefore S_r = S_s - 3 = 36 - 3 = 33$

     Rotor bar current, $I_b = 0.85 \dfrac{6T_sI_s}{S_r} = \dfrac{0.85 \times 6 \times 240 \times 12.183}{33}$

$$= 451.88 \ A$$

Let, current density in rotor bar, $\delta_b = 4 \ A/mm^2$

Area of cross-section of rotor bar $\Big\} a_b = \dfrac{I_b}{\delta_b} = \dfrac{451.88}{4} = 112.96 = 113 \ mm^2$

End ring current, $I_e = \dfrac{S_rI_b}{\pi p} = \dfrac{33 \times 451.88}{\pi \times 6} = 791.1 \ A$

Let, current density in end ring, $\delta_e = 4 \ A/mm^2$

Area of cross-section of end ring $\Big\} a_e = \dfrac{I_e}{\delta_e} = \dfrac{791.1}{4} = 197.775 \ mm^2$

Let, area of cross-section of end ring, $a_e = 200 \ mm^2$

RESULT

Diameter of stator bore,	D	= 0.275 m
Length of stator core,	L	= 0.12 m
Turns per phase,	T_s	= 240
Number of stator slots,	S_s	= 36
Number of rotor slots,	S_r	= 33
Area of cross-section of stator conductor,	a_s	= 4.061 mm^2
Area of cross-section of rotor bar,	a_b	= 113 mm^2
Area of cross-section of end ring,	a_e	= 200 mm^2

EXAMPLE 8.5
(PTU, May' 19, 5 M)

A 3-phase, 4 pole induction motor has 24 slots. Calculate the order of slot harmonics produced. It is desired to completely eliminate the higher order slot harmonic, find the angle through which the bars must be skewed. Find the effect of skewing on the lower harmonic.

GIVEN DATA

3-phase   4 pole $S_s = 24$

SOLUTION

Slots per pole per phase, $q = \dfrac{S_s}{3p} = \dfrac{24}{3 \times 4} = 2$

Order of slot harmonics, $n_{sh} = 6\,Aq \pm 1$

Let, $A = 1$, $\therefore n_{sh} = 6 \times 1 \times 2 \pm 1 = 13,\ 11$

Here 13th order harmonic is the higher order harmonic and so let us completely eliminate the 13th harmonic.

The angle of skew to eliminate n^{th} order harmonic is $2\pi/n$.

Let, θ_{sk_13} = Angle of skew to eliminate 13th order harmonic.

$\therefore \theta_{sk\_13} = \dfrac{2\pi}{13} = 0.4833$ electrical radian

Distribution factor for 13th order harmonic, $K_{d13} = \dfrac{\sin \dfrac{n\,\theta_{sk\_13}}{2}}{\dfrac{n\,\theta_{sk\_13}}{2}}$

$$= \dfrac{\sin \dfrac{13}{2} \times \dfrac{2\pi}{13}}{\dfrac{13}{2} \times \dfrac{2\pi}{13}} = \dfrac{\sin \pi}{\pi} = 0$$

Distribution factor for 11th order harmonic, $K_{d11} = \dfrac{\sin \dfrac{n\,\theta_{sk\_11}}{2}}{\dfrac{n\,\theta_{sk\_11}}{2}}$

$$= \dfrac{\sin \dfrac{11}{2} \times \dfrac{2\pi}{13}}{\dfrac{11}{2} \times \dfrac{2\pi}{13}} = \dfrac{\sin \dfrac{11\pi}{13}}{\dfrac{11\pi}{13}} = 0.1748$$

Therefore, the 13th order harmonic is completely eliminated and the 11th harmonic emf is reduced to 17.48% of value obtained without skewing.

EXAMPLE 8.6
(KTU, Feb' 18, 10 M) (MU, May' 19, 10 M) (AU, Nov' 18, 13 M)

Find the main dimensions, number of radial ventilating ducts, number of stator slots and number of turns per phase of a 3.7 kW, 400 V, 3-phase, 4 pole, 50 Hz, squrriel cage induction motor to be started by a star delta starter. Work out the winding details. Assume average flux density in the air-gap equal to 0.45 Wb/m², Ampere conductors per meter = 23000, η = 0.85, powerfactor = 0.84. Choose main dimension to achieve cheap design. Winding factor = 0.955, iron stacking factor = 0.9.

GIVEN DATA

3.7 kW	50 Hz	$\cos\phi = 0.84$
400 V	ac = 23000 amp.con./m.	$k_{ws} = 0.955$
3-phase	$B_{av} = 0.45$ Wb/m²	Stacking factor = 0.9
4 pole	$\eta = 0.85$	

SOLUTION

For a cheap design choose the ratio, $L/\tau = 1.5$

$$\therefore L = 1.5\,\tau = 1.5 \times \frac{\pi D}{p} = \frac{1.5 \times \pi}{4} D = 1.178\,D \qquad \boxed{\tau = \frac{\pi D}{p}}$$

$\therefore$ Length of stator core, $L = 1.178\,D$ $\qquad$ ....(1)

kVA input, $Q = \dfrac{kW}{\eta \cos\phi}$

$$= \frac{3.7}{0.85 \times 0.84} = 5.18 \ kVA$$

Output coefficient, $C_o = 11\,k_{ws}\,B_{av}\,\textbf{ac} \times 10^{-3}$

$$= 11 \times 0.955 \times 0.45 \times 23000 \times 10^{-3}$$

$$= 108.7 \ kVA/m^3\text{-}rps$$

Synchronous speed, $n_s = \dfrac{2f}{p} = \dfrac{2 \times 50}{4} = 25 \ rps$

Also, kVA input, $Q = C_o\,D^2\,L\,n_s$

$$\therefore D^2 L = \frac{Q}{C_o\,n_s}$$

$$= \frac{5.18}{108.7 \times 25} = 1.906 \times 10^{-3} \ m^3$$

$$\therefore D^2 L = 1.906 \times 10^{-3} \ m^3$$

$$\therefore D^2 \times 1.178\,D = 1.906 \times 10^{-3} \qquad \boxed{\text{Using equation (1)}}$$

$\therefore$ Diameter of stator bore, $D = \left(\dfrac{1.906 \times 10^{-3}}{1.178}\right)^{1/3} = 0.117 \ m \approx 0.12 \ m$

$\therefore$ Length of stator core, $L = 1.178\,D$ $\qquad \boxed{\text{Using equation (1)}}$

$$= 1.178 \times 0.12 = 0.141 \ m$$

Radial ventilating ducts are provided if length exceeds 0.125 m.

Since the length of core is 0.141 m, one radial duct of 10 mm wide is provided.

Flux per pole, $\phi m = B_{av}\,L\,\tau = B_{av}\,L\left(\dfrac{\pi D}{p}\right)$

$$= 0.45 \times 0.141 \times \frac{\pi \times 0.12}{4}$$

$$= 5.98 \times 10^{-3} \ Wb$$

Stator turns per phase, $T_s = \dfrac{E_s}{4.44\, f\, \phi_m K_{ws}}$

Stator is delta connected.

$\therefore E_s = V_L$

$$= \dfrac{400}{4.44 \times 50 \times 5.98 \times 10^{-3} \times 0.955} = 316$$

The stator slot pitch should lie between 15 to 25 mm.

Stator slot, $S_s = \dfrac{\pi D}{y_{ss}}$

When, $y_{ss} = 15$ mm, $S_s = \dfrac{\pi \times 0.12}{15 \times 10^{-3}} = 25$

When, $y_{ss} = 25$ mm, $S_s = \dfrac{\pi \times 0.12}{25 \times 10^{-3}} = 15$

The stator slots, S_s should lie between 15 to 25.

The stator slots be multiple of q, where q is slots per pole per phase.

Stator slots, S_s = Number of phase × Poles × q

When   q = 2, $S_s = 3 \times 4 \times 2 = 24$

When   q = 3, $S_s = 3 \times 4 \times 3 = 36$

The S_s value of 24 lie in the range of 15 to 25.

$\therefore S_s = 24$

Check for slots loading

Stator conductor per phase, $I_s = \dfrac{kVA \times 10^3}{3 \times E_s}$

Stator is delta connected.

$\therefore E_s = V_L$

kVA = 3 E_s I_s × 10^{-3}

$$= \dfrac{5.18 \times 10^3}{3 \times 400} = 4.3167\ A$$

Conductors per slot, $Z_{ss} = \dfrac{6\, T_s}{S_s}$

$$= \dfrac{6 \times 316}{24} = 79$$

Slot loading = $Z_{ss} I_s$ = 79 × 4.3167 = 341 $amp.cond.$

When S_s = 24, slot loading 500 $amp.cond.$ is not exceeded.

$\therefore$ Total stator conductors, $Z = S_s \times Z_{ss} = 24 \times 79 = 1896$

New value of turns per phase, $T_s = \dfrac{Z_{ss} S_s}{6}$

$$= \dfrac{79 \times 24}{6} = 316$$

Let, current density in stator conductors, $\delta_s = 3$ A/mm^2

$\therefore$ Area of cross-section of stator conductors, $a_s = \dfrac{I_s}{\delta_s}$

$$= \dfrac{4.3167}{3} = 1.4389\ mm^2 \approx 1.4\ mm^2$$

RESULT

Diameter of stator bore, $\qquad$ D = 0.12 m

Length of stator core, $\qquad$ L = 0.141 m

Number of radial ventilating ducts $\qquad$ = 1

Conductors per slot, $\qquad$ Z_{ss} = 79

Turns per phase, $\qquad$ T_s = 316

Total stator conductors, $\qquad$ Z = 1896

Stator slots, $\qquad$ S_s = 24

Area of cross-section of stator conductor, a_s = 1.4 mm².

EXAMPLE 8.7

(PU, Nov' 19, 8 M) (AU, Nov' 18, 7 M)

A 11 *kW*, 3-phase, 6 pole, 50 *Hz*, 220 *V*, star connected induction motor has 54 stator slots, each containing 9 conductors. Calculate the value of bar and end ring currents. The number of rotor bars is 64. The machine has an efficiency of 86% and a power factor of 0.85. The rotor mmf may be assumed to be 85% of stator mmf. Also find the bar and the end ring section if the current density is 5 *A/mm²*.

GIVEN DATA

3-phase	$V_L = 220\ V$	$\delta_b = \delta_e = 5\ A/mm^2$	Star connected
11 *kW*	$S_S = 54$	$S_r = 64$	
50 *Hz*	pf = 0.85	$Z_{ss} = 9$	
6 pole	$\eta = 0.86$	Rotor mmf = 0.85 × Stator mmf	

SOLUTION

$$\text{kVA input, } Q = \frac{P_o}{\eta\ pf} \qquad\qquad .....(1)$$

Also, kVA input, $Q = \sqrt{3}\ I_L\ V_L \times 10^{-3}$

$$\text{Stator current per phase, } I_s = \frac{Q \times 10^{-3}}{\sqrt{3}\ V_L}$$

$$= \frac{P_o \times 10^3}{\sqrt{3}\ V_L\ \eta\ pf}$$

> Star connected.
> $\therefore I_s = I_L$

> Using equation (1)

$$= \frac{11 \times 1000}{\sqrt{3} \times 220 \times 0.86 \times 0.85} = 39.49\ A$$

Number of stator conductors, Z = 54 × 9 = 486

$\therefore$ Stator turns per phase, $T_s = \dfrac{Z}{6} = \dfrac{486}{6} = 81$

Stator mmf = $3\ I_s\ T_s$ = 3 × 39.49 × 81 = 9596 *AT*

$\therefore$ Rotor mmf = 0.85 × Stator mmf

$= 0.85 \times 9596 = 8157\ AT$

Also, rotor mmf = $\dfrac{S_r\ I_b}{2} = \dfrac{64\ I_b}{2} = 32\ I_b$

On equating the above two equations of rotor mmf we get,

$32 I_b = 8157$

$\therefore$ Current in rotor bars, $I_b = \dfrac{8157}{32} = 254.91\ A$

End ring current, $I_e = \dfrac{S_r I_b}{\pi p}$

$$= \dfrac{64 \times 254.91}{\pi \times 6} = 865.5\ A$$

$\therefore$ Area of cross-section of each bar, $a_b = \dfrac{I_b}{\delta_e}$

$$= \dfrac{254.91}{5} = 51\ mm^2$$

Area of cross-section each end ring, $a_e = \dfrac{I_e}{\delta_e}$

$$= \dfrac{865.5}{5} = 173.1 = 173\ mm^2$$

RESULT

Rotor bar current,	I_b	$= 254.91\ A$
End ring current,	I_e	$= 865.5\ A$
Area of cross-section of each rotor bar,	a_b	$= 51\ mm^2$
Area of cross-section of each end ring,	a_e	$= 173\ mm^2$.

EXAMPLE 8.8   *(AU, Apr' 18, 8 M)*

Design a cage rotor for a 40 *HP*, 3-phase, 400 *V*, 50 *Hz*, 6 pole, delta connected induction motor having a full load efficiency, η of 87% and a full load pf of 0.85. Take, D = 33 *cm* and L = 17 *cm*. Stator slots = 54, conductors per slot = 14. Assume suitably the missing data if any.

GIVEN DATA

3-phase	$f = 50\ Hz$	$S_s = 54$
40 *HP*	$p = 6$	$Z_{ss} = 14$
400 *V*	$\eta = 0.87$	$D = 33\ cm$
Delta connected	$pf = 0.85$	$L = 17\ cm$

SOLUTION

We know that, $(S_s - S_r)$ cannot be, $0, \pm p,\ \pm 2p,\ \pm 3p,\ \pm 5p,$
$\pm 1,\ \pm 2,\ \pm(p \pm 1),\ \pm(p \pm 2)$

Here $p = 6$, $\therefore (S_s - S_r)$ cannot be, $0, \pm 6,\ \pm 12,\ \pm 18,\ \pm 30,$
$\pm 1,\ \pm 2,\ \pm 7,\ \pm 5,\ \pm 8,\ \pm 4.$

$(S_s - S_r)$ can be ± 3 or ± 9

Let, $S_s - S_r = 3$

$\therefore$ Rotor slots, $S_r = S_s - 3 = 54 - 3 = 51$

Total stator conductors, $Z = S_r Z_{ss}$

$$= 54 \times 14$$

$$= 756$$

Also, Total stator conductors, $Z = 6 T_s$

∴ Turns per phase, $T_s = \dfrac{Z}{6}$

$$= \dfrac{756}{6} = 126$$

kVA input, $Q = \dfrac{HP \times 0.746}{\eta \times pf}$

$$= \dfrac{40 \times 0.746}{0.87 \times 0.85} = 40.352 \, kVA$$

Also, kVA input, $Q = 3 E_s I_s \times 10^{-3}$

∴ $I_s = \dfrac{Q}{3 E_s \times 10^{-3}}$

$\boxed{\text{Stator is delta connected, } \therefore \ E_s = V_L}$

$$= \dfrac{40.352}{3 \times 400 \times 10^{-3}} = 33.63 \, A$$

Stator mmf $= 6 \, T_s I_s$

Let, Rotor mmf $= 0.85 \times$ Stator mmf

$$= 0.85 \times 6 \, T_s I_s$$

Also, rotor mmf $= S_r I_b$

On equating the above two equations of rotor mmf we get,

$$S_r I_b = 0.85 \times 6 \, T_s I_s$$

∴ Rotor bar current, $I_b = 0.85 \times \dfrac{6 T_s I_s}{S_r} = \dfrac{0.85 \times 6 \times 126 \times 33.63}{51}$

$$= 423.7 \, A$$

Let, $\delta_b = 4 \, A/mm^2$

∴ $a_b = \dfrac{I_b}{\delta_b} = \dfrac{423.7}{4} = 105.9 = 106 \, mm^2$

∴ Area of cross-section of each rotor bar, $a_b = 106 \, mm^2$

End ring current, $I_e = \dfrac{S_r I_b}{\pi p} = \dfrac{51 \times 423.7}{\pi \times 6} = 1146.4 \, A$

Let, $\delta_e = 4 \, A/mm^2$

$a_e = \dfrac{I_e}{\delta_e} = \dfrac{1146.4}{4} = 286.6 = 287 \, mm^2$

∴ Area of cross-section of each end ring, $a_e = 287 \, mm^2$

In induction motors the length of rotor core is same as that of stator core.

∴ Length of rotor core, $L_r = 17 \, cm = 0.17 \, m$

Length of air-gap, $l_g = 0.2 + 2 \times \sqrt{DL} = 0.2 + 2 \times \sqrt{0.33 \times 0.17}$

$$= 0.67 \ mm = 0.7 \ mm$$

Diameter of rotor, $D_r = D - 2 l_g$

$$= 0.33 - 2 \times 0.7 \times 10^{-3}$$

$$= 0.3286 \ m$$

RESULT

Length of rotor,	L_r	=	$0.17 \ m$
Diameter of rotor,	D_r	=	$0.3286 \ m$
Length of air-gap,	l_g	=	$0.7 \ mm$
Area of cross-section of each rotor bar,	a_b	=	$106 \ mm^2$
Area of cross-section of each end ring,	a_e	=	$287 \ mm^2$

EXAMPLE 8.9

A 3-phase induction motor has 54 stator slots with 8 conductors per slot and 72 rotor slots with 4 conductors per slot. Find the number of stator and rotor turns. Find the voltage across the rotor slip rings, when rotor is open circuited and at rest. Both stator and rotor are star connected and a voltage of 400 V is applied across the stator terminals.

GIVEN DATA

3-phase	$Z_{ss} = 8$	Star connected stator
$S_s = 54$	$Z_{sr} = 4$	Star connected stator
$S_r = 72$	$V_L = 400 \ V$	

SOLUTION

Total stator conductors, $Z = 6 \ T_s$

Also, total stator conductors, $Z = Z_{ss} \ S_s$

On equating the above two equations we get,

$$6 \ T_s = Z_{ss} \ S_s$$

∴ Stator turns per phase, $T_s = \dfrac{Z_{ss} \times S_s}{6} = \dfrac{8 \times 54}{6} = 72$

Similarly,

Rotor turns per phase, $T_r = \dfrac{Z_{rs} \times S_r}{6} = \dfrac{4 \times 72}{6} = 48$

The turns ratio of induction motor is,

$$\dfrac{E_r}{E_s} = \dfrac{K_{wr} T_r}{K_{ws} T_s} \quad \Rightarrow \quad \dfrac{E_r}{E_s} = \dfrac{T_r}{T_s}$$

$\boxed{\text{Let, } K_{ws} = K_{wr}}$

∴ Rotor emf per phase at standstill, $E_r = E_s \dfrac{T_r}{T_s} = \dfrac{400}{\sqrt{3}} \times \dfrac{48}{72}$

$$= 153.96 \ V \approx 154 \ V$$

Rotor emf between slip rings (line value), $E_{rL} = \sqrt{3} \, E_r = \sqrt{3} \times 154 = 266.7 \ V$

RESULT

Stator turns per phase,	T_s	= 72
Rotor turns per phase,	T_r	= 48
Rotor emf between slip rings at standstill,	E_{rL}	= 266.7 V

EXAMPLE 8.10

Estimate the rotor current, area of rotor conductor, width of rotor tooth and depth of rotor core for a 100 kW, 500 V, 50 Hz, 4 pole, star connected slip ring induction motor. Assume flux per pole = 0.13 Wb, rotor slots = 63, stator turns per phase = 72, rotor turns per phase = 48, stator current per phase = 17.8, diameter of rotor = 0.45, depth of rotor slot = 0.16, rotor slot pitch = 0.0034, flux density in rotor core = 0.35, net iron length of stator core = 0.7.

GIVEN DATA

100 kVA	$\phi_m = 0.13\ Wb$	$I_s = 17.8\ A$	$B_{cr} = 0.35$
5000 V	$S_r = 63$	$D_r = 0.45$	$L_i = 0.7$
4 pole	$T_s = 72$	$D_{sr} = 0.16$	
50 Hz	$T_r = 48$	$Y_{sr} = 0.0034$	

SOLUTION

Rotor current per phase, $I_r = \dfrac{0.85\ I_s\ T_s}{T_r}$

> Rotor ampere turns is 85% of stator ampere turns.

$$= \frac{0.85 \times 17.8 \times 72}{48} = 22.695\ A$$

Let current density in rotor, $\delta_r = 5\ A/mm^2$.

$\therefore$ Area of cross-section of rotor conductors $\Big\}\ a_r = \dfrac{I_r}{\delta_r} = \dfrac{22.695}{5} = 4.539 = 5\ mm^2$

Minimum width of rotor tooth, $W_{tr\_min} = \dfrac{\pi\,(D_r - 2d_{sr})}{S_r} - y_{sr}$

$$= \frac{\pi\,(0.45 - 2 \times 0.16)}{63} - 0.0034$$

$$= 0.0031\ mm$$

Depth of rotor core, $d_{cr} = \dfrac{\phi_m}{2\ B_{cr}\ L_i} = \dfrac{0.13}{2 \times 0.35 \times 0.7} = 0.2653\ m$

RESULT

Rotor current per phase,	I_r	= 22.695 A
Cross-section of rotor conductor,	a_r	= 5 mm²
Minimum width of rotor tooth,	W_{tr_min}	= 0.0031 mm
Depth of rotor core,	d_{cr}	= 0.2653 m

EXAMPLE 8.11   *(JNTKU, Apr' 19, 15 M) (AU, Apr' 17, 10 M)*

A 90 kW, 500 V, 50 Hz, 3-phase, 8 pole induction motor has a star connected stator winding accommodated in 63 slots with 6 conductors per slot. If the slip ring voltage on open circuit is not to exceed 400 V, find a suitable rotor winding by estimating number of slots, number of conductors per slot, coil span, slip ring voltage on open circuit, approximate full load current per phase in rotor. Assume $\eta = 0.9$ and $pf = 0.86$.

GIVEN DATA

90 kW	$V_L = 500\ V$	Star connected stator
50 Hz	$S_s = 63$	Voltage between slip rings $\leq 400\ V$
3-phase	$Z_{ss} = 6$	$pf = 0.86$
8 pole	$\eta = 0.9$	

SOLUTION

Let rotor slots per pole per phase = q_r

Rotor slot should be multiple of q_r for integral slot winding.

$\boxed{\text{Number of stator and rotor poles are same.}}$

Number of rotor poles, p = 8

∴ Rotor slots, S_r = Number of phase × Poles × q_r = 3 p q_r

For, $q_r = 2$, $S_r = 3 \times 8 \times 2 = 48$

For, $q_r = 3$, $S_r = 3 \times 8 \times 3 = 72$

For, $q_r = 4$, $S_r = 3 \times 8 \times 4 = 96$

To eliminate harmonics, $(S_s - S_r)$ should not be equal to

$$0, \pm p, \pm 2p, \pm 3p, \pm 5p, \pm 1, \pm 2,$$
$$\pm(p \pm 1), \pm(p \pm 2).$$

For p = 8, $(S_s - S_r)$ should not be equal to

$$0, \pm 8, \pm 16, \pm 24, \pm 40, \pm 1, \pm 2, \pm 9, \pm 7, \pm 10, \pm 6.$$

Here, $(S_s - S_r)$ can be equal to, $\pm 3, \pm 4, \pm 5, \pm 11, \pm 12, \pm 13, \pm 14, \pm 15$, etc.

For $S_r = 48$, $S_s - S_r = 63 - 48 = 15$.

Hence the rotor slots can be 48, which also results in integral slot winding.

∴ $S_r = 48$

Let line voltage between slip rings, $E_{rL} = 400\ V$

Rotor emf per phase, $E_r = 400/\sqrt{3} = 231\ V$

Stator emf per phase, $E_s = 500/\sqrt{3} = 289\ V$

Stator turns per phase, $T_s = \dfrac{Z_{ss} \times S_s}{6} = \dfrac{6 \times 63}{6} = 63$

The turns ratio for induction motor is given by,

$$\frac{K_{wr}\,T_r}{K_{ws}\,T_s} = \frac{E_r}{E_s}$$

∴ Rotor turns per phase, $T_r = T_s \dfrac{E_r}{E_s}$

$$\boxed{K_{wr} = K_{ws}}$$

$$= 63 \times \frac{231}{289} \approx 50$$

Rotor conductors per slot, $Z_{rs} = \dfrac{6T_r}{S_r} = \dfrac{6 \times 50}{48} = 6.25 = 6$

Let rotor conductors per slot, $Z_{rs} = 6$

New value of rotor turns per phase, $T_r = \dfrac{S_r Z_{rs}}{6} = \dfrac{48 \times 6}{6} = 48$

New value of rotor emf per phase, $E_r = E_s \times \dfrac{T_r}{T_s} = 289 \times \dfrac{48}{63} = 220\ V$

Emf between slip rings, $E_{rL} = \sqrt{3}\,E_r = \sqrt{3} \times 220 = 381\ V$

Rotor coil span for full pitch coils $= \dfrac{\text{Slots}}{\text{Number of poles}} = \dfrac{48}{8} = 6\ \text{slots}$

kVA rating, $Q = \dfrac{kW}{\eta \times pf}$

$$= \frac{90}{0.9 \times 0.86} = 116.28\ kVA$$

Stator current per phase, $I_s = \dfrac{kVA}{3E_s \times 10^{-3}}$

$$\boxed{kVA = 3\,E_s\,I_s \times 10^{-3}}$$

$$= \frac{116.28}{3 \times 289 \times 10^{-3}} = 134.12\ A$$

Rotor current per phase, $I_r = \dfrac{0.85\,I_s\,T_s}{T_r}$

$$\boxed{\begin{array}{l}\text{Rotor ampere turns}\\\text{is 85\% of stator}\\\text{ampere turns.}\end{array}}$$

$$= \frac{0.85 \times 134.12 \times 63}{48} = 149.63\ A$$

Let current density in rotor, $\delta_r = 5\ A/mm^2$.

∴ Area of cross-section } $a_r = \dfrac{I_r}{\delta_r} = \dfrac{149.63}{5} = 29.926 = 30\ mm^2$
of rotor conductors

RESULT

Number of stator slots,	S_s =	63
Number of rotor slots,	S_r =	48
Emf between slip rings,	E_{rL} =	381 V
Rotor turns per phase,	T_r =	48
Rotor conductor per slot,	T_{rs} =	6
Rotor current per phase,	I_r =	149.63 A
Area of cross-section of rotor conductor,	a_r =	30 mm^2

EXAMPLE 8.12

(AU, Apr' 19, 13 M)

A 20 *HP*, 440 *V*, 4 pole, 50 *Hz*, 3-phase induction motor is built with a stator bore of 0.25 *m* and core length of 0.16 *m*. The specific electric loading is 23000 *amp. cond./m*. Find the specific magnetic loading of the machine. Assume full load efficiency of 84% and a power factor of 0.82. Using the data of the above machine determine the main dimensions, number of stator slots and stator conductors for a 15 *HP*, 460 *V*, 6 pole, 50 *Hz* motor. Take K_{ws} = 0.955.

GIVEN DATA

	Machine-I		**Machine-II**
	20 *HP*	4 pole	15 *HP*
	440 *V*	50 *Hz*	460 *V*
	D = 0.25 *m*	3-phase	50 *Hz*
	L = 0.16 *m*		6 pole
	ac = 23000 *amp.cond./m*		K_{ws} = 0.955

SOLUTION

Machine-I

$$\text{kVA input, } Q_1 = \frac{HP_1 \times 0.746}{\eta \times pf} = \frac{20 \times 0.746}{0.84 \times 0.82} = 21.66 \; kVA$$

$$\text{Synchronous speed, } n_{s1} = \frac{2f}{p} = \frac{2 \times 50}{4} = 25 \; rps$$

Also, kVA input, $Q_1 = C_o \, D^2 \, L \, n_s$

where, $C_o = 11 \, B_{av} \, ac \, K_{ws} \times 10^{-3}$

$$\therefore Q_1 = 11 \, B_{av} \, ac \, K_{ws} \times 10^{-3} D^2 \, L \, n_s$$

$$\therefore B_{av} = \frac{Q_1}{11 \, ac \, K_{ws} \times 10^{-3} D^2 \, L \, n_s}$$

$\boxed{K_{ws} = 0.955}$

$$= \frac{21.66}{11 \times 23000 \times 0.955 \times 10^{-3} \times 0.25^2 \times 0.16 \times 25} = 0.3586 \; Wb/m^2$$

Machine-II

$$\text{kVA input, } Q_2 = \frac{HP_2 \times 0.746}{\eta \times pf} = \frac{15 \times 0.746}{0.84 \times 0.82} = 16.246 \; kVA$$

$$\text{Synchronous speed, } n_{s2} = \frac{2f}{p} = \frac{2 \times 50}{6} = 16.667 \; rps$$

The value of **ac** (specific electric loading) decreases when voltage rating is increased. Hence the ratio of specific electric loading can be expressed as shown below.

$$\frac{ac_2}{ac_1} = \frac{V_{L2}}{V_{L1}}$$

$$\text{The ratio of voltage rating} = \frac{V_{L2}}{V_{L1}} = \frac{460}{440} = 1.0455$$

$$\therefore ac_2 = ac_1 \times \frac{V_1}{V_2} = \frac{ac_1}{V_2 / V_1}$$

$$= \frac{ac_1}{1.0455} = \frac{23000}{1.0455} = 21999.044 \approx 22000 \; amp.cond./m$$

Let us assume same L/τ ratio for both the machines.

For machine-I, $\quad \dfrac{L_1}{\tau_1} = \dfrac{L_1}{\dfrac{\pi D_1}{p_1}} = \dfrac{L_1\, p_1}{\pi\, D_1} = \dfrac{0.16 \times 4}{\pi \times 0.25} = 0.8149$

For machine-II, $\quad \dfrac{L_2}{\tau_2} = 0.8149$

$\boxed{\tau = \dfrac{\pi D}{p}}$

$\therefore L_2 = 0.8142 \ \tau_2 = 0.8149 \times \dfrac{\pi D_2}{p_2} = \dfrac{0.8149 \times \pi}{6}\, D_2 = 0.4267\, D_2$

$\therefore$ Length of stator core, $L_2 = 0.4267\, D_2$ .....(1)

kVA input, $Q_2 = C_{o2}\, D_2^{\,2}\, L_2\, n_{s2}$

where, $C_{o2} = 11\, B_{av}\, ac_2\, K_{ws} \times 10^{-3}$

Taking same B_{av} for both machines.

$\therefore Q_2 = 11\, B_{av}\, ac_2\, K_{ws} \times 10^{-3} \times D_2^{\,2}\, L_2\, n_{s2}$

$\therefore D_2^{\,2}\, L_2 = \dfrac{Q_2}{11\, B_{av}\, ac_2\, K_{ws} \times 10^{-3}\, n_{s2}}$

$\boxed{K_{ws} = 0.955}$

$\qquad = \dfrac{16.246}{11 \times 0.3586 \times 22000 \times 0.955 \times 10^{-3} \times 16.667} = 0.0118\, m^3$

$\therefore D_2^{\,2}\, L_2 = 0.0118\, m^3$

$D_2^{\,2}\, (0.4267\, D_2) = 0.0118$

$\therefore$ Diameter of stator bore, $D_2 = \left(\dfrac{0.0118}{0.4267}\right)^{1/3} = 0.3024\, m = 0.3\, m$

$\therefore$ Length of stator core, $L_2 = 0.4267 \times 0.3 = 0.128\, m$

Maximum flux per pole, $\phi_m = \dfrac{B_{av}\, \pi\, D_2\, L_2}{p}$

$\qquad = \dfrac{0.3586 \times \pi \times 0.3 \times 0.128}{4} = 0.011\, Wb$

Stator turns per phase, $T_s = \dfrac{E_s}{4.44\, f\, \phi_m\, K_{ws}}$

Delta connected motor.
$\therefore E_s = V_L$

$\qquad = \dfrac{440}{4.44 \times 50 \times 0.011 \times 0.955} = 188$

The stator slot pitch should be lie between 15 mm to 25 mm.

Stator slot, $S_s = \dfrac{\pi D}{y_{ss}}$

When, $y_{ss} = 15\ mm, \quad S_s = \dfrac{\pi \times 0.3}{15 \times 10^{-3}} = 62$

When, $y_{ss} = 25\ mm, \quad S_s = \dfrac{\pi \times 0.3}{25 \times 10^{-3}} = 38$

The stator slot should be lie between 38 to 62.

The stator slot be multiple of q, where q is slots per pole per phase.

Stator slot, S_s = Number of phase $\times$ Poles $\times$ q = 3 p q

When, q = 2　　　$S_s = 3 \times 4 \times 2 = 24$

When, q = 3　　　$S_s = 3 \times 4 \times 3 = 36$

When, q = 4　　　$S_s = 3 \times 4 \times 4 = 48$

The S_s value of 48 lie in the range of 38 to 64.

Conductors per slot, $Z_{ss} = \dfrac{6\,T_s}{S_s}$

$$= \frac{6 \times 188}{48} = 23.5 \simeq 24$$

∴ Total stator conductors, $Z = S_s \times Z_{ss} = 48 \times 24 = 1152$

∴ New value of turns per phase, $T_s = \dfrac{Z_{ss} \times S_s}{6}$

$$= \frac{48 \times 24}{6} = 192$$

RESULT

Machine-I	Specific magnetic loading,	B_{av}	=	0.3586 Wb/m^2
Machine-II	Diameter of stator bore,	D_2	=	0.2972 m
	Length of stator core,	L_2	=	0.127 m
	Conductors per slot,	Z_{ss}	=	24
	Stator slot,	S_s	=	48
	Total stator conductors,	Z	=	1152
	Stator turns per phase,	T_s	=	192

EXAMPLE 8.13

Calculate the specific electric and magnetic loading of 100 HP, 3000 V, 3-phase, 50 Hz, 8 pole, star connected, flame proof induction motor having stator core length = 0.5 m and stator bore = 0.66 m. Turns per phase = 286. Assume full load efficiency as 0.938 and pf as 0.86.

GIVEN DATA

100 HP	η = 0.938	Star connected
3000 V	L = 0.5 m	pf = 0.86
3-phase	D = 0.66 m	T_{ph} = 286
50 Hz	8 Pole	

SOLUTION

kVA input, $Q = \dfrac{HP \times 0.746}{\eta \times pf} = \dfrac{100 \times 0.746}{0.938 \times 0.86} = 92.48\ kVA$

Also, kVA input, $Q = \sqrt{3}\,V_L\,I_L \times 10^{-3}$

∴ Line current, $I_L = \dfrac{Q}{\sqrt{3}\,V_L \times 10^{-3}}$

$$= \frac{92.48}{\sqrt{3} \times 3000 \times 10^{-3}} = 17.8\ A$$

∴ Stator current per phase, $I_s = I_L = 17.8\ A$

$\boxed{\text{Star connected stator}}$

∴ Current through each conductor, $I_z = I_s = 17.8\ A$

Number of armature conductors, $Z =$ Number of phase $\times\ 2\ \times\ T_{ph}$

$$= 3 \times 2 \times 286 = 1716$$

Specific electric loading, $\mathbf{ac} = \dfrac{I_z Z}{\pi D} = \dfrac{17.8 \times 1716}{\pi \times 0.66} = 14731.38 \approx 15000\ amps.cond./m$

Synchronous speed, $n_s = \dfrac{2f}{p} = \dfrac{2 \times 50}{8} = 12.5\ rps$

Also, kVA input, $Q = C_o\ D^2\ L\ n_s$ and $C_o = 11\ B_{av}\ \mathbf{ac}\ K_{ws} \times 10^{-3}$

$\boxed{K_{ws} = 0.955}$

∴ $Q = 11\ B_{av}\ \mathbf{ac}\ K_{ws} \times 10^{-3} \times D^2\ L\ n_s$

∴ Specific magnetic loading, $B_{av} = \dfrac{Q}{11\ \mathbf{ac}\ K_{ws} \times 10^{-3} \times D^2\ L\ n_s}$

$$= \dfrac{92.48}{11 \times 15000 \times 0.955 \times 10^{-3} \times 0.66^2 \times 0.5 \times 12.5}$$

$$= 0.22\ Wb/m^2$$

RESULT

Specific electric loading, $\mathbf{ac}\ =\ 15000\ amp.cond./m$

Specific magnetic loading, $B_{av}\ =\ 0.22\ Wb/m^2$

EXAMPLE 8.14

A 70 kW, 400 V, 4 pole, 50 Hz, 3-phase, star connected slip ring induction motor has the following data: stator bore = 0.68 m, stator core length = 0.52 m, number of stator slots = 96, number of rotor slots = 72, number of stator turns per phase = 288, specific permeance due to stator slot = 2.2 μ_o and due to to rotor slot = 1.6 μ_o, harmonic leakage reactance per phase = 0.8 Ω, magnetizing current = 6 A. Estimate the total standstill leakage reactance of motor referred to stator. Take K_w = 0.955 for stator and rotor.

GIVEN DATA

70 kW	50 Hz	D = 0.68 m	K_w = 0.955
400 V	S_s = 96	L = 0.52 m	X_h = 0.8 Ω
3-phase	S_r = 72	λ_{ss} = 2.2 μ_o	I_m = 6 A
4 pole	T_s = 288	λ_{sr} = 1.6 μ_o	

SOLUTION

Stator slots per pole per phase, $q_s = \dfrac{S_s}{\text{Number of phase} \times \text{Poles}} = \dfrac{96}{3 \times 4} = 8$

Rotor slots per pole per phase, $q_r = \dfrac{S_r}{\text{Number of phase} \times \text{Poles}} = \dfrac{72}{3 \times 4} = 6$

Stator slot leakage reactance, $X_{ss} = 8\ \pi\ f\ T_s^2\ L\left(\dfrac{\lambda_{ss}}{pq_s}\right)$

$\boxed{\mu_o = 4\pi \times 10^{-7}\ H/m}$

$$= 8\pi \times 50 \times 288^2 \times 0.52 \times \left(\dfrac{2.2 \times 4\pi \times 10^{-7}}{4 \times 8}\right) = 4.68\ \Omega$$

Rotor slot leakage reactance, $\lambda_{sr}' = \left(\dfrac{K_{ws}}{K_{wr}}\right)^2 \dfrac{S_s}{S_r} \lambda_{sr}$

$$= \left(\dfrac{0.955}{0.955}\right)^2 \times \dfrac{96}{72} \times 1.6\,\mu_0 = 2.13\,\mu_0$$

Rotor slot leakage reactance referred to stator $\Big\}$ $X_{sr}' = 8\,\pi\,f\,T_s^2\,L\left(\dfrac{\lambda_{sr}'}{pq_s}\right)$

$$= 8\pi \times 50 \times 288^2 \times 0.52 \times \left(\dfrac{2.13 \times 4\pi \times 10^{-7}}{4 \times 8}\right) = 4.53\,\Omega$$

Stator voltage per phase, $E_s = \dfrac{400}{\sqrt{3}} = 231\,V$

Star connected motor.
$\therefore E_s = V_L / \sqrt{3}$

Magnetizing reactance, $X_m = \dfrac{E_s}{I_m} = \dfrac{231}{6} = 38.5\,\Omega$

Zigzag leakage reactance, $X_z = \dfrac{5}{6}\dfrac{X_m}{m_s^2}\left(\dfrac{1}{q_s^2 + q_r^2}\right)$

$m_s = 3$ = Number of stator phase

$$= \dfrac{5}{6} \times \dfrac{38.5}{3^2} \times \left(\dfrac{1}{8^2} + \dfrac{1}{6^2}\right) = 0.15\,\Omega$$

Pole pitch, $\tau = \dfrac{\pi D}{p} = \dfrac{\pi \times 0.68}{4} = 0.53\,m$

Stator slot pitch, $y_{ss} = \dfrac{\pi D}{S_s} = \dfrac{\pi \times 0.68}{96} = 0.02\,m$

$$L_0\lambda_0 = \mu_0 \dfrac{k_s\,\tau^2}{\pi\,y_{ss}}$$

Assume, $k_s = 1$

$$= \mu_0 \times \dfrac{1 \times 0.53^2}{\pi \times 0.02} = 4.47\,\mu_0$$

Overhang leakage reactance, $X_0 = 8\,\pi\,f\,T_s^2\,L_0\left(\dfrac{\lambda_0}{pq_s}\right) = 8\,\pi\,f\,T_s^2 \times \dfrac{4.47\,\mu_0}{p\,q_s}$

$$= 8\pi \times 50 \times 288^2 \times \dfrac{4.47 \times 4\pi \times 10^{-7}}{4 \times 8} = 18.3\,\Omega$$

Let, X_s = Total standstill leakage reactance of motor referred to stator

$\therefore X_s = X_{ss} + X_{sr}' + X_z + X_0 + X_h$

$$= 4.68 + 4.53 + 0.15 + 18.3 + 0.8 = 28.5\,\Omega$$

RESULT

Total standstill leakage reactance, $X_s = 28.5\,\Omega$

EXAMPLE 8.15

Find the leakage reactance of a 7 kW, 400 V, 3-phase, 50 Hz, 4 pole cage type induction motor. The stator bore is 0.19 m and the core length is 0.15 m. The stator has 36 slots and the rotor 31 slots. The stator slot is parallel sided and rotor slot is circular as shown in Figs. 1 and 2 respectively. The length of air-gap is 0.8 mm and the number of stator turns is 278. The length of overhang on one side is 0.26 m. The stator winding factor may be assumed as 0.955. The stator winding is mush type for which the constant k = 0.37 × 10⁻⁶ to calculate overhang leakage permeance. Neglect harmonic leakage. Stator slot dimensions in mm: h_1 = 23, h_2 = 2, h_3 = 2, h_4 =1, y_{ss} = 9.5 and W_{os} = 3. Rotor slot dimensions in mm: y_{sr} = 9.5, W_{or} = 3.

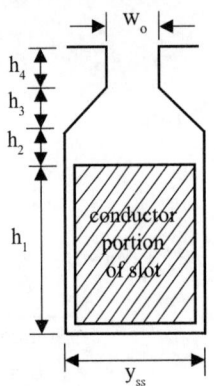

Fig. 1: *Parallel sided slot.*

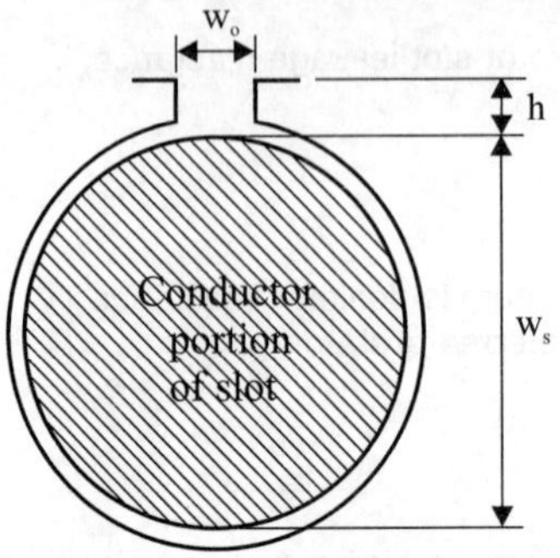

Fig. 2: *Circular slot.*

GIVEN DATA

7 kW	$l_g = 0.8$ mm	$k = 0.37 \times 10^{-6}$	$y_{ss} = 9.5$ mm	$h_1 = 23$ mm
400 V	$D = 0.19$ m	$T_s = 278$	$W_{os} = 3$ mm	$h_2 = 2$ mm
3-phase	$L = 0.15$ m	$L_o = 0.26$ m	$y_{sr} = 9.5$ mm	$h_3 = 2$ mm
4 pole	$S_s = 36$	$K_{ws} = 0.955$	$W_{or} = 3$ mm	$h_4 = 1$ mm
50 Hz	$S_r = 31$			

SOLUTION

Stator slots per pole per phase, $q_s = \dfrac{S_s}{\text{Number of phase} \times \text{Poles}}$

$$= \frac{36}{3 \times 4} = 3$$

Stator slot specific permeance, $\lambda_{ss} = \mu_o \left[\dfrac{h_1}{3W_o} + \dfrac{h_2}{y_{ss}} + \dfrac{2h_3}{y_{ss} + W_o} + \dfrac{h_4}{W_o} \right]$

$$= \mu_o \left[\frac{23}{3 \times 9.5} + \frac{2}{9.5} + \frac{2 \times 2}{9.5 + 3} + \frac{1}{3} \right] = 1.67 \, \mu_o$$

Rotor slot specific permeance, $\lambda_{sr} = \mu_o \left[0.66 + \dfrac{h}{W_o} \right]$

$$= \mu_o \left[0.66 + \frac{1}{3} \right] = \mu_o$$

Rotor slot specific permeance referred to stator $\Big\}$ $\lambda_{sr}' = \left(\dfrac{K_{ws}}{K_{wr}} \right)^2 \dfrac{S_s}{S_r} \lambda_{sr}$

$$= \left(\frac{0.955}{0.955} \right)^2 \times \frac{36}{31} \times \mu_o = 1.16 \, \mu_o$$

Stator slot pitch, $y_{ss} = \dfrac{\pi D}{S_s} = \dfrac{\pi \times 0.19 \times 10^3}{36} = 16.6$ mm

Width of stator tooth, $W_{ts} = y_{ss} - W_{os}$

$$= 16.6 - 3 = 13.6 \text{ mm}$$

Rotor slot pitch, $y_{sr} = \dfrac{\pi D}{S_r} = \dfrac{\pi \times 0.19 \times 10^3}{31} = 19.3$ mm

Width of rotor tooth, $W_{tr} = y_{sr} - W_{or}$

$$= 19.3 - 3 = 16.3 \; mm$$

Specific permeance due to zigzag leakage $\left.\right\} \lambda_z = \mu_0 \dfrac{W_{ts} \; W_{tr} \; (W_{ts}^2 + W_{tr}^2)}{12 \, l_g \, y_{ss}^2 \; y_{sr}}$

$$= \mu_0 \dfrac{13.6 \times 16.3 \times (13.6^2 + 16.3^2)}{12 \times 0.8 \times 16.6^2 \times 19.3} = 1.97 \; \mu_0$$

Specific permeance due to slot and zigzag leakage $\left.\right\} \lambda = \lambda_{ss} + \lambda_{sr}{}' + \lambda_z$

$$= \mu_o \; (1.67 + 1.35 + 1.97) = 4.99 \; \mu_o$$

Leakage reactance due to slot and zigzag leakage $\left.\right\} X_s + X_z = 8\pi \, f \, T_s^2 \, L \left(\dfrac{\lambda}{pq_s} \right) = 8 \, \pi \, f \, T_s^2 \, L \, \dfrac{4.99 \; \mu_0}{p \, q_s}$

$$= 8\pi \times 50 \times 278^2 \times 0.15 \times \left(\dfrac{4.99 \times 4\pi \times 10^{-7}}{4 \times 3} \right) = 7.61 \, \Omega$$

Specific permeance due to overhang leakage $\left.\right\} \lambda_o = \dfrac{k \, L_o}{2\sqrt{2} \; y_{ss}}$

$$= \dfrac{0.37 \times 10^{-6} \times 0.26}{2 \times \sqrt{2} \times 16.6 \times 10^{-3}} = 2.05 \times 10^{-6}$$

Leakage reactance due to overhang leakage $\left.\right\} X_o = 8 \, \pi \, f \, T_s^2 \, L_o \left(\dfrac{\lambda_o}{pq_s} \right)$

$$= 8\pi \times 50 \times 278^2 \times 0.26 \times \left(\dfrac{2.05 \times 10^{-6}}{4 \times 3} \right) = 4.31 \, \Omega$$

Let, X_s = Total leakage reactance of motor referred to stator

$$\therefore \; X_s = X_s + X_z + X_o$$

$$= 7.61 + 4.31 = 11.92 \; \Omega$$

RESULT

Total leakage reactance, $X_s = 11.92 \; \Omega$.

EXAMPLE 8.16   *(PU, Nov' 19, 6 M) (UTU, Dec' 13, 10 M)*

A 16 kW, 400 V, 3-phase, 50 Hz, 6 pole induction motor has a diameter of 0.3 m and the length of core 0.12 m. The number of stator slot is 72 with 20 conductors per slot. The stator is delta connected. Calculate the value of magnetizing current per phase if the length of air-gap is 0.6 m. The gap contraction factor is 1.2. Assume the mmf required for the iron parts to be 35 percent of the air-gap mmf. Take stator winding factor as 0.955.

GIVEN DATA

16 kW	6 pole	$Z_{ss} = 20$	mmf for iron parts = $\dfrac{35}{100} \times AT_g$
400 V	D = 0.3 m	$l_g = 0.6 \; mm$	
3-phase	L = 0.12 m	$K_g = 1.2$	
50 Hz	$S_s = 72$	$K_{ws} = 0.955$	

SOLUTION

Stator turns per phase, $T_s = \dfrac{Z_{ss} \, S_s}{6} = \dfrac{20 \times 72}{6} = 240$

$\therefore$ Flux per pole, $\phi_m = \dfrac{E_p}{4.44 \, f \, T_s \, K_{ws}}$

> Delta connected stator.
> $\therefore E_p = V_L$

$\qquad = \dfrac{400}{4.44 \times 50 \times 240 \times 0.955} = 7.9 \times 10^{-3} \, Wb$

Area per pole, $A_p = \dfrac{\pi D L}{p} = \dfrac{\pi \times 0.3 \times 0.12}{6} = 18.85 \times 10^{-3} \, m^2$

Average air-gap density, $B_{av} = \dfrac{\phi_m}{A_p}$

$\qquad = \dfrac{7.9 \times 10^{-3}}{18.85 \times 10^{-3}} = 0.419 \, Wb/m^2$

Let, B_{g60} = Gap flux density at 60° from pole axis

$B_{g60} = 1.36 \, B_{av} = 1.36 \times 0.419 = 0.57 \, Wb/m^2$

mmf required for air-gap, $AT_g = 800{,}000 \times B_{g60} \times K_g \times l_g$

$\qquad = 800{,}000 \times 0.57 \times 1.2 \times 0.6 \times 10^{-3} = 328.32 \, AT$

mmf for iron parts $= \dfrac{35}{100} \times AT_g$

$\qquad = 0.35 \times 328.32 = 114.91 \, AT$

$\therefore$ Total mmf, AT_{60} = mmf for air-gap + mmf for iron parts

$\qquad = 328.32 + 114.91 = 443.23 \, AT$

Magnetizing current per phase, $I_m = \dfrac{0.427 \times p \times AT_{60}}{K_{ws} \, T_s}$

$\qquad = \dfrac{0.427 \times 6 \times 443.23}{0.955 \times 240} = 4.95 \, A$

RESULT

Magnetizing current per phase, $I_m = 4.95 \, A$.

EXAMPLE 8.17

(PU, Nov' 19, 6 M)

A 75 *kW*, 3300 *V*, 50 *Hz*, 8 pole, 3-phase, star connected induction motor has a magnetizing current which is 40% of full load current. Calculate the value of stator turns per phase if the mmf required for flux density at 60° from the pole axis is 500 *AT*. Winding factor = 0.95, efficiency = 0.94 and power factor = 0.86. Assume suitable data if required.

GIVEN DATA

$P_o = 75 \, kW$ 3-phase $\eta = 0.94$

$V_L = 3300 \, V$ $I_m = \dfrac{40}{100} I_L = 0.4 \, I_L$ pf = 0.86

50 *Hz* $AT_{60} = 500 \, AT$

8 pole $K_{ws} = 0.95$

SOLUTION

Power input, $P_i = \dfrac{\text{Power output}}{\text{Efficiency}} = \dfrac{P_o}{\eta}$ .....(1)

kVA input, $Q = \dfrac{\text{Power input}}{\text{Power factor}}$

$\boxed{\text{Using equation (1)}}$

$= \dfrac{P_i}{pf} = \dfrac{P_o}{\eta\, pf}$

Also, kVA input, $Q = \sqrt{3}\ V_L\ I_L \times 10^{-3}$

On equating two equations of kVA input we get,

$\sqrt{3}\ V_L\ I_L \times 10^{-3} = \dfrac{P_o}{\eta\, pf}$

Full load current, $I_L = \dfrac{P_o}{\sqrt{3}\ V_L\ \eta\, pf}$

$= \dfrac{75 \times 1000}{\sqrt{3} \times 3300 \times 0.94 \times 0.86} = 16.2316\ A$

∴ Magnetizing current, $I_m = 0.4 \times 16.2316 = 6.4926\ A$

We know that,

Magnetizing current per phase, $I_m = \dfrac{0.427\ p\ AT_{60}}{K_{ws}\ T_s}$

∴ Stator turns per phase, $T_s = \dfrac{0.427\ p\ AT_{60}}{K_{ws}\ I_m}$

$= \dfrac{0.427 \times 8 \times 500}{0.95 \times 6.4926} = 276.9 \approx 278$

RESULT

Stator turns per phase, $T_s = 278$

EXAMPLE 8.18

Calculate the equivalent resistance of rotor per phase in terms of stator, current in each bar and end ring and total rotor copper loss for 4 pole, 3-phase, 50 Hz, 400 V cage motor having 48 slots in stator with 35 conductors per slot. Each conductor carries a current of 12 A. The rotor has 57 slots, each slot has a bar of 0.14 m length and 50 mm² area. The mean diameter of each ring is 0.3 m and area is 176 mm². Resistivity is 0.02 × 10⁻⁶ Ω-m, winding pitch factor is 0.955.

GIVEN DATA

400 V	$S_s = 48$	$L_b = 0.14\ m$	$a_b = 50\ mm^2$
50 Hz	$Z_{ss} = 35$	$D_e = 0.3$	$k_{ws} = k_{wr} = 0.955$
3-phase	$I_s = 12\ A$	$a_e = 176\ mm^2$	
4 pole	$S_r = 57$	$\rho = 0.02 \times 10^{-6}\ \Omega\text{-}m$	

SOLUTION

Stator turns per phase, $T_s = \dfrac{Z_{ss} \times S_s}{6}$

$$= \frac{35 \times 48}{6} = 280$$

Current in each bar, $I_b = 0.85 \times \dfrac{6 \, I_s \, T_s}{S_r}$

> Rotor ampere-turns is 85% of stator ampere-turns.

$$= 0.85 \times \frac{6 \times 12 \times 280}{57} = 300.63 \text{ A}$$

Current in each ring, $I_e = \dfrac{S_r \, I_b}{\pi p}$

$$= \frac{57 \times 300.63}{\pi \times 4} = 1364 \text{ A}$$

Resistance of each bar, $r_b = \dfrac{\rho \, L_b}{a_b}$

$$= \frac{0.02 \times 10^{-6} \times 0.14}{50 \times 10^{-6}} = 56 \times 10^{-6} \, \Omega$$

Copper loss in rotor bars $= S_r \, I_b^2 \, r_b$

$$= 57 \times (300.63)^2 \times 56 \times 10^{-6} = 288.5 \ W$$

Resistance of each ring, $r_e = \dfrac{\rho \, \pi \, D_e}{a_e}$

$$= \frac{0.02 \times 10^{-6} \times \pi \times 0.3}{176 \times 10^{-6}} = 107 \times 10^{-6} \, \Omega$$

Copper loss in two end rings $= 2 \, I_e^2 \, r_e$

$$= 2 \times 1364^2 \times 107 \times 10^{-6} = 398 \ W$$

Total rotor copper loss = Copper loss in rotor bars + Copper loss in two end rings

$$= 288.5 + 398 = 228.83 \ W$$

Rotor resistance per phase, $r_r = \dfrac{\text{Total rotor copper loss}}{I_b^2}$

$$= \frac{228.83}{300.63^2} = 2.53 \times 10^{-3} \, \Omega$$

Rotor resistance referred to stator per phase, $r_r' = \left(\dfrac{k_{ws} \, T_s}{k_{wr} \, T_r} \right) r_r$

> $k_{ws} = k_{wr}$

$$= \left(\frac{T_s}{S_r} \right)^2 r_r$$

> $T_r = S_r$

$$= \left(\frac{280}{57} \right)^2 \times 2.53 \times 10^{-3} = 0.061 \, \Omega$$

RESULT

Rotor resistance referred to stator per phase, $r_r' = 0.061 \ \Omega$.

EXAMPLE 8.19

The following data refer to a 11 kW, 50 Hz, 3 pole, 440 V, slip ring induction motor with 3-phase delta connected stator winding. Magnetizing current = 4.47 A, iron loss = 510 W, friction and windage loss = 110 W, total leakage reactance = 6.52 Ω, stator resistance per phase = 1.41 Ω, rotor resistance referred to stator = 1.35 Ω. Draw the circle diagram and deduce therefrom the line current, efficiency, power factor and slip at full load. Find also the maximum output and the pullout (or maximum) torque.

GIVEN DATA

11 kW	440 V	$r_r' = 1.35\ W$
50 Hz	$I_m = 4.47\ A$	Iron loss = 510 W
3 pole	$X_s = 6.52\ \Omega$	Friction and windage loss = 110 W
3-phase	$r_s = 1.41\ \Omega$	

SOLUTION

Total no-load loss = No-load stator copper loss + Iron loss + Friction and windage loss

$$= 3\,I_s^2\,r_s + 510 + 110 \approx 3\,I_m^2\,r_s + 510 + 110 \qquad \boxed{I_s \approx I_m}$$

$$= (3 \times 4.47^2 \times 1.41) + 510 + 110 = 704.5\ W$$

Loss component of no-load current, $I_l = \dfrac{\text{Total no-load loss}}{3\,E_S}$

$$\boxed{\begin{array}{l}\text{Delta connected motor.}\\ \qquad \therefore\ E_s = V_L\end{array}}$$

$$= \dfrac{704.5}{3 \times 440} = 0.53\ A$$

No-load current, $I_o = \sqrt{I_m^2 + I_l^2}$

$$= \sqrt{4.47^2 + 0.53^2} = 4.5\ A$$

Phase angle of no-load current, $\phi_o = \tan^{-1} \dfrac{I_m}{I_l}$

$$= \tan^{-1} \dfrac{4.47}{0.53} = 83^\circ$$

Total motor resistance referred to stator, $R_s = r_s + r_r'$

$$= 1.41 + 1.35 = 2.76\ \Omega$$

Total impedence, $Z_s = \sqrt{R_s^2 + X_s^2}$

$$= \sqrt{2.76^2 + 6.52^2} = 7.08\ \Omega$$

Phase angle of short circuit current, $\phi_{sc} = \tan^{-1} \dfrac{X_s}{R_s}$

$$= \tan^{-1} \dfrac{6.52}{2.76} = 67^\circ$$

Short circuit current, $I_{sc} = \dfrac{E_S}{Z_s}$

$$\boxed{\begin{array}{l}\text{Delta connected motor.}\\ \qquad \therefore\ E_s = V_L\end{array}}$$

$$= \dfrac{440}{7.08} = 62\ A$$

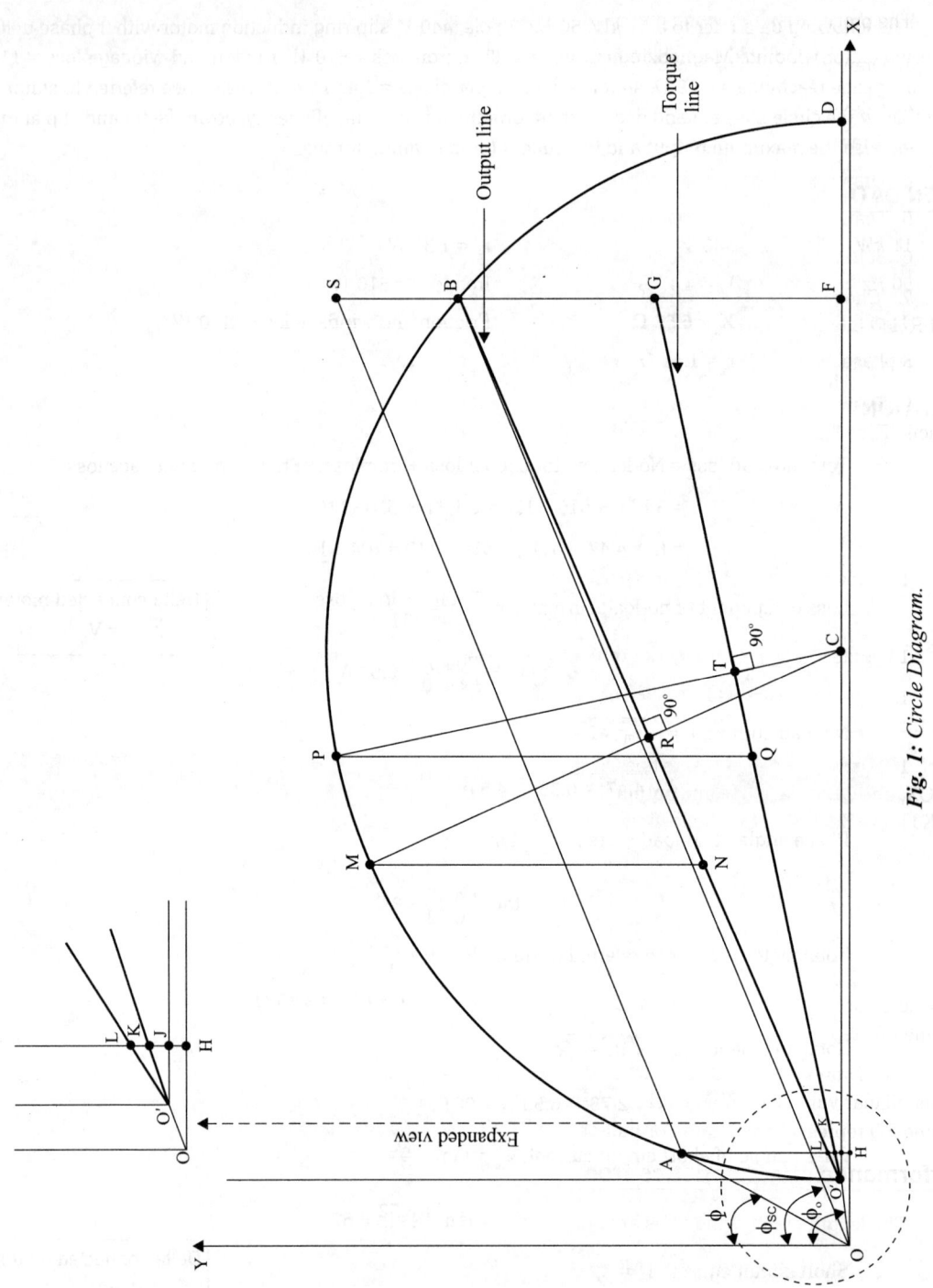

Fig. 1: Circle Diagram.

Procedure to draw circle diagram

The procedure for drawing circle diagram is given below (Refer Fig. 1).

1. Draw two lines OX and OY perpendicular to each other.

2. Choose a current scale, 1 cm = 4 A.

3. Draw line OO' whose length is equal to the no-load current per phase, I_o at an angle ϕ_o with line OY.

4. Draw a line O'D passing through O' and parallel to line OX.

5. Draw line OB whose length is equal to short circuit current per phase, I_{sc} at an angle, ϕ_{sc} with line OY.

6. Join the point O' and B, now the line O'B is the output line.

7. Choose mid-point, R in line O'B and construct the perpendicular bisector of O'B passing through point R and intersecting the line O'D at point C, where point C is the centre of circle having radius O'C.

8. Draw the circle O'BD, with centre at point C.

9. Draw a perpendicular to O'D such that it passes through point B and meeting the line O'D at point F. Then divide the line BF in such a way that

$$\frac{BG}{GF} = \frac{\text{Rotor resistance referred to stator}}{\text{Stator resistance}} = \frac{r_r'}{r_s}$$

10. Join O' to G. Now line O'G is known as torque line.

11. Now, Power scale, 1 cm = Current scale × E_s

$$= 4 \times 440 = 1760 \ W$$

Delta connected motor.
$\therefore E_s = V_L$

12. Extend line FB and mark point S such that line BS is full load power output per phase.

13. Draw a line parallel to output line O'B passing through point S and cutting the circle at point A. Now the point A is the full load operating point and line OA is full load current per phase.

14. Draw a perpendicular to line OX passing through point A. Let the crossing point of this perpendicular line OX, line OD, torque line and output line be respectively H, J, K and L. Now the line segments AH, AK, AL, LK, KJ and JH represent the following.

AH = Power input	LK = Rotor copper loss
AK = Rotor input	KJ = Stator copper loss
AL = Power output	JH = Constant loss

15. Draw a perpendicular to output line passing through point C. Let the crossing point of this perpendicular with circle be point M, Draw a line parallel to line FS crossing the output line at point N. Now the line MN represents maximum output.

16. Draw a perpendicular to torque line passing through point C. Let the crossing point of this perpendicular with circle be point P. Draw a line parallel to line FS crossing the torque line at point Q. Now the line PQ represents maximum or pull-out torque.

Performance characteristics from circle diagram

The length of following lines are measured from circle diagram

OA = 3 cm	AK = 2.4 cm	PK = 6.4 cm
AH = 2.7 cm	AL = 2.3 cm	PQ = 6.4 cm
LK = 0.1 cm	MN = 5.1 cm	

The following performance characteristics are calculate using the above measured values.

Full load stator phase current, I_s = OA × Current scale

$$= 3 \times 4 = 12 \, A$$

∴ Line current, $I_L = \sqrt{3} \, I_s = \sqrt{3} \times 12 = 20.78 \, A$ | Delta connected stator. |

Stator power factor at full load, $\cos \phi = \dfrac{AH}{OA} = \dfrac{2.7}{3} = 0.9$

% Slip, s = $\dfrac{\text{Rotor copper loss}}{\text{Rotor input}} = \dfrac{LK}{AK} \times 100$

$$= \dfrac{0.1}{2.4} \times 100 = 4.2 \, \%$$

% Efficiency, $\eta = \dfrac{AL}{AH} \times 100$

$$= \dfrac{2.3}{2.7} \times 100 = 85.2 \, \%$$

Maximum output = 3 × MN × Power scale

$$= 3 \times 5.1 \times 1760 = 26928 \, W = 26.928 \, kW$$

Maximum torque = 3 × PQ × Power scale

$$= 3 \times 6.4 \times 1760 = 33792 \, Synchronous \, Watt = 33.792 \, synchronous \, kW$$

$$\dfrac{\text{Maximum torque}}{\text{Full load torque}} = \dfrac{PQ}{AK} = \dfrac{6.4}{2.4} = 2.667$$

$$\dfrac{\text{Maximum output}}{\text{Full load output}} = \dfrac{MN}{AL} = \dfrac{5.1}{2.3} = 2.22$$

RESULT

Full load line current,	I_L	= 20.78 A
Full load power factor,	$\cos \phi$	= 0.9
Full load efficiency,	η	= 85.2 %
Full load slip,	s	= 4.2 %
Maximum output		= 26.928 kW
Maximum torque		= 33.792 synchronous kW.

8.13 COMPUTER PROGRAMS

PROGRAM 8.1

Write a C Program to estimate the stator and rotor turns and rotor emf between slip ring of a 3-phase induction motor for the specifications in Example 8.9.

Method-1

```
/* Program to estimate rotor and stator turns and rotor emf between slip
   ring */
```

```
/* VARIABLE DECLARATION
..............................................
Ss = Number of stator slots.
Sr = Number of rotor slots.
Zss = Stator conductor per slot.
Zsr = Rotor conductor per slot.
VL = Voltage.
.........................................*/
#include<stdio.h>
#include<conio.h>
#include<math.h>
#include<iostream.h>
void main()
{
int Ss,Sr,Zss,Zsr,Ts,VL,Tr;
float Es,Er,ErL;
clrscr();
printf("ÏNPUT DATA\n");
printf("Enter number of stator slots, Ss = ");
scanf("%d", &Ss);
printf("Enter number of rotor slots, Sr =  ");
scanf("%d", &Sr);
printf("Enter stator conductor per slot, Zss = ");
scanf("%d", &Zss);
printf("Ënter rotor conductor per slot, Zsr = ");
scanf("%d", &Zsr);
printf("Ënter Voltage, VL = ");
scanf("%d", &VL);
Ts = Zss*Ss/(6);        /* Ts = Stator turns per phase. */
Tr = Zsr*Sr/(6);        /* Tr = Rotor turns per phase. */
Es = VL/(1.732);        /* Es = Stator phase voltage. */
Er = Es*Tr/Ts;          /* Er = Rotor phaes voltage. */
ErL = 1.732*Er;         /* ErL = Rotor emf between slipring.*/
printf("OUTPUT\n");
printf("Stator turns per phase, Ts = %8.4f \n", Ts);
printf("Rotor turns per phase, Tr = %8.4f \n", Tr);
printf("Rotor emf between slipring, ErL = %8f V\n", ErL);
getch();
}
```

Input data

Enter number of stator slots,	Ss	= 54
Enter number of rotor slots,	Sr	= 72

```
Enter stator conductor per slots,   Zss =  8
Enter rotor conductor slots,        Zsr =  4
Enter voltage,                      VL  =  400
```

Output

```
Stator turns per phase,        Ts  = 72
Rotor turns per phase,         Tr  = 48
Rotor emf between slip ring, ErL = 266.66 V
```

Method-2

```
/* Program to estimate rotor and stator and emf between slip ring */

#include<stdio.h>
#include<conio.h>
#include<math.h>
#include<iostream.h>
void main()
{
clrscr();
int Ss,Sr,Zss,Zsr,Ts,VL,Tr;
float Es,Er,ErL;
Ss = 54;                /* Ss = Number of stator slots. */
Sr = 72;                /* Sr = Number of rotor slots. */
Zss = 8;                /* Zss = Stator conductor per slot. */
Zsr = 4;                /* Zsr = Rotor conductor per slot. */
VL = 400;               /* VL = voltage. */

Ts = Zss*Ss/(6);        /* Ts = Stator turns per phase. */
Tr = Zsr*Sr/(6);        /* Tr = Rotor turns per phase. */
Es = VL/(1.732);        /* Es = Stator phase voltage. */
Er = Es*Tr/Ts;          /* Er = Rotor phaes voltage. */
ErL = 1.732*Er;         /* ErL = Rotor emf between slipring.*/
printf("OUTPUT\n");
printf("Stator turns per phase, Ts = %8.4f \n", Ts);
printf("Rotor turns per phase, Tr = %8.4f \n", Tr);
printf("Rotor emf between slipring, ErL = %8f V\n", ErL);
getch();
}
```

Output

```
Stator turns per phase,        Ts  = 72
Rotor turns per phase,         Tr  = 48
Rotor emf between slip ring, ErL = 266.66 V
```

PROGRAM 8.2

Write a C program to estimate the rotor current, area of rotor conductor, width of rotor tooth and depth of rotor core of a star connected slipring rotor for the specifications in Example 8.10.

```
/* Program to  estimate rotor current, area of rotor conductor, width
of rotor tooth and depth of rotor core. */

#include<stdio.h>
#include<conio.h>
#include<math.h>
#include<iostream.h>
void main()
{
clrscr();
int Ts,Tr,Sr,DELTAr;
float ar,Dr,Dsr,Ysr,Wtr,Bcr,Li,dcr,PHIm,Is,Ir;
Sr = 63;                  /* Sr = Number of rotor slots.*/
Ts = 72;                  /* Ts = Stator turns per phase.*/
Tr = 48;                  /* Tr = Rotor turns per phase. */
Is = 17.8;                /* Is = Stator current per phase. */
DELTAr = 5;               /* DELTAr = Current density.*/
Dr = 0.45;                /* Dr = Diameter of rotor.*/
Dsr = 0.16;               /* Dsr = Depth of rotor slot.*/
Ysr = 0.0034;             /* Ysr = Rotor slot pitch.*/
Bcr = 0.35;               /* Bcr = Flux density in rotor core. */
Li = 0.7;                 /* Li = Net iron length of stator core. */
PHIm = 0.13;              /* PHIm = Maximum flux. */

Ir = (0.85*Is*Ts)/Tr;            /* Ir = Rotor current per phase. */
ar = Ir/DELTAr;                  /* ar = Area of rotor conductor. */
Wtr = 3.14*(Dr-2*Dsr)/(Sr)-Ysr; /* Wtr = Width of rotor tooth. */
dcr = PHIm/(2*Bcr*Li);           /* dcr = Depth of rotor core.*/

printf("OUTPUT\n");
printf("Rotor current, Ir = %8.4f A\n", Ir);
printf("Area of rotor conductor, ar = %8.4f mm2\n", ar);
printf("Width of rotor tooth, Wtr = %8.4f mm\n", Wtr);
printf("Depth of rotor core dcr = %8.4f m\n", dcr);
getch();
}
```

Output

```
Rotor current,              Ir  = 22.6950 A
Area of rotor conductor,    ar  = 4.5390 mm2
Width of rotor tooth,       Wtr = 0.0031 mm
Depth of rotor core,        dcr = 0.2653 m
```

Program 8.3

Write a C program to estimate the stator slot leakage reactance, rotor slot leakage reactance, overhang leakage reactance and Zig zag leakage reactance of a star connected slipring inductioon motor for the specifications in Example 8.14.

```c
/* Program to estimate  stator slot leakage reactance, Rotor slot leakage
reactance,overhang lekage reactance and Zigzag leakage reactance. */

#include<stdio.h>
#include<conio.h>
#include<math.h>
#include<iostream.h>
void main()
{
clrscr();
int Ss,Sr,ph,p,Qs,Qr,f,Ts,Ms,VL;
float L,LAMDAss,Xss,Xsr,Xo,LoLAMDAo,LAMDAsr,Xz,Xm,Es,Im;
Ss = 96;          /* Ss = Number of stator slots. */
Sr = 72;          /* Sr = Number of rotor slots. */
ph = 3;           /* ph = number of phase.*/
p = 4;            /* p = number of poles. */
f = 50;           /* f = Frequency.*/
Ts = 288;         /* Ts = Stator turns per phase.*/
L = 0.52;         /* L = Stator length. */
LAMDAss = 2.2;    /* LAMDAss = Specific permeance due to stator slot. */
LAMDAsr = 2.13;   /* LAMDAsr = Specific permeance due to rotor slot. */
LoLAMDAo = 4.47;  /* LoLAMDAo = Length and specific permeance overhang. */
Im = 6;           /* Im = Magnetizing current.*/
Ms = 3;           /* Ms = Number of stator phase. */

Qs = Ss/(ph*p);              /* Qs = Stator slots per poles. */
Qr = Sr/(ph*p);              /* Qr= Rotor slots per poles. */
Es = VL/(1.732);             /* Es = Stator voltage per phase.*/
Xm = Es/Im;                  /* Xm = Magnetizing reactance. */

Xss = (8*3.14*f*Ts*Ts*L)*(LAMDAss*4*3.14*10e-8)/(p*Qs);
                       /* Xss = Stator slot leakage reactance. */
```

```
Xsr = (8*3.14*f*Ts*Ts*L)*LAMDAsr*4*3.14*10e-8/(p*Qs);
                        /* Xsr = Rotor slot leakage reactance. */
Xo = 8*3.14*f*Ts*Ts*(LoLAMDAo*4*3.14*10e-8)/(p*Qs);
                        /* Xo = Overhang leakage reactance. */
Xz = 0.833*Xm/(Ms*Ms)*1/(Qs*Qs)+1/(Qr*Qr);
                        /* Xz = Zigzag leakage reactance. */
printf("OUPUT\n");
printf("Stator slot leakage reactance, Xss = %8.4f ohm\n", Xss);
printf("Rotor slot reactance referred to stator, Xsr = %8.4f ohm\n",Xsr);
printf("Overhang leakage reactance, Xo = %8.4f ohm\n", Xo);
printf("Zigzag lekage reactance, Xz = %8.4f ohm\n", Xz);
getch();
}
```

Output

```
Stator slot leakage reactance,                   Xss =  4.6778 ohm
Rotor slot leakage reactance referred to stator, Xsr =  4.5289 ohm
Overhang leakage reactance,                      Xo  = 18.2777 ohm
Zigzag lekage reactance,                         Xz  =  0.2502 ohm
```

PROGRAM 8.4

Write a C program to estimate the Rotor bar current, endring current, Area of croos-section of each rotor bar, Area of cross-section of each end ring of a star connected slipring induction motor for the specifications in Example 8.7.

```
/* Program to estimate Rotor bar and end ring current, Area of cross-
section of rotor bar and end ring. */


#include<stdio.h>
#include<conio.h>
#include<math.h>
#include<iostream.h>
void main()
{
clrscr();
int Po,VL,Ss,Zss,f,Z,Ts,Sr,p,Q;
float pf,Eta,Rm,Is,Ib,Ie,ab,ae,Sm,Dele;
Po = 11;                /* Po = Power output of induction motor */
VL = 220;               /* VL = Line voltage. */
Ss = 54;                /* Ss = Number of stator slots. */
Sr = 64;                /* Sr = Number of rotor slots. */
```

```c
pf = 0.85;              /* pf = Power factor. */
Eta = 0.86;             /* Eta = Efficiency. */
Zss = 9;                /* Zss = Conductors per stator slot. */
Dele = 5;               /* Dele = Current density in end ring */
f =50;                  /* f = frequency.*/
p = 6;                  /* p = Number of poles. */


Q = Po/(Eta*pf);           /* Q = kVA rating. */
Is = Q*1000/(1.732*VL);    /* Is = Stator current per phase.*/
Z = Ss*Zss;                /* Z = Number of stator conductor. */
Ts = Z/6;                  /* Ts = Stator turns per phase. */
Sm = 3*Is*Ts;              /* Sm = Stator mmf. */
Rm = 0.85*Sm;              /* Rm = Rotor mmf.*/
Ib = Rm/(Sr/2);            /* Ib = Rotor bar current.*/
Ie = Sr*Ib/(3.14*p);       /* Ie = End ring current. */
ab = Ib/(Dele);            /* ab = Area of cross-section of rotor bar. */
ae = Ie/(Dele);            /* ae = Area of cross-section of end ring. */

printf("OUTPUT\n");
printf("Rotor bar current, Ib = %8.2f A\n", Ib);
printf("End ring current, Ie = %8.2f A\n", Ie);
printf("Area of cross-section of each rotor bar, ab = %d mm2\n", ab);
printf("Area of cross-section of each end ring, ae = %d mm2\n", ae);
getch();
}
```

Output

```
Rotor bar current, Ib =    254.89 A
End ring current, Ie =    865.88 A

Area of cross-section of each rotor bar, ab = 50 mm2
Area of cross-section of each end ring, ae = 173 mm2
```

PROGRAM 8.5

Write a C program to estimate the Specific electric loading and magnetic loading of a 3-phase star connected slipring induction motor for the specifications in Example 8.13.

```
/* Program to estimate the Specific electric loading and Specific Magnetic
loading. */
```

```c
#include<stdio.h>
#include<conio.h>
#include<math.h>
void main()
{
int HP,VL,f,ph,Tph,p,Z,Q;
float Eta,L,D,pf,Iz,Bav,ns,Kws,ac;
clrscr();
HP = 100;          /* HP = Power output in horse power rating.*/
VL = 3000;         /* VL = Line voltage. */
L = 0.5;           /* L = Length of stator core. */
D = 0.66;          /* D = Diameter of  stator bore. */
pf = 0.86;         /* pf = Power factor. */
Eta = 0.938;       /* Eta = Efficiency. */
p = 8;             /* p = Numer of poles.     */
ph = 3;            /* ph = Number of phase. */
f =50;             /* f = frequency.*/
Tph = 286;         /*Tph = Turns per phase. */
Kws =  0.955;      /* Kws = Stator winding factor.*/

Z = ph*2*Tph;                      /*Z = Number of armature
                                         conductors.*/
Q = HP*0.746/(Eta*pf);             /* Q = kVA rating.*/
Iz = Q/(1.732*VL*10e-3);           /* Iz = Current through a
                                         stator conductor */
ac = Iz*Z/(3.14*D);                /* ac = Specific electric loading. */
ns = 2*f/p;                        /* ns = Synchronous speed. */
Bav = Q/(11*ac*Kws*10e-3*D*D*L*ns); /* Bav = Specific magnetic loading.*/

printf("OUTPUT\n");
printf("Specific electric loading, ac = %8.4f amp.cond./m\n", ac);
printf("Specific magnetic loading, Bav = %8.2f Wb/m2\n", Bav);
getch();
}
```

Output

```
Specific electric loading, ac  = 15000 amp.cond./m
Specific magnetic loading, Bav = 0.23 Wb/m2
```

PROGRAM 8.6

Write a C program to estimate the Value of stator turns per phase of a 3-phase star connected slipring induction motor for the specifications in Example 8.17.

```
/* Program to estimate value of stator turns per phase. */
```

```c
#include<stdio.h>
#include<conio.h>
#include<math.h>
void main()
{
int Po,VL,f,ph,p,Ts;
float Eta,pf,Kws,IL,Im,AT60;
clrscr();
Po = 75;            /* HP = Power output in horse power rating.*/
VL = 3300;          /* VL = Line voltage. */
pf = 0.86;          /* pf = Power factor. */
Eta = 0.94;         /* Eta = Efficiency. */
p = 8;              /* p = Number of poles.    */
ph = 3;             /* ph = Number of phase. */
f =50;              /* f = frequency.*/
Kws =  0.95;        /* Kws = Stator winding factor.*/
AT60 = 500;         /* AT60 =  mmf for Im using flux density 60degree from
                                interpolar axis. */

IL = Po*1000/(1.732*VL*Eta*pf);  /* IL = Line current.*/
Im = 0.4*IL;                     /* Im = Magnetizing current.*/
Ts = 0.427*p*AT60/(Kws*Im);      /* Ts = Stator turns per phase.*/

printf("OUTPUT\n");
printf("Stator turns per phase, Ts = %d\n", Ts);
getch();
}
```

Output

```
Stator turns per phase, Ts = 277
```

8.14 SHORT-ANSWER QUESTIONS

Q8.1   What are the different types of induction motor? How they differ from each other?

The two different types of induction motor are squirrel cage and slip ring type. The stator is identical for both types but they differ in the construction of rotor.

The squirrel cage rotor has copper or aluminium bars mounted on rotor slots and short circuited at both ends by end rings.

The slip ring rotor carries a three-phase winding. One end of each phase is connected to a slip ring and other ends are star connected.

Q8.2 *List the constructional elements of squirrel cage induction motor.*

The constructional elements of squirrel cage induction motor are,

Stator - Frame   **Rotor** - Rotor core
        - Stator core   - Rotor bars
        - Stator winding   - End ring

Q8.3 *List the constructional elements of slip ring induction motor.*

The constructional elements of slip ring induction motor are,

Stator - Frame   **Rotor** - Rotor core
        - Stator core   - Rotor winding
        - Stator winding   - Slip rings

Q8.4 *Define specific magnetic loading of induction motor.*

The specific magnetic loading is defined as the average flux density over the air-gap of a machine.

$$\text{Specific magnetic loading, } B_{av} = \frac{\text{Total flux around the air-gap}}{\text{Area of flux path at the air-gap}} = \frac{p\phi}{\pi DL}$$

Q8.5 *Define specific electric loading of induction motor.*

The specific electric loading is defined as the number of armature (or stator) ampere conductors per metre of armature (or stator) periphery at the air-gap.

$$\text{Specific electric loading, } \mathbf{ac} = \frac{\text{Total armature ampere conductors}}{\text{Armature periphery at air-gap}} = \frac{I_z Z}{\pi D}$$

Q8.6 *Give typical values of specific electric and magnetic loading of induction motor.*

Specific magnetic loading, $B_{av} = 0.3$ to 0.6 *Wb/m²*

Specific electric loading, **ac** $= 5000$ to 45000 *amp.cond./m.*

Q8.7 *What is output equation of induction motor?*   *(PTU, May' 19, 2 M)*

The equation which relates the kVA input to the main dimensions (D and L), Specific loadings (B_{av} and **ac**) and speed (n) of a machine is known as output equation.

The output equation of induction machine is, $Q = C_o D^2 L n_s$

where, Q = kVA rating or input of induction machine.

        C_o = Output coefficient $= 11\ B_{av}\ \mathbf{ac}\ k_{ws} \times 10^{-3}$

Q8.8 *Write the expression for output coefficient of induction machine.*   *(AU, Apr' 19, 2 M)*

For induction machine, Output coefficient, $C_o = 11\ B_{av}\ \mathbf{ac}\ K_{ws} \times 10^{-3}$

The unit of C_o is *kVA/m³-rps.*

Q8.9 *What is the significance of the ratio of core length and pole pitch in induction motor?*

In induction motors the operating characteristics are mainly influenced by the ratio of core length and pole pitch, L/τ. The factors influencing this choice are,

For minimum cost, $L/\tau = 1.5$ to 2 | For good efficiency, $L/\tau = 1.5$
For good power factor, $L/\tau = 1.0$ to 1.25 | For good overall design, $L/\tau = 1.0$

Q8.10 What is the equation that is used to design an induction motor for best power factor?

For best power factor the separation of D and L is performed using the equation,

Pole pitch, $\tau = \sqrt{0.18\,L}$

where, $\tau = \pi D/p$ *(HTU, Dec' 18, 10 M)*

Q8.11 What are the factors that decide the choice of specific magnetic loading in induction motor?

The value of specific magnetic loading in induction motor is determined by,

1. Maximum flux density in iron parts of machine   3. Core losses
2. Magnetizing current and power factor   4. Over load capacity.

Q8.12 The effect of magnetizing current is considered as important in case of induction motor. Why?

In induction motor the magnetizing current decides the power factor of the motor. When the magnetizing current is high, the power factor is low and vice versa. If power factor is low, then to deliver the same power output the current rating will be higher, which increases the cost of winding and motor.

Q8.13 What are the factors that decide the choice of specific electric loading in induction motor?

The choice of specific electric loading in induction motor depends on the following factors.

1. Permissible temperature rise and copper loss   3. Voltage rating and size of machine
2. Over load capacity   4. Current density.

Q8.14 Give typical values for specific electric and magnetic loading for a 3.7 kW, 1440 rpm squirrel cage induction motor.

One of the typical choice of specific loadings for a 3.7 *kW*, 1440 *rpm* squirrel cage induction motor are,

Specific electric loading, **ac** = 28,000 *amp.cond./m.*

Specific magnetic loading, B_{av} = 0.56 *Wb/m²*

Q8.15 Smaller machines have low specific magnetic loadings. Why?   *(AKTU, Dec' 19, 3 M)*

A higher value of specific magnetic loading results in increased core loss and higher temperature rise. Consequently the efficiency of the machine will be low. In small machines, the losses has to be kept low in order to get higher power output and so a low value of specific magnetic loading is preferred in small machines.

Q8.16 Name the losses that occur in three-phase induction motors.   *(AU, Apr' 18, 2 M)*

There are four types of losses occur in three-phase induction motors,

1. Stator copper losses   3. Stator iron losses
2. Rotor losses   4. Friction and winding losses.

Q8.17 Why wound rotor construction is adopted?

The wound rotor has the facility of increasing the rotor resistance through slip rings. High value of rotor resistance is needed during starting to get a high value of starting torque.   *(AU, Nov' 18, 2 M)*

Q8.18 What is rotating transformer? (or) why induction machine is called a rotating transformer?

The principle of operation of induction motor is similar to that a transformer. The stator winding is equivalent to primary of a transformer and the rotor winding is equivalent to short circuited secondary of a transformer. In transformer the secondary is fixed but in induction motor it is allowed to rotate. Hence the induction motor is also called rotating transformer.

Q8.19 What type of starter cannot be used for squirrel cage motors?

The starter which cannot be used for squirrel cage motor is rotor resistance starter.

Q8.20 How the slip ring motor is started?

The slip ring motor is started by using rotor resistance starter. The starter consists of star connected variable resistances and protection circuits. The resistances are connected to slip rings. While starting the full resistance is included in the rotor circuit to get high starting torque. Once the rotor starts rotating, the resistances are gradually reduced in steps. At running condition the slip rings are shorted and so it is equivalent to squirrel cage rotor.

Q8.21 What type of connection is preferred for stator of induction motor?

Under running condition the stator of induction motor is normally connected in delta. (In delta connection the torque developed will be higher than the star connection). But for reducing the starting current, the stator can be connected in star while starting and then changed to delta.

Q8.22 Why fractional slot winding is not used for induction motor?   *(AU, Apr' 19, 2 M)*

Windings with fractional number of slots per pole per phase create asymmetrical mmf distribution around the air-gap and favour the creation of noise in the motor. Therefore, fractional slot windings are not used for induction motor stator and only integral slot windings are used.

Q8.23 What are the materials used for slip rings and brushes in induction motor?

The slip rings are made of brass or phosphor bronze. The brushes are made of metal graphite which is an alloy of copper and carbon.

Q8.24 What are the special features of the cage rotor of induction machine?

1. The cage rotor can adopt itself for any number of phases and poles
2. It is suitable for any type of starting method except using rotor resistance starter
3. It is cheaper and rugged
4. Rotor overhang leakage reactance is lesser which results in better power factor, greater pull out torque and overload capacity.

Q8.25 Describe the constructional features of three-phase slip ring induction motor.

The slip ring induction motor has two major parts and they are stator and rotor. The stator has a core and winding and they are covered by a frame. The rotor has core and winding. One end of each phase of rotor winding is terminated on a slip ring mounted on the shaft. The other ends of the winding are star connected.

Q8.26 What are the advantages of cage induction motor over slip ring induction motor?   *(AU, Apr' 18, 5 M)*

The advantages of squirrel cage motor are,

1. It is cheaper than slip ring motors
2. It does not have any wear and tear parts like slip rings, brush gear and short circuiting devices. Hence the construction will be rugged.
3. The rotor slots can be fully occupied by the conductor due to absence of insulation. Hence the rotor bars may have low resistance. Also there is no overhang (or small overhang) in rotor winding. Due to these two factors the rotor copper loss will be lesser than slip ring motors and so efficiency will be slightly higher.
4. Due to smaller overhang leakage reactance, the motor will have better power factor, a greater pull out torque and over load capacity.

Q8.27 Name the materials used to insulate the laminations of the core of induction motor.

The materials used to insulate the laminations are kaolin and varnish.

Q8.28 What are the main dimensions of induction motor?

The main dimensions of induction motor are stator core internal diameter (D) and stator core length (L).

Q8.29 *Comparison of squirrel cage induction motor and wound rotor induction motor.* *(JNTKU, Apr' 19, 6 M)*

Squirrel Cage Induction Motor		Wound Rotor Induction Motor	
1.	Rotor conductors are made of solid copper bars short circuited at both ends by end rings.	1.	Round conductors are made of winding copper wires wound on the rotor slots
2.	Solid rotor bars act as rotor winding.	2.	Rotor carries a three-phase winding with one end star connected and the other end terminated on slip rings.
3.	Additional resistance can not be added in rotor bars to improve starting torque.	3.	Additional resistance can be added through slip rings to improve starting torque.
4.	No wear and tear parts like brushes and slip rings.	4.	Carbon brushes are mounted using brush holders to make contact with slip rings. Due to wear and tear the slip rings and carbon brushes has to be replaced periodically.
5.	Rugged construction and cost lesser than wound rotor machine.	5.	Cost higher than squirrel cage rotor machine.
6.	Good power factor, higher pull-out torque, efficiency and overload capacity than wound rotor machine and absence of rotor overhand reactance make this motor performance superior to that of wound rotor machine.	6.	Poor power factor, lesser pull-out torque, efficiency, and overload capacity than squirrel cage motor and presence of rotor overhand reactance make this motor performance inferior to that of squirrel cage motor.

Q8.30 *What are the advantages of slip ring rotor over cage rotor?*

The advantages of slip ring rotor are,
1. The starting torque can be varied by adding resistance to rotor.
2. The speed of the machine can be varied by injecting an emf through slip rings to the rotor.

Q8.31 *What are the ranges of efficiency and power factor in induction motor?*

(AU, Apr' 17, 2 M)
(PTU, May' 17, 2 M)

Squirrel cage motors

Efficiency   = 0.72 to 0.91

Power factor = 0.66 to 0.9

Slip ring motors

Efficiency   = 0.84 to 0.91

Power factor = 0.7 to 0.92

The ISI specifications says that the product of efficiency and power factor shall be in the range of 0.83 to 0.88.

Q8.32 *How the induction motor can be designed for best power factor?*

For best power factor the pole pitch, τ is choesn such that, $\tau = \sqrt{0.18\,L}$.

Q8.33 *What are the different types of stator windings in induction motor?*

The different types of stator windings are mush winding, lap winding and wave winding.

Q8.34 *What is integral slot winding and fractional slot winding?*

In integral slot winding the total number of slots is chosen such that the slots per pole is an integer. The integer should also be a multiple of number of phases.

In fractional slot winding the total number of slots is chosen such that the slots per pole is not an integer.

Q8.35 *What is full pitch and short pitch or chording?*

When the coil span is equal to pole pitch (180 °e). The winding is called full pitched winding.

When the coil span is less than the pole pitch (180 °e). The winding is called short pitched or chorded.

Q8.36 Why short chorded windings are employed in induction motor?   *(HTU, Dec' 18, 2 M)*

For short chorded windings the length of mean turn will be lesser than the full pitch coils. Hence it results in reduction of copper. Also the short chorded windings eliminates certain harmonic magnetic fields.

Q8.37 Where mush winding is used?

The mush winding is used in small induction motors of ratings less than 5 *HP*.

Q8.38 Write the expression for length of mean turn of stator winding.

Length of mean turn of stator, $L_{mts} = 2L + 2.3\,\tau + 0.24$

where, L = Length of stator and τ = Pole pitch.

Q8.39 What type of slots are preferred in induction motor?

Semienclosed slots are preferred for induction motor. It results in less air-gap contraction factor giving a small value of magnetizing currents, low tooth pulsation loss and much quieter operation (less noise).

Q8.40 What is slot space factor?

The slot space factor is the ratio of conductor (or copper) area per slot and slot area. It gives an indication of the space occupied by the conductors and the space available for insulation. The slot space factor for induction motor varies from 0.25 to 0.4.

Q8.41 What is the minimum value of slot pitch of a three-phase induction motor?   *(AU, Nov' 17, 2 M)*

The minimum value of slot pitch in three-phase induction motor is 15 *mm*.   *(AKTU, Dec' 19, 5 M)*

Q8.42 What are the factors to be considered for selecting the number of slots in induction machines stator?

The factors to be considered for selecting the number of slots are tooth pulsation loss, leakage reactance, magnetizing current, iron loss and cost. Also the number of slots should be multiple of slots per pole per phase for integral slot winding.

Q8.43 What are the criteria used for the choice of number of slots of an induction machine?

The factors to be considered for the choice of number of stator slots of an induction machine are slot loading, slot pitch, type of winding and harmonic torques.

Slot loading:	Slot loading should not exceed 750 *amp.cond.*
Slot pitch:	The slot pitch should lie between 15 *mm* to 25 *mm*.
Type of winding:	1. For integral slot winding the stator slots should be a multiple of slots per pole per phase.
	2. For double layer winding, the conductors per slot should be even.
Harmonic Torque:	Certain combinations of stator and rotor slots give rise to harmonic torques which results in crawling and cogging. To avoid these undesirable effects the difference between stator and rotor slots should not be equal to 0, ±1, ±2, ±p, ±2p, ±3p, ±5p, ±(p ± 1), ±(p ± 2).

Q8.44 Which part of induction motor has maximum flux density? What is the maximum value of flux density in that part?

The teeth of the stator and rotor core will have maximum flux density. The maximum value of flux density in the teeth is 1.7 Wb/m^2.

                                                                                          (AKTU, Dec' 19, 5 M)

Q8.45 What are the advantages and disadvantages of large air-gap length in induction motor?

Advantage : A large air-gap length results in higher overload capacity, better cooling, reduction in noise and reduction in unbalanced magnetic pull.

Disadvantage : The disadvantage of large air-gap length is that it results in high value of magnetizing current and hence poor power factor.

Q8.46 What are the factors to be considered for estimating the length of air-gap in induction motor?

The following factors are to be considered for estimating the length of air-gap.

1. Power factor   4. Unbalanced magnetic pull
2. Overload capacity   5. Cooling
3. Pulsation loss   6. Noise

Q8.47 What happens if the air-gap of an induction motor is doubled?   *(AU, Apr' 18, 2 M)*

If the air-gap of an induction motor is doubled then the mmf and magnetizing current approximately doubles. (Because in induction motor the air-gap requires large mmf when compared to rest of the magnetic circuit).

Also the increase in air-gap length increases the overload capacity, offers better cooling, reduces noise and reduces unbalanced magnetic pull.

Q8.48 What is the effect of air-gap on the leakage reactance of an induction motor?

The length of air-gap affects the value of zigzag leakage reactance which forms a large part of total leakage reactance. If the length of air-gap is large then the zig zag leakage flux will be less and the leakage reactance will be less.

Q8.49 List out the methods to improve the power factor of induction motor.

The power factor of the induction motor can be improved by reducing the magnetizing current and leakage reactance.

The magnetizing current can be reduced by reducing the length of air-gap. The leakage reactance can be reduced by reducing the depth of stator and rotor slots, by providing short chorded winding and reducing the overhang in stator winding.

Q8.50 Why the air-gap of an induction motor is made as small as possible?

The mmf and the magnetizing current are primarily decided by length of air-gap. If air-gap is small then mmf and magnetizing current will be low, which in turn increase the value of power factor. Hence by keeping small air-gap, high power factor is achieved.

Q8.51 Write the formula for air-gap in case of three-phase induction motor in terms of length and diameter.

$$\text{The length of air-gap, } l_g = 0.2 + 2\sqrt{DL} \text{ in } mm$$

where D and L are expressed in metre.

Q8.52 Discuss the relative merits of open and closed slots for induction motor rotor.

The closed slots will not increase reluctance of air-gap and has lesser noise but it has difficulty in casting the rotor bars.

The open slots increase the reluctance of air-gap and has high noise but it offers flexibility in casting rotor bars.

Q8.53 List the undesirable effects produced by certain combination of rotor and stator slots.

The following problems may develop in induction motor with certain combination of rotor and stator slots.

1. The motor may refuse to start (cogging).
2. The motor may run at subsynchronous speed (crawling).
3. Severe vibrations may develop and the noise will be excessive.

Q8.54 What are the different types of windings used for the rotor of induction motor?

The different types of windings employed in induction motor rotor are mush winding and double layer bar type winding.

Q8.55 What is harmonic induction torque and harmonic synchronous torque?

Harmonic induction torques are torques produced by harmonic fields due to stator winding and slots.

Harmonic synchronous torques are torques produced by the combined effect of same order of stator and rotor harmonic fields.

Due to both the harmonic torques the machine may crawl, but there will be difference in the crawling speeds. In case of harmonic induction torque crawling speed is slightly lesser than the subsynchronous speed. But in case of harmonic synchronous torque the crawling speed is same as that of subsynchronous speed and this is also called synchronous cusps.

Q8.56 What is crawling and cogging?

Crawling is a phenomena in which the induction motor runs at a speed lesser than subsynchronous speed. Cogging is a phenomena in which the induction motor refuse to start.

Q8.57 Explain the phenomena of cogging. *(AU, Nov' 18, 2 M)*

When the number of stator and rotor slots are equal, the speeds of all the harmonics produced by stator slotting coincide with the speed of corresponding, rotor harmonics. Thus harmonics of every order would try to exert synchronous torques at their corresponding synchronous speeds and the machine would refuse to start. This is known as cogging.

Q8.58 What are the methods adopted to reduce harmonic torques?

The methods used for reduction or elimination of harmonic torques are chording, integral slot winding, skewing and increasing the length of air-gap.

Q8.59 What is skewing?

Skewing is twisting either the stator or rotor core. The motor noise, vibrations, cogging and synchronous cusps can be reduced or even entirely eliminated by skewing.

In order to eliminate the effect of any harmonic, the rotor bars should be skewed through an angle so that the bars lie under alternate harmonic poles of the same polarity or in other words the bars must be skewed through two harmonic pole pitches.

Q8.60 What will be the effects on the performance of a 3-phase squirrel cage induction motor when copper end rings of the cage rotor are replaced by aluminium?

The aluminium has higher resistance than copper and so the rotor resistance may increase. Due to increase in rotor resistance the starting torque and rotor copper loss will increase but the efficiency will decrease.

Q8.61 Define dispersion coefficient.

The dispersion coefficient is defined as the ratio of magnetizing current to short circuit current.

$$\text{Dispersion coefficient, } \sigma = \frac{I_m}{I_{sc}}$$

where, I_m = Magnetizing current ; I_{sc} = Short circuit current

Q8.62 List the leakage fluxes associated with three-phase induction motors.

The different types of leakage fluxes associated with induction motors are the following,

1. Slot leakage flux
2. Zigzag leakage flux
3. Harmonic or differential leakage flux
4. Peripheral leakage flux.
5. Tooth top leakage flux
6. Overhang leakage flux
7. Skew leakage flux

Q8.63 What is the condition for obtaining the maximum torque in case of 3-phase induction motor?

The maximum torque occurs in induction motor when rotor reactance is equal to rotor resistance.

Q8.64 How the dimensions of induction generator differs from that of an induction motor?

The construction of induction generator resembles that of a motor. The differences in construction are listed below.

1. The rotor is designed to have low loss and low reactance in order to achieve high efficiency and high power factor.

2. The rotor should be designed to withstand runaway speed, which will be three times the normal speed.

3. The diameter will be small and length will be large so as to make the rotor to withstand the stresses developed during runaway speeds.

4. The rotor bars are strongly wedged and special attention is given to end-ring connections, in order to withstand stresses during high speeds.

5. The stator may have heaters to keep the insulation dry when the machine is idle.

Q8.65 State the rules for selecting rotors slots of squirrel cage machines.   *(AU, Apr' 17, 2 M)*

The following general rules should be followed concerning the choice of rotor slots for squirrel cage machines.

1. Number of stator slots should never be equal to rotor slots. Satisfactory results are obtained when S_r is 15 to 30% larger or smaller than the S_s.

2. The difference $(S_s - S_r)$ should not be equal to $\pm p$, $\pm 2p$ or $\pm 5p$ to avoid synchronous cusps.

3. The difference $(S_s - S_r)$ should not be equal to $\pm 3p$ for 3-phase machine to avoid magnetic locking.

4. The difference $(S_s - S_r)$ should not be equal to $\pm 1, \pm 2, \pm(p \pm 1)$ or $\pm(p \pm 2)$ to avoid noise and vibrations.

Summarising, $(S_s - S_r)$ should not be equal to $0, \pm p, \pm 2p, \pm 3p, \pm 5p, \pm 1, \pm 2, \pm(p \pm 1), \pm(p \pm 2)$.

8.15 EXERCISES

I.   Fill in the blanks

1. The major parts of three-phase induction motor are .... and ....

2. The stator core is made of laminated sheet steel of thickness ....

3. The two types of rotor of three-phase induction motor are .... and .... rotor.

4. For best power factor the pole pitch, τ = ...

5. The range of slot pitch in induction motor is ....

6. The no-load current of three-phase induction motor is approximately .... of full load current.

7. The starting torque of squirrel cage induction motor can be increased by providing ....

8. The inner cage of a double cage induction motor will have .... reactance and .... resistance.

9. The cost of a squirrel cage induction motor is .... a slip ring motor of same rating.

10. If an induction motor runs at a stable speed less than the rated speed then it is called ....

11. If the motor refuse to start when the supply is switched ON, then it is called ....

12. The twist provided in the rotor is called ....

13. In induction motors, low power factor is due to .... in the value of magnetizing current.

14. If specific magnetic loading is increased then the core loss .... and efficiency ....

15. Space factor is the ratio of .... to total slot area.

16. In high .... machines, the space factor is less due to .... thickness of insulation.

17. In rotating machines if .... is the linear dimension then the output is proportional to ....

18. In rotating machines, if .... is the linear dimension then the losses are proportional to ....

19. The .... loss is proportional to frequency and .... loss is proportional to the square of the frequency.

Answers

1.	stator, rotor	11.	cogging
2.	0.5 *mm*	12.	skew
3.	squirrel cage, wound	13.	increase
4.	$\sqrt{0.18\,L}$	14.	increases, decreases
5.	15 to 25 *mm*	15.	bare conductor area
6.	25 to 40%	16.	voltage, higher
7.	double cage rotor (or deep bar rotor)	17.	x, x^4
8.	high, low	18.	x, x^3
9.	less than	19.	hysteresis, eddy current
10.	crawling		

II.   State whether the following statements are True/False

1. The cage rotor is electrically equivalent to slip ring rotor.

2. The starting torque of slip ring motor cannot be varied.

3. Semienclosed slots are popularly used in induction motors.

4. The air-gap of induction motor is kept higher to improve the power factor.

5. The overload capacity of induction motor increases with increase in air-gap length.

6. The cogging in induction motor can be reduced by skewing either the rotor or stator .

7. The size of three-phase induction motor will be less than single-phase motor of same rating.

8. The slip ring induction motor can be started only with rotor resistance starter.

9. The short chorded winding is employed in induction machine to reduce harmonics.

10. The cage rotor adopt itself for any number of phases and poles.

11. The winding factor is accounted only for induction machines.

12. In induction machines, higher value of magnetizing current is preferred in order to achieve good power factor.

13. Core loss is observed only when the iron parts are subjected to alternating magnetization.

14. Air-gap flux density is directly proportional to frequency of flux reversals.

15. Higher values of **ac** are used for machines having round conductors, because space factor is less for them.

Answers

1. True	6. True	11. True
2. False	7. True	12. False
3. True	8. False	13. True
4. False	9. True	14. False
5. True	10. True	15. False

III   Unsolved problems

E8.1 Determine the approximate diameter and length of stator core, the number of stator slots and the number of conductors for a 20 kW, 400 V, 3ϕ, 4 pole, 1200 rpm, delta connected induction motor. B_{av} = 0.5 Wb/m^2, η = 0.82, ac = 26000 $amp.cond./m$, p.f = 0.8, L/τ = 1, double layer stator winding.

$$(D = 0.22 \ m \ ; L = 0.18 \ m \ ; S_s = 36, T_s = 120, Z = 720).$$

E8.2 Estimate the main dimensions, air-gap length, stator slots, stator turns per phase and cross sectional area of stator and rotor conductors of a 3-phase, 110 kW, 3300 V, 50 Hz, 10 poles, star connected induction motor. B_{av} = 0.48 Wb/m^2, ac = 28000 $amp.cond./m$, L/τ = 1.25, η = 0.9, p.f. = 0.86.

$$(D = 0.635 \ m \ ; L = 0.25 \ m, S_s = 90, Z_{ss} = 24, S_r = 122, T_s = 360 \ ;$$
$$a_s = 6.22 \ mm^2 \ ; \ a_b = 62.35 \ mm^2, a_e = 242 \ mm^2)$$

E8.3 Design a cage rotor for a 18.8 HP, 3ϕ, 440 V, 50 Hz, 1000 rpm, induction motor having a full load η of 0.86 and a power factor of 0.86. D = 0.25 m, L = 0.14 m, Z_{ss} = 16, S_s = 54. Assume missing data if any.

$$(D_r = 0.2488 \ mm \ ; l_g = 0.6 \ mm \ ; L_r = 0.14 \ m,$$
$$S_r = 51, \delta_b = 5 \ A/mm^2 \ ; \ a_b = 72 \ mm^2 \ ; \ a_e = 194 \ mm^2)$$

E8.4 Design the main dimensions of 37 kW, 3-phase 50 Hz, 415 V, 720 rpm slip ring induction motor. B_{av} = 0.48 Wb/m^2, ac = 26000 $amp.cond./m$, full load efficiency = 85%, full load power factor = 0.82 and winding factor = 0.95. Take L/τ = 1.

$$(p = 8 \ ; D = 0.4353 \ m \ ; L = 0.171 \ m)$$

E8.5 Find the suitable diameter and length for a 3.7 kW, 440 V, 50 Hz, cage type induction motor to run at a full load speed of 1440 rpm. B_{av} = 0.45 Wb/m^2, ac = 440 $amp.cond./cm$, full load efficiency = 85% and full load power factor = 0.85. Take L/τ = 1.5.

$$(p = 4; D = 0.094 \ m \ ; L = 0.11 \ m)$$

E8.6 A 13 kW, 3-phase, 6 pole, 50 Hz, 300 V, star connected induction motor has 52 stator slots, each containing 11 conductors. Calculate the value of bar and end ring currents. The number of rotor bars is 62. The machine has an efficiency of 90% and a power factor of 0.8. The rotor mmf may be assumed to be 85% of stator mmf. Also find the bar and the end ring section if the current density is 5 A/mm^2.

$$(I_b = 27.40 \ A, I_e = 90.12 \ A, a_b = 6 \ mm^2, a_e = 18 \ mm^2)$$

E8.7 Calculate the specific electric and magnetic loading of 50 HP, 2000 V, 3-phase, 50 Hz, 6 pole, star connected, flame proof induction motor having stator core length = 0.52 m and stator bore = 0.6 m. Turns per phase = 262. Assume full load efficiency as 0.911 and pf as 0.9.

$$(ac = 11000 \ amp.cond./m, B_{av} = 0.13 \ Wb/m^2)$$

E8.8 A 11 kW, 300 V, 3-phase, 50 Hz, 4 pole induction motor has a diameter of 0.25 m and the length of core 0.18 m. The number of stator slot is 68 with 16 conductors per slot. The stator is delta connected. Calculate the value of magnetizing current per phase if the length of air-gap is 0.4 m. The gap contraction factor is 1.2. Assume the mmf required for the iron parts to be 25 percent of the air-gap mmf. Take stator winding factor as 0.955.

$$(I_m = 1.37 \ A)$$

> Chapter 9

DESIGN OF SYNCHRONOUS MACHINE

List of symbols

Symbol	Meaning	Unit
AT_a	Armature mmf per pole	AT
AT_{fo}	Field mmf on no-load	AT
AT_f	Field mmf required to induce voltage per phase on no-load	AT
AT_{fl}	Field mmf on load or Full load field mmf	AT
AT_{fl_TA}	Full load field mmf of turbo alternator	AT
AT_c	mmf for armature core	AT
AT_g	mmf for air-gap	AT
AT_t	mmf for armature teeth	AT
AT_p	mmf for poles	AT
AT_y	mmf for yoke	AT
A_{d_ring}	Area of cross-section of end ring of damper winding	mm^2
A_p	Area of cross-section of pole body	mm^2
A_d	Total area of damper bars per pole	mm^2
a_a	Cross-section of stator conductor	mm^2
a_f	Area of field conductor	mm^2
a_d	Area of cross-section of each damper bar	mm^2
ac	Specific electric loading	$amp.cond./m$
B_p	Flux density in pole body	Wb/m^2
B_g	Maximum flux density in the air-gap	Wb/m^2
B_{av}	Specific magnetic loading	Wb/m^2
b	Pole arc	m
b_p	Width of pole body	m
C_o	Output coefficient	$kVA/m^3\text{-}rps$

Symbol	Meaning	Unit
C_l	Leakage coefficient	$kVA/m^3\text{-}rps$
D	Diameter of stator bore	m
D_r	Diameter of rotor	m
d_d	Diameter of damper winding conductors	mm
d_f	Depth of field winding	mm
d_{ss}	Depth of stator slot	mm
E_f	Voltage across each field coil	V
E_o	Induced emf per phase on open circuit	V
E_s	Stator emf per phase	V
E_g	Induced emf on load	V
f	Frequency	Hz
h_p	Height of pole body	m
h_f	Height of field winding	m
h_{fc}	Height of conductor	m
h_s	Height of pole shoe	m
I_f	Field current	A
I_s	Armature or Stator current per phase	A
I_z	Current through a stator conductor	A
K_c	Chording factor	-
K_d	Distribution factor	-
K_f	Form factor	-
K_g	Total gap contraction factor	-
K_{wr}	Rotor winding factor	-
K_{ws}	Stator winding factor	-
L	Length of stator core	m
L_p	Length of pole body	m
L_d	Length of damper bars	m
L_{mtf}	Length of mean turn of field coil	m
l_g	Length of air-gap	mm
N_d	Number of damper bars over a pole	-
n_s	Synchronous speed	rps
P_d	Reduction factor to find field mmf equivalent to armature mmf	-
p	Number of ploes	-

Symbol	Meaning	Unit
Q	kVA rating	kVA
Q_f	Total loss dissipated in field winding	W
q	Slots per pole per phase	-
q_f	Loss per unit surface of field winding	W/m^2
R_a	Armature resistance per phase	Ω
R_f	Resistance of field coil	Ω
S_r	Number of rotor slots	-
S_s	Total stator slots	-
S_f	Stacking factor	-
S_{fc}	Copper space factor of field winding	-
SCR	Short Circuit Ratio	-
T_f	Number of field turns	-
T_r	Number of rotor turns per phase	-
T_s	Stator turns per phase	-
V_a	Peripheral speed	m/s
V_f	Exciter voltage	V
y_{sr}	Rotor slot width or pitch	mm
X_d	Direct axis reactance	Ω
X_s	Synchronous reactance per phase	Ω
y_{sd}	Slot pitch for damper winding	-
y_{ss}	Stator slot pitch	-
Z	Number of stator conductor	-
Z_f	Number of field conductor	-
Z_{rs}	Conductor per rotor slot	-
Z_{ss}	Conductors per stator slot	-
η	Efficiency	-
ρ	Resistivity of copper	Ω/m
ρ_d	Reduction factor to find field mmf equivalent to armature mmf	Ω/m
δ	Current density	A/mm^2
δ_a	Current density in armature conductor	A/mm^2
δ_d	Current density in damper bar	A/mm^2
ϕ	Flux per pole	Wb
ϕ_p	Flux in pole body	Wb
τ	Pole pitch	m
τ_c	Effective span of coils	m

9.1 TYPES OF SYNCHRONOUS MACHINE

The synchronous machines may be classified as salient pole machines and cylindrical rotor machines depending upon the type of construction used for the rotor.

Salient pole machines are driven by water wheels or diesel engines. They operate at low speeds and so large number of poles are required to produce desired frequency. This type of machine has projecting poles and field coils are mounted on the poles.

Cylindrical rotor machines are driven by steam turbines and gas turbines which run at very high speeds. They have slots on the outer periphery of smooth cylindrical rotor. The field conductors are placed on this slots.

Constructional elements of salient pole synchronous machine

Stator	- Frame	**Rotor**	- Field pole
	- Armature core		- Pole shoe
	- Armature winding		- Field winding
			- Damper winding

Constructional elements of cylindrical rotor synchronous machine

Stator	- Frame	**Rotor**	- Solid rotor
	- Armature core		- Field conductors or bars
	- Armature winding		

Synchronous machines operating on general power supply networks may be divided into the following categories.

Hydro-generators: The prime mover is a water wheel and driven at 100 to 1000 *rpm*. Available upto a capacity of 750 *MW*. The type of prime mover for this alternators depends on available water head. Pelton wheel is used for water heads of 400 *m* and above. Francis turbine is used for water heads upto 380 *m*. Kaplan turbine is used for water heads upto 50 *m*. The peripheral speed of the rotor is limited to 80 *m/s*.

Turbo-alternators: The prime mover is steam turbine or gas turbine which run at 3000 *rpm*, manufactured upto a capacity of *1000 MW*. Turbo alternators are characterised with long axial length and short diameter. The high speed of the rotor limits the diameter of the rotor to about 1.2 *m* giving a peripheral speed of about 175 *m/s*.

Engine driven: The prime mover is internal combustion engine (diesel or petrol) with speeds upto 1500 *rpm* and manufactured upto a rating of 20 *MW*.

Motors: Motors are manufactured in wide ranging capacity and they are provided with damper windings.

Compensators: Compensators are synchronous motors run at leading power factors to supply the reactive power to a transmission network. They are manufactured upto a rating of 100 *MVAR* and runs upto 3000 *rpm*.

Runaway speed of alternator

The runaway speed is defined as the speed which the prime mover would have, if it is suddenly unloaded, when working at its rated load.

Steam turbines are equipped with quick acting overspeed governors which are set to trip at 1.1 times the rated speed. Hence turbo alternators are designed for 1.25 times the rated speed. But in waterwheel the gate valves (valves which regulate water flow) are not fast acting and so the machine attains higher value of runaway speed. The following are the runaway speeds for various water wheels with full gate opening.

Pelton wheel   -   1.8 times rated speed

Francis turbine   -   2 to 2.2 times the rated speed

Kaplan turbine   -   2.5 to 2.8 times the rated speed

Thus the salient pole machines are designed to withstand mechanical stresses encountered at runaway speeds.

9.1.1 Main Dimensions of Synchronous Machine   *(KU, Dec' 19, 6 M) (VTU, Dec' 19, 6 M)*

The main dimensions of synchronous machine are diameter of stator bore, D and length of stator core, L. The main dimensions of the two types of synchronous machine are shown in Fig. 9.1.

In Fig. 9.1,

D_r = Diameter of rotor

$D_r = D - 2\, l_g$

where, l_g = Length of air-gap

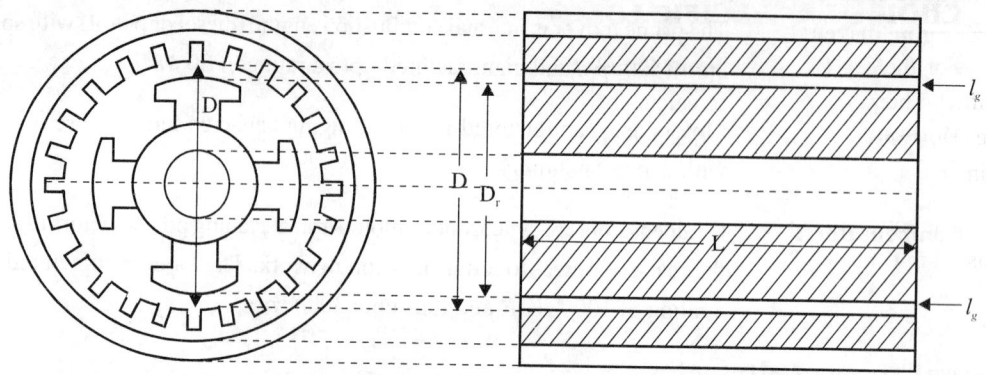

Fig. a: *Main dimensions of salient pole synchronous machine.*

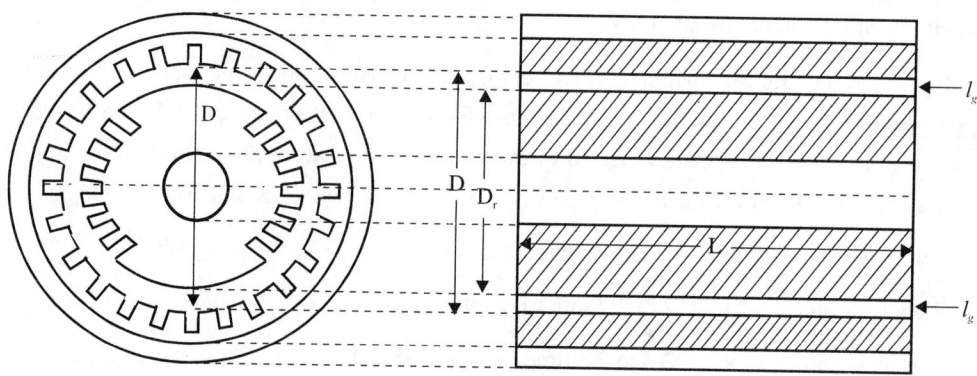

Fig. b: *Main dimensions of cylindrical rotor synchronous machine.*

Fig. 9.1: *Main dimensions of synchronous machine.*

9.2 OUTPUT EQUATION

In case of synchronous machine the equation for kVA rating is considered as output equation.

The kVA rating refer to kVA output for alternators and kVA input for synchronous motors.

kVA rating, $Q = C_o D^2 L n_s$ .....(9.1)

Output coefficient, $C_o = 11 B_{av} \, \textbf{ac} \, K_{ws} \times 10^{-3}$

where,   D   =   Diameter of stator bore B_{av} = Specific magnetic loading

   L   =   Length of stator core   **ac** = Specific electric loading

   n_s =   Synchronous speed K_{ws} = Stator winding factor

Note : *The output equation of induction machine and synchronous machine are same. Therefore, for derivation of output equation, refer chapter-8.*

Also, kVA rating can be calculated from line and phase voltage and current.

kVA rating, $Q = \sqrt{3} \, V_L I_L \times 10^{-3} = 3 \, E_s I_s \times 10^{-3}$

9.3 CHOICE OF SPECIFIC LOADINGS

For the design of synchronous machine, the kVA rating, speed, power factor and efficiency are specified. Therefore, in order to calculate the value of D^2L, we must evaluate the output coefficient. The value of output coefficient depends upon the choice of specific electric loading (**ac**) and specific magnetic loading (B_{av}).

The equations for total and specific loadings of synchronous machine are same as that of DC machines discussed in Chapter-6, Section 6.4.

$\therefore$ Total magnetic loading $= p\phi$

Total electric loading $= I_z Z$

Specific magnetic loading, $B_{av} = \dfrac{p\phi}{\pi DL}$

Specific electric loading, $\mathbf{ac} = \dfrac{I_z Z}{\pi D}$

9.3.1 Choice of Specific Magnetic Loading

The choice of magnetic loading depends on,

1. Iron loss
2. Voltage rating
3. Transient short circuit current
4. Stability
5. Parallel operation.

A higher value of magnetic loading results in increased iron loss with consequent decrease in efficiency and increase in temperature rise. A lower value of magnetic loading should be used in high voltage machines to avoid excessive values of flux density in teeth and core.

A high value of magnetic loading results in decrease in the leakage reactance, which results in higher short circuit currents. Hence to limit initial short circuit currents, a low value of magnetic loading should be chosen.

The maximum power which a cylindrical rotor machine can deliver under steady state conditions is $P_{max} = EV / X_s$ where E is the induced emf, V is the terminal voltage and X_s is the synchronous reactance. Hence the maximum power or steady state stability limit of a machine is inversely proportional to its synchronous reactance. A high value of flux density results in lesser turns per phase and so lesser reactance. Therefore, the use of high magnetic loading increases the steady state stability limit.

The satisfactory parallel operation of synchronous generators depends on the synchronising power. Higher values of synchronising power results in better stability of the machines in parallel. The synchronising power is inversely proportional to reactance. Therefore machines with higher magnetic loading operate satisfactorily in parallel.

The specific magnetic loading, B_{av} lies in the range of 0.52 to 0.65 Wb/m^2.

9.3.2 Choice of Specific Electric Loading   *(AKTU, Dec' 19, 4 M)*

The choice of specific electric loading depends on,

1. Copper loss
2. Temperature rise
3. Voltage rating
4. Synchronous reactance
5. Stray load losses.

A high value of **ac** gives higher copper loss resulting in lower efficiency and higher temperature rise. The value of **ac** used also depends on cooling coefficient. A higher value of **ac** can be used for low voltage machines since the space required for insulation is small. A high value of **ac** results in higher value of synchronous reactance. Hence a machine designed with high value of **ac** has the following characteristics.

1. Poor voltage regulation
2. Low current under short circuit conditions
3. Low value of steady state stability limit
4. Low value of synchronising power.

The stray load losses are also increases steeply with an increase in **ac**.

The value of **ac** lies in the range of 20000 to 40000 *amp.cond./m* for salient pole machines. For turbo alternator it is in the range of 50000 to 75000 *amp.cond./m*.

9.4 DESIGN OF SALIENT POLE MACHINES   *(AU, Apr' 17, 6 M) (AU, Apr' 18, 8 M)*

9.4.1 Design of Main Dimensions

The main dimensions of salient pole machines are D and L.

where, D refer to inner diameter of stator bore and L refer to length of stator.

The cross-section of salient pole synchronous machine is shown in Fig. 9.1.

The two types of poles used in salient pole machines are Round poles and Rectangular poles as shown in Fig. 9.2.

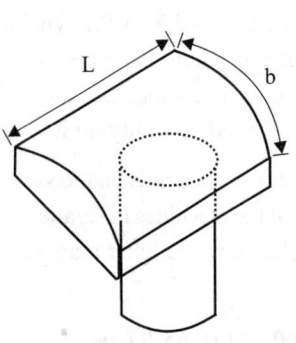 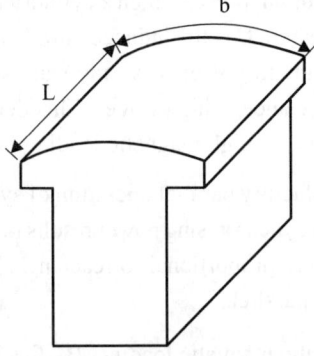

        ***Fig. a:** Round Pole.*   ***Fig. b:** Rectangular Pole.*

***Fig. 9.2:** Types of salient poles.*

The choice of diameter depends on the type of pole and the permissible peripheral speed. When round poles are used the ratio of pole arc to pole pitch, b/τ is between 0.6 to 0.7. For low values of b/τ it is possible to use round poles with square pole shoes. Hence for round poles with square pole shoes.

Length of pole, L = Width of pole shoe, b_s

∴ $b/\tau = L/\tau = 0.6$ to 0.7

Here, Length of pole is also equal to length of stator core.

The ratio of pole arc to pole pitch varies between 1 to 5 for rectangular poles. However, this ratio should not exceed 3 for normal machines otherwise the design of field system becomes uneconomical.

∴ For rectangular poles,

$L/\tau = 1$ to 5

The value of allowable peripheral speeds depends on the type of pole attachment. For bolted pole construction the peripheral speed is 50 *m/s*. For dovetail and T-head construction, the peripheral speed is 80 *m/s*.

The values of main dimensions D and L can be calculated using output equation and choice of the ratio, L/τ. The product $D^2 L$ is determined from kVA rating, specific electric and magnetic loadings.

kVA rating, $Q = C_o D^2 L n_s$

$$\therefore D^2 L = \frac{Q}{C_o n_s} \qquad \qquad .....(9.2)$$

where, $C_o = 11 B_{av}$ **ac** $K_{ws} \times 10^{-3}$

Assume a suitable value for L/τ.

Pole pitch, $\tau = \dfrac{\pi D}{p}$

Let, $\dfrac{L}{\tau} = k$

where, k = Value of L/τ.

$$\therefore L = k \tau = k \frac{\pi D}{p} \qquad \qquad .....(9.3)$$

where, k = 0.6 to 0.7 - For round poles

= 1 to 5 - For rectangular poles

The equations (9.2) and (9.3) are used to solve the values of D and L.

9.4.2 Short Circuit Ratio (SCR) *(UTU, Dec' 13, 10 M) (KU, Dec' 19, 8 M) (VTU, Dec' 19, 8 M)*

Short Circuit Ratio (SCR) is defined as the ratio of field current required to produce rated voltage on open circuit (OC) to field current required to circulate rated current at short circuit (SC).

$$SCR = \frac{\text{Field current for OC voltage}}{\text{Field current for SC current}} \qquad \qquad .....(9.4)$$

From the open circuit characteristics (OCC) and short circuit characteristics (SCC) shown in Fig. 9.3 we can write,

$$SCR = \frac{OF_o}{OF_s}$$

OF_o = Per unit (pu) field current required to develop rated voltage on open circuit (OC).

OF_s = Per unit (pu) field current required to develop rated current on short circuit (SC).

In Fig. 9.3, both the current axis are in per unit (pu). Hence, from the characteristics shown in Fig. 5.3 we can conclude that,

$$OF_o = CF_o$$

$$OF_s = BF_s = AF_o$$

Note : Here in pu current axis OF_s = 1 pu and BF_s = 1 pu. Hence OF_s = BF_s. Similarly OF_o and CF_o corresponds to same value of pu current when referred to current axis.

$$\therefore SCR = \frac{OF_o}{OF_s} = \frac{CF_o}{BF_s} = \frac{CF_o}{AF_o} = \frac{1}{AF_o / CF_o} = \frac{1}{\dfrac{\text{pu volt on open circuit}}{\text{pu SC current corresponding to pu volt}}}$$

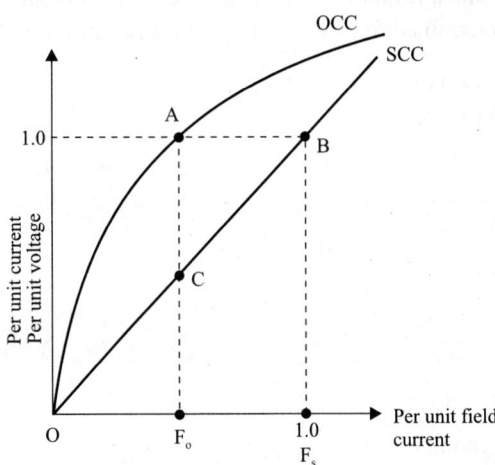

Fig. 9.3: OCC and SCC characteristics of alternators.

We know that,

$$\text{Direct axis reactance, } X_d = \frac{\text{pu volt}}{\text{pu SC current}}$$

$$\therefore SCR = \frac{1}{X_d} \qquad\qquad\qquad .....(9.5)$$

Thus the short circuit ratio (SCR) is defined as the reciprocal of synchronous reactance, X_d in per unit (pu). For turbo alternators SCR is normally between 0.5 to 0.7. For salient pole hydro electric generators SCR varies from 1.0 to 1.5.

The choice of SCR affects the performance of the synchronous machine. With higher values of SCR the machine has higher stability limit and low value of regulation. But it gives high value of short circuit current and results in longer air-gap. With long air-gap, the mmf is higher and hence field system will be large and the machines become costlier. The modern design trend is to design the machine with low SCR.

9.4.3 Armature or Stator Design

In AC machine the stator is also known as armature. The armature design of synchronous machines are almost similar to stator design of induction motors except the few differences in stator slot dimensions, stator core flux density and stator winding conductor current density.

Armature winding

The windings used in synchronous machines may be single layer or double layer type.

Modern practice, is to employ double layer wave or lap winding. The coil span for the winding are chosen such that harmonics are reduced. The highest amplitude harmonics in the flux distribution curve of salient pole generators are likely to be 5th and 7th harmonic. The maximum reduction of these harmonics is given by a coil span of 8.33% lesser than the pole pitch. Therefore, this coil span should not be equal to ratio of number of slots to number of poles. The coil span used should be as near to this ratio as possible.

Turns per phase

The flux per pole, $\phi = B_{av} \tau L$ .....(9.6)

Let, a = Number of parallel paths per phase

If a = 1, then

Turns per phase, $T_s = \dfrac{E_s}{4.44 \, \phi \, f \, K_{ws}}$ .....(9.7)

If a > 1 then,

Turns per phase, $T_s = \dfrac{E_s \times a}{4.44 \, \phi \, f \, K_{ws}}$ .....(9.8)

Length of mean turn

The approximation length of mean turn of the winding of synchronous machine can be estimated from following empirical relationships.

Length of mean turn of stator, $L_{mts} = 2L + 2.5\tau + 0.06kV + 0.2$

where, kV = kV voltage rating of synchronous machine

Here the values of L and τ are expressed in *m*.

Area of cross-section of armature conductor

Armature or stator current per phase, $I_s = \dfrac{kVA}{3 \, E_s \times 10^{-3}}$ $\boxed{kVA = 3 \, E_s \, I_s \times 10^{-3}}$ .....(9.9)

If number of parallel path per phase, a = 1, then,

Current through a conductor, $I_z = I_s$

If number of parallel path per phase, a > 1, then

Current through a conductor, $I_z = \dfrac{I_s}{a}$ .....(9.10)

The area of cross-section of armature conductor can be estimated by assuming a suitable current density.

The range of current density, $\delta_a = 3$ to $5\ A/mm^2$

$\therefore$ Area of cross-section of armature conductor, $a_a = \dfrac{I_z}{\delta_a}$ .....(9.11)

9.4.4 Armature Resistance

The resistance, R of a conductor of a length, l and area of cross-section, a is given by,

$$R = \frac{\rho l}{a}$$

where, R = Resistivity of the material of the conductor.

In armature of synchronous machine the total length of armature winding is given by product of number of turns per phase, T_s and length of mean turn of the armature winding, L_{mts}.

$\therefore$ DC resistance of armature conductor per phase $\left.\right\}$ $R_{a\_dc} = \dfrac{\rho\, T_s\, L_{mts}}{a_a}$

where, ρ = Resistivity of copper

a_a = Area of cross-section of armature conductor

The resistance of armature calculated by the above equation is called DC resistance, because the equation does not account the eddy current losses in the conductor. The AC resistance of the armature can be obtained from DC resistance by multiplying the DC resistance by average eddy current loss factor, k_{e_av}. The value of k_{e_av} for critical depth of slot is 1.33.

$\therefore$ AC resistance of armature conductor per phase $\left.\right\}$ $R_{a\_ac} = k_{e\_av} \dfrac{\rho\, T_s\, L_{mts}}{a_a}$

$$= \frac{1.33\, \rho\, T_s\, L_{mts}}{a_a}$$

Per unit value of AC resistance $\left.\right\}$ $= \dfrac{I_s\, R_{a\_ac}}{E_s}$

9.4.5 Armature Leakage Reactance

In synchronous machine the armature leakage reactance has two components. Slot leakage reactance and overhang leakage reactance. The slot leakage reactance in AC machine is discussed in Chapter-3, Section 3.15.

$\therefore$ Slot leakage reactance, $X_s = \dfrac{8\pi\, f\, T_s^2\, L\, \lambda_s}{pq}$ | From equation (3.78) |

The equation for overhang leakage reactance can be obtained from above equation by replacing LT_s by $L_o\, \lambda_o$.

$\therefore$ Overhang leakage reactance, $X_o = \dfrac{8\pi\, f\, T_s^2\, L_o\, \lambda_o}{pq}$

where, $L_o\, \lambda_o = \dfrac{k\, L_o^2}{2\sqrt{2}\ T_s}$

$$k = 0.23 \times 10^{-6} \quad - \quad \text{For concentric winding}$$

$$= 0.29 \times 10^{-6} \quad - \quad \text{For barrel winding}$$

$$\left. \begin{array}{l} \therefore \text{Stator leakage} \\ \text{reactance per phase} \end{array} \right\} X_{LS} = X_s + X_o$$

$$= \frac{8\pi\, f\, T_s^2\, f\, \lambda_s}{pq} + \frac{8\pi\, f\, T_s^2\, L_o\, \lambda_o}{pq}$$

$$= \frac{8\pi\, f\, T_s^2}{pq} (L\, \lambda_s + L_o\, \lambda_o)$$

$$\left. \begin{array}{l} \text{Per unit value of} \\ \text{leakage reactance} \end{array} \right\} = \frac{I_s\, X_{LS}}{E_s}$$

9.4.6 Armature Winding Coil Dimensions

The coils of synchronous machine armature winding are made of copper stripes or bars. Therefore, the armature winding is also called bar winding (lap or wave connected bar winding). The two types of coils used in bar winding are Single turn coils and Multi-turn coils.

Single-Turn Coil

The synchronous machine armature windings are double layer windings and so in windings with single turn coil each slot will accommodate two coil sides: Top coil side and bottom coil side.

Each coil side is a bar type conductor made from rectangular copper strips or strands with thickness 3 *mm* and width 4 *mm* to 7 *mm*. The thickness may also depend on current density and size of slot. Each strand is insulated with treated asbestos or glass roving of thickness 0.5 *mm*. The main insulation between conductor and slot is provided with an insulation of thickness 2 *mm* to 6 *mm* depending on voltage rating of the machine. Also, a separator with 2 to 5 *mm* thickness is provided between top and bottom coil side. A typical arrangement of single turn bar type conductor is shown in Fig. 9.4.

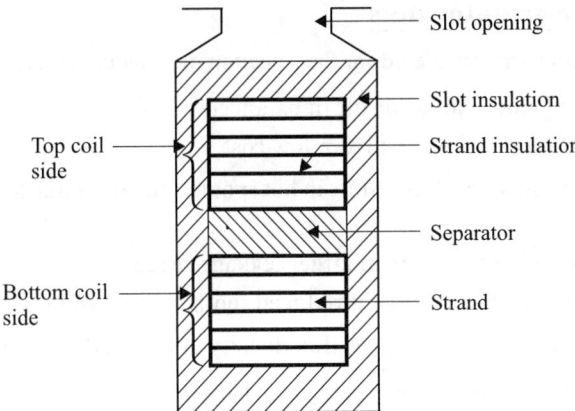

Fig. 9.4: Cross-section of slot showing arrangement of single-turn coil.

Multi-Turn Coil

In multi-turn coil, the coil sides will have multiple bar type conductor similar to that single turn coil side. Interturn insulation of thickness 2 *mm* will be provided between turns. Mostly multi-turn coil will have even number of coil sides for symmetrical arrangement of coil sides either width wise or depth wise in slots. A typical arrangement of two-turn bar type conductor is shown in Fig 9.5. In this arrangement both top and bottom coil sides have two turns.

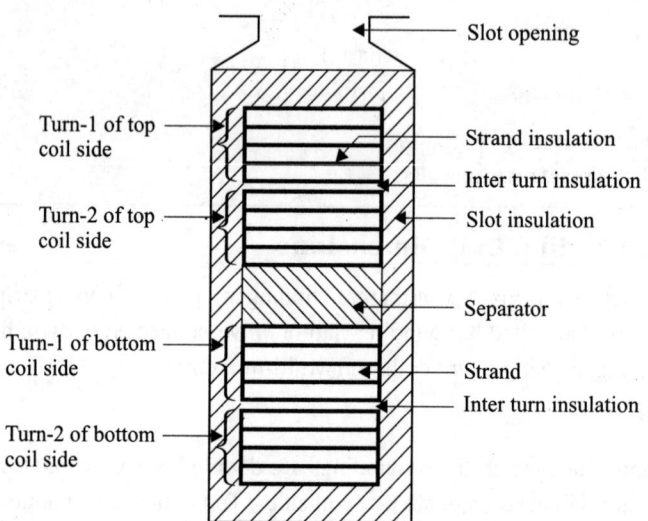

Fig. 9.5: Cross-section of slot showing arrangement of multi-turn coil.

9.4.7 Stator Core

The stator core is made of laminations of thickness 0.5 *mm*. The design of stator core involves selection of number of slots, estimation of dimensions of teeth and depth of stator core.

Number of armature or stator slots

The following factors are considered for the selection of armature slots.

1. The number of slots should result in balanced winding.
2. A smaller number of slots results in low cost.
3. But a smaller number of slots results in hot spot temperature due to bunching of conductors in fewer slots.
4. Small number of slots results in larger leakage reactance.
5. Larger number of slots results in reduced tooth ripples and losses.
6. Larger number of slots, increases flux density at the teeth and hence iron losses.

The usual values of stator slot pitch, y_{ss} are given below.

$y_{ss} \leq 25$ *mm* for low voltage machines

$y_{ss} \leq 40$ *mm* for 6 *kV* or low voltage machines

$y_{ss} \leq 60$ *mm* for machines upto 15 *kV*

In salient pole machines, the number of stator slots per pole per phase is usually between 2 to 4.

$$\text{Stator slot pitch, } y_{ss} = \frac{\text{Gap surface}}{\text{Total number of stator slots}}$$

Let, S_s = Total stator slots

πD = Gap surface

$$\text{Number of stator slots, } S_s = \frac{\text{Gap surface}}{\text{Stator slot pitch}} = \frac{\pi D}{y_{ss}} \quad \text{.....(9.12)}$$

Number of stator conductors, Z = Number of phase × Conductors per phase

$$= 3 \times 2\,T_s = 6\,T_s$$

where, T_s = Stator turns per phase

$$\text{Conductors per slot, } Z_{ss} = \frac{\text{Number of stator conductors}}{\text{Number of stator slots}} = \frac{6T_s}{S_s} \quad \text{.....(9.13)}$$

> *Note : Z_{ss} must be even for double layer winding.*

Area of stator slot

$$\text{Approximate area of each slot} = \frac{\text{Copper area per slot}}{\text{Space factor}}$$

$$= \frac{Z_{ss} \times a_s}{\text{Space factor}}$$

The space factors vary from 0.25 to 0.4. High voltage machines have lower space factors due to large thickness of insulation. After obtaining the area of the slot, the dimensions of the slot should be adjusted. The slot should not be too wide to give a thin tooth. The width of the slot should be so adjusted such that the mean flux density in the tooth lies between 1.3 to 1.7 *Wb/m²*. The width of tooth should not be too large as it results in narrow and deep slots. The deeper slots give a large value of leakage reactance. In general the ratio of slot depth to slot width should be between 3 and 6.

Stator teeth

The concept discussed for design of stator teeth in induction motor stator teeth are applicable for synchronous machine.

Therefore, from equation (8.27) of Chapter-8, we get,

$$\therefore \text{ Minimum width of stator tooth, } W_{ts\_min} = \frac{\phi_m\,p}{1.7\,S_s\,L_i}$$

Depth of stator core

The concept discussed for design of depth of stator core for induction motor are applicable for synchronous machine.

Therefore, from equation (8.28) and (8.29) of Chapter-8, we get,

$$\text{Depth of stator core, } d_{cs} = \frac{\phi_m}{2\,B_{cs}\,L_i}$$

$$\text{Outer diameter of stator core, } D_o = D + 2\,(d_{ss} + d_{cs})$$

9.4.8 Estimation of Length of Air-gap   *(AU, Apr' 17, 10 M) (KTU, Feb' 18, 4 M)*

The length of air-gap greatly influences the performance of synchronous machine. The advantages of large air-gap are,

  1. Reduction in armature reaction
  2. Small value of regulation
  3. Higher value of stability
  4. A higher synchronizing power which makes the machine less sensitive to load variations
  5. Better cooling
  6. Lower tooth pulsation loss
  7. Less noise
  8. Smaller unbalanced magnetic pull

The disadvantage of large value of air-gap is that the required field mmf will increase which in turn result in larger field winding and the machine becomes costlier.

In salient pole machines the length of the air-gap is not constant over the pole arc but minimum at centre of pole and increases from centre to outwards. An attempt is usually made to obtain sinusoidal distribution of flux by proper shaping and proportioning of the pole shoe.

Let, l_g = Length of air-gap at the centre of poles

For salient pole machines of normal construction and having open type slots, the length of air-gap can be determined from equation (9.14).

$$\frac{\text{Length of air-gap}}{\text{Pole pitch}} = \frac{l_g}{\tau} = 0.01 \text{ to } 0.015 \qquad\qquad .....(9.14)$$

For synchronous motors designed with maximum output equal to 1.5 times rated output, the length of air-gap can be determined from equation (9.15).

$$\frac{\text{Length of air-gap}}{\text{Pole pitch}} = \frac{l_g}{\tau} = 0.02 \qquad\qquad .....(9.15)$$

Estimation of Air-gap using SCR

The air-gap can also be estimated from short circuit ratio.

We know that,

mmf required for air-gap, $AT_g = 800,000 \ B_g \ K_g \ l_g$ $\qquad\qquad .....(9.16)$

Also, mmf required for air-gap is approximately equal to 80% of no-load field mmf.

Let, AT_{fo} = Field mmf on no-load.

∴ mmf required for air-gap, $AT_g = 0.8 \ AT_{fo}$

On equating the two equations of mmf for air-gap we get,

$$800{,}000 \, B_g \, K_g \, l_g = 0.8 \, AT_{fo}$$

$$\therefore \text{ Length of air-gap, } l_g = \frac{0.8 \, AT_{fo}}{800000 \, B_g K_g} = \frac{AT_{fo}}{B_g K_g \times 10^6} \qquad .....(9.17)$$

We know that,

$$B_g = \frac{B_{av}}{K_f}$$

$$AT_{fo} = AT_a \times SCR$$

On substituting the expression for AT_{fo} and B_g in the equation for l_g we get,

$$\text{Length of air-gap, } l_g = \frac{AT_a \, SCR \, K_f}{B_{av} \, K_g \times 10^6} \qquad .....(9.18)$$

$$\text{Armature mmf per pole, } AT_a = 2.7 \, \frac{I_s \, T_s \, K_{ws}}{p}$$

where, I_s = Stator current per phase K_{ws} = Winding factor of stator

   T_s = Turns per phase K_f = Form factor

   p = Number of poles

9.4.9 Design of Rotor (or) Design of Field System   *(AU, Nov' 18, 13 M) (AU, Apr' 19, 13 M)*

(AU, Apr' 18, 5 M)

The main dimensions of rotor are diameter of rotor, D_r and length of rotor. The length of rotor is same as that of length of stator, L.

$$\text{Diameter of rotor, } D_r = D - 2 \, l_g$$

where, l_g = Length of air-gap at centre of pole.

The design of D and L are discussed in Section 9.4.

The rotor of synchronous machine consists of poles mounted on outer periphery. The two types of poles employed in synchronous machine are poles with circular and rectangular cross-section. The cross-section and dimensions of salient pole are shown in Fig. 9.6.

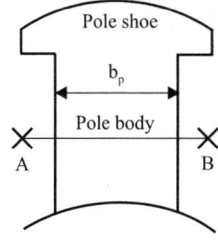

Fig. a: Cross-section of pole.

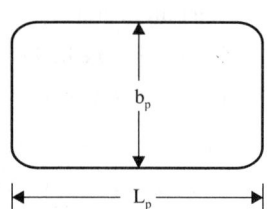

Fig. b: Rectangular cross-section at AB.

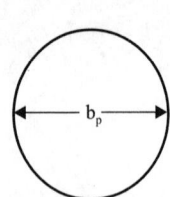

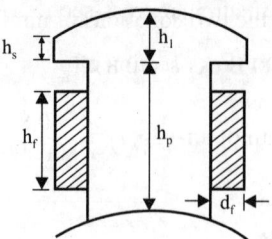

Fig. c: *Circular cross-section at AB.*   **Fig. d:** *Dimensions of pole.*

Fig. 9.6: *Pole of salient pole synchronous machine.*

Width of pole

Let, ϕ_p = Flux in pole body

B_p = Flux density in pole body

$$\boxed{B_p = \frac{\phi_p}{A_p}}$$

Area of cross-section of pole body, $A_p = \dfrac{\phi_p}{B_p}$ .....(9.19)

Flux density in pole body, B_p = 1.5 to 1.7 *Wb/m²*

Flux in pole body, ϕ_p = Leakage coefficient × Flux per pole

$$= C_l \, \phi \qquad\qquad\qquad .....(9.20)$$

The flux per pole can be estimated using the equation of induced emf.

Stator induced emf per phase, E_s = 4.44 f ϕ T_s K_{ws}

$\therefore$ Flux per pole, $\phi = \dfrac{E_s}{4.44 \, f \, T_s \, K_{ws}}$ .....(9.21)

where, f = Frequency

T_s = Turns per phase

K_{ws} = Stator winding factor

For poles with rectangular cross-section,

Area of cross-section of pole body, $A_p = S_f \, L_p \, b_p$ .....(9.22)

where, S_f = Stacking factor

L_p = Length of pole body

b_p = Width of pole body

In case of laminated pole the stacking factor is chosen as 0.98 and for solid pole construction, S_f= 1. The L_p is chosen same as length of rotor, L.

$\therefore$ Width of pole body, $b_p = \dfrac{A_p}{S_f \, L_p} = \dfrac{A_p}{S_f \, L}$ .....(9.23)

For poles with circular cross-section,

$$\text{Area of cross-section of pole body, } A_p = \frac{\pi}{4} b_p^2$$

$$\boxed{\text{Area of circle} = \frac{\pi}{4} d^2} \quad .....(9.24)$$

$$\therefore \text{ Width of pole body, } b_p = \sqrt{\frac{4}{\pi} \times A_p} \quad .....(9.25)$$

> *Note : Poles with circular cross-section are not laminated and will have solid construction and so no stacking factor, S_f*

Height of pole

The height of pole body can be estimated from the knowledge of field mmf.

Let, AT_{fl} = Full load field mmf

AT_a = Armature mmf

Choose, $AT_{fl} = k\, AT_a$

where, k = Constant in the range 2 to 2.5

$$AT_a = 2.7 \frac{I_s\, T_s\, k_{ws}}{p} \quad .....(9.26)$$

where, I_s = Armature current per phase

T_s = Armature turns per phase

k_{ws} = Stator winding space factor

p = Number of poles

$$\text{mmf per unit height of field winding} = \frac{AT_{fl}}{h_f} \quad .....(9.27)$$

where, h_f = Height of field winding

Also, mmf per unit height of field winding $= 10^4 \times \sqrt{S_{fc}\, d_f\, q_f} \quad .....(9.28)$

On equating the above two equations of mmf per unit height of field winding we get,

$$\frac{AT_{fl}}{h_f} = 10^4 \times \sqrt{S_{fc}\, d_f\, q_f}$$

$$\therefore \text{ Height of field winding, } h_f = \frac{AT_{fl}}{10^4 \times \sqrt{S_{fc}\, d_f\, q_f}} \quad .....(9.29)$$

where, S_{fc} = Copper space factor of field winding

q_f = Loss per unit surface of winding

d_f = Depth of field winding

The depth of field winding, d_f depends on the diameter of armature, D and can be chosen from Table 9.1.

Table 9.1: Depth of Field Winding

Armature diameter, D m	Winding depth, d_f mm
0.20	30
0.35	35
0.50	40
0.65	45
1.00	50
1.00 and above	55

The height of pole body, h_p is decided by the sum of heigth of field winding and clearance needed for insulation.

$\therefore$ Height of pole body, $h_p = h_f +$ Clearance for insulation   .....(9.30)

The height of pole shoe should be sufficient to accommodate the damper winding conductors.

$\therefore$ Height of pole shoe, $h_s = 2 d_d$ .....(9.31)

where, d_d = Diameter of damper winding conductors

The height h_1 is decided by the mechanical dimension of pole arc.

9.4.10 Design of Damper Winding
(UTU, Dec' 13, 10 M) (AU, Nov' 18, 7 M)

The damper winding is provided in salient pole synchronous generator to suppress the negative sequence field and to damp the oscillations during hunting. In salient pole synchronous motor the damper winding is provided to produce starting torque and to damp the oscillations during hunting.

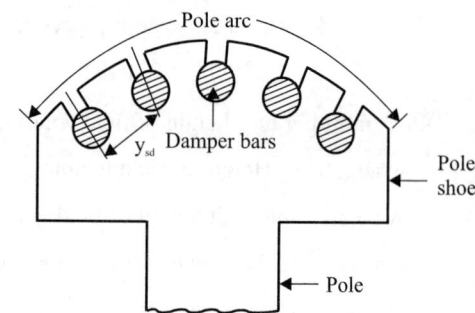

The damper windings are copper bars accommodated in the slots cut on pole shoe of field poles and short circuited at both ends by end rings. The cross-section of pole shoe showing cross-section of damper bars is shown in Fig. 9.7.

Fig. 9.7: *Cross-section of pole shoe with damper bars.*

The mmf for damper winding is estimated as 20% of specific electric loading over a pole pitch.

$\therefore$ mmf for damper winding = 0.2 **ac** τ

where, **ac** = Specific electric loading

τ = Pole pitch

Also, mmf for damper winding is given by product of total area of cross-section of damper bars under a pole, A_d and current density in damper bar, δ_d.

$\therefore$ mmf for damper winding = $A_d \delta_d$

On equating the above two equations of mmf for damper winding we get,

$$A_d \, \delta_d = 0.2 \, \text{ac} \, \tau$$

$$\left.\begin{array}{l}\therefore \text{ Total area of cross-section} \\ \text{of damper bars under a pole}\end{array}\right\} A_d = \frac{0.2 \, \text{ac} \, \tau}{\delta_d} \qquad \qquad .....(9.32)$$

The current density in the damper bars is usually taken as 3 to 4 *A/mm²*.

The damper winding consists of N_d damper bars accommodated in slots cut on the pole shoe. The slot pitch for damper winding is made smaller than stator slot pitch in order to reduce current induced in damper windings due to tooth ripples.

Let, y_{sd} = Slot pitch for damper winding

$\qquad y_{ss}$ = Stator slot pitch

Choose, $y_{sd} = 0.8 \, y_{ss}$

Now, number of damper bars over a pole, N_d is given by,

$$\text{Number of damper bars over a pole, } N_d = \frac{\text{Pole arc}}{y_{sd}} = \frac{b}{y_{sd}} = \frac{b}{0.8 \, y_{ss}} \qquad \qquad .....(9.33)$$

where, b = Pole arc

Let, a_d = Area of cross-section of each damper bar

Here, $N_d \, a_d = A_d$

$\qquad \therefore$ Area of cross-section of each damper bar, $a_d = \dfrac{A_d}{N_d} \qquad \qquad .....(9.34)$

When circular damper bars are employed, the diameter of damper bars can be estimated using the following equation.

$$a_d = \frac{\pi}{4} \, d_d^2$$

where, d_d = Diameter of damper bars

The damper bars are short circuited at both ends by using end rings. The area of end ring can be chosen as 80% to 100% of total area of damper bars under a pole, A_d so that current density in end rings will be almost same as that of damper bars.

$$\left.\begin{array}{l}\therefore \text{ Area of cross-section of} \\ \text{end ring of damper winding}\end{array}\right\} A_{d\_ring} = k \, A_d \qquad \qquad .....(9.35)$$

where, k = 0.8 to 1

The length of damper bars are chosen slightly larger than the length of stator core, in order to connect end rings of damper winding.

$$\therefore \text{ Length of damper bars, } L_d = 1.1 \times L \quad - \quad \text{For small machines}$$

$$= 0.1 + L \quad - \quad \text{For large machines}$$

where, L = Length of stator core

9.4.11 Determination of Full Load Field mmf

The full load field mmf can be estimated by vectorically adding the mmf required on no-load and mmf to overcome armature reaction.

Field mmf required on no-load

The field mmf required on no-load is given by sum of mmf required for air-gap, armature teeth, armature core, poles and yoke.

$$\text{mmf on no-load, } AT_{fo} = AT_g + AT_t + AT_c + AT_p + AT_y \qquad \qquad .....(9.36)$$

where, AT_g = mmf for air-gap

AT_t = mmf for armarure teeth

AT_c = mmf for armature core

AT_p = mmf for poles

AT_y = mmf for yoke

Armature mmf per pole

The field mmf should have a component to over come armature reaction and this component is called armature mmf per pole. The armature mmf can be estimated from the following equation.

$$\text{Armature mmf per pole, } AT_a = \frac{2.7 \, I_s \, T_s \, K_{ws}}{p} \, \rho_d \qquad \qquad .....(9.37)$$

where, ρ_d = Reduction factor to find field mmf equivalent to armature mmf

The usual value of ρ_d is 0.85.

Procedure to find full load field mmf

In synchronous machines it is difficult to numerically estimate field mmf on no-load and field mmf to overcome armature reaction and adding them vectorically to find full load field mmf. Therefore, a graphical procedure is developed to determine components of field mmf from open circuit characteristics (OCC) and then adding vectorically.

1. Determine the open circuit characteristies (OCC) shown in Fig. 9.8, by running the machine on no-load.

Here, E_o = Induced emf per phase on open circuit

AT_{fo} = Field mmf required to induce E_o on no-load

AT_{fo} = Number of poles × AT per pole

$$= p \times I_f T_f \qquad \qquad .....(9.38)$$

where, T_f = Number of turns in a field coil.

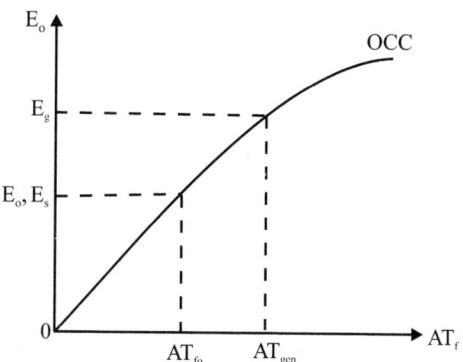

Fig. 9.8: OCC of synchronous machine.

2. From the OCC find the rated value of induced emf per phase, E_o on open circuit which is the emf corresponding to AT_o. This E_o is also stator terminal voltage per phase, E_s on load.

3. Estimate the induced emf on load, E_g by vectorically adding the induced emf on no-load to resistive and reactive drop of armature winding as shown in Fig. 9.9.

$$\therefore \ \overline{E}_g = \overline{E}_s + \overline{I}_s R_a + j\overline{I}_s X_s \qquad\qquad .....(9.39)$$

where, R_a = Armature resistance per phase

X_s = Synchronous reactance per phase

$\overline{I}_s$ = Armature current per phase

Note : Over bar on E_g, E_s and I_s indicates phasor value.

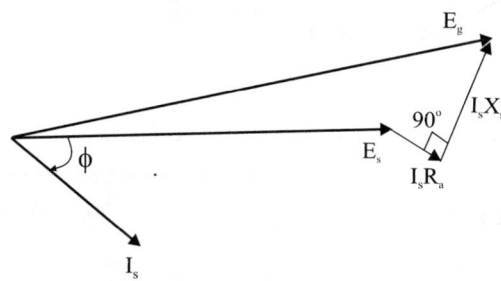

Fig. 9.9: Vector addition to find generated emf on load.

4. Determine the mmf required on load, AT_{gen} from OCC which is the AT corresponding to E_g.

5. The full load field mmf can be obtained by vectorically adding mmf generated on load, armature and leakage mmf as shown in Fig. 9.10.

The vector diagram is drawn by choosing a suitable scale for mmf, for example 1 *cm* = 200 *AT*.

In Fig. 9.10, the vectors representing various mmf are,.

od = mmf generated on load, AT_{gen}

de = Armature mmf equivalent to field mmf, AT_a

df = Magnetizing mmf due to leakage reactance

og = Field mmf on full load, AT_{fl}.

In order to construct vector diagram, first draw a horizontal line and mark od. Then draw the armature current, I_s at angle, ϕ lagging od.

Draw a perpendicular to armature current, I_s through point d.

From point d, mark de corresponding to AT_a estimated using the following equation.

$$AT_a = \frac{2.7\, I_s\, T_s\, K_{ws}}{p}\, \rho_d \qquad\qquad\qquad .....(9.40)$$

Then from point e mark ef in the line de, where ef is the mmf due to leakage reactance.

Now draw a line from point o through point f. Then draw a perpendicular to touch point e such that meeting point of perpendicular is point g.

Now the field mmf on load is given by vector og.

∴ Field mmf on load, AT_{fl} = Length of og × mmf scale

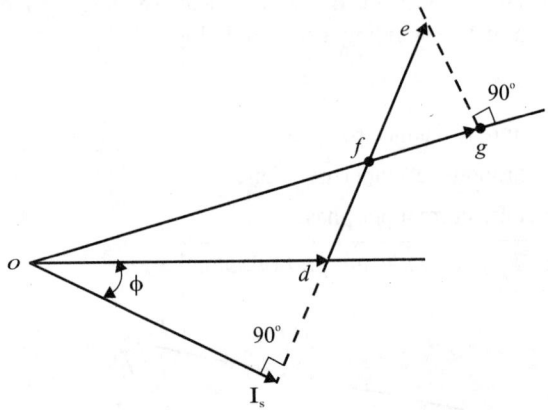

Fig. 9.11: *Determination of full load field mmf.*

9.4.12 Design of Field Winding   *(KTU, Feb' 18, 12 M)*   *(AU, Apr' 19, 15 M)*

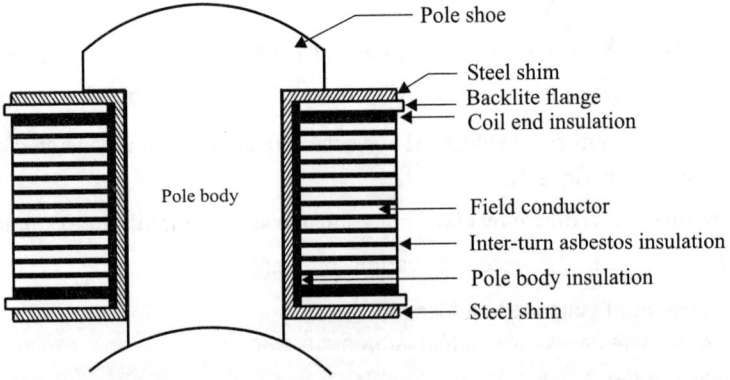

Fig. 9.11: *Cross-section of field coil.*

In case of low capacity synchronous machine the field winding is insulated copper wire wound coil and in case of large capacity machines the field winding is wound using bare copper strips with asbestos insulation between turns.

A steel shim is provided around pole body to hold the field coil. The field coil is also supported by bakelite flanges at the two ends and sides of field coils. The paper insulation is fixed with varnish or shellac.

The mean length of field coil turn can be estimated by considering the coil surrounding the pole as two sides of rectangle of length, L and two semi-circle of radius, R_{mean}.

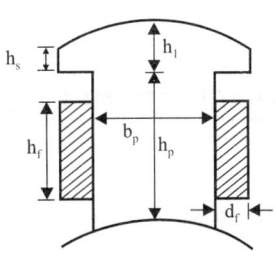

Fig. 9.12: *Dimension of pole.*

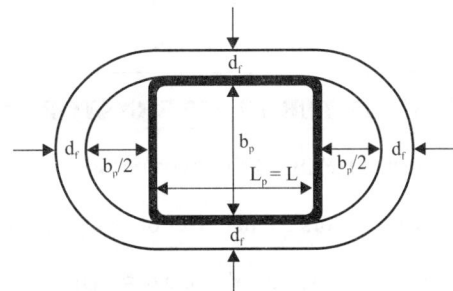

Fig. 9.13: *Plan of field coil to estimate length of mean turn.*

With reference to Fig. 9.12 we can write,

Length of mean turn of field coil, $L_{mtf} = 2L + 2\pi R_{mean}$ .....(9.41)

where, $R_{mean} = \dfrac{b_f}{2} + \text{Insulation thickness} + d_f$

Insulation thickness = 5 *mm*

In synchronous machine, each pole will carry a field coil and field coils of all the poles are connected in series to form field winding.

The voltage across the field winding is chosen as 80 to 85% of exciter voltage.

$\therefore$ Voltage across each field coil, $E_f = \dfrac{(0.8 \text{ to } 0.85) \, V_f}{p}$ .....(9.42)

where, V_f = Exciter voltage

$\quad\;\; p$ = Number of poles

Let, R_f = Resistance of field coil

$\therefore R_f = \dfrac{T_f \, \rho \, L_{mtf}}{a_f}$ .....(9.43)

Also, $R_f = \dfrac{E_f}{I_f}$

where, I_f = Field current

On equating the above two equations of field resistance we get,

$$\frac{E_f}{I_f} = \frac{T_f \, \rho \, L_{mtf}}{a_f} \qquad \qquad .....(9.44)$$

∴ Area of cross-section of field conductor, $a_f = \dfrac{I_f \, T_f \, \rho \, L_{mtf}}{E_f}$ $\boxed{AT_{fl} = I_f \, T_f}$

$$= \frac{AT_{fl} \, \rho \, L_{mtf}}{E_f}$$

where, T_f = Number of field turns

ρ = Resistivity of copper

9.5 DESIGN OF TURBO ALTERNATOR (or) NON-SALIENT POLE ALTERNATOR

9.5.1 Design of Main Dimensions   *(AU, Apr' 18, 15 M)*

In turbo alternators the diameter of rotor, D_r is limited by the maximum peripheral speed, V_a.

Peripheral speed, $V_a = \pi D_r n_s \approx \pi D n_s$ $\boxed{\text{Neglecting air-gap}}$

∴ Stator inner diameter, $D = \dfrac{V_a}{\pi n_s}$ .....(9.45)

The output equation of synchronous machine (equation 9.1) can be modified by using the above relation,

kVA rating, $Q = C_o \, D^2 \, L \, n_s$ .....(9.46)

where, $C_o = 11 \, B_{av} \, \mathbf{ac} \, K_{ws} \times 10^{-3}$ .....(9.47)

On substituting for D and C_o from equations (9.45) and (9.47) in equation (9.46) we get,

kVA rating, $Q = 11 \, B_{av} \, \mathbf{ac} \, K_{ws} \times 10^{-3} \left(\dfrac{V_a}{\pi n_s} \right)^2 L \, n_s$

$$= 1.11 \, B_{av} \, \mathbf{ac} \, K_{ws} \, \frac{V_a^2}{n_s} \, L \times 10^{-3} \qquad \qquad .....(9.48)$$

The length of the armature, L can be estimated from equation (9.48) and stator inner diameter, D can be estimated using equation (9.45).

The value of specific loading for conventionally cooled turbo alternators are,

B_{av} = 0.54 to 0.65 *Wb/m²*

ac = 50000 to 75000 *amp.cond./m*

The specific loadings used in large water cooled turbo alternators are,

B_{av} = 0.54 to 0.65 *Wb/m²*

ac = 180000 to 200000 *amp.cond./m*

9.5.2 Length of Air-gap

The length of air-gap for turbo alternator can be estimated from the ratio of length of air-gap, l_g and pole pitch, τ.

$$\frac{\text{Length of air-gap}}{\text{Pole pitch}} = \frac{l_g}{\tau} = 0.02 \text{ to } 0.025$$

Alternatively, the length of air-gap can be estimated from short circuit ratio (SCR) as shown below.

mmf for air-gap $= 800{,}000 \; K_g \, B_g \, l_g$

Also, mmf for air-gap $= 80\%$ of no-load field mmf, AT_{fo}

The value of SCR is 0.5 to 0.7 for turbo alternators.

On equating the above two equations of mmf for air-gap,

$$800{,}000 \; K_g \, B_g \, l_g = 0.8 \; AT_{fo}$$

$$800{,}000 \; K_g \, B_g \, l_g = 0.8 \times SCR \times AT_a$$

$$800{,}000 \; K_g \, B_g \, l_g = 0.8 \times SCR \times \mathbf{ac} \; \tau / 2$$

$\boxed{AT_{fo} = SCR \times AT_a}$

$\boxed{AT_a = \mathbf{ac} \times \dfrac{\tau}{2}}$

$$\therefore \text{ Length of air-gap}, \; l_g = \frac{0.8 \times SCR \times \mathbf{ac} \, (\tau / 2)}{800000 \; K_g B_g} = \frac{0.5 \; SCR \; \mathbf{ac} \; \tau}{K_g B_g \times 10^6}$$

$$= \frac{0.5 \; SCR \; \mathbf{ac} \; \tau \times 10^{-6}}{K_g B_{av} / K_f} = \frac{0.5 \; SCR \; \mathbf{ac} \; K_f \times 10^{-6}}{K_g B_{av}}$$

$\boxed{B_g = \dfrac{B_{av}}{K_f}}$

9.5.3 Stator Design   *(AU, Nov' 17, 16 M)*

The armature slot, winding, turns per phase and conductor designs of turbo alternator are same as that of salient pole alternator. However there is a slight difference in the following aspects of design.

1.   The number of stator slots per pole phase lies between 2 to 4 in salient pole machine but in the case of turbo alternators 8 or 9 slots per pole per phase may be used.

2.   The slot pitch is normally about 25 to 60 *mm* in salient pole machine but in the case of large turbo alternators it may be 75 to 90 *mm*.

3.   Single layer concentric or two layer short pitched windings may be used in turbo alternators. The advantage of single layer winding is that it can be easily clamped and easy for short chording. Two layer winding chorded by about 1/6 pole pitch is more common because it practically eliminates 5th and 7th as well as 17th and 19th harmonics.

4.   The stator conductors must be subdivided and transposed to reduce eddy current losses.

5.   In windings of large modern turbo alternators it is common practice to assemble two conductors per slot and two parallel circuits per phase.

6.   The current density in the stator windings of modern water-cooled generators is usually between 8 to 9.5 *A/mm²* as compared to about 4 *A/mm²* in the case of conventionally cooled machines.

7.   The stator winding of turbo alternators is deliberately put in deep slots in order to increase the leakage reactance and hence to reduce the forces under short circuit conditions.

9.5.4   Rotor Design (or) Design of Field System   *(RGPV, Jun' 20, 5 M) (AU, Apr' 18, 15 M)*

The rotor of turbo alternators are solid cylindrical structure with slots on outer periphery. The rotor is constructed using an alloy of chromium nickel steel or chromium molybdenum steel. The main dimensions of rotor are rotor diameter, D_r and rotor length, L. The length of rotor is same as length of stator, L.

Diameter of rotor, $D_r = D - 2\,l_g$

where, l_g = Length of air-gap

The design of main dimensions D and L are discussed in section 9.5.1.

Rotor winding

The rotor winding is made of solid or multi-turn conductors placed on the rotor slots. The rotor slots are made on two-third of pheriphery (Refer Fig. 9.1b) and the rotor conductors are distributed on these slots.

Concentric multi-turn coils are used for field winding as shown in Fig. 9.14. The number of wound slots should be an integer. The slot pitch is so chosen that undesirable harmonics are not introduced in the flux density wave.

Epoxy glass and absestos moulded resin glass and or synthetic rubberized glass are used as insulation for rotor winding. The insulation thickness varies from 0.25 to 0.33 *mm* per 100 *V* across the winding.

Rotor current density is 2.5 *A/mm²* for conventionally cooled machines and for modern direct cooled generators the rotor current densities may be 9.5 to 14 *A/mm²*.

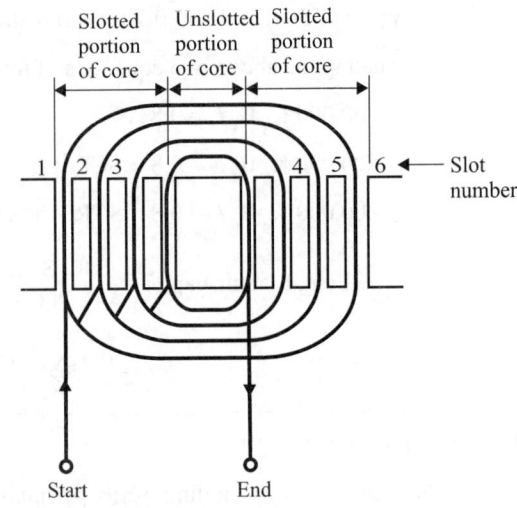

Fig. 9.14: Rotor winding.

Procedure for rotor winding design

1. Full load field mmf of turbo alternator can be taken as twice the armature mmf.

    $\therefore$ Full load field mmf, $AT_{fl\_TA} = 2\,AT_a$

    where, $AT_a = \dfrac{2.7\,I_s\,T_s\,k_w}{p}$

2. The voltage across the field winding is chosen as 80 to 85% of exciter voltage.

    $\therefore$ Voltage across each field coil, $E_f = \dfrac{(0.8 \text{ to } 0.85)\,V_f}{p}$

    where, V_f = Exciter voltage

    p = Number of poles

3. The length of mean turn of field winding can be estimated from following equation.

$$L_{mtf} = 2L + 2.3\tau_c + 0.24 \qquad .....(9.49)$$

where, τ_c = Effective span of coils

Let, R_f = Resistance of field coil

$$\therefore R_f = \frac{T_f \, \rho \, L_{mtf}}{a_f}$$

Also, $R_f = \dfrac{E_f}{I_f}$

where, I_f = Field current

On equating the above two equations of field resistance we get,

$$\frac{E_f}{I_f} = \frac{T_f \, \rho \, L_{mtf}}{a_f}$$

$$\therefore \text{Area of cross-section of field conductor, } a_f = \frac{I_f \, T_f \, \rho \, L_{mtf}}{E_f}$$

$$= \frac{AT_{fl} \, \rho \, L_{mtf}}{E_f} \qquad \boxed{AT_{fl} = I_f \, T_f}$$

where, T_f = Number of field turns

ρ = Resistivity of copper

4. Assume a suitable current density, δ_f for the field winding

$$\text{Total area of field conductors, } A_f = \frac{2p \, AT_{fl}}{\delta_f} \qquad .....(9.50)$$

$$\text{Number of field conductors, } Z_f = \frac{2p \, AT_{fl}}{\delta_f \, a_f} \qquad .....(9.51)$$

$$\text{Conductors per rotor slot, } Z_{rs} = \frac{2p \, AT_{fl}}{\delta_f \, a_f \, S_r} \qquad .....(9.52)$$

where, S_r = Number of wound slots in the rotor.

9.6 SUMMARY OF DESIGN EQUATIONS

Part 1: Design of Salient Pole Machines

1. Output equation of salient pole alternator

i) kVA rating, $Q = C_o \, D^2 \, L \, n_s$

ii) kVA rating, $Q = \sqrt{3} \, V_L \, I_L \times 10^{-3}$

iii) kVA rating, $Q = 3 \, E_s \, I_s \times 10^{-3}$

Note : Q is output kVA in alternator and input kVA in motor.

2.   Output coefficient, $C_o = 11\ B_{av}\ ac\ K_{ws} \times 10^{-3}$

3.   Choice of L/τ for salient pole synchronous machines

    i)   When round poles are employed,   L/τ = 0.6 to 0.7

    ii)   When rectangular poles are employed,   L/τ = 1 to 5

4.   Direct axis reactance, $X_d = \dfrac{\text{pu volt}}{\text{pu SC current}}$

5.   Short circuit ratio = $1/X_d$

6.   Ratio of length of air-gap and pole pitch

    i) $\dfrac{l_g}{\tau} = 0.01$ to 0.015   -   For normal construction

    ii) $\dfrac{l_g}{\tau} = 0.02$ -   For construction with higher maximum output

7.   i)   Length of air-gap, $l_g = \dfrac{AT_{fo}}{B_g K_g \times 10^6}$

    ii)   Length of air-gap, $l_g = \dfrac{AT_a\ SCR\ K_f}{B_{av}\ K_g \times 10^6}$

Armature Design

8.   Flux per pole, $\phi = \dfrac{E_s}{4.44\ f\ T_s\ K_{ws}}$

9.   i)   Turns per phase, $T_s = \dfrac{E_s}{4.44\ \phi\ f\ K_{ws}}$ -   When parallel path, a = 1

    ii)   Turns per phase, $T_s = \dfrac{E_s \times a}{4.44\ \phi\ f\ K_{ws}}$ -   When parallel path, a > 1

10.   Current per phase, $I_s = \dfrac{kVA}{3\ E_s \times 10^{-3}}$

11.   Current through a conductor, $I_z = \dfrac{I_s}{a}$

12.   Area of cross-section of armature conductor, $a_a = \dfrac{I_z}{\delta_a}$

Design of Rotor

13.   Diameter of rotor, $D_r = D - 2\ l_g$

14.   i)   Area of cross-section of pole body, $A_p = \dfrac{\phi_p}{B_p}$

    ii) Area of cross-section of pole body, $A_p = S_f L_p b_p$ -   For rectangular poles

    iii) Area of pole body, $A_p = \dfrac{\pi}{4} b_p^2$ -   For round poles

15.   Flux in pole body, $\phi_p = C_I \phi$

16.   Induced emf per phase, $E_s = 4.44 \, f \, \phi \, T_s \, K_{ws}$

17.   Flux per pole, $\phi = \dfrac{E_s}{4.44 \, f \, T_s \, K_{ws}}$

18.   i) Width of pole body, $b_p = \dfrac{A_p}{S_f L_p} = \dfrac{A_p}{S_f L}$ -   For rectangular poles

    ii) Width of pole body, $b_p = \sqrt{\dfrac{4}{\pi} \times A_p}$ -   For round poles

19.   i) mmf per unit height of field winding $= \dfrac{AT_{fl}}{h_f}$

    ii) mmf per unit height of field winding $= 10^4 \times \sqrt{S_{fc} \, d_f \, q_f}$

20.   Height of field winding, $h_f = \dfrac{AT_{fl}}{10^4 \times \sqrt{S_{fc} \, d_f \, q_f}}$

21.   Height of pole body, $h_p = h_f +$ Clearance for insulation

22.   Height of pole shoe, $h_s = 2 \, d_d$

Design of Damper winding

23.   i) mmf for damper winding $= A_d \delta_d$

    ii) mmf for damper winding $= 0.2 \, \mathbf{ac} \, \tau$

24. $\left. \begin{array}{l} \text{Total area of cross-section} \\ \text{of damper bars under a pole} \end{array} \right\} A_d = \dfrac{0.2 \, \mathbf{ac} \, \tau}{\delta_d}$

25.   Number of damper bars, $N_d = \dfrac{b}{y_{sd}} = \dfrac{b}{0.8 \, y_{ss}}$

26.   Area of cross-section of each damper bar, $a_d = \dfrac{A_d}{N_d}$

27. $\left. \begin{array}{l} \text{Area of cross-section of} \\ \text{end ring of damper winding} \end{array} \right\} A_{d\_ring} = k \, A_d$

Design of Field Winding

28.   mmf on no-load, $AT_{fo} = AT_g + AT_t + AT_c + AT_p + AT_y$

29.   Armature mmf per pole, $AT_a = \dfrac{2.7 \, I_s \, T_s \, K_{ws}}{p} \, \rho_d \, ;$ where, $\rho_d = 0.85$

30. Field mmf on load, $AT_{fl} = $ Length of $og \times$ mmf scale

where, og is length of vector representing vector sum of components of field mmf.

31. Length of mean turn of field coil, $L_{mtf} = 2L + 2\pi\, R_{mean}$

32. $R_{mean} = \dfrac{b_f}{2} + $ Insulation thickness $+ d_f$

33. Voltage across each field coil, $E_f = \dfrac{(0.8 \text{ to } 0.85)\, V_f}{p}$

34. Area of cross-section of field conductor, $a_f = \dfrac{I_f\, T_f\, \rho\, L_{mtf}}{E_f}$

Part 2 : Design of Turbo Alternators

1. Peripheral speed, $V_a = \pi\, D\, n_s$ $\boxed{\text{Neglecting air-gap}}$

2. kVA rating, $Q = 1.11\, B_{av}\, \textbf{ac}\, K_{ws}\, \dfrac{v_a^2}{n_s}\, L \times 10^{-3}$

3. No-load field mmf, $AT_{fo} = SCR \times AT_a$

4. Full load field mmf of turbo alternator, $AT_{fl\_TA} = 2\, AT_a$

5. Length of air-gap, $l_g = \dfrac{0.5\, SCR\, \textbf{ac}\, \tau\, K_f \times 10^{-6}}{K_g B_{av}}$

Rotor Winding Design

6. Voltage across each field coil, $E_f = \dfrac{(0.8 \text{ to } 0.85)\, V_f}{p}$

7. Length of mean turn of field coil, $L_{mtf} = 2L + 2.3\tau_c + 0.24$

where, $\tau_c = $ Effective span of coil

8. Area of cross-section of field conductor, $a_f = \dfrac{I_f\, T_f\, \rho\, L_{mtf}}{E_f}$

9. Total area of field conductors, $A_f = \dfrac{2p\, AT_{fl}}{\delta_f}$

10. Number of field conductors, $Z_f = \dfrac{2p\, AT_{fl}}{\delta_f\, a_f}$

11. Conductors per rotor slot, $Z_{rs} = \dfrac{2p\, AT_{fl}}{\delta_f\, a_f\, S_r}$

9.7 SOLVED PROBLEMS

EXAMPLE 9.1

(AU, Apr' 18, 15 M)

A 600 *rpm*, 50 *Hz*, 10000 *V*, 3-phase, synchronous generator has the following design data. $B_{av} = 0.48$ *Wb/m²*, $\delta = 2.7$ *A/mm²*, slot space factor = 0.35, number of slots = 144, slot size = 120 × 20 *mm*, D = 1.92 *m* and L = 0.4 *m*. Determine the kVA rating of the machine.

GIVEN DATA

3-phase	$B_{av} = 0.48$ *Wb/m²*	D = 1.92 *m*
600 *rpm*	$\delta = 2.7$ *A/mm²*	L = 0.4 *m*
10000 *V*	Slots = 144	Slot size = 120 × 20 *mm*
		Slot space factor = 0.35

SOLUTION

We know that,

Current density, $\delta = \dfrac{I_z}{a_z}$

∴ Current in each armature conductor, $I_z = \delta\, a_z$ .....(1)

Conductors area in a slot = Slot area × Slot space factor = 120 × 20 × 0.35 = 840 *mm²*

Total number of armature conductors $\Big\}$ $Z = \dfrac{\text{Conductors area in a slot} \times \text{Number of slots}}{\text{Area of cross-section of each conductor}}$

$$= \dfrac{840 \times 144}{a_z} \qquad .....(2)$$

Specific electric loading, $ac = \dfrac{I_z Z}{\pi D}$

> Using equations (1) and (2).

$$= \dfrac{\delta a_z \dfrac{840 \times 144}{a_z}}{\pi D} = \dfrac{\delta \times 840 \times 144}{\pi D}$$

$$= \dfrac{2.7 \times 840 \times 144}{\pi \times 1.92} = 54144 \; amp.cond./m$$

kVA rating, $Q = C_o\, D^2\, L n_s$

$$= 11\, B_{av}\, ac\, K_{ws} \times 10^{-3}\, D^2\, L\, n_s$$

> $C_o = 11\, B_{av}\, K_{ws} \times 10^{-3}$

$$= 11 \times 0.48 \times 54144 \times 0.955 \times 10^{-3} \times 1.92^2 \times 0.4 \times \dfrac{600}{60}$$

$$= 4025 \; kVA$$

RESULT

kVA rating, Q = 4025 *kVA*

EXAMPLE 9.2

The output coefficient of 1250 kVA, 300 rpm, synchronous generator is 200 kVA/m^3-rps. Determine the main dimensions D and L for following three cases.

Case (i): The ratio of length to diameter is 0.2.

Case (ii): Specific loadings are decreased by 10% each with speed remaining the same as in case (a).

Case (iii): Speed is decreased to 150 rpm with specific loadings remaining the same as in case (a).

Assume the same ratio of length to diameter in all the three cases. Comment upon the results.

GIVEN DATA

$C_o = 200$ kVA/m^3-rps $\qquad$ $Q = 1250$ kVA

$\dfrac{L}{D} = 0.2$ $\qquad$ $N_s = 300$ rpm

SOLUTION

Case (i)

Given that, $\dfrac{L}{D} = 0.2$

$\therefore$ Length of stator core, L = 0.2 D $\qquad\qquad$ .....(1)

Synchronous speed, $n_s = \dfrac{300}{60} = 5$ rps

kVA rating, $Q = C_o D^2 L n_s$

$\therefore D^2 L = \dfrac{Q}{C_o n_s} = \dfrac{1250}{200 \times 5} = 1.25$ m^3

$\therefore D^2 L = 1.25$

$D^2 (0.2D) = 1.25$ $\qquad$ $\boxed{\text{Using equation (1)}}$

$\therefore$ Diameter of stator bore, $D = \left(\dfrac{1.25}{0.2}\right)^{1/3} = 1.842$ m

$\therefore$ Length of stator core, L = 0.2 D = 0.2 × 1.842 = 0.368 m $\qquad$ $\boxed{\text{Using equation (1)}}$

Case (ii)

Let, Output coefficient of machine-II, $C_{o2} = \pi^2 B_{av2}$ $ac_2 \times 10^{-3}$

In case (ii) the specific loadings are reduced (decreased) by 10%.

$\therefore B_{av2} = 0.9$ B_{av} and $ac_2 = 0.9$ ac

$\left. \begin{array}{l} \therefore \text{Output coefficient} \\ \text{of machine – II} \end{array} \right\}$ $C_{o2} = 11 B_{av2}$ $ac_2 \times 10^{-3}$

$\qquad\qquad = 11 \times 0.9$ $B_{av} \times 0.9$ $ac \times 10^{-3}$

$\qquad\qquad = 0.9 \times 0.9 \times (11$ B_{av} $ac \times 10^{-3}) = 0.9 \times 0.9 \times C_o$

$\qquad\qquad = 0.9 \times 0.9 \times 200 = 162$ kVA/m^3-rps

kVA rating of machine-II, $Q_2 = C_{o2} D_2^2 L_2 n_s$

$$\therefore D_2^2 L_2 = \frac{Q_2}{C_{o2} n_s}$$

$$= \frac{1250}{162 \times 5} = 1.543 \, m^3$$

$\boxed{Q_2 = 1250 \, kVA \text{ (same as case (a))}}$

$\therefore D_2^2 L_2 = 1.543$

$D_2^2 (0.2 \, D_2) = 1.543$

$\boxed{\begin{array}{c} \text{Using equation (1)} \\ L_2 = 0.2 \, D_2 \end{array}}$

$\therefore$ Diameter of stator bore, $D_2 = \left(\dfrac{1.543}{0.2}\right)^{1/3} = 1.976 \, m$

$\therefore$ Lenght of stator core, $L_2 = 0.2 \, D_2 = 0.2 \times 1.976 = 0.395 \, m$

Comment : Here $D_2^2 L_2 = 1.543 \, m^3$ but $D^2 L = 1.25 \, m^3$, therefore volume of machine-II is more than machine-I and so the size of the machine increases with decrease in specific loadings.

Case (iii)

In case (iii) the speed is reduced to 150 *rpm*.

$\therefore$ Synchronous speed in rps, $n_{s3} = \dfrac{150}{60} = 2.5 \, rps$

kVA rating of machine-III, $Q_3 = C_o D_3^2 L_3 n_{s3}$

$\boxed{\begin{array}{l} Q \text{ and } C_o \text{ same as case (a)} \\ \therefore Q_3 = 1250 \, kVA \text{ and} \\ \therefore C_o = 200 \, kVA/m^3\text{-}rps. \end{array}}$

$$\therefore D_3^2 L_3 = \frac{Q_3}{C_o n_{s3}} = \frac{1250}{200 \times 2.5} = 2.5 \, m^3$$

$\therefore D_3^2 L_3 = 2.5$

$D_3^2 (0.2 \, L_3) = 2.5$

$\boxed{\begin{array}{c} \text{Using equation (1)} \\ L_3 = 0.2 \, D_3 \end{array}}$

$\therefore$ Diameter of stator bore, $D_3 = \left(\dfrac{2.5}{0.2}\right)^{1/3} = 2.32 \, m$

$\therefore$ Length of stator core, $L_3 = 0.2 \, D_3 = 0.2 \times 2.32 = 0.464 \, m$

Comment : Here $D_3^2 L_3 = 2.5 \, m^3$ but $D^2 L = 1.25 \, m^3$, therefore volume of machine-III is more than machine-I and so the size of machine increases with decrease in operating speed.

RESULT

Case (i)	Case (ii)	Case (iii)
$D = 1.842 \, m$	$D_2 = 1.976 \, m$	$D_3 = 2.32 \, m$
$L = 0.368 \, m$	$L_2 = 0.395 \, m$	$L_3 = 0.464 \, m$

EXAMPLE 9.3   *(KTU, Feb' 18, 12 M) (AU, Apr' 17, 16 M) (AU, Nov' 17, 16 M) (AU, Apr' 18, 13 M) (AU, Nov' 18, 13 M)*

For a 250 *kVA*, 1100 *V*, 12 pole, 500 *rpm*, 3-phase alternator, determine air-gap diameter, core length, number of stator conductors, number of stator slots and cross-section of stator conductors. Assuming average gap density as 0.6 *Wb/m²* and specific electric loading of 30,000 *amp.cond./m*. L/τ = 1.5.

GIVEN DATA

250 *kVA*	L/τ = 1.5
12 pole	V_L = 1100 *V*
3-phase	B_{av} = 0.6 *Wb/m²*
500 *rpm*	**ac** = 30,000 *amp.cond./m*

SOLUTION

Output coefficient, $C_o = 11\,B_{av}\,\textbf{ac}\,K_{ws} \times 10^{-3}$

$\boxed{K_{ws} = 0.955}$

$$= 11 \times 0.6 \times 30000 \times 0.955 \times 10^{-3}$$

$$= 189.09 \; kVA/m^3\text{-}rps$$

Synchronous speed, $n_s = \dfrac{2f}{p} = \dfrac{2 \times 50}{12} = 8.33 \; rps$

Given that, $L = 1.5\,\tau = 1.5 \times \dfrac{\pi D}{p} = \dfrac{1.5 \times \pi}{12}\,D = 0.3927\,D \approx 0.4\,D$

$\boxed{\tau = \dfrac{\pi D}{p}}$

∴ Length of stator core, L = 0.4 D   .....(1)

kVA rating, $Q = C_o\,D^2\,L\,n_s$

∴ $D^2 L = \dfrac{Q}{C_o\,n_s} = \dfrac{250}{189.09 \times 8.33} = 0.1587 \; m^3$

∴ $D^2 L = 0.1587$

$D^2(0.4\,D) = 0.1587$

∴ Diameter of stator bore, $D = \left(\dfrac{0.1587}{0.4}\right)^{1/3} = 0.7348 \; m$

$\boxed{\text{Using equation (1)}}$

∴ Length of stator core, L = 0.4 D = 0.4 × 0.7348 = 0.2939 *m*

Flux per pole, $\phi = B_{av}\,\tau L$

$$= B_{av}\,\dfrac{\pi D}{p}\,L = 0.6 \times \dfrac{\pi \times 0.7348}{12} \times 0.2939 = 0.0339 \; Wb$$

Turns per phase, $T_s = \dfrac{E_s}{4.44\,f\,\phi\,K_{ws}}$

$\boxed{\begin{array}{c}\text{Assume star connected stator.}\\ \therefore\; E_s = V_L / \sqrt{3}\end{array}}$

$$= \dfrac{1100 / \sqrt{3}}{4.44 \times 50 \times 0.0339 \times 0.955} = 88.36 \approx 90$$

Total armature conductors, $Z = 6\,T_s = 6 \times 90 = 540$

Let, q = Slots per pole per phase

S_s = Stator slots (Armature slots)

The stator slots should be multiple of q, where q is slots per pole per phase.

$\therefore$ Stator slot, S_s = Number of phase × Poles × q

When, q = 2, S_s = 3 × 12 × 2 = 72

When, q = 3, S_s = 3 × 12 × 3 = 108

Let, S_s = 72

$\therefore$ Slot pitch, $y_{ss} = \dfrac{\pi D}{S_s} = \dfrac{\pi \times 0.7348}{72} = 0.032\ m = 32\ mm$

Here, y_{ss} < 60 mm, hence the choice of 72 is valid.

$\therefore$ Conductor per slot, $Z_{ss} = \dfrac{Z}{S_s} = \dfrac{540}{72} = 7.5 \approx 8$

$\therefore$ New value of total stator conductors, $Z = S_s \times Z_{ss}$

New value of turns per phase, $T_s = \dfrac{Z_{ss}\ S_s}{6} = \dfrac{8 \times 72}{6} = 96$

Current per phase, $I_s = \dfrac{kVA}{\sqrt{3}\ V_L \times 10^{-3}}$

$\boxed{\begin{array}{c} kVA = \sqrt{3}\ V_L\ I_L \times 10^{-3} \\ \text{Since star connected,} \\ I_s = I_L \end{array}}$

$= \dfrac{250}{\sqrt{3} \times 1100 \times 10^{-3}} = 131.21\ A$

Let, current density, δ = 3.5 A/mm²

$\left.\begin{array}{l}\therefore \text{Area of cross-section} \\ \text{of armature conductors}\end{array}\right\}\ a_a = \dfrac{I_s}{\delta} = \dfrac{131.21}{3.5} = 37.48\ mm^2$

RESULT

Diameter of stator bore (Air-gap diameter),	D	=	0.7348 m
Length of stator core (Core length),	L	=	0.293 m
Number of stator conductor,	Z	=	576
Number of stator slots,	S_s	=	72
Cross-section of stator conductor,	a_a	=	37.48 mm²

EXAMPLE 9.4

For a 250 kVA, 1000 V, 12 pole, 50 Hz, 3-phase alternator, determine number of stator conductors, number of stator slots and area of cross-section of stator conductors. Assuming flux per pole = 0.1192 Wb, winding factor = 0.955, slot per pole per phase = 3.

GIVEN DATA

250 kVA	3-phase
1000 V	ϕ = 0.1192 Wb
12 pole	q = 3
50 Hz	K_w = 0.955

SOLUTION

Synchronous speed, $n_s = \dfrac{2f}{p} = \dfrac{2 \times 50}{12} = 8.33 \; rps$

Turns per phase, $T_s = \dfrac{E_s}{4.44 \, f \, \phi \, K_{ws}}$

> Assume star connected stator.
> $\therefore E_s = V_L / \sqrt{3}$

$\qquad = \dfrac{1000 / \sqrt{3}}{4.44 \times 50 \times 0.1192 \times 0.955} = 22.84 \approx 23$

Total armature conductors, $Z = 6 \, T_s = 6 \times 23 = 138$

Total number of slots, $S_s =$ Number of phase × Number of poles × Slots per pole

$\qquad = 3 \times 12 \times 3 = 108$

Conductor per slot, $Z_{ss} = \dfrac{Z}{S_s} = \dfrac{138}{108} = 1.27 \approx 1$

$\therefore$ New value of total stator conductors, $Z = S_s \times Z_{ss}$

New value of turns per phase, $T_s = \dfrac{Z_{ss} \, S_s}{6} = \dfrac{1 \times 108}{6} = 18$

Current per phase, $I_s = \dfrac{kVA}{\sqrt{3} \; V_L \times 10^{-3}}$

> $kVA = \sqrt{3} \; V_L \, I_L \times 10^{-3}$
> Since star connected,
> $I_s = I_L$

$\qquad = \dfrac{250}{\sqrt{3} \times 1000 \times 10^{-3}} = 144.34 \; A$

Let, current density, $\delta = 3.5 \; A/mm^2$

$\left. \begin{array}{l} \therefore \text{Area of cross-section} \\ \text{of armature conductors} \end{array} \right\} a_a = \dfrac{I_s}{\delta} = \dfrac{144.34}{3.5} = 41.24 \; mm^2$

RESULT

Total stator conductor,	Z	$= 138$
Number of stator slots,	S_s	$= 108$
Cross-section of stator conductor,	a_a	$= 41.24 \; mm^2$

EXAMPLE 9.5   *(RGPV, Jun' 20, 10 M) (JNTU, Apr' 19, 15 M) (VTU, Dec' 19, 8 M) (VTU, Dec' 19, 10 M) (AU, Apr' 19, 13 M)*

Determine the output coefficient for a 1500 *kVA*, 2200 *V*, 3-phase, 10-pole, 50 *Hz*, star connected alternator with sinusoidal flux distribution. The winding has 60° phase spread and full pitch coils. **ac** = 30000 *amp.cond./m*, B_{av} = 0.6 *Wb/m²*. If the peripheral speed of the rotor must not exceed 100 *m/s* and the ratio pole pitch to core length is to be between 0.6 and 1, find D and L. Assume an air-gap length of 6 *mm*. Find also the approximate number of stator conductors.

GIVEN DATA

1500 *kVA*	Star connected	**ac** = 30000 *amp.cond./m*
2200 *V*	60° phase spread	B_{av} = 0.6 *Wb/m²*
3-phase	Full pitch coils	τ/L = 0.6 to 1
10 pole	50 *Hz*	l_g = 6 *mm*
		Peripheral speed ≤ 100 *m/s*

SOLUTION

For a phase spread of 60°, K_d = 0.955

For full pitched coils, K_c = 1

∴ Stator winding factor, K_{ws} = $K_d K_c$ = 0.955 × 1 = 0.955

Output coefficient, C_o = 11 B_{av} **ac** K_{ws} × 10^{-3}

$$= 11 × 0.6 × 30000 × 0.955 × 10^{-3}$$

$$= 189 \ kVA/m^3\text{-}rps$$

Synchronous speed, $n_s = \dfrac{2f}{p} = \dfrac{2 × 50}{10} = 10 \ rps$

kVA rating, $Q = C_o D^2 L n_s$

$$∴ D^2 L = \frac{Q}{C_o n_s} = \frac{1500}{189 × 10} = 0.793 \ m^3 \qquad \qquad .....(1)$$

Let, D_r = Diameter of rotor

∴ D = $D_r + 2l_g$

Let, V_a = 80 m/s	Let V_a = 40 m/s
∴ $D_r = \dfrac{V_a}{\pi n_s} = \dfrac{80}{\pi × 10}$	∴ $D_r = \dfrac{V_a}{\pi n_s} = \dfrac{40}{\pi × 10}$
$= 2.546 \ m$	$= 1.2732 \ m$
$D = D_r + 2l_g$	$D = D_r + 2l_g$
$= 2.546 + 2 × 6 × 10^{-3}$	$= 1.2732 + 2 × 6 × 10^{-3}$
$= 2.558 \ m$	$= 1.2852 \ m$
From equation (1),	From equation (1),
$L = \dfrac{0.793}{D^2} = \dfrac{0.793}{2.558^2} = 0.1212 \ m$	$L = \dfrac{0.793}{D^2} = \dfrac{0.793}{1.2852^2} = 0.4801 \ m$
$\tau = \dfrac{\pi D}{p} = \dfrac{\pi × 2.558}{10} = 0.8036 \ m$	$\tau = \dfrac{\pi D}{p} = \dfrac{\pi × 1.2852}{10} = 0.4038 \ m$
$\dfrac{\tau}{L} = \dfrac{0.8036}{0.1212} = 6.630$	$\dfrac{\tau}{L} = \dfrac{0.4038}{0.4801} = 0.8411$
Since τ / L is not between 0.6 and 1 the calculated values of D and L are not suitable	In this case τ / L is between 0.6 and 1 Hence the calculated values of D and L are acceptable.

∴ Diameter of stator bore, D = 1.2852 m

∴ Length of stator core, L = 0.4801 m

Flux per pole, $\phi = B_{av} \tau L = 0.6 × 0.4038 × 0.4801 = 0.1163 \ Wb$

Turns per phase, $T_s = \dfrac{E_s}{4.44 \, f\phi K_{ws}} = \dfrac{2200/\sqrt{3}}{4.44 \times 50 \times 0.1163 \times 0.955} = 51.5 \approx 52$

Star connected,
$\therefore E_s = V_L / \sqrt{3}$

Total armature conductors, $Z = 6 \, T_s = 6 \times 52 = 312$

Alternate method

Current through each conductor, $I_z = I_s = \dfrac{kVA}{3E_s \times 10^{-3}}$

$kVA = \sqrt{3} \, V_L \, I_L \times 10^{-3}$
Since star connected,
$I_s = I_L$

$$= \dfrac{1500}{\sqrt{3} \times 2200 \times 10^{-3}} = 393.65 \, A$$

Specific electric loading, $\mathbf{ac} = \dfrac{I_z \, Z}{\pi \, D}$

$\therefore$ Number of stator conductors, $Z = \dfrac{\mathbf{ac} \, \pi \, D}{I_z} = \dfrac{\mathbf{ac} \, \pi \, D}{I_s}$

$$= \dfrac{30000 \times \pi \times 1.285^2}{393.65} = 307.7 \approx 308$$

Turns per phase, $T_s = \dfrac{Z}{6} = \dfrac{308}{6} = 51.3 \approx 52$

New value of total stator conductors, $Z = 6 \, T_s = 6 \times 52 = 312$

RESULT

Output coefficient,	C_o =	189 kVA/m^3-rps
Diameter of stator bore,	D =	1.2852 m
Length of stator core,	L =	0.4801 m
Total number of stator conductors,	Z =	312

EXAMPLE 9.6　　　　　　　　　　　　　　　　　　　　　　　　　*(AU, Apr' 19, 13 M)*

Calculate the MMF required for the air-gap of a salient pole synchronous machines having core length of 0.2 m including 4 ducts of 10 mm each, pole arc = 0.19 m, slot pitch = 65.4 mm, slot opening = 5 mm, air-gap length = 5 mm, flux per pole = 52 mWb, carter's coefficient for slot opening = 0.18, carter's coefficient for ducts = 0.28.

GIVEN DATA

L = 0.2 m	y_{ss} = 65.4 mm	ϕ = 52 mWb
l_j = 5 mm	W_o = 5 mm	b = 0.19 m
n_d = 4	W_d = 10 mm	

Carter's coefficient for slot opening, K_{cs} = 0.18

Carter's coefficient for ducts, K_{cd} = 0.28

SOLUTION

Flux density at centre of pole, $B_g = \dfrac{B_{av}}{\psi} = \dfrac{p\phi/\pi DL}{b/\tau}$

$\qquad\qquad\qquad\qquad B_{av} = \dfrac{p\phi}{\pi DL}$

$$= \dfrac{p\phi}{\pi DL} \times \dfrac{\pi D}{bp} = \dfrac{\phi}{Lb}$$

$\qquad\qquad\qquad \psi = \dfrac{b}{\tau} \qquad \tau = \dfrac{\pi D}{p}$

$$= \dfrac{52 \times 10^{-3}}{0.2 \times 0.19} = 1.368 \; Wb/m^2$$

Gap contraction factor for slot opening, $K_{gs} = \dfrac{y_{ss}}{y_{ss} - K_{cs} w_o}$

$\qquad\qquad\qquad\qquad$ Refer Chapter-1

$$= \dfrac{65.4}{65.4 - (0.18 \times 5)} = 1.014$$

Gap contraction factor for ducts, $K_{gd} = \dfrac{L}{L - K_{cd} \, n_d \, w_d}$

$\qquad\qquad\qquad\qquad$ Refer Chapter-1

$$= \dfrac{0.2}{0.2 - (0.28 \times 4 \times 10 \times 10^{-3})} = 1.059$$

$\therefore$ mmf required for air-gap, $AT_g = 800,000 \; K_{gs} \, K_{gd} \, B_g \, l_g$

$$= 800,000 \times 1.014 \times 1.059 \times 1.368 \times 5 \times 10^{-3}$$

$$= 5876 \; AT$$

RESULT

mmf required for air-gap, AT_g = 5876 AT

EXAMPLE 9.7

A 1500 kVA, 300 rpm, 3-phase, 50 Hz, 2400 V, star connected salient pole alternator has the following design data: Stator bore = 2.2 m, core length = 0.4 m, slot per pole per phase = 4, conductors per slot = 4, leakage factor = 1.4, winding factor = 0.955. The flux density in pole core is 1.5 Wb/m^2, the winding depth is 30 mm, the ratio of full load field mmf to armature mmf is 2.1, field winding space factor is 0.84 and the field winding dissipated 1800 W/m^2 of inner and outer surface without the temperature rise exceeding the permissible limit. Leave 30 mm for insulation, flanges and height of pole shoe along the height of pole. Find the flux per pole, length and width of pole, winding height and pole height.

GIVEN DATA

1500 kVA	D = 2.2 m	K_{ws} = 0.955	B_p = 1.5 Wb/m^2
300 rpm	L = 0.4 m	d_f = 30 mm = 0.03 m	Conductors per slot = 4
3-phase	q = 4	q_f = 1800 W/m^2	AT_{fl}/AT_a = 2.1
V_L = 2400 V	C_l = 1.4	S_{fc} = 0.84	Clearance for insulation = 30 mm
			= 0.03 m

SOLUTION

Number of poles, $p = \dfrac{120\, f}{N}$

$$= \dfrac{120 \times 50}{300} = 20$$

Total number of slots, S_s = Number of phase × Number of poles × Slots per pole

$$= 3 \times 20 \times 4 = 240$$

Total number of conductors, Z = Slots × Conductors per slots

$$= 240 \times 4 = 960$$

Turns per phase, $T_s = \dfrac{Z}{6} = \dfrac{960}{6} = 160$

Voltage per phase, $E_s = \dfrac{2400}{\sqrt{3}} = 1385.6\ V$

> Star connected alternator.

Flux per pole, $\phi = \dfrac{E_s}{4.44\, f\, T_s\, K_{ws}}$

$$= \dfrac{1385.6}{4.44 \times 50 \times 160 \times 0.955}$$

$$= 0.0408\ Wb = 40.8 \times 10^{-3}\ Wb$$

Flux in pole body, $\phi_p = C_l\, \phi$

$$= 1.4 \times 0.0408$$

$$= 0.0571\ Wb = 57.1 \times 10^{-3}\ Wb$$

Area of cross-section of pole body, $A_p = \dfrac{\phi_p}{B_p}$

$$= \dfrac{57.1 \times 10^{-3}}{1.5} = 0.038\ m^2$$

Length of pole body, L_p = Length of armature core, $L = 0.4\ m$

Width of pole body, $b_p = \dfrac{A_p}{L_p}$

> $kVA = \sqrt{3}\ V_L\, I_L \times 10^{-3}$
> Since star connected,
> $\quad I_s = I_L$

$$= \dfrac{0.038}{0.4} = 0.095\ m$$

Current in each phase, $I_s = \dfrac{kVA}{\sqrt{3}\ V_L \times 10^{-3}}$

$$= \dfrac{1500}{\sqrt{3} \times 2400 \times 10^{-3}} = 360.84\ A$$

Armature mmf per pole, $AT_a = 2.7\, \dfrac{I_s\, T_s\, k_{ws}}{p}$

$$= \dfrac{2.7 \times 360.84 \times 160 \times 0.955}{20} = 7443\ AT$$

Given that, $\dfrac{AT_{fl}}{AT_a} = 2.1$

$\therefore$ Field mmf at full load, $AT_{fl} = 2.1 \times AT_a$

$$= 2.1 \times 7443 = 15630.3 \; AT$$

Heigth of field winding, $h_f = \dfrac{AT_{fl}}{10^4 \times \sqrt{S_{fc} \, d_f \, q_f}}$

$$= \dfrac{15630.3}{10^4 \times \sqrt{0.84 \times 0.03 \times 1800}} = 0.232 \; m$$

Height of pole body, $h_p = h_f +$ Clearance for insulation

$$= 0.232 + 0.03 = 0.262 \; m$$

RESULT

Turns per phase,	T_s	= 160
Voltage per phase,	E_s	= 1385.6 V
Flux per pole,	ϕ	= 40.8 × 10⁻³ Wb
Width of pole body,	b_p	= 0.095 m
Heigth of field winding,	h_f	= 0.232 m
Height of pole body,	h_p	= 0.262 m

EXAMPLE 9.8

A 1000 kVA, 3-phase, 6600 V, salient pole synchronous machine has following design data. Design a damper winding for the synchronous machine. Diameter of stator bore = 1.55 m, length of stator = 0.44 m, number of poles = 18, specific electric loading = 28000 $amp.cond./m$. Ratio of pole arc to pole pitch, b/τ = 0.69, stator slot pitch = 30 mm. Damper bar current density = 3 A/mm^2.

GIVEN DATA

1000 kVA	L = 0.44 m	ac = 28000 $amp.cond./m$.
3-phase	D = 1.55 m	$\dfrac{b}{\tau} = 0.69$
6600 V	p = 18	y_{ss} = 30 mm
		δ_d = 3 A/mm^2

SOLUTION

Pole pitch, $\tau = \dfrac{\pi D}{p}$

$$= \dfrac{\pi \times 1.55}{18} = 0.2705 \; m$$

Number of damper bars over a pole, $N_d = \dfrac{b}{0.8 \, y_{ss}} = \dfrac{0.69 \, \tau}{0.8 \, y_{ss}}$

$$\boxed{\dfrac{b}{\tau} = 0.69}$$

$$= \dfrac{0.69 \times 0.2705}{0.8 \times 30 \times 10^{-3}} = 7.7769 \simeq 8$$

$$\left.\begin{array}{l}\text{Total area of damper} \\ \text{bars per pole}\end{array}\right\} A_d = \frac{0.2 \, \mathbf{ac} \, \tau}{\delta_d}$$

$$= \frac{0.2 \times 28000 \times 0.2705}{3} = 505 \; mm^2$$

Area of each bar, $a_d = \dfrac{A_d}{N_d}$

$$= \frac{505}{8} = 63.125 \; mm^2$$

Diameter of damper bars, $d_d = \sqrt{\dfrac{4}{\pi} \times a_d}$

$$\boxed{a_d = \frac{\pi}{4} \, d_d^2}$$

$$= \sqrt{\frac{4}{\pi} \times 63.125} = 8.965 \; mm$$

Length of damper bar, $L_d = 1.1 \times L$

$$= 1.1 \times 0.44 = 0.484 \; m$$

Area of end ring, $A_{d\_ring} = 0.85 \times A_d$

$$= 0.85 \times 505 = 429.25 \; mm^2$$

RESULT

Number of damper bars over a pole,	N_d	= 8
Diameter of damper bars,	d_d	= 8.965 mm
Length of damper bar,	L_d	= 0.484 m
Area of end ring,	A_{d_ring}	= 429.25 mm^2

EXAMPLE 9.9

The field coils of a salient pole alternator are wound with a single layer winding of bare copper strip 32 mm deep, with separating insulation 0.17 mm thick. Determine a suitable winding length, number of turns and thickness of conductor to develope an mmf of 10000 AT with a potential difference of 5 V per coil and with a loss of 1200 W/m^2 of total coil surface. The mean length of turn is 1.1 m. The resistivity of copper is 0.023 × 10⁻⁶ Ω-m.

GIVEN DATA

$AT_{fl} = 10000 \; AT$ $\rho = 0.023 \times 10^{-6} \; \Omega\text{-}m$

$L_{mtf} = 1.1 \; m$ $d_f = 32 \; mm = 0.032 \; m$

$E_f = 5 \; V$ Insulation thickness = 0.17 mm

$q_f = 1200 \; W/m^2$

SOLUTION

Area of field conductor, $a_f = \dfrac{AT_{fl} \, \rho \, L_{mtf}}{E_f}$

$$= \frac{10000 \times 0.023 \times 10^{-6} \times 1.1}{5}$$

$$= 50.6 \times 10^{-6} \; m^2 = 50.6 \; mm^2$$

Let the cross-section of field conductor be rectangle.

$\therefore$ Area of cross-section, $a_f = h_{fc} \times d_f$

$\therefore$ Thickness or height of conductor, $h_{fc} = \dfrac{a_f}{d_f}$

$$= \frac{50.6}{32} = 1.58 \simeq 1.6 \ mm$$

$\therefore$ New value of area of conductor, $a_f = d_f \times h_{fc}$

$$= 32 \times 1.6 = 51.2 \ mm^2$$

Height occupied by each conductor $= h_{fc} +$ Insulation thickness

$$= 1.6 + 0.17 = 1.77 \ mm$$

$\therefore$ Height winding, $h_f = T_f \times$ Hight occupied by each conductor $= T_f \times 1.77 \times 10^{-3}$ \qquad .....(1)

Total heat dissipating surface, $S = 2 \ L_{mtf} \ (h_f + d_f)$

$$= 2 \times 1.1 \times (h_f + 0.032)$$

$$= 2.2 \ h_f + 0.0704$$

Total loss dissipated, $Q_f = q_f \ S = 1200 \ (2.2 \ h_f + 0.0704)$

$$= 2640 \ h_f + 84.48$$

$\therefore$ Field current, $I_f = \dfrac{Q_f}{E_f}$

$$= \frac{2640 \ h_f + 84.48}{5} = 528 \ h_f + 16.896 \qquad .....(2)$$

Given that,

Field mmf, $AT_{fl} = 10000 \ AT$

Also, Field mmf, $AT_{fl} = I_f \ T_f$

$\therefore I_f \ T_f = 10000 \ AT$

$(528 \ h_f + 16.896) \ T_f = 10000$ \hfill $\boxed{\text{Using equation (2)}}$

$(528 \times T_f \times 1.77 \times 10^{-3} + 16.896) \ T_f = 10000$ \hfill $\boxed{\text{Using equation (1)}}$

$0.935 \ T_f^2 + 16.896 \ T_f = 10000$

$\therefore 0.935 \ T_f^2 + 16.896 \ T_f - 10000 = 0$

The root of above quadratic equation are,

$$T_f = \frac{-16.896 \pm \sqrt{16.896^2 + 4 \times 0.933 \times 10000}}{2 \times 0.935} = \frac{-16.896 \pm 194.127}{2 \times 0.935}$$

$\boxed{\text{Taking only positive root.}}$

$$= 94.78 \approx 95$$

Number of turns in field coil, $T_f = 95$

Height of winding, $h_f = T_f \times 1.77 \times 10^{-3} = 95 \times 1.77 \times 10^{-3} = 0.168 \ m = 168 \ mm$

RESULT

Height of field winding,	h_f	=	168 mm
Number of turns in field coil,	T_f	=	95
Thickness of field conductor,	h_{fc}	=	1.6 mm

EXAMPLE 9.10

(AKTU, Dec' 19, 6 M)

A 3000 *rpm*, 50 *Hz*, 3-phase turbo alternator has a core length of 0.92 *m*. The average gap density is 0.44 *Wb/m²* and the ampere conductors per metre are 23000. The peripheral speed of rotor is 95 *m/s* and the length of air-gap is 18 *mm*. Find the kVA output of the machine when the coils are (i) full pitch (ii) chorded by 1/3 pole pitch. The winding can be taken as infinitely distributed with a phase spread of 60°.

GIVEN DATA

3000 *rpm*	l_g = 18 *mm* = 0.018 *m*
50 *Hz*	V_a = 95 *m/s*
3-phase	B_{av} = 0.44 *Wb/m²*
L = 0.92 *m*	**ac** = 23000 *amp.cond./m.*

SOLUTION

Synchronous speed, $n_s = \dfrac{N}{60}$

$$= \frac{3000}{60} = 50 \text{ rps}$$

Peripheral speed, $V_a = \pi \, D_r \, n_s$

$\therefore$ Diameter of rotor, $D_r = \dfrac{V_a}{\pi \, n_s}$

$$= \frac{95}{\pi \times 50} = 0.6048 \ m$$

Diameter of stator bore, $D = D_r + 2l_g$

$$= 0.6048 + (2 \times 0.018) = 0.6408 \ m$$

With infinite distribution and 60° phase spread,

Distribution factor, $K_d = 0.955$

Case (i)

With full pitch coils,

Pitch factor, $K_p = 1$

$\therefore$ Winding factor, $K_{ws1} = K_d \times K_p$

$$= 0.955 \times 1 = 0.955$$

kVA output, $Q_1 = C_o D^2 L n_s$

$$\boxed{C_o = 11 B_{av} \ ac \ K_{ws} \times 10^{-3}}$$

$$= 11 B_{av} \ ac \ K_{ws1} \times 10^{-3} \times D^2 L n_s$$

$$= 11 \times 0.44 \times 23000 \times 0.955 \times 10^{-3} \times 0.6408^2 \times 0.92 \times 50$$

$$= 2008 \ kVA$$

Case (ii)

Angle of chording, $\alpha = 180/3 = 60°$

$\therefore$ Pitch factor, $K_p = \cos \alpha/2 = \cos 60/2 = 0.866$

$\therefore$ Winding factor, $K_{ws2} = K_d \times K_p$

$$= 0.955 \times 0.866 = 0.827$$

Since, specific loadings, diameter, length and speed remaining the same, the output of the machine is directly propotional to the winding factor.

$$\therefore \frac{Q_2}{Q_1} = \frac{K_{ws2}}{K_{ws1}}$$

$\therefore$ kVA output, $Q_2 = Q_1 \times \dfrac{K_{ws2}}{K_{ws1}}$

$$= 2008 \times \frac{0.827}{0.955} = 1739 \ kVA$$

RESULT

Diameter of rotor,	D_r	=	$0.6048 \ m$
Diameter of stator bore,	D	=	$0.6408 \ m$
kVA output with full pitched winding,	Q_1	=	$2008 \ kVA$
kVA output with short pitched winding,	Q_2	=	$1739 \ kVA$

EXAMPLE 9.11

A 3-phase, 1800 kVA, 3.3 kV, 50 Hz, 250 rpm, salient pole alternator has the following design data. Stator bore diameter = 230 cm, gross length of stator bore = 38 cm, number of stator slot = 216, number of conductors per slot = 4 and sectional area of stator conductor = 86 mm^2. Calculate flux per pole, flux density in the air-gap, current density and size of stator slot.

GIVEN DATA

1800 kVA	$D = 230 \ cm = 2.3 \ m$
3.3 kV	$L = 38 \ cm = 0.38 \ m$
50 Hz	$S_s = 216$
250 rpm	$Z_s = 4$
3-phase	$a_a = 86 \ mm^2$

SOLUTION

Total number of conductors, Z = Slots × Conductor per slot

$$= 216 \times 4 = 864$$

Turns per phase, $T_{ph} = \dfrac{Z}{6} = \dfrac{864}{6} = 144$

Emf per phase, $E_{ph} = \dfrac{3300}{\sqrt{3}} = 1905\,V$

Flux per pole, $\phi = \dfrac{E_{ph}}{4.44\,T_{ph}\,f\,K_w}$ $\boxed{\text{Let, } K_w = 0.955}$

$\qquad = \dfrac{1905}{4.44 \times 144 \times 50 \times 0.955} = 0.0624\,Wb = 62.4\,mWb$

Poles, $p = \dfrac{120f}{N} = \dfrac{120 \times 50}{250} = 24$

Pole pitch, $\tau = \dfrac{\pi D}{p} = \dfrac{\pi \times 2.3}{24} = 0.3011\,m$

Flux density in air-gap, $B_{av} = \dfrac{\phi}{\tau\,L} = \dfrac{0.0624}{0.3011 \times 0.38} = 0.5454\,Wb/m^2$

$\boxed{\begin{array}{c} kVA = \sqrt{3}\;V_L\,I_L \times 10^{-3} \\ \text{In star connection,} \\ I_s = I_L \end{array}}$

Current per phase, $I_s = \dfrac{kVA \times 1000}{\sqrt{3} \times V_L} = \dfrac{1800 \times 1000}{\sqrt{3} \times 3300} = 315\,A$

Current density, $\delta = \dfrac{I_s}{\text{Area of cross-section of conductor}} = \dfrac{315}{86} = 3.7\,A/mm^2$

Let us choose double layer lap winding with two turns per coil side or conductor.

Let, each conductor is made of two strips in parallel and with dimensions of each strip as $3 \times 6\ mm$.

$\quad\therefore$ Area of cross-section, $a_a = 2 \times 2 \times 3 \times 7 = 84\ mm^2$

New current density, $\delta_a = \dfrac{I_s}{a_a} = \dfrac{315}{84} = 3.75\,A/mm^2$

There are 4 conductors per slot and so it is possible to arrange 2 conductors in top layer and 2 conductors in bottom layer. Therefore, there will be 2 turns per coil.

Let 2 mm thick insulation strips between layers and 0.5 mm thick conductor insulation. The thickness of main wall insulation is 2.5 mm.

Slot width estimation:

Bare conductor,		$= 7\ mm$
Conductor insulation,	$2 \times 0.5 =$	$1\ mm$
Main slot insulation,	$2 \times 2.5 =$	$5\ mm$
Slot width,		$w_s = 13\ mm$

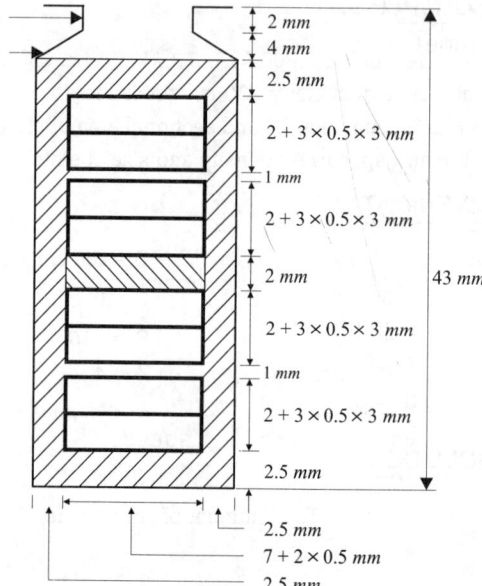

Fig. 1: *Conductors arrangement in slot.*

Slot depth estimation:

Bare conductor, 3×8 = 24 *mm*

Insulation between layer, 1×2 = 2 *mm*

Conductor insulation, 12×0.5 = 6 *mm*

Main slot insulation, 2×2.5 = 5 *mm*

Slot wedge,   = 4 *mm*

Slot lip,   = 2 *mm*

Slot depth, d_s = 43 *mm*

RESULT

Flux per pole, ϕ = 62.4×10^{-3} *Wb*

Flux density in air-gap, B_{av} = 0.5454 *Wb/m²*

Current density, δ_a = 3.5 *A/mm²*

Size of stator slot, $w_s \times d_s$ = 13×43 *mm*

9.8 COMPUTER PROGRAMS

PROGRAM 9.1

Write a C program to estimate the output coefficient, number of stator conductor, diameter and length of stator bore of a 3-phase star connected alternator for the specifications in Example 9.5.

Method-1

```
/* Program to estimate output coefficient, Number of stator conductor,
diameter and length of stator bore*/
    /* VARIABLE DECLARATION
    ......................................
    Kd = Distribution factor for a phase spread of 60 degree.
    Kc = Chording factor for full pitched coils.
    Bav = Specific magnetic loading
    ac = Specific electric loading.
    p = Number of poles.
    f= Frequency.
    Q = Kva rating.
    Va = Peripheral speed.
    lg = Air-gap length.
    ......................................*/
    #include<stdio.h>
    #include<conio.h>
    #include<math.h>
```

```
#include<iostream.h>
void main()
{
int Kc,p,f,Q,Va,Z;
float Kd,Kws,Bav,ac,D2L,Dr,D,L,Co,lg,ns;
clrscr();
printf("Enter Distribution factor for a phase spread of 60 degree, Kd = ");
scanf("%f", &Kd);
printf("Enter Chording factor for full pitched coils, Kc = ");
scanf("%d", &Kc);
printf("Enter Specific Magnetic loading, Bav =");
scanf("%f", &Bav);
printf("Enter Specific electric loading, ac =");
scanf("%d", &ac);
printf("Enter number of poles, p =");
scanf("%d", &p);
printf("Enter frequency, f =");
scanf("%d", &f);
printf("Enter kVA rating, Q =");
scanf("%d", &Q);
printf("Enter peripheral speed, Va = ");
scanf("%d", &Va);
printf("Enter Air-gap length, lg =");
scanf("%d", &lg);

Kws = Kd*Kc;                /* Kws = Stator winding factor.*/
Co = 11*Bav*ac*Kws*10e-4;   /* Co = Output coefficient.*/
ns = 2*f/p;                 /* ns = Synchronous speed.*/
D2L = Q/(Co*ns);            /* D2L = Diameter with length of stator core.*/
Dr = Va/(3.14*ns);          /* Dr = Diameter of rotor.*/
D = Dr+2*lg*10e-4;          /* D = Diameter of stator bore.*/
L = D2L/(D*D);              /* L = Length of stator bore.*/
Z = 6*Ts;                   /* Z = Number of stator conductor.*/
printf("OUTPUT\n");
printf("Output coefficient, Co = %8.1f KVA/m3\n", Co);
printf("Diameter of stator bore, D = %8.4f m\n", D);
printf("Length of stator bore, L = %8.4f m\n", L);
printf("Number of stator conductor, Z = %d\n",Z);
getch();
}
```

Input data

Enter Distribution factor for a phase spread of 60 degree, Kd = 0.955
Enter Chording factor for full pitched coils , Kc = 1
Enter Specific magnetic loading,   Bav = 0.6
Enter Specific electric loading,   ac = 30000
Enter number of poles,   p   = 10
Enter frequency,   f   = 50
Enter kVA rating,   Q   = 1500
Enter Peripheral speed,   Va = 40
Enter Air-gap length,   lg = 6

Output

Output coefficient,   Co =   189.1 KVA/m3
Diameter of stator bore,   D =   1.2859 m
Length of stator bore,   L =   0.4798 m
Number of stator conductor, Z =   312

Method-2

/* Program to estimate output coefficient, number of stator conductor,
diameter and length of stator bore */

```
#include<stdio.h>
#include<conio.h>
#include<math.h>
#include<iostream.h>
void main()
{
clrscr();
int Kc,p,f,Q,Va,Z,Ts;
float Kd,Kws,Bav,ac,D2L,Dr,D,L,Co,lg,ns;
Kd = 0.955;              /* Kd = Distribution factor for a phase spread of
                            60 degree.*/
Kc = 1;                  /* Kc = Chording factor for full pitched coils.*/
Bav = 0.6;               /* Bav = Specific magnetic loading.*/
ac = 30000;              /* ac = Specific electric loading.*/
p =10;                   /* p = Number of poles.*/
f= 50;                   /* f= Frequency.*/
Q = 1500;                /* Q = Kva rating.*/
Va= 40;                  /* Va = Peripheral speed.*/
lg = 6;                  /* lg = Air-gap length.*/
Ts = 52;                 /* Ts = Turns per phase. */
```

```
Kws = Kd*Kc;                        /* Kws = Stator winding factor.*/
Co = 11*Bav*ac*Kws*10e-4;           /* Co = Output coefficient.*/
ns = 2*f/p;                         /* ns = Synchronous speed.*/
D2L = Q/(Co*ns);                    /* D2L = Diameter with length of stator
                                           core.*/
Dr = Va/(3.14*ns);                  /* Dr = Diameter of rotor.*/
D = Dr+2*lg*10e-4;                  /* D = Diameter of stator bore.*/
L = D2L/(D*D);                      /* L = Length of stator bore.*/
Z = 6*Ts;                           /* Z = Number of stator conductor.*/

printf("OUTPUT\n");
printf("Output coefficient, Co = %8.1f KVA/m3\n", Co);
printf("Diameter of stator bore, D = %8.4f m\n", D);
printf("Length of stator bore, L = %8.4f m\n", L);
printf("Number of stator conductor, Z = %d\n",Z);
getch();
}
```

Output

```
Output coefficient,           Co = 189.1 KVA/m3
Diameter of stator bore,      D  = 1.2859 m
Length of stator bore,        L  = 0.4798 m
Number of stator conductor, Z   = 312
```

PROGRAM 9.2

Write a C program to estimate the number of stator conductor, number of stator slots and area of cross-section of stator conductor of a 3-phase star connectrd alternator for the specifications in Example 9.4.

```
/*Program to estimate number of stator conductor, slots and conductor
area of cross-section.*/
#include<stdio.h>
#include<conio.h>
#include<math.h>
#include<iostream.h>
void main()
{
clrscr();
int f,p,ph,q,Ss,Q,VL,Ts;
float PHI,Kws,Z,Is,DEL,Aa;
f = 50;                             /* f = Frequency.*/
PHI = 0.1192;                       /* PHI = Flux per pole.*/
Kws = 0.955;                        /* Kws = Stator winding factor.*/
p = 12;                             /* p = Number of poles.*/
```

```
ph = 3;                        /* ph = Number of phase.*/
q = 3;                         /* q = Slots per pole per phase.*/
Q = 250;                       /* Q = kVA rating.*/
VL = 1000;                     /* VL = Voltage.*/
DEL = 3.5;                     /* DEL = Current density.*/

Ts = VL/1.732/(4.44*f*PHI*Kws);   /* Ts = Turns per phase.*/
Z = 6*Ts;                      /* Z = Number of stator conductor.*/
Ss= ph*p*q;                    /* Ss = Number of stator slots.*/
Is = Q/(1.7328*VL*10e-4);      /* Is = Current per phase.*/
Aa = Is/DEL;

printf("OUTPUT\n");
printf("Total number of stator conductor, Z = %8.1f\n", Z);
printf("Number of stator slots, Ss = %d\n", Ss);
printf("Area of crosssection of stator conductor Aa = %8.4f mm2\n", Aa);
getch();
}
```

Output

```
Total number of stator conductor,          Z  = 138
Number of stator slots,                    Ss = 108
Area of cross-section of stator conductor, Aa = 41.24 mm2
```

PROGRAM 9.3

Write a C program to estimate the diameter and length of synchronous generator for the specifications in Example 9.2.

```
/*Program to determine diameter and length for 3 cases

1. Ratio of length to diameter.
2. Specific magnetic loading reduced by 10%
3. Speed is decreased to 150 rpm. */

#include<stdio.h>
#include<conio.h>
#include<math.h>
#include<iostream.h>
void main()
{
clrscr();
int Q;
float LD,D2L,D,L,Bav,ns3,Co,ns;
LD = 0.2;                      /* LD = Ratio of length to diameter.*/
Q = 1250;                      /* Q = Kva rating. */
```

```c
Co = 200;                        /* Co = Output coefficient.*/
ns = 5;                          /* ns = synchronous speed.*/
Bav = 0.9;                       /* Bav = Specific loading reduced by 10% */
ns3 = 2.5;                       /* ns3 = Speed is reduced to 150 rpm. */

D2L = Q/(Co*ns);                 /*  D2L = Square of diameter with length.*/
D = pow(D2L/(LD),0.33);          /* D = Diameter of stator bore.*/
L = LD*D;                        /* L = Length of stator core. */

printf("OUTPUT\n");
printf("CASE1\n");
printf("Diameter of stator bore, D = %8.4f m\n", D);
printf("Length of stator core, L = %8.4f m\n", L);
printf("................................\n");

D = pow(Q/(Bav*Bav*Co*ns)/(LD) , 0.33);   /* D = Diameter of stator bore.*/
L = LD*D;                                   /* L = Length of stator core. */


printf("CASE2\n");
printf("Diameter of stator bore, D = %8.4f m\n",D);
printf("Length of stator core, L = %8.4f m\n", L);
printf("................................\n");
D = pow(Q/(Co*ns3)/(LD),0.33);            /* D = Diameter of stator bore.*/
L = LD*D;                                   /* L = Length of stator core. */

printf("CASE3\n");
printf("Diameter of stator bore, D = %8.4f m\n", D);
printf("Length of stator core, L = %8.4f m\n", L);
printf("................................\n");
getch();
}
```

Output

```
CASE 1
Diameter of stator bore, D = 1.8421 m
Length of stator core,   L = 0.3402 m
................................
CASE 2
Diameter of stator bore, D = 1.9626 m
Length of stator core,   L = 0.3925 m
................................
CASE 3
Diameter of stator bore, D = 2.3013 m
Length of stator core,   L = 0.4603 m
```

PROGRAM 9.4

Write a C program to estimate the diameter of rotor, diameter of stator bore, kVA output with full pitched winding and kVA output with short pitched winding of a 3-phase turbo-alternator for the specifications in Example 9.10.

```
/* Program to estimate diameter of rotor and stator bore, kVA output
with full and short pitched winding.*/
        #include<stdio.h>
        #include<conio.h>
        #include<math.h>
        #include<iostream.h>
        void main()
        {
        clrscr();
        int N,Va,ph,Q1,Q2;
        float L,Dr,lg,D,Kd,Kwsf,Bav,Kps,Kwss,ns,Kpf,ac,ALPHA;
        L = 0.92;                /* L = Length.*/
        N = 3000;                /* N = Speed in rpm.*/
        Va = 95;                 /* Va = Peripheral speed of rotor.*/
        Lg = 0.018;              /* lg = Length of air gap.*/
        Kpf = 1;                 /* Kpf = Pitch factor for full pitch winding.*/
        Kd = 0.955;              /* Kd = Distribution factor.*/
        Bav = 0.44;              /* Bav = Average gap density.*/
        ac = 23000;              /* ac = Specific electric loading.*/
        ph = 3;                  /* ph = Number of phase.*/
        Kps = 0.866;             /* Kps = Pitch factor for short pitch winding.*/

        ns = N/60;                         /* ns = Synchronous speed.*/
        Dr = Va/(3.14*ns);                 /* Dr = Diameter of rotor.*/
        D = Dr+(2*Lg);                     /* D = Diameter.*/
        Kwsf = Kd*Kpf;                     /* Kwsf = Winding factor for full
                                                     pitch winding.*/

        Q1 = 11*Bav*ac*Kwsf*10e-4*D*D*L*ns; /* Q1 = kVA output for full pitch
                                                     winding.*/

        ALPHA = 180/ph;                    /* ALPHA = Angle of chording.*/
        Kwss = Kd*Kps;                     /* Kwss = Winding factor for short
                                                     pitch winding.*/

        Q2 = Qf*(Kwss/Kwsf);               /* Q2 = kVA output for short pitch
                                                     winding.*/
        printf("OUTPUT\n");
        printf("Diameter of rotor, Dr = %8.4f m\n", Dr);
        printf("Diameter of stator bore, D = %8.4f m\n", D);
        printf("Kva output for full pitch winding, Q1 = %d KVA\n", Qf);
```

```
printf("Kva output for short pitch winding, Q2 = %d KVA\n", Qs);
getch();
}
```

Output

Diameter of rotor,	Dr = 0.6051 m
Diameter of stator bore,	D = 0.6411 m
kVA output for full pitch winding,	Q1 = 2009 kVA
kVA output for short pitch winding,	Q2 = 1740 kVA

PROGRAM 9.5

Write a C program to estimate the kVA rating of the machine for a 3-phase synchronous generator for the specifications in Example 9.1.

```
/* Program to estimate kVA rating of machine. */

#include<stdio.h>
#include<conio.h>
#include<math.h>
void main()
{
int N,S,Ssi,Q;
float DEL,D,L,az,Kws,Bav,IzZ,ac,ns;
clrscr();
N = 600;       /* N = Speed in rpm.*/
DEL = 2.7;     /* DEL = Current density.*/
D = 1.92;      /* D = Diameter of stator bore.*/
L = 0.4;       /* L = Length of stator core.*/
S = 144;       /* S = Number of stator slots.*/
az = 0.35;     /* az = Slot space factor.*/
Ssi = 2400;    /* Ssi = Slot size.*/
Bav = 0.48;    /* Bav = Specific magnetic loading.*/
Kws = 0.955;   /* Kws = Stator winding factor.*/

IzZ = DEL*az*S*Ssi;
ac = IzZ/(3.1415*D);   /* ac = Specific electric loading.*/
ns = N/60;             /* ns = Speeed in rps.*/
Q = 11*Bav*ac*Kws*10e-4*D*D*L*ns;   /* Q = kVA rating.*/

printf("OUTPUT\n");
printf("kVA rating, Q = %d kVA\n", Q);
getch();
}
```

Output

kVA rating, Q = 4025 kVA

Program 9.6

Write a C program to estimate the mmf required for air-gap of a salient pole synchronous machine for the specifications in Example 9.6.

```
/* Program to estimate mmf required for air-gap.*/

#include<stdio.h>
#include<conio.h>
#include<math.h>
void main()
{
float L,yss,b,Kcs,Kcd,Kgs,Kgd,Bg,ATg,PHI,lg,nd,Wo,Wd;
clrscr();
L = 0.2;          /* L = Length of stator core.*/
yss = 65.4;       /* yss = Stator slot pitch.*/
Wo = 5;           /* Wo = Slot opening.*/
Wd = 10;          /* Wd = Slot ducts.*/
b =0.19;          /* b = Pole arc.*/
PHI = 52;         /* PHI = Flux per pole.*/
lg = 5;           /* lg = Length of air-gap.*/
nd = 4;           /* nd = Number of ducts.*/
Kcs = 0.18;       /* Kcs = Carter's coefficient for slot opening.*/
Kcd = 0.28;       /* Kcd = Carter's coefficient for ducts.*/

Bg = PHI*10e-3/(L*b);     /* Bg = Maximum flux density in air-gap.*/
Kgs = yss/(yss-(Kcs*Wo)); /* Kgs = Gap contraction factor for slot
                                    opening.*/
Kgd = L/(L-Kcd*nd*Wd*10e-3);      /* Kgd = Gap contraction factor for
                                          ducts.*/
ATg = 800000*Kgs*Kgd*Bg*lg*10e-3; /* ATg = Mmf required for air-gap.*/

printf("OUTPUT\n");
printf("mmf required for air-gap, ATg = %d AT\n", ATg);
getch();
}
```

Output

```
mmf required for air-gap, ATg = 5876 AT
```

PROGRAM 9.7

Write a C program to estimate the Number of damper bars, Diameter and length of damper bars, Area of end ring of a salient pole synchronous machine for the specifications in Example 9.8.

```c
/*Program to estimate the Number of damper bars, diameter and length of
damper bars, Area of end ring. */

#include<stdio.h>
#include<conio.h>
#include<math.h>
void main()
{
int p,Nd;
float L,D,ac,Dow,Ad,ad,dd,Ld,Ad_ring,Nd,DELd,yss;
clrscr();
L = 0.44;    /* L = Length of stator core.*/
D = 1.55;    /* D = Diameter of stator bore.*/
p = 18;      /* p = Number of poles.*/
yss = 30;    /* yss = Stator slot pitch. */
DELd = 3;    /* DELd = Current density in damper bar.*/
ac = 28000;  /* ac = Specific electric loading.*/

Dow = 3.1415*D/(p);              /* Dow = Pole pitch.*/
Nd = 0.69*Dow/(0.8*yss*10e-3);   /* Nd = Number of damper bars over a
                                         pole.*/
Ad = 0.2*ac*Dow/(DELd);    /* Ad = Total area of damper bars per pole.*/
ad = Ad/(Nd);              /* ad = Area of each bar.*/
dd = sqrt(4/(3.1415)*ad);  /* dd = Diameter of damper bars.*/
Ld = 1.1*L;                /* Ld = Length of damper bar.*/
Ad_ring = 0.85*Ad;         /* Ad_ring = Area of end ring.*/

printf("OUTPUT\n");
printf("Number of damper bars over a pole, Nd = %8.4f\n", Ad);
printf("Diameter of damper bars, dd = %8.3f mm\n",dd);
printf("Length of damper bars, Ld = %8.3f m\n", Ld);
printf("Area of end ring, Ad_ring = %8.2f mm2\n", Ad_ring);
getch();
}
```

Output

Number of damper bars,	Nd	= 7.7769
Diameter of damper bars,	dd	= 8.965 mm
Length of damper bars,	Ld	= 0.484 m
Area of end ring,	Ad_ring	= 429.22 mm2

9.9   SHORT-ANSWER QUESTIONS

Q9.1   *Name the two types of synchronous machines.*

Based on construction the synchronous machines may be classified as,

1. Salient pole machines
2. Cylindrical rotor machines.

Q9.2   *What are the two types of poles used in salient pole machines?*

The two types of poles used in salient pole machines are Round poles and Rectangular poles.

Q9.3   *What is runaway speed?*   *(AU, Nov' 18, 2 M) (AU, Nov' 17, 2 M)*

The runaway speed is defined as the speed which the prime mover would have, if it is suddenly unloaded, when working at its rated load.

Q9.4   *List the constructional elements of salient pole synchronous machine.*

The various constructional elements of salient pole synchronous machine are,

Stator	- Frame	**Rotor**	- Field pole
	- Armature core		- Pole shoe
	- Armature winding		- Field winding
			- Damper winding

Q9.5   *What are the constructional elements of cylindrical rotor synchronous machine?*

The constructional elements of cylindrical rotor synchronous machine are,

Stator	- Frame	**Rotor**	- Solid rotor
	- Armature core		- Field conductors or bars
	- Armature winding		

Q9.6   *Define specific magnetic loading of synchronous machine.*

The specific magnetic loading is defined as the average flux density over the air-gap of a machine.

$$\text{Specific magnetic loading, } B_{av} = \frac{\text{Total flux around the air-gap}}{\text{Area of flux path at the air-gap}} = \frac{p\phi}{\pi DL}$$

Q9.7   *Define specific electric loading of synchronous machine.*

The specific electric loading is defined as the number of armature (or stator) ampere conductors per metre of armature (or stator) periphery at the air-gap.

$$\text{Specific electric loading, } \mathbf{ac} = \frac{\text{Total armature ampere conductors}}{\text{Armature periphery at air-gap}} = \frac{I_z Z}{\pi D}$$

Q9.8   Give typical values of specific electric and magnetic loading of synchronous machine.

Machine	Specific magnetic loading, B_{av} in Wb/m^2	Specific electric loading, ac in $amp.cond./m.$
Salient pole machine	0.52 to 0.65	20,000 to 40,000
Turbo-alternator (conventionally cooled)	0.52 to 0.65	50,000 to 75,000
Turbo alternator water cooled	0.52 to 0.65	180,000 to 200,000

Q9.9   What is output equation of synchronous machine?

The equation which relates the kVA rating to the main dimensions (D and L), Specific loadings (B_{av} and **ac**) and speed (n) of a machine is known as output equation.

$$\text{kVA rating, } Q = C_0 \, D^2 \, L \, n_s$$

where, C_0 = Output coefficient = $11 \, B_{av} \, \textbf{ac} \, K_{ws} \times 10^{-3}$, in $kVA/m^3\text{-}rps$

Q9.10 Write the expression for output coefficient in synchronous machine.

Output coefficient, $C_0 = 11 \, B_{av} \, \textbf{ac} \, K_{ws} \times 10^{-3}$, in $kVA/m^3\text{-}rps$

Q9.11 What is the significance of core length/pole arc in synchronous machine?

In synchronous machines the ratio L/τ influences the peripheral speed, short circuit ratio, number of poles.

Q9.12 What are the factors to be considered for the separation of D and L of synchronous machine?

In synchronous machine the separation of D and L depends on pole proportions peripheral speed, number of poles and short circuit ratio.

Q9.13 List the various values of L/τ used for separation of D and L in synchronous machine.

In salient pole synchronous machines, when round poles are used the ratio of L/τ is between 0.6 to 0.7, and when rectangular poles are employed the ratio of L/τ is between 1 to 5.

> **Note :** *In cylindrical rotor synchronous machines, the ratio L/τ is not used for separation of D and L. In this type of machine, the permissible peripheral speed is used to calculate D. Then L is estimated using the value of D in the output equation.*

Q9.14 What is peripheral speed? Write the expression for peripheral speed of a rotating   *(PU, May' 19, 2 M)*
machine.

The peripheral speed is a translational speed that may exist at the surface of the rotor, while it is rotating. (It is a translational speed equivalent to the angular speed at the surface of the rotor).

$$\text{Peripheral speed, } V_a = \pi D_r \, n$$

where, D_r = Diameter of rotor

   n = Speed of the rotor.   *(AKTU, Dec' 19, 5 M)*

Q9.15 High speed alternators have very long armature. Why?

In high speed alternators, the peripheral speed will be high and so the diameter has to be kept low to limit the peripheral speed. For a given volume of active part, if the diameter is kept low then the length has to be increased. Therefore the high speed alternators have very long armature.

Q9.16 With a sketch indicate the location of damper windings in a synchronous machine.

In salient pole synchronous machines, dampers are placed on the pole shoes. (In cylindrical rotor synchronous machines, dampers are not used). Fig. Q9.16 shows the location of dampers on the pole shoes.

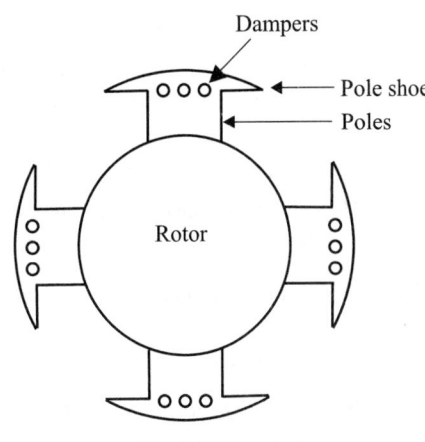

Fig. Q9.16.

Q9.17 What are the prime movers used for (a) Salient pole alternator (b) Non-salient pole alternator.

(a) The prime movers used for salient pole alternators are water wheels like Kaplan turbine, Francis turbine, Pelton wheel, etc., and diesel or petrol engines.

(b) The prime movers used for non-salient pole alternators are steam turbines and gas turbines.

Q9.18 State three important features of turbo alternator rotors.   *(AU, Nov' 17, 2 M)*
                                                                                        (AU, Apr' 17, 2 M)

1. The rotors of turbo alternators have large axial length and small diameters.

2. Damping torque is provided by the rotor itself and so there is no necessity for additional damper winding.

3. They are suitable for high speed operations and so number of poles is usually 2 or 4.

Q9.19 Sketch the shape of rotor of a turbo alternator.

The shape of turbo alternator rotor is cylindrical. The slots are cut on the outer surface of the rotor.

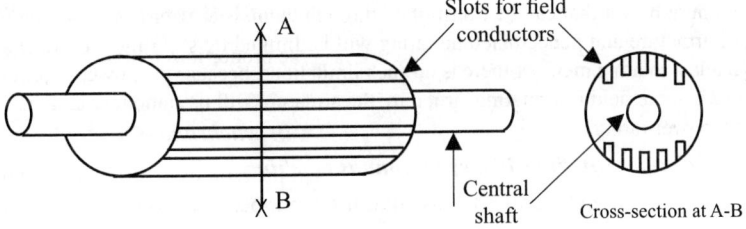

Fig. Q9.19.

Q9.20 Sketch the shape of salient pole rotor for synchronous machine.

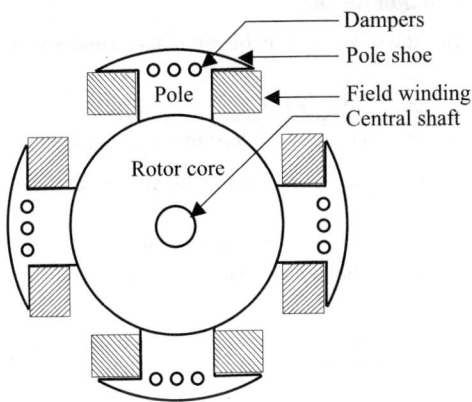

Fig. Q9.20.

Q9.21 Distinguish between cylindrical pole and salient pole construction.
                                                                                          (AU, Nov' 18, 2 M)

In cylindrical pole construction the rotor is made of solid cylinder and slots are cut on the outer periphery of the cylinder to accommodate field conductors.

In salient pole construction, the circular or rectangular poles are mounted on the outer surface of a cylinder. The field coils are fixed on the pole.

The cylindrical pole construction is suitable for high speed operations, whereas the salient pole construction is suitable for slow speed operations.

Q9.22 Salient pole alternators are not suitable for high speeds. Why?

The salient pole rotors cannot withstand the mechanical stresses developed at high speeds. The projecting poles may be damaged due to mechanical stresses.

Q9.23 What is critical speed of alternator?

When the rotor of the alternator has an eccentricity, it may have a deflection while rotating. This deflection will be maximum (or infinite) at a speed called critical speed. When a rotor with eccentricity pass through critical speed, severe vibrations are developed.

Q9.24 What type of prime movers are used in hydroelectric stations depending on the head?

The type of water turbine used in hydroelectric stations depends on water head. Pelton wheel is used for water heads of 400 *m* and above. Francis turbine is used for water heads upto 380 *m*. Kaplan turbine is used for water heads upto 50 *m*.

Q9.25 Why the field structure is a rotating member in 3-phase synchronous machine?

The electrical connection to the rotating member is made through slip ring and brush arrangement which has a limitation of maximum power that can be transmitted through them. If armature is a rotating member then the power rating of armature and hence machine rating will be limited by slip ring and brush arrangement. When the field is made a rotating member there is no such limitations, because the power requirement of field is very low. Hence when the field is a rotating structure, the armature will be stationary and the armature can be designed for any power rating.

Q9.26 Mention the uses of damper windings in a synchronous machine.   *(AU, Apr' 18, 2 M)*

1. The damper winding is used to reduce the oscillations developed in the rotor of alternator when it is suddenly loaded.
2. The damper winding is used to start the synchronous motor as an induction motor.

Q9.27 Define pitch factor and distribution factor.

The pitch factor is defined as the ratio of vector sum of emf induced in a coil to arithmetic sum of emf induced in the coil.

$$\text{Pitch factor, } K_c = \frac{\text{Vector sum of emf induced in a coil}}{\text{Arithmetic sum of emf induced in the coil}}$$

The distribution factor is defined as the ratio of vector sum to arithmetic sum of emf induced in the conductors of one phase spread.

$$\text{Distribution factor, } K_d = \frac{\text{Vector sum of emfs induced in the conductors of a phase under a pole}}{\text{Arithmetic sum of emfs induced in the conductors of a phase under a pole}}$$

Q9.28 Mention the advantages of fractional slot winding.

1.   In low speed machines with large number of poles, the fractional slot winding will reduce tooth harmonics.

2.   A range of machines with different speeds can be designed with a single lamination.

3.   The fractional slot winding reduces the harmonics in mmf and the leakage reactance of the windings.

4.   The fractional slot winding allows only short chorded winding. Therefore the length of mean turn (L_{mts}) of a stator winding coil reduces which results in shorter end connections and so saving in copper.

Q9.29 Why alternators are rated in kVA?

The *kVA* rating of alternators depends on power factor of the load. The power factor in turn depends on the operating conditions. The operating conditions differ from place to place. Therefore the *kVA* rating is specified for all alternators.

Q9.30 What are the factors to be considered for the choice of specific magnetic loading in synchronous machines ?   *(AU, Apr' 18, 2 M) (AU, Apr' 19, 2 M)*

The factors to be considered for the choice of specific magnetic loading are,

1. Iron loss   4. Stability

2. Voltage rating   5. Parallel operation

3. Transient short circuit current.

Q9.31 List the factors to be considered for the choice of specific electric loading in synchronous machines.

The factors to be considered for the choice of specific electric loading are,

1. Copper loss   4. Synchronous reactance

2. Temperature rise   5. Stray load losses

3. Voltage rating.

Q9.32 What is Short Circuit Ratio (SCR)?   *(AU, Apr' 19, 2 M)*

The Short Circuit Ratio (SCR) is defined as the ratio of field current required to produce rated voltage on open circuit to field current required to circulate rated current at short circuit.

It is also given by the reciprocal of synchronous reactance, X_d in p.u (per unit). For turbo-alternators SCR is normally between 0.5 to 0.7. For salient pole alternator SCR varies from 1.0 to 1.5.

Q9.33 How the value of SCR affects the design of alternator.   *(AU, Apr' 17, 2 M)*

For high stability and low regulation, the value of SCR should be high, which requires large air-gap. When the length of air-gap is large, the mmf requirement will be high and so the field system will be large. Hence the machine will be costlier.

Q9.34 What are the advantages of large air-gap in synchronous machines?

The advantages of large air-gap are,

1.   Reduction in armature reaction   5.   Better cooling

2.   Small value of regulation   6.   Lower tooth pulsation loss

3.   Higher value of stability   7.   Less noise

4.   A higher synchronizing power   8.   Smaller unbalanced magnetic pull
     which makes the machine less
     sensitive to load variation.

Q9.35 Write the expressions for length of air-gap in salient pole synchronous machine?

i) Length of air-gap, $l_g = \dfrac{AT_{fo}}{B_g \, K_g \times 10^6}$

ii) Lenght of air-gap, $l_g = \dfrac{AT_a \, SCR \, K_f}{B_{av} \, K_g \times 10^6}$

Q9.36 List the influence of length of air-gap on the performance of the synchronous machine.

The performance characteristics that are influenced by length of air-gap are armature reaction, regulation, stability, synchronising power, cooling, tooth pulsation loss, noise and unbalanced magnetic pull.

The synchronous machines usually have large air-gap. With increase in air-gap length the following factors decreases.

1. Armature reaction   4. Regulation
2. Noise   5. Tooth pulsation loss
3. Unbalanced magnetic pull   6. Sensitivity to load variations

Also the higher value of air-gap length improves the stability, and cooling of machines.

Q9.37 List the factors to be considered for the choice of number of slots in synchronous machine.

The factors to be considered for the choice of number of slots are

1. Balanced winding   4. Leakage reactance
2. Cost   5. Tooth losses
3. Hot spot temperature in winding   6. Tooth flux density.

Q9.38 Determine the total number of slots in the stator of an alternator having 4 poles, 3-phase, 6 slots per pole for each phase?

Total number of slots = Slots per pole per phase × Number of poles × Number of phase

$$= 6 \times 4 \times 3 = 72 \text{ slots.}$$

Q9.39 What is the limiting factor for the diameter of synchronous machine?

The limiting factor for the diameter of synchronous machine is the peripheral speed. The limiting value of peripheral speed is 175 *m/s* for cylindrical rotor machines and 80 *m/s* for salient pole machines.

Q9.40 Write the expression for air-gap length in cylindrical rotor machine.

Length of air-gap, $l_g = \dfrac{0.5 \, SCR \, \mathbf{ac} \, \tau \, K_f \times 10^{-6}}{K_g \, B_{av}}$

Q9.41 What are the factors to be considered for selecting the number of poles in an alternator?

The number of poles depends on the speed of the prime mover and frequency of generated emf.

Q9.42 Discuss how the ventilation and cooling of a large high speed alternator is carried out.

For high speed alternator two cooling methods are available and they are conventional cooling and direct cooling.

In conventional cooling methods, radial and axial ventilating ducts are provided in the core. Cooling is performed by forced circulation of air or hydrogen at a pressure higher than atmosphere.

In direct cooling methods, cooling ducts are provided in the stator and rotor slots or the conductor itself will be in the form of tubes. Coolants like water or oil or hydrogen are circulated in the ducts to remove the heat directly from the conductors.

Q9.43 Mention the factors that govern the design of field system of alternator.

The following are the factors to be considered for the design of field system in alternator.

1. Number of poles and voltage across each field coil

2. Amp-turn per pole (or mmf per pole)

3. Copper loss in field coil

4. Dissipating surface of field coil

5. Specific loss dissipation and allowable temperature rise.

9.10 EXERCISES

I.   Fill in the blanks

1.   The range of specific magnetic loading in alternator is ....

2.   The range of specific electric loading in salient pole alternator is ....

3.   The range of specific electric loading in turbo-alternator is ....

4.   The .... is equal to inverse of synchronous reactance.

5.   In alternators when length of air-gap is increased then stability ....

6.   The .... in alternator is decided by the frequency and speed of operation.

7.   The material used for the construction of rotor of turbo alternator is ....

8.   .... is provided to reduce hunting in alternators.

9.   The value of specific magnetic loading in synchronous machines is .... than DC machines and .... than induction machines.

Answers

1. 0.52 to 0.65 Wb/m^2	6. number of poles
2. 20,000 to 40,000 $amp.cond./m$	7. chromium nickel steel
3. 50,000 to 75,000 $amp.cond./m$	8. Damper winding
4. short circuit ratio	9. lower, higher
5. increases	

II.   State whether the following statements are True/False

1.   The hunting occurs in alternator when it is suddenly loaded.

2.   The short circuit ratio is directly proportional to synchronous reactance.

3.   The synchronous reactance decreases with increase in specific electric loading.

4.   For cylindrical rotor machines the limiting value of peripheral speed is 175 m/s.

5.   In salient pole alternators the air-gap length will not be uniform.

6.   In cylinder rotor machines, the pheripheral speed limits the diameter of rotor.

Answers

1. True	4. True
2. False	5. True
3. False	6. True

III   Unsolved problems

E9.1   Determine for a 15 *MVA*, 11 *kV*, 50 *Hz*, 2 pole, star connected turbo alternator (i) air-gap diameter, (ii) core length, (iii) number of stator conductors, from the given data, B_{av} = 0.55 *Wb/m²*, ac = 36000 *amp.cond./m*, δ = 5 *A/mm²*, synchronous speed, n_s = 50 *rps*, K_{ws} = 0.98, peripheral speed = 160 *m/s*.

$$(D = 1 \ m \ ; L = 1.41 \ m \ ; Z = 144, T_s = 24)$$

E9.2   Calculate the main dimension of 3000 *kVA*, 6600 *V*, 3ϕ, 50 *Hz*, 187.5 *rpm*, star connected alternator, given p.f = 0.8. Choose L/τ = 1.5, V_a = 30 *m/s*, l_g/τ = 0.0165.

$$(l_g = 5 \ mm, D = 3.065 \ m \ ; L = 0.45 \ m, p = 32)$$

E9.3   Determine the main dimensions of a 20 *MVA*, 3000 *rpm*, alternator with the following data. B_{av} = 0.5 *Wb/m²*, ac = 56000 *amp.cond./m*, $V_a \leq$ 150 *m/s*, length of air-gap = 30 *mm* and winding factor = 0.955.

$$(D = 1.015 \ m \ ; L = 1.32 \ m)$$

E9.4   Determine the main dimensions for a three-phase star connected 8 pole alternator rated at 300 *kVA*, 3300 *V*, 50 *Hz*, B_{av} = 0.6 *Wb/m²*, ac = 280 *amp.cond./cm* and pole arc = 0.65τ.

$$(D = 0.81 \ m \ ; L = 0.207 \ m)$$

E9.5   Determine suitable stator dimension for a 500 *kVA*, 3-phase alternator to run at 375 *rpm*. B_{av} = 0.55 *Wb/m²*, ac = 25000 *A/m*. The peripheral speed should not exceed 35 *m/s*.

$$(V_a = 30 \ m/s \ ; D = 1.528 \ m \ ; L = 0.237 \ m)$$

E9.6   A 800 *rpm*, 50 *Hz*, 8000 *V*, 3-phase, synchronous generator has the following design data. B_{av} = 0.55 *Wb/m²*, δ = 3.5 *A/mm²*, slot space factor = 0.43, number of slots = 144, slot size = 120 × 20 *mm*, D = 2.3 *m* and L = 0.5 *m*. Determine the kVA rating of the machine.

$$(Q = 14667 \ kVA)$$

E9.7   For a 400 *kVA*, 1600 *V*, 12 pole, 500 *rpm*, 3-phase alternator, determine air-gap diameter, core length, number of stator conductors, number of stator slots and cross-section of stator conductors. Assuming average gap density as 0.8 *Wb/m²* and specific electric loading of 30,000 *amp.cond./m*. L/τ = 1.9.

$$(D = 0.7249 \ m, L = 0.3625 \ m, Z = 480, S_s = 72, a_a = 41.24 \ mm²)$$

E9.8   A 500 *kVA*, 3-phase, 6600 *V*, salient pole synchronous machine has following design data. Design a damper winding for the synchronous machine. Diameter of stator bore = 0.9 *m*, length of stator = 0.4 *m*, number of poles = 14, specific electric loading = 28000 *amp.cond./m*. Ratio of pole arc to pole pitch, b/τ = 0.5, stator slot pitch = 24 *mm*. Damper bar current density = 2.3 *A/mm²*.

$$(N_d = 6, d_d = 10.2 \ mm, L_d = 0.44 \ m, A_{d\_ring} = 418.2 \ mm²)$$

E9.9   A 1000 *rpm*, 50 *Hz*, 3-phase turbo alternator has a core length of 0.7 *m*. The average gap density is 0.32 *Wb/m²* and the ampere conductors per metre are 21000. The peripheral speed of rotor is 85 *m/s* and the length of air-gap is 16 *mm*. Find the kVA output of the machine when the coils are (i) full pitch (ii) chorded by 1/3 pole pitch. The winding can be taken as infinitely distributed with a phase spread of 60°.

$$(D_r = 0.5411 \ m, D = 0.5731 \ m, Q_1 = 276 \ kVA, Q_2 = 239 \ kVA)$$

ANNA UNIVERSITY (AU)

B.E/ B.TECH. DEGREE EXAMINATION, APRIL/ MAY 2019

Sixth Semester

Electrical and Electronics Engineering

EE 6604 - DESIGN OF ELECTRICAL MACHINES

(Regulation 2013)

(Common to PTEE 6604-Design of Electrical Machines for B.E.(Part-time)-Fifth semester-Electrical and Electronics Engineering)

(Regulation 2014)

Time: 3 Hours   **Maximum:** 100 Marks

Answer all questions

PART A - (10 × 2 = 20 Marks)

1. Define space factor.

    Chapter 6, SA - Q6.52 *[Page No - 6.80]*

2. What are the electrical properties of insulating materials?

    Chapter 2, SA - Q2.4 *[Page No - 2.12]*

3. Write down the carter's coefficient of DC Machine.

    Chapter 3, SA - Q3.25 *[Page No - 3.49]*

4. What are the factors to be considered in the design of commutator of a DC Machine?

    Chapter 6, SA - Q6.71 *[Page No - 6.82]*

5. Define window space factor.

    Chapter 7, SA - Q7.16 *[Page No - 7.81]*

6. What are the methods of cooling of transformer?

    Chapter 7, SA - Q7.44 *[Page No - 7.85]*

7. Write down the equation for output coefficient in an induction motor.

    Chapter 8, SA - Q8.8 *[Page No - 8.81]*

8. Why fractional slot winding is not used for induction motor?

    Chapter 8, SA - Q8.22 *[Page No - 8.83]*

9. What are the factors that influence the choice of specific magnetic loading in a synchronous machine?

    Chapter 9, SA - Q9.30 *[Page No - 9.63]*

10. Define short circuit ratio of a synchronous machine.

    Chapter 9, SA - Q9.32 *[Page No - 9.63]*

PART - B (5 × 13 = 65 Marks)

11. (a) Discuss about the factors that influence the choice of specific electric and magnetic loadings in the design of rotating machines.   (13)

Chapter 6, Section 6.4 *[Page No - 6.12]*

(OR)

(b) Describe the methods of measurements of temperature rise in various parts of an electrical machine.   (13)

Chapter 2, Section 2.8.3 *[Page No - 2.10]*

12. (a) (i) Derive the output equation of DC machine.   (6)

Chapter 6, Section 6.2 *[Page No - 6.5]*

(ii) A 5 *kW*, 250 *V*, 4 pole, 1500 *rpm* DC shunt generator is designed to have a square pole face. The specific magnetic loading and specific electric loading are 0.42 *Wb/m²* and 15000 *ac/m* respectively. Find the main dimensions of the machines. Assume full load efficiency 87% and poles arc to pole pitch ratio is 0.66.   (7)

Chapter 6, Example 6.6 *[Page No - 6.46]*

(OR)

(b) Calculate the MMF required for the air-gap of a salient pole synchronous machines having core length of 0.2 *m* including 4 ducts of 10 *mm* each; pole are = 0.19 *m*. slot pitch = 65.4 *mm*; slot opening = 5 *mm*. Air gap length = 5 *mm*. Flux per pole = 52 *m/Wb*; carter's coefficient is 0.18 for opening/gap = 1; carter's coefficient is 0.28 for opening/gap = 2.   (13)

Chapter 9, Example 9.6 *[Page No - 9.40]*

13. (a) Explain the different methods of cooling of transformers.   (13)

Chapter 7, Section 7.8.1 *[Page No - 7.34]*

(OR)

(b) A single-phase, 400 *V*, 50 *Hz*, transformer is built from stampings having a relative permeability of 1000. The length of the flux path is 2.5 *m*. The area of cross section of the core is 2.5 × 10⁻³ *m²* and the primary winding has 800 turns. Estimate the maximum flux and no-load current of the transformer. The iron loss at the working flux density is 2.6 *W/kg*. Iron weight 7.8 ×10³ *kg/ m³*, stacking factor is 0.9.   (13)

Chapter 7, Example 7.10 *[Page No - 7.55]*

14. (a) Write short notes on: (i) Design of rotor bars and slots. (ii) Design of end rings.   (7+6)

Chapter 8, Section 8.7.2 *[Page No - 8.18]*

Chapter 8, Section 8.7.3 *[Page No - 8.19]*

(OR)

(b) A 15 *kW*, 440 *V*, 50 *Hz*, 3-phase induction motor is built with a stator bore 0.25 *m* and a core length of 0.16. The specific electric loading is 23000 *amp. cond./m*. Using the data of this machine, determine the core dimensions number of stator slots and number of stator conductors for a 11 *kW*, 460 *V*, 6 pole, 50 *Hz* motor. Assume a full load efficiency of 84% and power factor of 0.82 for each machine. The winding factor is 0.955. (13)

Chapter 8, Example 8.12 *[Page No - 8.59]*

15. (a) Briefly discuss the step by step procedure involved in the design of rotor in salient pole synchronous machine. (13)

Chapter 9, Section 9.4.9 *[Page No - 9.17]*

(OR)

(b) Determine the output for a 1500 *kVA*, 2200 *V*, 3-phase, 10 pole, 50 *Hz*, star connected alternator with sinusoidal flux distribution. The winding has 60° phase spread full pitch coils. **ac** = 30000 *amps.Conductor/m*, B_{av} = 0.6 *Wb/m²*. If the peripheral speed of the rotor must not exceed 100 *m/s* and the ratio of pole pitch to core length is to be between 0.6 and find D and L. Asssume an air-gap length of 6 *mm*. Find also the approximate number of stator conductors. (13)

Chapter 9, Example 9.5 *[Page No - 9.38]*

PART - C (1 × 15 = 15 Marks)

16. (a) Explain the various steps involved in the design of armature winding of DC machines. (15)

Chapter 6, Section 6.6 *[Page No - 6.16]*

(OR)

(b) Explain the step by step procedure for the design of field winding of synchronous machines. (15)

Chapter 9, Section 9.4.12 *[Page No - 9.24]*

ANNA UNIVERSITY (AU)

B.E/ B.TECH. DEGREE EXAMINATION, NOVEMBER/DECEMBER 2018

Sixth Semester

Electrical and Electronics Engineering

EE 6604 - DESIGN OF ELECTRICAL MACHINES

(Regulations 2013)

(Also common to PTEE 6604 - Design of electrical machines for B.E.(Part-time) fifth semester - Electrical and Electronics Enginering / Regulation 2014)

Time: 3 Hours   **Maximum**: 100 Marks

Answer All Questions

PART - A (10 × 2 = 20 Marks)

1. Mention the various duty cycles of motor.

   Chapter 4, SA - Q4.18 *[Page No - 4.47]*

2. What are the different conducting materials used in rotating machines?

   Chapter 2, SA - Q2.2 *[Page No - 2.12]*

3. What are the constituents of magnetic circuits in a DC machine.

   Chapter 6, SA - Q6.11 *[Page No - 6.74]*

4. Define specific electric and magnetic loading.

   Chapter 6, SA - Q6.12 *[Page No - 6.75]*

5. What are the advantages of stepped core in transformer?

   Chapter 7, SA - Q7.20 *[Page No - 7.81]*

6. How is iron loss reduced in transformer?

   Chapter 7, SA - Q7.22 *[Page No - 7.82]*

7. Why induction machine is called a rotating transformer?

   Chapter 8, SA - Q8.18 *[Page No - 8.82]*

8. What is cogging? How is it avoided in induction motor?

   Chapter 8, SA - Q8.57 *[Page No - 8.87]*

9. Define run away speed of an alternator.

   Chapter 9, SA - Q9.42 *[Page No - 9.3]*

10. Distinguish salient pole and smooth cylindrical rotor construction of alternators.

   Chapter 9, SA - Q9.21 *[Page No - 9.45]*

PART - B (5 × 13 = 65 Marks)

11. (a) Discuss in detail the desirable properties and classification of insulating materials used in rotating machines. (13)

Chapter 2, Section 2.8 *[Page No - 2.8]*

(OR)

(b) Discuss in detail the factors affecting the choice of a specific electric and magnetic loading in rotating machines. (13)

Chapter 6, Section 6.4 *[Page No - 6.12]*

12. (a) (i) Derive an exprssion for the mmf for air-gap of a slotted armature with ducts. (6)

Chapter 3, Section 3.8 *[Page No - 3.15]*

(ii) Determine the air-gap length of a DC machine from the following data. gross core length = 0.12 m, number of ducts = 1 of 10 mm width, slot pitch = 25 mm, slot width = 10 mm, carters coefficient for slots and ducts = 0.32, gap density at pole center = 0.7 T. Field mmf per pole = 3900 AT, mmf required for iron parts of magnetic circuit = 800 AT. (7)

Chapter 3, Example 3.3 *[Page No - 3.41]*

(OR)

(b) Calculate the diameter and length of armature for a 55 kW, 110 V, 1000 rpm, 4 pole DC shunt generator assuming specific electric and magnetic loading as 26000 $amp.condn/ m$ and 0.5 Wb/ m^2 respectively. The Pole arc should be about 70% of the pole pitch and length of core about 1.1 times the pole arc. Allow 10 A for the field current and assume a voltage drop of 4 V for the armature circuit. Specify the winding used and also determine suitable values for the number of conductors and number of armature slots. (13)

Chapter 6, Example 6.9 *[Page No - 6.51]*

13. (a) (i) Derive the output equation of three-phase transformer. (6)

Chapter 7, Section 7.3 *[Page No - 7.8]*

(ii) Determine the dimensions of core and yoke for a 200 kVA, 50 Hz single-phase core type transformer. A cruciform core is used with distance between centers of adjacent limbs equal to 1.6 times the width of largest lamination. Assume voltage per turn is 14 V, Maximum flux density, B_m = 1.1 Wb/m^2. Window Space factor 0.32, Current density = 3 A/mm^2. Stacking factor = 0.9. The net iron area is 0.56 d^2 for cruciform core. width of largest stamping = 0.85 d. (7)

Chapter 7, Example 7.9 *[Page No - 7.54]*

(OR)

(b) A 250 *kVA*, 6600/400 *V*, three-phase core type transformer has a total loss of 4800 *W* at full load. Transformer tank is 1.25 *m* in height and 1 *m* 0.5 *m* in plan (top view). Design a suitable scheme for cooling tubes if the temperature rise is to be limited to 35°C .The diameter of the tubes is 50 *mm* and are placed 75 *mm* from each other. The average height of the tube is 1.05 *m*. Sp. heat of dissipstion through radiation and convection are 6 and 6.5 *W/m²-°C*. Assume that the convection is improved by 35% due to the provision of tubes. (13)

Chapter 7, Example 7.19 *[Page No - 7.65]*

14. (a) Find the main dimensions, number of radial ventialting ducts, number of stator slots and number of turns per phase of a 3.7 *kW*, 400 *V*, 3-phase, 4 pole, 50 *Hz*, squirrel cage induction motor to be started by a star delta starter. Work out the winding details. Assume average flux density in the air-gap equal to 0.45 *Wb/m²*, *amp. cond./m.* = 23000, η = 0.85, power factor = 0.84. choose main dimension to achieve cheap design. Winding factor = 0.955, Iron stacking factor = 0.9. (13)

Chapter 8, Example 8.6 *[Page No - 8.49]*

(OR)

(b) (i) Derive an expression for the end ring current in three-phase induction motor. (6)

Chapter 8, Section 8.7.2 *[Page No - 8.18]*

(ii) A 11 *kW*, three-phase 6 pole, 50 *Hz*, 220 *V* star connected induction motor has 54 stator slots, each containing conductors. Calculated the value of bar and end ring currents. The number of rotor bars is 64. The machine has an efficiency of 86 percent and a power factor of 0.85. The rotor MMF may be assumed to be 85 percent of stator MMF. Also find the bar and the end ring sections if the current density is 5 *A/mm²* . (7)

Chapter 8, Example 8.7 *[Page No - 8.52]*

15. (a) Determine the main dimensions of a 75000 *kVA*, 13.8 *kV*, 50 *Hz*, 62.5 *rpm*, 3 phase star connected alternator. The peripheral speed should be about 40 *m/s*. Assume average gap density = 0.65 *Wb/m²*, *amp.cond./m* = 40,000, current density = 4 *A/mm²*. Also find the number of stator slots, conductors per slot and conductor area. Assume slot pitch = 55 *mm*. (13)

Chapter 9, Example 9.3 *[Page No - 9.36]*

(OR)

(b) Explain the design procedure for the field system of a salient pole alternator. (13)

Chapter 9, Section 9.4.9 *[Page No - 9.17]*

PART - C (1 × 15 = 15 Marks)

16. (a) The armature of a 10 pole 1000 kW, 500 V, 300 rpm DC generator has a diameter of 1.6 m. There are 450 coils. Determine suitable length and diameter of the commutator, giving details of brushes having regard to commutation conditions and temperature rise. The design limitations are: Peripheral speed of commutator ≤ 20 m/s, pitch of segments ≥ 4, current per brush ≤ 70 A, Temperature rise ≤ 40°C. The other data given are: The brushes span three segments approximately, brush contact drop 1.5 V, coefficient of friction 1.5, brush pressure 20 kN/m^2.

cooling coefficient = $\dfrac{0.012}{1 + 0.1\,V}$. Make suitable assumptions for clearance between brushes, staggering of brushes and end play. (15)

Chapter 6, Example 6.15 *[Page No - 6.61]*

(OR)

(b) (i) Explain the design of damper winding in synchronous machine. (7)

Chapter 9, Section 9.4.10 *[Page No - 9.20]*

(ii) A 250 kVA, 3-phase, 6600 V, salient pole alternator has the following data. Air-gap diameter = 1.6 m; length of core = 0.45 m; number of poles = 20; a_c = 28000; pole arc to pole pitch ratio = 0.68; stator slot pitch = 28 mm; current density in damper winding = 3 A/mm^2. Design a suitable damper winding for the machine. (8)

Chapter 9, Example 9.3 *[Page No - 9.36]*

ANNA UNIVERSITY (AU)

B.E/ B.TECH. DEGREE EXAMINATION, APRIL/ MAY 2018

Sixth Semester

Electrical and Electronics Engineering

EE 6604 - DESIGN OF ELECTRICAL MACHINES

(Regulation 2013)

Time: 3 Hours   **Maximum:** 100 Marks

Answer all questions

PART A - (10 × 2 = 20 Marks)

1. What are the major considerations to evolve a good design of electrical machine ?

    Chapter 1, SA - Q1.8 *[Page No - 1.37]*

2. What are the different types of heat transfer methods found in electrical machines?

    Chapter 7, SA - Q7.2 *[Page No - 7.80]*

3. Mention the two types of armature winding used in DC machine and compare.

    Chapter 5, SA - Q5.3 *[Page No - 5.112]*

4. What factor decides the minimum number of armature coils?

    Chapter 6, SA - Q6.55 *[Page No - 6.81]*

5. Why stepped core are generally used in transformers?

    Chapter 7, SA - Q7.31 *[Page No - 7.83]*

6. The voltage per turn of a 500 *kVA*, 11 *kV*, Δ/Y 3-phase transformer is 8.7 *V*. Calculate the number of turns per phase of LV and HV windings.

    Chapter 7, SA - Q7.39 *[Page No - 7.84]*

7. What happens if the air-gap length of induction motor is doubled ?

    Chapter 8, SA - Q8.47 *[Page No - 8.86]*

8. Name the losses that occur in 3-phase induction motors.

    Chapter 8, SA - Q8.16 *[Page No - 8.82]*

9. What are the factors to be considered for the choice of specific magnetic loading in synchronous machines?

    Chapter 9, SA - Q9.30 *[Page No - 9.63]*

10. Give the purpose of providing damper windings in synchronous machines.

    Chapter 9, SA - Q9.26 *[Page No - 9.62]*

PART - B (5 × 13 = 65 Marks)

11. (a) (i) Classify the insulating materials based on thermal consideration.   (5)

Chapter 2, Section 2.8.1 [Page No - 2.9]

      (ii) A 350 *kW*, 500 *V*, 450 *rpm*, 6 pole DC generator is built with an armature diameter of 0.87 *m* and core length of 0.32 *m*. The lap wound armature has 660 conductors. Calculate the specific electric and magnetic loadings.   (8)

Chapter 1, Example 1.1 [Page No - 1.28]

(OR)

  (b) Derive the heating and cooling curve of an electrical machine.   (13)

Chapter 4, Section 4.4 and 4.5 [Page No - 4.10]

12. (a) (i) Draw the magnetic circuit of DC machine.   (5)

Chapter 6, Section 6.3.1 [Page No - 6.9]

      (ii) Find the main dimensions and the number of poles of a 37 *kW*, 230 *V*, 1400 *rpm* shunt motor so that a square pole face is obtained. The average gap density is 0.5 *Wb/m²* and the amphere conductors per meter are 22000. The ratio of pole arc to pole pitch is 0.7 and the full load efficiency is 90 percent.   (8)

Chapter 6, Example 6.6 [Page No - 6.46]

(OR)

  (b) (i) What are the advantages and disadvantages of large number of poles in a DC machine? (5)

Chapter 6, Section 6.5 [Page No - 6.14]

      (ii) Design a suitable commutator for a 350 *kW*, 600 *rpm*, 440 *V*, 6 pole DC generator having an armature diameter of 0.75 *m*. The number of coils is 288. Assume suitable values wherever necessary.   (8)

Chapter 6, Example 6.15 [Page No - 6.61]

13. (a) (i) What are the salient features of distribution transformer?   (5)

Chapter 7, SA Q7.6 [Page No - 7.80]

      (ii) Estimate the main dimensions including winding conductor area of a 3-phase, Δ-Y core type transformer rated at 300 *kVA*, 6600/440*V*, 50 *Hz*. A suitable core with 3-steps having a circumscribing circle of 0.25 *m* diameter and a leg spacing of 0.4 *m* is available. Emf/turn = 8.5 $V, \delta = 2.5 \, A/mm^2, K_w = 0.28, S_f = 0.9$.   (8)

Chapter 7, Example 7.2 [Page No - 7.44]

(OR)

  (b) (i) List the different method of cooling of transformers.   (5)

Chapter 7, Section 7.8.1 [Page No - 7.34]

(ii) The tank of 1250 kVA, natural oil cooled transformer has the dimensions length, width and height as $1.55 \times 0.65 \times 1.85$ m respectively. The full load loss = 13.1 kW, loss dissipations due to radiations = 6 $W/m^{2o}C$, improvement in convection due to provision of tubes = 40.1 temperature rise = 40 ^oC length of each tube = 1 m, diameter of tube = 50 mm, loss dissipation due to provision of tubes = 50 mm. Find the number of tubes for this transformers. Neglect the top and bottom surface of the tank as a regards the cooling.   (8)

Chapter 7, Example 7.18 *[Page No - 7.63]*

14. (a) (i) What are the advantages of squirrel cage induction motors over slip ring induction motor?
                                                                                                 (5)

Chapter 8, SA - Q8.26 *[Page No - 8.83]*

(ii) Determine the D and L of a 70 HP, 415 V, 3-phase, 5 Hz, star connected, 6 pole induction motor for which **ac** = 30000 $amp.cond/m$ and B_{av} = 0.51 Wb/m^2. Take η = 90% and pf = 0.91. Assume τ = L. Estimate the number of stator conductors required for a winding in which the conductors are connected in 2-parallel paths. Choose a suitable number of conductors/slots, so that the slot loading does not exceed 750 $amp.cond$.   (8)

Chapter 8, Example 8.3 *[Page No - 8.44]*

(OR)

(b) (i) List the rules for selecting rotor slots.   (5)

Chapter 8, Section 8.7.1 *[Page No - 8.18]*

(ii) Design a cage rotor for a 40 HP, 3-phase, 400 V, 50 Hz, 6 pole, delta connected induction motor having a full load η =87% and a full load pf of 0.85. Take D = 33 cm and L = 17 cm, stator slots = 54, conductors/slots = 14. Assume suitable values wherever necessary.   (8)

Chapter 8, Example 8.8 *[Page No - 8.53]*

15. (a) Determine for a 250 kVA, 1100 V, 12 pole, 500 rpm, 3-phase alternator.

1) Air gap diameter
2) Core length
3) Number of stator conductors
4) Number of stator slots and
5) Cross-section of stator conductors.

Assuming average gap density as 0.6 Wb/m^2 and specific electric loading of 30000 $amp. cond/m$, L/τ = 1.5.   (13)

Chapter 9, Example 9.3 *[Page No - 9.36]*

(b) (i) Mention the factors that govern the design of field system alternator   (5)

Chapter 9, Section 9.4.9 *[Page No - 9.17]*

(ii) Sketch the shape of salient pole rotor and cylindrical rotor. What are the constructional difference between salient pole types alternator and cylindrical rotor type alternators? (8)

Chapter 9, Section 9.4 *[Page No - 9.8]*

PART - C (1 × 15 = 15 Marks)

16. (a) A 600 *rpm*, 50 *Hz*, 3-phase, synchronous generator has the following design data. B_{av} = 0.48 *Wb/m²*, δ = 2.7 *amp/mm²*, slot space factor = 0.35, number of slots = 144, slot size = 120 × 20 *mm*, D = 1.092 *m* and L = 0.4 *m*. Determine the kVA rating of the machine.   (15)

Chapter 9, Example 9.1 *[Page No - 9.33]*

(OR)

(b) Show the design procedure of field system of non-salient pole alternator.   (15)

Chapter 9, Section 9.5.4 *[Page No - 9.28]*

ANNA UNIVERSITY (AU)

B.E./B.Tech. DEGREE EXAMINATION, NOVEMBER/DECEMBER 2017

Sixth Semester

Electrical and Electronics Engineering

EE6604 - DESIGN OF ELECTRICAL MACHINES

(Regulations 2013)

Time: 3 Hours   **Maximum:** 100 Marks

Answer All question

PART - A (10 × 2 = 20 Marks)

1. Define short time rating.

    Chapter 4, SA - Q4.8 *[Page No - 4.46]*

2. List the standard specifications of transformer.

    Chapter 1, SA - Q1.11 *[Page No - 1.37]*

3. Define specific electric and magnetic loading.

    Chapter 6, SA - Q6.12 *[Page No - 6.75]*

4. State the advantages of having larger number of poles in DC machines.

    Chapter 6, SA - Q6.30 *[Page No - 6.77]*

5. How does a distribution transformer differ from a power transformer in design aspects?

    Chapter 7, SA - Q7.12 *[Page No - 7.80]*

6. Why stepped cores are used in transformers?

    Chapter 7, SA - Q7.31 *[Page No - 7.83]*

7. What are the factors to be considered in selecting the number of stator slots in induction machines?

    Chapter 8, SA - Q8.42 *[Page No - 8.85]*

8. What is unbalanced magnetic pull in induction machines?

    Chapter 3, SA - Q3.66 *[Page No - 3.55]*

9. What are the design features of turbo alternators.

    Chapter 9, SA - Q9.18 *[Page No - 9.61]*

10. Define run way speed of an alternator.

    Chapter 9, SA - Q9.4 *[Page No - 9.59]*

PART - B (5 × 16 = 80 Marks)

11. (a)   Discuss in detail the factors affecting the choice of specific electric and magnetic loading in rotating machines.   (16)

Chapter 6, Section 6.4 *[Page No - 6.12]*

(OR)

(b)   Derive an expression for the heating and cooling curve in electrical machines.   (16)

Chapter 4, Section 4.4 and 4.5 *[Page No - 4.10]*

12. (a)   (i)   Derive an expression for the mmf of air-gap of a machine with slotted armature and ventilating ducts.   (8)

Chapter 3, Section 3.6 *[Page No - 3.13]*

(ii)   A 15 *kW*, 230 *V*, 4 pole DC machine has the following data: Armature diameter = 0.125 *m* ; armature core length = 0.125 *m*; length of air-gap at pole center = 2.5 *mm*; flux density pole = 11.7 × 10^{-3} *Wb*; pole arc to pole pitch ratio = 0.66. calculate the mmf required for the air-gap (i) if the armature surface is treated as smooth (ii) if the armature is slotted and the gap contraction factor is 1.18.   (8)

Chapter 3, Example 3.2 *[Page No - 3.41]*

(OR)

(b)   Determine the diameter and length of armature core for a 55 *kW*, 110 *V*, 1000 *rpm*, 4 pole shunt generator assuming specific electric and magnetic loading of 26000 *amp.cond./m.* and 0.5 *Wb/m^2* respectively. The pole arc should be about 70% of pole pitch and length of core about 1.1 times the pole arc. Allow 10 ampere for the field current and assume a voltage drop of 4 *V* for the armature circuit. Specify the winding used and also determine the values of the number of armature conductors and number of armature slots.   (16)

Chapter 6, Example 6.9 *[Page No - 6.51]*

13. (a)   Estimate the main dimensions including winding conductor area of a 3-phase delta star core type transformer rated at 300 *kVA*, 6600/40 *V*, 50 *Hz*. A suitable core with three steps having a circumscribing circle of 0.25 *m* diameter and a leg spacing of 0.4 *m* is available. The emf per turn is 8.5 *V*. Assume a current density of 2.5 *A/mm^2*, a window space factor of 0.28 and a stacking factor of 0.9   (16)

Chapter 7, Example 7.2 *[Page No - 7.44]*

(OR)

(b)   A 1000 kVA, 6600/440 V, 50 Hz, 3-phase delta star core type oil immersed natural cooled (ON) transformer has the following design data ; Distance between centers of adajacent limbs = 0.47 m; outer diameter of HV winding = 0.44 m; Height of frame = 1.24 m; core loss = 3.7 kW and I^2R loss = 10.5 kW. Design a suitable tank for the transformer. The average temperature rise of the oil should not exceed 35°C. The specific heat dissipation for the tank walls is 6 W/m^2-°C and 6.5W/m^2-°C due to radiation and convection respectively. Assume that the convection is improved by 35% due to the provision to tubes.   (16)

Chapter 7, Example 7.20 *[Page No - 7.67]*

14. (a)   Determine the stator bore and core length of a 70 HP, 415 V, 3-phase, 50 Hz star connected, 6 pole induction motor for which the specific electric and magnetic loading are 3200 A/m and 0.51 Wb/m^2 respectively. Take the efficiency as 90% and power factor as 0.91. Assume pole pitch = core length. Estimate the number of stator conductors required for a winding in which the conductors are connected in two parallel paths.choose a suitable number of conductor per slot so that in two parallel paths. choose a suitable number of conductors per slot so that the slot loading does not exceed 750 ampere conductors.   (16)

Chapter 8, Example 8.3 *[Page No - 8.44]*

(OR)

(b)   Estimate the main dimensions, air-gap length, stator slots, stator turns per phase and cross sectional area of stator and rotor conductors for 3-phase, 15 HP, 400 V, 6 pole, 50 Hz, 975 rpm, induction motor. The motor is suitable for star delta starting, B_{av} = 0.45 Wb/m^2, **ac** = 20000 $amp.$ $cond./m$. L/T ratio = 0.85, efficiency = 0.9 and power factor = 0.85.   (16)

Chapter 8, Example 8.4 *[Page No - 8.46]*

15. (a)   Determine the main dimensions of a 75000 kVA, 13.8 kV, 50 Hz, 62.5 rpm, 3-phase, star connected alternator, also find the number of stator slots, conductors per slot, conductor area and work out winding details. The peripheral speed is about 40 m/s. Assume average gap density = 0.65 $Wb/$ m^2, $amp.cond./m.$ = 40000 and current density 4 A/mm^2.   (16)

Chapter 9, Example 9.3 *[Page No - 9.36]*

(OR)

(b)   Explain the design procedure for stator and rotor of turbo alternators.   (16)

Chapter 9, section 9.5.4 *[Page No - 9.27]*

ANNA UNIVERSITY (AU)

B.E./B.Tech. DEGREE EXAMINATION, APRIL/MAY 2017

Sixth Semester

EE 6604 - DESIGN OF ELECTRICAL MACHINES

(Regulations 2013)

Time: 3 Hours   **Maximum:** 100 Marks

Answer All Questions

PART - A (10 × 2 = 20 Marks)

1. What are the electrical properties of insulating materials?

    Chapter 2, SA - Q2.4 *[Page No - 2.12]*

2. Mention the different types of duties of a machine.

    Chapter 4, SA - Q4018 *[Page No - 4.47]*

3. Distinguish between real and apparent flux densities in the tooth section of slot.

    Chapter 3, SA - Q3.38 *[Page No - 3.51]*

4. Write down the expression for brush friction losses.

    Chapter 6, SA - Q6.13 *[Page No - 6.75]*

5. What is window space factor in the design of transformer?

    Chapter 7, SA - Q7.16 *[Page No - 7.81]*

6. How magnetic curves are used for calculating the no-load current of a transformer?

    Chapter 7, SA - Q7.60 *[Page No - 7.87]*

7. State the rules for selecting rotors slots of squirrel cage machines.

    Chapter 8, SA - Q8.65 *[Page No - 8.88]*

8. What are the range of efficiency and power factor in induction motor?

    Chapter 8, SA - Q8.31 *[Page No - 8.84]*

9. What are the factors that are affected due to SCR?

    Chapter 9, SA - Q9.33 *[Page No - 9.63]*

10. State three important features of turbo-alternator rotors.

    Chapter 9, SA - Q9.18 *[Page No - 9.61]*

PART - B (5 × 16 = 80 Marks)

11. (a) (i)   Classify the insulating materials based on thermal consideration.   (8)

                            Chapter 2, Section 2.8.1 [Page No - 2.9]

        (ii)   What are the major considerations to evolve a good design of electrical machine?   (8)

                            Chapter 1, Section 1.1.5 [Page No - 1.6]

(OR)

   (b) (i)   List the methods used for determine the motor rating for variable load drives. Explain any one method.   (8)

                            Chapter 4, Section 4.10.2 [Page No - 4.26]

        (ii)   Write a short notes an standard specifications. List the indian standard specifictions for transformer and induction motor.   (8)

                            Chapter 1, Section 1.1.4 [Page No - 1.4]

12. (a) (i)   Derive the expressions for reluctance of air-gap in machines with smooth armature and slotted armature.   (6)

                            Chapter 3, Section 3.4 [Page No - 3.10]

        (ii)   Determine the air-gap length of a DC machine from the following particulars : gross-length of core = 0.12 m, number of ducts = one and is 10 mm wide, slot pitch = 25 mm, slot width = 10 mm, carter's coefficient for slots and ducts = 0.32, gap density at pole centre = 0.7 Wb/m^2 ; field mmf/pole = 3900 AT, mmf required for iron parts of magnetic circuit = 800 AT.   (10)

                            Chapter 3, Example 3.3 [Page No - 3.41]

(OR)

   (b) (i)   Determine the main dimensions of a 80 kW, 4 pole, 600 rpm DC shunt generator, the full load terminal voltage being 220 V. The maximum gap density is 0.75 Wb/m^2 and ampere conductors per meter are 27000. Assume a square pole face.   (8)

                            Chapter 6, Example 6.4 [Page No - 6.44]

        (ii)   Give the expression for the torque developed by a DC motor in terms of main dimensions of the armature.   (8)

                            Chapter 6, Section 6.2 [Page No - 6.7]

13. (a) (i)   Differentiate the design features of power and distribution type transformers.   (6)

                            Chapter 7, Section 7.1.4 [Page No - 7.5]

        (ii)   Estimate the main dimensions including winding conductor area of a 3-phase, Δ-Y core type transformer rated at 300 kVA, 6600/440 V, 50 Hz. A suitable core with 3-steps having a circumscribing circle of 0.25 m diameter and a leg spacing of 0.4 m is available. Emf per turn = 8.5 V, δ = 2.5 A/mm^2, K_w = 0.28, S_f = 0.9.   (10)

                            Chapter 7, Example 7.2 [Page No - 7.44]

(OR)

(b) (i) List and explain the different methods of cooling of transformers.   (6)

Chapter 7, Section 7.8.1 [Page No - 7.34]

(ii) The tank of a 500 *kVA*, 1-phase, 50 *Hz*, 6600/400 *V* transformer is 110 *cm* × 65 *cm* ×155 *cm*. If the load loss is 6.2 *kW*, find and show the suitable arrangements for the cooling tubes to limit the temperature rise to 35 °C. Take the diameter of the cooling tubes as 5 *cm* and average length of the tubes as 110 *cm*.   (10)

Chapter 7, Example 7.18 [Page No - 7.63]

14. (a) (i) Derive the expression for output equation of induction motor.   (6)

Chapter 8, Section 8.2 [Page No - 8.5]

(ii) Estimate the stator core dimensions, number of stator slots and number of stator conductors per slot for a 100 *kW*, 3300 *V*, 50 *Hz*, 12 pole, star connected slip ring niduction motor. B_{av} = 0.4 *Wb/m²*, ac = 25000 *amp.cond./m*, η = 0.9, pf = 0.9. Choose main dimensions tⁿ give best power factor. The slot loading should not exceed 500 *amp.conductors*.   (10)

Chapter 8, Example 8.2 [Page No - 8.41]

(OR)

(b) (i) What are the factors to be considered for estimating the length of air-gap in induction motor?   (6)

Chapter 8, Section 8.6 [Page No - 8.13]

(ii) A 90 *kW*, 500 *V*, 50 *Hz*, 3-phase, 8 pole induction motor has a star connected stator winding accommodated in 63 slots with 6 conductors per slot. If the slip ring voltage on open circuit is not to exceed 400 *V*, find a suitable rotor winding by estimating number of slots, number of conductors per slot, coil span, slip ring voltage on open circuit, approximate full load current per phase in rotor. Assume η = 0.9 and pf = 0.86.   (10)

Chapter 8, Example 8.11 [Page No - 8.57]

15. (a) (i) Sketch the shape of salient pole rotor for synchronous machine.   (6)

Chapter 9, Section 9.4 [Page No - 9.8]

(ii) What are the factors to be considered for fixing the air-gap length for synchronous machine.   (10)

Chapter 9, Section 9.4.8 [Page No - 9.16]

(OR)

(b) For a 250 *kVA*, 1100 *V*, 12 pole, 500 *rpm*, 3-phase alternator. Determine air-gap diameter, core length, number of stator conductors, number of stator slots and cross-section of stator conductors, Assuming average gap density as a 0.6 *Wb/m²* and specific electric loading of 30000 *amp.cond./m*. L/τ = 1.5.   (16)

Chapter 9, Example 9.3 [Page No - 9.36]

KARNATAKA TECHNICAL UNIVERSITY (VTU)

Sixth Semester, B.E. Degree Examination, Dec.2019/Jan.2020

15EE64 - ELECTRICAL MACHINE DESIGN

Time: 3 Hours   **Maximum:** 80 Marks

Note: 1. Answer any Five full questions,choosing ONE full question from each module.
2. Assume any missing data suitably.

Module-1

1. a. What are limitations involved in design of electrical machines?   (6)

        Chapter 1, Section 1.8 *[Page No - 1.21]*

    b. What are the desirable properties of conducting materials?   (5)

        Chapter 2, SA - Q2.1 *[Page No - 2.12]*

    c. What are ferromagnetic materials and solid core materials?   (5)

        Chapter 2, Section 2.7.1 *[Page No - 2.6]*

OR

2. a. Compare aluminium and copper wires.   (4)

        Chapter 2, Section 2.2.2 *[Page No - 2.2]*

    b. What are the desirable properties of insulating materials? Give the classification of insulating materials based on thermal consideration with two examples in each class.   (8)

        Chapter 2, Section 2.8 *[Page No - 2.8]*

    c. What is cold rolled grain oriented silicon steel? What are advantages of using these materials in electrical machines?   (4)

        Chapter 2, Section 2.7.2 *[Page No - 2.7]*

Module-2

3. a. Define "specific magnetic loading" and "specific electric loading". What are advantages and disadvantages of using higher specific loadings?   (8)

        Chapter 1, SA - Q1.18 and Q1.19 *[Page No - 1.38]*

    b. Find the main dimensions and number of poles of a 50 *HP*, 230 *V*, 1400 *rpm* shunt motor so that a square pole is obtained. Specific magnetic loading in the air-gap is 0.5 *weber/m²* and the ampere conductors per arc 22000. The ratios of pole arc to pole pitch is 0.7. Assume the efficiency as 90%. Check that the obtained values are within permissible limits. Take 1 *HP* = 0.7355 *kW*.

        Chapter 6, Example 6.6 *[Page No - 6.46]*   (8)

OR

4   a. What are the advantages and disadvantages of large number of poles in DC machines?   (6)

        Chapter 6, Section 6.5 *[Page No - 6.14]*

    b. Calculate the diameter and length of armature for a 7.5 *kW*, 4 pole, 1000 *rpm*, 220 *V* shunt
        motor. Given that the full load efficiency is 83%. Maximum air gap flux density is 0.9 *webers/*
        m^2. Specific electric loading is 30,000 ampere conductors per meter, field form factor is 0.7
        Assume that the maximum efficiency occurs at full load and field current is 2.5% of rated current.
        The pole face square.   (10)

        Chapter 6, Example 6.6 *[Page No - 6.46]*

Module-3

5. a. Prove that EMF /turn of a single-phase transformer is $K\sqrt{Q}$ where Q = output KVA rating of
        transformer per phase.   (6)

        Chapter 7, Section 7.3.1 *[Page No - 7.10]*

    b. The tank of a 1250 *kVA* natural oil colled transformer has the following dimensions, length width
        and height as 0.65 *m*, 1.55 *m*, and 1.85 *m* respectively. The full load loss is 13.1 *kW*. Assume
        heat dissipiation due to convection as 6.5 *W/m^2 °C* and due to radiation as 6.0 *W/m^2 °C*. Improvement
        in convection due to provision of tubes is 40%. Limit for temperature rise is 40°C. Length of each
        tube is 1.0 *m* and diameter of each tube is 50 *mm*. Find number of tube to be provided for the
        transformer. Neglect top and bottom surfaces of the tank as regards cooling.   (10)

        Chapter 7, Example 7.18 *[Page No - 7.63]*

OR

6. a. Derive the output equation of a three-phase core type transformer.   (6)

        Chapter 7, Section 7.3 *[Page No - 7.8]*

    b. Determine the dimensions of core, the number of turns, the cross section area of conductors of
        primary and secondary windings of a 100 *kVA*, 2200/480 *V* single phase core type of transformer
        to operater at a frequency of 50 *Hz* by assuming the following data:

        Approximate volts per turn = 7.5 *V*, maximum flux density is 1.2 *weber/m^2*. Ratio of effective
        cross section area of core to the square of the diameter of circumscribing circle is 0.6, Ratio of
        height to width of window is 2.0, window space factor K_w = 0.28, current density δ = 2.5 *A/mm^2*.

        Chapter 7, Example 7.6 *[Page No - 7.49]*   (10)

Module-4

7. a. Derive the output equation of a three-phase induction motor.   (6)

        Chapter 8, Section 8.2 *[Page No - 8.5]*

b. Determine the main dimensions, turns per phase number of slots, conductor cross section and slot area of a 250 *HP*, 3-phase, 50 *Hz*, 400 *V*, 1410 *rpm*, slip ring induction motor. Assume specific magnetic loading, B_{av}=0.5 T, specific electric loading, **ac** = 30,000 ampere conductors per meter, effciency is 90%, winding factor is 0.955, current density = 3.5 A/mm^2. The slot space factor is 0.4 and ratio of core length to pole pitch is 1.2. The machine is delta connected. Take 1 *HP* = 0.7355 *kW*.   (10)

> Chapter 8, Example 8.2 *[Page No - 8.41]*

OR

8.   a. What are the factors to be considered for estimating the length of air-gap for induction motors? Explain them.   (6)

> Chapter 8, Section 8.6 *[Page No - 8.13]*

b. Estimate the stator dimensions, number of stator slots, and number of staor conductor per slot for a 100 *kW*, 3300 *V*, 50 *Hz*, 12 pole, star connected slip ring induction motor. Assume an average flux density of 0.4 *webers/m²* in the air gap, ampere conductors per main dimensions to give best power factor. The slot loading must not exceed 500 ampere conductors.   (10)

> Chapter 8, Example 8.2 *[Page No - 8.41]*

Module-5

9.   a. Derive the output equation of a synchronous machine in terms of its main dimensions and specific loadings.   (6)

> Chapter 9, Section 9.1.1 *[Page No - 9.5]*

b. Find the main dimensions of a 100 *MVA*, 11 *kV*, 50 *Hz*, 150 *rpm*, 3-phase, water wheel driven alternator. The average air-gap flux density is 0.65 *webers/m²* and ampere conductors per meter are 40,000. The peripheral speed should not exceed 65 *m/sec* at normal running speed in order to limit the runaway speed. Assume a winding factor K_{ws} = 0.955.   (10)

> Chapter 9, Example 9.5 *[Page No - 9.38]*

OR

10.   a. Define Short Circuit Ratio (SCR) of a synchronous machine and discuss its effects on the machine performance.   (8)

> Chapter 9, Section 9.4.2 *[Page No - 9.9]*

b. Determine the main dimensions of a 1000 *kVA*, 50 *Hz*, 3-phase, 375 *rpm* alternator. The average air-gap flux density is 0.55 *webers/m²* and the ampere conductors per meter are 28,000. Use rectangular poles and assume a winding factor K_{ws} = 0.955. Boiled on pole construction is used for which the maximum permissible peripheral speed is 50 *m/sec*. The runaway speed is 1.8 times the synchronous speed.   (8)

> Chapter 9, Example 9.5 *[Page No - 9.38]*

PUNJAB TECHNICAL UNIVERSITY (PTU)

B.Tech (Electronics Engg) (E-1 2012 Onwards) (Sem-6) - May 2019

ELECTRICAL MACHINE DESIGN

Subject code : BTEEE-603A

M.Code : 72842

Time: 3 Hours   **Maximum:** 60 Marks

SECTION-A

1. **Answer briefly:**   (10 × 2 = 20)

   a. What are the "main dimensions" in machine design?]

       Chapter 1, Section 1.21 *[Page No - 1.7]*

   b. What are the factors that affect volume of a three-phase induction motor?

       Chapter 8, Section 8.2 *[Page No - 8.6]*

   c. Define short time and intermittent rating of induction motor.

       Chapter 4, SA - Q4.8 and Q4.9 *[Page No - 4.45]*

   d. Why machines with large dimensions are more efficient?

       Chapter 1, SA - Q1.54 *[Page No - 1.43]*

   e. Write the differences between core and shell type transformer.

       Chapter 7, Section 7.1.2 *[Page No - 7.4]*

   f. What is output equation of induction motor?

       Chapter 8, SA - Q8.7 *[Page No - 8.81]*

   g. Write the function of frame of a three phase induction motor.

       Chapter 4, SA - Q4.15 *[Page No - 4.47]*

   h. Which design factors influence the power factors of induction motor?

       Chapter 8, SA-Q8.31 *[Page No - 8.84]*

   i. What is peripheral speed of rotating machine?

       Chapter 9, SA - Q9.14 *[Page No - 9.60]*

   j. How the mmf of a magnetic circuit is determined?

       Chapter 3, SA-Q3.18 *[Page No - 3.49]*

SECTION-B   (4 × 5 = 20)

2. What are the cooling methods applied for reduction of temperatur rise of a transformer?

    Chapter 7, Section 7.8 *[Page No - 7.34]*

3. Derive the output equation for design of a three phase transformer.

    Chapter 7, Section 7.3 *[Page No - 7.8]*

4. Derive the condition for design of minimum cost in a transformer.

5. Discuss the steps of calculations for designing number of stator slots and area of stator slots of a three-phase induction motor.

    Chapter 8, Section 8.5.2 *[Page No - 8.10]*

6. A 3-phase, 4 pole induction motor has 24 slots. Calculate the order of slot harmonics produced. It is desired to completely eliminate the higher order slot harmonic, find the angle through which the bars must be skewed. Find the effect of skewing on the lower order harmonic.

    Chapter 8, Example 8.5 *[Page No - 8.49]*

SECTION-C $(2 \times 10 = 20)$

7. Determine the core dimensions, number of stator slots, number of stator conductors for a 10 kW, 415 V, 3-phase, 4 pole, 50 Hz motor. Assume winding factor 0.955, full load efficiency 0.85, power factor 0.8. For the machine B_{av} = 0.35 Wb/m^2, **ac** = 20000 A/m. Output coefficient, C_o = 87.2, L/τ = 0.83.

    Chapter 8, Example 8.2 *[Page No - 8.41]*

8. Calculate the approximate overall dimensions for a 100 kVA, 6000/440 V, 50 Hz. 3-phase core type transformer. Volts/turn, E_t = 10 V, Flux density B_m = 1.1 Wb/m^2, current density δ = 2.5 A/m^2, Winding factor K_w = 0.3, Overall height = overall width, stacking factor 0.9. Use a 3-Stepped core.

    Chapter 7, Example 7.2 *[Page No - 7.44]*

9. Derive the equation of temperature rise with time in electric machines. What is heating time constant? Briefly discuss about induced, forced, radial and axial ventilation. Write the advantages of hydrogen cooling in turbo alternators.

    Chapter 4, Section 4.7 *[Page No - 4.16]*

JNT KAKINADA UNIVERSITY (JNTKU)

III B.Tech II Semester Supplementary Examinations, April/May - 2019

ELECTRICAL MACHINE DESIGN

(Electrical and Electronics Engineering)

Subject code : R32021

Time: 3 Hours   **Maximum:** 75 Marks

Answer any FIVE questions
All Questions carry equal marks

1. a) Briefly discuss the modern trends in the design of electric machines.   (6)

        Chapter 1, Section 1.10 *[Page No - 1.23]*

    b) List and explain various conducting materials used in electrical machines.   (9)

        Chapter 2, Section 2.1 *[Page No - 2.1]*

2. a) How the dimensions of induction generator differs from that of an induction motor?   (4)

        Chapter 8, SA - Q8.64 *[Page No - 8.88]*

    b) A 4 pole, simplex lap wound DC armature has 64 slots and 1152 conductors. The number of commutator segments is 192. Dtermine the number of coil sides per slot, number of turns per coil and the winding pitches. Draw the winding table. Specify whether the winding is symmetrical or not.   (11)

        Chapter 5, Example 5.2 *[Page No - 5.29]*

3. a) Discuss the main parts of a DC commutator machine.   (3)

    b) Derive the output equation of a DC machine.   (5)

        Chapter 6, Section 6.2 *[Page No - 6.5]*

    c) Find the main dimensions of a 200 *kW*, 250 *V*, 6 pole, 100 *rpm* generator. The maximum value of flux density in the gap is 0.87 *Wb/m²* and the ampere conductors per meter of armature periphery are 31000. The ratio of pole arc to pole pitch is 0.67 and the efficiency is 91%. Assume the ratio of length of core to pole pitch = 0.75.   (7)

        Chapter 6, Example 6.1 *[Page No - 6.40]*

4. a) Draw and explain the constructional details of a three-phase core type transformer.   (7)

        Chapter 7, Section 7.3 *[Page No - 7.8]*

    b) What is the need for cooling in transformers? List and briefly discuss various cooling schemes for transformers.   (8)

        Chapter 7, Section 7.8.1 *[Page No - 7.34]*

5. a) Show that the output of a 3-phase core type transformer is: $Q = 5.23 \, f \, B_m H d^2 H_w \times 10^{-2} \, kVA$, where f = frequency, Hz; B_m = maximum flux density, Wb/m^2; d = effective diameter of the core, m; H = magnetic potential gradient in limb, A/m; H_w = height of window, m. (7)

Chapter 7, Example 7.10 *[Page No - 7.56]*

b) The current densities in the primary and secondary windings of a transformer are 2.2 and 2.1 A/mm^2 respectively. The ratio of transformation is 10:1 and the length of mean turn of the primary 10% greater than that of the secondary. Calculate the resistance of the secondary winding given that the primary winding resistance is 10 Ω.

Chapter 7, Section 7.1.2 *[Page No - 7.4]*

6. a) Compare the squirrel cage induction motor with wound rotor machine. (6)

Chapter 8, SA-Q8.29 *[Page No - 8.84]*

b) What are the main dimensions of the induction motor? What are the desired values of L/τ, peripheral speed and width of ventilation ducts. (9)

Chapter 8, Section 8.3 *[Page No - 8.7]*

7. a) A 90 kW, 500 V, 50 Hz, 3-phase, 8 pole induction motor has a star connected stator winding accomdated in 63 slots with 6 conductors per slot. If the slip ring voltage on open circuit is to be about 400 V, find a suitable rotor winding, stating: a) number of slots b) number of conductors per slot c) coil span d) slip ring voltage on open circuit e) approximate full load current per phase in rotor. Assume efficiency = 0.9, power factor = 0.86. (15)

Chapter 8, Example 8.11 *[Page No - 8.57]*

8. Find the main dimension of a 100 MVA 11 kV, 50 Hz, 150 rpm, 3-phase waterwheel generator. The average gap density is 0.65 Wb/m^2 and ampere conductors perimeter are 40000. The peripheral speed should not exceed 65 m/s at normal running speed in order to limit run away peripheral speed. (15)

Chapter 9, Example 9.5 *[Page No - 9.38]*

RAJIV GANDHI TECHNICAL UNIVERSITY (RGPV)
EE-603(GS)
B.E. VI Semester Examination, June 2020
Grading System (GS)
ELECTRICAL MACHINE DESIGN

Time: 3 Hours   **Maximum:** 70 Marks

1. Find the suitable number of poles and the diameter of the core of a 400 *kW,* 550 *V,* 180 *rpm* DC generator having 92% efficiency. Assume an average flux density in the air gap of about 0.6 *Wb/m²* and ampere conductor per meter to be 35000.   (10)

    Chapter 6, Example 6.1 *[Page No - 6.40]*

2. Write an algorithm to calculate the following for a 3-phase IM.
    i) Stator core dimensions
    ii) Flux/pole
    iii) No.of stator slots.   (10)

    Chapter 8, Section 8.5.2 *[Page No - 8.10]*

3. a) Give the criteria for the selection of stator slots and rotor slots in 3ϕ induction motor.   (5)

        Chapter 8, Section 8.5.2 *[Page No - 8.11]*

        Chapter 8, Section 8.7.1 *[Page No - 8.18]*

    b) Discuss crawling and cogging in three-phase induction motor.   (5)

        Chapter 8, Section 8.7.1 *[Page No - 8.15]*

4. What do you understand by reliablity of a machine? What are different factor affecting reliablity? Explain whether these factors can be taken in design of a machine.   (10)

5. Differentiate between the Non linear optimization and linear optimization techniques. How these can be programmed through computer.   (10)

                                                OR

    Derive the output equation in case of single-phase transformer and three-phase transformer.   (10)

    Chapter 7, Section 7.3 *[Page No - 7.6]*

6.   Find the main dimension of a 100 *MVA*, 11 *kV*, 50 *Hz*, 150 *rpm,* three phase water wheel generator. The average gap density of 0.65 *Wb/m²* and ampere conductors per meter are 40000. The peripheral speed should not exceed 65 *m/s* at normal running speed in order to limit the run away peripheral speed.   (10)

Chapter 9, Example 9.5 *[Page No - 9.38]*

OR

How do you calculate the following for an three-phase induction motor?   (10)

i)   Area of stator slots.   Chapter 8, Section 8.5.2 *[Page No - 8.11]*

ii) Length of mean turn.   Chapter 8, Section 8.5 *[Page No - 8.9]*

7.   In a 500 *kW*, 440 *V*, 375 *rpm*, 8 pole compound generator, the external diameter of armature is 1.1 *mt,* gross armature length is 0.3 *mt,* armature slot pitch is 2.5 *cm* and flux/pole is 0.0875 *wb*.   (10)

Determine:

i)   No.of conductors in armature winding.

ii)   No.of slots in armature

iii) Resistance of the armature winding.

Chapter 6, Example 6.13 *[Page No - 6.59]*

8.   Write short notes on any two of the following.   (10)

i)   Design of the rotor of a turbo alternator.

Chapter 9, Section 9.5.4 *[Page No - 9.28]*

ii)   Design of field winding of a DC machine.

Chapter 6, Section 6.8 *[Page No - 6.25]*

iii) Optimal design of induction motor.

Chapter 8, Section 8.3 *[Page No - 8.7]*

MUMBAI UNIVERSITY (MU)

Paper/Subject Code: 42203/Electrical Machine Design-May 2019

IT00817 - B.E.(Electrical)(Sem VII)(CBSGS)/42203 - Electrical Machine Design 68741

Time : 3 Hours   **Total Marks : 80**

Q1. Answer the following questions.

    a. Explain different types of magnetic materials.   (5)

        Chapter 2, Section 2.5 *[Page No - 2.6]*

    b. Discuss the choice of flux density for designing of transformer.   (5)

        Chapter 7, Section 7.5.5 *[Page No - 7.19]*

    c. Explain in brief the methods of cooling of a transformer.   (5)

        Chapter 7, Section 7.8.1 *[Page No - 7.34]*

    d. Discuss various insulating properties of transformer oil.   (5)

        Chapter 2, SA - Q2.3 *[Page No - 2.11]*

Q2. a. Derive an output equation of single-phase and three-phase transformer.   (10)

        Chapter 7, Section 7.2 and 7.3 *[Page No - 7.8]*

    b. Determine the main dimensions and number of turns of a 100 kVA, 6600/440 V, delta star connection, 50 Hz, 3-phase core type transformer with data, three step core arrangement, Emf per turn = 10 V maximum flux density: 1.3 wb/m^2, current density : 2.5 A/mm^2, window space factor = 0.3, stacking factor : 0.9 over all height = overall width.   (10)

        Chapter 7, Example 7.2 *[Page No - 7.44]*

Q3. a. Discuss designing of cooling tanks and tubes in a transformer.   (10)

        Chapter 7, Section 7.9 *[Page No - 7.36]*

    b. A 100 kVA, 2000/400 V, 50 Hz, 1-phase, Shell type transformer, has sandwich coils. There are two full hv coils, one full lv coil and 2 half lv coils. Calculate the value of leakage reactance referred to hv side. The data given is : depth of hv coil= 40 mm, depth of lv coils= 36 mm, depth of duct between hv and lv= 16 mm, width of winding= 0.12 m, length of mean turn= 1.5 m, the no of turns in hv winding are 200.   (10)

        Chapter 7, Example 7.15 *[Page No - 7.60]*

Q4. a. Discuss the various mechanical forces developed in transformer with sketches. Explain how they are taken care while fabrication.   (10)

        Chapter 7, Section 7.6.6 *[Page No - 7.28]*

b. Derive the equation for leakage reactance calculation for a two winding core type transformer
(10)

Chapter 7, Section 7.6.4 *[Page No - 7.23]*

Q5. a. Derive the output equation of a three-phase induction motor in terms of main dimensions.   (10)

Chapter 8, Section 8.2 *[Page No - 8.5]*

b. Determine the main dimensions, total conductors and number of slot, area of slot, conductor per slot for minimum cost design, for a 3.7 *kW,* 400 *V,* 1410 *rpm*, 3 phase, 4 pole, 50 *Hz*, delta connected, squirrel cage induction motor with the data average flux density in the air-gap = 0.45 *Wb/m²*· ampere conductors: 23,000 *A/m,* efficiency : 0.85, power factor : 0.84, winding factor 0.955, current density : 3.5 *A/mm²,* stacking factor = 0.9, slot space factor = 0.4.   (10)

Chapter 8, Example 8.6 *[Page No - 8.49]*

Q6. a. Discuss various steps to be followed while designing a rotor of induction motor for main dimensions.
(10)

Chapter 8, Section 8.3 *[Page No - 8.7]*

b. Discuss the design modifications in a stator and rotor of an energy efficient motor.   (10)

KERELA TECHNICAL UNIVERSITY (KTU)

Eighth Semester, B.Tech. Degree Examination, Feburary - 2018

(2013 Scheme)

ELECTRICAL MACHINE DESIGN

Time: 3 Hours   **Maximum:** 100 Marks

PART - A

Answer all questions from Part - **A**.   (5 × 4 = 20 Marks)

1. Draw and explain the temperature rise-time curve of electrical machines when put on a constant load.

    Chapter 2, Section 2.8.3 *[Page No - 2.10]*

2. What is window space factor? On what factors does it depend?

    Chapter 7, SA Q7.16 *[Page No - 7.81]*

3. What are commutating poles and why are they used? Why are these poles wound with series turns?

    Chapter 6, Section 6.8.3 *[Page No - 6.35]*

4. Enumerate the advantages and disadvantages of having a large air-gap in synchronous machines.

    Chapter 9, Section 9.4.8 *[Page No - 9.16]*

5. Explain the different computer aided design methods.

    Chapter 1, Section 1.11 *[Page No - 1.23]*

PART - B

Module - I

6. a) Derive output equation of a three-phase core type transformer in terms of design constants and hence deduce expression for emf per turn.   (8)

        Chapter 7, Section 7.3 *[Page No - 7.8]*

    b) Determine the main dimensions of the core and window for a 500 *kVA*, 6600/400 *V*, 50 *Hz*, single-phase, core type, oil immersed, self cooled transformer. Assume : flux density = 1.2 T, current density = 2.5 *A/mm²*, window space factor = 0.32, volt\turn = calculate the number of turns and cross-sectional area of the conductors used for the primary and secondary windings.
                                                                                             (12)

        Chapter 7, Example 7.8 *[Page No - 7.52]*

7. a) Explain about the different cooling methods adopted in transformers.   (8)

        Chapter 7, Section 7.8.1 *[Page No - 7.34]*

b)   Determine the main dimensions and winding details for a 125 kVA, 2000/400 V, 50 Hz, single-phase, shell type transformer with volt\turn = 11.2, flux density = 1.0 T, current density = 2.2 A/mm^2, window space factor = 0.33.   (12)

$$\frac{\text{Depth of stacked core}}{\text{Width of central limb}} = 2.6 \qquad\qquad \frac{\text{Height of window}}{\text{Width of window}} = 3$$

Chapter 7, Example 7.8 *[Page No - 7.52]*

Module - II

8.   a)   Discuss the various factors that govern choice of number of poles in a DC machine.   (10)

Chapter 6, Section 6.5 *[Page No - 6.14]*

    b)   A 350 kW, 500 V, 450 rpm, 6 pole DC generator is built with an armature diameter of 0.87 m and core length of 0.32 m. The lap wound armature has 660 conductors. Calculate the specific electric and magnetic loadings.   (10)

Chapter 1, Example 1.1 *[Page No - 1.28]*

9.   a)   What are the factors to be considered while fixing the number of poles in a DC machine?   (8)

Chapter 6, Section 6.5 *[Page No - 6.15]*

    b)   The core diameter and length of an armature of a 250 kW, 500 V, 600 rpm, DC generator are 75 cm and 30 cm respectively. The lap connected armature gas 720 conductors. Using the data obtained from this machine, obtain the main dimensions and the number of poles are a 350 kW, 500 V, 720 rpm, DC generator. Assume the full load efficiency as 0.85.   (12)

Chapter 6, Example 6.3 *[Page No - 6.42]*

Module - III

10.   a)   Differentiate between real and apparent flux densities in a rotating electrical machine. Obtain a relationship between the two.   (8)

Chapter 3, Section 3.10 *[Page No - 3.19]*

    b)   Find the main dimensions of a 100 MVA, 11 kV, 50 Hz, 150 rpm, 3-phase, water wheel driven alternator. The average air-gap flux density is 0.65 $weber/m^2$ and ampere conductors per meter are 40,000. The peripheral speed should not exceed 65 m/sec at normal running speed in order to limit the runaway speed. Assume a winding factor, $K_{ws} = 0.955$.   (12)

Chapter 9, Example 9.5 *[Page No - 9.38]*

11.   a)   Explain the procedure for design of field winding in salient pole synchronous machines.   (8)

Chapter 9, Section 9.4.12 *[Page No - 9.24]*

b) Determine the dimensions of stator core, number of stator conductors and number of slots for a 3-phase star connected 300 kVA, 3.3 kV, 8 pole, 50 Hz alternator. Assume specific loadings as 28,000 A/m and 0.6 Wb/m^2 And square pole faces with pole arc = 0.65 of pole pitch.   (12)

Chapter 9, Example 9.3 *[Page No - 9.36]*

Module - IV

12. a) Explain the factors to be considered while choosing length of air-gap in induction machines.   (10)

Chapter 8, Section 8.6 *[Page No - 8.13]*

b) Find the diameter and length of stator core of a 7.5 kW, 220 V, 50 Hz, 4 pole induction motor for the best power factor. Given, specific electric loading = 22,000 ampere conductors per metre, specific electric loading = 0.4 wb/m^2, efficiency = 0.86 and power factor = 0.87.   (10)

Chapter 8, Example 8.2 *[Page No - 8.41]*

13. a) Derive the output equation of a 3ϕ induction motor.   (8)

Chapter 8, Section 8.2 *[Page No - 8.5]*

b) Determine the main dimensions, total conductors and number of slots and the number of turns per phase of a 3.7 kW, 400 V, 3-phase, 4 pole, 50 Hz, squrriel cage induction motor to be started by a star delta starter. Assume average flux density in the air-gap = 0.45 wb/m^2, ampere conductors per meter = 23,000 A/m, Efficiency = 85%, power factor = 0.84, winding factor = 0.955, stacking factor = 0.9. Machines rated at 3.7 kW, 4-poles are sold at a competitive price and therefore choose the main dimensions to give a cheap design.   (10)

Chapter 8, Example 8.6 *[Page No - 8.49]*

APJ ABDUL KALAM TECHNOLOGICAL UNIVERSITY (AKTU)

Seveth Semester, B.Tech, Degree examination (R&S), December - 2019

Course Code: EE409

ELECTRICAL MACHINE DESIGN

Maximum: 100 Marks   **Time:** 3 Hours

PART A

1. Briefly explain the different types of enclosures used in electrical machines.   (5)

    Chapter 4, Section 4.6 *[Page No - 4.15]*

2. Give two differences between power transformer and distribution transformer.   (5)

    Chapter 7, Section 7.1.4 *[Page No - 7.5]*

3. Explain in steps the design of series field winding for a DC machine.   (5)

    Chapter 6, Section 6.8.2 *[Page No - 6.33]*

4. Salient pole alternators are not suitable for high speeds. why?   (5)

    Chapter 9, SA Q9.15 *[Page No - 9.60]*

5. State and explain the factors considered for selection of air-gap length in induction motors.   (5)

    Chapter 8, SA-Q8.42 *[Page No - 8.85]*

6. List and justify the advantages of a larger air-gap in induction motor.   (5)

    Chapter 8, SA Q8.45 *[Page No - 8.77]*

7. List out and explain the features of three finite element based softwares for analysis of electrical machines.   (5)

8. Explain the basic concept of computer aided design.   (5)

    Chapter 1, Section 1.11 *[Page No - 1.23]*

PART B

9. Explain the procedure to calculate mmf for air-gap and teeth in an electrical machine.   (10)

    Chapter 3, Section 3.9 *[Page No - 3.17]*

10. a) Write the design equations to find the area of cross-section of conductor for both primary and secondary of a transformer.   (3)

    Chapter 7, Section 7.2 *[Page No - 7.6]*

    b) Determine the dimensions of core and window for a 5 *kVA*, 50 *Hz*, 1-phase, core type transformer. A rectangular core is used with long side twice as long as short side. The window height is 3 times the width. Voltage per turn = 1.8 *V*, space factor = 0.2, Current density is 1.8 *A/mm²*, Maximum flux density is 1*Wb/m²*.   (7)

    Chapter 7, Example 7.4 *[Page No - 7.47]*

11. a) A 15 kW, 230 V, 4 pole DC machine has the following data :

Armature diameter = 0.25 m, armature core length = 0.125 m, length of air-gap at pole centre = 2.5 mm, flux per pole = 11.7 × 10^{-3} Wb, Ratio of pole arc/pole pitch = 0.66. Calculate the mmf required for air-gap (i) if the armature surface is treated as smooth (ii) if the armature is slotted and the gap contraction factor is 1.18.   (7)

Chapter 3, Example 3.2 *[Page No - 3.41]*

b) Examine the factors that influence the choice of flux density of a transformer.   (3)

Chapter 7, Section 7.5.5 *[Page No - 7.19]*

PART C

12. a) Smaller machines have low specific magnetic loadings. Why?   (3)

Chapter 8, SA - Q8.15 *[Page No - 8.82]*

b) A 4 pole, 25 HP, 500V, 600 rpm DC series motor has an efficiency of 82%. The pole faces are square and the ratio of pole arc to pole pitch is 0.67. Take B_{av} = 0.55 Wb/m^2 and **ac** = 17000 $amp.cond./m$. Obtain the main dimensions, number of slots, and conductors per slot. Assume it to be wave winding.   (7)

Chapter 6, Example 6.8 *[Page No - 6.48]*

13. a) List the factors to be considered for the choice of specific electric loading in synchronous machines.   (4)

Chapter 9, Section 9.3.2 *[Page No - 9.8]*

b) A 3000 rpm, 50 Hz, 3-phase turbo alternator has a core length of 0.92 m. The average gap density is 0.44 Wb/m^2 and the ampere conductors per metre are 23000. The peripheral speed of rotor is 95 m/s and the length of air-gap is 18 mm. Find the kVA output of the machine when the coils are (i) full pitch (ii) chorded by 1/3 pole pitch. The winding can be taken as infinitely distributed with a phase spread of 60°.   (6)

Chapter 9, Example 9.10 *[Page No - 9.46]*

14. a) What are the advantages and disadvantages of higher number of poles in DC machine?   (5)

Chapter 6, Section 6.5 *[Page No - 6.14]*

b) Explain any 3 methods of cooling for turbo alternators.   (5)

Chapter 4, Section 4.9 *[Page No - 4.19]*

PART D

15. Estimate the stator core dimensions, number of stator slots and number of stator conductors per slot for a 100 kW, 3300 V, 50 Hz, 12 pole, star connected slip ring induction motor. B_{av} = 0.4 Wb/m^2, **ac** = 25000 $amp.cond./m$, η = 0.9, pf = 0.9. Choose main dimensions to give best power factor. The slot loading should not exceed 500 $amp.conductors$.   (10)

Chapter 8, Example 8.2 *[Page No - 8.41]*

16. a) Explain on synthesis method of solving electrical machine using CAD with a flow chart.   (7)

    Chapter 1, Section 1.11.2 *[Page No - 1.12]*

    b) What are the advantages of hybrid methods?   (3)

    Chapter 1, SA-Q1.61 *[Page No - 1.44]*

17. a) Explain how finite element method is used for analysis of electrical machines.   (5)

    b) Derive from first principles, the output equation of a 3-phase induction motor. Explain each term used.   (5)

    Chapter 8, Section 8.2 *[Page No - 8.5]*

PUNE UNIVERSITY (PU)

T.E.(ELECTRICAL)

DESIGN OF ELECTRICAL MACHINES - Nov 2019

(2015 Pattern) (Semester - II)

Time: 2 1/2 Hours   **Maximum:** 70 Marks

Q1. a) Derive the output equation for a three-phase core type transformer.   (6)

         Chapter 7, Section 7.3 *[Page No - 7.8]*

     b) Explain the different modes of heat dissipiation.   (6)

         Chapter 4, Section 4.1.1 *[Page No - 4.3]*

     c) Explain the mechanical forces developed under short circuit conditions. Also state the measures
         to overcome this effect.   (8)

         Chapter 7, Section 7.6.6 *[Page No - 7.28]*

OR

Q2. a) What are the advantages and disadvantages of higher number of poles in DC machine?   (6)

         Chapter 6, Section 6.5 *[Page No - 6.14]*

     b) Explain the procedure to calculate the no-load current in case of a three-phase transformer. (6)

         Chapter 7, Section 7.7.2 *[Page No - 7.33]*

     c) Explain the procedure for the design of tank with tubes and derive the relation for the number
         of tubes.   (8)

         Chapter 7, Section 7.9 *[Page No - 7.36]*

Q3. a) Define specific electrical and magnetic loadings. Explain the factors to be considered for the
         choice of specific electrical loading and specific magnetic loading.   (8)

         Chapter 3, Section 1.3.1 and 1.3.3 *[Page No - 1.9]*

     b) Derive the Output equation for three-phase induction motor with usual notations.   (8)

         Chapter 8, Section 8.2 *[Page No - 8.5]*

OR

Q4. a) Compare the squirrel cage induction motor with wound rotor machine.   (8)

         Chapter 8, SA-Q8.29 *[Page No - 8.84]*

     b) What are the main dimensions of the induction motor? What are the desired values of L/τ,
         peripheral speed and width of ventilation ducts.   (8)

         Chapter 8, Section 8.3 *[Page No - 8.7]*

Q5. a) What are the suitable combinations of designing stator and rotor slots?   (8)

        Chapter 8, Section 8.7.1 *[Page No - 8.18]*

  b) Derive the equation for end ring current for the rotor of squirrel cage induction motor.   (8)

        Chapter 8, Section 8.7.3 *[Page No - 8.19]*

Q6. a) Explain the various factors which affect the length of air-gap in an induction motor.   (8)

        Chapter 8, Section 8.6 *[Page No - 8.13]*

  b) A 11 kW, 220 V, 3-phase, 6 pole, star connected squirrel cage induction motor has the follwoing data : number of stator slots = 54, number of conductor in each stator slot = 9, number of rotor bars = 64, efficiency = 0.86, power factor = 0.85, current density = 5 A/mm^2. Find bar current, end ring current area of bar, area of end ring. Assume Rotor mmf as 85% of stator mmf. Assume suitable data if required.   (8)

        Chapter 8, Section 8.7 *[Page No - 8.52]*

Q7. a) Explain

      i)   Slot leakage

      ii)   Tooth top leakage

      iii)   Zig-zag leakage

      iv)   Overhang, leakage with the help of necessary diagrams.   (6)

        Chapter 3, Section 3.12 *[Page No - 3.23]*

  b) Explain the effects of ducts on calculation of magnetizing diagrams.   (6)

        Chapter 3, Section 3.6 *[Page No - 3.13]*

  c) A 75 kW, 3300 V, 50 Hz, 8 pole 3 phase, star connected induction motor has a magnetizing current which is 40% of full load current. Calculate the value of stator turns per phase if the mmf required for flux density at 60° from the pole axis is 500 A, winding factor 0.95, efficiency = 0.94 and power factor = 0.86. Assume suitable data if required.   (6)

        Chapter 8, Example 8.16 *[Page No - 8.65]*

                                               OR

Q8. a) Explain the mmf calculation for air-gap, stator teeth, stator core, rotor teeth and rotor core. (6)

        Chapter 8, Section 8.5 *[Page No - 8.9]*

  b) Explain how the no-load current is calculated in case of induction motor.   (6)

        Chapter 8, Section 8.10 *[Page No - 8.27]*

  c) Calculate the magnetizing current for 415 V, 3-phase, 50 Hz, 4 pole induction motor which has the following dimensions: air-gap length is 0.5 mm; flux density at 60° is 0.478 wb/m^2. The stator winding is delta connected with 4 slots/pole/phase and 28 conductors/slot. The ampere turns for the iron path is equal to 45% of the air-gap ampere turns. Assume the gap contraction factor as 1.2, stator winding factor = 0.955. Assume suitable data if required.   (6)

        Chapter 8, Example 8.16 *[Page No - 8.65]*

HIMACHAL UNIVERSITY (HTU)
B.Tech. 7th Semester Examination - Dec 2018
Electrical Machine Design (OS)
EE - 7001

Section-A

1. a) Derive the equation of temperature rise with time for electrical machine. What is heating time constant? (10)

    Chapter 4, Section 4.4 *[Page No - 4.10]*

    b) Discuss briefly about properties of insulating material. (10)

    Chapter 2, Section 2.8 *[Page No - 2.8]*

2. a) Explain temperature rise of transformer. Derive expression for "number of tubes' in a three-phase transformer. (10)

    Chapter 7, Section 7.8 *[Page No - 7.34]*

    b) Explain hydrogen cooling of electrical machines. What are the advantages of hydrogen cooling?

    Chapter 4, Section 4.9.2 *[Page No - 4.20]* (10)

Section-B

3. a) Derive total specific slot permeance of parallel sided slots. (10)

    Chapter 3, Section 3.14.1 *[Page No - 3.28]*

    b) What are the factors affecting choice of specific electric and magnetic loadings? (10)

    Chapter 1, Section 1.3.2 and 1.3.3 *[Page No - 1.9]*

4. a) What types of mechanical forces are developed in transformer windings? (10)

    Chapter 7, Section7.6.6 *[Page No - 7.28]*

    b) Explain the procedure for the design of tank with tubes and derive the relation for the number of tubes. (10)

    Chapter 7, Section 7.9 *[Page No - 7.36]*

Section-C

5. a) Develop the output equation for a single-phase as well as three-phase transformer. (10)

    Chapter 7, Section 7.2 and 7.3 *[Page No - 7.6]*

    b) Determine the dimensions for core and yoke for a 5 *kVA*, 50 *Hz*, single phase core type transformer. A rectangular core is used with long side twice as long as short side. Thre window height is 3 times the width. Voltage per turn is 1.8 *V*; space factor 0.2; current density 1.8 *A/mm²*; flux density 1 *Wb/m²*. (10)

    Chapter 7, Example 7.4 *[Page No - 7.47]*

6.   a)   Calculate the no-load current and magnetizing volt ampere for a single-phase transformer. (10)

           Chapter 7, Section 7.7.1 *[Page No - 7.31]*

     b)   Draw the flow chart for overall design of a transformer. The design must include

           i)   Efficiency

           ii)   Cost

           iii) Main dimension, core and yoke dimension.

           iv)   Design of tank and cooling system.   (10)

Section-D

7.   a)   Which factors should be considered when estimating the length of the air-gap of induction motor? why the air-gap should be as small as possible?   (10)

           Chapter 8, Section 8.6 *[Page No - 8.13]*

     b)   Find the values of diameter and length of a 7.5 kW, 200 V, 50 Hz, 4 pole, 3 phase induction motor for best power factor. Given: specific magnetic loading = 0.4 Wb/m^2; specific electric loadings = 22000 A/m; efficiency = 0.88 and power factor = 0.87. Also find the main dimension if the ratio of core length to pole pitch is unity.   (10)

           Chapter 8, Example 8.2 *[Page No - 8.41]*

8.   a)   Explain methods to reduce the harmonic torque in induction motor.   (10)

           Chapter 8, Section 8.7 *[Page No - 8.15]*

     b)   Draw the flow chart for overall design of 3-phase induction motor. The design must include

           (i)   Winding design

           (ii)   Losses and efficiency

           (iii)   Conductor size

           (iv)   Temperature rise.   (10)

Section-E   (10 × 2)

9.   a)   Define gap contraction for slots.

           Chapter 3, SA Q3.21 *[Page No - 3.49]*

     b)   List the methods used for estimating the mmf for teeth.

           Chapter 3, SA - Q3.37 *[Page No - 3.51]*

     c)   Why short time rating of an electrical machine is much higher than the continous rating?

           Chapter 4, SA - Q4.10 *[Page No - 4.46]*

     d)   Define stacking factor.

           Chapter 7, SA Q7.28 *[Page No - 7.82]*

     e)   Why stepped core are generally used for transformer?

           Chapter 7, SA Q7.31 *[Page No - 7.83]*

     f)   What is window space vector?

           Chapter 7, SA Q7.16 *[Page No - 7.81]*

g) Name the factors to be considered to choose the type of winding for a core type transformer.

  Chapter 7, SA - Q7.34 [Page No - 7.84]

h) What are the factors to be considered for the choice of specific magnetic loading in induction motor design?

  Chapter 8, SA Q8.11 [Page No - 8.82]

i) Why short chorded windings are employed in induction motor?

  Chapter 8, SA Q8.36 [Page No - 8.85]

j) Define unbalanced magnetic pull.

  Chapter 3, SA Q3.66 [Page No - 3.55]

UTRAKHAND TECHNICAL UNIVERSITY (UTU)

B.Tech (SEM. VI) (EVEN SEM.) EXAMINATION, DEC - 2013

ELECTRICAL MACHINE DESIGN

Time : 3 Hours   **Total Marks : 100**

                                                                              (5 × 4 = 20)

1. Attempt any FOUR of the following :

    (a) How materials are classified according to their degree of magnetism?

    (b) Find the apparent tooth density at a section of the tooth in the following case when the real tooth density at that section is 2.15 Wb/m^2. Gross armature length is 32 cm, number of ventilating ducts is 4, each 1 cm wide, teeth width at the section is 1.2 cm, slot width with parallel sides is 1 cm. Permeablity of the teeth correspondings to real tooth density is 35.8 × 10^{-6}.

    Chapter 3, Example 3.5 *[Page No - 3.42]*

    (c) Derive the equation for end ring current for the rotor of squirrel cage induction motor.

    Chapter 8, Section 8.7.3 *[Page No - 8.19]*

    (d) Define the term 'Thermal resistivity' which is used for calculating the heat dissipiation by conduction. Derive the basic equations of heat dissipiation by conduction, convection and Radiation.

    Chapter 4, Section 4.1.1, 4.1.2 and 4.1.3 *[Page No - 4.3]*

    (e) Estimate the stator core dimensions, number of stator slots and number of stator conductors per slot for a 100 kW, 3300 V, 50 Hz, 12 pole, star connected slip ring induction motor. B$_{av}$ = 0.4 Wb/m^2, **ac** = 25000 $amp.cond./m$, η = 0.9, pf = 0.9. Choose main dimensions to give best power factor. The slot loading should not exceed 500 $amp.conductors$.

    Chapter 8, Example 8.2 *[Page No - 8.41]*

    (f) Explain methods of cooling.

    Chapter 7, Section 7.7.1 *[Page No - 7.24]*

3. Attempt any TWO of the following :   **(10 × 2 = 20)**

    (a) Explain the following:

        (i) Field form co-efficient

            Chapter 3, SA-Q3.35 *[Page No - 3.51]*

        (ii) Carter's co-efficient

            Chapter 3, SA-Q3.25 *[Page No - 3.49]*

        (iii) Gap contraction factor

            Chapter 3, SA-Q3.23 *[Page No - 3.49]*

A DC machine has the following data: Pole arc = 32 cm, length of armature, L = 40 cm, length of air-gap, I_g=0.8 cm, slot pitch = 2.6 cm, width of slot = 1.2 cm, No of ventilating ducts in armature = 5, width of each ventilating duct = 1 cm, flux per pole = 0.75 × 10^{-3} Wb. Find the mmf required for the air-gap. Given:

$$\frac{\text{Slot width}}{\text{Gap length}} = 1.5, \quad \text{Carter's co efficient for slot} = 0.21$$

$$\frac{\text{Duct width}}{\text{Gap length}} = 1.25, \quad \text{Carter's coefficient for duct} = 0.18$$

Chapter 3, Example 3.7 [Page No - 3.44]

(b)   Derive the output equation of a three-phase transformer. What are the factors affecting the choice of flux density of core in a transformer?

Chapter 7, Section 7.3 [Page No - 7.8]

(c)   A 100 kVA, 2000/400 V, 50 Hz, single phase, shell type transformer has the following particulars: B_{max} = 1.1 Wb/m^2, current density δ = 2.2 A/mm^2, k_w = 0.33, volts per turn = 11, core is rectangular and stampinsare 7 cm wide. Length of window = 2(width of window). Obtain (i) A_i (ii)A_w (iii) Dimensions and weight of core. Specific gravity of iron = 7.8 gm/cm^3. Sketch the core and show how LV and LV windings are arranged.

Chapter 7, Example 7.12 [Page No - 7.57]

3.   Attempt any TWO of the following :   **(10 × 2 = 20)**

(a)   Differentiate between harmonic induction torque and harmonic synchronous torque developed in an induction motor. Describe their effects.

Chapter 8, Section 8.17 [Page No - 8.15]

(b)   A 15 kW, 400 V, 3-phase, 50 Hz, 6 pole, Delta connected squirrel cage induction motor has a diameter of 0.3 m and a core length of 0.12 m. There are 72 stator slots with 20 conductors/slot. Calculate the magnetizing current per phase if the air-gap length is 0.55 mm and the gap contraction factor is 1.2. Assume mmf required for iron parts to be 35% of the mmf for the air gap and the coil span is 11 slots.

Chapter 8, Example 8.14 [Page No - 8.14]

(c)   Explain methods to reduce the harmonic torque in induction motor.

Chapter 8, Section 8.7 [Page No - 8.15]

4.   Attempt any TWO of the following :   **(10 × 2 = 20)**

(a)   What is the role of damper windings in (i) Synchronous generator (ii) synchronous motor?

Chapter 9, Section 9.4.10 [Page No - 9.20]

(b)   What is the meant by SCR of an alternator? Discuss its significance in relation to stablity, voltage regulation and parallel operation of synchronous generators.

Chapter 9, Section 9.4.2 [Page No - 9.9]

(c)   A 500 *kVA*, 3.3 *kV*, 50 Hz, 600 *rpm,* 3-phase, and salient pole alternator has 180 turns/phase. Estimate the length of air-gap if the average flux density is 0.54 *Wb/m²*. The ratio of pole arc to pole pitch = 0.66, the short circuit ratio = 1.2, and the gap extension coefficient = 1.15. The mmf required for gap is 80% of no-load field mmf. Winding factor = 0.955.

Chapter 8, Example 8.16 *[Page No - 8.65]*

5.   Attempt any Two of the following: (10 × 2 = 20)

(a)   Explain the following approaches used for the machine design with the help of flow charts:

(i)   Analysis method

Chapter 1, Section 1.11.1 *[Page No - 1.23]*

(ii) Synthesis method.

Chapter 1, Section 1.11.2 *[Page No - 1.25]*

(b)   Explain in detail the concept of optimization in electrical machine design.

Chapter 1, Section 1.11.4 *[Page No - 1.26]*

(c)   Discuss the various mechanical forces developed in transformer with sketches. Explain how they are taken care while fabrication.

Chapter 7, Section 7.6.6 *[Page No - 7.28]*

INDEX